Pavement Management 2018

AA005500

Transportation Research Record: Journal of the Transportation Research Board Volume 2672, Issue 40

ISBN: 978-1-5108-8248-5

Printed from e-media with permission by:

Curran Associates, Inc.
57 Morehouse Lane
Red Hook, NY 12571

Some format issues inherent in the e-media version may also appear in this print version.

Published by Sage Publications USA

ISBN (Print) 978-1-5108-8248-5
ISBN 2018 Printed Set (All Issues) 978-1-5108-7735-1

Printed by Curran Associates, Inc. (2019)

Additional copies of this publication are available from:

Curran Associates, Inc.
57 Morehouse Lane
Red Hook, NY 12571 USA
Phone: 845-758-0400
Fax: 845-758-2633
Email: curran@proceedings.com
Web: www.proceedings.com

Volume 2672 Number 40 December 2018

TRR
JOURNAL OF THE TRANSPORTATION RESEARCH BOARD

Contents

Pavement Management 2018

Articles

Transportation Officials, Washington, D.C.
K. Jane Williams, *Acting Administrator, Federal Transit Administration, U.S. Department of Transportation*
Paul Zukunft *(Admiral, U.S. Coast Guard), Commandant, U.S. Coast Guard, Washington, D.C.*

---

* Membership as of February 2019.

TRB

TRANSPORTATION RESEARCH BOARD

CORE SPONSORS

Transportation Departments of the 50 States and the District of Columbia

Federal Government

U.S. Department of Transportation
- Federal Aviation Administration (FAA)
- Federal Highway Administration
- Federal Motor Carrier Safety Administration (FMCSA)
- Federal Railroad Administration (FRA)
- Federal Transit Administration
- National Highway Traffic Safety Administration (NHTSA)
- Office of the Assistant Secretary of Transportation for Research and Technology (OST-R)

U.S. Air Force Civil Engineer Center (AFCEC)
U.S. Coast Guard (USCG)
U.S. Department of Energy
U.S. Department of the Interior (DOI)
U.S. Environmental Protection Agency (EPA)

Nongovernmental Organizations

American Public Transportation Association (APTA)
Association of American Railroads (AAR)
California Air Resources Board

The National Academies of
SCIENCES · ENGINEERING · MEDICINE

The **National Academy of Sciences** was established in 1863 by an Act of Congress, signed by President Lincoln, as a private, nongovernmental institution to advise the nation on issues related to science and technology. Members are elected by their peers for outstanding contributions to research. Dr. Marcia McNutt is president.

The **National Academy of Engineering** was established in 1964 under the charter of the National Academy of Sciences to bring the practices of engineering to advising the nation. Members are elected by their peers for extraordinary contributions to engineering. Dr. C. D. Mote, Jr., is president.

The **National Academy of Medicine** (formerly the Institute of Medicine) was established in 1970 under the charter of the National Academy of Sciences to advise the nation on medical and health issues. Members are elected by their peers for distinguished contributions to medicine and health. Dr. Victor J. Dzau is president.

The three Academies work together as the National Academies of Sciences, Engineering, and Medicine to provide independent, objective analysis and advice to the nation and conduct other activities to solve complex problems and inform public policy decisions. The Academies also encourage education and research, recognize outstanding contributions to knowledge, and increase public understanding in matters of science, engineering, and medicine.

Learn more about the National Academies of Sciences, Engineering, and Medicine at **www.national-academies.org.**

The **Transportation Research Board** is one of seven major programs of the National Academies of Sciences, Engineering, and Medicine. The mission of the Transportation Research Board is to increase the benefits that transportation contributes to society by providing leadership in transportation innovation and progress through research and information exchange, conducted within a setting that is objective, interdisciplinary, and multimodal. The Board's varied activities annually engage about 7,000 engineers, scientists, and other transportation researchers and practitioners from the public and private sectors and academia, all of whom contribute their expertise in the public interest. The program is supported by state transportation departments, federal agencies including the component administrations of the U.S. Department of Transportation, and other organizations and individuals interested in the development of transportation.

Learn more about the Transportation Research Board at **www.TRB.org**.

TRR
JOURNAL OF THE TRANSPORTATION RESEARCH BOARD

Article

Transportation Research Record
2018, Vol. 2672(40) 1–6

DOI: 10.1177/0361198118767415
journals.sagepub.com/home/trr
SAGE

Time-Series Review of Highway Performance Monitoring System Data

Amy L. Simpson[1], James M. Bryce[1], Gonzalo R. J. Rada[1], and Jonathan L. Groeger[1]

Abstract

The passing of the Moving Ahead for Progress in the 21st Century Act required the FHWA to adopt pavement performance measures for evaluating the condition of the interstate highway system (IHS) and the national highway system (NHS). In January 2015, the FHWA issued a notice of proposed rulemaking (NPRM) to establish performance measures to assess pavement conditions on the IHS and NHS, with the final rule published in January 2017. The four measures were 1) Percentage of pavements on the IHS in good condition; 2) Percentage of pavements on the IHS in poor condition; 3) Percentage of pavements on the NHS (excluding IHS) in good condition; and 4) Percentage of pavements on the NHS (excluding IHS) in poor condition. Pavement condition is determined by the following metrics in the Highway Performance Monitoring System (HPMS): International Roughness Index, cracking percent, rutting, and faulting. FHWA commissioned a study to collate a statistically significant sample of the IHS to evaluate the efficacy of the HPMS and to recommend improvements in data collection and reporting processes to meet NPRM requirements. As part of the referenced study, a time-series review was conducted of the HPMS to identify changes in the data from 2013 to 2015 and to determine how these data compare with the study-collected data. This paper discusses the review of HPMS data, identifying changes in the pavement condition metrics, and overall condition. In general, the HPMS database exhibited improvements, with more complete data and fewer anomalies, in 2015 compared with 2013.

The Moving Ahead for Progress in the 21st Century Act (MAP-21) required performance measures to be established for the interstate highway system (IHS) and the succeeding bill—the Fixing America's Surface Transportation Act—which maintains support of the performance management elements established under MAP-21. Consequently, the FHWA issued a notice of proposed rulemaking (NPRM) to establish performance measures to assess the condition of the pavements on the IHS and the national highway system (NHS). The NPRM established four pavement performance measures to assess pavement condition as follows: percentage of pavements on the IHS in good and poor condition, and percentage of pavements on the NHS (excluding IHS) in good and poor condition. These four measures were confirmed in the final rule issued in January 2017(*1*).

As a result, the FHWA undertook the Interstate Pavement Condition Sampling study. The objectives of this study were to:

- Collect an unbiased baseline data set for a statistically significant sample of the IHS, and produce a report indicating condition of IHS nationally and in each state where data were collected.
- Determine if the Highway Performance Monitoring System (HPMS) is an unbiased representation of pavement condition on the IHS.
- Recommend improvements to HPMS data collection and reporting to either make HPMS unbiased or improve its precision.

Simpson et al. presented findings from the initial phases of the Interstate Pavement Condition Sampling study to answer such questions as: Is two-way data collection necessary? Does data need to be collected in more than one lane in a direction? What is the optimum reporting length? For what percentage of the network should data be collected? Should data be collected annually? (*2*).

The originally stated objectives were completed in July 2016 and were published in the Interstate Pavement Condition Sampling final report (*3*). In addressing these objectives, several items of interest to FHWA surfaced. Of greatest interest to the FHWA were the following specific data analyses and evaluations:

[1]Amec Foster Wheeler, Beltsville, MD

Corresponding Author:
Address correspondence to Amy L. Simpson: amy.simpson@amecfw.com

- Comparison of measured data to the most recently submitted 2015 HPMS data.
- Temporal analysis of 2013, 2014, and 2015 HPMS data sets.
- Evaluation of whether regional conditions affect interstate pavement conditions.
- Evaluation of metrics that are driving classification of condition ratings, using HPMS and project data.
- Data management and quality evaluation.

The objective of this paper is to review the time-series trends in the HPMS data from 2013 to 2015, which is the second of the above stated data analyses.

HPMS Data Discussion

The first step in the temporal analysis was to perform a review of the data contained in the HPMS data sets for each year – 2013, 2014 and 2015. Table 1 provides a review of the summary statistics for the HPMS data sets by year. The HPMS data condition metrics other than the international roughness index (IRI) are provided on a sample of the network and not on the full network. In each year of data collection, the extent of sampling has expanded. For example, percent cracking is provided for 25,690 mi on the IHS in the year 2013, 25,690 mi in 2014, and 26,574 mi in 2015. A similar trend is observed for each of the condition metrics.

Reviews were then performed to identify the trends in data over time using data contained in the 2013, 2014, and 2015 data sets. Figure 1 presents the cumulative distributions of the IRI for each of the 3 years in the comparison, along with the best fit normal distribution for comparison purposes. Similar figures were developed for percent cracking, rut depth, and faulting. Observations from this analysis include the following:

- Very little change was observed in the distributions of IRI among the 3 years of data.
- Distributions of IRI did not follow a normal distribution, but more closely matched a generalized extreme value distribution.
- Average percent cracking decreased in each year, but the proportion of values of percent cracking above 50% was higher in 2014 and 2015.
- Some agencies rounded cracking data to the nearest multiple of five, whereas others did not; therefore, a smooth distribution cannot be fit to the data.
- Rutting in 2013 was generally higher, whereas rutting in 2014 was generally lower than rutting in 2015.
- Two percent of the 2013 data reported rutting higher than 1 in., whereas zero percent of the 2014 and 2015 data reported rutting greater than 1 in.
- Although rutting data appear to follow a log-normal distribution, a smooth distribution could not be fit to the data because some agencies rounded values to the nearest 0.1 in., whereas other agencies provided unrounded values.
- Faulting was observed to generally decrease over time, with a significant portion of the data reported as zero.

Table 1. Summary Statistics of Data

Condition metric	Statistic	HPMS data 2013	2014	2015
IRI, in./mi	Mean	81	74	79
	Standard deviation	43	38	43
	Minimum	21	6	18
	Maximum	982	700	974
Cracking, %	Mean	4.9	4.7	3.4
	Standard deviation	11.6	9.2	10.3
	Minimum	0.0	0	0
	Maximum	100	100	100
Rut depth, in.	Mean	0.25	0.12	0.12
	Standard deviation	0.81	0.10	0.35
	Minimum	0.0	0.0	−100
	Maximum	16.9	1.8	1.45
Faulting, in.	Mean	0.12	0.05	0.04
	Standard deviation	0.16	0.14	0.11
	Minimum	0	0	0
	Maximum	2.9	9	2.8

Note: in = inches; mi = mile.

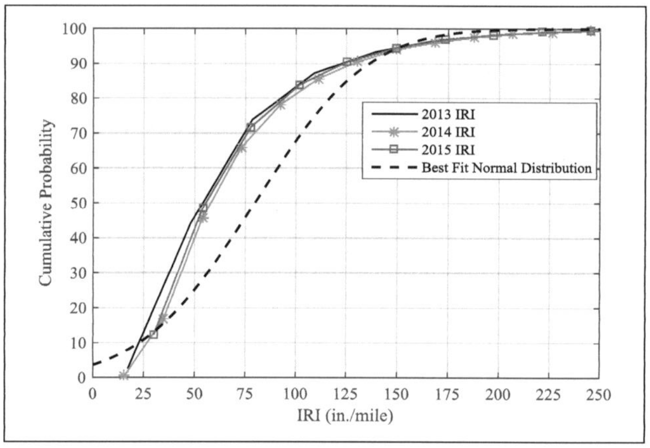

Figure 1. Comparison of IRI in 2013, 2014, and 2015 HPMS data.

Comparing distributions of the data over time gives the opportunity to review for significant biases in the data between years. However, this approach does not provide the opportunity to review changes in specific segments over time. In order to complete that type of analysis, it is necessary to identify the same pavement segment in each of the three data sets. A set of 205,633 segments were identified as occurring in each of the three data sets (2013, 2014, and 2015).

Once the segments were paired over time, the differences in their performance metrics were investigated. A histogram of differences was developed for each metric between 2013

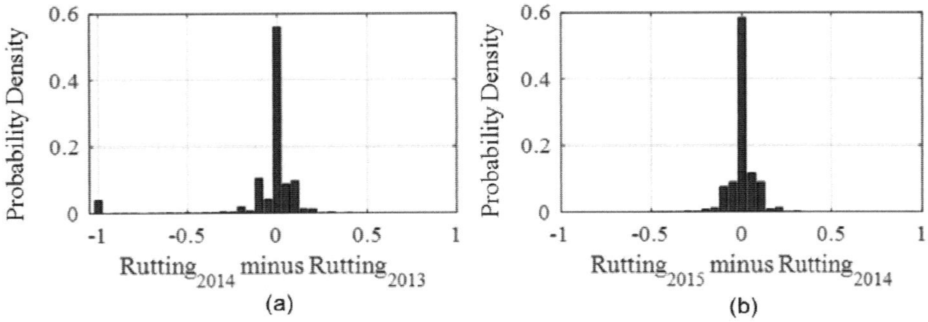

Figure 2. Paired differences in the 2013 and 2014 rut depth data (*a*) and the 2014 and 2015 rut depth data (*b*).

and 2014 data, as well as between 2014 and 2015 data. Figure 2 provides an example of these histograms as developed for rut depth. The following observations were made from these histograms:

- For IRI, the differences are evenly distributed around zero with an average change in IRI from 2013 to 2014 of 0.85 in./mi, and an average change between 2014 and 2015 of −1.49 in./mi.
- One percent of the IRI values for the paired segments do not record any change in IRI between the 3 years of data reviewed, and no State has more than 3% of its reported IRI data (limited to only the paired segments) indicating no change in IRI over the 3 years reviewed.
- For cracking, the average change is slightly negative, which indicates a small decrease on average.
- For rutting, the mean change between the 2013 and 2014 data is −0.19 in., which indicates a significant decrease in rutting between the 2 years.
- Of segments that recorded rutting of less than 1 in., the change between 2013 and 2014 data is −0.01 in., which is a very small decrease, indicating that the very high rutting values recorded in 2013 influenced the temporal comparison significantly.
- The change between the 2014 and 2015 rutting for the paired segments is less than 0.01 in., which can be interpreted as effectively no change.
- For rutting, approximately half of the segments indicated no change between the years.
- For faulting, the mean change between the 2013 and 2014 measurements is −0.04 in., and the mean change between the 2014 and 2015 measurements is −0.03 in., indicating a decrease in both yearly intervals.

The next investigation was to review the mean of the measurements in comparison to the median for each paired segment. For IRI, the mean IRI tended to be larger than the median of the three measurements, indicating that significant deviations in the measurements are more likely to result in an increase in reported roughness than a decrease. Additionally, the results indicate that significant deviations from the mean of percent cracking are expected to be caused by a measurement reading much higher than the other two cracking values over the range that encompasses the majority of cracking measurements (less than 35%). This result was expected given that the average cracking for all segments was very low and cracking is truncated such that it cannot fall below zero or above 100. Similar results were found for rutting and faulting, where, over the range that encompasses the majority of the data, median values tended to be less than mean values. This indicates that, in many cases, measurement errors tend to result in larger (worse) values for each metric.

The next analysis performed using these paired data included determining the average growth over the 3 years for each segment. A histogram of the calculated slopes was then developed separately by surface type. Figure 3 provides an example of these histograms for IRI. Observations from this analysis include:

- The distributions of growth rate for IRI for each pavement type are similar.
- It is assumed that a portion of the negative changes over time came from pavements that received work during the analysis period, and another portion of the negative changes are due to measurement errors.
- The cracking growth on asphalt concrete (AC) pavements is evenly distributed around zero, whereas cracking growth on continuously reinforced concrete pavement (CRCP) and jointed concrete pavement (JCP) pavements is much more likely to be negative. In this case, the growth rates for CRCP are much more likely to be negative, indicating that a reduction in cracking was seen on the majority of CRCP.
- For rutting, the peak of the distribution was observed at −0.2 in./year due to the high rutting measurements recorded for many segments in the year 2013.
- The average faulting growth rate is negative (mean value equal to −0.05 in./year) indicating that faulting generally declined over the 3 years.

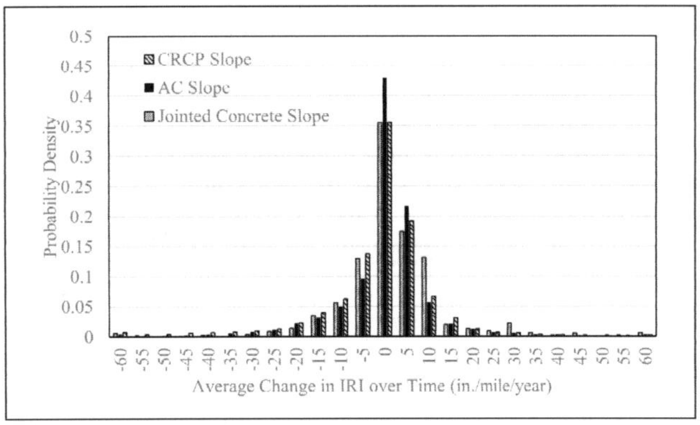

Figure 3. Average IRI growth rate for PCC- and AC-surfaced pavements.

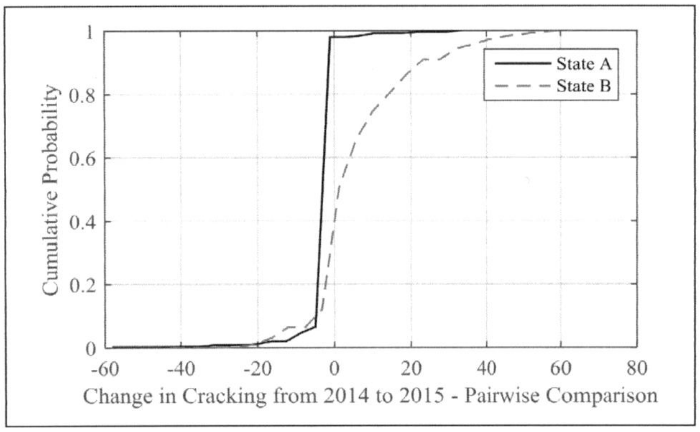

Figure 4. Comparison of changes in cracking for two states.

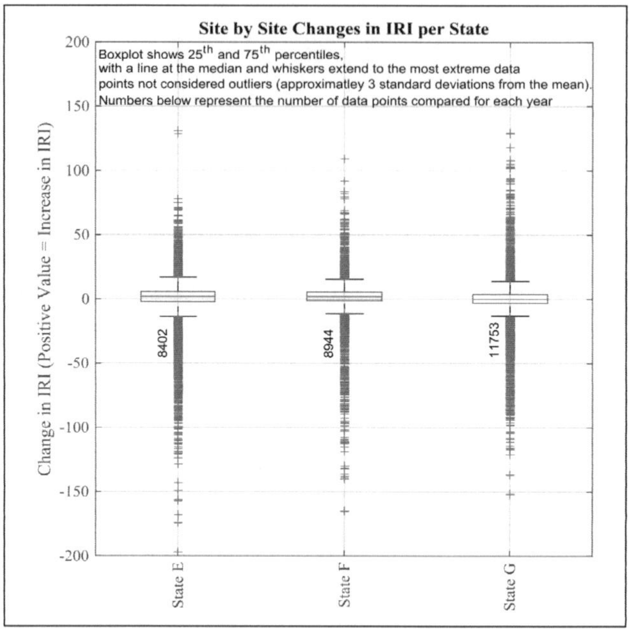

Figure 5. Paired changes in 2014/2015 IRI measurements for select states.

While investigating the changes in the metrics throughout the years, it was noted that many States had certain metrics that did not change over time. As an example, Figure 4 shows that a large majority of segments in State A recorded no change in cracking between 2014 and 2015, whereas the data from State B shows a significant range of changes between the 2 years. The range of changes for each metric based on comparing the same pavement segments in both years (i.e., performing a pairwise comparison) was reviewed for each State.

As part of this review, it was important to note the required frequency of data submissions to the HPMS. Whereas the IRI data are required to be submitted every year, data for cracking, rutting, and roughness can be submitted on a 2-year cycle. This could explain why some metrics have zero change for some States. However, some agencies do submit annual data for all metrics.

Several cases were investigated in more detail. Three States with relatively little change in IRI over the 2 years are shown in Figure 5, along with the data considered as outliers. The median for each of these three sites is slightly positive (i.e., small growth in IRI). At the opposite end of the spectrum is the data from State A, which indicates that a significant portion of the pavement segments had a sizable decrease in IRI.

Figure 6 shows the cumulative distributions for the IRI data from State A (not paired); as can be seen, the 2015 data are consistently lower than the 2014 data by a considerable amount. Some States were observed to have data that do not change over the 2 years. Using the pairwise comparison, 96% of the segments in State C had no change reported in the IRI between the 2 years. Although many factors can be used to explain the large range of changes in the condition metrics, it is clearly seen from the information provided that the quality of IRI data reported is inconsistent from one State to another.

Similar to the IRI results, several States report very little change in cracking, whereas others have a large range. An interesting case is the consistent reduction in cracking exhibited by the State C data. Figure 7 shows the cumulative distribution of the cracking data for 2014 and 2015 as reported by State C; as can be seen, the 2015 data contained significantly more data than the 2014 data. However, the 2014 data contained no segments with zero cracking, whereas 97% of the 2015 data had no cracking reported. Another case that was further investigated is the consistent cracking reduction shown in the State D data. Figure 8 shows the cumulative distributions of cracking for all segments from State D. As shown, the 2015 data contains more than twice the amount of segments with no cracking reported as the 2014 data. Although several factors may be used to explain the differences in the reported data, it

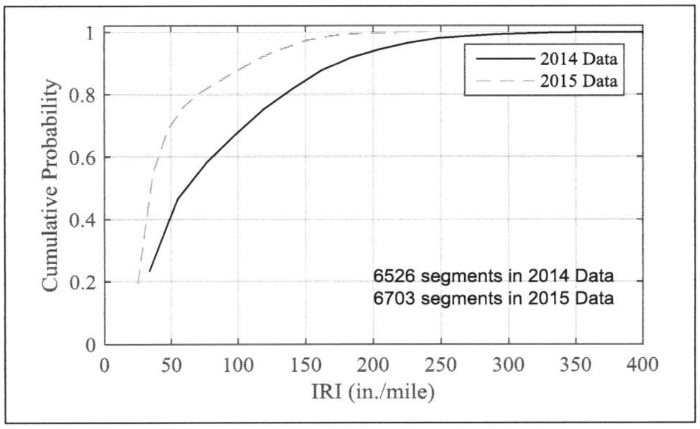

Figure 6. Comparison of 2014 and 2015 IRI data from State A.

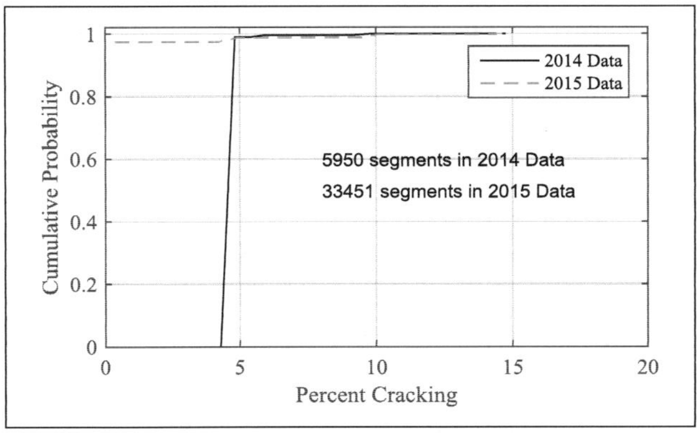

Figure 7. Comparison of 2014 and 2015 cracking data from State C.

is clear that the changes in each metric are highly dependent on the State that is reporting the data.

Similar to the IRI and cracking data, the rutting data from some States showed no changes, whereas many states showed a wide range of changes. For example, State H showed practically no changes in rutting over the 2 years, whereas the State I data indicate an average decrease in rutting with a large range of year-to-year changes. Figure 9 shows the cumulative distributions using data from State H. As shown, very little change occurs on average. The step-wise nature of the plot in Figure 9 is due to rounding of the reported data. Figure 10 shows the cumulative distribution of rutting data from State I. As shown, the 2015 data are much lower on average than the 2014 data. Whereas practically no pavement had zero rutting in the 2014 State I data, 42% of the 2015 AC segments had rutting reported.

The year-to-year changes in faulting data from State J and State K were investigated in more detail. State J reported very little change in faulting between the 2 years and the State K data showed a significant increase in faulting on

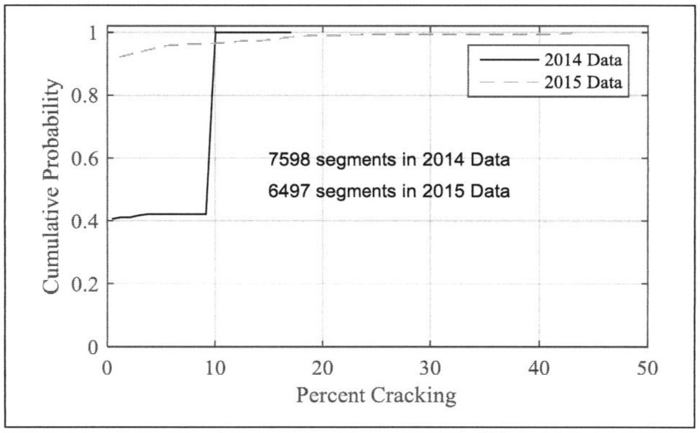

Figure 8. Comparison of 2014 and 2015 cracking data from State D.

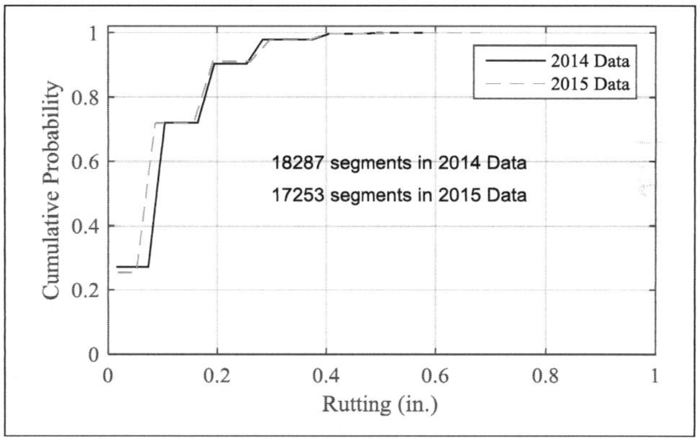

Figure 9. Comparison of 2014 and 2015 rutting data from State H.

average. Figure 11 shows the cumulative distribution of faulting from State J for the 2 years. As shown, the 2015 data reported practically no faulting (only 8 of the 5587 segments had faulting reported). Figure 12 shows the cumulative distributions of the faulting data from State K; as can be seen, many segments in State K have faulting values greater than 1 in. (3% of the 2014 data and 28% of the 2015 data).

Conclusions

This paper presented the results of a temporal analysis conducted to assess changes in the 2013, 2014, and 2015 HPMS data. This assessment of temporal changes in the HPMS data revealed several interesting results:

- The quality of the data appears to be improving in terms of the amount and values reported.
- In the case of faulting, the 2013 data is significantly different when compared with 2014 and 2015 data. This holds true to a lesser extent when evaluating the rutting data, which includes a high

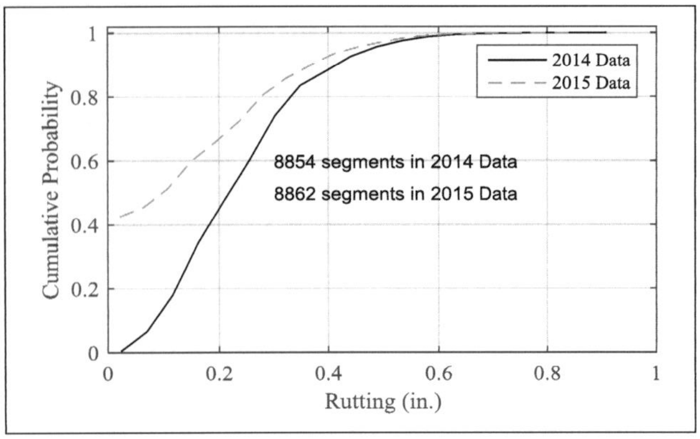

Figure 10. Comparison of 2014 and 2015 rutting data from State I.

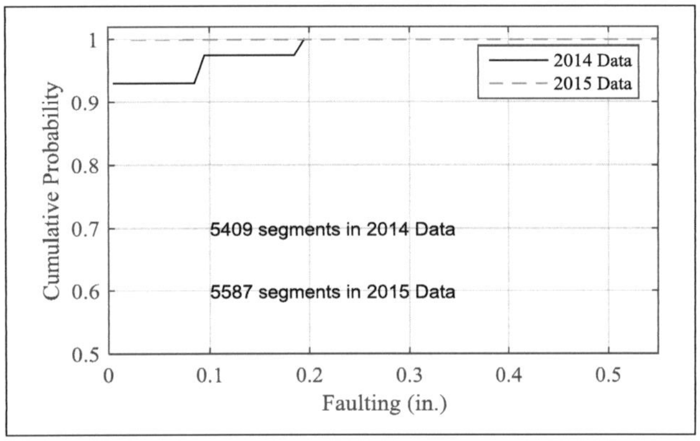

Figure 11. Comparison of 2014 and 2015 faulting data from State J.

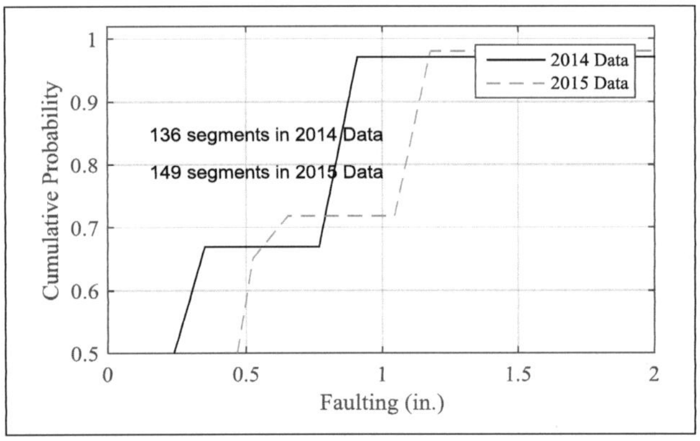

Figure 12. Comparison of 2014 and 2015 faulting data from State K.

number of unreasonably large rut measurements in 2013. However, the cracking matched reasonably well and the IRI data matched very well over the 3 years.

- The measurements for each metric on a given pavement segment had considerable variance, and thus the estimation of an average growth rate using the data produced highly varied results. For example, nearly 20% of the segments reported an absolute value of average IRI growth greater than or equal to 10 in./mi/year, and the growth rates were fairly evenly distributed around zero. It is expected that, as more data becomes available for each segment, the effect of this variance on estimating growth rates of the metrics will be mitigated.
- The range of changes in the metrics for a given pavement segment between 2 years is dependent on the state that reports the metrics. For example, some agencies report changes in IRI on the same pavement segment between −15 in./mi and 15 in./mi, whereas other report changes of the order of −80 in./mi to 80 in./mi between 2 years.

Based on the results presented, it can be concluded that, although significant measurement errors exist that contribute to high levels of variance for metrics on any given pavement segment, the sample size of the data in the HPMS is large enough to offset the effects of these errors. Effectively, a single pavement segment (or small group of pavements) can be expected to exhibit temporal characteristics not typically associated with pavement condition data (e.g., large improvements in the metrics in the absence of maintenance), but the effects of this can be mitigated over a large sample size.

References

1. Federal Register. *National Performance Management Measures; Assessing Pavement Condition for the National Highway Performance Program and Bridge Condition for the National Highway Performance Program.* 23 CFR Part 490, 2017.
2. Simpson, A. L., G. R. Rada, B. A. Visintine, and J. L. Groeger, *Interstate Pavement Condition Sampling.* Federal Highway Administration, Washington, D.C., 2016.
3. Simpson, A. L., J. M. Bryce, B. A. Visintine, G. R. Rada, and J. L. Groeger. *Interstate Pavement Condition Sampling: Report Addendum*, Federal Highway Administration, Washington, D.C., 2017.

The Standing Committee on Pavement Management Systems peer-reviewed this paper (18-00261).

Article

TRR
JOURNAL OF THE TRANSPORTATION RESEARCH BOARD

Transportation Research Record
2018, Vol. 2672(40) 7–13

DOI: 10.1177/0361198118768522
journals.sagepub.com/home/trr
SAGE

International Roughness Index Prediction Model for Thin Hot Mix Asphalt Overlay Treatment of Flexible Pavements

Jinsong Qian[1], Chen Jin[1], Jiake Zhang[1], Jianming Ling[1], and Chao Sun[1]

Abstract

Pavement performance prediction after maintenance and rehabilitation is important to pavement management. A two-parameter exponential international roughness index (IRI) regression model for thin hot mix asphalt overlay was developed based on information from the U.S. Long Term Pavement Performance (LTPP) database. The model influence parameters α and β, which represent the initial IRI as the thin overlay completion and shape factor of IRI deterioration curve, were statistically analyzed. The results suggested that the IRI deterioration trends in high-temperature and low-temperature regions are different. This is because β was mainly affected by the structural strength and equivalent single axle loads in the high and medium temperature region and mainly affected by the average annual precipitation in low temperature region. In-situ data from LTPP database was used to verify the IRI prediction model, and it was found that the predicted IRI and measured IRI exhibited similar trends.

Pavement performance modeling is important to pavement management, which allows engineers and department of transportation (DOT) to plan for maintenance and rehabilitation. Thus, DOT can budget for future expenses and minimize the cost (*1*). Current pavement performance prediction models highly rely on the performance data, which does not account for the influence of maintenance and rehabilitation activities. However, several maintenance and rehabilitation activities are conducted during the service life of pavement, and the pavement performance prediction model for the rehabilitated road may be different for predicting the performance of pavement after rehabilitation.

Several researchers developed treatment-specific pavement performance models (*2–5*) or emphasized the needs for such models (*6*). Current literature shows that the influence of treatment in performance prediction can be classified into two types: one is the performance jump (PJ) which represents the performance change after treatment and the other is the performance deterioration curve after treatment. Rajagopal and George (*2*) developed a mechanistic-empirical model to predict performance jump and deterioration rate as rehabilitation was finished. Lu et al. (*7*) found that the pavement treatment effectiveness in international roughness index (IRI) follows a polynomial relationship with the pavement pretreatment condition. Also, a pavement deterioration rate prediction model after slurry seal and chip seal was developed. A number of deterioration models to predict the IRI change for slurry seal, chip seal, and creak seal for thin overlay were proposed as well (*4, 5, 8*). Then, a performance development model was also proposed for the thin overlay. Kerali et al. (*9*) invented performance models with the variables of pavement thickness, cumulative equivalent standard axle, and California Bearing Capacity Ratio (CBR). Based on the in-serve pavement performance data in Indiana DOT, Gulen et al. (*10*) proposed a post-overlay IRI prediction model with the function of time and Annual Average Daily Traffic (AADT). In general, the prediction models on PJ are relatively well-developed but limited research on the post-treatment performance deterioration curve was conducted.

The objective of this paper is to develop a post-thin-overlay IRI prediction model using a unified model function form, which can be used for different regions. The influence of thin overlay was characterized by several factors such as structural strength, annual precipitation, etc. An IRI prediction method was proposed based on the analysis of influence factors, and findings from this work can provide the guideline for other performance indictors or treatments.

Selection of the Functional Form for Post-Overlay IRI Prediction Model

Since the World Bank proposed the concept of IRI during the late 1970s, many IRI prediction models were developed

[1]Key Laboratory of Road and Traffic Engineering of the Ministry of Education, Tongji University, Shanghai, P.R. of China

Corresponding Author:
Address correspondence to Jiake Zhang: zhjiake@tongji.edu.cn

Table 1. Investigated Functional Form of Post-Treatment IRI Prediction Model

Model type	Functional form
Exponential model (I)	$IRI = e^{(\alpha+\beta\cdot AATA\cdot t+\gamma\cdot ANDX\cdot t)}$
Exponential (II)	$IRI = \alpha\cdot e^{\beta\cdot Age}$
Power function	$IRI = \alpha\cdot Age^{\beta}$
Polynomial	$IRI = \alpha+\beta\cdot Age+\gamma\cdot Age^2$

using linear or nonlinear regression method (*11*). With the development of mathematics and computer-related technologies in recent years, more methods were used for pavement performance prediction. For instance, Texas Department of Transportation established a pavement performance prediction using an artificial neural network method. Chen (*12*) developed an IRI prediction model using grey prediction model, and Amador-Jimenez (*13*) applied time series methods to predict pavement roughness.

However, current IRI prediction models have geographical limitations, and it is difficult for them to be widely accepted or applied in different regions. In this paper, a unified functional form of post-treatment IRI prediction model was selected for different regions, structures, and traffic loads. The variability for different IRI deterioration in pavement was described by the difference of parameters in IRI prediction model. A variety of functional forms were investigated using statistical software, and then models which can more precisely predict the pavement performance were selected. These model functional forms are shown in Table 1 (*14*).Previous research has shown that the exponential model (I) and (II) can better fit the IRI curves under different conditions (*15, 16*). However, the average annual truck traffic (AATA) data and average annual freeze index (ANDX) data, which are the input parameters of pavement age and exponential model (I), are not widely available in pavement management database. Thus, the exponential model (I) is not suitable for IRI prediction due to lack of these input data. So, the exponential model (II) was used to predict the IRI with time after the overlay competition.

Data Preparation for IRI Prediction Model

Data from the U.S. Long Term Pavement Performance (LTPP) database was used in this work (*17*). The climate information, structural characterization, traffic data, and performance dataset from Specific Pavement Study (SPS3) were used, and the SPS3 locations are shown in Figure 1.

IRI data after thin overlay from 79 test sections with more than five years was selected as the sample space, which was randomly divided into two parts. One part with 70 test sections was used to develop the post-overlay IRI prediction model and the other one was used as case study to verify the validity of post-overlay IRI prediction model.

The completion of thin overlay was selected as the timeline zero, and a regression analysis using an exponential functional form was conducted to obtain the regression parameters for each section. The histogram frequency distribution diagram of R^2 in IRI regression model is shown in Figure 2. The result indicates that 70% of R^2 values, which suggests the parameter of determination between regression model and the actual data, were higher than 0.8 and all the R^2 values were higher than 0.5. Therefore, it is reasonable to assume that post-overlay IRI curve can be represented by exponential function form (II).

Model Parameters Determination

Besides the equational form, the post-overlay IRI prediction model also needs a determining method for the model parameters α and β.

Parameter α

Considering the completion of thin overlay as timeline zero, the IRI value at timeline zero is parameter α. Many researchers studied the differences between IRI values before and after thin overlay (Performance Jump, PJ). Relevant research results revealed that the PJ value relates to the IRI before thin overlay (IRI_{pre}) and the PJ prediction model was developed based on this finding. Figure 3 shows two representative PJ prediction models. As illustrated in Figure 3, the predicted results between different PJ prediction models are similar, and this is because IRI_{pre} ranges between 1 and 3. This paper uses the following formula to calculate PJ by considering large sample space of Irfan's PJ prediction model (*14*) as

$$PJ = 1.11\ln\left(IRI_{pre}\right) + 0.377 \quad (1)$$

and the parameter α can be calculated as following

$$\alpha = IRI_{pre} - PJ = IRI_{pre} - 1.11\ln(IRI_{pre}) - 0.377 \quad (2)$$

where, α = calculated model parameter, PJ = the change of IRI after thin overlay, and IRI_{pre} = the value of IRI before thin overlay.

Parameter β

Parameter β indicates the shape of IRI curve and deterioration rate, and it is essential to propose a calculation method for determining parameter β. However, parameter β is affected by both external and internal factors. The external factors include climate, traffic, and the internal factors include layer thickness, overlay thickness, and structural strength.

Analyses of influence factors which include climate, traffic, layer thickness, overlay thickness, IRI_{pre}, and deflection were conducted in this work. Since the test sections in the LTPP database were paved with flexible pavement, the deflection can reflect the structural strength of the pavement.

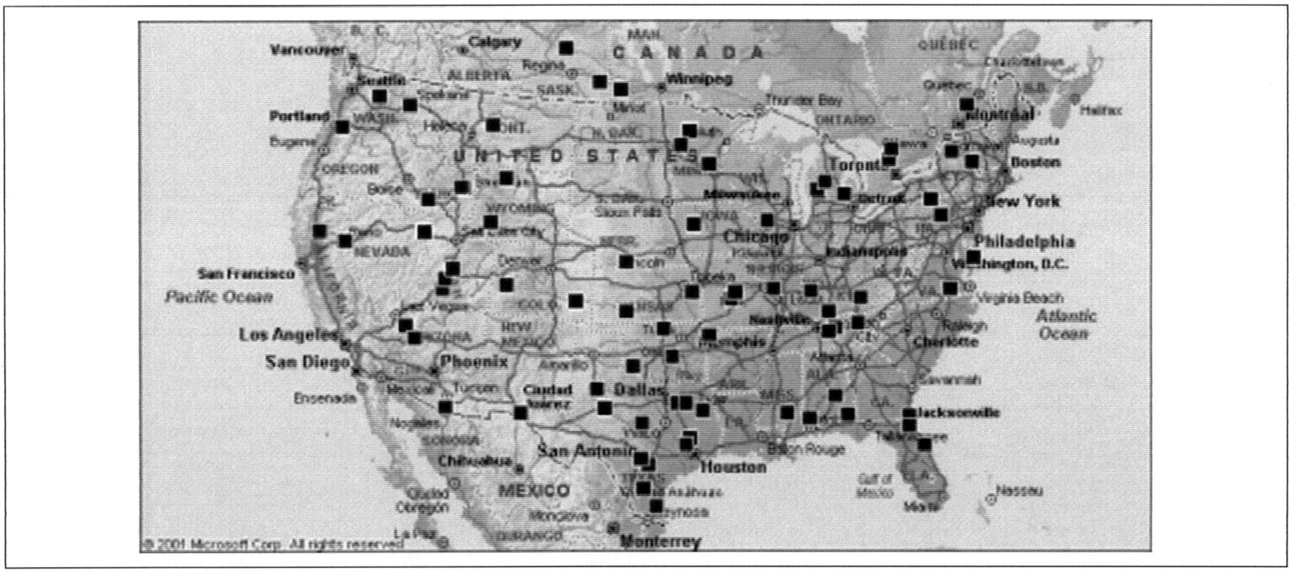

Figure 1. SPS3 locations.

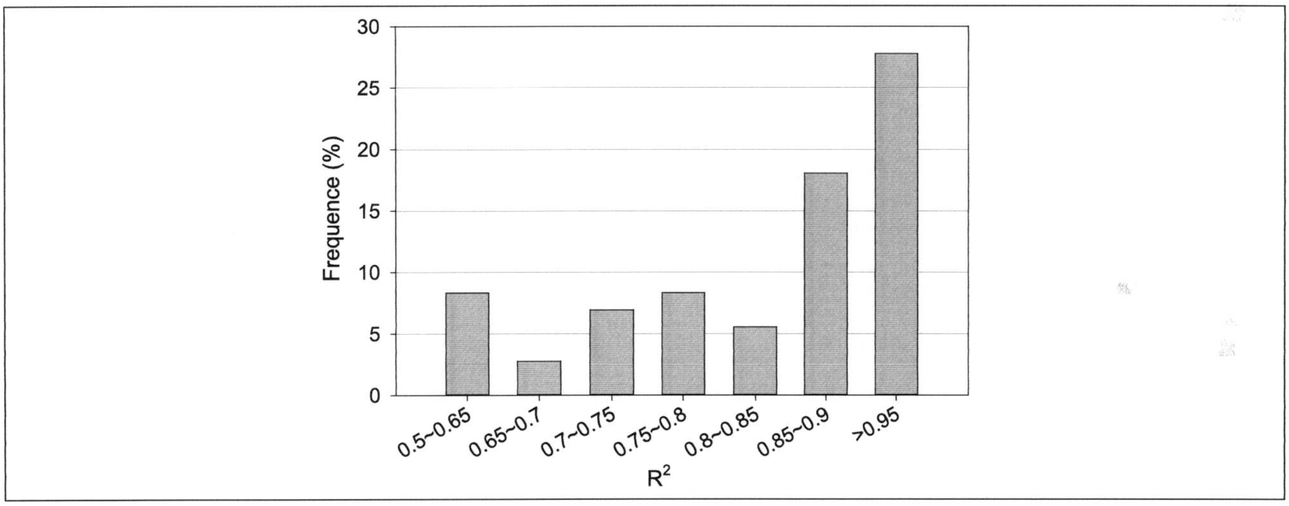

Figure 2. Frequency distribution of R^2 in IRI regression model.

Analysis of Influence Factors

Climate

Climate significantly affects the IRI value for flexible pavement. As demonstrated from previous research, temperature is the most important factor affecting the IRI reduction (*8*), and an IRI prediction model with AAFI (Accumulated Annual Freezing Index) as independent variable was proposed. In this paper, considering existing research results and temperature data provided by LTPP database, the average annual temperature was selected as the temperature input data.

The influence of temperature on IRI deterioration was mainly due to the softening of asphalt concrete. However, parameter β, which indicates the deterioration rate, decreases with temperature as shown in Figure 4. This is because thawing settlement in low temperature regions would accelerate IRI deterioration. It indicates that different deterioration mechanisms exist between high temperature regions and low temperature regions.

IRI before Thin Overlay (IRI_{pre})

Different application conditions can lead to different IRI deterioration trends even with the same treatment (*18*). IRI_{pre} was grouped into five categories in this work as shown in Table 2 for evaluating the influences of initial IRI.

Overlay Thickness

The thickness of thin overlay is an important design parameter which affects the shape parameter β. This work divided

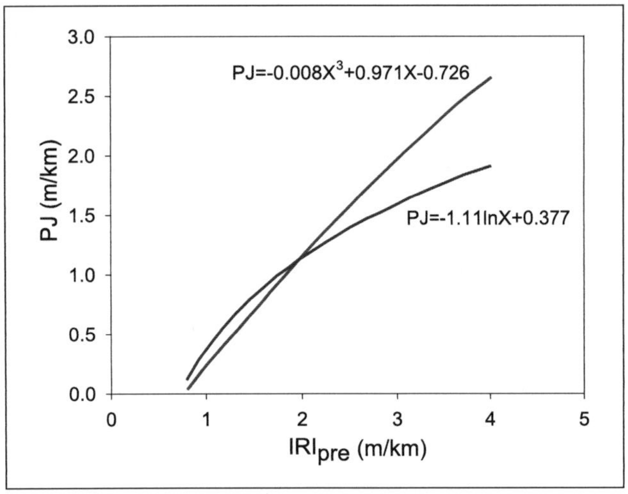

Figure 3. Relationship between PJ and IRI_{pre} after thin overlay.

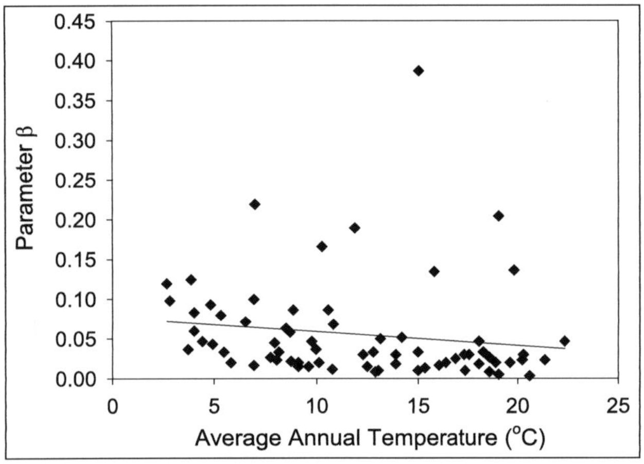

Figure 4. Relationship between parameter β and average annual temperature.

the overlay thickness into four categories: 1)2–2.5 cm; 2) 2.5–3 cm; 3) 3–3.5 cm; 4) 3.5–4 cm, as shown in Table 3. It suggested that the variation of β increases with the overlay thickness, and thicker overlay results in parameter β being more widely distributed.

Deflection

The initial deflection l_0 before thin overlay was used to characterize the pavement structural strength in this analysis. This is because deflection gives the comprehensive influence of pavement material, structure, base thickness, modulus of subgrade and other factors, and won't be affected by preventive maintenance.

The initial pavement deflection is between 0–80 mm, and β increases with the initial deflection. In this paper, the initial deflection was divided into five categories, as shown in Table 4. It illustrates that the sample mean and the maximum β increases with the deflection in general, which suggested a positive correlation between β and the initial deflection again. Also, this indicates that β didn't follow the rule mentioned above when deflection higher than 50, which may be due to the small sample space.

Table 2. Statistics Analysis of β for Different IRI_{pre} Levels before Thin Overlay

IRI_{pre}	No.	Average	Standard deviation	Min	Max
0.5–1	13	0.065	0.068	0.008	0.220
1–1.2	17	0.042	0.028	0.004	0.100
1.2–1.5	17	0.064	0.024	0.014	0.387
1.5–2	14	0.033	0.022	0.011	0.098
>2	11	0.078	0.051	0.012	0.189

Table 3. Statistics Analysis Results of Parameter β for Different Thin Overlay Thickness

Overlay thickness (cm)	No.	Average	Standard deviation	Min	Max
2–2.5	7	0.041	0.042	0.004	0.120
2.5–3	20	0.059	0.049	0.008	0.189
3–3.5	19	0.051	0.051	0.009	0.221
3.5–4	12	0.077	0.101	0.009	0.387

Table 4. Statistics Analysis Results of Parameter β for Different Deflection

Deflection: 0.01 mm	No	Average	Standard deviation	Min	Max
<20	15	0.021	0.041	0.004	0.043
20–30	18	0.048	0.117	0.006	0.167
30–40	15	0.063	0.047	0.008	0.189
40–50	10	0.092	0.040	0.015	0.387
>50	10	0.051	0.041	0.012	0.135

Table 5. Statistics Analysis Results of Parameter β under Different ESAL

ESAL	No.	Average	Standard deviation	Min	Max
<100	19	0.040	0.037	0.004	0.167
100–200	19	0.045	0.044	0.009	0.189
200–400	13	0.061	0.039	0.019	0.135
400–1000	14	0.051	0.054	0.019	0.221
>1000	7	0.120	0.128	0.006	0.387

Equivalent Single Axle Loads (ESAL)

Traffic load significantly impacts the pavement performance. In this article, ESAL was used to represent traffic load. The larger ESAL results in IRI decreasing rapidly and increasing β value. To find the relationship between β and ESAL, ESAL was divided into five categories based on the characteristics of the sample: 1) below 100; 2) 100–200; 3) 200–400; 4) 400–1000; 5) above 1000, as shown in Table 5. Here β

Table 6. Regression Model of Parameter β in IRI Prediction Model

Temperature	Variables	Coefficient	Standard error	P[|Z|>z]	R^2
High	Constant	-1.03×10^{-2}	0.0102	0.334	0.964
	ESAL	6.23×10^{-5}	0.0004	0.000	
	Deflection	6.01×10^{-4}	0.0003	0.030	
Medium	Constant	1.07×10^{-3}	0.0240	0.965	0.254
	ESAL	2.8×10^{-5}	0.0001	0.007	
	Deflection	1.16×10^{-3}	0.0009	0.083	
Low	Constant	-9.00×10^{-2}	0.0370	0.030	0.669
	ESAL	1.19×10^{-4}	0.0004	0.012	
	Deflection	2.16×10^{-3}	0.0009	0.007	
	Average annual precipitation	1.20×10^{-3}	0.0030	0.008	

changes with similar trend to the qualitative analysis, which increases with ESAL increases.

Parameter β Calculation Formula

As mentioned in the discussion of the influence factors, different temperature regions have different IRI deterioration mechanisms. The temperature range was divided into three intervals in this work: 1) 18°C above; 2) 10–18°C; 3) 10°C or less. This paper provides parameter β calculation formula for different temperature regions.

Linear Regression Model

Ordinary least square method was applied to calculate the coefficients of linear regression model for parameter β. Significant variables were selected by stepwise regression method and the statistical analysis result is shown in Table 6.

If the value of significance (P[|Z|>z]) is less than 0.05, the influence factors are considered as the significant variables. It should be noted that the significance of deflection in medium temperature region is 0.083 and that also could be accepted when a higher significance level 0.1 is considered.

R^2 values of the regression equation for predicting β were 0.964, 0.669 and 0.254 in high, low and medium temperature regions, respectively. Thus, linear regression model can be used to predict β in high and low temperature regions, but this is difficult in the medium temperature area, so a nonlinear regression model might be needed.

Nonlinear Regression Model

This work compared the polynomial model and exponential regression model, and found that the polynomial model gives higher accuracy. However, the polynomial model has many regression parameters and complex form, and the exponential model gives reasonable accuracy as well. Thus, this work chose the exponential model during analysis with simplified forms, and the regression results are shown below. The

Table 7. Coefficient β in Regression Model

Region	a_0	a_1	a_2	a_3	R^2
High	1.34×10^{-4}	0.871	0.1	-	0.905
Medium	6.55×10^{-6}	0.575	1.58	-	0.743
Low	0.046	0.333	0.968	0.585	0.788

exponential regression model of β in the high and medium temperature region is shown in Equation 3 as

$$\beta=a_0 \mathrm{ESAL}^{a_1} l_0^{a_2} \tag{3}$$

The exponential regression model of β in the low temperature region is shown in Equation 4 as

$$\beta=a_0 \mathrm{ESAL}^{a_1} l_0^{a_2} q^{a_3} \tag{4}$$

where, β = the calculated parameter, l_0 = deflection before overlay, q = average annual precipitation, and a_0, a_1, a_2, a_3 = regression coefficients.

The regression results of β in high, medium and low temperature regions are shown in Table 7.

The Determination of Shape Factor β

The linear model is more suitable for high temperature regions and the nonlinear model for medium and low temperature regions by comparing the R^2 values. Based on that, the equations to determine the shape factor β in IRI deterioration can be developed, as shown in Table 8.

Case Study

Test sections were randomly selected in high, medium and low temperature regions from the LTPP database to evaluate the post-thin-overlay IRI prediction model. The climate information and traffic data of test sections are shown in Table 9.

Table 8. Calculation Formula of Parameter β in IRI Prediction Model

Region	Temperature	Calculation formula	R^2
High	>18°C	$\beta=-1.03\times10^{-2}+6.23\times10^{-5}ESAL+6.01\times10^{-4}l_0$	0.964
Medium	10°C~18°C	$\beta=6.55\times10^{-6}ESAL^{0.575}l_0^{1.58}$	0.743
Low	<10°C	$\beta=4.6\times10^{-5}ESAL^{0.333}l_0^{0.968}q^{0.585}$	0.788

Table 9. Climate Information and Traffic Data of the Test Sections

Region	Temperature	ESAL	Deflection	Precipitation: cm
Arizona	19.8°C	2137	40.0	16.7
Missouri	12.3°C	60.1	41.4	123.9
Illinois	8.0°C	43.6	18.5	76.8

Table 10. Calculated Parameters α and β from IRI Prediction Models

Region	α	β	IRI prediction models
Arizona	0.53	0.146	$IRI = 0.53e^{0.146Age}$
Missouri	0.82	0.025	$IRI = 0.82e^{0.025Age}$
Illinois	0.97	0.034	$IRI = 0.97e^{0.034Age}$

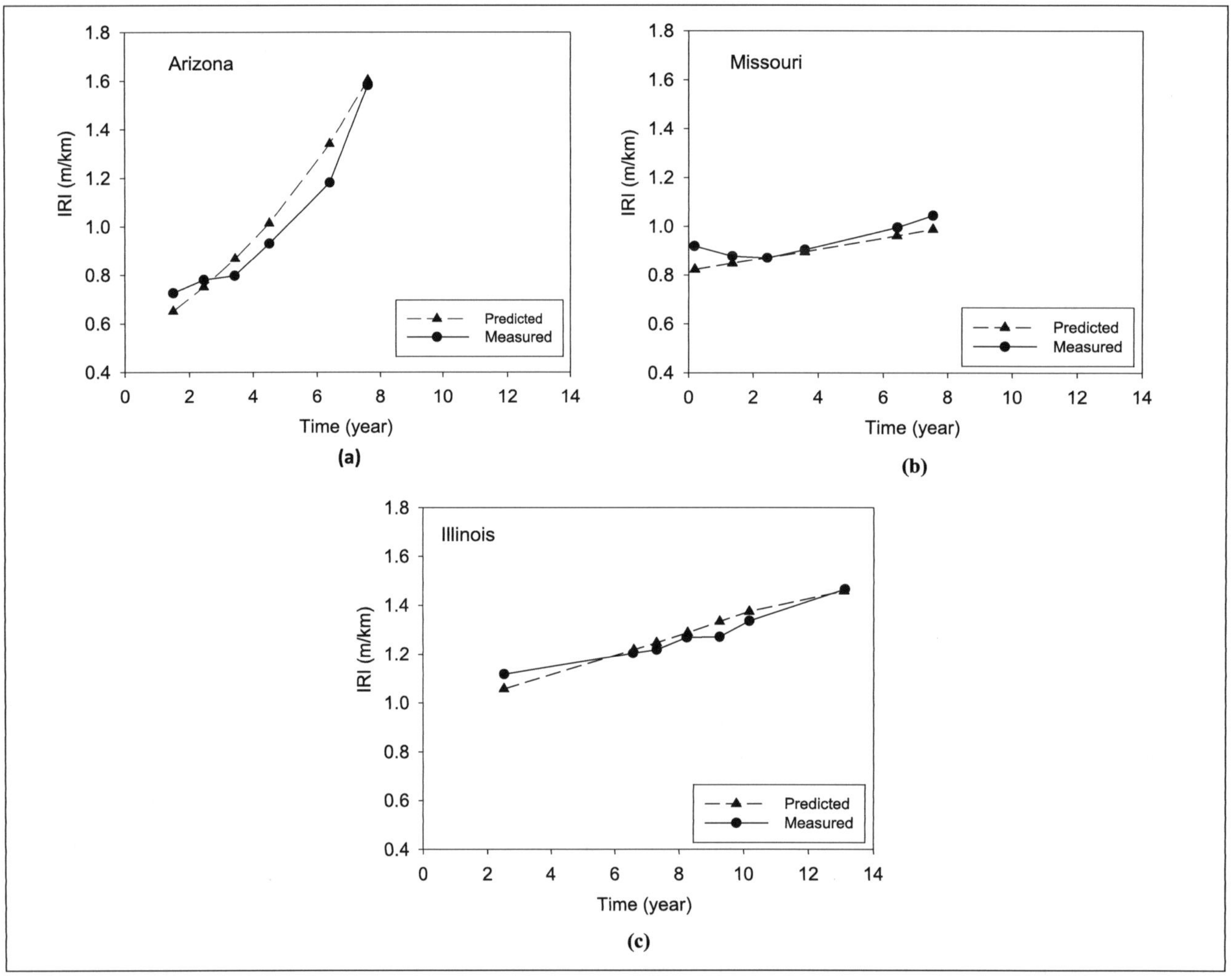

Figure 5. Comparison of measured IRI and predicted IRI for three locations: (*a*) Arizona; (*b*) Missouri; and (*c*) Illinois.

Based on results from the tested sections from the three locations, the developed IRI prediction models were used to calculate the parameters α and β as shown in Table 10.

Relationships between the measured IRI and predicted IRI are shown in Figure 5, and it can be noted that the predicted IRI and measured IRI exhibit similar trend. Both measured IRI and predicted IRI increase with time, and the predict model for the low temperature region fits the measured IRI best, which indicates that the IRI prediction model has acceptable accuracy and reliability.

Summary and Conclusions

In this study, an IRI regression model for asphalt pavement after thin overlay was developed. The model parameter estimation method was proposed based on analyzing the data from the LTPP database. The following conclusions can be drawn from this work:

(1) The two-parameter exponential model can match with the IRI curve very well after thin overlay.
(2) Different post-overlay IRI deterioration mechanisms in high-temperature regions and low-temperature regions were observed based on analyzing the influence factors. It was observed that deflection and ESAL are the significant factors in high and medium temperature regions. Whereas deflection, ESAL, and average annual precipitation are the significant factors in low-temperature regions.
(3) The shape factor β determination method was proposed based on statistical analysis of the influencing factors, multiple linear and nonlinear regressions for high, medium and low temperature regions. It was found that the linear model is more suitable for high temperature regions and the nonlinear model for medium or low temperature regions by comparing the R^2 values.
(4) Information from LTPP database was randomly selected from the high, medium and low temperature regions to compare the measured IRI and predicted IRI. The results show that the predicted IRI matches well with the measured IRI, which indicates the reliability of the IRI prediction model.

Acknowledgments

The authors would like to thank Shanxi Department of Transportation and Key Laboratory of Road and Traffic Engineering of the Ministry of Education at Tongji University (Project No. 1600219286) for supporting this research.

References

1. Smith, R. E., B. Mukherjee, M. Zulyaminayn, C. C. Pilson, T. Dossey, and B. F. McCullough. *Integration of Network- and Project-Level Performance Models for TxDOT PMIS*. Final Report. Texas Department of Transportation, Austin, 2001.
2. Rajagopal, A. S., and K. P. George. Pavement Maintenance Effectiveness. *Transportation Research Record: Journal of the Transportation Research Board*, 1991. 1276: 62–66.
3. Livneh, M. Deterioration Modeling for Un-laid and Overlaid Pavements. *Transportation Research Record: Journal of the Transportation Research Board*, 1996. 1524: 177–184.
4. Lamptey, G. *Optimal Scheduling of Pavement Preventive Maintenance using Life Cycle Cost Analysis*. M.S. thesis. Purdue University, 2004.
5. Irfan, M., M. B. Khurshid, S. Labi, and W. Flora. Evaluating the Cost-effectiveness of Flexible Rehabilitation Treatments using Different Performance Criteria. *Journal of Transportation Engineering*, Vol. 135, 2009, pp. 753–763.
6. Zimmerman, K. A., and D. G. Peshkin. Integrating preventive maintenance and pavement management practices. *Proc., 2003 Mid-Continent Transportation Research Symposium*, Center for Transportation Research and Education, Iowa State University, Ames, 2003.
7. Lu, P., and D. Tolliver. Pavement Treatment Short-Term Effectiveness in IRI Change Using Long-Term Pavement Program Data. *Journal of Transportation Engineering*, Vol. 138, No. 11, 2012, pp. 1297–1302.
8. Labi, S., and K. C. Sinha. *The Effectiveness of Maintenance and Its Impact on Capital Expenditures*. Report FHWA/IN/JTRP-(2002)-27. Purdue University, West Lafayette, 2003.
9. Kerali, H. R., A. J. Lawrance, and K. R. Awad. Data Analysis Procedures for Long-Term Pavement Performance Prediction. *Transportation Research Record: Journal of the Transportation Research Board*, 1996. 1524: 152–159.
10. Gulen, S., K. Zhu, J. Weaver, J. Shan, and W. Flora. *Development of Improved Pavement Performance Model*. Indiana Department of Transportation, Indianapolis, 2001.
11. Park, K., N. E. Thomas, and K. W. Lee. Applicability of the International Roughness Index as a Predictor of Asphalt Pavement Condition. *Journal of Transportation Engineering*, Vol. 133, No. 12, 2007, pp. 706–709.
12. Chen, C. T., C. T. Hung, C. C. Chou, Z. Chiang, and J. Lin. The Predicted Model of International Roughness Index for Drainage Asphalt Pavement. In *Advanced Intelligent Computing Theories and Applications, With Aspects of Theoretical and Methodological Issues*. Springer, Berlin, Heidelberg, 2008, pp. 937–945.
13. Amador-Jimenez, L. E., and D. Mrawira. Capturing Variability in Pavement Performance Models from Sufficient Time-Series Predictors: A Case Study of the New Brunswick Road Network. *Canadian Journal of Civil Engineering*, Vol. 38, No. 2, 2011, pp. 210–220.
14. Irfan, M. *A Framework for Developing Optimal Pavement Life-Cycle Activity Profiles*. Dissertation, Purdue University. 2010.
15. Hajj, E. Y., L. Loria, P. E. Sebaaly, C. M. Borroel, and P. Leiva. Optimum Time for Application of Slurry Seal to Asphalt Concrete Pavements. *Transportation Research Record: Journal of the Transportation Research Board*, 2011. 2235: 66–81.
16. Kamram, M., V. Bezhad, and C. K. James. Pavement Management System to Maximize Pavement Investment and Minimize Cost. *Transportation Research Record: Journal of the Transportation Research Board*, 1990. 1272: 65–73.
17. Hall, K. T., A. L. Simpson, and C. E. Correa. *LTPP Data Analysis: Effectiveness of Maintenance and Rehabilitation Options*. National Cooperative Highway Research Program, 2002.
18. Haider, S. W., and M. B. Dwaikat. Estimating Optimum Timing for Preventive Maintenance Treatment to Mitigate Pavement Roughness. *Transportation Research Record: Journal of the Transportation Research Board*, 2011. 2235: 43–53.

The Standing Committee on Pavement Management Systems (AFD10) peer-reviewed this paper (18-00933).

Article

TRR
JOURNAL OF THE TRANSPORTATION RESEARCH BOARD

Transportation Research Record
2018, Vol. 2672(40) 14–21

DOI: 10.1177/0361198118758693
journals.sagepub.com/home/trr
SAGE

Estimation of Gravel Roads Ride Quality Through an Android-Based Smartphone

Waleed Aleadelat[1], Cameron H. G. Wright[2], and Khaled Ksaibati[1]

Abstract

This study demonstrated the ability of smartphone sensors in evaluating gravel roads conditions. Seventy gravel roads with various conditions, surface materials, and geometric features were included in this study. The analysis was based on signal demodulation and wavelet transformation to reduce the effect of many external factors (i.e., speed dependency, engine vibrations, and suspension system) that may affect the obtained measurements. It was found that the acquired signals from a smartphone accelerometer can reflect the actual conditions of a gravel road. In addition, the location and the severity of surface deteriorations such as potholes could be identified. A regression model (R^2 = 0.78) based on the acquired signals from smartphones was developed to predict the overall rating of the gravel road condition according to the Riding Quality Rating Guide (RQRG) system. An initial validation analysis, conducted on 35 new gravel roads, showed that this model was able to return reasonable ratings. Also, the statistical analysis showed that any difference between the predicted and the actual ratings of <1.3 was not significant. The proposed methodology can be considered as a baseline for building a low cost crowdsourcing platform that helps local agencies in managing their inventory of gravel roads.

The local governments in the state of Wyoming own a large inventory of gravel roads which formulates 90% of the local roads network in the state (*1*). These gravel roads carry frequent additional heavy traffic because of the different mineral and drilling activities in the state. This heavy traffic can greatly affect the structural capacity of gravel roads and change their deterioration patterns enormously. Therefore, the Wyoming Technology Transfer Center (WYT^2/LTAP) is currently in the process of developing a new gravel roads management system (GRMS). Such a GRMS is intended to provide feasible practices to help local agencies in dealing with the different challenges associated with maintaining gravel roads.

Gravel roads are usually managed by small local agencies with limited resources. Normally, the available local expertise is more proficient in performing maintenance and rehabilitation works than managing them. Therefore, the developed GRMS must be tailored to suit the needs of these agencies.

Currently, the WYT^2/LTAP utilizes inexpensive, less labor intensive windshield surveys to evaluate gravel road conditions. Most of these surveys are modifications to the Pavement Surface Evaluation and Rating (PASER) guide developed by the Wisconsin Transportation Information Center (*2, 3*). One of these modifications is the Ride Quality Rating Guide (RQRG) which rates the roads on a scale from 1 (failed) to 10 (excellent) (*2*). The RQRG reflects the perceptions of road users with regards to the driving quality of the gravel road. For any gravel road, the overall rating given by the RQRG is considered as the most important factor in identifying maintenance and rehabilitation needs for that road. Representing the overall condition of a gravel road by a single figure can highly reduce the data input of any GRMS to be in line with the limited resources of the different rural counties in the Midwest region (*4*).

Collecting feasible gravel road condition data is considered one of the main challenges of building any GRMS (*4*). Gravel roads are very dynamic and their conditions vary quickly based on different traffic and weather conditions (*2, 4–7*). This leaves any plans for scheduled condition monitoring without any significant value (*7*). Therefore, the WYT^2/LTAP is considering many approaches to facilitate and automate the collection of gravel road condition data. One of these approaches is considering the utilization of smartphone sensors to estimate the RQRG ratings of gravel roads. The RQRG is similar to the Present Serviceability Index (PSI) of

[1]Department of Civil and Architectural Engineering, University of Wyoming, Laramie, WY
[2]Department of Electrical and Computer Engineering, University of Wyoming, Laramie, WY

Corresponding Author:
Address correspondence to Khaled Ksaibati: Khaled@uwyo.edu

the paved roads in evaluating the ride quality based on the public perceptions. Therefore, it is expected that the roughness of the gravel roads will be the main factor in defining the final ratings (*8*). Consequently, there should be merit in utilizing the accelerometer of a smartphone to predict the RQRG ratings. The proposed approach is intended to establish a baseline for a crowdsourcing platform where the residents of the rural areas can provide local agencies with gravel road condition data. This way local agencies can control the highly changeable nature of these roads by obtaining an automated frequent data feed from the actual road users. This low cost data feed can be used to establish an alarming system, in which the value of triggered maintenance scheduling can be appreciated. Also, this leaves the procurement process of condition data in the hands of the locals, which could enhance the quality of the data. Locals will be concerned about the quality of their everyday commuting routes more than anyone else.

Background

Gravel road assessment methods are mostly manual and dependent on visual surveys or actual measurements of road surface conditions (*2*). The U.S. Army Corps of Engineers assessment system (USACE) is considered to be one of the earliest efforts to develop a measurable consistent index for gravel roads (*2, 3, 9*). This index is known as the Unsurfaced Road Condition Index (URCI) and depends on the actual measurement of the level and extent of each distress related to gravel roads. Then, using a deduct value for each distress type (e.g., potholes and corrugations), an overall URCI, on a scale from 1 to 100, can be calculated (*9*). In another study conducted by Chamorro et al. in Chile, a new Gravel Roads Condition Index (UPCI) was developed (*10*). Through linear regression equations a correlation was established between objective distresses measures (e.g., rutting depth) and a given subjective rating (UPCI). The developed equations gave the UPCI ratings the required level of objectivity, which may improve the quality of data and reduce required survey times (*10*). Nonetheless, any trial of developing objective based indices is still impractical for local agencies. Estimating these types of indices requires substantial resources to objectively measure each type of distress, especially when dealing with large networks. In addition, manually measuring the level and extent of each distress still implies some degree of subjectivity. Therefore, these indices are more practical for research purposes than road management (*2*). This encourages local agencies to deal with more practical assessment methods, such as visual surveys.

Currently, there are many available gravel road visual assessment systems such as, the South African Council for Scientific and Industrial Research (CSIR) system, the Federal Highway Administration (FHWA) Central Federal Lands Highway Division method, and the Wisconsin PASER system. The PASER rating system is considered to be the most popular rating system used for gravel road evaluation in the U.S (*2*). The PASER system rates the overall gravel road condition on a scale from 1 (failed) to 5 (excellent) by incorporating distresses related to the following aspects of the road:

- Crown
- Drainage
- Gravel layer
- Surface deformation:
  - Washboard
  - Potholes
  - Ruts
- Surface defects:
  - Dust
  - Loose aggregate

In the PASER system, the gravel road is evaluated from a decision-maker view point (*2, 11*). This explains the inclusion of gravel properties and drainage conditions within the rating system. However, the PASER's short scale limits the rater's ability to make consistent judgments. In such short scales the rater tends to make judgements closer to the end points (error of leniency). In other occasions, raters avoid making extreme judgements, or what is known as the central tendency error (*12*). Therefore, the WYT^2/LTAP modified the current PASER system to a different rating system (RQRG) that uses an expanded scale from 1 to 10. This expanded rating scale is intended to reduce the possibility of errors within the rating process. Moreover, the modified scale reflects the perceptions of road users regarding their comfortability while driving over a specific road segment (*2*). Therefore, RQRG is affected more by surface deformation modes such as potholes, washboards, and rutting. These types of deformations are considered to be the main failure mode for gravel roads (*5*). Thus, the RQRG system has an advantage over the PASER system by its ability to alleviate possible rating errors and represent the actual road conditions at the same time, which leads to a better decision making process (*2*).

A few studies have tried to introduce a sort of platform for automating the gravel road assessment process (*13–15*). For example, the Forest Engineering Research Institute of Canada (FERIC) developed the Opti-Grade road management system. This system depends on accelerometers and a global positioning system (GPS) to detect and calculate the roughness of the Canadian forest roads. It was found that this system was helpful in improving the effectiveness of maintenance works and reducing their costs (*13*). In another indirect approach, Zhang and Elaksher used an unmanned aerial vehicle (UAV) based imaging system to identify the different surface distresses present in gravel roads. Using aerial imagery, a perceived three dimensional (3D) surface of gravel roads was created. Then, using image processing algorithms, they were able to identify the extent and severity of surface

Figure 1. Test vehicle and smartphone orientation.

distresses. The derived measurements from the perceived 3D road surface were in good agreement with the actual measurements (*14*). In 2015, Alhasan et al. used a terrestrial laser scanner to quantify the roughness of gravel roads (*15*). The acquired data points using the laser scanner were used to construct 3D representation maps for the road surface. Then, using the quarter car model, Fast Fourier transform (FFT), and statistical analysis, they were able to estimate the international roughness index (IRI) values, locate corrugations within road segments, and infer some information related to surface material characteristics (e.g., gradation). Regarding the use of smartphones in evaluating roads conditions, most of the previous work was related to paved roads, in particular estimating the IRI and locating anomalies within the pavement surface (*16–23*).

In conclusion, previous efforts made to automate the data collection for gravel roads are still impractical for application and require an analysis level that may supersede the capabilities of small local agencies. In addition, these methods are not able to control the highly changeable conditions of gravel roads. Therefore, an investigation is required to evaluate the ability of smartphones in predicting gravel road conditions. The proposed approach is anticipated, in addition to reducing the cost, to deal with the dynamic conditions of gravel roads and enhance the quality of the collected data through the automation process.

Methodology

A smartphone internal accelerometer was used to capture the vertical vibrations while driving over gravel roads. The acquired measurements using the smartphone accelerometer formulate different signals that can be considered as a reflection of the actual gravel road conditions. These acquired signals were then analyzed using the Frequency Demodulation and the Discrete Wavelet Transform (DWT). Later on, these analyzed signals were correlated with the RQRG ratings. The following subsections describe this process.

Experimental Design

A Samsung Galaxy S III smartphone was selected to collect the vertical acceleration data. The smartphone was glued onto the testing vehicle's dashboard. This location was selected as it facilitates the monitoring process of the smartphone. In addition, it would help to capture the relative vibrations generated directly on both wheel paths. A 2010 Ford (F-150) pickup truck was selected as the testing vehicle for this study. This truck represents the majority of vehicles that are used by locals in Wyoming. Figure 1 shows the test vehicle and the smartphone arrangement that were used in this study. Seventy gravel roads were selected from Laramie County in Wyoming. These roads were randomly selected as part of the new GRMS building process. The selected roads have different conditions and surface materials with various geometric features. Table 1 shows a summary of statistics for the gravel roads that were included in this study.

Smartphone Data Processing

As the smartphone was fixed horizontally in the X and Y directions, the accelerometer readings obtained along the Z axis were the focus of this study. On gravel roads, it is very hard for a driver to maintain a constant speed. Therefore, the acquired signals are speed dependent and mixed with different signals because of the vehicle suspension system, engine rotation, tire movements, road slope, and many other external factors (*22*). To overcome this speed dependency and to attenuate the effect of all other factors related to the vehicle and road geometry, it was assumed that the

Table 1. Summary of Gravel Roads Included in the Study

Number of gravel roads	Parameter	RQRG	Length (miles)	Speed (mph)
70	Mean	6.40	1.66	31.91
	Median	7.00	1.05	30.00
	Standard deviation	1.83	1.46	9.25
	Max	9.00	7.50	55.00
	Min	1.00	0.10	08.00

acquired accelerometer signal acts as a carrier signal for broadcasting the radio waves, or what is known as the frequency modulated (FM) signals. In radio technology, a carrier signal is a main high frequency signal that is modulated (i.e., modified) with different low frequency input signals. In gravel roads, a carrier signal with high frequency represents the actual signal resulting from the road surface and the low frequency input signals represent the signals resulting from all other factors (i.e., speed dependency and engine vibrations). Therefore, using frequency demodulation helps in reducing the effect of low frequency components that may be present in the acquired signals. This method was used by Seraj et al. in 2014 to reduce the effect of external factors while they were trying to detect road anomalies using smartphones (*22*).

After demodulating the acquired signals by smartphones, DWT was used to analyze the demodulated signals. The DWT is a comprehensive tool for analyzing random signals in both frequency and time domains (*24–26*). In the DWT the acquired signal is decomposed to a series of wavelet functions. This way the different frequencies that are present in the acquired signals can be discriminated. In addition, the power of the signal can be calculated along the time domain, which allows the identification of any abnormal events (e.g., potholes). The DWT is more suitable for analyzing non-stationary signals, which is the case when dealing with accelerometer data acquired from a road surface. Therefore, DWT can supersede the power spectrum density (PSD) analysis, which is more suitable for stationary signals, and for discriminating the different deterioration modes within the road surface (*22, 24, 25*). The DWT is defined according to the following equation (*25, 26*):

$$W_{\psi}(s,\tau) = \int_{-\infty}^{\infty} X(t)\psi_{s,\tau}(t)dt \qquad (1)$$

where:

$W_{\psi}(s,\tau)$ represents wavelet transform coefficients. At any specific time, a higher obtained coefficient means more energy present in the signal at that time.

s and τ are continuous variables related to the scale and translation parameters, respectively,

$\psi_{s,\tau}(t)$ denotes the wavelet analyzing function, in the DWT, this wavelet analyzing function is given by Equation 2, where $s = 2^j$, $\tau = k*2^j$, j is an integer that designates the level of decomposition, and k is an integer from minus infinity to infinity:

$$\psi_{s,\tau}(t) = \frac{1}{\sqrt{s}}\psi(\frac{t-\tau}{s}) \qquad (2)$$

The level of decomposition (*j*) can be used to define the required level of analysis. In this study, the DWT decomposition was performed at five levels ($j = 5$). Using these multiple levels will allow for analyzing signals into finer bands. This allows the discrimination of the various abnormalities within the raw signals. Selecting a higher decomposition level will lead to a decomposition with a higher resolution. For the wavelet function, there are many available wavelet families such as Haar, Daubecheies, and Coiflets. However, there is no single known procedure for deciding which family is suitable for certain types of signals (*25, 26*). Therefore, a wavelet analyzing function from the Symlets family, which is Sym6, was selected based on trial and error to perform the analysis. Finally, the analyzed signals were examined for the possible identification of any distresses that may be present in the gravel road surface. Also, a correlation between the RQRG ratings and the estimated DWT coefficients was established.

Data Collection

The smartphone accelerometer data was collected using an Android application called "AndroSensor." This application is available at the Google Play store as a free download. In addition to the ability of AndroSensor to capture the accelerometer data, it can be used to record data from most of the available sensors in any Android smartphone. The used sampling frequency for capturing accelerometer data was 200 Hz (200 samples/s).

For the RQRG ratings, one team of two persons (driver and rater) drove all over the selected gravel roads and rated them according to the RQRG system. The rating process was conducted as part of the new GRMS data collection process. The beginning and the ending GPS coordinates for each gravel road were provided by the Wyoming Department of Transportation (WYDOT). By uploading these coordinates into Microsoft Streets and Trips software, the rating team was able to identify the exact location of each gravel road considered in this study. As gravel roads have a dynamic nature, the driving speed was variable from road to road and even within any road itself. However, raters did not exceed the posted speed limits. The accelerometer data and the RQRG ratings were both collected in the summer of 2017.

Data Analysis

Accelerometer data was extracted from the smartphone, uploaded to a computer in Microsoft Excel format (*.CSV), and imported into MATLAB for further analysis. The analysis was started by performing the frequency demodulation on the obtained signals to minimize the effect of low frequency inputs. By trial and error, it was determined that a frequency of 98 Hz can represent the potential frequency for the carrier signal. Figure 2 shows a plot for a gravel road segment with an overall rating of 8 before and after performing the frequency demodulation.

It can be noted that demodulating the signal can highly isolate the carrier signal from the unwanted components. Also, the demodulated signal can clearly show any possible abnormalities within the road surface (vertical peaks). The gaps between the vertical peaks represent parts of the road surface that can be considered smooth compared with the rest of the road. Therefore, to get a better understanding of the different deterioration patterns along the road surface, DWT analysis was performed. Figure 3 shows the 3rd level coefficients of the DWT for the same gravel road shown in Figure 2. It can be noted that the DWT can provide a better image of the deterioration patterns along the road surface. For example, between mile posts of 0.1 and 0.2, high frequent peaks can be considered as an indication for washboards. The same behavior can be noted between the mile posts of 0.6 and 0.7, but with less severity. The less frequent peaks that are close to the mile posts of 0.6, 0.8, and 1 can be considered as a sign of potholes. The magnitude of the DWT coefficients can be considered as a way to identify the severity of these deterioration patterns. In other words, a higher coefficient means a higher deterioration level. Therefore, most of the previously identified deterioration patterns are of less severity (below the red line). This threshold of 2.5 was determined after comparing signals with different overall ratings between 1 and 9. The 3rd level coefficients were selected as it was found to best represent the actual road conditions and gave the best results compared to the other levels of the decomposition.

Figure 4 shows the 3rd level coefficients of the DWT for a gravel road with an overall rating of 4. As seen from this figure, the density and magnitude of the 3rd level coefficients are very high compared to Figure 3. Also, by comparing both figures, it can be noted that the coefficients with an absolute value >2.5 are the dominant indicators in identifying the overall rating level (above the red line). From the same figure, potholes can be clearly located around the mile posts of 0.4, 1, 1.5, 2.9, 3.5, 4 and 4.2. In addition, severe washboards can be clearly identified between mile posts of 2.5 and 3. The same washboards with less extent can be identified between the mile posts of 0.2 and 0.5. However, further investigations are required to associate between the severity

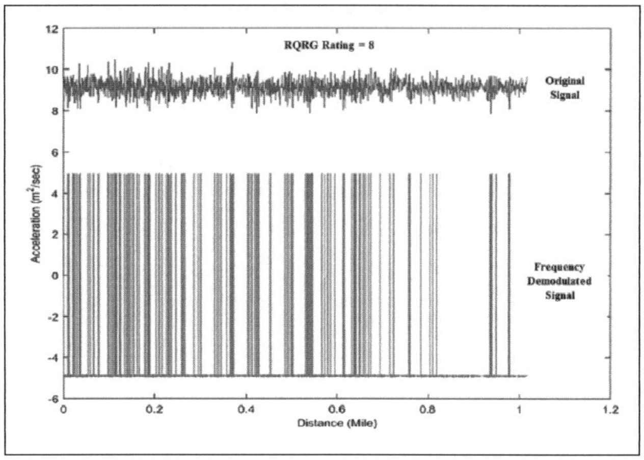

Figure 2. Original versus demodulated signal.

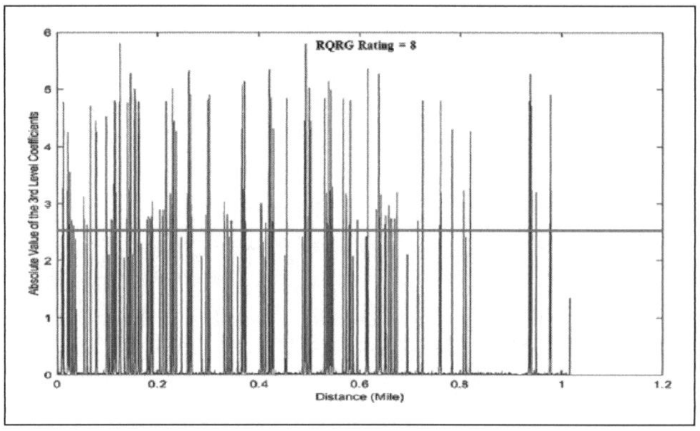

Figure 3. 3rd level coefficients of DWT (RQRG rating of 8).

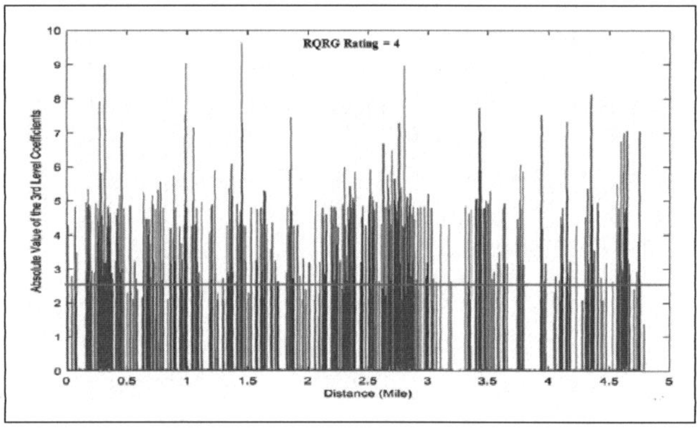

Figure 4. 3rd level coefficients of DWT (RQRG rating of 4).

of these deterioration patterns and the measured signals. This could include searching for explanatory variables measured using a smartphone that can correlate well with a subjective rating system specific to potholes or washboards, such as the Gravel Roads Rating Standards (GRRS) or the Wisconsin PASER system.

To evaluate the ability of the smartphone in predicting RQRG ratings, different attempts led to finding an explanatory variable measured using the smartphone which can correlate well with the RQRG ratings. This variable is the sum of the local maxima points (SLM) within the absolute value of the 3rd level coefficients. Figure 5 shows the established correlation between the RQRG ratings and the SLM points. It can be noted that there is an exponential relationship between the given ratings and the SLM points.

Equation 3 shows the RQRG rating prediction model using the smartphone. This model reveals the worth of using the frequency demodulation during the analysis by having this significant relationship. The found relationship can explain 78% of the variations in the predicted ratings, regardless of the different driving speeds. In addition, the obtained results reflect the actual conditions of the road in a rational way as the ratings decrease with the increase of the SLM. These found results are very promising, especially when dealing with such dynamic roads where the surface conditions change very often.

$$RQRG = 7.5e^{-0.003*SLM}; R^2 = 0.78 \quad (3)$$

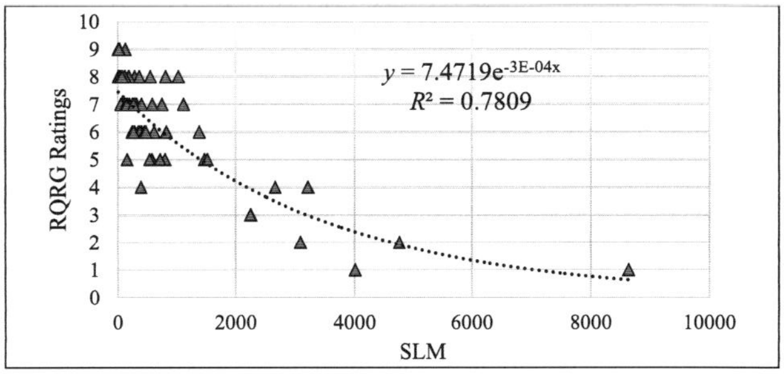

Figure 5. RQRG ratings versus the sum of the local maxima points.

Table 2. Summary of Validation Gravel Roads

Number of gravel roads	Parameter	RQRG	Length (miles)	Speed (mph)
35	Mean	6.74	2.62	38.29
	Median	7.00	2.00	35.00
	Standard deviation	1.59	2.33	9.63
	Max	9.00	10.66	60.00
	Min	1.00	0.50	25.00

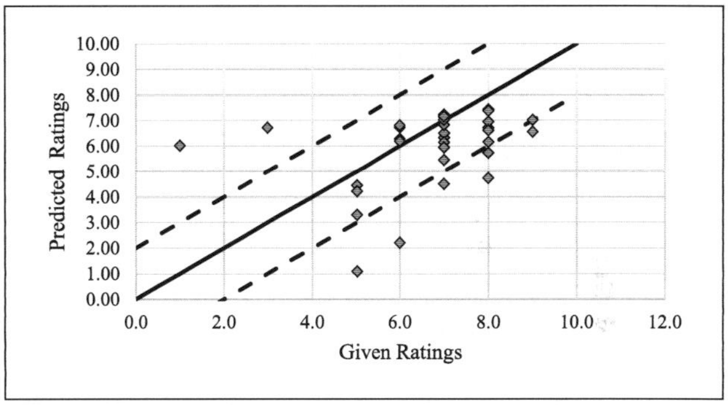

Figure 6. Predicted versus actual ratings.

Validation of the RQRG Model

To validate the ability of the smartphone in predicting the overall ratings of gravel roads, 35 new gravel roads were selected to perform the experiment again using the same combination of the smartphone, testing vehicle, and the rating team. Table 2 shows a summary of statistics for the selected gravel roads for the validation process.

Figure 6 shows a plot for the predicted ratings using Equation 3 and the actual given ratings. It can be noted that the predicted and the given ratings are in good agreement. The majority of the predicted ratings are within the two units offset band (dashed lines). In fact, only eight gravel roads fall outside the offset band. This means Equation 3 was able to provide reasonable predictions for 77% of the given ratings. Nonetheless, the paired t-test results showed that a difference of <1.3 units between the predicted and the actual ratings was not significant at a 95% confidence level [t-stat(34) = −1.94; p-value = .06].

The obtained results can be improved by using more than 1 rating team to perform the rating process. This way more agreement on the actual ride quality can be achieved. In addition, dividing gravel roads into short sub-segments (i.e., 0.1 mile) for the rating purposes can highly enhance the harmony between the subjective ratings and the smartphone measurement. Suggesting that the overall gravel road condition can be represented by the average of the sub-segment ratings.

Conclusions

This study validated the ability of using smartphones accelerometers in evaluating gravel road conditions. The data was collected on 105 gravel roads with various conditions and geometric features that may exist under any GRMS. One rating team drove over the test roads and rated them according to the RQRG system. Using MATLAB, the acquired signals from a smartphone accelerometer were demodulated to reduce the effect of possible external factors. Then, using DWT analysis, the behavior of the acquired signals was examined. The analyzed signals were able to specify the main patterns of deterioration such as washboards and potholes. Also, a significant model was developed to predict the overall RQRG ratings of gravel roads. Later on, a validity analysis was conducted to verify the reliability of the RQRG model. The validation results showed that the new model was able to produce ratings in good agreement with the actual ones. Also, any difference of <1.3 units between the predicted and the actual ratings was not statistically significant. The main conclusions drawn from this study are as follows:

1. The acquired signals from the smartphone accelerometer can be used as a reflection of the actual conditions of gravel roads. Therefore, it can be used to identify the locations and the severity of deterioration patterns along the road surface.
2. Frequency demodulating the acquired signals from the smartphone accelerometer can greatly help in reducing the effects of external factors that may affect the response of the accelerometer.
3. The DWT is a very valuable tool for discriminating the different deterioration modes within the gravel road surface. It is more suitable for non-stationary signals, which is the case when dealing with gravel roads.
4. The SLM can be used as a good explanatory variable to predict the RQRG ratings.
5. At this stage, and using the same combinations of vehicle and smartphone, Equation 3 can be used to reasonably predict the RQRG ratings of gravel roads.

The RQRG rating represents a figure that can describe the general condition of any gravel road. Therefore, predicting the RQRG rating using a smartphone can highly support any GRMS within small local agencies. Also, identifying the location and the severity of the various deterioration patterns within the road surface can be vital to any triggered maintenance scheduling plans. In reality, having a frequent condition data feed can play a significant role in controlling the highly changeable conditions of gravel roads. The proposed approach can be considered as a baseline for a crowdsourcing platform where the residents take the responsibility for providing gravel road condition data to the local agencies. It provides an inexpensive solution toward automating gravel road data collection. Future research will focus on addressing the effect of the different external variables on this approach. For example, the actual profile of a non-treated gravel road can be determined manually using a dipstick profiler to establish a reference frame for calibration (*27*). This measured profile can then be analyzed to identify the different vehicle characteristics that may affect the acquired signals from a smartphone. For treated gravel roads, a profilometer may be used directly to measure the actual road profile. In addition, different types of smartphones and fixing locations (e.g., drivers' pocket) will be tested using different vehicles to simulate real time circumstances.

Acknowledgments

The authors would like to thank Mountain Plains Consortium (MPC), Wyoming Counties, and WYDOT for supporting this research study.

References

1. Huntington, G., and K. Ksaibati. Annualized Road Works Cost Estimates for Gravel Roads. *Journal of Transportation Engineering*, Vol. 135, No. 10, 2009, pp. 702–710.
2. Huntington, G., and K. Ksaibati. Visual Assessment System for Rating Unsealed Roads. *Transportation Research Record: Journal of the Transportation Research Board*, 2015. 2474: 116–122.
3. Walker, D. *Gravel PASER Manual: Pavement Surface Evaluation and Rating*. Wisconsin Transportation Information Center, Madison, 1989.
4. Huntington, G., and K. Ksaibati. *Gravel Roads Management*. Report prepared for the Wyoming Department of Transportation and the Mountain-Plains Consortium, Wyoming Technology Transfer Center (T2/LTAP). Laramie, 2011.
5. Huntington, G., and K. Ksaibati. Gravel Roads Surface Performance Modeling. *Transportation Research Record: Journal of the Transportation Research Board*, 2007. 2016: 56–64.
6. Huntington, G., and K. Ksaibati. Improvement Recommendations for Unsealed Gravel Roads. *Transportation Research Record: Journal of the Transportation Research Board*, 2011. 2205: 165–172.
7. Huntington, G., and K. Ksaibati. Implementation Guide for the Management of Unsealed Gravel Roads. *Transportation Research Record: Journal of the Transportation Research Board*, 2011. 2205: 189–197.
8. Carey, W. N., Jr., , and P. E. Irick. The Pavement-Serviceability Concept. *Highway Research Board Bulletin 250*, HRB, National Research Council, Washington, D.C., 1960, pp. 40–58.
9. Eaton, R. A., and R. E. Beaucham. *Unsurfaced Road Maintenance Management*. CRREL Special Report 92-26. Cold Regions Research and Engineering Laboratory, U.S. Army Corps of Engineers, Hanover, N.H., 1992.
10. Chamorro, A., H. de Solminihac, M. Salgado, and E. Barrera. Development and Validation of a Methodology to Evaluate Gravel Road Condition by Using Objective Distress Measures. *Transportation Research Record: Journal of the Transportation Research Board*, 2009. 2101: 3–9.
11. Huntington, G., A. Pearce, N. Stroud, J. Jones, and K. Ksaibati. *Mitigating Impacts of Oil and Gas Traffic on Southeastern Wyoming County Roads*. University of Wyoming, Laramie, 2013. wwweng.uwyo.edu/wyt2/ [Special Projects]. Accessed June 20, 2017.
12. Nair, S. K., and W. R. Hudson. Serviceability Prediction from User-Based Evaluations of Pavement Ride Quality. *Transportation Research Record : Journal of the Transportation Research Board,* 1986. 1084: 66–75.
13. Brown, M., S. Mercier, and Y. Provencher. Road Maintenance with Opti-Grade®: Maintaining Road Networks to Achieve the Best Value. *Transportation Research Record: Journal of the Transportation Research Board*, 2003. 1819: 282–286.
14. Zhang, C., and A. Elaksher. An Unmanned Aerial Vehicle-Based Imaging System for 3D Measurement of Gravel Road

Surface Distresses. *Computer-Aided Civil and Infrastructure Engineering*, Vol. 27, No. 2, 2012, pp. 118–129.

15. Alhasan, A., D. J. White, and K. D. Brabanter. Quantifying Roughness of Gravel Roads by Terrestrial Laser Scanning. *Transportation Research Record: Journal of the Transportation Research Board*, 2015. 2523: 105–114.

16. Strazdins, G., A. Mednis, G. Kanonirs, R. Zviedris, and L. Selavo. Towards Vehicular Sensor Networks with Android Smartphones for Road Surface Monitoring. *Proc., 2nd International Workshop on Networks of Cooperating Objects*, Chicago, 2011. www.dee.srv1.eu/refs/files/4/105.pdf. Accessed June 15, 2017.

17. Douangphachanh, V., and H. Oneyama. A Study on the Use of Smartphones for Road Roughness Condition Estimation. *Journal of the Eastern Asia Society for Transportation Studies*, Vol. 10, 2013, pp. 1551–1564.

18. Jiménez, L. A., and N. Matout. A Low *Cost Solution to Assess Road's Roughness Surface Condition for Pavement Management*. 93rd Annual Meeting of the Transportation Research Board, Washington, D.C., 2014.

19. Hanson, T., C. Cameron, and E. Hildebrand. Evaluation of Low-Cost Consumer-Level Mobile Phone Technology for Measuring International Roughness Index (IRI) Values. *Canadian Journal of Civil Engineering*, Vol. 41, No. 9, 2014, pp. 819–827.

20. Islam, S., W. G. Buttlar, R. G. Aldunate, and W. R. Vavrik. Measurement of Pavement Roughness Using Android-Based Smartphone Application. *Transportation Research Record: Journal of the Transportation Research Board*, 2014. 2457: 30–38.

21. Aleadelat, W., and K. Ksaibati. Estimation of Pavement Serviceability Index Through Android-Based Smartphone Application for Local Roads. *Transportation Research Record: Journal of the Transportation Research Board*, 2017. 2639: 129–135.

22. Seraj, F., B. J. van der Zwaag, A. Dilo, T. Luarasi, and P. Havinga. RoADS: A Road Pavement Monitoring System for Anomaly Detection Using Smart Phones. In *Big Data Analytics in the Social and Ubiquitous Context*, Springer, Cham, 2014, pp. 128–146.

23. Mohan, P., V. N. Padmanabhan, and R. Ramjee. Nericell: Rich Monitoring of Road and Traffic Conditions Using Mobile Smartphones. *Proc., 6th ACM Conference on Embedded Network Sensor Systems*, Raleigh, NC, USA, ACM, New York, 2008, pp. 323–336.

24. Hesami, R., and K. J. McManus. Signal Processing Approach to Road Roughness Analysis and Measurement. In *TENCON 2009-2009 IEEE Region 10 Conference, Singapore*, Singapore, IEEE, New York, 2009, pp. 1–6.

25. Burrus, C. S., R. A. Gopinath, and H. Guo. *Introduction to Wavelets and Wavelet Transforms: A Primer*. Prentice Hall, New Jersey, 1997.

26. Haddadi, R., E. Abdelmounim, M. El Hanine, and A. Belaguid. Discrete Wavelet Transform Based Algorithm for Recognition of QRS Complexes. In *2014 International Conference on IEEE Multimedia Computing and Systems (ICMCS)*, Marrakech, Morocco, IEEE, New York, 2014, pp. 375–379.

27. Gillespie, T. D., and M. W. Sayers. *NCHRP Report 228: Calibration of Response-Type Road Roughness Measuring Systems*. TRB, National Research Council, Washington, D.C., 1980.

The Standing Committee on Pavement Management Systems peer-reviewed this paper (18-04288).

Article

TRR
JOURNAL OF THE TRANSPORTATION RESEARCH BOARD

Comparing Traffic Speed Deflectometer and Falling Weight Deflectometer Data

Transportation Research Record
2018, Vol. 2672(40) 22–31

DOI: 10.1177/0361198118768524
journals.sagepub.com/home/trr
SAGE

Eyal Levenberg[1], Matteo Pettinari[2], Susanne Baltzer[2], and Britt Marie Lekven Christensen[3]

Abstract

In recent years the pavement engineering community has shown increasing interest in shifting from a stationary falling weight deflectometer (FWD) to moving testing platforms such as the traffic speed deflectometer (TSD). This paper dealt with comparing TSD measurements against FWD measurements; it focused on the comparison methodology, utilizing experimental data for demonstration. To better account for differences in loading conditions between the two devices a new FWD deflection index was formulated first. This index served as reference/benchmark for assessing the corresponding TSD measurements. Next, a Taylor diagram was proposed for visualizing several comparison statistics. Finally, a modern agreement metric was identified and applied for ranking comparison results across different datasets. Overall, the suggested methodology is deemed generic and highly applicable to future situations, especially for assessing the worth of emerging device upgrades or improved interpretation schemes (or both).

The most common testing device for nondestructive evaluation of pavement condition is the falling weight deflectometer (FWD). For a chosen test location, the FWD generates a vertical stress-pulse at the pavement surface about 30 milliseconds in duration; this is achieved by dropping a mass and then blocking its fall in a controlled manner (*1*). The device records the time-history of the loading as well as the time-history of the resulting vertical surface velocities at several offset distances from the load center (*2,3*). The velocity signals are internally integrated with respect to time to yield deflection time-histories. FWD measurements are usually employed to assess mechanical layer properties by means of backcalculation (*4–6*). For this purpose, a pavement model is assumed in which layers are treated as continuous having a priori known thicknesses.

Pavement management activities entail rapid inspection of structural condition on a network level. For such wide-area application backcalculation cannot always be performed. One reason is that layers are often distressed to a point that violates the continuity assumption in the modeling; another reason is that layer thicknesses are not readily available; finally, interpretation time is prohibitive, as ample data of drop-experiments are collected in field surveys. In light of these reasons the accepted FWD interpretation approach for pavement management is based on some index derived directly from the deflection peaks. The index is chosen such that it exhibits good statistical correlation with some definition of "structural capacity" or "remaining life" (*7*). The Surface Curvature Index SCI_{300} is one such a parameter; it has units of length and denotes the difference between the central FWD peak deflection and the peak deflection at an offset of 300 mm (*8,9*). It is calculated for a peak FWD load of 50 kN applied over a load-plate with a radius of 150 mm. The SCI_{300} is roughly related to pavement fatigue life because, for a given system layering and loading, it correlates to the tensile strain at the bottom of the asphalt layer (*10–12*).

As means of increasing FWD testing safety, testing efficiency, and spatial coverage of pavement condition, and as means of minimizing traffic disruption during deflection measurements, the pavement engineering community has been pushing for the development of moving deflection-measuring platforms. One promising device in this category is the traffic speed deflectometer (TSD). This device consists of a custom-built truck equipped with a linear array of Doppler lasers (*13,14*). The lasers, mounted on a stiff beam, are oriented downward toward the pavement surface with a small incidence angle; they measure instantaneous velocities in the direction of the laser rays. These readings embody the truck's travel speed as well as the pavement deflections caused by the truck's loading. After normalization w.r.t. travel speed (measured independently), deflection-slopes are obtained.

[1]Department of Civil Engineering, Technical University of Denmark, Kgs. Lyngby, Denmark
[2]The Danish Road Directorate, Hedehusene, Denmark
[3]Norwegian State Road Administration - East Region, Oslo, Norway

Corresponding Author:
Address correspondence to Eyal Levenberg: eylev@byg.dtu.dk

Device		Step 1	Step 2	Step 3	Step 4
TSD	20 km/h	1 2 3	-	-	-
	40 km/h	1 2 3	-	4 5	-
	60 km/h	1 2 3	-	4 5	-
	70 km/h	1 2 3	-	4 5	-
FWD		-	0 m → 80 m @ 10 m (9 points) 85 m → 350 m @ 5 m (54 points)	-	355 m - 435 m @ 5 m (17 points)

Figure 1. Testing campaign for generating TSD and FWD comparison data.

Given that FWDs have been in service for a long time whereas the TSD is newer and keeps evolving, there is a growing interest within the pavement community to quantify the agreement level between measurements performed by the two devices. The possibility of accessing FWD-based indices from a moving measurement platform offers considerable savings in all aspects of data collection for pavement condition evaluation. This ability is especially important for road authorities and pavement managers because maintenance decisions are often based on past experience gained with FWD deflection data. Therefore, at least until the new technologies mature, there is a need to relate back and carry out comparisons against FWD results.

Such comparisons have been recently carried out by several research groups (*15–17*). The main approach taken in these studies included: (i) measuring the same pavement section by both devices; (ii) calculating an index directly from the measurements of the two devices—such as the SCI_{300}; and (iii) quantifying the differences between the two datasets by application of established/classical statistical tools such as Pearson's correlation coefficient, Deming regression (*18*), and Bland–Altman plots (*19*).

This paper also deals with contrasting TSD measurements against FWD data. Similar to other studies, it describes and compares data collected by both devices over the same pavement section. However, focus here is not placed on the comparison results but rather on the comparison methodology, so that the work remains relevant when newer devices or improved technologies emerge. Subsequently, the objective here is to suggest and demonstrate three methodology advances: (i) the formulation of a new FWD deflection index that better accounts for the differences in loading conditions between the two devices, and is therefore better suited to serve as basis for statistical comparison (*20*); (ii) the graphical display of several statistics in a Taylor diagram to facilitate and enrich the comparative assessment (*21*); and (iii) the application of a modern agreement metric that embodies both correlation information and information on deviation magnitudes (*22*).

Field Data

A newly constructed asphalt road, located close to the city of Slagelse (Denmark), was chosen for a testing campaign that included both TSD and FWD measurements. The designed pavement layering included (top to bottom): 190 mm asphalt concrete, 200 mm unbound granular base course (gravely material), and about 400 mm drainage subbase (sandy material). The Subgrade was visually classified as gravelly boulder clay, with 100% passing the 63 mm sieve, and about 50% passing the 0.063 mm sieve. The road was in pristine condition during testing, before any traffic was allowed, and before the final 50 mm wearing course was paved. All measurements were performed within 5 hours during a single day; the pavement surface temperature, as monitored by the TSD, was 17.6°C on average, with a general increasing trend, fluctuating in the range of 14.4°C and 20.1°C.

A line was painted over the asphalt base layer along a stretch of 430 m to facilitate and guide the testing. Both the TSD and FWD device operators were instructed to measure along this line. Data collection was done in four steps according to the scheme shown in Figure 1, in which TSD runs and run numbers are indicated by the shaded square markers. In Step 1, the TSD was driven at four different speeds: 20, 40, 60, and 70 km/h; this was repeated three times (Runs #1, #2, and #3). In Step 2, FWD measurements were performed at either 5 m or 10 m intervals over 63 different stations, covering a distance of 350 m. Step 3 included more TSD measurements, with the device running at three different speeds: 40, 60, and 70 km/h; this was repeated twice (Runs #4 and #5). In the final and fourth Step, FWD measurements were resumed at 5 m intervals to complete the coverage of the 430 m road stretch, for a total of 80 test locations.

Measurements from Steps 2, 3, and 4 were chosen for interpretation here; these include all 80 FWD test locations, and TSD Runs #4 and #5 at 40, 60, and 70 km/h. A wider variation in pavement temperature was the main motivation for excluding Step 1. Nonetheless, so doing is inconsequential, as the work focuses on the comparison methodology.

TSD Measurements

The TSD device employed for the testing was a first-generation model. The device had a heavy rear single-axle equipped with dual-wheels on each side. Under stationary conditions this axle exerted a total load of 100 kN. Four Doppler lasers (Polytec model OFV-353, December 2003) were incorporated for reading deflection-slopes; these were fixed to a stiff

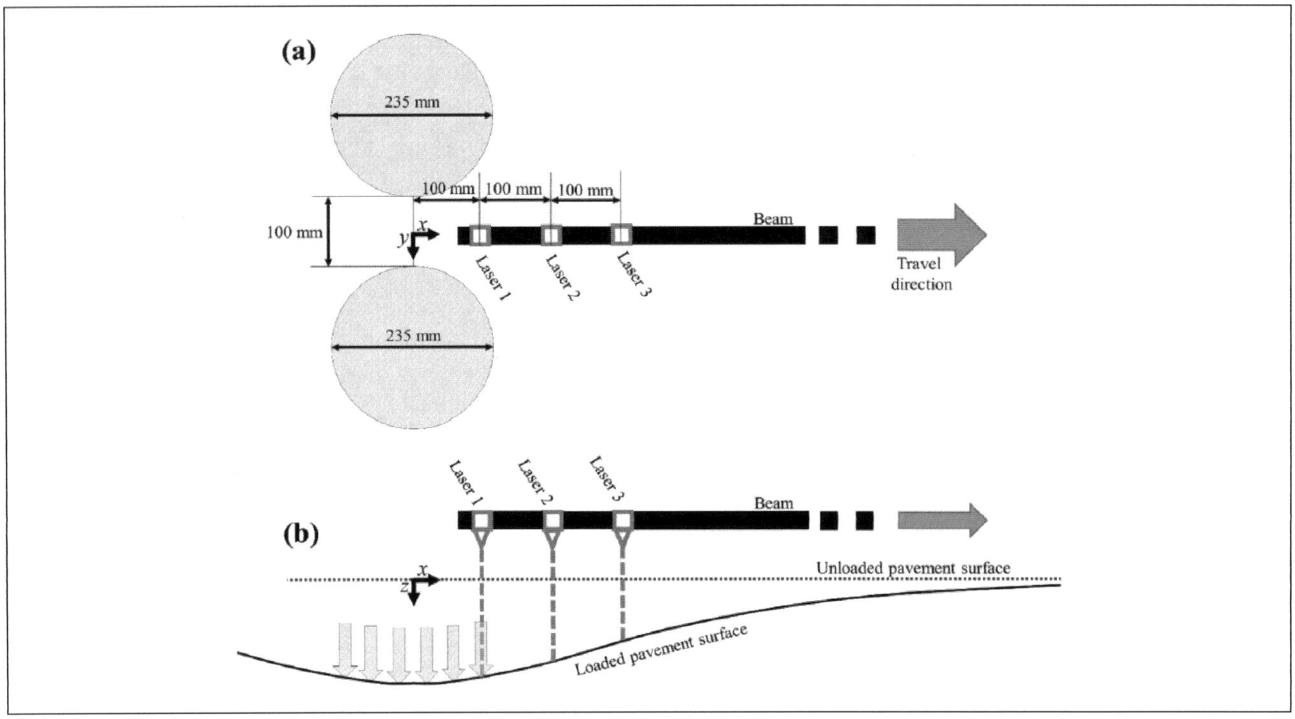

Figure 2. TSD loading and measurement setup: (*a*) plan view and (*b*) side view.

beam that was oriented in the travel direction and positioned between the dual-wheels on the right-hand side. Three out of the four lasers were placed relatively close to the dual-wheels with 100 mm spacing. The fourth laser served as reference and placed 3.6 m away, where the pavement deflection is assumed negligibly small.

Figure 2 presents a sketch of the TSD loading and measurement setup, focusing on the loaded zone and therefore displaying only three lasers. The tire–pavement contact areas are represented by shaded circles in Figure 2*a* and by shaded arrows in Figure 2*b*. The shown dimensions were obtained by direct measurement of the dual-tire assembly, with the diameter taken as the tire contact width. A Cartesian coordinate system is also included in this figure, positioned between the two tires, with its origin located at the unloaded (undeformed) pavement surface (see Figure 2*b*). The x -axis points in the travel direction, the y -axis points in the transverse direction, and the z-axis points downward into the pavement medium.

Each time the TSD traversed the 430 m stretch of road, deflection-slope measurements from Lasers 1, 2, and 3 were collected, averaged over 10 m intervals, and ascribed to the interval center location. Based on these measurements the device provided an internally computed index, here named TSD_{300} . This index has length units and resembles SCI_{300} ; it denotes the difference between the deflection between the dual-wheels, that is, at the origin of the coordinate system, and the deflection under Laser 3, that is, at $x = 300\,\mathrm{mm}$.

For computing TSD_{300} an analytic two-dimensional model is employed by the device (*23–25*). According to this model the pavement system is represented by an infinitely long and weightless Euler–Bernoulli (EB) beam supported on a Winkler foundation. A vertical point-force is employed to represent the loading of the dual-wheels assembly. Considering a rectangular coordinate system such that the x -axis coincides with the undeformed EB beam and the z -axis points downward toward the support, the point-force appears acting at the origin. Based on this description, the EB beam deflection-slope u_z' at any offset distance x from the applied force is given by

$$u_z' = -A\sin(Bx)e^{-Bx} \qquad (x \geq 0) \qquad (1)$$

in which the constants A (unitless) and B (units of length^{-1}) are both positive, representing model parameters, such as: EB beam properties, Winkler foundation properties, and loading intensity. The numerical values of these constants are found by best-matching the three measured TSD deflection-slopes with calculated EB beam slopes at three offset distances: 100, 200, and 300 mm.

Once A and B are obtained, the EB beam deflection u_z at any offset distance x is given by the expression:

$$u_z = \frac{A}{2B}e^{-\mathrm{Bx}}\left(\sin(Bx) + \cos(Bx)\right) \qquad (x \geq 0) \quad (2)$$

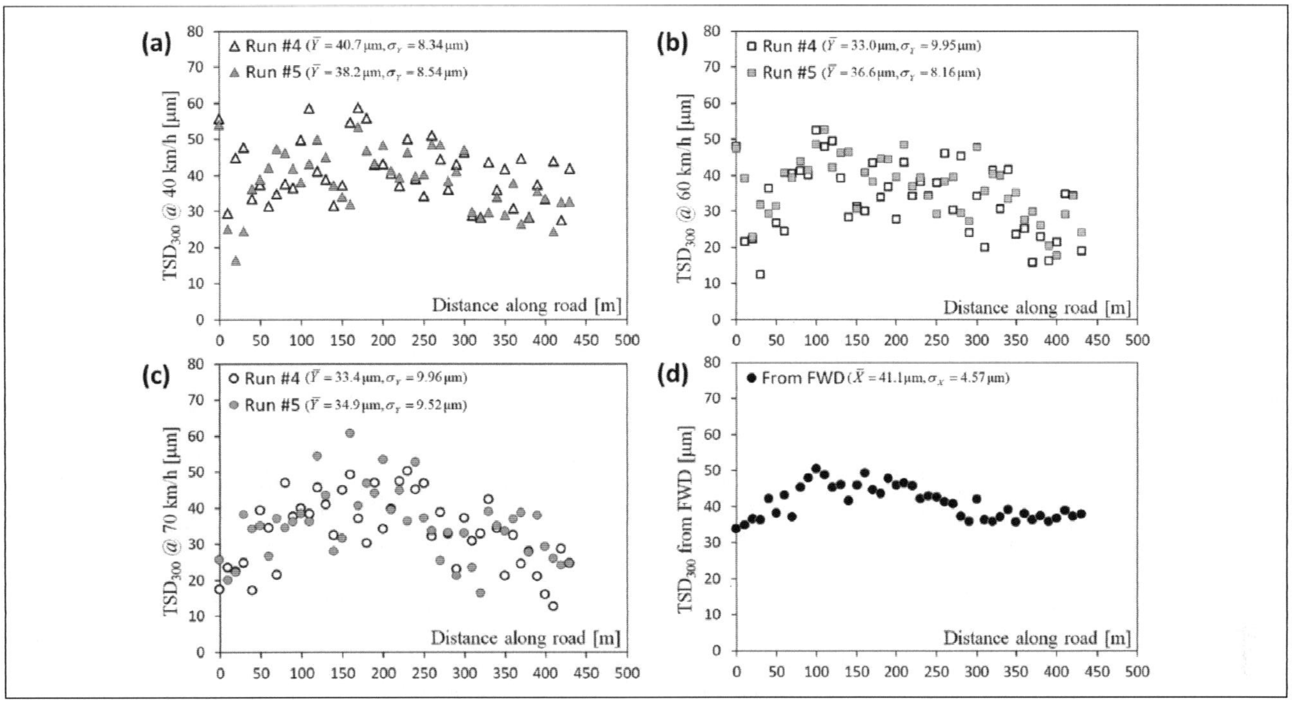

Figure 3. Values of TSD_{300} index reported at 10 m intervals for the same tested road: (*a*) acquired from TSD device at 40 km/h, (*b*) acquired from TSD device at 60 km/h, (*c*) acquired from TSD device at 70 km/h, and (*d*) acquired from interpretation of FWD deflections.

from which the TSD_{300} index is derived as follows:

$$TSD_{300} = u_z(x=0) - u_z(x=300\,\text{mm}) \qquad (3)$$

This expression, consistent with all abovementioned mechanistic assumptions and modeling simplifications, is based on the premise that maximal deflection and zero slope occur at $x = 0$. In actuality, the deflection bowl generated by a moving load over a viscoelastic pavement system is not symmetric, and the maximal vertical surface displacement is expected to take place behind the axle, that is, where $x < 0$ (*26,27*).

Figure 3 shows TSD_{300} values measured during Step 3 of the field testing campaign (refer to Figure 1). Figure 3, *a, b,* and *c* present, respectively, measurements collected at 40, 60, and 70 km/h. For each speed there are 44 data points from Run#4 (hollow markers) and 44 data points from Run#5 (shaded markers). Included in the charts are average TSD_{300} values (denoted as $\bar{Y}$) and corresponding standard deviations (denoted as σ_Y).

FWD Measurements

As indicated by Figure 1, the FWD device was operated at intervals of 5 m or 10 m along the 430 m test road. The device used was Dynatest Model 8002 equipped with a load-plate radius of 150 mm and nine geophones at the following offsets: 0, 200, 300, 450, 650, 900, 1200, 1500, and 1800 mm. Four separate FWD drops were executed in every test location, with only the last three deflection basins employed for analysis. In actuality, peak applied loads were very close to 50 kN; nonetheless, deflection peaks were linearly normalized to correspond to a peak load of exactly 50 kN. The admissibility of this procedure is based on the premise that nonlinear effects w.r.t. loading level are small.

Comparison Methodology

The purpose here is to compare the level of similarity between TSD and FWD measurements. Since the same exact road stretch was tested, results are expected to be similar, but only if a common pavement index is employed for the comparison. Subsequently, given that the TSD device reports TSD_{300} index values at 10 m intervals, the comparison approach commences by constructing/calculating TSD_{300} values from FWD measurements, also at 10 m intervals. Once this is completed, statistical tools become applicable to judge closeness of the datasets.

Analysis of FWD Deflections

The analysis presented hereafter aims at calculating TSD_{300} from FWD measurements. To achieve this there is a need to consider and account for the dissimilarity in loading configuration between the two devices. FWD deflections are the result of loading a single circular plate with a peak force of 50 kN. On the other hand, the TSD involves a dual-wheels assembly loaded (nominally) to 50 kN, with a center-to-center tire spacing of 335 mm (see Figure 2).

Subsequently, the idea advocated here is to virtually place a second FWD plate with a spacing of 335 mm (center-to-center) as means of imitating the TSD loading configuration. This is done assuming a linear pavement response for which superposition is applicable. With reference to Figure 2, such an analysis requires accessing FWD peak deflections at two nonstandard offsets. The first offset is 167.5 mm (=335/2), corresponding to the midpoint between the dual-tires. The second offset is 343.6 mm ($=\sqrt{167.5^2+300^2}$) corresponding to the point under Laser 3. The difference between these two deflections is deemed equivalent to the TSD_{300} index.

For each FWD drop, nine FWD deflection peaks $d_0 \ldots d_8$ were measured, corresponding to nine offset distances $x_0 \ldots x_8$, with $x_0 = 0$ and $x_{i+1} > x_i$. A continuous and smooth deflection basin function $w(x)$ was defined by means of a piecewise cubic interpolant passing through the $[x_i, d_i]$ pairs:

$$w(x) = \begin{cases} w_1(x) & x_0 \le x < x_1 \\ w_2(x) & x_1 \le x < x_2 \\ \vdots & \vdots \\ w_8(x) & x_7 \le x \le x_8 \end{cases} \tag{4}$$

in which x is the lateral (or offset) coordinate with origin under the load-plate center, and:

$$w_i(x) = d_{i-1} + B_i(x - x_{i-1}) + C_i(x - x_{i-1})^2 + D_i(x - x_{i-1})^3 \tag{5}$$

The constants B_i s, C_i s, and D_i s are spline constants, determined according to the usual requirements, that is, $w_i(x_i) = d_i$, $w_i'(x_i) = w_{i+1}'(x_i)$, $w_i''(x_i) = w_{i+1}''(x_i)$, and $w_8''(x_8) = 0$. To enforce zero deflection-slope under the load, that is, $w_1'(x_0) = 0$, a condition that $B_1 = 0$ was also be included in the formulation. After defining $\Delta d_i = d_i - d_{i-1}$ and $\Delta x_i = x_i - x_{i-1}$, the resulting equation set for obtaining the spline constants is

$$\begin{bmatrix}
\Delta x_1^2 & \Delta x_1^3 & & & & & & & & & & & \\
2\Delta x_1 & 3\Delta x_1^2 & -1 & & & & & & & & & & \\
2 & 6\Delta x_1 & & -2 & & & & & & & & & \\
& & \Delta x_2 & \Delta x_2^2 & \Delta x_2^3 & & & & & & & & \\
& & 1 & 2\Delta x_2 & 3\Delta x_2^2 & -1 & & & & & & & \\
& & & 2 & 6\Delta x_2 & & -2 & & & & & & \\
& & & & \ddots & \ddots & \ddots\ddots & \ddots & & & & & \\
& & & & & & & \Delta x_7 & \Delta x_7^2 & \Delta x_7^3 & & & \\
& & & & & & & 1 & 2\Delta x_7 & 3\Delta x_7^2 & -1 & & \\
& & & & & & & & 2 & 6\Delta x_7 & & -2 & \\
& & & & & & & & & & \Delta x_8 & \Delta x_8^2 & \Delta x_8^3 \\
& & & & & & & & & & & 2 & 6\Delta x_8
\end{bmatrix}
\begin{Bmatrix} C_1 \\ D_1 \\ B_2 \\ C_2 \\ D_2 \\ \vdots \\ B_7 \\ C_7 \\ D_7 \\ B_8 \\ C_8 \\ D_8 \end{Bmatrix}
=
\begin{Bmatrix} \Delta d_1 \\ 0 \\ 0 \\ \Delta d_2 \\ 0 \\ 0 \\ \vdots \\ \Delta d_7 \\ 0 \\ 00 \\ \Delta d_8 \\ 0 \end{Bmatrix} \tag{6}$$

For each measured deflection basin in Steps 2 and 4 of the field testing campaign (see Figure 1), Equation 6 was solved, spline constants were determined, and a continuous deflection function was generated according to Equation 4. Then, the two sought nonstandard deflections were found by evaluating $w(x = 167.5\,\text{mm})$ and $w(x = 343.6\,\text{mm})$, and TSD_{300} was calculated as the difference between the two. The above description is graphically presented in Figure 4 for a randomly chosen deflection set.

The aforementioned calculations were carried out for all test locations using the last three drops out of the four executed FWD drops. The three calculated TSD_{300} values were then averaged to arrive at a representative index value for the test location. Ultimately, TSD_{300} from FWD were reported at intervals of 10 m along the experimental road stretch to correspond to the reported TSD results. For this purpose, FWD tests taken 5 m before and after the location of interest were factored-in with a relative weight of 25%. For example, the TSD_{300} from FWD at location 180 m was composed of 50% of the index measured at location 180 m, plus 25% of the index measured at location 175 m plus 25% of the index measured at location 185 m. Doing so somewhat imitates the TSD device, in which continuous data were averaged over 10 m long road sections. The final set of TSD_{300} values calculated from FWD

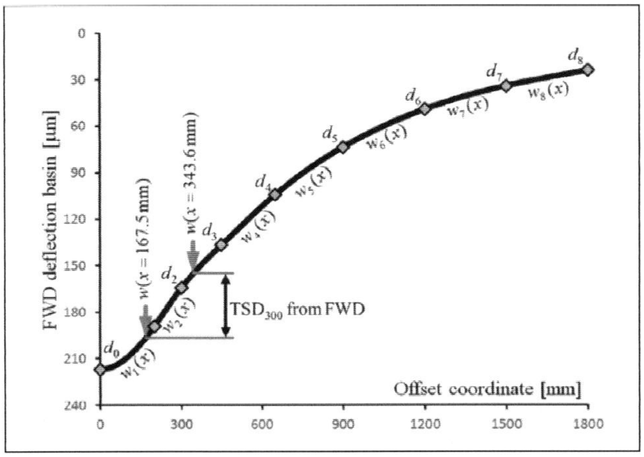

Figure 4. Calculation of TSD_{300} from FWD by means of spline interpolation; the d_i s (rotated square markers) are measured FWD peak deflections normalized to a peak force of 50 kN; the w_i s (solid line), are components of a fitting spline curve with x as the offset coordinate (see Equation 4).

according to the above-described procedure is shown in Figure 3*d*. This chart includes a total of 44 values serving as reference for the corresponding TSD_{300} values measured by the TSD device during any run (with a given speed). Also included in the Figure are the average (denoted as $\bar{X}$) and standard deviation (denoted as σ_X). It is clear from Figure 3 that the TSD results exhibit a larger spread (higher variability) as compared with the FWD derived results. If both devices had identical measurement resolution and accuracy, and given that the TSD provides readings averaged over 10 m intervals, higher variability would be expected in the FWD measurements.

Visualization of Comparison Statistics

In what follows, X_n s denote values of TSD_{300} derived from FWD (Figure 3*d*) with $n = 1...N$ and $N = 44$. Denoted as Y_n s are values of TSD_{300} directly reported by the TSD device (Figure 3, *a–c*), under a given measurement speed (40, 60, or 70 km/h) and for a given Run (#4 or #5).

A popular statistical tool for comparing two datasets is the Pearson product-moment correlation coefficient:

$$r = \frac{\sum_{n=1}^{N}(X_n - \bar{X})(Y_n - \bar{Y})}{N\sigma_X\sigma_Y} \quad (7)$$

in which $\bar{X}$ and $\bar{Y}$ are the respective means of X_n and Y_n, while σ_X and σ_Y are the corresponding standard deviations. The r statistic is a dimensionless metric that indicates the degree or strength of the linear dependence between the X_n s and the Y_n s. Values of r range between -1 and 1, and describe both the linear relationship strength and the linear relationship direction.

Scatterplots of X_n versus Y_n are included in Figure 5 for the different TSD runs, along with the associated Pearson correlation coefficient. As can be seen, in all considered cases a positive correlation was attained, with r values in the range of 0.335–0.698. All r values were subjected to a statistical significance test to assess whether they are larger than zero. This significance test was carried out assuming r follows a Student's t-distribution with $N-2$ degrees of freedom. Accordingly, t-values were calculated in each case with the expression: $r\sqrt{N-2}/\sqrt{1-r^2}$ and then the significance level was obtained from the associated t-distribution based on a one-tailed check. All r values for the data shown in Figure 5 were confirmed to be positive, with a statistical significance that is better than 1.3%. The verbal descriptions appearing in the charts (i.e., weak, moderate, and strong) are based on the categorization suggested in Evans (*28*). This categorization, see Table 1, is context dependent; it is offered here as one possible means for interpreting correlation magnitudes.

The Pearson correlation coefficient is scale-blind as it does not convey information on the deviation magnitudes. A popular agreement metric that overcomes this drawback, and considers dimensional information, is

$$E = \sqrt{\frac{\sum_{n=1}^{N}(X_n - Y_n)^2}{N}} \quad (8)$$

in which E is the root-mean-square error (RMSE); it essentially aggregates the squared (individual) differences between X_n and Y_n across the population. In effect, two different error types are represented in E, namely: pattern mismatch (E') and mean mismatch ($\bar{E}$). These two error types may be decomposed according to the formula $E^2 = \bar{E}^2 + E'^2$ in which $\bar{E} = \bar{Y} - \bar{X}$ and E' is the centered RMSE metric:

$$E' = \sqrt{\frac{\sum_{n=1}^{N}\left((X_n - \bar{X}) - (Y_n - \bar{Y})\right)^2}{N}} \quad (9)$$

The standard deviations (σ_X and σ_Y) and the Pearson product-moment correlation coefficient (r) are related to the centered RMSE metric (E') according to the expression $E'^2 = \sigma_Y^2 + \sigma_X^2 - 2r\sigma_Y\sigma_X$. This expression bears similarity to the trigonometric law of cosines; it led Taylor (*21*) to propose a diagram that simultaneously depicts these different statistics in a single chart. Such a Taylor diagram is shown in Figure 6, representing data from Figures 3 and 5.

In this diagram, standard deviations are depicted on both the abscissa and ordinate while the Pearson correlation coefficient r is represented by azimuthal positions with regard to the coordinate origin such that the abscissa coincides with $r = 1$ (i.e., perfect linear correlation) and the ordinate coincides with $r = 0$ (i.e., no linear correlation). Points

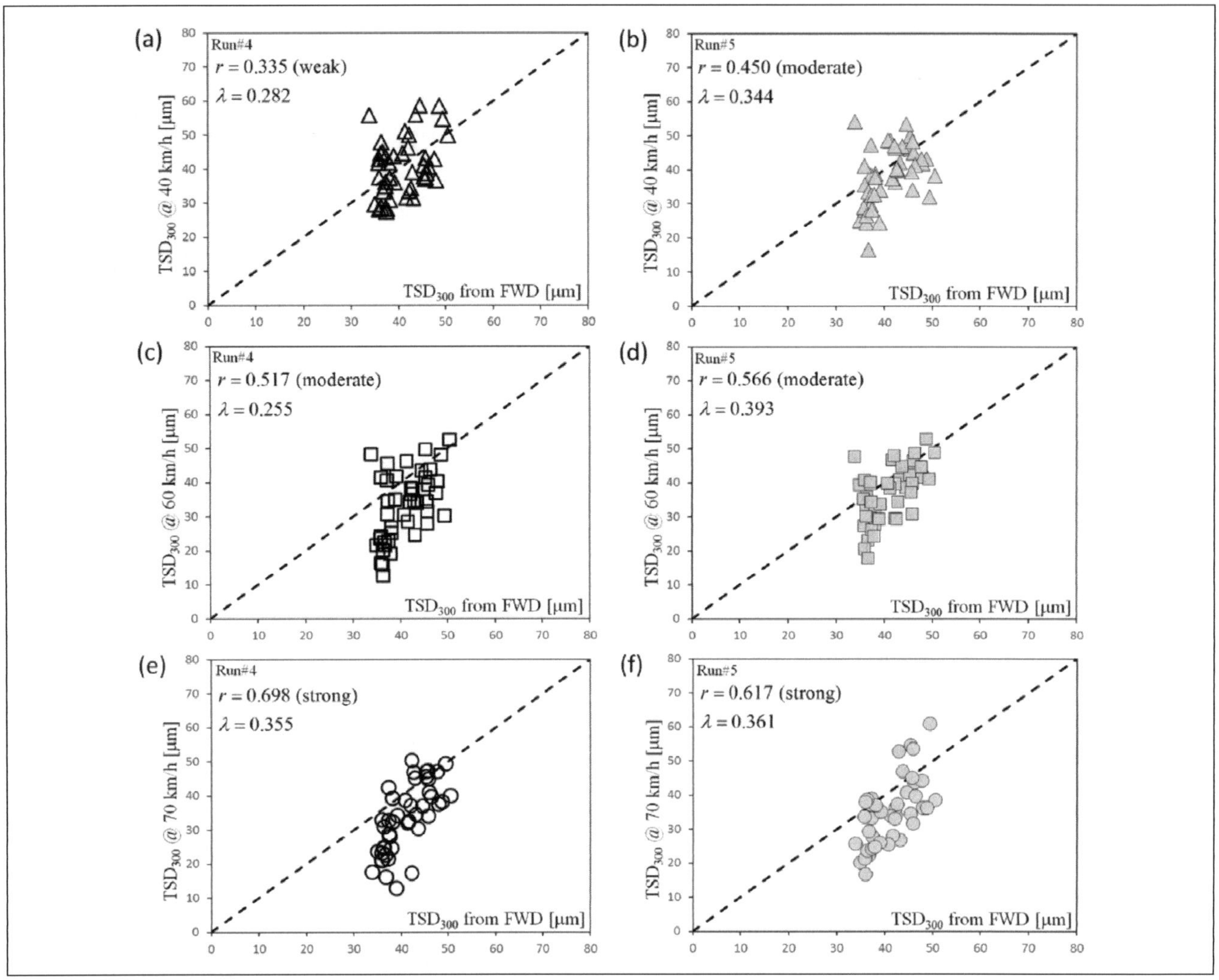

Figure 5. Cross-plots of measured TSD_{300} data during Runs #4 and #5 versus TSD_{300} obtained from FWD testing: (a) and (b) measurements at 40 km/h, (c) and (d) measurements at 60 km/h, and (e) and (f) measurements at 70 km/h.

Table 1. Verbal Interpret ation of Pearson Correlation Offered for Comparing TSD and FWD Data

Correlation range	Verbal description (*28*)
0.00 ↔ 0.19	Very weak
0.20 ↔ 0.39	Weak
0.40 ↔ 0.59	Moderate
0.60 ↔ 0.79	Strong
0.80 ↔ 1.00	Very strong

representing the different TSD runs are shown in the chart: triangular markers represent measurements collected at 40 km/h, square markers represent a measurement speed of 60 km/h, and measurement speed of 70 km/h is represented by circular markers. The distance of these points from the coordinate origin represents their standard deviation and is indicated by dashed arcs. The reference FWD dataset is depicted with a solid circular marker on the abscissa at a standard deviation $\sigma_X = 4.57\,\mu m$ (refer to Figure 3*d*). Radial distances from the reference point are indicated by dotted arcs and represent centered RMSE metric values (Equation 9).

Overall, the Taylor diagram makes it easy and intuitive to assess similarity levels between TSD and FWD measurements. In the specific examples shown, three features are immediately apparent from the graphical depiction: (i) the standard deviations of the TSD_{300} values (dashed arcs), residing between 8 µm and 10 µm, exhibit a slight tendency to increase with increasing measurement speed; (ii) the centered RMSE metric values (dotted arcs) appear insensitive to TSD measurement speed; and (iii) the correlation between TSD and FWD (azimuthal positions) somewhat improves as TSD measurement speed increases.

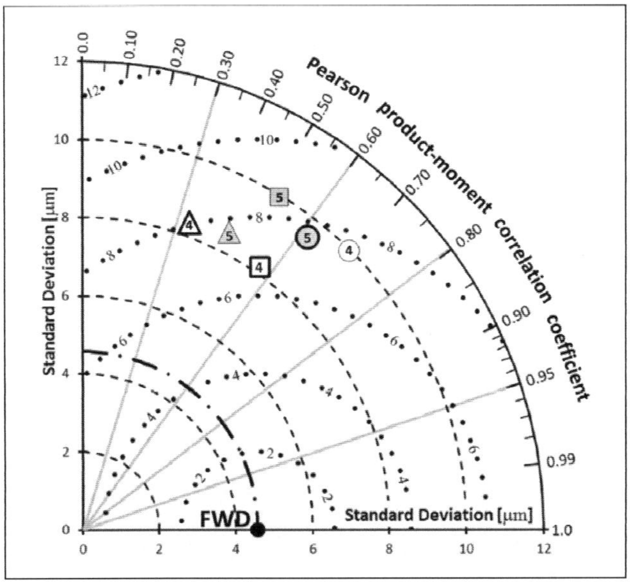

Figure 6. Taylor diagram of TSD_{300} index statistics for comparing FWD and TSD measurements at three different speeds: 40 km/h (triangular markers), 60 km/h (square markers), and 70 km/h (circular markers).

Modern Agreement Metric

The Taylor diagram depicts several centered pattern error statistics. It does not, however, offer a single similarity metric needed for performing inter-comparison or for ranking different cases. To this end, it is instructive to introduce a modern agreement metric λ, suggested originally in the field of biometrics for evaluating method reproducibility (*29*), and highlighted recently by Duveiller et al. (*22*). The λ metric summarizes the closeness of two datasets in a single index, considering both correlation and bias. It is dimensionless, symmetric, and relatively "cheap" to compute. For positive Pearson correlation coefficient values it is given by the expression:

$$\lambda = \frac{2r}{\frac{\sigma_X}{\sigma_Y} + \frac{\sigma_Y}{\sigma_X} + \frac{(\bar{X} - \bar{Y})^2}{\sigma_X \sigma_Y}} \qquad (r \geq 0) \qquad (10)$$

in which the purpose of the denominator is to "penalize" the value of r for increasing additive bias or multiplicative bias (or both) between the datasets. As can be seen, if two datasets X_n and Y_n have identical means and standard deviations, then the λ metric becomes equal to the Pearson correlation coefficient. In all other cases, for which the means or standard deviations (or both) are dissimilar, the λ metric is smaller than r. Therefore, λ in Equation 10 ranges from zero to unity.

Calculated λ values are included in the charts in Figure 5 just beneath the r values. As expected, they are different and always smaller than r. Based on these λ values Table 2 ranks the agreement level between FWD and TSD across the different datasets collected in the field experiment; it also presents agreement rankings based on the Pearson correlation coefficient.

As can be seen in Table 2, the rankings differ. The highest level of λ is reported for TSD Run#5 at 60 km/h ($\lambda = 0.393$). If only correlation level was taken as an agreement criterion then Run#4 at 70 km/h ($r = 0.698$) would be ranked as best. The lowest level of λ is reported for TSD Run#4 at 60 km/h ($\lambda = 0.255$). If only correlation level was taken as an agreement criterion then Run#4 at 40 km/h ($r = 0.335$) would be ranked as worst. Given that λ embodies both correlation and bias, rankings based on this metric are deemed superior and more appropriate than rankings based on r. It may also be noticed from Table 2 that the three top-ranked agreement levels refer to the higher measurement speeds.

Summary and Comments

The introduction of new pavement evaluation devices require comparison against accepted methods so that existing knowledge and experience can remain useful, at least during some transition period. In this connection, the paper was concerned with comparing TSD and FWD measurements taken along the same road stretch. The work focused on the analysis approach rather than on the comparison results. This is because comparison outcomes depend on the specific TSD and FWD devices and operational modes employed during data collection. Comparison outcomes also depend on the tested pavement conditions because the quality of measured values, especially if small, may be influenced by finite sensing resolution.

Here, comparison was based on a deflection index named TSD_{300} which is internally calculated, and subsequently reported, when operating the TSD device. The first interpretation step involved estimating TSD_{300} from FWD, with the results serving as a reference/benchmark dataset for subsequent evaluation. This was done as an initial attempt to account for the differences in loading configuration between the devices. The estimation of TSD_{300} from FWD presumed linear quasi-static pavement response, and considered only load and deflection peaks.

The second interpretation step employed here involved a Taylor diagram as a tool for visualizing dissimilarity between datasets. Finally, a modern metric λ was proposed for quantifying the closeness of two datasets and therefore rank-calculated agreement levels across different cases. Both statistical tools are generic and therefore applicable for evaluating other/different datasets. In particular, they appear to be highly suited for displaying trends in agreement resulting from employing better/improved interpretation algorithms. Similarly, they should also be effective in demonstrating the merits of changing operational settings or modifying data-collection procedures. Moreover, although not the focus here, the λ metric is well suited for

Table 2. Ranking of TSD and FWD Agreement Level Based on the λ Metric (Equation 10) and Based on the Pearson Correlation Coefficient r (Equation 7)

TSD Run #	Speed [km/h]	Agreement ranking based on λ	Agreement ranking based on r
4	40	5	6
	60	6	4
	70	3	1
5	40	4	5
	60	1	3
	70	2	2

quantifying device reproducibility (*30*), for example, the similarity level between two TSD run results. Therefore, the proposed methodology is deemed highly suited for the task of quantifying the value of forthcoming device upgrades.

In future work, more sophisticated efforts may be envisioned to enhance comparison validity in which systematic differences between devices are further reduced before any statistical assessment. The following aspects related to loading conditions should be considered as candidate for interpretation improvement: (i) area of loaded zones, that is, diameter of FWD load-plate versus TSD tire imprint size; (ii) contact stress distribution uniformity; (iii) type of loading, that is, FWD impact versus TSD rolling wheels; and (iv) oscillation of TSD load caused by vehicle dynamics in combination with pavement roughness. Moreover, the following pavement modeling aspects should be considered: (i) employ identical mechanical model for FWD and TSD measurements when estimating TSD_{300}, for example, abandon the EB beam model; and (ii) employ more faithful mechanical representation of the pavement layers, for example, nonlinear response, time-dependence, thermal-sensitivity, and inertia effects. The above list of proposed ideas for reducing systematic differences (before agreement testing) implicitly advocates using mechanical layer properties as basis for comparison, given that they are device independent.

Acknowledgments

The authors would like to acknowledge the Measurement Division of the Danish Road Directorate for their help in carrying out the FWD and TSD measurements. Also acknowledged is Mr. Alexander Schimmer Molin for his contribution in identifying relevant literature on comparative statistics.

References

1. Madsen, S. S., and E. Levenberg. Dynamic Backcalculation with Different Load-Time Histories. *Road Materials and Pavement Design*, 2017. DOI: 10.1080/14680629.2017.1307263.
2. ASTM D4694-09. *Standard Test Method for Deflections with a Falling-Weight-Type Impulse Load Device*. ASTM International, West Conshohocken, Pa., 2015.
3. ASTM D4695-03. *Standard Guide for General Pavement Deflection Measurements*. ASTM International, West Conshohocken, Pa., 2015.
4. Bush, A. J., and G. Y. Baladi. *Nondestructive Testing of Pavements and Backcalculation of Moduli*. ASTM STP 1026. ASTM International, West Conshohocken, Pa., 1989.
5. Von Quintus, H. L., A. J. Bush, and G. Baladi. *Nondestructive Testing of Pavements and Backcalculation of Moduli: Second Volume*. ASTM STP 1198. ASTM International, West Conshohocken, Pa., 1994.
6. Tayabji, S., and E. Lukanen. *Nondestructive Testing of Pavements and Backcalculation of Moduli: Third Volume*. ASTM STP 1375. ASTM International, West Conshohocken, Pa., 2000.
7. Kim, Y. R., Y. C. Lee, and S. R. Ranjithan. Flexible Pavement Condition Evaluation using Deflection Basin Parameters and Dynamic Finite Element Analysis Implemented by Artificial Neural Networks. In *Non-destructive Testing of Pavements and Backcalculation of Moduli: Third Volume* (S. D. Tayabji, and E. O. Lukanen), ASTM STP 1375, ASTM International, West Conshohocken, Pa., 2000, pp. 514–530.
8. Horak, E. Benchmarking the Structural Condition of Flexible Pavements with Deflection Bowl Parameters. *Journal of the South African Institution of Civil Engineering*, Vol. 50, No. 2, 2008, pp. 2–9.
9. Horak, E., A. Hefer, S. Emery, and J. Maina. Flexible Road Pavement Structural Condition Benchmark Methodology Incorporating Structural Condition Indices Derived from Falling Weight Deflectometer Deflection Bowls. *Journal of Civil Engineering and Construction*, Vol. 4, No. 1, 2015, pp. 1–14.
10. Kilareski, W. P., and B. A. Anani. Evaluation of In-Situ Moduli and Pavement Life from Deflection Basins. *Proc., 5th International Conference on the Structural Design of Asphalt Pavements*, Delft, the Netherlands, Vol. 1, 1982, pp. 349–366. University of Michigan, Ann Arbor.
11. Molenaar, A. A. A. *Structural Performance and Design of Flexible Pavements and Asphalt Concrete Overlays*. PhD dissertation. Delft University of Technology, Delft, The Netherlands, 1983.
12. Thyagarajan, S., N. Sivaneswaran, K. Petros, and B. Muhunthan. Development of a Simplified Method for Interpreting Surface Deflections for In-Service Flexible Pavement Evaluation. Presented at 8th International Conference on Managing Pavement Assets, Santiago, Chile, 2011.
13. Hildebrand, G., and S. Rasmussen. *Development of a High Speed Deflectograph*. Danish Road Institute, Report 117, Road Directorate, Roskilde, Denmark, 2002.
14. Simonin, J. M., D. Lièvre, S. Rasmussen, and G. Hildebrand. Assessment of the Danish High Speed Deflectograph in France. Proc., 7th International Conference on Bearing Capacity of Roads, Railways and Airfields, Trondheim, Norway, 2005. Horvli, Ivar (ed), Norwegian University of Science and Technology, Trondheim, Norway, ISBN 8292506241.
15. Muller, W. B., and J. Roberts. Revised Approach to Assessing Traffic Speed Deflectometer Data and Field Validation of Deflection Bowl Predictions. *International Journal of Pavement Engineering*, Vol. 14, No. 4, 2013, pp. 388–402.
16. Flintsch, G., S. Katicha, J. Bryce, B. Ferne, S. Nell, and B. Diefenderfer. *Report S2-R06F-RW-1: Assessment of Continuous Pavement Deflection Measuring Technologies*. Transportation Research Board of the National Academies, Washington, D.C.,2013.
17. Chai, G., S. Manoharan, A. Golding, G. Kelly, and S. Chowdhury. Evaluation of the Traffic Speed Deflectometer Data using Simplified Deflection Model. *Transportation Research Procedia*, Vol. 14, 2016, pp. 3031–3039.

18. Deming, W. E. *Statistical Adjustment of Data*. Dover, New York, 1943.
19. Bland, J. M., and D. G. Altman. Statistical Methods for Assessing Agreement between Two Methods of Clinical Measurement. *The Lancet*, Vol. 327, No. 8476, 1986, pp. 307–310.
20. Christensen, B. M. L. *A Comparison of Traffic Speed Deflectometer and Falling Weight Deflectometer Data – Different Approaches for Comparing the Two Devices*. Civil Engineering Bachelor thesis. Technical University of Denmark (DTU), Kongens Lyngby, Denmark, 2016.
21. Taylor, K. E. Summarizing Multiple Aspects of Model Performance in a Single Diagram. *Journal of Geophysical Research*, Vol. 106, 2001, pp. 7183–7192.
22. Duveiller, G., D. Fasbender, and M. Meroni. Revisiting the Concept of a Symmetric Index of Agreement for Continuous Datasets. *Scientific Reports*, Vol. 6, 2016, pp. 19401.
23. Dias, K., J. Gravesen, P. G. Hjorth, P. Larsen, C. Please, N. Radulovic, L. Wang, and P. L. Aagaard. *Beneath the Wheel - Greenwood Engineering*. 54th European Study Group with Industry (ESGI-54), Final Report, University of Southern Denmark, Odense, Denmark, 2005.
24. Krarup, J., S. Rasmussen, L. Aagaard, and P. Hjorth. *Output from the Greenwood Traffic Speed Deflectometer. Paper presented at the 22nd Australian Road Research Board (ARRB) Conference*, Canberra, Australia, 2006.
25. Pedersen, L., P. G. Hjorth, and K. Knudsen. *Viscoelastic Modelling of Road Deflections for use with the Traffic Speed Deflectometer*. Report IMM-PHD-2013, No. 310. Technical University of Denmark, Kgs. Lyngby, 2013.
26. Levenberg, E. Inverse Analysis of Viscoelastic Pavement Properties using Data from Embedded Instrumentation. *International Journal for Numerical and Analytical Methods in Geomechanics*, Vol. 37, No. 9, 2013, pp. 1016–1033.
27. Levenberg, E. Viscoelastic Pavement Modeling with a Spreadsheet. *Proc., 8th International Conference on Maintenance and Rehabilitation of Pavements (Mairepav8), Research Publishing*, Singapore, 2016, pp. 746–755. 27-29 July 2016, Singapore.
28. Evans, J. D. *Straightforward Statistics for the Behavioral Sciences, Brooks/Cole Publishing*. Pacific Grove, Calif., 1996.
29. Lin, L. A Concordance Correlation Coefficient to Evaluate Reproducibility. *Biometrics*, Vol. 45, 1989, pp. 255–268.
30. Lin, L., A. S. Hedayat, B. Sinha, and M. Yang. Statistical Methods in Assessing Agreement. *Journal of the American Statistical Association*, Vol. 97, No. 457, 2002, pp. 257–270.

The Standing Committee on Pavement Condition Evaluation (AFD20) peer-reviewed this paper (18-00936).

Article

TRR
JOURNAL OF THE TRANSPORTATION RESEARCH BOARD

Transportation Research Record
2018, Vol. 2672(40) 32–40

DOI: 10.1177/0361198118792117
journals.sagepub.com/home/trr
SAGE

Issues in the Prediction of the Mechanical Properties of Open Graded Mixes

Hong Zhang[1], Kumar Anupam[1], Athanasios Scarpas[1,2], and Cor Kasbergen[1]

Abstract

Within the pavement engineering community, open graded mixes (OGM) are regarded as mixes capable of reducing noise and improving wet skid resistance. However, during their design life, these asphalt mixes are known to suffer from a particular distress type known as raveling. This results in a premature failure of a road network. In order to study the propensity of OGM to raveling, homogenization-based approaches are considered to be accessible and effective. One of the most widely accepted homogenization models for asphalt concrete is proposed by Christensen et al. Several studies related to homogenization techniques have been conducted in the past; however, to the best of the authors' knowledge not a lot of attention has been paid to the study of OGM by means of homogenization models. The other limitation of the Christensen model is that some parameters are difficult to physically understand. Under the above realization, the objective of the paper is twofold: (1) to propose a modification of the Christensen model for OGM; and (2) to verify the modified model's capability in predicting the mechanical properties of OGM. In general, it was found that once the proposed factor is calibrated for a given OGM by laboratory tests the obtained results are accurate.

Open graded mixes (OGM) are increasingly used because of their effectiveness in reducing noise caused by tire–pavement interaction and improving skid resistance, especially during wet seasons (*1*). However, because of the high air voids characteristic, OGM always suffer from a particular distress known as raveling, which occurs as individual aggregate particles dislodge from the pavement surface (*2*). Normally, there are two types of damage modes for raveling distress: cohesive damage and adhesive damage (Figure 1) (*2*). Cohesive damage is the failure of mastic properties, whereas adhesive damage is the loss of bond between aggregate particles and mastic.

The methods proposed by researchers (*3*) at the Delft University of Technology to evaluate the characteristics of cohesive damage and adhesive damage in the laboratory are shown in Figure 1). A mastic column, the size of which is 6 mm in diameter and 12 mm in height, and a stone column, which consists of two stone columns and a mastic sample (8 mm in diameter and 2 mm in thickness) between them, are used to perform dynamic shear stress or strain to obtain fatigue life curves for analyzing cohesive damage and adhesive damage, respectively.

Although the characteristics of cohesive damage and adhesive damage can be evaluated by laboratory tests, the occurrences of these two damage modes in the field are affected by various factors, such as repetition of loads, moisture, temperature, and so forth (*4*). Therefore, it is a big challenge to analyze the propensity of these two phenomena during the service life of OGM pavement.

In the recent past, computational models based on finite element methods (FEM) and/or discrete element methods (DEM) have been proposed as a mean to obtain the stresses and/or strains at mix component level of OGM, which are further used to analyze the raveling damage (*5*, *6*). Although FEM/DEM-based models are able to handle complex compositions and almost realistic mix component material properties, they require the development of very large FEM meshes (usually by means of post-processing of the results of CT scans) and very large-scale computational facilities. For example, for a specimen of 60 mm in height and 150 mm in diameter, about 6.8 million elements are needed to carry out a reliable analysis and a running time of several days (Figure 2*a*) (*6*). Computational tools and facilities of such scale are not typically available in engineering practice.

[1]Section of Road Engineering, Faculty of Civil Engineering & Geosciences, Delft University of Technology, Delft, the Netherlands
[2]Khalifa University of Science and Technology, Abu Dhabi, UAE

Corresponding Author:
Address correspondence to Hong Zhang: h.zhang-4@tudelft.nl

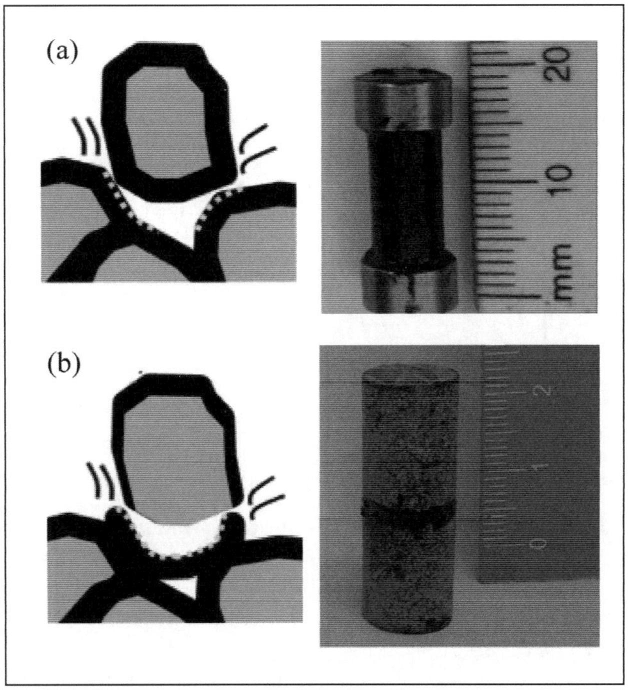

Figure 1. Two types of damage modes of raveling distress. (*a*) Cohesive damage; (*b*) Adhesive damage.

Homogenization methods (*7*) offer an attractive alternative to the above issues (Figure 2*b*). By means of a homogenization technique, the effective micromechanical properties of the composite are determined from the properties of its different phases, which further allows the calculation of the stress/strain field of the different phases based on the applied loading conditions. This information, in combination with the laboratory tests mentioned above for evaluating the characteristics of cohesive damage and adhesive damage, can be utilized to investigate the propensity of a given OGM for raveling.

On the basis of the homogenization theory, many models have been developed for estimating the mechanical properties of asphalt mixes (*8*). Among these models, the Hirsch model is one of the most commonly used semi-empirical models because of its similarity and requirement of few constituent properties (*8*). It was originally developed by Hirsch (*9*) to investigate the modulus of concrete as determined by the moduli of cement and aggregate particles. In the Hirsch model, different phases are assumed to be in a combination of parallel and series arrangements. The modulus of the composite is controlled by the volume fractions and the moduli of all its phases. Further, Christensen et al. (*10*) modified the original Hirsch model to make it applicable for asphalt concrete, and the Christensen model has been independently evaluated by several researchers (*11*, *12*) with generally good results.

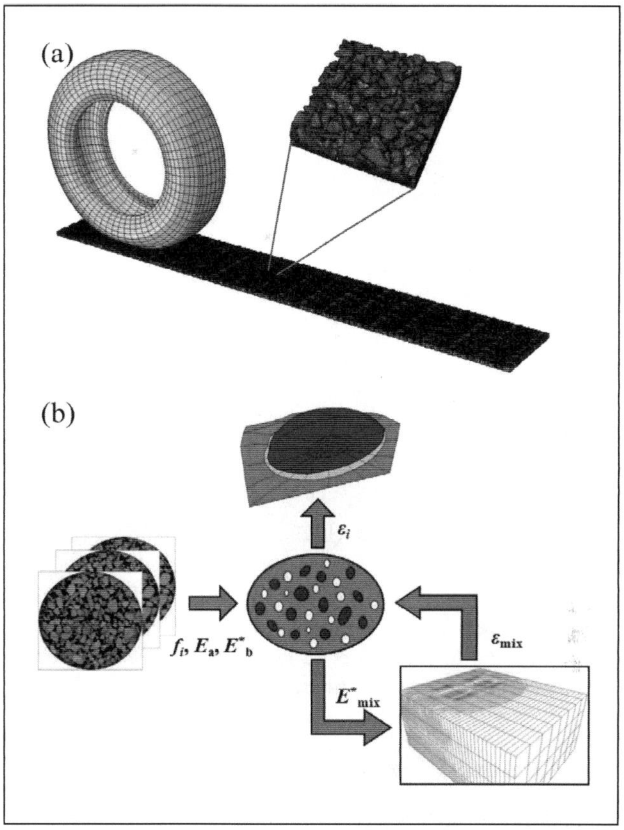

Figure 2. Different approaches for OGM analysis. (*a*) FEM approach; (*b*) Homogenization approach.

From the above analysis, it can be concluded that in comparison with the FEM and DEM methods, the homogenization approach is a more accessible and effective way for obtaining the mechanical properties of OGM. It can also be summarized that the Hirsch model, which relates the properties of each phase to the overall properties of mixes, is a simple but effective homogenization model. Based on this realization, the objective and research scopes will be proposed in the following section.

Objective and Research Scopes

The objective of this paper is to outline the homogenization methodology for estimating the mechanical properties of OGM via the Hirsh model.

The scopes of the proposed study include:

- To modify the expression of the Christensen model and propose an aggregates organization factor to describe the frequency/temperature-dependent contribution of the aggregate phase to the overall behavior of OGM.

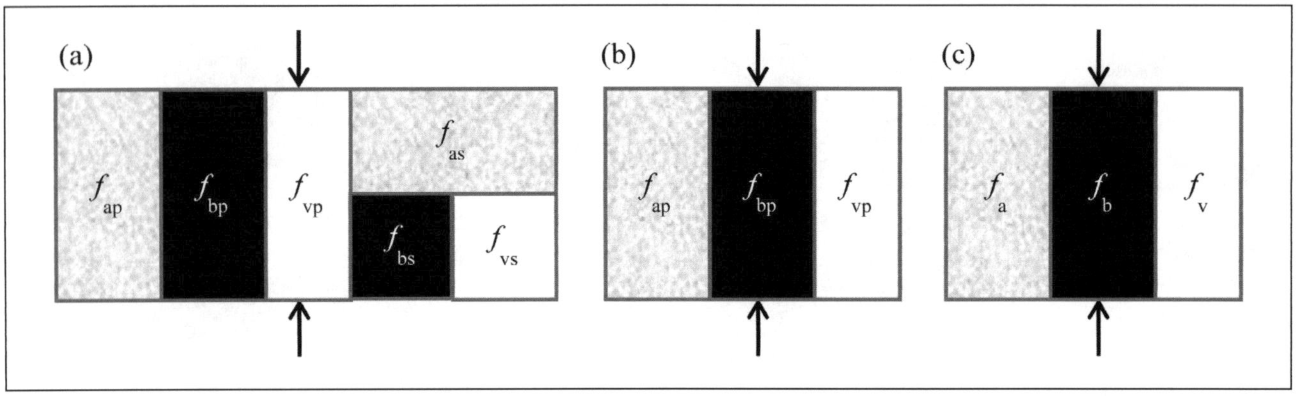

Figure 3. Arrangement of the Christensen model for asphalt mixes. (*a*) Original arrangement (*10*); (*b*) Simplified arrangement (*13*); (*c*) Proposed revised arrangement.

- To determine and validate the expression for the aggregates organization factor from laboratory tests of OGM.
- To analyze the contribution made by each phase to the modulus of OGM based on the modified expression of the Christensen model.

Background Knowledge

Application of the Hirsch Model on Asphalt Mixes

On the basis of the research needs described earlier, descriptions about some of the aspects related to the Hirsch model will be presented in this section. In the early research of Christensen et al. (*10*), various phase arrangements were proposed to modify the original Hirsch model to make it applicable for asphalt mixes. After thorough investigation of several proposed versions of the modified Hirsh model, the researchers came to a conclusion that the version in which the parallel and series sub-units composed of asphalt binder, aggregates, and air voids are arranged in parallel provided the most accurate predicted results of the modulus of asphalt concrete (Figure 3*a*).

In this version, the dynamic Young's modulus of an asphalt mix $|E^*|_{mix}$ is obtained from the volume fractions and moduli of asphalt binder, aggregates and air voids, Equation 1.

$$|E^*|_{mix}(f) = f_{ap}E_a + 3f_{bp}|G^*|_b(f) + (f_{as} + f_{bs} + f_{vs})^2 \left[\frac{f_{as}}{E_a} + \frac{(f_{bs} + f_{vs})^2}{3f_{bs}|G^*|_b(f)}\right]^{-1} \quad (1)$$

where the subscripts p and s represent the parallel portion and the series portion, respectively; f_a and E_a are the volume fraction and Young's modulus of the aggregate phase, respectively; f_b and $|G^*|_b$ are the volume fraction and the dynamic shear modulus of the asphalt binder phase, respectively; and f_v is the volume fraction of the air voids phase.

However, it was observed that the effects of the series sub-unit on the estimated modulus of an asphalt mix are much less significant than the effects of the parallel sub-unit (*13*), which indicates that the characteristics of an asphalt mix are similar to a parallel arrangement of individual phases. Under this realization, the authors (*13*) further simplified the original Christensen model to a simple parallel arrangement, as shown in Figure 3*b*, Equation 2.

$$|E^*|_{mix}(f) = f_{ap}E_a + 3f_{bp}|G^*|_b(f) \quad (2)$$

where f_{ap}, f_{bp}, E_a and $|G^*|_b$ are described in Equation 1.

Equation 2 has two unknown parameters f_{ap}, f_{bp}, which makes it difficult to be determined by using laboratory/field tests. Researchers (*10*) have proposed further simplification to solve this issue by introducing factors such as contact factor P_c.

Contact Factor P_c

P_c is the share of the parallel component of the total volume of the composite and therefore varies between 0 and 1. As can be seen from Equation 2, if the parallel part occupies the same proportion in each phase, Equation 2 can be further reduced to the following equation, Equation 3.

$$|E^*|_{mix}(f) = P_c(f)(f_aE_a + 3f_b|G^*|_b(f)) \quad (3)$$

where f_a, f_b, E_a and $|G^*|_b$ are described in Equation 1. However, it is highlighted here that the underlying

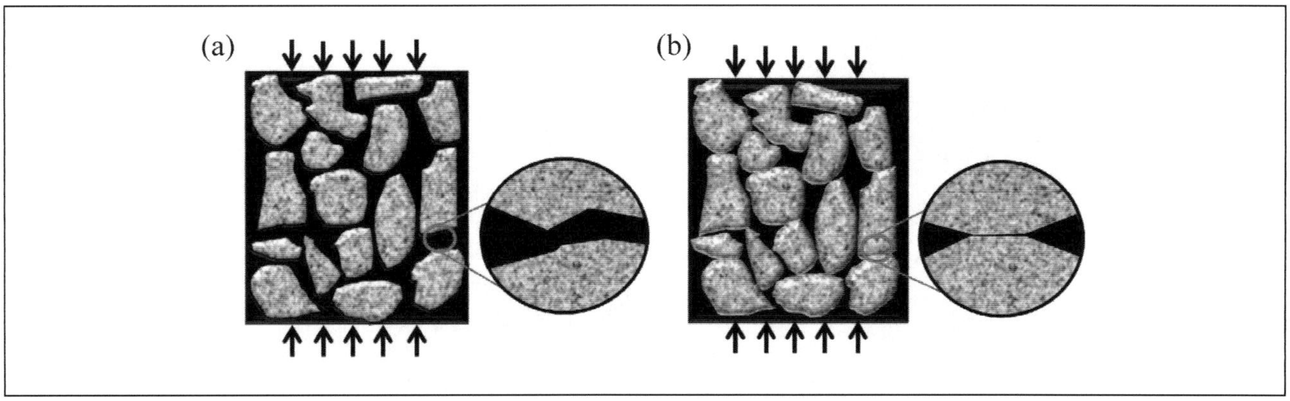

Figure 4. Aggregate contacts at different frequencies/temperatures. (*a*) High frequency/low temperature; (*b*) Low frequency/high temperature.

assumption is that the value of P_c does not change with different phases but remains the same for all the phases of the composite.

It is obvious that P_c is one of the critical parameters in the calculation of $|E^*|_{mix}$. As the behavior of asphalt concrete depends on frequency/temperature, the relative proportions of the series part and parallel part are also frequency/temperature dependent. As proposed by the researchers (*10*), the values of P_c can be determined from the laboratory tests. The researchers concluded that the proposed factor P_c is frequency/temperature sensitive, as expected.

In line with the above research work, researchers (*10*, *13*) showed that in general a good agreement exists between the predicted results of the Christensen model and the laboratory tests. However, to the best of the authors' knowledge, it can be stated that the physical representation of P_c is difficult. In their pioneering studies, researchers (*10*) hypothesized that P_c represents the aggregate contact factor and interpreted it as the contribution from the portion of aggregate particles in intimate contact with each other. They also noted that high values of P_c at high frequencies/low temperatures indicate more contact among aggregate particles. However, this interpretation does not comply with the physical situation where fewer aggregate particles are expected to be in intimate contact at high frequencies/low temperatures (Figure 4*a*). This behavior is expected because of the existence of stiff asphalt binder at high frequencies/low temperatures. Similarly, at low frequencies/high temperatures when the modulus of asphalt binder is soft, it is expected that the aggregate would find it easier to move, which will result in more pronounced contact (Figure 4*b*).

Based on the above analysis, it can be concluded that the interpretation of P_c does not consider the aggregate contact interaction aptly. Thus, in the following section, a modified expression for the Christensen model will be proposed.

Modified Expression of the Christensen Model

As mentioned above, the effects of the series element in the Christensen model on the estimated modulus of asphalt concrete are negligible as compared with the effects of the parallel element. Therefore, in the revised arrangement, it is proposed that the total volume of asphalt binder, aggregates, and air voids are arranged in parallel (Figure 3*c*). This arrangement is the same as the arrangement proposed by the original researchers (*13*) except that in the revised arrangement the part of the volume is replaced by the total volume. The relationship between $|E^*|_{mix}$ and the properties of individual phases is shown in Equation 4.

$$|E^*|_{mix}(f) = f_a E_a + 3 f_b |G^*|_b(f) \qquad (4)$$

where f_a, f_b, E_a and $|G^*|_b$ are described in Equation 1.

The authors proposed to modify Equation 4 by introducing a factor P_a, which describes the contribution to $|E^*|_{mix}$ from the arrangement of aggregate particles at different frequency/temperature conditions, Equation 5. A detailed description about P_a will be presented in the following subsection.

$$|E^*|_{mix}(f) = P_a(f) f_a E_a + 3 f_b |G^*|_b(f) \qquad (5)$$

where f_a, f_b, E_a and $|G^*|_b$ are described in Equation 1.

Aggregates Organization Factor, P_a

An asphalt mix has a higher value of f_a than f_b. Moreover, the value of E_a is also much higher than $|G^*|_b$. Therefore, the contributions made by the aggregate phase to the modulus of the asphalt mix are expected to be higher than other phases. If a non-frequency/temperature-dependent factor was introduced for evaluating the contribution from the aggregate phase to

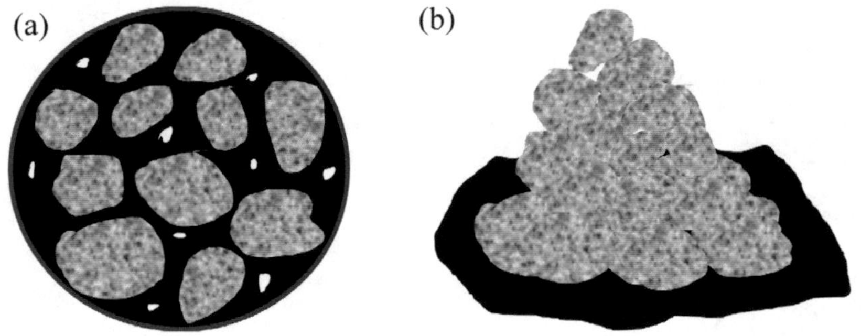

Figure 5. Organization of aggregates at different frequencies/temperatures. (*a*) High frequency/low temperature; (*b*) Low frequency/high temperature

the overall response of the mix, $|E^*|_{mix}$ would also be primarily frequency/temperature independent. This can be clearly deduced from Equation 5, where the aggregate phase contribution ($P_a f_a E_a$) becomes temperature/frequency independent as a result of the temperature/frequency-independent P_a.

In order to physically understand the concept of P_a, if hypothetically two extreme conditions of low temperatures (or high frequencies) and high temperatures (or low frequencies) are considered, then in the case of low temperatures the asphalt binder would be stiff and it would be able to bond the particles together well (Figure 5*a*), whereas, at high temperatures, the binder would be too soft to bind the particles (Figure 5*b*). In the former case, the whole structure will act together in the load-bearing capacity; on the contrary, in the latter case, the asphalt binder will not take part in the load-bearing capacity and the mix would slowly collapse.

From the above discussions, it can be stated that the introduction of the factor, P_a, is logical and meaningful. It is also expected to capture the frequency/temperature-dependent contribution of the aggregate phase in predicting $|E^*|_{mix}$. P_a, which is termed the "aggregates organization factor," has been determined and validated by the laboratory tests in this study. The description of the sample preparation, laboratory tests, test results and analysis will be presented in the later sections.

Material Properties

Two OGM specimens and one dense asphalt concrete (DAC) specimen, denoted as "OGM-1," "OGM-2," and "DAC", were prepared. Specimen OGM-1 was utilized for carrying out necessary tests to calibrate the values of P_a, whereas specimens OGM-2 and DAC were used for the subsequent validation of the proposed model. The properties of different used materials are shown in Table 1.

These specimens with a height of 150 mm and a diameter of 100 mm were prepared by using a gyratory compactor in accordance with the AASHTO T 342-11 standard method (*14*). The laboratory test methods to measure the mechanical properties of these specimens will be introduced in the next section.

Laboratory Tests

Dynamic Shear Rheology Test

The values of $|G^*|_b$ was measured by using the parallel plate configuration of dynamic shear rheology equipment. Frequency sweep tests were conducted in a frequency range of 50–0.1 Hz, at four different temperatures, –10°C, 4°C, 21°C and 37°C. Two geometrically different asphalt binder specimens 8 mm and 25 mm in diameter, and 2 mm and 1 mm in height, were used.

Uniaxial Compression Test

The values of $|E^*|_{mix}$ were measured under cyclic sinusoidal uniaxial compression load by universal testing machine (UTM) at different frequencies and temperatures. In accordance with the AASHTO T 342-11 standard method (*14*), four temperatures, –10°C, 4°C, 21°C and 37°C, were applied during the test and for each temperature, six different frequencies, which were 20 Hz, 10 Hz, 5 Hz, 1 Hz, 0.5 Hz, and 0.1 Hz, were performed. The load applied on the specimens was under stress-controlled mode and three linear variable differential transformer sensors were installed to measure their displacements.

Results and Discussion

According to the time–temperature superposition principle, the measured results of $|G^*|_b$ and the phase angle δ_b of asphalt binder at different temperatures were shifted

Table 1. Properties of Materials

Type of materials	Gradation of OGM (% passing)	Gradation of DAC (% passing)	Density (kg/m^3)	Volume fraction OGM-1	Volume fraction OGM-2	Volume fraction DAC
16	98.3	97	2686	0.735	0.777	0.831
11.2	77.3	85	2686			
8	43.8	74.5	2678			
5.6	22.1	57.5	2670			
2	15	40	2673			
0.063	4.1	7	2658			
Filler	0	0	2638			
Asphalt 70/100	4.5	6.4	1032	0.086	0.091	0.137
Air voids	-	-	-	0.179	0.132	0.032

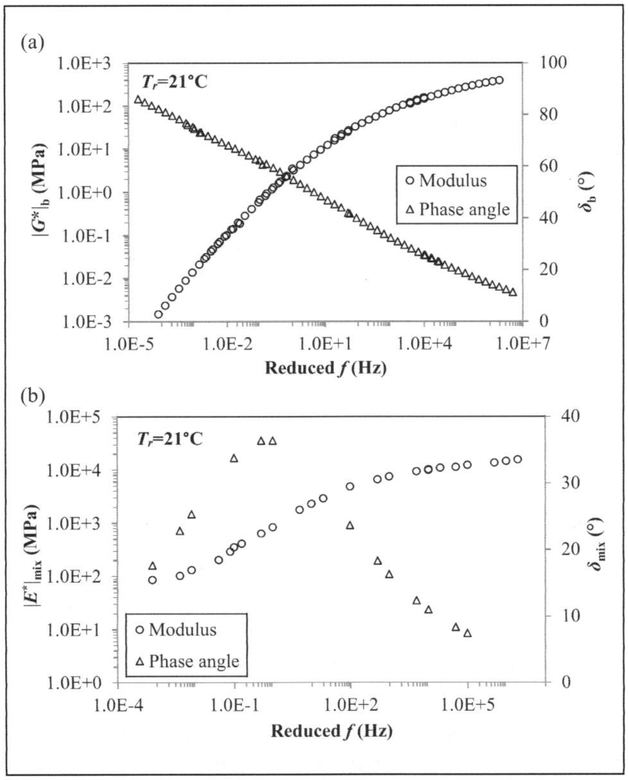

Figure 6. Laboratory test results of asphalt binder and OGM-1. (*a*) Master curve of $|G^*|_b$ and δ_b; (*b*) Master curve of $|E^*|_{mix}$ and δ_{mix}.

to obtain master curves over a large range of frequencies at a reference temperature (T_r) of 21°C (Figure 6*a*). The plots tend to follow the power law curve in the test frequency range, which is consistent with the results from other research (*15*).

From the UTM tests, the master curves of $|E^*|_{mix}$ and phase angle δ_{mix} of OGM-1 specimen were obtained at the same reference temperature (Figure 6*b*). It can be observed that the value of $|E^*|_{mix}$ decreases from high frequencies to low frequencies. The plot of $|E^*|_{mix}$ reaches a stable plateau at high/low frequencies beyond which there is only a slight change in the value of $|E^*|_{mix}$. This trend further solidifies the hypothesis made regarding P_a in the earlier section that weaker bond characteristics among aggregate particles are expected with decreasing frequencies.

The value of δ_{mix} reaches a peak and starts to further decline as a result of the decrease in frequency. This indicates that the mix behaves elastically both at higher and lower frequencies. This behavior can also be explained in terms of P_a where at low temperatures the binder itself starts to behave elastically thus the overall mix response is elastic; on the other hand, at high temperatures P_a is too low to contribute in the load-bearing capacity of the mix, and that is why the bulk of loading is taken over by the aggregates themselves, which are elastic in nature. A laboratory-based approach to determine P_a will be presented in the following paragraphs.

Determination of P_a from Laboratory Tests

Equation 5 can be rearranged in the form of the following equation, Equation 6:

$$P_a(f) = \frac{|E^*|_{mix}(f) - 3f_b|G^*|_b(f)}{f_a E_a} \tag{6}$$

Where f_a, f_b, E_a and $|G^*|_b$ are described in Equation 1; the values of f_a and f_b can be found in Table 1 and the value of E_a is assumed as 53000 MPa.

The values of P_a can be calculated with the mix properties and Equation 6. A typical curve as obtained for specimen OGM-1 is shown in Figure 7. The shape of the plot shows that P_a is indeed a frequency-dependent factor as discussed before. It can also be observed that P_a increases with the increase in frequency and more or less follows a sigmoidal curve for the tested specimen. This in

Table 2. Fitting Results of Constant Parameters

Parameters	*a*	*b*	*c*	*d*
Values	0.0017	0.62	0.72	–0.17

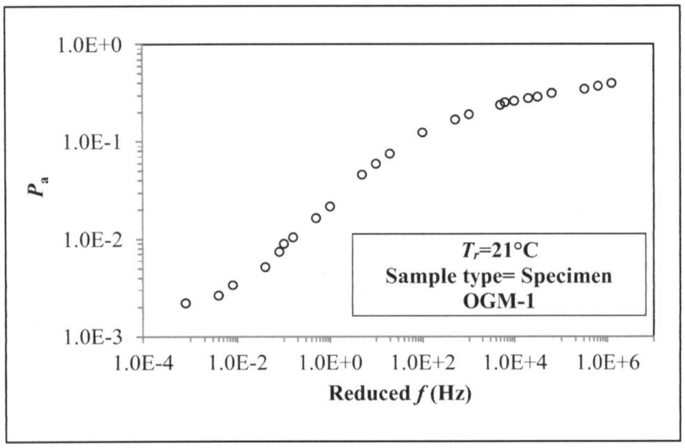

Figure 7. Calculation results of P_a.

terms of physical meaning implies that the contribution made by the aggregate phase to the modulus of OGM increases with the frequencies.

By assuming P_a to follow a sigmoidal curve, Equation 7 can be deduced. It is noted here that Equation 7 is similar in nature to the P_c function as described elsewhere (*10*). After fitting the test results in Equation 7, constant parameters of the equation can be obtained, as shown in Table 2.

$$P_a = a + (1 - a)\frac{\exp(b + c\ln(f_b/(f_b + f_v) \times |G^*|_b) + d(f_b + f_v))}{1 + \exp(b + c\ln(f_b/(f_b + f_v) \times |G^*|_b) + d(f_b + f_v))} \quad (7)$$

where f_a, f_b, f_v and $|G^*|_b$ are described in Equation 1.

It is important that the proposed model is validated. In order to judge the suitability of P_a, a two-step validation approach was adopted. In the first step the overall response of the specimen OGM-1 was compared against the predicted results, and in the second step, predictions were made for the specimen OGM-2 and the specimen DAC on the basis of parameters obtained in Table 2.

Validation of the Modified Expression of the Christensen Model

Figure 8, *a–c*, shows the predicted results of $|E^*|_{mix}$ for specimen OGM-1, specimen OGM-2 and specimen DAC, respectively. It can be observed that the predicted results of $|E^*|_{mix}$ for specimen OGM-1 are in a good agreement with the laboratory tests, which provides a good check for the calibration procedure.

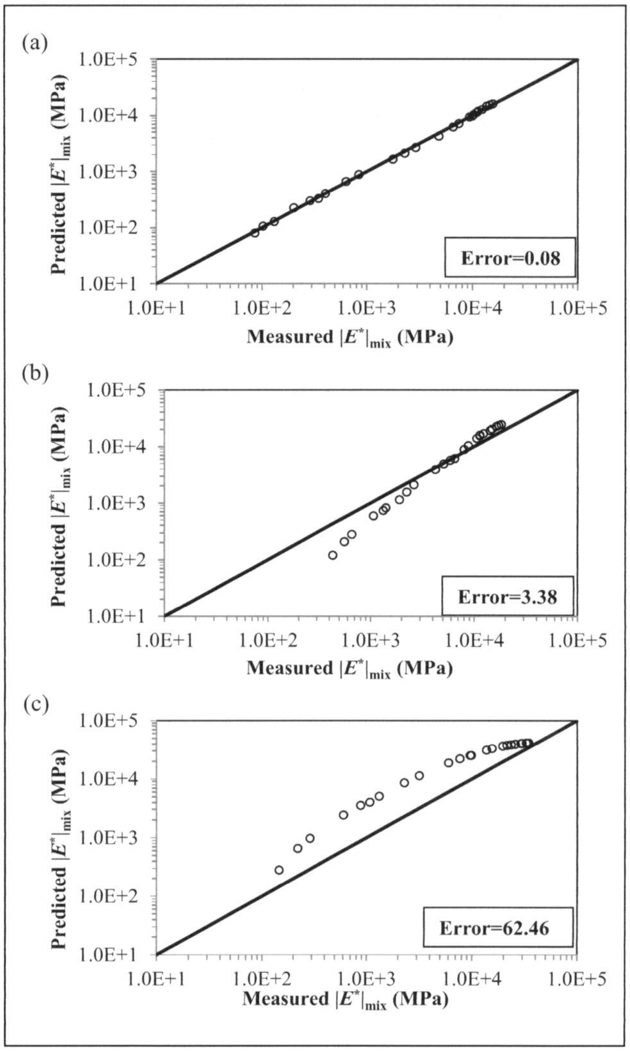

Figure 8. Comparison between predicted results of $|E^*|_{mix}$ to the laboratory tests. (*a*) Specimen OGM-1; (*b*) Specimen OGM-2; (*c*) Specimen DAC.

The predicted results of $|E^*|_{mix}$ for specimen OGM-2 on the basis of parameters obtained from specimen OGM-1 fit with reasonable accuracy. These differences were already expected because the parameters used for both OGM specimens in Equation 7, although with different properties (see Table 1), are the same. Despite these differences, the predicted and the test results for practical purposes match quite well for OGM.

A deliberate comparison between two altogether different types of mixes was further made to judge the suitability of the proposed model. As one would expect, it is

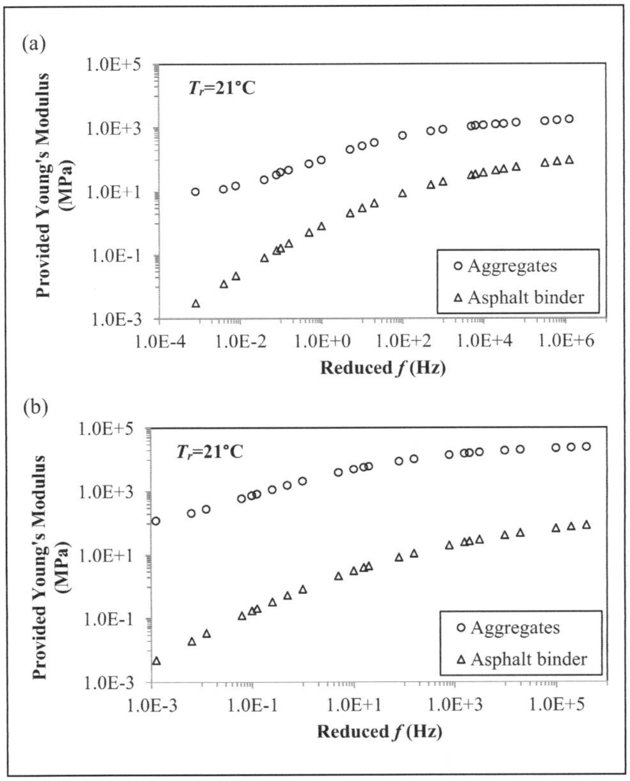

Figure 9. Modulus provided by aggregates and asphalt binder in OGM. (*a*) Specimen OGM-1; (*b*) Specimen OGM-2.

clear that the differences between these two results are substantial (Figure 8*c*).

Overall, it can be concluded that if the parameters are obtained for one type of mix, the prediction could be made with fair accuracy, and if the type of mix is completely changed then new parameters would need to be calibrated.

Contributions of Each Phase in the $|E^*|_{mix}$ *Prediction*

According to Equation 5, the contribution of the aggregate phase to the total mix modulus is obtained by $P_a f_a E_a$, whereas the contribution of the asphalt binder phase could be obtained by $3f_b|G^*|_b$. Figure 9, *a* and *b*, shows the plots of the relative contribution of each phase in the overall response of the OGM specimens. It can be seen that both phases are frequency dependent as explained earlier.

As expected it is found that the aggregate phase makes a more significant contribution to the modulus of OGM. It can be further deduced that in OGM the aggregate phase provides the load-bearing capacity while the role of the asphalt binder phase is to bind the aggregate particles together ensuring the structural organization of the composite material.

Conclusions

Homogenization methodology makes the analysis of OGM accessible for determining the propensity of raveling. In order to describe the frequency/temperature-dependent contribution of each phase to the overall response of OGM, the expression of the Christensen model was modified and a temperature/frequency-dependent factor P_a was proposed. A function for calculating P_a was determined and verified against test results. Furthermore, the contributions made by each phase to the modulus of OGM were analyzed. The following conclusions were drawn:

- By means of P_a, the effects of the frequency/temperature-dependent contribution of the aggregate phase on the overall mix response can be accounted for, and the modified expression of the Christensen model can produce the shape and values of the frequency/temperature-dependent modulus of asphalt mixes.
- The calibrated function for describing P_a can only obtain accurately predicted results of the material which it is calibrated for.
- In the whole range of frequencies/temperatures, the contribution made by the aggregate phase to the overall modulus of OGM is much more significant than the contribution of the asphalt binder phase.

Current Research

The basic limitation of the presented work is the need to experimentally determine P_a. In future research, the current work will be extended to enable the prediction of asphalt mix properties exclusively on the basis of the mechanical properties of its constituents.

Author Contributions

The authors confirm contribution to the paper as follows: study conception and design: Athanasios Scarpas, Kumar Anupam, Hong Zhang; data collection: Hong Zhang; analysis and interpretation of results: Hong Zhang, Kumar Anupam, Athanasios Scarpas, Cor Kasbergen; draft manuscript preparation: Kumar Anupam, Hong Zhang. All authors reviewed and the results and approved the final version of the manuscript.

References

1. Hagos, E. T. *The Effect of Aging on Binder Properties of Porous Asphalt Concrete*. Delft University of Technology, Delft, 2008.
2. Mo, L., M. Huurman, S. Wu, and A. A. A. Molenaar. Ravelling Investigation of Porous Asphalt Concrete Based on Fatigue Characteristics of Bitumen–Stone Adhesion

and Mortar. *Materials & Design*, Vol. 30, No. 1, 2009, pp. 170–179.
3. Khedoe, R. N., and J. Moraal. *Lifetime Optimisation Tool: Sample Preparation and Laboratory Testing for the LOT*. Delft University of Technology, Delft, 2007.
4. Kringos, N., and A. Scarpas. Raveling of Asphaltic Mixes Due to Water Damage: Computational Identification of Controlling Parameters. *Transportation Research Record: Journal of the Transportation Research Board*, 2005. 1929: 79–87.
5. Kringos, N., and A. Scarpas. Development of a Finite Element Tool for Simulation of Ravelling of Asphaltic Mixes. *Proc., International Conference on Computational and Experimental Engineering and Sciences (ICCES)*, Madeira, 2004, pp. 1211–1216.
6. Anupam, K., S. K. Srirangam, A. Varveri, C. Kasbergen, and A. Scarpas. Microstructural Analysis of Porous Asphalt Concrete Mix Subjected to Rolling Truck Tire Loads. *Transportation Research Record: Journal of the Transportation Research Board*, 2016. 2575: 113–122.
7. Ghossein, E. *Numerical Validation of Analytical Homogenization Models for the Case of Randomly Distributed and Oriented Ellipsoidal Fibers Reinforced Composites*. Université de Montréal, Montreal, 2014.
8. Eberhardsteiner, L. *Extension of a Multiscale Model for Hot Mix Asphalt Stiffness Prediction to Introduce Binder Aging into Pavement Design*. Vienna University of Technology, Wien, 2014.
9. Hirsch, T. J. Modulus of Elasticity of Concrete Affected by Elastic Moduli of Cement Paste Matrix and Aggregate. *Proceeding of the American Concrete Institute*, Vol. 59, No. 3, 1962, pp. 427–452.
10. Christensen, D. W., J. T. Pellinen, and R. F. Bonaquist. Hirsch Model for Estimating the Modulus of Asphalt Concrete. *Journal of the Association of Asphalt Paving Technologists*, 2 Vol. 72, 2003, pp. 97–121.
11. Dongre, R., L. Myers, J. D'Angelo, C. Paugh, and J. Gudimettla. Field Evaluation of Witczak and Hirsch Models for Predicting Dynamic Modulus of Hot-Mix Asphalt. *Journal of the Association of Asphalt Paving Technologists*, Vol. 74, 2005, pp. 381–442.
12. Pellinen, T., A. Zofka, M. Marasteanu, and N. Funk. Asphalt Mix Stiffness Predictive Models. *Journal of the Association of Asphalt Paving Technologists*, Vol. 76, 2007, pp. 575–625.
13. Christensen, D. W., and R. Bonaquist. Improved Hirsch Model for Estimating the Modulus of Hot-Mix Asphalt. *Road Materials and Pavement Design*, Vol. 16, 2015, pp. 254–274.
14. American Association of State Highway Transportation Officials. *Determining Dynamic Modulus of Hot-Mix Asphalt Concrete Mixes*. Washington, D.C.: AASHTO T342-11, 2011.
15. Underwood, B. S. *Multiscale Constitutive Modelling of Asphalt Concrete*. ProQuest Dissertations and Theses, Raleigh, N.C., 2011.

The Standing Committee on Pavement Condition Evaluation (AFD20) peer-reviewed this paper (18-03260).

Article

TRR
JOURNAL OF THE TRANSPORTATION RESEARCH BOARD

Transportation Research Record
2018, Vol. 2672(40) 41–49

DOI: 10.1177/0361198118759951
journals.sagepub.com/home/trr
SAGE

An Area-Based Faulting Measurement Method Using Three-Dimensional Pavement Data

Georgene M. Geary[1], **Yichang (James) Tsai**[1,2], **and Yiching Wu**[1]

Abstract

Faulting is one of the three performance measures recognized by the Code of Federal Regulations (CFR) for jointed concrete pavements. U.S. Federal Highway Administration rules developed for the CFR require the state Departments of Transportation to use the AASHTO Standard R 36, Standard Method for Evaluating Faulting of Concrete Pavements, for measurement of faulting. R 36 allows both manual testing using a faultmeter and automatic testing using a high-speed inertial profiler (HSIP). There is a concern in the literature that the HSIP methods are not accurate enough for network level testing of faulting. Recent studies have also shown that three-dimensional (3-D) laser technology is showing promise for providing more accurate faulting measurements. Still, there currently is not a specific method in R 36 that utilizes 3-D data for gathering faulting measurements. This paper proposes a new method that takes full advantage of the full-coverage capabilities of 3-D to measure faulting more accurately and consistently. This method uses the comparison of smoothed areas on both sides of a joint instead of a single longitudinal profile to measure elevation differences. Field tests were performed to compare the 3-D method with manual readings taken with a Georgia faultmeter on two sections of interstate in Georgia. In addition, the impact of various footprints and measuring locations of faulting measurements using this 3-D technology were quantitatively evaluated using the proposed method. Based on this case study the new proposed 3-D method appears to be appropriate to be considered to be added as a 3-D automatic method in AASHTO R 36.

Faulting, cracking, and international roughness index (IRI) are the three performance measures recognized in the Code of Federal Regulations (23 CFR Part 490) for jointed concrete pavements. State Departments of Transportation (DOTs) must report these pavement performance measures to the U.S. Federal Highway Administration (FHWA) starting in 2018. As part of the recent regulation changes, the percentage of roadways in Good and Poor condition will be monitored by the federal government, and each state DOTs use of federal funding will be impacted by these measurements. All three pavement measurements must fall into a good performance region for a section of road to be rated Good (*1*). Therefore, each of these measurements is critical for recognizing good performing jointed concrete pavements.

IRI measurements are well defined and several well-vetted standards are used for IRI such as the American Association of State Highway and Transportation Officials (AASHTO) Standards M 328, R 54, and R 56. Certification programs are also available for the high-speed inertial profiler (HSIP) equipment used to measure IRI (*2*).

Cracking, even though somewhat more nebulous than IRI, does now have a clear standard definition in the latest Highway Performance Monitoring System (HPMS) field manual. The HPMS field manual considers it to be transverse cracking of a slab that extends at least one-half the width of the lane (*3*). Cracking is not measured per se, but is considered as yes or no per slab for classification purposes, so the measurement of cracking is not subject to much measurement uncertainty.

Faulting, on the other hand, even though it has been performed by the long-term pavement performance program (LTPP) and many state DOTs for over 20 years, and although a simple concept, is not as defined a measurement as IRI or cracking. FHWA rules developed for the CFR require the state DOTs to use the AASHTO Standard R 36, Standard Method for Evaluating Faulting of Concrete Pavements, for measurement of faulting (*1*). AASHTO R 36 provides both manual and automatic methods to compute faulting (*4*). However, the manual method is not realistically a viable option because of the FHWA requirement of measuring every joint on the network, every year. The manual method described in the R 36 standard is also not equivalent to the

[1]School of Civil and Environmental Engineering, Georgia Institute of Technology, Atlanta, GA
[2]Changjiang Scholar, Chang'an University, Xi'an, China

Corresponding Author:
Address correspondence to Georgene M. Geary: ggeary@gatech.edu

manual faultmeter equipment that has been used for over 20 years. As reported in a 2012 ASTM international report on pavement performance measures "Currently, no such manual devices have been built to meet the requirements in AASHTO R 36-12"(*5*). Florida DOT recently constructed and tested a faultmeter that meets the R 36 requirements; their effort will be described later in this paper. AASHTO R 36 also includes two automatic methods that use the longitudinal profile from the same HSIP equipment that is used for IRI. Even though the HSIP have been widely used and validated for IRI measurements, they have not been as extensively tested for faulting measurements, and when they have there have been accuracy concerns. A 2016 report developed for FHWA (*6*) noted a concern with collecting reliable and repeatable faulting data with HSIP equipment alone and suggested that 3-D methods be pursued. Some states that use HSIP for faulting, such as Illinois DOT, specifically refer to using manual faulting equipment in the field to check on unusual measurements because of the concern for "inaccuracy and misrepresentation" of automatic faulting measurements (*7*).

This paper is organized first with a review of faulting and the faulting measurement literature. Next the proposed 3-D method is defined. Following this, the field test is described and the data are presented. Finally, the paper concludes with a summary of the results and recommendations for future analysis.

Review of Faulting and Faulting Measurements

Before discussing the proposed 3-D method it is appropriate to review the concept of faulting in jointed concrete pavements.

What is Faulting?

Faulting is caused by a change in elevation of a concrete slab near a joint. It is reported as the difference in elevation of an approach slab as compared to the elevation of a leave slab at a joint or crack. Based on the amount of movement and type of joint, it also involves loss of aggregate interlock and/or movement or distress of dowels in the case of doweled pavements. In colloquial terms faulting is the "thumpity-thump" you sometimes hear and feel on concrete pavements as you cross over the joints. Beyond noise, increases in faulting have been correlated to increases in IRI in previous studies (*8*). Faulting can be a major problem in undoweled pavements on erodible bases. As many states have gone to doweled jointed concrete pavements as standard practice, faulting of doweled pavements does not seem to be as big a concern with these newer pavements (*8, 9*). Even the original AASHO Road Test, which used dowels at all the transverse joints for the jointed concrete pavements, did not experience faulting at the joints, only at the cracks (*10*). Nonetheless, because of heavier loads and as the doweled sections age considerably, and as more concrete overlays (typically composed of undoweled panels) are utilized, faulting will continue to be an important pavement distress measure into the future.

How is Faulting Measured?

Manual faulting measurements have long been performed using a Georgia Faultmeter (GFM), which was first built by the Georgia DOT in 1987. Georgia DOT also built the first modified version used by SHRP for LTPP testing in the late 1980s, termed in this paper the SHRP faultmeter (SFM). Figure 1 shows how a faulting measurement is taken with the GFM. The legs are placed on the leave slab and the probe measures the faulting from the approach slab. The GFM reads out in positive or negative integer readings (i.e., −2, −1, 0, 1, 2) that are equivalent to 1/32 in. (0.03 in. or 0.8 mm) measurements. It can measure positive and negative faulting. Positive faulting is a drop in elevation along the direction of travel and negative faulting is a rise in elevation in the direction of travel, as shown in the figure. Positive faulting is considered the expected change because of traffic effects. Negative faulting, though still possible, has also been linked to data errors such as measuring faulting at cracked or repaired areas, excessive joint sealant, placing the GFM in the wrong direction, or just because of the accuracy of the measurement itself when the actual elevation difference is near 0 (*8*). The range of the GFM reading is from −20 to +20 or ±20/32 in. (±0.6 in. or ±16 mm) and it has a clear space between the closest leg to the probe and the probe of 50 mm (2 in.). The SFM is basically the same as the GFM except it was modified to read out in millimeters to the nearest millimeter (0.04 in. or 1 mm) and it was also modified to have 100 mm (4 in.) clear spacing between the closest leg and the probe, so it could also be used for shoulder drop-off readings (*11*). Minnesota DOT also made changes to the footprint of the faultmeter they used for the MNROADs test sections to improve repeatability. They found that a three leg "bolt" system was more stable than the four long legs used in the GFM, but they also found that marking where the exact location measurements were taken along the joint was also crucial to repeatable measurements (*12*). Of course, manual faulting measurements are time consuming, labor intensive, and can be a safety concern. Therefore, automatic methods have been developed using longitudinal profiles from HSIPs and these will be discussed in the next section in relation to the current faulting standard.

Current Faulting Standard. AASHTO R 36, Evaluating Faulting of Concrete Pavements, is the only current standard for faulting measurements. R 36-17 currently provides three methods to measure the faulting value: manual, automatic method A, and automatic method B. The schematic for manual measurements in R 36 is different than the current GFM or SFM. The dimensions for the distance between the probe

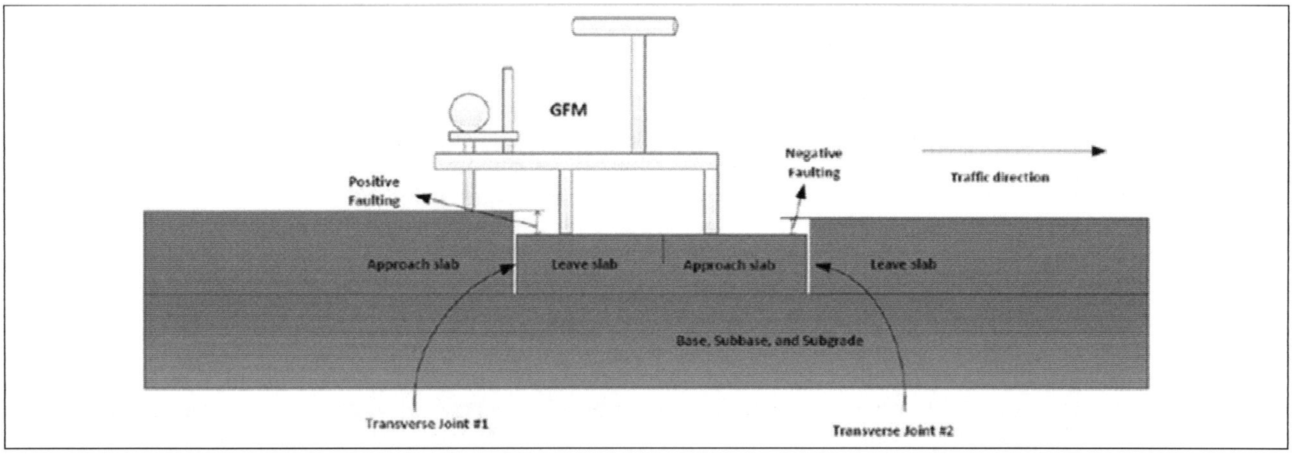

Figure 1. LTPP faultmeter operation (*9*).

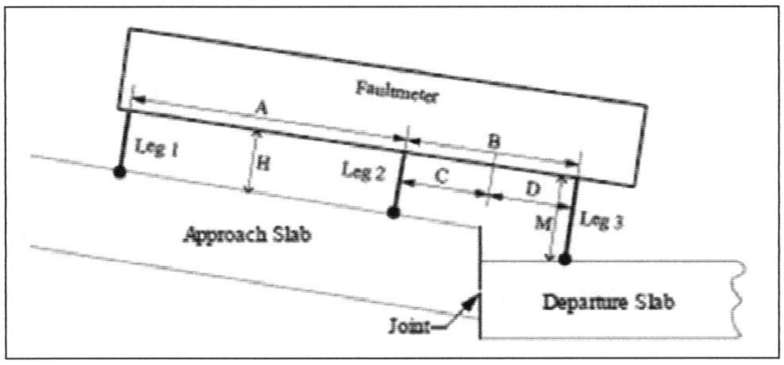

Figure 2. Generic faultmeter from AASHTO R 36 Standard (*4*).

and the joint and the joint and the front legs (C and D as shown in Figure 2) are each noted to be between 76 and 226 mm (3 to 8.9 in). Therefore, the R 36 generic faultmeter has a clear space between the closest leg to the probe and the probe of 152 to 452 mm (6 to 17.8 in.) as compared to 50 mm (2 in.) for the GFM and 100 mm (4 in.) for the SFM. Also, the probe in R 36 is on the departure slab and the legs are on the approach slab whereas the probe for the GFM and SFM (see Figure 1) is placed on the approach slab and the legs are placed on the departure slab. The automatic methods both use one longitudinal profile from a HSIP to compute faulting.

Automatic method A uses a regression line for the approach and departure slab to identify elevation measurements on each side of the joint. The regressed length is 1219 mm (4 ft) on each side of the joint. The difference in elevations for a section consisting of a distance of 76 to 226 mm (3 to 8.9 in.) from the joint are averaged.

Method B is similar in that it averages the values between 76 and 226 mm (3 and 8.9 in.), but it uses the actual profile points and measures the elevation difference at 300 mm (11.8 in.) horizontal locations. Both methods have inherent issues. Using a 1219 mm (4 ft) regression will include the effect of profile roadway slope. Measuring elevation differences like method B will provide different results for a roadway on a constant slope as compared to a section that is changing slope, even if the faulting is the same.

Recently, comparison of manual and automatic faulting measurements accuracy using R 36 equipment and longitudinal profile data were documented by Florida DOT. They used their newly developed faultmeter that read out to 0.01 mm and met R 36 requirements, and they concluded that their faultmeter was repeatable to 0.42 mm in the field and that their automatic method was repeatable to 0.6 mm when the same HSIP was used, although the results increased to 0.9 mm when different HSIPs were used. Nonetheless, they experienced issues with joint identification, in which of the 93 joints marked and manually tested only 39 were consistently picked up with the HSIP profilers, and therefore they were the only joints compared for faulting, putting somewhat of an unknown bias into the analysis (*13*). They also identified a bias in the automatic procedure as compared to the manual faulting with a confidence limit of 0.2 mm (0.01 in.) to 0.7 mm (0.03 in.). It should also be noted that 36 of the 39 manual readings that were used measured less than a 4 mm elevation difference (0.125 in.).

Concerns with Current Faulting Methods. GFMs are historically considered to only be accurate to ± one reading (1 mm or 1/32 in.). Minnesota DOT made modifications to the original GFM to improve repeatability of their faulting measurements. By replacing the four legs with three bolt feet and adding an offset rod they noticed a change in the faulting trends. They further marked the actual locations on the pavement when they felt that repeatability because of surface texture could be an issue (*12*). Although these improvements are applicable to a static accelerated test facility like MNRoads, they cannot be incorporated into actual network level surveys.

Manual GFM readings are time consuming and as it is necessary to place the equipment in the travel lane to get a measurement, they can also impede traffic. There is always a safety issue when physically encroaching into a travel lane. Automatic methods are faster, and by their dynamic nature, can be much safer as they are performed at highway speeds.

Automated faulting methods can use a single longitudinal profile or 3-D data that provide both longitudinal and transverse information. Automated methods that use a single longitudinal profile can be affected by surface texture as the depth of typical tining is of the order of 4 mm (0.125 in.) (*14*), and therefore measuring faulting to less than 1 mm (0.03 in.) is a challenge using a longitudinal profile alone.

A recent study performed for FHWA on collecting network level data for the Highway Performance Monitoring System (HPMS) noted that the HSIP they used found an average of 0.4 joints as compared to 3.4 joints for a laser crack measurement system (LCMS) equipped vehicle. They noted that the LCMS (which is a form of 3-D data collection) count of 3.4 joints was much closer to the actual value of 3.6 joints per 0.01 mile section (*6*). As noted earlier, Florida DOT experienced a similar issue with their HSIP readings not picking up every joint using a longitudinal profile. This is a concern with HSIP and longitudinal profiles, and is a known challenge (*9, 15*). Even though this paper does not address automatic joint location, only faulting measurements, 3-D pavement data naturally would lend itself to easier identification of joints as compared to a single longitudinal profile.

Proposed Method

A recent (2015) survey as part of an NCHRP Synthesis 501 on Putting Data to Work (*16*) noted that out of the 41 respondents, 18 were already using or in the act of implementing 3-D automated data collection for pavement condition surveys. Additionally, 17 respondents noted that they plan to go to 3-D data within two years. If this occurs, more than one-half of the state DOTs will have access to 3-D type data in the near future.

3-D analysis of faulting data has the benefit of being able to identify joints and also provides a 3-D view of the edge of the pavement that can be used to smooth out irregularities, such as spalling, and is also able to measure as close to the joint as possible to remove curl, warp, and longitudinal profile aspects. Preliminary studies of faulting measurements using 3-D data performed by Tsai (*17, 18*) and others (*19*) have shown potential improved results over HSIP methods.

Description of Proposed Method

The placement and functionality of a GFM was considered in developing a method to measure faulting using 3-D data. The GFM places a probe on the approach side as shown previously in Figure 1. It rests on the departure side, essentially creating a plane. With manual placement the operator can observe spalling or other irregularities and place the device such that it does not include those areas. With an automatic method that is not possible, but to address that issue a rectangular area was chosen, instead of a point location, to represent both the approach and departure slab. In this manner, any irregularities can be smoothed out, or identified by also averaging the data from several points within the area and identifying the standard deviation of the data. Therefore, the proposed faulting measurement method using 3-D data involves measuring within a smoothed rectangular shaped area on each side of the joint as shown in Figure 3*a*. The rectangular box shown in the figure is located 120 mm from the lane line and 10 mm from the center of the joint and the box is 20 mm longitudinally by 200 mm transversely. The concept is similar to method B in R 36 in which the values are averaged, although in contrast to R 36 in which the values are averaged along the longitudinal profile, the elevation differences are computed at five points within the box and then are averaged transversely along the joint. Five measurements are taken within each smoothed box and each elevation difference is computed (X2 – X1) and then the average is computed for each joint as shown in Figure 3*b*. The standard deviation is also computed for each joint, which can be checked for anomalies. A program was developed in MATLAB to read the 3-D files and allow for the change of the size and location of the measurement box, and also for computing the elevation difference at several points within the area. The area is located in reference to the center of the joint and the lane marking. The results were compared to ground truth performed by a Georgia Faultmeter and the field test is described in the next section.

Field Test of Proposed Method

A field test was performed on February 27, 2017 on Interstate 16 in central Georgia to compare manual GFM measurements to faulting from 3-D pavement data. Faulting at 20 joints (joints 1 to 20) was measured between Eastbound MP 6-7 and faulting at 20 joints (joints 21 to 40) was measured between Westbound MP 15-14. MP 6-7 consists of 30 ft slabs on 90 degree joints. MP 15-14 consists of slabs at a random spacing (17 ft, 23 ft, 22 ft, 16 ft) with joints skewed at 10 degrees. Both sections were originally constructed in the 1970s with transverse tining, but now have a diamond ground surface. The actual joint locations within the 2 miles were chosen to get the widest range of faulting possible. The GFM readings varied from −9 to +17 (negative 9/32 in. to positive 17/32 in.) or −7 mm to +13.4 mm. A GFM in use by the Georgia DOT was used and Georgia DOT personnel that normally do the manual surveys for Georgia GDOT operated the GFM. The GFM was checked for calibration at the beginning of each mile section. The GFM reads out in whole numbers. It was identified that the GFM truncates readings instead of rounding (i.e. an actual reading of 2.9 would be seen as 2 and recorded as 2 instead of 3 for the GFM). Three readings were taken at locations 8 to 40, whereas one reading

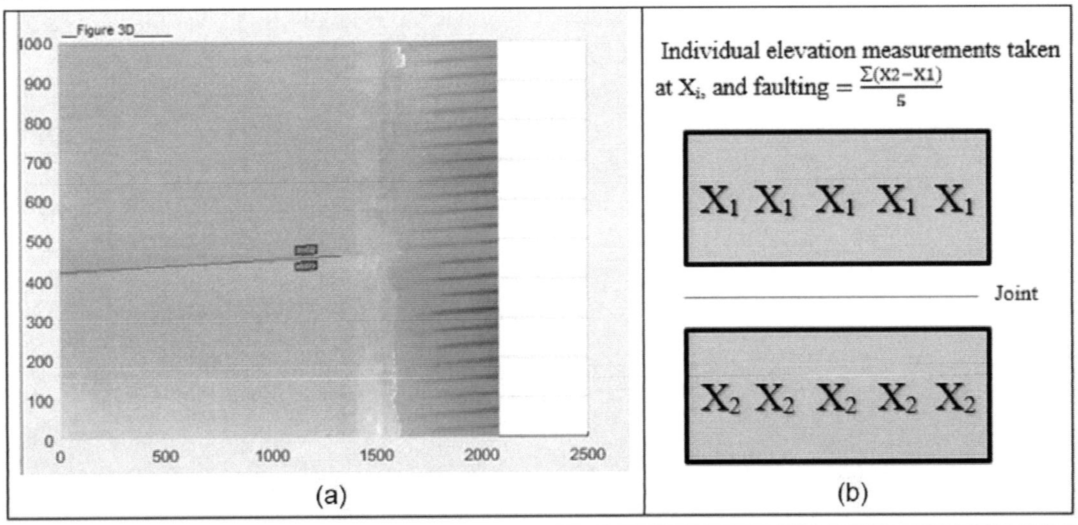

Figure 3. (a) Location of smoothed areas; (b) Averaging method used to calculate faulting.

Figure 4. Template used to mark faulting locations in the field.

was taken at locations 1 to 7. The joint locations were marked with a template to allow for a consistent repeat reading with the GFM, and to locate the exact joints tested, as shown in Figure 4. The Georgia Tech Sensing Van collected 3-D pavement data over the sections the same day, after all the manual readings were taken.

Field Test Results. As noted, three replicates were taken at 33 of the 40 joints. Although a template was used, and the faultmeter was placed back within the template markings each time, the GFM readings for each set of the faultmeter did vary. Of the 33 readings with three replicates, only 14 consistently read the same GFM value each of the three times. This could be because of inherent variability attributable to the measurement increment (0.8 mm) and pavement irregularities, or potentially because of rounding of the GFM value as noted earlier. The 3-D faulting data were first computed by identifying the location of the probe marking in the image (the black dot inside the template as shown in Figure 4) from the distance from the shoulder, and using a 20 mm by 200 mm rectangular box with five measurements taken in the box. The box was located 10 mm off the joint on each side, which provided an area measurement located between 10 and 30 mm off the joint, to match the 25 mm probe to joint measurement used by the GFM. The comparisons of the GFM and 3-D elevation differences derived this way are shown in Figure 5 (note that the GFM values were converted to millimeters). The GFM readings are shown by the dots and the 3-D readings are shown by the X. When more than one GFM reading was recorded at a location then a vertical line is shown depicting the range of the GFM readings. An apparent bias between the GFM and 3-D readings are displayed. 90% of the 3-D readings are lower than the average GFM readings. An average bias of 0.6 mm was observed.

The size of the rectangular box was increased but the best results were found when the box size was maintained at 20 mm by 100 mm. Next the location of the box was varied. The box was located at a consistent distance of 120, 220, 320, 420, and 640 mm off the lane line (shoulder). The box was also moved away from the joint at distances of 10, 20, 40, 60, 80, 100, 200, and 300 mm. Figure 6 shows an example of the locations of the boxes for the different measurements. The photo shows only approximately half of the lane. The shoulder is on the right-hand side of the photo and the outside lane marking is directly to the left of the shoulder. The approach slab is in the bottom of the picture and the departure slab in the top. The joint is mirrored by the boxes representing the different locations of the tests for the sensitivity analysis. The values shown are in millimeters. The results of this sensitivity analysis are further described below.

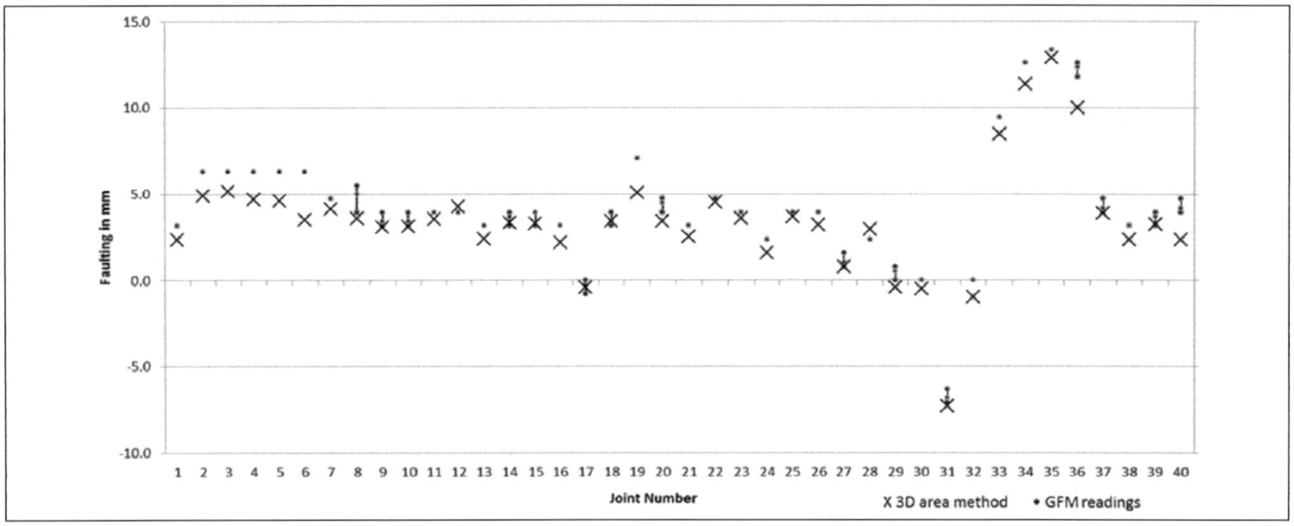

Figure 5. GFM readings versus the 3-D method for 40 joint field test.

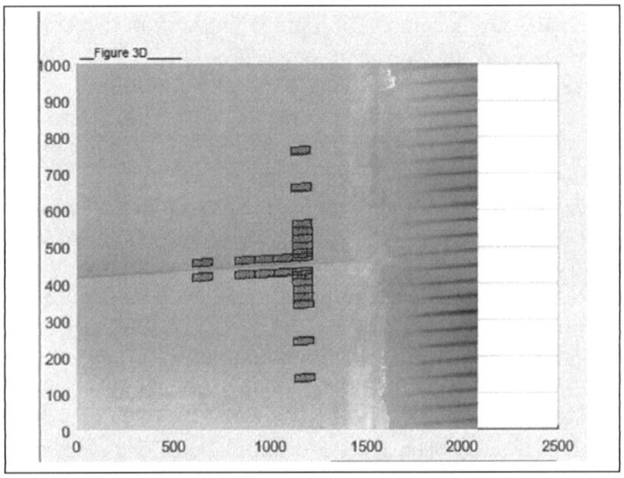

Figure 6. Varied measurement locations considered.

Sensitivity Analysis—Distance from the Shoulder. The elevation differences when placing the box 120 mm, 220 mm, 320 mm, 420 mm, and 640 mm from the lane line were calculated for all the joints. For clarity, only six joints are shown in Figure 7. The figure shows some of the joints (joints 2, 20, 10, 6, 1, and 24) that had the highest and least positive and negative difference between the 120 mm reading and the 640 mm reading. 120 mm and 640 mm were chosen as the GFM is typically measured around 120 mm from the shoulder, and the wheelpaths are considered between 300 mm and 750 mm from the shoulder. Some of the joints showed a higher faulting value at the edge and some showed a higher value farther from the edge. Joints not shown here were similar to the ones shown; they varied in values but no consistent trend was identified overall. As can be seen, the values did not vary by more than 1 mm either way for most of the joints shown. This was also the case for all the joints, as the average absolute difference between the elevation difference at 120 mm and 640 mm was 0.57 mm. This result is consistent with comparison of SFM readings at the edge and wheelpath performed using LTPP data in a faulting study published by FHWA in 2000. In that study, it was found that 90% of the joints that they compared from the LTPP database had less than a ±1 mm difference in faulting readings between the edge and the wheelpath of the same joint. The authors felt that the variation was more related to the accuracy of the equipment used to measure faulting than true elevation differences (*8*).

The joints with higher measured GFM faulting at the edge (joints 33 to 36 in Figure 5) consistently had the highest differences between the edge and wheelpath readings, and in all cases the joints with the highest GFM values (GFM readings over 5 mm) went down as the readings were taken farther from the shoulder, as shown in Figure 8. This would tend to indicate that measuring faulting at the edge instead of the wheelpath should provide more conservative faulting values. The benefit of 3-D pavement data are that this hypothesis could be checked for each section of roadway relatively easy, as both the edge and wheelpaths are covered in the data.

Sensitivity Analysis—Distance from the Joint. As expected, the elevation difference computed at different locations away from the joint varied more than in the case of transverse distance from the shoulder. Changes in elevation related to distance from the joint can come from curl and warp of the slabs, profile elevation changes or cracking of the slabs. For the first section (first 19 joints) the elevation difference computed at each location of the box is shown in Figure 9*a*. Figure 9*b* is a close-up of the area showing only the distance between 10 mm and 60 mm. The figures show

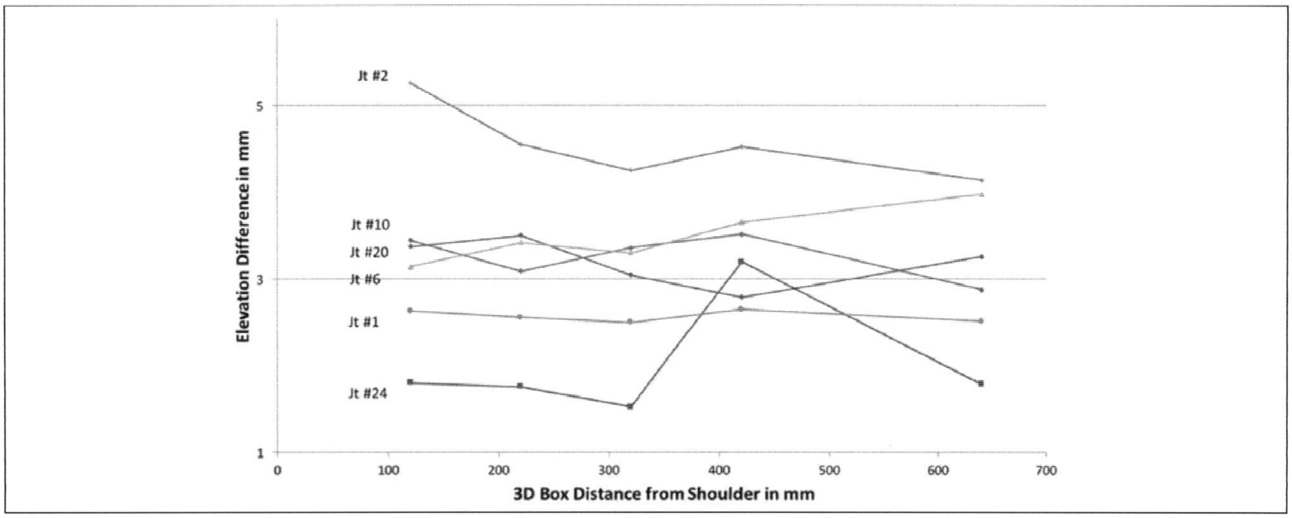

Figure 7. Variability in faulting reading based on distance from shoulder.

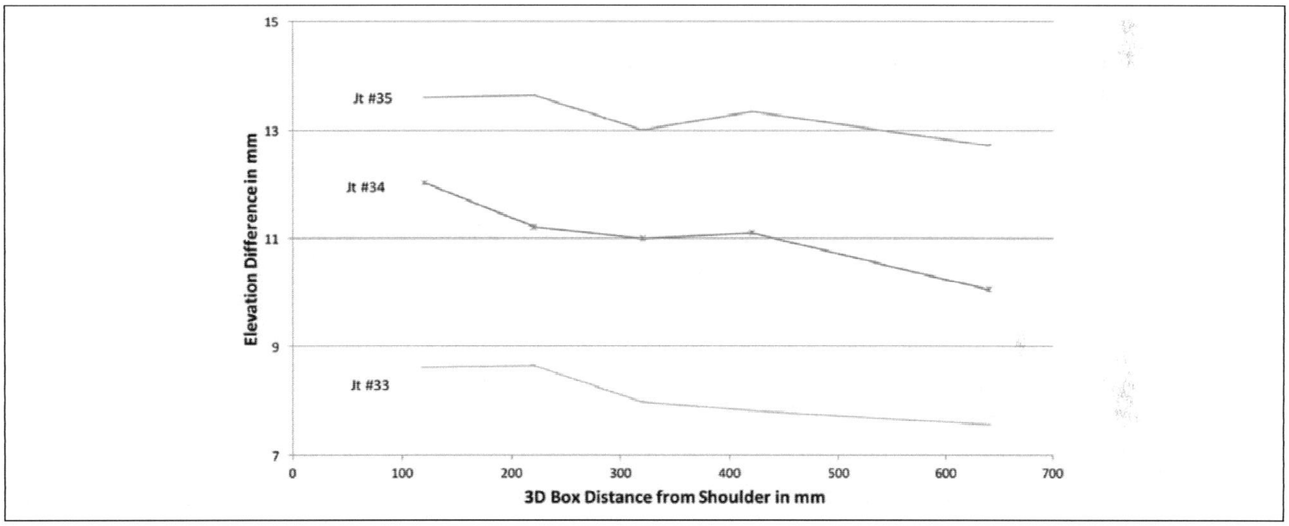

Figure 8. Reduction in faulting measurement when measured farther from the shoulder.

an almost consistent trend of the measured elevation difference becoming smaller as the measurement is taken farther from the joint. Once again, this would tend to indicate that measuring faulting closer to the joint should provide more conservative faulting values. A 3-D depiction of joint 6 is shown in Figure 10 to illustrate the differences more clearly.

Figure 10 illustrates why the elevation difference is getting smaller and even negative, as the measurement is taken farther from the joint. At the joint (located at a horizontal distance of 300 mm in the figure) the approach slab clearly is higher than the departure slab, but if readings are taken farther from the joint the departure slab is at a higher elevation than the approach slab, providing a negative elevation difference. The three axes shown are in millimeters.

Conclusion

Even though this study was performed on a limited sample, at only two locations with similar texture, it did showcase a wide range of faulting values and two different joint orientations. The 3-D pavement data method described in this paper provides a different way to analyze faulting on jointed concrete pavements. Relatively consistent results were gathered when compared to manual GFM readings. Analysis of different locations for testing showed that the most comparable faulting measurement to the GFM is between 10 and 20 mm from the joint. The measurements closest to the joint also provided the most conservative values of faulting. AASHTO R 36 currently does not address 3-D data for faulting and this procedure could be developed into an alternative method C.

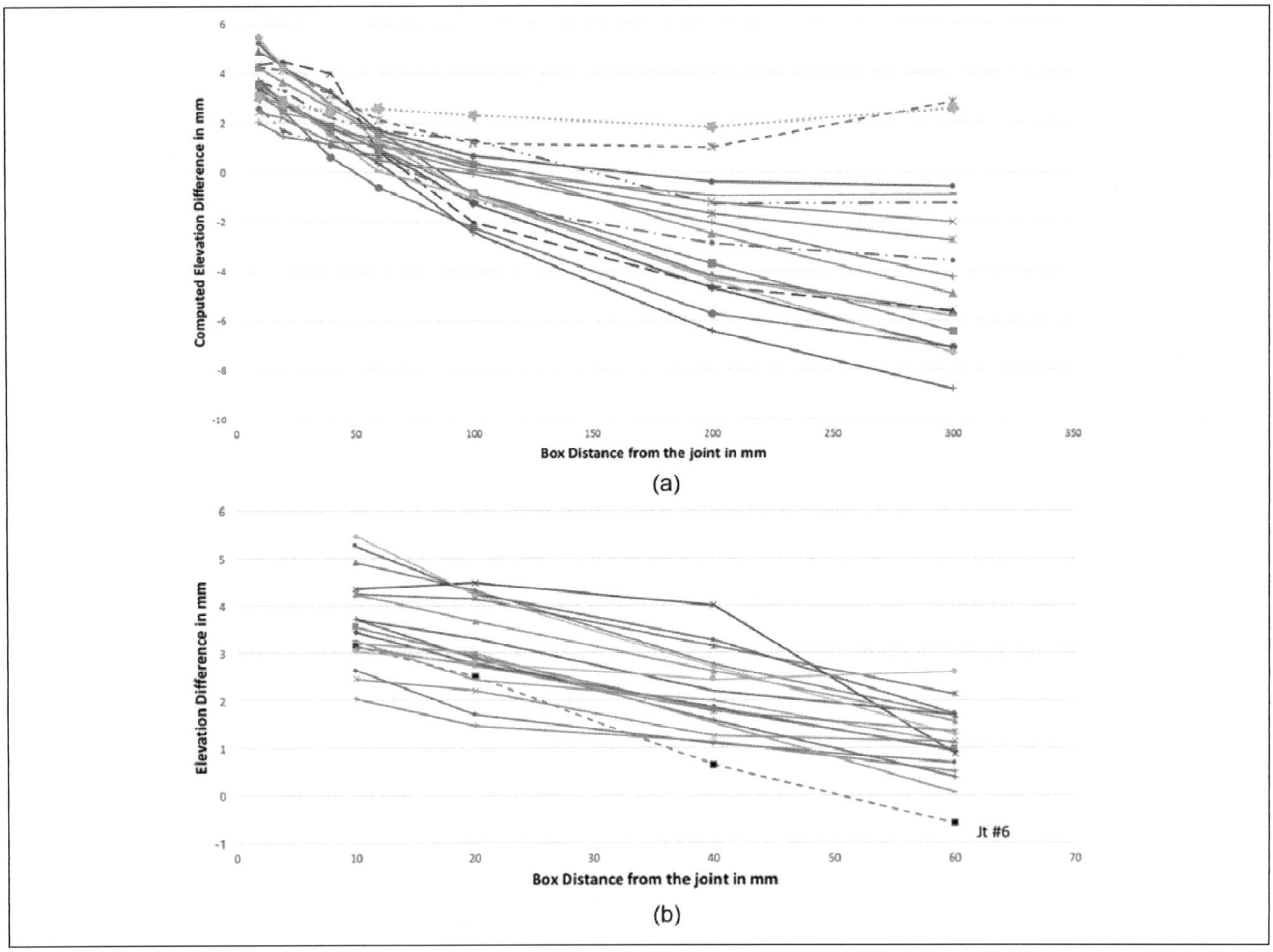

Figure 9. (*a*) Elevation differences determined at locations farther from the joint (first 19 joints at 10, 20, 40, 60, 80, 100, 200, and 300 mm). (*b*) A closer view of the elevation differences for joint 6 at 10 to 60 mm.

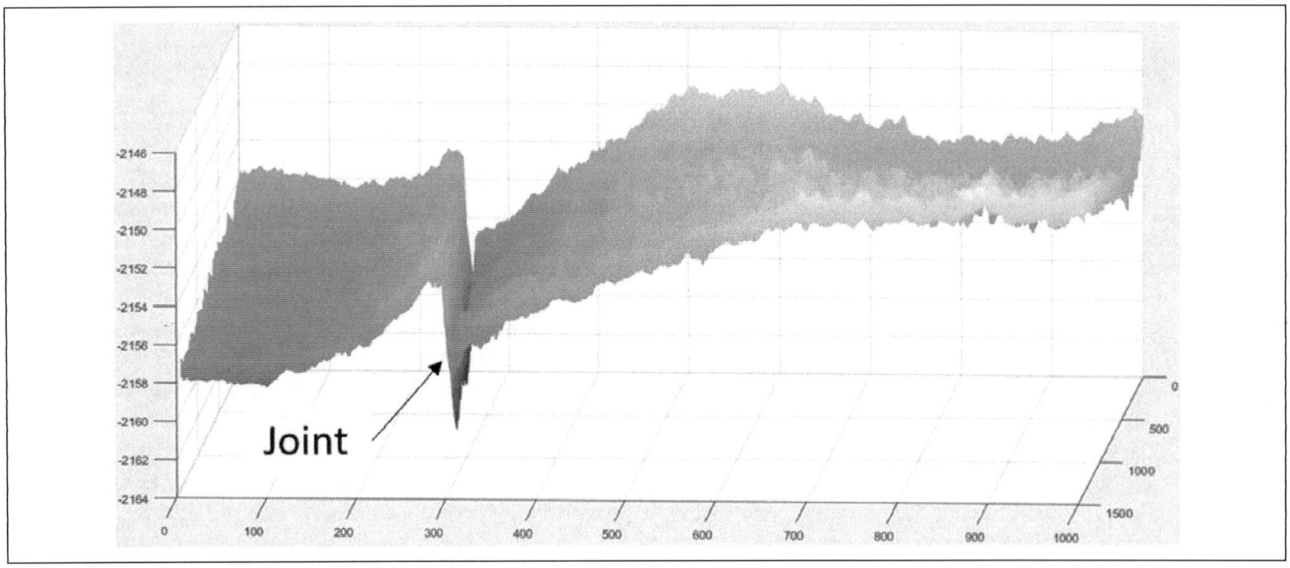

Figure 10. 3-D view of joint 6: joint at 300 mm, approach slab to the left, and departure slab to the right.

Future Research Needs

Further use on different textures, additional joint spacings and different joint widths would be beneficial. Also the 3-D data needs to be collected at different temperatures to check repeatability under conditions of curl and warp.

Acknowledgment

This material is based upon work supported by the National Science Foundation Graduate Research Fellowship under Grant No. DGE1650044, USDOT project DTOS59-10-H-0003, and GDOT RES PROJ 13-19.

References

1. U.S. Department of Transportation, Federal Highway Administration. *Transportation Performance Management Website: Two Performance Management Final Rules Take Effect*. U.S. DOT, FHWA, Washington. D.C., 2017. https://www.fhwa.dot.gov/tpm/rule.cfm. Accessed July 2017.
2. CalTrans. *Smoother Pavements: Certification Location*. CalTrans, Calif., 2017. http://www.dot.ca.gov/hq/maint/Pavement/Offices/Pavement_Engineering/Certification_Location.shtml. Accessed July 2017.
3. United States Federal Highway Administration. *Highway Performance Monitoring System Field Manual*. OMB No. 2125-0028, FHWA, U.S. Department of Transportation, Washington, D.C., 2014.
4. AASHTO R36-17. *Evaluating Faulting of Concrete Pavements*. Standard Specifications for Transportation Material and Methods of Sampling and Testing, AASHTO, 2017.
5. ASTM International. *Pavement Performance: Current Trends, Advances, and Challenges*. STP1555-EB. ASTM International, West Conshohocken, Pa., 2012.
6. Simpson, A. L., G. R. Rada, B. Visintine, and J. Groeger. *Interstate Pavement Condition Sampling Final Report*. FHWA-HIF-17-022. Federal Highway Administration, Washington, D.C., 2016.
7. ILDOT. *Chapter 53- Pavement Rehabilitation*, Bureau of Design and Environment Manual, Illinois DOT, Springfield, IL, 2010.
8. Selezneva, O., J. Jiang, and S. D. Tayabji. *Preliminary Evaluation and Analysis of LTPP Faulting Data - Final Report*. FHWA-RD-00-076. Federal Highway Administration, McLean, Va., 2000.
9. Agurla, M., and S. Lin. *LongTerm Pavement Performance Automated Faulting Measurement*. US DOT, FHWA-HRT-14-092. Federal Highway Administration, McLean, Va., 2015.
10. Highway Research Board. *The AASHO Road Test, Report 5, Pavement Research*. Special Report 61E, National Academy of Sciences-National Research Council, Washington, D.C., 1962.
11. Stone, J. *Design, Operation, and Maintenance Manual for Georgia Digital Faultmeter*. FHWA Report No. GA-91-SP9010, Georgia DOT, Forest Park, GA, 1991.
12. Worel, B., and T. Clyne. *MnROAD Faultmeter*. Minnesota Road Research, Maplewood, MN, 2009.
13. Mraz, A., A. Nazef, H. Lee, C. Holzschuher, and B. Choubane. Precision of Florida Methods for Automated and Manual Faulting Measurements. *Transportation Research Record: Journal of the Transportation Research Board*, 2012. 2306: 131–137.
14. Rasmussen, R. O., R. Sohaney, P. Wiegand, and D. Harrington. *Measuring and Analyzing Pavement Texture*. Tech Brief. National Concrete Pavement Technology Center, Ames, Iowa, 2011.
15. Chang, G., A. Nazef, J. Watkins, and S. M. Karamihas. Automated Fault Measurement (AFM) in Proval. *Proc., Pavement Evaluation 2010 Conference*, Roanoke, Virginia, 2010. https://www.vtti.vt.edu/impact/pe-2010.html. Accessed July 2017.
16. Zimmerman, K. *NCHRP Synthesis Report 501: Pavement Management Systems: Putting Data to Work*. Transportation Research Board, Washington, D.C., 2017.
17. Tsai, Y. C., Y. C. Wu, and C. Ai. *Feasibility Study of Measuring Concrete Joint Faulting Using 3D Continuous Pavement Profile Data*. Presented at 90th Annual Meeting of the Transportation Research Board, Washington, D.C., 2011.
18. Tsai, Y., Y. Wu, C. Ai, and E. Pitts. Critical Assessment of Measuring Concrete Joint Faulting Using 3D Continuous Pavement Profile Data. *Journal of Transportation Engineering*, Vol. 138, No. 11, 2012, pp. 1291–1296.
19. Wang, K. C. P., L. Li, and J. Q. Li. Automated Joint Faulting Measurement Using 3D Pavement Texture Data at 1 mm Resolution. In *T&DI Congress 2014: Planes, Trains, and Automobiles*, Orlando, FL, ASCE, 2014, pp. 498–510.

The Standing Committee on Pavement Condition Evaluation peer-reviewed this paper (18-03389).

Any opinions, findings, and conclusions or recommendations expressed in this material are those of the authors and do not necessarily reflect the views of the National Science Foundation.

Article

TRR
JOURNAL OF THE TRANSPORTATION RESEARCH BOARD

Transportation Research Record
2018, Vol. 2672(40) 50–58

DOI: 10.1177/0361198118798474
journals.sagepub.com/home/trr
SAGE

A Cost-Effective and Objective Full-Depth Patching Identification Method using 3D Sensing Technology with Automated Crack Detection and Classification

Yi-Chang (James) Tsai[1,2], Yi-Ching Wu[1], and Geoffrey Price[1]

Abstract

Full-depth patching is one of the commonly used asphalt pavement maintenance and rehabilitation methods in which deteriorated base and surface layers are repaired to restore strength and improve ride quality. During resurfacing projects, areas requiring full-depth patching are identified and quantified as construction priorities because of the high costs associated with the labor and materials for the procedure. Currently, the manual surveys conducted to identify these locations are time-consuming and labor-intensive. Thus, large projects often cannot easily quantify the full-depth patching need because of the significant labor that would be required. This paper proposes a method that uses emerging 3D laser technology to identify the full-depth patching need by processing and analyzing the pavement distresses automatically extracted from 3D laser images. The proposed method consists of five steps: (1) 3D data acquisition, calibration, and validation, (2) crack detection, (3) crack classification, (4) rutting detection and measurement, and (5) determination of image-based patching need using the established decision tree. A case study of one mile of 3D pavement images, collected from US 80/S.R. 26, was conducted to demonstrate the use and feasibility of the proposed method. Results show the proposed method is capable of correctly classifying 95.4% of the images that show pavements requiring patching and 84.2% of the images showing pavements not requiring patching for a combined accuracy of 94.1%. The method shows promise for identifying patch locations in a cost-effective manner and will save money and time for transportation agencies.

Full-depth patching is a commonly used flexible pavement rehabilitation strategy in which deteriorated base and surface layers are repaired to restore strength and improve ride quality. As asphalt pavements deteriorate, cracks allow water and other foreign material to enter the pavement and wash away pavement material. This accelerates deterioration and reduces load-carrying capacity. Full-depth patching involves removing the defective pavement layers down to a stable, distress-free surface and backfilling the location with new material before compaction. The procedure improves strength and restores a smooth surface; however, it is costly, time-consuming, labor-intensive, and dangerous. In addition, it requires experienced pavement engineers to conduct this task. Because it is expensive, it is typically a localized treatment applied before resurfacing for the entire project. Many resurfacing projects require all patching locations to be identified and fixed before resurfacing can begin. This is because, if any of these full-depth patching locations go unidentified, it could severely affect the time, cost, and labor needed for the project.

In addition to costing additional time, labor, and materials, unidentified full-depth patch locations can severely affect future construction costs. If a patching location is resurfaced before patching can be performed, the new surface is likely to not last long because the base on which the new surface is laid is crumbling and unstable. This means that transportation agencies would require more work in the future to fix the problem. This is why experienced engineers are tasked with identifying these locations so that the need for full-depth patching

[1]School of Civil and Environmental Engineering, Georgia Institute of Technology, Atlanta, GA
[2]Chang'an University, Xi'an, China

Corresponding Author:
Address correspondence to Geoffrey Price: gprice7@gatech.edu

can be determined. Through performing a manual survey, engineers can outline the full-depth patching locations by hand for rehabilitation.

The two types of surveys primarily used are walking surveys and windshield surveys. Walking surveys involve noting pavement distresses while an engineer walks along the pavement shoulder. Windshield surveys involve noting distresses while an engineer rides in a vehicle (*1*). In addition to being time-consuming and labor-intensive, these methods may also be subjective, showing bias based on surveyor experience. Although many transportation agencies have rules for classifying different distresses, there is often large variability between surveyor results when they identify distress type and severity (*2*). Surveyor experience and training have been shown to influence the identification and quantification of distresses. In particular, less experienced surveyors tend to underestimate distresses and have trouble identifying and quantifying more complex distresses such as alligator cracking (*3*). Because full-depth patching locations typically exhibit alligator cracking and severe rutting, these locations may not be identified completely, which may result in unexpected costs. To obtain more consistent results, new technologies can be utilized to obtain more objective results from pavement condition surveys.

Using 3D Sensing Technology for Pavement Distress Detection

In recent years, many emerging technologies including line lasers, GPS, and distance measuring instruments (DMIs) have been used in remote sensing vehicles capable of collecting high-resolution pavement surface data while driving at traffic speeds. This method allows for more efficient and cost-effective collection of detailed surface information than traditional survey methods. The 3D laser technology is capable of measuring high-resolution depth (range) data and reflectivity (intensity) of the pavement. The range data are useful for the identification and measurement of pavement cracks and the calculation of rut depth (*4*, *5*).

Pavement cracks are identified using both automatic and semi-automatic methods. Automatic methods use range information to determine crack presence while semi-automatic methods use a "path-finding" algorithm to detect the crack path between manually entered points along the crack (*6*). Similarly, the range data in the 3D laser files can be used to measure rutting in the pavement wheel paths using a straight-line method (*7*). Severe cracking and rutting are the two pavement distresses most representative of pavements requiring full-depth patching. Using 3D laser technology, the distresses can be quickly measured from the 3D laser files, significantly reducing the amount of time needed to survey in the field. Additionally, the distress detection algorithms can help reduce the bias that is inherent with manual survey methods, allowing for more complete, objective pavement distress data.

To facilitate best practices in a pavement management system, identifying all full-depth patching locations is crucial to properly estimating labor and material costs. The objective of this report is to propose a method for the identification of full-depth patching need by using 3D sensing technology and semi-automated crack detection and classification. The paper is structured as follows. The first section identifies the research need and objective. The second section presents the proposed method. The third section presents a case study using the proposed method. Finally, the fourth section summarizes the findings from the case study and gives recommendations for future research.

Proposed Method

This section presents the proposed method for the identification and quantification of full-depth patching locations using 3D pavement images with semi-automatic crack detection, crack classification, and rutting measurement. The proposed method is composed of the following processes: (1) 3D data acquisition, calibration, and validation, (2) crack detection, (3) crack classification, (4) rutting detection and measurement, and (5) determination of image-based patching need using the decision tree established in this study.

3D Data Acquisition

For this paper, data were collected by the Georgia Tech Sensing Vehicle (GTSV) (Figure 1*a*). The vehicle is equipped with two laser crack measurement systems (LCMS), GPS, a DMI, high-resolution cameras, and other remote sensing technologies. The LCMS systems are used to collect laser profiles while driving at 60 mph (~100 km/h). At this speed, the system can capture a laser profile every 5 mm in the driving (x) direction. These profiles are measured at 1 mm resolution in the transverse (y) direction and 0.5 mm resolution in the z-direction (*4*) (Figure 1*b*).

The high-resolution 3D data are segmented into pavement images that correspond to a 5 m × 4.16 m pavement section. Each of these images contains over 4.1 million data points, each with a corresponding x, y, and z value. To maintain consistent precision, the LCMS and DMI are calibrated and validated regularly to observe any changes in LCMS measuring capabilities. Vehicle vibration, tire pressure, and sensor configuration can all significantly affect the line-laser measurements of the LCMS. To minimize the effects of these parameters, repeatability

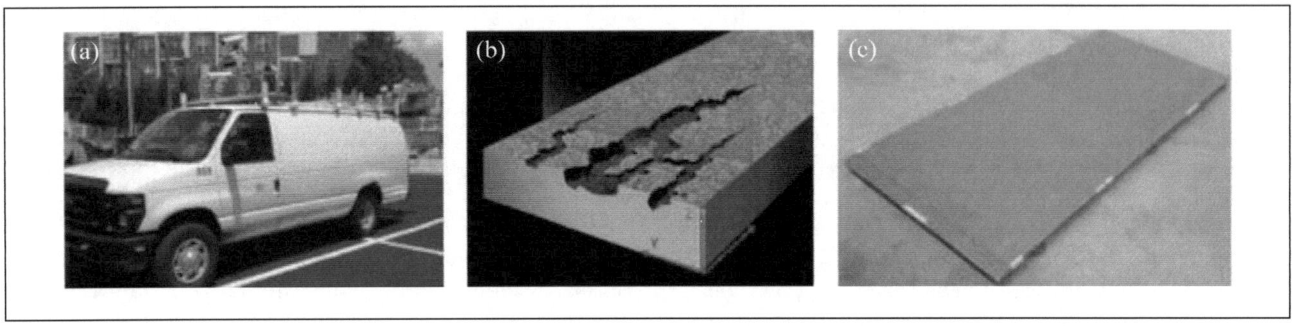

Figure 1. GTSV laser scanning and calibration: (*a*) GTSV; (*b*) pavement scan; and (*c*) calibration board.

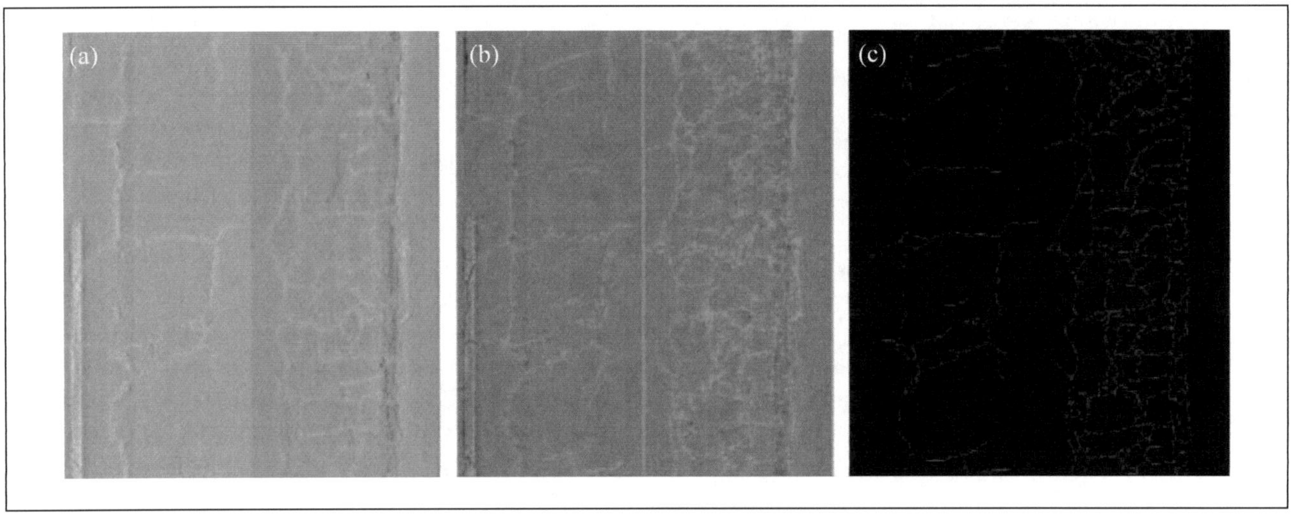

Figure 2. Crack detection process: (*a*) threshold crack detection; (*b*) semi-automatic detection; and (*c*) binary crack map.

studies are run periodically to ensure the laser is configured properly and the measurements are consistent.

The precisely machined metal plate (Figure 1*c*) is scanned at different orientations for both sensors before data collection. Because the precise dimensions of the metal plate are known, scans from different days can be compared with each other to identify external parameters that may have affected the LCMS units. If one of the sensors has suffered damage or is not working properly, the 3D images from the left and right sensors will show subtle changes in the 3D scan. These changes may result from changes in tire pressure or LCMS configuration from constant driving. With properly calibrated systems, distress detection can be performed on the collected laser scans.

Crack Detection

After the 3D data are collected, the images are processed to identify the cracks in the pavement surface. Automatic crack detection is performed using a thresholding crack detection algorithm that identifies the darker, deeper locations in the image that correspond to cracks (*8*). With the potential cracks identified, the algorithm uses a voting system to identify crack pixels from non-crack pixels. This algorithm, however, does not perform as well on severe alligator cracking (Figure 2).

Although most cracks are properly detected at lower cracking severities, detection accuracy is reduced as the cracks become denser. This incomplete detection of cracks may hinder the performance of the crack classification algorithm, so cracks are adjusted using manual input for semi-automated crack detection. Using this method, the original crack detection results are corrected to obtain accurate crack maps (Figure 2*c*). These crack maps are processed to measure crack properties such as crack length, average crack width, and crack orientation, which are indicative of cracking severity.

Crack Classification using Crack Fundamental Element

After the cracks are fully detected, the images need to be classified based on the severity of cracking within the

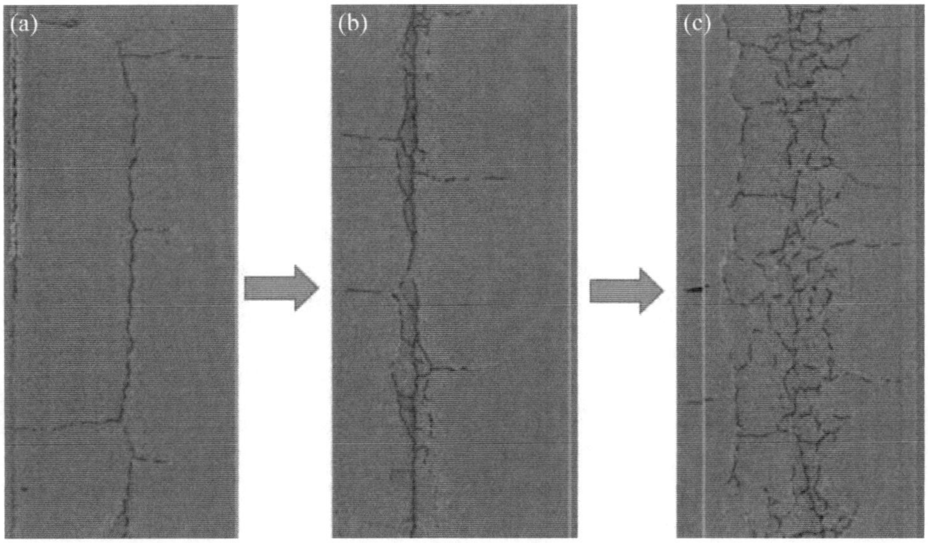

Figure 3. Load cracking progression: (*a*) Severity Level 1; (*b*) Severity Level 2; and (*c*) Severity Level 3.

image. To classify images by severity, the proposed method uses a classification algorithm based on the Georgia Department of Transportation's (GDOT's) Pavement Condition Evaluation System (PACES) (*9*), which identifies the extent and severity of load/fatigue cracking and block/transverse cracking in asphalt pavements. This paper focuses on load/fatigue cracking, which is caused by repeated traffic loads and is typically present in the pavement wheel paths (*9*). This type of cracking is important because it is used by GDOT engineers to identify locations that require full-depth patching. The GDOT PACES manual defines the progression of load cracking from a minor distress where no rehabilitation is required, to a severe distress where full-depth patching is necessary (Figure 3).

Severity Level 1 cracking (Figure 3*a*) presents as a single, longitudinal crack in the wheel path. As the cracking progresses to Severity Level 2 (Figure 3*b*), more longitudinal cracks begin to form, and transverse cracks begin to extend from the longitudinal cracks. Severity Level 3 cracking (Figure 3*c*) occurs when the three or more longitudinal cracks have formed and the transverse cracks increase, forming polygons in the pavement surface that resemble alligator hide. This is also known as "alligator cracking." Crack classification uses these changes in crack topology to determine the severity. For this paper, a crack classification methodology developed by Jiang (*8*) is used to classify cracks by type and severity according to PACES specification developed by GDOT; however, the method can be extended to other state DOT distress protocols with minor modification. Classification is performed using the crack fundamental element (CFE) segmentation method developed by Tsai et al which groups crack segments at different scales to identify crack clusters (Figure 4) (*10*).

Once the CFE clusters are identified, the images are classified into different load cracking severities based on the following eight cracking characteristics: (1) length of longitudinal crack segments, (2) longitudinal ratio, (3) number of crack lines, (4) initial number of CFEs, (5) number of crack intersect points, (6) average crack width, (7) maximum crack width, and (8) width of the CFE at final scale. Because it is possible to have multiple severities of load cracking in one image, the CFE returns the predominant classification of load cracking in the image. Research shows this method achieves 95.2% accuracy for identifying images with alligator cracking (*10*). After crack classification, the images are processed to measure rut depth and width.

Rutting Detection and Measurement

Ruts are defined in the PACES manual as "...longitudinal depressions that form under traffic in the wheel paths." Rutting is a permanent deformation that can be caused by a variety of factors including heavy traffic loading, improper compaction during construction, an unstable base, and so forth. Using 3D laser technology, rut depth can be measured using the straightedge method and classified according to severity. For this paper, rutting is measured from the transverse profiles taken by the LCMS systems. Each laser profile is smoothed, and the depth of the rut is measured from the rotated pavement surface to the imaginary horizontal plane that

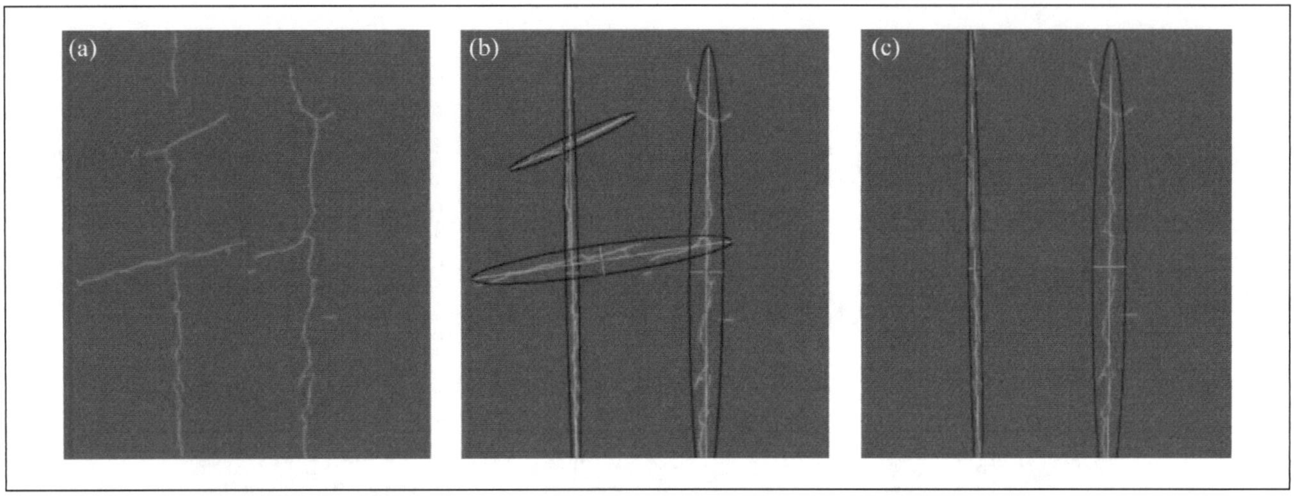

Figure 4. Crack fundamental element: (*a*) detected cracks; (*b*) final CFE clusters; and (*c*) CFEs for load cracking classification (*10*).

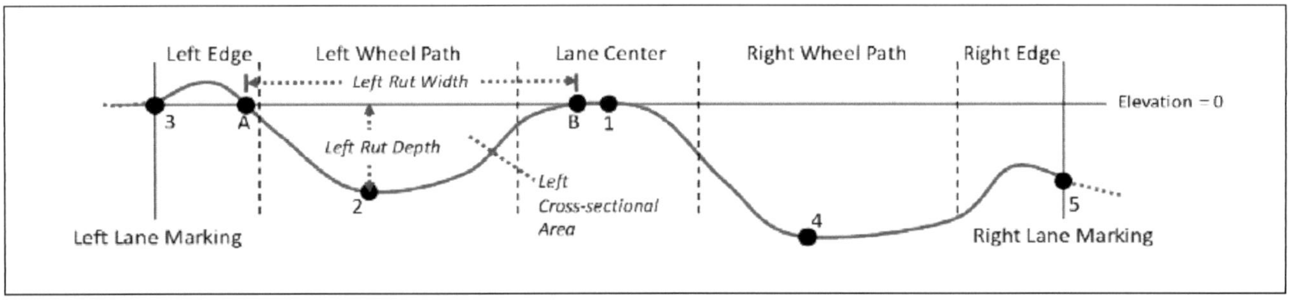

Figure 5. Smoothed and rotated laser profile for left wheel path rut measurement (7) (not to scale).

connects the lane center and the respective left or right lane marking (Figure 5).

Rut depth measurements are taken using the straight-edge method. In this method, the profile is smoothed and rotated until the lane center is horizontal with the pavement marking. From this horizontal line, rut depth is measured as the largest difference in elevation to the pavement surface. In addition to rut depth, the width of the rut is measured as the distance between the two points where the pavement surface elevation crosses the horizontal pavement line (*7*). By rotating the laser profile about the lane center until the right lane marking is at an elevation of zero, the rut depth and width for the right wheel path are measured by the same method.

A significant rut in the wheel path is one of the signs used by pavement experts to confirm the need for full-depth patching. A significant rut indicates the possible movement of base material. If the base material is loose enough to move around and cause the surface to sink, it has likely crumbled beyond the point of simple repair and is in need of full-depth patching. By assessing the rutting characteristics of pavements in different conditions, a rut depth threshold is determined to help classify pavement images based on the need for full-depth patching.

Determination of Image-Based Full-Depth Patching Need

After measuring and classifying the cracking and rutting within each image, the results were combined to determine if patching is needed. Typically, pavement engineers use high-severity (Levels 3 and 4) load cracking and rutting $\geqslant 1/4$ in. as general rules to identify pavements needing patching; however, these rules do not seem consistent with the results obtained from a visual inspection by a GDOT pavement expert. To better understand the characteristics of pavements requiring patching, the classification results are compared with manual classification performed by the pavement expert Ritchie Swindell of GDOT.

Mr. Swindell performs a visual assessment in which he determines the need for patching in a series of range images. His results, based on his experience, give good indications about which pavements should/should not be patched. For the image-based validation, images requiring patches ⩾2 ft in length were classified as "patching" images, while all other images were classified as "non-patching." Once the patching images were classified from non-patching images, the rutting in the wheel paths for each image was measured using 3D laser technology. For each image, the rut depth was measured in each of the 1,000 transverse profiles in the image. The image rut depth was recorded as the median depth obtained from the 1,000 transverse profiles. The calculated rut depths and assessed load cracking severity level for each image are compared with the ground truth established by GDOT pavement experts.

Typically, all locations showing alligator cracking will require full-depth patching; however, limited funding often reduces the amount of full-depth patching that can be performed for a project. Because of this, surveyors often need to prioritize patch locations, allowing other locations to deteriorate further. Because some locations will be left to deteriorate, engineers must use their knowledge to select the most severe patching locations. This is why it is crucial to have experienced surveyors when determining the need for full-depth patching; however, the distress severity and surveyor experience have been shown to affect distress rating results (*11*). In an effort to mimic the decision process of engineers in the field, the load cracking and rutting characteristics of images with known classification are used.

By using images with known classification, the distress characteristics of images requiring patching can be observed to develop thresholds for classifying images with unknown classification. This also helps remove the objectivity of surveyor classification. The threshold values for rut depth and load cracking severity, as well as additional cutoff values for total crack length (TCL) and length-weighted width (LWW) are shown in Table 1. These threshold values are used to classify images to determine patching need.

These minimum acceptable values are determined by comparing the distress values of patching images with no-patching images. The primary classification parameters are selected to maximize the correct classification of patching images while minimizing the number of non-patch images incorrectly classified. However, there are several cases in which patching images could not be correctly identified as shown in Figure 6.

Table 1. Minimum Distress Values for Full-Depth Patching

Pavement distress indicator	Primary classification	Adjacent image classification
Minimum TCL	11.5 m (37.8 ft)	17. m (55.8 ft)
Minimum LWW	14.0 mm (0.55 in.)	16.0 mm (0.63 in.)
Minimum rut depth	5 mm (~3/16 in.)	na
Minimum load cracking severity	2	1
Cracking extent	N/A	4.50 m (14.8 ft)

Note: na = not applicable.

The five images shown in Figure 6 represent the three major issues encountered when attempting to classify pavement images based on cracking and rutting characteristics. In Figure 6*a*, two adjacent images are shown and the need for patching in each is highlighted; however, when looking at the images separately, the need for patching of the pavement in the image on the right is not evident. This is because, although the proposed method is based on individual images, the need for patching of the pavement in any image may depend on the need for patching shown in adjacent images. Similar to the second image in Figure 6*a, b* contains a patch that is too small to significantly change the crack characteristics shown in the image. In cases like this, the images are classified as Severity Level 1 by the CFE method.

In Figure 6*c*, the image shows predominantly Severity Level 1 load cracking. In terms of full-depth patching, the patch and pothole highlighted in the image should be patched before resurfacing is performed. Although there are potholes present, the predominant load cracking severity is Severity Level 1, leading to the pavement in the image being classified as pavement not requiring a patch. Similarly, because the cracking in Figure 6*d* is primarily on the edge of the pavement and not in the wheel path, the image is classified as Severity Level 1 load cracking, which does not require full-depth patching according to the proposed method. Although the classification is correct, the edge of the pavement in this image would require full-depth patching because of the significant deterioration that may soon spread to the wheel path. Cases like this are important to consider when identifying full-depth patching locations because they can severely affect the accuracy of classification for the few patching images in a set.

The cases mentioned above account for a significant number of the patch images that are not classified correctly using only the primary classification threshold values given in Table 1. To correct for this, an additional classification is used to determine the need for patching in a given image using the cracking condition of adjacent images. A diagram of the process is given in Figure 7.

First, the cracking and rutting statistics of the original image are compared with the primary classification statistics. If the criteria are met, the image is classified as showing the pavement requires patching. If the criteria are not met, the cracking and rutting statistics of the adjacent images are checked against the adjacent image classification threshold values given in Table 1. If either of the adjacent images shows that patching is required and meets the secondary criteria, the original image is

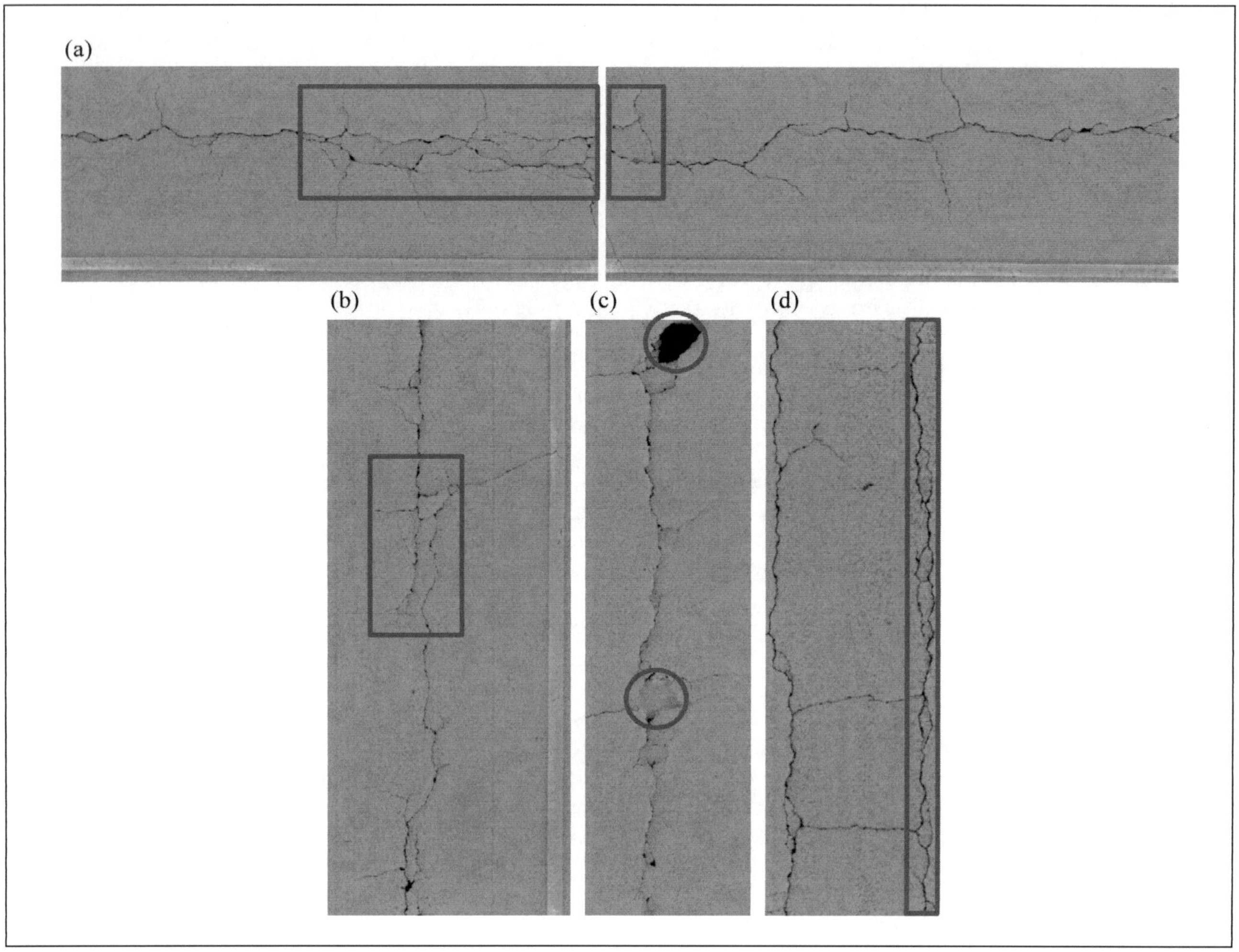

Figure 6. Classification special cases.

classified to show that the pavement requires patching. If the original image does not meet the primary criteria and the adjacent images do not meet the secondary criteria, the original image is classified to show that the pavement does not require patching.

Case Study: US80/SR26 Near Savannah, GA

The proposed method is validated using the 3D data collected in June 2015 on SR. 26/US Highway 80 near Savannah, Georgia. Laser surface images and high-resolution video log images were taken in the westbound direction at milepost 7–8. This pavement shows extensive high-severity load cracking as well as extensive rutting in the wheel paths. The data for the 1-mi length of roadway were collected as 322 separate 4 × 5 m pavement images, each of which was split in half to assess the left and right wheel paths separately, resulting in 644 total pavement images (322 × 2).

Once the 3D data were collected, it was processed using automatic and semi-automatic crack detection to obtain accurate crack maps. These maps were then processed to determine crack length, width, and severity. After severity classification, the images were reviewed to ensure the precision of the crack classification algorithm, and image classifications were adjusted to show the predominant severity in each image (Table 2).

Once the images were classified into the different load cracking severities, the rut depth for each wheel path was measured for the 1,000 transverse profiles in each image, and the median rut depth was taken for each image. Although the rut depth may vary for the different transverse profiles in each image, the median value was used as a representative rut depth value for the image.

The measurements from the individual images were compared with these values. If TCL, LWW, rut depth,

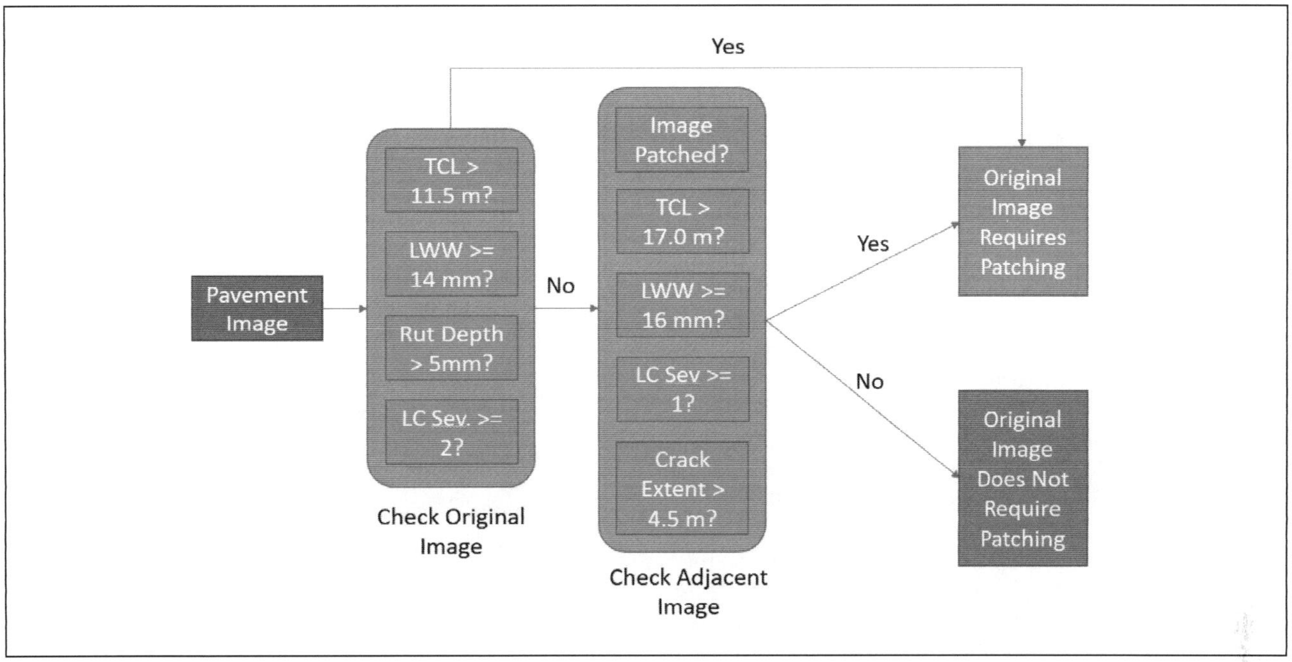

Figure 7. Full-depth patch decision tree.

Table 2. Pavement Cracking Classification

Load cracking severity	Number of images
0	74
1	511
2	28
3+	31

and load cracking severity levels for the images were greater than or equal to the default values from Table 1, the image was considered a patching image; otherwise, the image was considered a non-patch image. With this analysis complete, the results were compared with those given by GDOT personnel to measure the performance of the method. The results of this analysis and the validation with expert opinion are given in the next section.

Analysis Results

For analysis, the 644 pavement images representing one mile of pavement were randomly split into two equal-sized training and testing sets and the need for patching within each test image was confirmed by the GDOT pavement expert. Of the 322 pavement images used in the test data set, 38 were identified as pavements requiring full-depth patching based on what was visually observed from the range image. After classification according to the proposed method, 84.2% of the patching images were correctly identified in addition to 95.4% of non-patching images. This gives an overall classification of 94.1% for the entire set of test images.

This method provides an objective and cost-effective means for transportation agencies to identify full-depth patching locations. Through this method, the time spent performing field inspections can be significantly reduced. Pavement locations that require full-depth patching are quickly and objectively identified without performing roadside surveys. With these candidate patching locations identified, the total length of pavement requiring inspection for full-depth patching is significantly reduced, reducing time spent in the dangerous roadway environment while giving objective full-depth patching locations that can be confirmed in the field.

Summary and Conclusion

Full-depth patching is a time-consuming and expensive method of roadway repair that requires the careful identification of distressed pavement that is likely to have a base problem. Using 3D laser technology, the pavement surface can be mapped at high-resolution and the pavement distresses, such as cracking and rutting, can be measured. In this way, the distresses used to identify locations with base problems can be automatically identified and categorized based on severity. The proposed method uses automatically and semi-automatically detected pavement cracks to determine the cracking extent, TCL, length-weighted crack width, and load cracking severity for pavement images collected from one mile of asphalt pavement on US80 near Savannah, Georgia. For these same images, the rut depth is also measured. Using these measurements, the images were classified based on their showing the pavement's need for full-depth patching.

A set of 644 continuous images was reviewed to determine the need for full-depth patching in the pavement shown in each image. Of the 644 images, 87 were determined to show pavement requiring patching. The images were then randomly split into equally sized training and testing sets. Using statistical analysis, the cracking and rutting characteristics of training images showing pavements requiring patching were compared with images showing pavements that do not require patching. Through this analysis, a set of classification threshold values was determined. In some cases, the need for patching shown in a given image is dependent on the adjacent images so an additional set of classification parameters was developed to help classify images by also looking at adjacent images. The test set of images was then classified according to the proposed method.

Through this classification process, the test set comprising 38 patching images and 284 non-patching images was classified using the proposed method. After classifying according to the developed decision tree, 95.4% of pavement images that showed no patching was needed and 84.2% of images showing pavements requiring patching were correctly classified. This results in an overall classification of 94.1% for the set of test images. The given proposed method shows a strong ability to identify the deep patching and can be used to significantly reduce the manual work required when assessing a project for pavement requiring full-depth patching.

Author Contributions

The authors confirm contribution to the paper as follows: Y. Tsai conceived of the presented idea and designed the framework. Y. Tsai, Y. Wu, and G. Price contributed to the design and implementation of the research, to the analysis of the results and to the writing of the manuscript. All authors provided critical feedback and helped shape the research, analysis and manuscript. All authors reviewed the results and approved the final version of the manuscript.

References

1. Timm, D. H., and J. M. McQueen. *A Study of Manual vs. Automated Pavement Condition Surveys*. Alabama, EE. UU: Auburn University, 2004.
2. Federal Highway Administration. TechBrief: *Variability of Pavement Distress Data from Manual Surveys*. Publication FHWA-RD-00-160. Federal Highway Administration, U.S. Department of Transportation, 2000.
3. Prakash, A., B. N. Sharma, and T. J. Kazmierowski. Investigation into Observational Variations in Pavement Condition Survey. *Proc., 3rd International Conference on Managing Pavements*. San Antonio, Texas, 1994.
4. Tsai, Y. J., C. Jiang, and Z. Wang. Pavement Crack Detection using High-resolution 3D Line Laser Imaging Technology. *Proc., 7th RILEM International Conference on Cracking in Pavements. Springer*, Netherlands, 2012.
5. Tsai, Y. J., Z. Wang, and Feng Li. Assessment of Rut Depth Measurement using Emerging 3D Continuous Laser Profiling Technology. *Journal of Marine Science and Technology*, Vol. 23, No. 3, pp. 322–330, 2015.
6. Kaul, V., A. Yezzi, and Y. Tsai. Detecting Curves with Unknown Endpoints and Arbitrary Topology using Minimal Paths. *IEEE Transactions on Pattern Analysis and Machine Intelligence*. Vol. 34, No. 10, 2012, pp. 1952–1965.
7. Wang, C.. *A Spatiotemporal Methodology for Pavement Rut Characterization and Deterioration Analysis Using Long-term 3D Pavement Data*. PhD dissertation. Georgia Institute of Technology, 2016.
8. Jiang, C. *A Crack Detection and Diagnosis Methodology for Automated Pavement Condition Evaluation*. PhD dissertation. Georgia Institute of Technology, 2015.
9. Georgia Department of Transportation. Pavement Condition Evaluation System. Georgia Department of Transportation, Office of Maintenance, Atlanta, GA, 2007.
10. Tsai, Y.-C., C. Jiang, and Y. Huang. Multiscale Crack Fundamental Element Model for Real-world Pavement Crack Classification. *Journal of Computing in Civil Engineering*. Vol. 28, No. 4, 04014012, 2014.
11. Prakash, A., B. N. Sharma, and T. J. Kazmierowski. Investigation into Observational Variations in Pavement Condition Survey. *Proceedings of the 3rd International Conference on Managing Pavements*, Vol. 1, No. 2, 2003, pp. 290–301.

The Standing Committee on Pavement Condition Evaluation (AFD20) peer-reviewed this paper (18-05950).

Article

TRR
JOURNAL OF THE TRANSPORTATION RESEARCH BOARD

Transportation Research Record
2018, Vol. 2672(40) 59–67
© National Academy of Sciences:
Transportation Research Board 2018
Article reuse guidelines:
sagepub.com/journals-permissions
DOI: 10.1177/0361198118756655
journals.sagepub.com/home/trr
SAGE

Evaluation of a Line Laser Scanner to Improve the Measurement of Average Least Dimension in Chip Seal Design Methods

Sareh Kouchaki[1], Hossein Roshani[2], Jorge A. Prozzi[1], Cristina Cordoba[1], and Joaquin Bernardo Hernandez[3]

Abstract

Chip seal, as the most widespread pavement preventive treatment, is regularly applied on existing pavements that are still in good structural condition to increase the pavement serviceability. The most effective key parameter on chip seal performance is binder application rate, which directly governs chip seal distresses such as bleeding and raveling. How this rate is calculated mainly depends on the value of least dimension (LD) of aggregate particles. However, the available measuring methods of LD value are slow, laborious, and subjective. This study presents the development of a new high-speed line laser scanner (LLS) prototype to measure the LD value of particles more quickly and accurately. The LD values of aggregate particles were also measured using a digital caliper and considered as control data. The repeatability and reliability of the developed LLS prototype were evaluated, as well as the speed of the prototype in calculating the LD values of 100 aggregate particles. The findings indicate that the measurements of the developed prototype are highly correlated with those of the caliper. In addition, it was found that the developed prototype is efficient and capable of calculating the LD values of several particles simultaneously.

Formerly, no maintenance treatment was performed on pavements until severe cracks emerged on surfaces and pavements deteriorated structurally (*1*). Over time, highway agencies have changed their strategies; rather than waiting until the pavement fails, a series of preservation techniques are applied during the life of the pavement to keep its condition at an acceptable level. These preservation practices are essential to correct minor distresses, delay pavement decay, maximize pavement service life, and enhance safety, and thereby increase drivers' satisfaction (*2*). Highway agencies around the world have pursued various preventive techniques. Among these preventive maintenance techniques, chip seal is the most commonly used surface treatment in the world (*3*). In 1999, Hickes et al. compared the cost and expected life of maintenance treatments. Chip seal was the best surface treatment technique because of its low cost and high durability (*4*). Similar results have been obtained by Chen et al. and Chen and Daleiden: chip seal outperforms other preventive treatments such as slurry seal and overlay in terms of cost and performance life (*5, 6*).

Chip seals, if designed properly, seal non-structural cracks to block water penetration, maintain the integrity of a cracked pavement, enrich macro-texture, enhance skid resistance, provide abrasion-resistant layers against traffic and climate, and correct light reflection of pavements during the night (*7–11*). On the other hand, poorly designed chip seals can cause significant distresses such as raveling and bleeding (*12*). These distresses pose a serious threat to drivers' safety. For instance, loose aggregates on the surface of the chip seal can be disturbed under traffic movement and could damage windshields and headlights. In addition, bleeding reduces the

[1]Department of Civil, Architectural, and Environmental Engineering, The University of Texas at Austin, Austin, TX
[2]City of San Antonio, Streets Engineering & Infrastructure Management, Transportation and Capital Improvements (TCI), San Antonio, TX
[3]Center for Transportation Research, The University of Texas at Austin, Austin, TX

Corresponding Author:
Address correspondence to Sareh Kouchaki: sareh.kouchaki@utexas.edu

skid resistance of chip seal and increases the potential for accidents (*8, 13*).

A chip seal consists of two materials: asphalt binder and chip (aggregate). The asphalt binder is first sprayed on the existing pavement surface, effectively sealing cracks and acting as a waterproof coating. This is immediately followed by spraying aggregates before the asphalt binder hardens (*13–15*). Determination of the asphalt binder and aggregate application rates is an important part of a chip seal design. At first, there was no rational design method determining the exact amount of asphalt binder and aggregate within the chip seal. Instead, trial and error was used to find the approximately correct amount of those materials. In 1935, Hanson pioneered the development of a mathematical method determining the ratio of aggregate to asphalt binder. This method was later adjusted and employed by different countries around the world. Some countries, such as the United Kingdom, South Africa, Australia, and New Zealand, have developed their own chip seal design methods that resulted from more comprehensive engineering principles. The following subsection provides a summary of chip seal design methods.

Chip Seal Design Methods

Hanson Method

The earliest seal coat design method was developed in New Zealand in 1935 by F.M. Hanson. The main factor in the calculations for the aggregate and asphalt binder spreading rates is the Average Least Dimension (ALD) of chip seals. The ALD refers to the overall average of the least dimension (LD) of the aggregate particles. Hanson measured the LD manually with the use of a caliper (*13*). These measurements provide information to determine the volume of voids in the aggregate covering the asphalt binder. Hanson determined that the voids in one-sized aggregate are at 50% after the aggregate is initially poured by the truck. After initial rolling, the voids are reduced to 30%. Finally, after compaction due to traffic loads, the voids should ultimately be reduced to 20% (*16*). Based on these findings, Hanson concluded that the rate of asphalt binder application should be determined such that 20% of the voids' volume after traffic compaction should become 60–75% filled, depending on the type of aggregate used and the volume of traffic (*13*). Therefore, the equation for asphalt binder application is $ALD*0.2*0.7$ (*16*).

Kearby Method

The Kearby method, designed by J. P. Kearby in 1953, was one of the first methods of determining chip seal application rates in the United States. Unlike the Hanson method, Kearby emphasized the importance of visual inspection and the use of judgment as essential parts of the design of chip seal (*13*). In addition, this method requires a series of laboratory tests in order to determine the aggregate and asphalt binder spreading rates. The laboratory tests determine the dry loose unit weight and the bulk specific gravity of the aggregate. The results of the laboratory measurements are used in a calculation that determines the aggregate spreading rate, given in cubic yards of aggregate per square yard area. Following that, a calculation is made for the asphalt binder spreading rate (gallons per square yard) which factors in the dry loose unit weight, dry bulk specific gravity of the aggregate, ALD, and embedment depth of aggregate particles into the asphalt (*17*). Originally, Kearby did not include any numerical correction factors for traffic levels and existing surface conditions. Instead, he visually inspected these characteristics and suggested increasing or decreasing design values according to variations observed (*18*). The modified Kearby method developed later accounts for these factors in the calculations (*13*).

Modified Kearby Method

The modified Kearby design method is a chip seal design method commonly used in the United States and Canada (*13*). This design method consists of laboratory tests, calculations, and good engineering judgment to find the most appropriate asphalt binder and aggregate spreading rates for optimal performance. Similar to the original Kearby method, the aggregate application rate is determined using dry loose unit weight and the bulk specific gravity of the aggregate found by laboratory tests. Once the aggregate properties and the existing roadway conditions are known, a series of calculations is performed to determine the asphalt application rate for asphalt cement. Factors in the calculation include the aggregate application rate, embedment depth, and dry bulk specific gravity. Percent embedment depth is obtained by calculating the average height of aggregates (*11, 19*). Correction factors for traffic and surface conditions are then included in the calculations to obtain the final result.

McLeod Method

The McLeod design method is another chip seal design method used in the United States and Canada (*13*). In this method, the aggregate application rate depends on the particle gradation, shape, and specific gravity (*19*). This rate is determined by computing the voids in the loose aggregate, ALD, bulk specific gravity of the aggregate, and the traffic whip-off wastage factor (*19*). Traffic whip-off accounts for the aggregate that is thrown to the side of the road by passing vehicles during the early stages of the seal coat (*19*). The asphalt binder application rate is determined by computing the ALD, traffic correction factor, voids in the loose aggregate, existing surface condition factor, aggregate absorption factor, and percent residual asphalt in the emulsion (*19*). A larger asphalt binder application rate is used for dry, pocketed, porous pavement surfaces because these have a greater absorption rate. Measurements of the dry loose unit weight of the cover aggregate and the bulk specific gravity of the aggregate are used to determine the percentage of voids. According to McLeod's practice, these should be, at most, 30 percent voids. In areas where there is low traffic volume, there would be a larger percentage of voids. Therefore, it is recommended to increase the asphalt binder application rate in order to assure good adhesion to the aggregate (*19*).

TRH3: South African Method

The existing pavement conditions play a large role in the South African design method TRH3. Existing conditions can be measured by visual assessment of defects and by measurements of surface texture depth, unevenness, permeability, and the expected embedment of aggregate (*20*). TRH3 specifies that a minimum of 42% voids should be filled with asphalt binder for chip seals. The design process begins with determining the equivalent light vehicles value for the most probable traffic scenario. The ball penetration test is used to measure potential embedment of aggregates into the existing surface. The ALD and the texture depth of the existing surface road are used to calculate the minimum and maximum asphalt binder application rates (*20*). Finally, adjustments are made to account for existing texture, climate, and traffic factors. A sensitivity analysis to the minimum and maximum application rate is undertaken, where the minimum consists of the highest expected traffic, highest ball penetration, smoothest texture, lowest design ALD, and the maximum consists of the lowest expected traffic, lowest ball penetration, coarsest texture, and highest design ALD (*20*).

Road Note 39: United Kingdom's Method

The chip seal design guide used by the United Kingdom not only describes a detailed design process but also recommends good engineering judgment, particularly for variations due to heavy traffic at high speeds. The parameters that go into selecting the spreading rates include surface temperature, road hardness, traffic flows/speeds and medium/heavy vehicles, geometric features (bends, gradients, junctions, parking, pedestrian crossings), and chipping properties (size(s), type, polished stone value) (*21*). Some adjustments are made to the asphalt binder spreading rate accounting for factors such as season and shade. For instance, shaded areas result in cooler road conditions, making it effectively harder, so it is recommended to use more asphalt binder (*21*).

Austroads: Australia's Method

The Austroads design method specifies two main factors that are important in chip seal design: ALD and average annual daily traffic (AADT). The ALD can be measured for nominal sizes of 10 mm or larger using the sieving and slot gauge test method. In addition, the design must take into account the rate of reorientation of aggregate particles dependent on traffic volume. For example, low traffic volumes will cause less reorientation resulting in random orientation of aggregate and larger void volume (*22*). The design traffic, which is determined from the AADT, must take into account the number of carriageways, direction of traffic, number of lanes, and percentage of total traffic traveling in each lane (*22*).

Adjustments are made to the aggregate spreading rate using measurements of material finer than 75 μm (0.0029 in.), flakiness index, particle size distribution on a dry sample, particle shape by proportional caliper, apparent particle density, particle density on a dry basis, and water absorption (*22*). Adjustments are made to the asphalt binder spreading rate accounting for the texture of the existing surface, potential embedment of the particles into the base, and potential absorption of the asphalt binder into the pavement and aggregate (*22*).

ALD Measurement

The review of design methods showed that the ALD of aggregate is essential to calculate the application rate of asphalt binder. Under traffic movement, aggregate particles are rotated until they sit on their flattest sides; hence, the LD value of aggregates is critical to assure proper coverage of the asphalt binder and prevent bleeding. The ALD is the overall average of the measured LD value of the representative particles. Test methods to determine the ALD vary among highway agencies. Mulry et al. grouped the existing ALD computation methods as follows (*24*).

1. A Vernier caliper, a dial gauge, or a slotted gauge

These devices directly measure the LD of aggregate samples. Although these apparatuses are simple and low cost, their measurement processes are slow and involve human error.

2. ALD machine developed in South Africa

The ALD machine (Figure 1) is used to calculate the LD of aggregate particles manually. The advantage of this device compared with previous devices is that the LD values of samples are stored on a computer. However, similar to the previous devices, the ALD machine measures the LD of aggregate particles one by one, which is time-consuming.

3. Jackson Nomograph and Equation

Jackson developed a nomograph (Figure 2) based on nominal size and flakiness index of aggregate particles to calculate the ALD (*25*). He stated that the nominal size of aggregate particles alone is not sufficient to calculate the ALD because flaky particles and cubical particles might have the same nominal size, but their LD values are different. To calculate ALD using the nomograph, first, the median

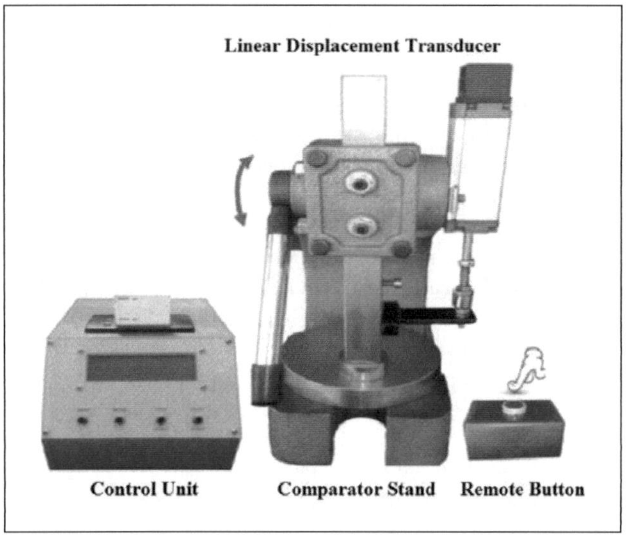

Figure 1. ALD machine developed in South Africa (*23*).

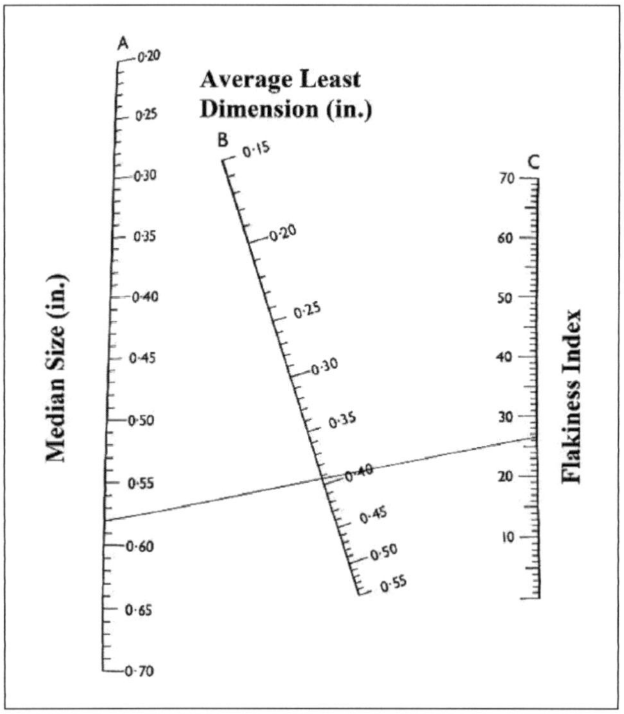

Figure 2. Jackson ALD nomograph (*27*).

size of particles and flakiness index are obtained through laboratory tests. Then, the values of these parameters are marked in the nomograph. The ALD is obtained by drawing a line between these two values (*25*). Later, the following equation was formulated by Jackson in 1963 based on the ALD nomograph (*26*). Two factors, median particle size and flakiness index, are required to determine the ALD:

$$\mathrm{ALD} = \frac{M}{1.139285 + 0.011506\,\mathrm{FI}} \tag{1}$$

where:
M = median particle size from sieve analysis; and
FI = flakiness index (%).

4. Dumas Method

In 2004, Dumas from South Africa criticized the model developed by Jackson (ALD nomograph and equation), claiming that the median size cannot fully represent the particle size distribution and instead proposed using five sieves corresponding to 10, 25, 50, 75, and 90% passing of particles rather than using only the median size. In this method, the ALD is measured by performing a sieve analysis and complex calculations (*24, 27*).

Automated Methods

Given the impact of the application rates of aggregate and the asphalt binder on chip seal performance, it is very important that highway agencies take careful consideration in quantifying the amount of the binder and aggregate. In most of the chip seal design methods, the ALD of aggregate particles is the key factor in the calculation of asphalt binder application rate. However, existing test methods used to measure the ALD are subjective and laborious. A rational method is required to measure the LD values of aggregate particles more quickly and accurately. In recent years, the use of image and laser scanning techniques in engineering surveys has gained increasing interest due to the advantages of being non-contact, rapid, and highly accurate.

Several researchers have used image-based techniques to improve methods for measuring aggregate properties. In one work, Masad developed the Aggregate Imaging System (AIMS), which was capable of analyzing the shape properties of coarse and fine aggregates (*28*). It consisted of a camera, video microscope, lighting systems, aggregate tray, and computer automated data acquisition system and processing software. AIMS was able to capture an image of 640 × 480 pixels of 6–10 fine aggregate particles. For coarse aggregate, only one particle could be captured per image. The drawback of using the AIMS was that it was only capable of taking 2D measurements. Therefore, there was some uncertainty in measuring the LD values of aggregates.

For more extensive measurements of the aggregates, it is preferable to obtain 3D data of the particles. Anochie-Boateng et al. have conducted several studies on the use of 3D laser scanner technologies to measure aggregate properties such as shape, surface area, and angularity (*29, 30*). The results of the studies showed that the laser could provide reliable and precise measurements. However, the 3D laser scanner used by Anochie-Boateng et al. has a chamber with the capacity for only one aggregate for scanning and measurement.

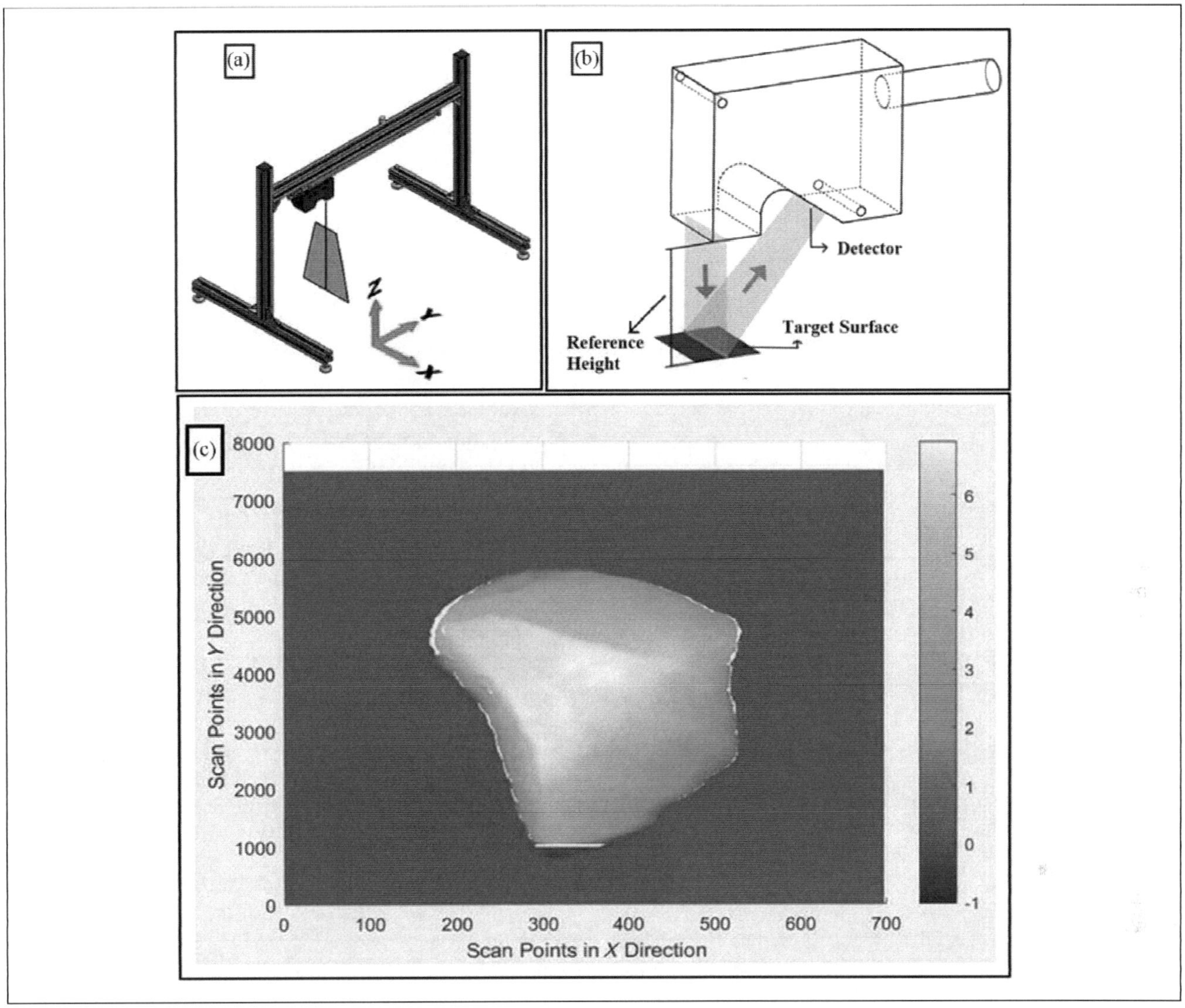

Figure 3. (*a*) 3D schematic view of the developed LLS prototype, (*b*) triangulation system of the LLS, and (*c*) 3D display of an aggregate surface.

Research Motivation and Scope

The existing methods for measuring ALD are slow and subjective. A logical and fast way is required for precise determination of ALD. In recent years, studies have turned to laser technology to measure aggregate shape properties. The goal of this study is to evaluate a new device, called the line laser scanner (LLS) prototype, which was developed to automatically and more precisely characterize the ALD of aggregates. The LLS used in this study is capable of scanning 100 particles simultaneously at high speed with a resolution of 0.025 mm, resulting in 3D data from the scan. Therefore, the limitations of the 2D image-based techniques can be overcome by using the LLS and a more rigorous measurement of aggregates can be achieved. The repeatability, reliability, and performance of the LLS prototype were evaluated by comparing its results to those from a caliper.

Methodology

Developed LLS Prototype

In this study, a 3D high-speed scanner system was developed to speed up the measurement of the LD values of aggregate particles and provide reliable results. The prototype (as shown in Figure 3*a*) consists of a LLS. Blue light with a wavelength of 405 nm is emitted from a source in the LLS prototype, and the reflection of the light is captured by a detector. Using the triangulation technique depicted in Figure 3*b*, the laser obtains the height of the points along the

Table 1. Common Aggregate Gradations for Texas Chip Seal (*32*)

Size (mm)	Grade 3	Grade 4	Grade 5
19	0	–	–
16	0–2	0	–
12.5	20–40	0–2	0
9.5	80–100	20–35	0–5
6.3	95–100	–	–
4.75	–	95–100	40–85
2	99–100	99–100	98–100
0.85	–	–	99–100

Note: Values are percent retained on sizes shown.

Table 2. Side-by-Side Comparison of the Measurement Results of LLS and Digital Caliper

Descriptive Statistic	LLS	Caliper
Mean	6.931 mm	7.01 mm
Standard Deviation	0.009 mm	0.065 mm
Minimum	6.917 mm	6.88 mm
Maximum	6.941 mm	7.09 mm
Range	0.024 mm	0.21 mm

laser strip with respect to the system's reference line. In this prototype, the LLS is mounted on a linear motion controller to enable the LLS to travel over a target and capture 3D data. Successive scans in the moving direction capture height points all over the target surface. These measurements are in millimeters. Both the motion controller and the laser are connected to a computer, and the scanned data are collected on the computer in an Excel file. By combining the collected height data, the 3D map of the target surface can be created. For example, a 3D display of an aggregate surface is shown in Figure 3*c*. As the laser mechanism is based on projecting light and capturing its reflection, all tests were performed in a laboratory with a constant light condition to avoid any further signal noise associated with light variation. Along with the light, the room temperature and humidity were also kept constant during the experiments. In a parallel research study conducted by Kouchaki et al. it was reported that the developed LLS prototype showed promising results in terms of scanning the surface micro-texture of aggregates (*31*).

Aggregate Selection and Preparation

Table 1 shows the master gradation ranges of the most common chip seal aggregates used in Texas. According to Table 1, aggregate retained on top of the sieve size 12.5 mm to aggregate retained on top of sieve size 2 mm might be included in chip seal applications given the selected gradation. A sample of particles retained on top of sieves of sizes 12.5, 9.5, 6.3, 4.75, and 2 mm (0.49, 0.37, 0.25, 0.19, and 0.08 in.) was selected to evaluate the capability of the new system in measuring variably sized particles. Aggregates used for this study were provided by Texas Department of Transportation (TxDOT) Construction Division in Austin, TX. The aggregates were first washed to remove any dust. The aggregate particles were then oven-heated to 160°C for 24 h followed by 4 h at room temperature to reach air-dry condition.

Measurement Processes and Results

Within the scope of this study, the reliability of the developed prototype was tested in three steps including (i) testing the repeatability, (ii) evaluating the effect of particle's orientation angle on measurements, and (iii) measuring the LD values of particles of variable sizes all together. The results of the developed LLS prototype were compared with those from a digital caliper.

Repeatability of the Developed LLS Prototype

One aggregate particle was put in the scanning area of the prototype and scanned 10 times in the same conditions to check the repeatability of the scanning system. Scanned data collected by the developed LLS prototype were stored on the computer for processing using MATLAB software. There are some points with very high or low values around the edge of the particle where the laser cannot see the reflection of light or due to the particle's shadow, where the light might reflect several times and cause noise in data processing. These invalid data need to be dealt with before calculating the LD value of particle. The scanned data were first converted to a binary image which has a black background and a white object. Following this, a boundary is created around the white object to separate the data associated with the aggregate from the background. It is to be noted that the LD value of the aggregate when sitting on its flattest side is a key factor in chip seal design. Therefore, in all tests, aggregate particles were arranged manually with their flattest side down. Accordingly, the elevation of the highest point within the boundary provides the LD value of that particle relative to the reference line.

The LD value of the selected particle was also measured 10 times using a digital caliper with a specified accuracy of 0.01 mm, and the measurements results were then compared with the LLS results. Table 2 provides a side-by-side comparison of the two measurements. As can be seen in the table, the mean value and the standard deviation of the 10 measurements captured by the developed LLS prototype is 6.931 and 0.009 mm (0.273 and 0.0004 in.), respectively. The mean value and the standard deviation of the 10 measurements captured by the caliper are 7.01 and 0.065 mm (0.276 and 0.0026 in.), respectively. This table also shows the maximum, minimum, and the range of measurements captured by the two devices. These results show that the measurements by the LLS have smaller standard deviation and range compared with the caliper. This suggests that the developed LLS prototype provides better repeatability.

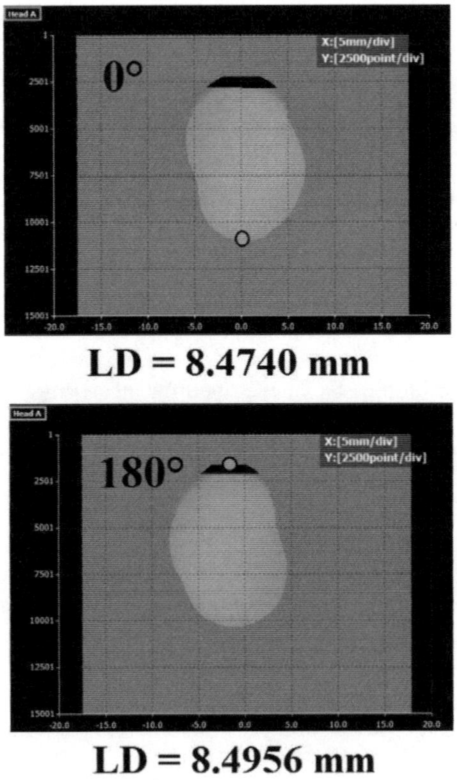

Figure 4. LD values of a particle at different orientations.

Effect of Aggregate Orientations on ALD Measurement

An aggregate particle tends to lie randomly on the asphalt binder layer; however, its LD value does not change with rotation. To evaluate the effect of orientations on the accuracy of the developed LLS prototype, an aggregate particle was rotated in four different directions ranging from 0° to 270° with an interval of 90° with respect to the laser line and then scanned. The result and the figure associated with each direction are provided in Figure 4. The black areas in the figures represent the parts very close to the edge of the aggregate where the laser was not able to see the light projection and missed the data. Since this area is related to the edge of the aggregate, the missing data did not impact the LD value measurement. The variability between measurements is calculated by the standard deviation, which is equal to 0.011 mm (0.0004 in.) and shows that the effect of the orientation is negligible.

Measurement of Particles with Variable Sizes

Devices such as calipers measure the LD values of representative particles separately. In addition, appropriate sieve analysis tests are required in approaches such as the Jackson and Dumas methods to measure the LD values of a sample of particles altogether. To evaluate the speed of the developed LLS prototype and its feasibility to measure the LD values of numerous particles, 100 variable-sized particles (in the range of 2–12.5 mm) were arranged manually in the scanning area. The developed LLS prototype has a scanning area of 160 mm × 600 mm (6.3 in. × 23.62 in.) within which particles must be spread. The particles should be in reasonable proximity to each other, as shown in Figure 5*a*, to reduce the error in analysis. Particles were scanned, and the collected data were stored in the computer. As explained in the previous step, after converting the data to a binary image, aggregate particles were discriminated from the background by finding a boundary around each particle and labeling them. The height points within each boundary were exported to a separate Excel sheet to determine the LD value of the corresponding particle. To evaluate the accuracy of the process, as shown in Figure 5*b*, each particle was also calipered several times to find the LD value of that particle, and the results compared with the developed LLS prototype measurements.

Figure 6 shows the plot of measurements captured by the developed LLS prototype versus those of the caliper. The best-fit line to the plotted data with correlation coefficient, R, of 0.99 (the square root of R^2) was calculated, which

Figure 5. (*a*) A total of 100 particles spread in the developed LLS prototype scanning area; (*b*) taking measurement using a caliper.

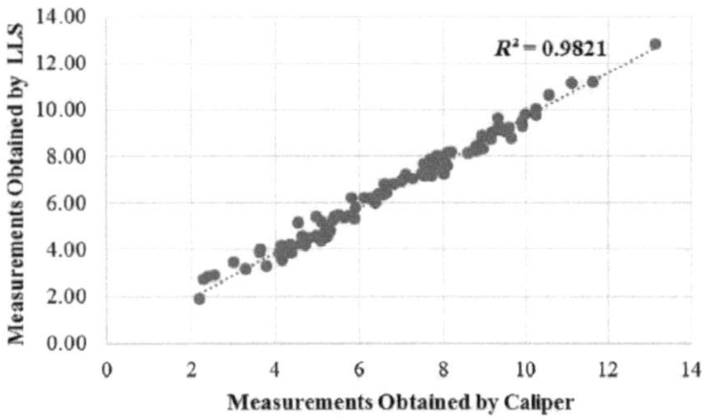

Figure 6. Plot of LD values of 100 particles captured by the developed LLS prototype and digital caliper.

presents a strong relationship between developed LLS prototype measurements and the caliper outcomes. As a result, the developed LLS prototype is capable of calculating the LD values of variable-sized particles simultaneously. Although the measurement of 100 particles by caliper took about 4 h, the measurement processes by LLS prototype—capturing data, data processing, and attaining results—were completed in less than 10 min. Accordingly, the developed LLS prototype helps in saving time and cost. In addition, more particle samples can be measured to obtain a precise ALD estimation.

Conclusions

Preventive maintenance is required when the pavement is in good structural condition in order to postpone distresses and extend service life. Chip seal is the most commonly used preventive treatment. This study reviewed the available chip seal design methods from within and outside the United States to highlight the effective parameters in calculating the asphalt binder application rate. The review highlighted several parameters such as the traffic type and volume, existing surface conditions, and ALD. It is observed that in most of the design methods, the ALD of aggregate particles is the key factor in the calculation of asphalt binder application rate. In this study, a new LLS prototype was developed for measuring the ALD of aggregate. Several tests were carried out to evaluate the repeatability, accuracy, reliability, and the speed of the developed LLS prototype. In all tests, the LD values of aggregate particles measured by the LLS were compared with measurements from a digital caliper.

The results are very promising. It was found that the LD values calculated by the LLS were similar to those of the caliper, which shows the reliability of the developed LLS prototype for measuring the LD values of aggregate particles. The repeatability and standard deviation test results showed that the new system outperforms the digital caliper in terms of repeatability. The accuracy of the LLS prototype was also investigated by measuring the LD value of an aggregate particle at various orientation angles. The standard deviation analysis implies that the missing data at the edge of the aggregate have a negligible impact on the LD value measurements. Therefore, highway agencies can consider the new prototype as an efficient alternative to LD value measurement methods.

The results also showed that the developed LLS prototype is able to improve on existing methods by scanning a lot of particles in a significantly shorter time, providing a higher data resolution, and calculating the LD values of a number of particles simultaneously. Accordingly, the LLS enables transportation agencies to measure the ALD more quickly, efficiently, and precisely as compared with conventional approaches, saving time and cost for those agencies not only by reducing human resources expenditure, but also by enhancing the chip seal design and its performance.

The next step in this study is to test the capability of the LLS prototype in measuring the other dimensions of aggregate particles, length, and width, as well as their surface area. Other potential applications of the developed LLS could include measuring the texture of aggregates. In addition, testing the applicability of the developed LLS prototype for determining the gradation of aggregate particles is recommended for further research.

References

1. Beatty, T. L., D. C. Jackson, D. A. Dawood, R. A. Ford, J. S. Moulthrop, G. D. Taylor, F. L. Danchetz, D. A. Blanck, J. M.

Fay, D. S. Keough, L. M. Rodriguez, M. D. Voth, and Z. L. Webb. *Pavement Preservation Technology in France, South Africa, and Australia*. Publication FHWA-PL-3-001, FHWA, U.S. and American Trade Initiatives, Alexandria, VA, 2002.
2. Scott, J. Seal coating practice in Saskatchewan. *Transportation Research Record: Journal of the Transportation Research Board*, 1986. 1096: 140–146.
3. Rahman, F., M. S. Islam, H. Musty, and M. Hossain. Aggregate retention in chip seal. *Transportation Research Record: Journal of the Transportation Research Board*, 2012. 2267: 56–64.
4. Hicks, R. G., S. B. Seeds, and D. G. Peshkin. *Selecting a Preventive Maintenance Treatment for Flexible Pavements*. Research Report, Foundation for Pavement Preservation, Washington, D.C., 2000.
5. Chen, D. H., D. F. Lin, and H. L. Luo. Effectiveness of Preventative Maintenance Treatments Using Fourteen SPS-3 Sites in Texas. *Journal of Performance of Constructed Facilities*, Vol. 17, No. 3, 2003, pp. 136–143.
6. Chen, D., and J. Daleiden. Lessons Learned from the Long-Term Pavement Performance Program and Several Recycled Sections in Texas. *E-Circular E-C078: Roadway Pavement Preservation 2005*. Transportation Research Board of the National Academies, Washington, D.C., pp.70–84.
7. Zoghi, M., A. Ebrahimpour, and V. Pothukutchi. *Performance Evaluation of Chip Seals in Idaho*. Publication FHWA-ID-10-190, Idaho Transportation Department, 2007.
8. Shuler, S. Chip seals for high traffic pavements. *Transportation Research Record: Journal of Transportation Research Board*, 1990. 1259: 24–34.
9. Elmore, W. E., M. Solaimanian, R. B. McGennis, C. Phromson, and T. W. Kennedy. *Performance-based seal coat specifications*. Research Report No. 1367-1, Center for Transportation Research, Austin, T.X., 1995.
10. Einarsson, I. T. *Chip Seals Examination of design and construction in two countries*. MSc Thesis, University of Washington, Seattle, W.A., 2009.
11. Kutay, M. E., U. Ozdemir, D. Hibner, Y. Kumbargeri, and M. Lanotte. *Development of an Acceptance Test for Chip Seal Projects*. Report No. SPR-1649, Michigan State University, East Lansing, M.I., 2016.
12. Pierce, L. M., and N. Kebede. *Chip seal performance measures_ Best practices*. Report No. WA-RD 841.1, Washington State Department of Transportation, Olympia, W.A., 2015.
13. Gransberg, D. D., and D. M. B. James. *NCHRP Synthesis 342: Chip Seal Best Practices*. Transportation Research Board of the National Academies, Washington, D.C., 2005.
14. Gransberg, D. D., S. Senadheera, and I. Karaca. *Analysis of Statewide Seal Coat Constructability Review*. Research Report TX-98/ 0-1787-1R, Texas Tech University, Lubbock, T.X., 1998.
15. Islam, M. D. S. *Evaluation of Lightweight Aggregates in Chip Seal*. MSc Thesis, Department of Civil Engineering, Kansas State University, Manhattan, Kansas, 2007.
16. Patrick, J. The Evolution of the New Zealand Chip seal Design Methodology. *1st Sprayed Sealing Conference*, Adelaide, Australia, 2008.
17. Epps, J. A., B. M. Gallaway, and C. H. Hughes. *Field Manual on Design and Construction of Seal Coats*. Research Report No. 214-25. Texas A&M Transportation Institute, College Station, Texas, 1981.
18. Senadheera, S., D. Gransberg, and T. Kologlu. *Statewide Seal Coat Constructability Review*. Report No. TX-97/0-1787-3. Texas Tech University, Lubbock, Texas, 2000.
19. *Seal Coat and Surface Treatment Manual* (2010). Texas Department of Transportation. http://onlinemanuals.txdot.gov/txdotmanuals/scm/scm.pdf. Accessed July 10, 2017.
20. *Technical Recommendations for Highways*; *Design and Construction of Surfacing Seals*. The South African National Roads Agency. www.nra.co.za/content/Blogin/04.pdf, Accessed July 10, 2017.
21. Roberts, C., and J. C. Nicholls. *Design Guide for Road Surface Dressing (Sixth edition)*. TRL Publication. United Kingdom, 2008.
22. *Update of the Austroads Sprayed Seal Design Method*. Austroads Technical Report, Report No. AP-T68/06, Austroads Inc., 2006.
23. Average least dimension Device, Reliance Laboratory Equipment, http://www.reliancelab.co.za/average-least-dimension-ald-device-reliance. Accessed Nov. 12, 2017.
24. Mulry, B., K. Feighm, and T. Casey. The use of average least dimension in surface dressing design. Civil Engineering Research in Ireland, 2016.
25. Jackson, G. P. *Surface dressing*. Shell International Petroleum Co., London, 1963.
26. Shuler, S., A. Lord, A. Epps-Martin, and D. Hoyt. *NCHRP Report 680: Manual for Emulsion-Based Chip Seals for Pavement Preservation*. HRB, National Research Council, Washington, D.C., 2011.
27. Dumas, B. The determination of the average least dimension of surfacing aggregates by computation. *20th South African Transport Conference*, South Africa, 2001.
28. Masad, E. *The development of a computer controlled Image Analysis System for measuring aggregate shape properties*. Final report, NCHRP-IDEA Project 77. Transportation Research Board of the National Academies, Washington, D.C., 2003.
29. Anochie-Boateng, J. K., J. J. Komba, and G. M. Mvelase. Three-dimensional laser scanning technique to quantify aggregate and ballast shape properties. *Journal of Construction and Building Materials*, Vol. 43, 2013, pp. 389–398.
30. Anochie-Boateng, J. K., J. J. Komba, and E. Tutumluer. Aggregate surface area quantified through laser measurements for South Africa asphalt mixtures. *Journal of Transportation Engineering*, Vol. 138, 2010, pp. 1006–1015.
31. Kouchaki, S., H. Roshani, J.A. Prozzi, and J.B Hernandez. Evaluation of aggregates surface micro-texture using spectral analysis. *Journal of Construction and Building Materials*, Vol. 156, 2017, pp. 944–955.
32. *Standard Specifications for the Construction and Maintenance of Highways, Streets, and Bridges*, Texas Department of Transportation, 1995.

The Standing Committee on Pavement Condition Evaluation peer-reviewed this paper (18-06373).

Article

TRR
JOURNAL OF THE TRANSPORTATION RESEARCH BOARD

Transportation Research Record
2018, Vol. 2672(40) 68–77

DOI: 10.1177/0361198118758392
journals.sagepub.com/home/trr
SAGE

Protocol for FWD Data Collection at Network-Level Pavement Management in Iran

Narges Matini[1], Nader Tabatabaee[1], and Mojtaba Abbasghorbani[2]

Abstract

The objective of this study was to develop an approach for incorporating techniques used to interpret and evaluate deflection data for network-level pavement management system applications. A national pavement management system is being developed in Iran and the use of falling weight deflectometers (FWDs) at the network level was deemed necessary to compensate for the lack of vital construction history data in the pavement inventory. Because FWD measurements disrupt traffic flow and are a potential safety hazard, it is imperative to increase the interval between FWD testing points as much as possible to allow scanning of the entire 51,000 km network of freeways, highways, and major roads in a reasonable time span with the least traffic disruption. A project-level dataset at 0.2 km intervals in different environments and diverse traffic categories was selected for analysis. In addition, data from continuous ground-penetrating radar was collected concurrently and compared with a limited number of cores. The overall analysis included evaluation of interval variation, segmentation, the structural condition index (SCI), and layer moduli calculated using the AASHTO and ELMOD methods. The analysis was done to determine the optimum interval between test points. Analysis showed that the collection intervals could be increased from 0.2 to 0.6 km. Subsequently, the applicability and time efficiency of the network-level intervals were verified by calculating overlay thickness and time required.

Nondestructive tests (NDTs) are important for the structural evaluation of pavements. The falling weight deflectometer (FWD) is a commonly-used device for NDTs. The impulse loads applied by a FWD models a typical tire of a heavy vehicle which is moving at a velocity of approximately 60 km/h (*1*). After the dynamic load has been applied, the vertical displacement or deflection of the pavement structure is measured using a series of geophones commonly spaced 30 cm (12 in) apart. Even though FWD tests are in use by most highway agencies, the most effective way of using FWD data for structural evaluation of pavements remains an important subject of research (*2*).

The high cost of FWD data collection for determining the structural condition of a pavement at the network level is a major issue in pavement management. Other major concerns are operational costs associated with FWDs, traffic control, and safety, especially on high-speed highways given the stop-and-go nature of FWD testing (*3*). Traffic levels should be considered when determining the optimum test spacing. High-traffic facilities require expensive lane closures and moderate traffic requires at least a sign truck and a crash attenuator mounted on a large vehicle. Ironically, low-traffic facilities typically rank lowest in priority in PMS data collection, but afford the opportunity to collect the most data (*4*).

Background

Many U.S. state Departments of Transportation have performed network surveys with an FWD to develop structural data for their pavements (*5*). To test the more than 2000 lane-miles of Virginia's interstate system, it was necessary to determine the number of test points per mile and the number of load levels at the network level. Comprehensive statistical analysis was performed on the subgrade moduli, effective structural number (SN) and layer moduli using the AASHTO and the evaluation of layer moduli and overlay design (ELMOD) methods (*6*). The results of this study indicated that network-level testing at three points per mile could be done instead of at 10 points per mile and the four FWD drop

[1]Sharif University of Technology, Tehran, Iran
[2]Technical and Soil Mechanics Laboratory, Tehran, Iran

Corresponding Author:
Address correspondence to Nader Tabatabaee: nader@sharif.edu

Table 1. Protocols for Collecting FWD Data in Different Agencies/Countries (*4, 7, 8*)

Agency/country	Spacing	Time intervals	Details
Virginia DOT	3 points per mile (530 m)	Every 4 or 5 years	One drop and two deflection basins at each drop level
Indiana DOT	3 points per mile (530 m)	Every 5 years	One drop and one deflection basin at each drop level
Texas DOT	3 points per mile (530 m), for more reliable results, 4 points per mile (400 m)	N/A	N/A
Kansas DOT	3 points per mile (530 m), for more reliable results, 5 points per mile (320 m)	One-third of network per year; however, one-fifth of network may be adequate	N/A
Alaska DOT	10 points per mile (160 m)	After repaving	N/A
Australia	200–1,000 m	N/A	N/A
Hong Kong	50 m	N/A	N/A
South Africa	500 m	N/A	N/A

levels could be reduced to only one drop level, at which two deflection basins were collected, without statistically compromising the quality of the data (*7*).

To determine the required FWD test spacing needed for implementing the structural condition index (SCI) at the network level of the Texas Department of Transportation (TxDOT), risk-based and bootstrap methods were employed. During testing, the variance of the SCI at the network level was assessed by randomly selecting a small representative sample of the pavement sections (*8*).

Protocols for VDOT, TxDOT and other states and countries have been reviewed. Table 1 summarizes the protocols currently in use for collecting network-level FWD data at agencies around the world (*4*). Based on ASTM D4695-03, deflection testing at the network level is typically performed at 100–500 m (or 250–1,000 ft) intervals. A minimum of seven tests per uniform pavement section is recommended to ensure a statistically significant sample.

Before this study, no proper protocol existed for collecting FWD data at the network level in Iran and most previous collections were at the project level. Most decision-making for PMS in Iran has been based on functional or composite indices such as the international roughness index (IRI) and pavement condition index (PCI). In some cases, FWD data at the project level have been used for maintenance and rehabilitation (M&R) decision-making.

Research Objective

A national pavement management system is being developed in Iran and the use of FWDs at the network level was deemed necessary to compensate for the lack of vital construction history data in the pavement inventory. Because of concerns for operational costs associated with FWD data collection and traffic control and safety, it was imperative to increase the interval between FWD testing points as much as possible to allow scanning of the entire 51,000 km network of freeways, highways, and major roads within a reasonable time without jeopardizing the usefulness of the data.

Methodology

Data Collection

The FWD tests were conducted on six routes on the Iranian road network. All six had flexible pavement structures. The routes totaled 300 km along the center-line and were located in southeastern, southwestern, eastern and central Iran as shown in Figure 1. These routes have diverse climates and traffic levels. FWD data was collected in the outer lane in both directions. The FWD data, air temperature, and temperature of the asphalt on the surface and at depth were collected. The thickness of the pavement layers at each measurement point was determined using ground penetrating radar (GPR) and coring. FWD testing was conducted using two or three drop levels. The measurements were recorded at approximately 1,300 points at 0.2 km intervals. Table 2 summarizes the collected data.

The characteristics of each route—traffic level, average temperature, and average pavement thickness—are presented in Table 3. The data represents diverse traffic levels and pavement thicknesses. It was collected from different regions in Iran having different temperatures. It should be noted that the specific data for each point, and not the overall averages, were used in the backcalculation.

Data Preparation

The deflection data was filtered using root mean square (RMS) and data points for which RMS >5 were eliminated from the database. RMS values were calculated using ELMOD 5.0 software for the filtering database. Because there was a linear relationship between deflections obtained at different load levels, all deflection data was normalized to a 9000-lb load (half of a standard axle load) and the sensor deflection data was first corrected to a temperature of 20°C (68°F) using the AASHTO correction factor. The asphalt temperature for the Jandagh-Moaleman and Sirjan-Shahre Babak routes were calculated using the BELLS3 method (*9*).

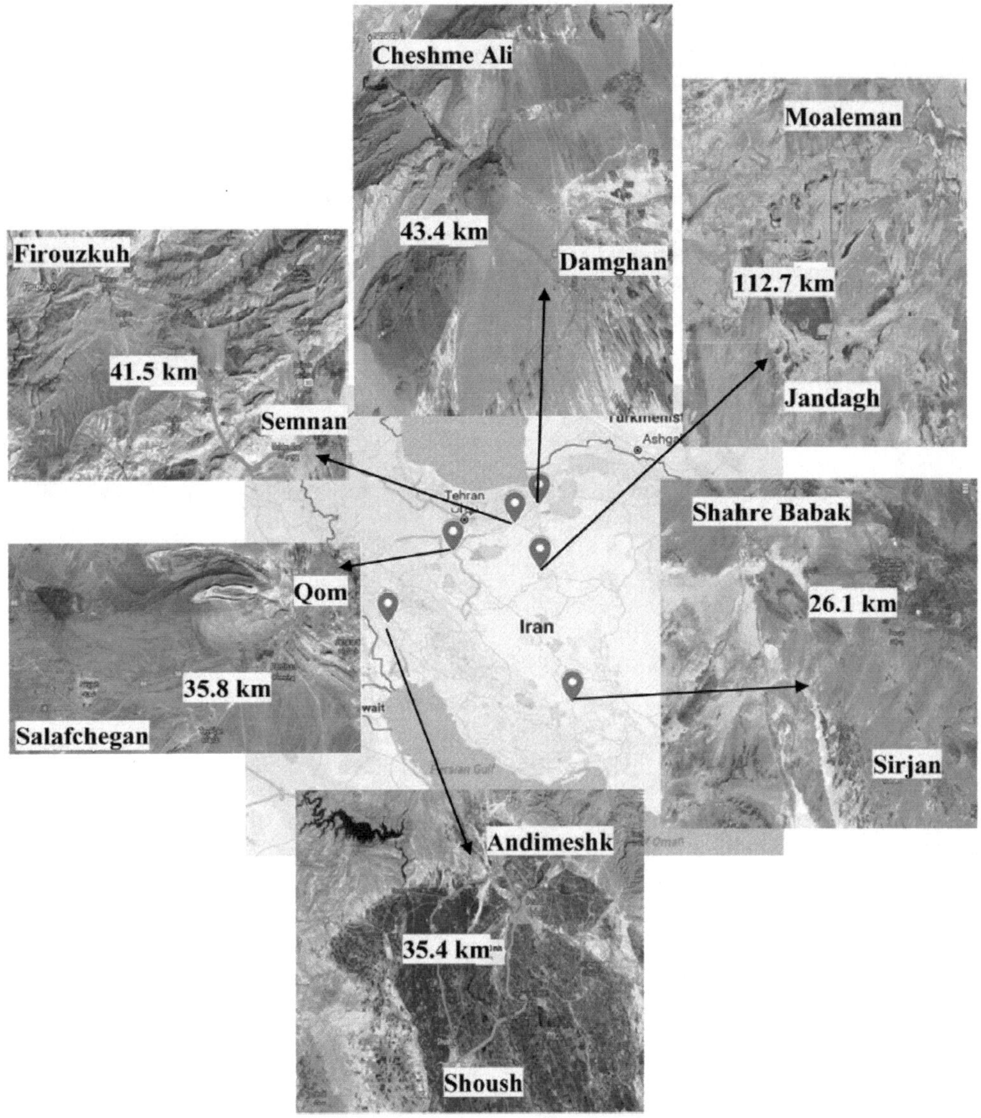

Figure 1. Location of routes in Iran.

Indicator Indices

SCI has been commonly used for pavement evaluation analysis in previous studies and was, thus, chosen as the main index for this study (*3*) and is calculated as:

$$SCI = \frac{SN_{eff}}{SN_{req}} \tag{1}$$

where SN_{eff} is the effective structural number and is calculated using the Rohde method and SN_{req} is the required structural number and is determined based on the 20-year ESAL and M_r (psi) (*3*).

Other indices, such as the structural strength index (SSI), structural health index (SHI) and normalized area ratio (A_r') were used for the purposes of comparison and to validate the initial results based on SCI. Note that in previous studies only SCI has been used as an indicator index for determining network-level FWD collection intervals and protocols (*3, 10, 11*). SSI is the result of a study to develop a structural condition-based index using more than 10,000 one-mile sections in Indiana and includes weather data, distress surveys, pavement type, and FWD measurements (*12*). It is calculated as:

$$SSI = 100\{1 - \alpha e^{\frac{-\beta}{\sigma\gamma}}\} \tag{2}$$

Table 2. Collected FWD Data

Parameter/route	Jandagh-Moaleman	Sirjan-Shahre Babak	Cheshme Ali-Damghan	Shoush-Andimeshk	Semnan-Firouzkuh	Qom-Salafchegan
Stress level (kPa)	660-930-1,360	580-880	570-900	570-700-900	570-900	560-860-1,260-1,870
Load plate radius (mm)	150	150	150	150	150	150
Number of drops	3 times	4 times	4 times	8 times	4 times	4 times
Number of geophones	7	9	9	9	9	9
Geophone positioning (cm)	0-20-30-45-60-90-120	0-20-30-45-60-90-120-150-180	0-20-30-45-60-90-120-150-180	0-20-30-45-60-90-120-150-180	0-20-30-45-60-90-120-150-180	0-20-30-45-60-90-120-150-180
Testing intervals (m)	200	200	200	200	200	200
Testing lane	Right-hand wheel path	Right-hand wheel path	Right-hand wheel path	Right-hand wheel path	Right-hand wheel path	Right-hand wheel path
Route length (km)	112.7	43.4	26.1	35.4	41.5	35.8
Number of points (before filtering*)	555	218	128	168	161	179
Number of points (after filtering)	494	211	105	165	142	143

*Data with RMS >5 was eliminated from the database.

Table 3. Overall Characteristics of Routes

	Asphalt thickness[a] (cm)	Total pavement thickness[a] (cm)	Measured asphalt temperature (°c)	Air temperature (°c)	Asphalt surface temperature (°c)	Million $ESAL_0$[b]
Shoush-Andimeshk	22.3	46	45	42.5	45	3.4
Cheshme Ali-Damghan	11.3	25.7	20.9	24.5	31	0.26
Jandagh-Moaleman	5.8	24.5	N/A[c]	13.9	16.7	0.5
Sirjan-Shahre Babak	11	37.2	N/A[c]	36.6	45.8	2.4
Semnan-Firouzkuh	16.6	31.2	27.3	25.3	29.8	0.5
Qom-Salafchegan	25	54	39.8	38.7	45.6	3.9

[a]Average of each point.
[b]Equivalent single axle load for first year.
[c]BELLS3 (9) used to determine asphalt temperature.

where σ is the first sensor deflection and α, β and γ are regression coefficients.

SHI is a SCI developed using 52 in-service pavement sections with a total length of approximately 320 miles in Louisiana. It uses a scale of 0–100 to describe the structural integrity of the pavement structure based on the backcalculated layer moduli of in-service pavements and a sigmoidal function (*13*) as:

$$\mathrm{SHI} = \frac{100}{1 + e^{0.15(\mathrm{SNloss\%} - 30)}} \tag{3}$$

where SNloss% is the loss of structural capacity and is quantified as the difference between the pavement SN at the time of construction and at the time of FWD testing.

A_r' is based on AREA and first deflection values and was developed using data collected by FWD and data generated by Circly computer simulation (*14*) as:

$$A_r' = \frac{\mathrm{AREA}}{\mathrm{Stiffest\ AREA} * D_0} \tag{4}$$

where AREA is the area of the deflection basin, stiffest AREA is the area of the deflection basin of the stiffest pavement (equal to 900 mm^2/mm) and D_0 is the first sensor deflection.

The application of the best approach with a high level of accuracy for the calculation of each parameter was of great importance. To calculate SCI, an accurate estimation of the SN was needed. Methods used to estimate SN include the Rohde, Wimsatt, and Noureldin methods (*15–17*). Based on the literature review and analyses conducted by the authors, the Rhode method was found to be the most appropriate to calculate SCI in this study. To validate the SCI and determine if this index was an appropriate indicator of the pavements under study, it was necessary to plot the SCI and rutting or fatigue ratio of each point and examine the correlation between them (*3*). Kenlayer was executed for each point in

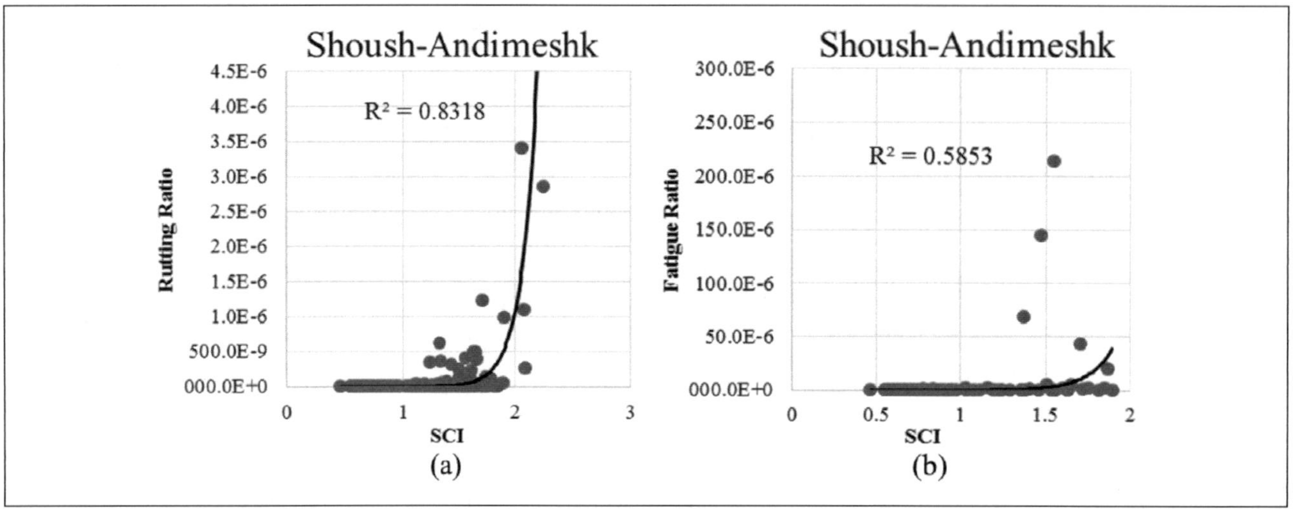

Figure 2. SCI validation: (*a*) correlation between rutting ratio and SCI values; (*b*) correlation between fatigue ratio and SCI values.

Table 4. Results of Validation of SCI Values

	R^2		Number of points (n)	
	Rutting ratio	Fatigue ratio	Rutting ratio	Fatigue ratio
Shoush-Andimeshk	0.83	0.58	164	118
Cheshme Ali-Damghan	0.80	0.53	105	73
Jandagh-Moaleman	0.67	0.50	101	102
Sirjan-Shahre Babak	0.70	0.50	200	158
Semnan-Firouzkuh	0.62	0.50	140	90
Qom-Salafchegan	0.63	0.55	143	143

MATLAB to calculate the rutting and fatigue ratio at every point as:

$$\text{Fatigue Ratio} = \frac{\textit{Number of ESALs to fatigue failure}}{\textit{Estimated } 20 - \textit{year EESAL}} \quad (5)$$

$$\text{Rutting Ratio} = \frac{\textit{Number of ESALs to rutting failure}}{\textit{Estimated } 20 - \textit{year EESAL}} \quad (6)$$

Figure 2 shows that the coefficient of determination (R^2) for the Shoush-Andimeshk route (35.4 km in length) was >0.5. This R^2 value is acceptable for this type of data; therefore, the SCI was considered to be an acceptable index for this study. Supplementary information about other routes is presented in Table 4.

SCI and SHI are based on SN; SSI and A_r' are based on pavement deflections. Because SN better correlates with pavement design approaches, SCI and SHI can be considered better indicator indices. Also, SCI seems to be a better pavement structural indicator than the SHI based on its validation analysis.

Segmentation and Determination of Representative Section Parameters

The accumulated difference and cumulative sum (CUMSUM) methods were used to divide each route into uniform sections. To determine the representative value for each section, two approaches have been suggested in previous studies. The first approach considers the closest value to the mean of each section as the representative value and the second considers the closest value to the 85% confidence interval (CI) of each section (*18*).

Another approach used in this paper was percentile based on the importance of the route. This method was suggested by the Asphalt Institute (AI) for determining the resilient modulus based on traffic levels. AI recommends the 87.5th percentile of the subgrade resilient moduli (M_r) of the section as the design M_r for heavy traffic conditions (*19*). A comparison of all three approaches indicates that the second method underestimates the pavement conditions and the third overestimates it. The estimation of the first method fell between those of the second and third methods: therefore, for the sake of brevity, results presented in this paper are based on the first method.

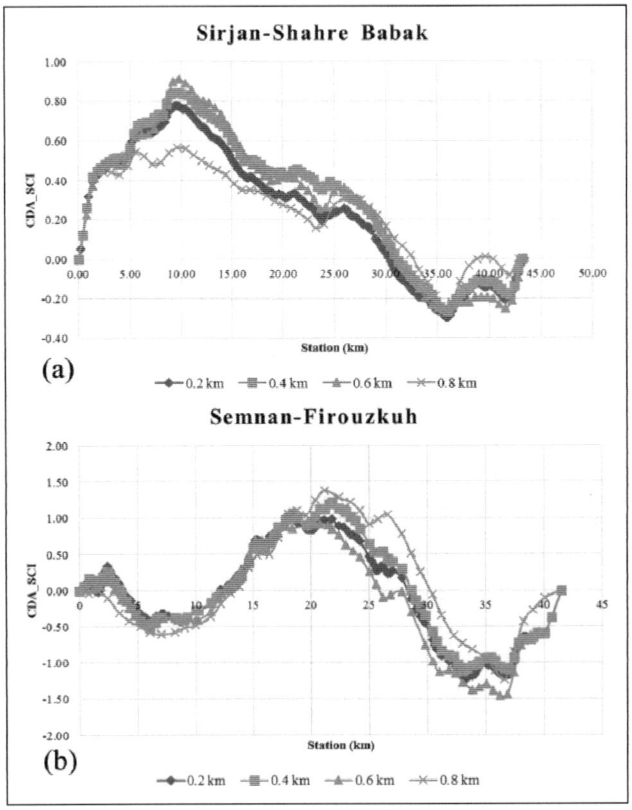

Figure 3. CDA of SCI values at different intervals: (*a*) Sirjan-Shahre Babak route; (*b*) Semnan-Firouzkuh route.

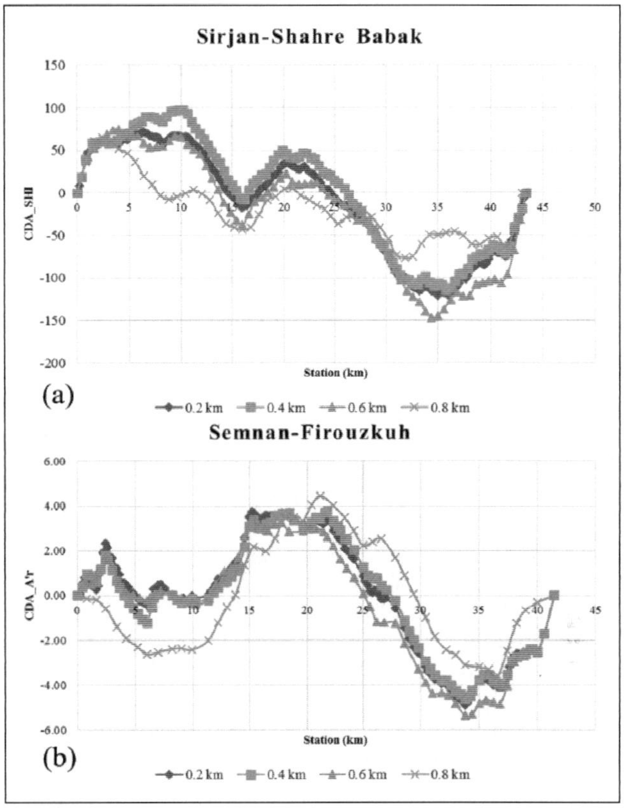

Figure 4. (*a*) CDA of SHI values at different intervals on Sirjan-Shahre Babak route; (*b*) CDA of A_r' values at different intervals on Semnan-Firouzkuh route.

Statistical Analysis and Results

The statistical phase of this study was conducted to determine the network-level FWD testing protocol using the existing project-level FWD database.

Network Level Intervals

The literature review revealed that one of the best approaches for determining network-level spacing is the use of project-level data. By eliminating the use of data points at specific intervals, the spacing was increased until the trend of the particular data index was similar to the initial spacing (*3*). Network-level intervals were determined using this approach. To determine changes in the indicator indices, the cumulative difference approach (CDA) was implemented.

The initial spacing interval for data collection was 0.2 km. To examine other possible spacing intervals, a new dataset with 0.4, 0.6, 0.8 and 1 km intervals was created by eliminating the middle points. For each spacing interval, the CDA of the SCI and the SSI, SHI and A_r' of other indices were calculated and plotted. Figure 3 shows the plots for two routes, Sirjan-Shahre Babak (43.4 km in length) and Semnan-Firouzkuh (41.5 km in length). As seen, changes in the SCI as shown by the CDA for the 0.4 and 0.6-km intervals were similar to that of the 0.2-km interval, but not for the 0.8-km interval. This means that the SCI for the 0.8-km interval no longer provided a reasonably accurate assessment of the pavement section compared with a complete set of data. For this particular route, the 0.6-km interval was found to be most appropriate for network-level FWD measurement. A similar analysis was conducted for the other indices and the plots for SHI and A_r' are presented in Figure 4. Table 5 shows the appropriate spacing for all routes tested. It was found that 0.6-km spacing was sufficient for most routes and there was a good correlation between the SCI and other indices.

Overlay Design

The required overlay for each section on each route was calculated at the project and network levels using the different methods. AASHTO equations and the SCI score-based, ELMOD and SHI-based overlay design methods were used. ELMOD overlay design requires an estimation of the material properties of each layer; thus, the FWD loading and layer moduli for each layer were first plotted. Then, depending on the use of stress softening or hardening, the material properties of each layer were estimated (*7*).

Table 5. Optimum Network-Level Intervals Based on Different Indices

Indices/route	Optimum network-level intervals (km)			
	SCI	SSI	A_r'	SHI
Shoush-Andimeshk	0.6	0.8	0.8	0.6
Cheshme Ali-Damghan	0.4	0.4	0.4	0.4
Sirjan-Shahre Babak	0.6	0.6	0.6	0.6
Jandagh-Moaleman	0.6	0.6	0.6	0.6
Semnan-Firouzkuh	0.6	0.6	0.6	0.6
Qom-Salafchegan	0.6	0.8	0.6	0.6

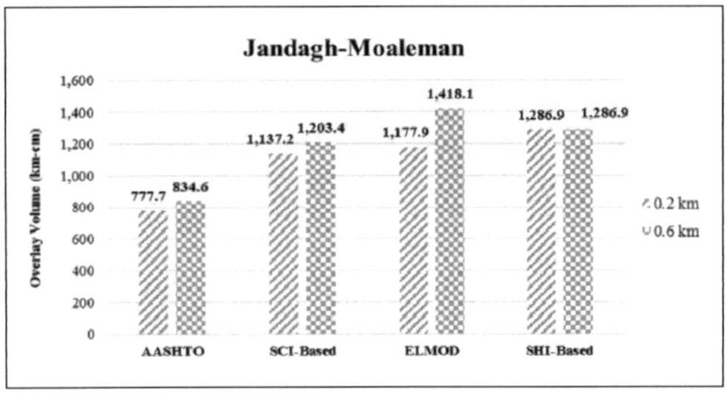

Figure 5. Overlay required for 0.2 and 0.6 km intervals using different methods.

Using the SCI-score approach, the required overlay thickness was determined based on the range of SCI scores used. For scores below 50, the required overlay was 15.2–20.3 cm (6–8 in) (*3*). Studies have shown that both functional and structural conditions should be considered for M&R decision-making (*5, 20*); therefore, SHI was used to consider both conditions when determining the overlay thickness (*13*). The thickness of the required overlay at each point and the average thickness of each uniform section were calculated. The thickness of the required overlay for each approach was reported as km-cm, which is the multiplication of the average required overlay thickness in centimeters by the length of the uniform section in kilometers for each uniform section.

Discussion

Statistical analysis was conducted to determine the effect of increasing the intervals at the network level on the evaluation of pavement conditions that may have an effect on M&R decisions. In previous studies, statistical analysis was restricted to the investigation of changes in indices and parameters. The effects of new intervals on network-level M&R decisions, such as overlays required and their application at the project level or initial spacing, was not considered.

Three types of analysis were conducted to validate the applicability of new intervals at the network level:

1. Percentage of change of overlay required for 0.2 and 0.6 km intervals
2. Percentage of change in SCI between the 0.2 km and 0.6 km intervals
3. Percentage of change in time required for testing for 0.2 and 0.6 km intervals

In the first analysis, the overlay thicknesses required for 0.2 and 0.6 km intervals were determined using the different methods. Figure 5 shows the overlay for 0.2 and 0.6 km intervals for one of the routes, Jandagh-Moaleman (112.7 km in length). As seen, the overlay required for the 0.6 km interval was very similar to that for the 0.2 km interval. The percentage of change for these intervals is presented in Table 6. In almost all methods, the changes were <10%. This can be interpreted as confirming that the evaluation of the pavement condition at 0.6 km intervals would be very similar to that at the 0.2 km interval. Increasing the spacing from 0.2 to 0.6 km would not change the required overlay considerably; therefore, instead of the 0.2 km interval, the FWD test can be conducted at the 0.6 km interval.

The second analysis showed that increasing the intervals and eliminating some points in-between changed the SCI <15%. This means that the SCI assessed the pavement conditions at the 0.6 and 0.2 km intervals as having similar values. Table 7 shows that, except for two sections for each overlay design method, all of the percentages of change were <15% and in most sections were <10%.

As indicated above, between 0.2 and 0.6 km, the SCI decreased and the overlay required increased. This demonstrates that increasing the intervals up to the specified values at the network level was safe in terms of the structural requirements.

In the third analysis, the time efficiency of the 0.6 km interval testing was investigated. The time required included the net testing time at a given point plus the travel time to the next point. Table 8 shows that, for almost all routes, the percentage of change in the time required for the 0.6 km interval testing decreased more than 45% compared with the 0.2 km interval, which increased the efficiency at the network level.

Summary and Conclusion

This study was conducted to develop an approach for incorporating techniques used to interpret and evaluate deflection data for network-level pavement management system applications. FWD deflection data collected from different regions in Iran were used. Different indicator indices were used to determine the network-level intervals. The effect of using new intervals on the overlay required, indicator indices and time required were determined to verify the applicability and efficiency of the network-level intervals.

Table 6. Overlay Required and Percentage of Change for 0.2 and 0.6 km Intervals

	Route length (km)	Overlay design method	Required overlay (km-cm)		Percentage of change
			0.2 km interval	0.6 km interval	
Shoush-Andimeshk	35.4	AASHTO	174	196.8	13.1
		SCI	97.8	113	15.6
		ELMOD	111.1	145.3	30.8
		SHI	240	258.3	7.6
Jandagh-Moaleman	112.7	AASHTO	777.7	834.6	7.3
		SCI	1137.2	1203.4	5.8
		ELMOD	1177.9	1418.1	20.4
		SHI	1286.9	1286.9	0
Cheshme Ali-Damghan	26.1	AASHTO	236.6	261.9	10.7
		SCI	280.4	288.8	3
		ELMOD	221.9	237.4	7
Sirjan-Shahre Babak	43.4	AASHTO	952.5	947.9	−0.5
		SCI	731.1	724.1	−0.9
		ELMOD	690.2	700.6	1.5
Semnan-Firouzkuh	41.5	AASHTO	103.2	115.7	12.1
		SCI	179.1	198.1	10.6
		ELMOD	135.1	131.9	−2.4
Qom-Salafchegan	35.8	AASHTO	329.5	320.8	−2.6
		SCI	240.4	248.5	3.4
		ELMOD	115.7	125.2	8.3
Overall	294.8	AASHTO	85.3	87	2
		SCI	96.4	99.4	3.2
		ELMOD	88.9	99.9	12.4
		SHI	128.7	129.3	0.5

Table 7. Percentage Change of SCI Values Between 0.2 and 0.6 km Intervals

	Section number	Section length (km)	Percentage change of SCI values	
			Mean	85% confidence
Shoush-Andimeshk	1	3.4	−3.76	−6.13
	2	10.6	−4.20	−10.99
	3	4.8	4.43	−6.42
	4	16.6	1.41	0.0
Cheshme Ali-Damghan	1	2.2	0.0	13.25
	2	5.4	−3.65	0.00
	3	2.1	28.44	0.00
	4	5.2	−14.19	0.00
	5	2.2	0.00	0.00
	6	4.4	3.60	0.00
	7	4.6	−1.19	−7.03
Sirjan-Shahre Babak	1	3	−6.46	0.00
	2	6.2	3.43	10.49
	3	8.1	0.00	−4.24
	4	6.2	−2.41	0.00
	5	14.2	−1.26	0.00
Jandagh-Moaleman	1	3.6	−19.79	−17.12
	2	10.4	5.81	0.00
	3	20.1	−0.99	−20.37
	4	14.7	13.92	0.00

(continued)

Table 7. (continued)

	Section number	Section length (km)	Percentage change of SCI values	
			Mean	85% confidence
	5	13.8	−3.44	9.63
	6	12.3	0.00	0.00
	7	8.2	0.00	−6.44
	8	8.2	0.19	4.86
	9	7.4	−2.54	0.00
	10	11.4	−1.91	0.00
	11	8.2	10.60	0.00
Semnan-Firouzkuh	1	8.4	−14.25	−11.88
	2	11.4	−6.87	4.26
	3	3.7	0.00	0.00
	4	12.3	−1.2	−5.79
	5	2.6	−8.19	−5.2
	6	3.6	4.38	0.00
Qom-Salafchegan	1	15.4	−4.76	0.00
	2	12.8	15.6	11.06
	3	2.8	−3.53	−6.29
	4	4.3	−4.75	1.27
Overall	37*	294.8	−0.94	−2.0

*Total number of sections.

Table 8. Time Required for FWD Testing at 0.2 and 0.6 km Intervals

	Type	0.2 km interval time (h)	0.6 km interval time (h)	Percentage of change
Shoush-Andimeshk	I	7.27	3.62	−50.28
	II	6.87	2.59	−62.22
	III	6.87	2.50	−63.64
Cheshme Ali-Damghan	I	4.80	3.30	−31.25
	II	4.35	2.65	−39.12
	III	4.35	2.56	−41.12
Sirjan-Shahre Babak	I	9.20	4.57	−50.35
	II	8.62	3.32	−61.43
	III	8.62	3.20	−62.87
Jandagh-Moaleman	I	21.70	11.84	−45.46
	II	19.87	8.60	−56.75
	III	19.87	8.27	−58.36
Semnan-Firouzkuh	I	6.81	4.24	−37.71
	II	5.97	3.02	−49.47
	III	5.97	2.9	−51.37
Qom-Salafchegan	I	6.56	3.85	−41.33
	II	5.93	2.84	−52.11
	III	5.93	2.74	−53.86

Assumptions

Type I
1. Speed of truck between points at 0.2 km interval = 20 km/h
2. Speed of truck between points at 0.6 km = 20 km/h
3. Acceleration of truck between points for both levels = 0

Type II
1. Speed of truck between points at 0.2 km interval = 20 km/h
2. Speed of truck between points at 0.6 km interval = 20 km/h
3. Acceleration of the truck between points for both intervals = 7,200 km/h^2

Type III
1. Speed of truck between points at 0.2 km interval = 20 km/h
2. Speed of truck between points at 0.6 km interval = 20 km/h
3. Acceleration of truck between points for both intervals = 9,600 km/h^2

The study examined 0.2, 0.4, 0.6, and 0.8 km intervals between the FWD data collection points with the aim of optimizing spacing without compromising the accuracy. The salient findings of the study are as follows:

- There was good agreement between the network-level intervals determined based on the indicator indices of SCI, SSI, SHI, and A_r'.
- There was an acceptable agreement between the network protocol in the study and intervals determined according to ASTM standard.
- Increasing the intervals at the network level caused <15% change in the SCI of each uniform section of each route.
- Increasing the intervals at the network level caused <10% change in the overlay thickness required. In

comparison with the initial spacing, this is appropriate for network-level PMS.

- The time required for data collection at the 0.6 km intervals was about 55% of that for the 0.2-km interval.
- Changing the data collection interval from 0.2 km to 0.6 km caused a 2, 3.2, 12.4, and 0.5% increase for the overall network overlay thickness using AASHTO equations, SCI score, ELMOD and SHI, respectively. In addition, this change caused a 0.94 and 2% decrease in the overall SCI for all routes based on the mean and 85% CI, respectively. These findings indicate that increasing the intervals to the specified values at the network level was safe in terms of the structural requirements.

Based on these findings and the results of analysis, it can be concluded that increasing the spacing interval from 0.2 to 0.6 km at the network level for FWD data collection will provide satisfactory results. There is a similarly acceptable correlation between the network spacing determined in this study and those of previous studies in other countries and agencies around the world. This study indicates that SCI is a more reliable and practical index than other indicator indices examined because it is based on the existing and required SNs and more directly relates to pavement design.

Acknowledgments

The authors would like to thank the Technical and Soil Mechanics Laboratory for providing the data needed to complete this study.

References

1. Collop, A. C., R. J. Armitage, and N. H. Thom. Assessing Variability of In Situ Pavement Material Stiffness Moduli. *Journal of Transportation Engineering*, Vol. 127, No. 1, 2001, pp. 74–81.
2. Hossain, M., and J. P. Zaniewski. Variability in Estimation of Structural Capacity of Existing Pavements from Falling Weight Deflectometer Data. *Transportation Research Record: Journal of the Transportation Research Board*, 1992. 1355: 17–26.
3. Peddibhotla, S. S. S., M. Murphy, and Z. Zhang. *Validation and Implementation of the Structural Condition Index (SCI) for Network-Level Pavement Evaluation*. Publication FHWA/TX-11/5-4322-01-1. FHWA, U.S. Department of Transportation, 2011.
4. Carvalho, R., R. Stubstad, R. Briggs, O. Selezneva, E. Mustafa, and A. Ramachandran. *Simplified Techniques for Evaluation and Interpretation of Pavement Deflections for Network-Level Analysis*. Publication FHWA-HRT-12-023. FHWA, U.S. Department of Transportation, 2012.
5. Chowdhury, T., R. A. Shekharan, and B. K. Diefenderfer. Implementation of Network Level Falling Weight Deflectometer Survey. *Transportation Research Record: Journal of the Transportation Research Board*, 2012. 2304: 3–9.
6. Dynatest Brochure, ELMOD 5.0. *Dynatest International A/S*, 2011. ftp://ftp.dynatest.com/Downloads/Elmod5/ELMOD%205.pdf. Accessed November 3, 2017.
7. Alam, J., K. A. Galal, and B. K. Diefenderfer. Statistical Determination of Minimum Testing Intervals and Number of Drop Levels for Network Level FWD Testing on Virginia's Interstate System. *Transportation Research Record: Journal of the Transportation Research Board*, 2007. 1990: 111–118.
8. Damnjanovic, I., and Z. Zhang. Determination of Required Falling Weight Deflectometer Testing Frequency for Pavement Structural Evaluation at the Network Level. *Journal of Transportation Engineering*, Vol. 132, No. 1, 2006, pp. 76–85.
9. ASTM D7228. *Standard Test Method for Prediction of Asphalt-Bound Pavement Layer Temperatures*. American Society for Testing and Materials, USA, 2012.
10Zhang, Z., G. Claros, L. Manuel, and I. Damnjanovic. Evaluation of the Pavement Structural Condition at Network Level Using Falling Weight Deflectometer (FWD) Data. Presented at 82nd Annual Meeting of the Transportation Research Board, Washington, D.C., 2003.
11. Zhang, Z., L. Manuel, I. Damnjanovic, and Z. Li. *Development of a New Methodology for Characterizing Pavement Structural Condition for Network-Level Applications*. Publication FHWA/TX-04/0-4322-1. FHWA, U.S. Department of Transportation, 2002.
12. Flora, W. F. *Development of a Structural Index for Pavement Management: An Exploratory Analysis*. Master of Science in Engineering thesis. Purdue University, West Lafayette, IN, 2009.
13Elbagalati, O., M. Elseifi, K. Gaspard, and Z. Zhang. Development of the Pavement Structural Health Index Based on Falling Weight Deflectometer Testing. *International Journal of Pavement Engineering*, Vol. 19, No. 1, 2016, pp. 1–8. http://dx.doi.org/10.1080/10298436.2016.1149838.
14Saleh, M. Simplified Approach for Structural Capacity Evaluation of Flexible Pavements at the Network Level. *International Journal of Pavement Engineering*, Vol. 17, No. 5, 2016, pp. 440–448.
15. Rohde, G. T. Determining Pavement Structural Number from FWD Testing. *Transportation Research Record: Journal of the Transportation Research Board*, 1994. 1448: 61–68.
16. Wimsatt, A. J. Direct Analysis Methods for Falling Weight Deflectometer Deflection Data. *Transportation Research Record: Journal of the Transportation Research Board*, 1999. 1655: 135–144.
17. Noureldin, A. S. New Scenario for Backcalculation of Layer Moduli of Flexible Pavements. *Transportation Research Record: Journal of the Transportation Research Board*, 1993. 1384: 23–28.
18. Molenaar, A. A. A. *Structural Design of Pavements, PART VI Structural Evaluation and Strengthening of Flexible Pavements Using Deflection Measurements and Visual Condition Surveys*. TU Delft, 2009.
19Huang, Y. H. *Pavement Analysis and Design*. Pearson Education, Inc., Upper Saddle River, New Jersey, 2004.
20. Bryce, J., G. Flintsch, S. Katicha, and B. K. Diefenderfer. Enhancing Network-Level Decision Making Through the Use of a Structural Capacity Index. *Transportation Research Record: Journal of the Transportation Research Board*, 2013. 2366: 64–70.

The Standing Committee on Pavement Condition Evaluation (AFD20) peer-reviewed this paper (18-06643).

Article

TRR
JOURNAL OF THE TRANSPORTATION RESEARCH BOARD

Transportation Research Record
2018, Vol. 2672(40) 78–87

DOI: 10.1177/0361198118757194
journals.sagepub.com/home/trr
SAGE

Effect of Methodological Choices on Pavement Life-Cycle Assessment

Rebekah Yang[1], Imad L. Al-Qadi[2], and Hasan Ozer[2]

The use of life-cycle assessment (LCA) to assess the environmental impacts of pavement systems has become more prevalent in recent years. When performing an LCA study, a series of methodological choices must be defined. As these decisions can change from study to study, it is important to understand the significance or insignificance of the methodological choices relevant to pavement LCA. This paper evaluated the sensitivity of five choices commonly made in pavement LCA; cut-off criteria, end-of-life (EOL) allocation, asphalt binder allocation, traffic growth, and type of energy reported. Eight case studies and four environmental indicators, that is, global warming potential, primary energy as fuel, total primary energy, and a unitless single score, were considered in the sensitivity analyses. Varying the cut-off criteria and asphalt binder allocation only had a significant impact on the environmental indicators when the use stage of the life-cycle is excluded and only the materials and construction, maintenance and rehabilitation, and EOL stages are considered. Using different EOL allocations, traffic growths, and types of energy reported had significant effects on the overall life-cycle results. Common methodological choices made in a pavement LCA are expected to have an impact on LCA results and subsequent interpretation, with the magnitude of the impact dependent on the scope of the analysis.

The application of life-cycle assessment (LCA) in evaluating the environmental impact of pavement systems has increased over the past two decades. The ability to quantify environmental impact over the lifespan of a pavement gives valuable insight to stakeholders in the transportation industry when making decisions about pavement planning, design, construction, and management. As a methodology, LCA is governed by a set of rules and preferences that are uniquely decided in the goal and scope of each study. Some of these choices refer to general LCA methodological elements such as those pertaining to the system boundary, cut-off criteria, or data quality requirements. Other choices are specific to the product system, such as the treatment of traffic growth or asphalt binder allocation for pavement LCA. When conducting an LCA, it is important to understand how rules and choices affect the results of the study.

In this paper, common pavement LCA choices are evaluated with respect to the magnitude of impact that they have on the overall life-cycle results. The outcome of this work is important for both pavement LCA practitioners and stakeholders. Sensitivity analyses can identify which methodological choices have a significant effect on pavement LCA and should be the subject of further scrutiny and research. By understanding the gravity of different choices, guidelines can be developed to further standardize pavement LCA practices by coming to a consensus on high-impact choices or requiring sensitivity analyses for these choices. If these high-impact choices are not consistent or acknowledged across LCA studies, stakeholders may be presented with misleading information when viewing results from studies with different methodological decisions.

The International Organization for Standardization (ISO) 14040 guidelines require sensitivity analyses as part of a larger evaluation of LCA results and interpretation (*1*). To establish confidence in a study's findings and conclusions, LCA practitioners are required to conduct a series of checks for completeness, sensitivity, and consistency as described in ISO 14044 (*2*). In addition to the general ISO standards, FHWA's Pavement LCA Framework also discusses requirements for LCA interpretation, including the use of sensitivity analyses (*3*). Recommendations for when to apply sensitivity analysis for the pavement system are mentioned throughout the document, such as in the goal and scope (e.g., cut-off criteria, system boundaries, allocation procedures, analysis period), inventory analysis (e.g., sources, data quality), pavement performance (e.g., metrics

[1]CNA, Arlington, VA
[2]Department of Civil and Environmental Engineering, University of Illinois at Urbana-Champaign, Urbana, IL

Corresponding Author:
Address correspondence to Rebekah Yang: rebekah.y.yang@gmail.com

used, future maintenance and rehabilitation), use (e.g., variables and models, changes in traffic growth and composition over time), end-of-life (EOL, e.g., highly recycled materials), and impact assessment (e.g., impact categories chosen, normalization reference data, weighting factors). It is not common practice yet to fully address all evaluation checks in a pavement LCA study, and the few studies that contain sensitivity analyses mainly addressed issues related to allocation methods or inventory sources. It should be noted that other pavement LCA studies not discussed in this paper use sensitivity analyses for comparative purposes to assess the environmental impacts of, for example, alternative design variations (*4*) or the increased use of recycled materials (*5*).

Sensitivity analyses related to allocation methods have been conducted for pavement LCA, but systematic analysis of other methodological issues such as analysis period, system boundaries, and cut-off criteria are not found in the current literature. In pavement LCA, allocation is especially relevant to co-products and EOL recycling. In 2010, Chen et al. found significant discrepancies in environmental impacts when using different allocation parameters (i.e., none, mass, economic) to distribute input flows from coal and steel production to fly ash and blast furnace slag (BFS), respectively (*6*). A study by Sayagh et al. in 2010 found a difference of 60% in the impact of greenhouse gas (GHG) over the 30-year life-cycle of a pavement when using 0% and 20% allocation factors for BFS as an aggregate base (*7*). Blomberg et al. in 2012 evaluated the environmental burdens from producing asphalt binder, finding higher inventory outputs when using mass versus economic allocation (*8*). In 2013, Huang et al. found that using a mass versus economic allocation for BFS as well as for asphalt binder resulted in higher impacts for a cradle-to-ground pavement reconstruction (*9*). The study also found that cradle-to-gate environmental benefits from using recycled material in asphalt concrete (AC) varied widely for different types of impacts depending on the EOL allocation used, reaching a 24% reduction in impacts for the cut-off method and 70% for substitution (*9*). Another study by Yang et al. in 2016 on asphalt binder found that environmental impacts were highest when using an allocation for distributing refinery processes based on mass, followed by energy content, volume, and then economic value as the lowest (*10*).

The effects of using different inventory sources in pavement LCA have also been evaluated using sensitivity analyses. Meil collected regional data for the production of AC and Portland cement concrete (PCC) in Canada (*4*). The results were performed using different mix designs typical of each region considered, preventing a sensitivity analyses of the inventory sources. In 2007, Huang performed an LCA for the production and construction of an airport asphalt pavement (*11*). A sensitivity check on data sources for AC mixing, asphalt binder production, and aggregate production concluded that there were no significant differences in overall results with different data sources. In 2012, Wang et al. conducted LCA on the full life-cycle of high- and low-volume roads (*12*). A sensitivity analyses of inventory sources for asphalt binder and Portland cement production showed no significant effect on the overall life-cycle results due to the large influence of the use stage. Yang et al. compiled regional processes for the production of various petroleum products, including diesel and asphalt binder, which showed significant differences in environmental impact of material production based on regional representation (*10*).

Existing pavement LCA studies have also included sensitivity analyses on pavement-specific items such as those related to hauling distances, use stage models, and traffic parameters. Meil included a sensitivity analyses of different hauling distances (as well as shoulder design) in that study (*4*). The study found that increases in GHGs were proportional to increases in transportation, and the differences were not significant unless the distances increase dramatically. As use stage models and traffic parameters tend to have a high level of uncertainty, many of these studies use sensitivity analyses to assess aspects of uncertainty (*13–15*), which are not discussed in this paper.

Although sensitivity analyses have been conducted for pavement LCA in the past, many of these analyses are limited to allocation and inventory or were used to assess areas of uncertainties or to compare design features. There is a gap in quantifying the effect of additional methodological choices such as cut-off criteria or choice of impact categories. Many existing studies only focus on one or two issues, so that the significant or insignificant discrepancies found in the sensitivity analyses are isolated and cannot be ranked against each other. The objective of the study is to conduct a series of systematic sensitivity analyses on various pavement LCA case studies. The five parameters assessed are related to methodological choices made in the goal and scope of an LCA, including some choices which have not yet been formally considered in the context of pavement LCA. Significant variation in results based on the scenarios considered in the analyses will identify parameters that can have a large impact on the overall LCA results.

Methods

The sensitivity analyses in this paper were performed on a set of case studies based on eight contracts from Illinois State Toll Highway Authority (ISTHA) between the years 1999 and 2013. An LCA software, developed specifically for ISTHA, was used to run the case studies with the software and original results described elsewhere (*16*). The case studies used in the analyses include both AC and PCC pavements as well as various reconstruction, widening, new

Table 1. Summary of Eight Case Study Contracts

ID	Year	Length (miles)	Design Life (years)	2013 AADT	% Truck	Pavement Type	Contract Type
AC1	2008	5.0	75	24,120 EB	29.4	FDHMA	Widening and reconstruction
AC2	2009	7.6	75	25,340 WB	29.4	FDHMA	Widening and reconstruction
NEW1	2006	1.42	62	31,720 NB; 30,960 SB	13.1	JPCP	New construction
NEW2	2006	1.72	62	34,275 NB; 33,325 SB	13.1	JPCP	New construction
PCC1	1999	5.1	44	51,475 EB; 50,720 WB	11.2	JPCP	Widening and reconstruction
PCC2	2004	4.6	78	67,005 EB; 67,655 WB	9.1	CRCP	Widening and reconstruction
PCC3	2007	2.6	62	61,270 SB	13.8	JPCP	Widening and reconstruction
RHB1	2001	12.8	58	70,864 NB; 70,968 SB	17.1	HMA overlay	Resurfacing and rehabilitation

construction, and resurfacing projects. A summary of the projects is given in Table 1, including details concerning the average annual daily traffic (AADT) in each direction [i.e., northbound (NB), southbound (SB), eastbound (EB), westbound (WB)] and pavement type [i.e., jointed plain concrete pavement (JPCP), continuously reinforced concrete pavement (CRCP), and full-depth hot-mix asphalt pavement (FDHMA)].

The system boundaries of the LCA considered for this paper include cradle-to-grave processes related to the pavement system of the contract including: materials and construction (M&C), maintenance and rehabilitation (M&R), use, and EOL. The M&C stage includes all processes regarding the acquisition, production, processing, and transportation of materials and equipment used for work activities as well as the actual construction itself, excluding any effects from work zone and resultant congestion. The work activities include work site preparation and evacuation, construction of the subgrade, base, and upper layers of the mainline and shoulders, and installation of markings and guardrails. The M&R stage includes both minor repairs, for example, crack sealant and patching, and major rehabilitations, for example, overlays and microsurfacing, required for the mainline and shoulders of the pavement based on typical ISTHA practices. The use stage considers impacts from vehicle fuel consumption throughout the life of the pavement as well as impacts from carbonation and radiative forcing related to surface albedo. To calculate additional vehicle fuel consumption attributed to the pavement, empirical models (*17*) are used to predict pavement performance as measured by the progression of international roughness index (IRI) and mean profile depth (MPD) over time combined with other models (*18*) correlating the effect of IRI and MPD on fuel consumption for passenger cars and trucks. The EOL stage considers impacts from the demolition, transportation, and/or landfilling of the pavement assuming a 5% landfilling rate.

A total of five sensitivity analyses were conducted to evaluate the effects of varying choices made regarding cut-off criteria, EOL allocation, asphalt binder allocation, traffic growth, and type of energy reported. The main four environmental indicators considered in the sensitivity analyses include global warming potential (GWP), primary energy as fuel (PEF), total primary energy (TPE), and a unitless single score (SS) based on the U.S. Environmental Protection Agency Tool for Reduction and Assessment of Chemicals and Other Environmental Impacts (TRACI) 2.1 method. PEF is defined as a subset of TPE that includes only energy values from fuel, whereas TPE also includes energy values from materials that are not used as fuel but have substantial energy content (e.g., lumber or non-fuel petroleum products). SS is a normalized (*19*) and weighted (*20*) metric based on 10 factors (i.e., ozone depletion, global warming, smog, acidification, eutrophication, carcinogenics, noncarcinogenics, respiratory effects, ecotoxicity, and fossil fuel depletion) characterized by TRACI. This metric is subjective and was developed specific to the U.S. context in 1999.

Results and Discussion

Cut-Off Criteria

The cut-off criteria is defined in ISO 14040 as the "specification of the amount of material or energy flow or the level of environmental significance associated with unit processes or product system to be excluded from a study" (*1*). In the case studies, the cut-off criteria for the M&C stage of the pavement system was uniquely defined to include pay items that satisfied at least one of the following criteria: existing in the top 95% of the total cost of all pavement-related items; or contributing at least 0.5% of the total cost of all pavement-related items. The criteria was based on cost to reduce the number of pay items analyzed to a manageable set. The chosen method relies on the assumption that economic cut-off

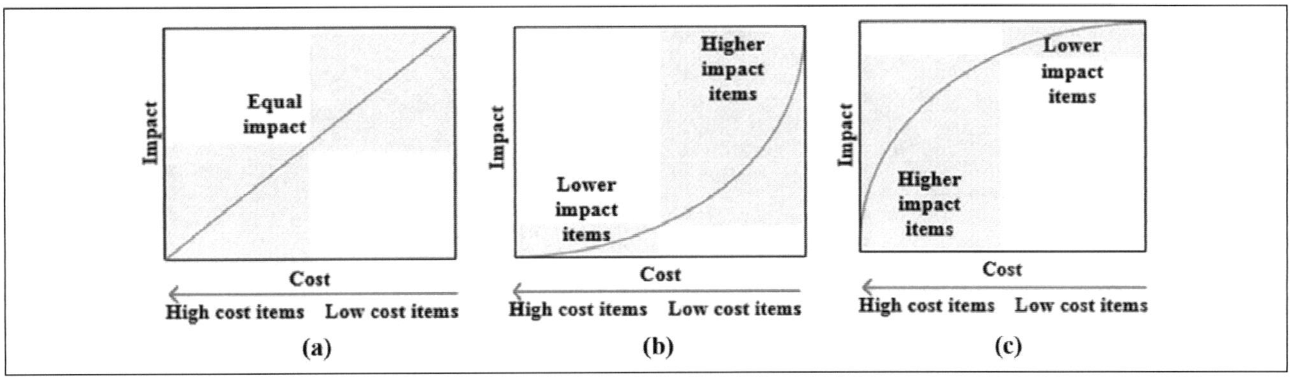

Figure 1. Explanation of cut-off criteria significance for (*a*) equality, (*b*) convex, and (*c*) concave curves.

criteria can be used for an environmental analysis because the two components are related.

To investigate the sensitivity of the cut-off criteria, the relationships between the economic and environmental contributions of pay items in the projects were examined. The reduction in environmental impacts from systematically excluding M&C pay items one at a time, in order of lowest to highest economic contribution, from the original set of M&C pay items were determined. The three graphs in Figure 1 demonstrate the types of behavior that can be expected. If the curves are closer to the equality line (Figure 1*a*), the economic and environmental impacts are proportional, so including more of any pay item is beneficial. If the curves are convex (Figure 1*b*), the low-cost items contribute to higher-impact items; thus, including more of the supporting or lower-costing pay items is important to capturing environmental impacts. If the curves are concave (Figure 1*c*), the higher-cost items contribute to higher environmental impacts, so including the most costly items is important.

The graphs in Figure 2 show the relationship between cost and environmental impacts for the M&C stage for each of the eight case studies. It can be seen that the curves are all relatively close to the equality line, implying that cost and environmental impact are related. The curves are slightly concave, implying that it is more important to include the top cost pay items. This observation is consistent with the 95%/0.5% cut-off criterion, which ensures that the items of highest environmental significance within established cost limits are included in the LCA.

Having established that including the top cost pay items is important, the sensitivity to the percentage of included pay items can also be investigated. For all projects, the average reduction in impacts when considering 90%, 80%, and 70% of the top cost pay items versus 95% can be seen in Figure 3*a*. When including M&C pay items falling within the top 70% of the total cost of the contract, the corresponding loss in environmental impacts from using the top 95% cut-off criteria is in the range of 7%–12% (i.e., less than the proportional 15%). This supports the conclusion made previously that the pay items contributing to the greatest environmental impacts are also those that contribute the greatest cost. A loss of 12% (or even 5% from the 90% cut-off) in the environmental impacts is quite high for the M&C stage, so it is not recommended to lower the cut-off criteria. This suggestion may be reconsidered when looking at the effect of changing the cut-off criteria on the entire life-cycle in Figure 3*b*. In this case, the excluded impacts when using a 70% cut-off constitutes approximately 2%–3% of the total life-cycle impacts. When considering the entire life-cycle, it may be possible for some LCA applications to lower the cut-off criteria to reduce the time and information needed for the LCA without compromising the evaluation of life-cycle environmental impacts.

EOL Allocation

Allocation deals with the "partitioning the input or output flows of a process or a product system between the product system under study and one or more other product systems," as defined in ISO 14040 (*1*). Although the sensitivity effect of allocation has been addressed by other pavement-related LCA studies in a more limited scope (*6–9, 18*), EOL allocation and asphalt binder allocation are discussed in this paper in the context of the entire pavement life-cycle.

Specifying an allocation procedure to use at EOL is especially important for road systems because a significant amount of AC and PCC material may be recycled or reused at the end of the system's life. Depending on the allocation method chosen, the benefits from using the subsequent recycled materials may be attributed to the original system, the future system or split between the two systems. A number of allocation methods have been established to attribute environmental impacts between chained product systems (Figure 4), where a portion of the product in System 1 is incorporated into System 2 (*22*). In this sensitivity analysis, the methodological choice of using the cut-off method (used in the original case studies) versus the substitution method for allocation was evaluated as they

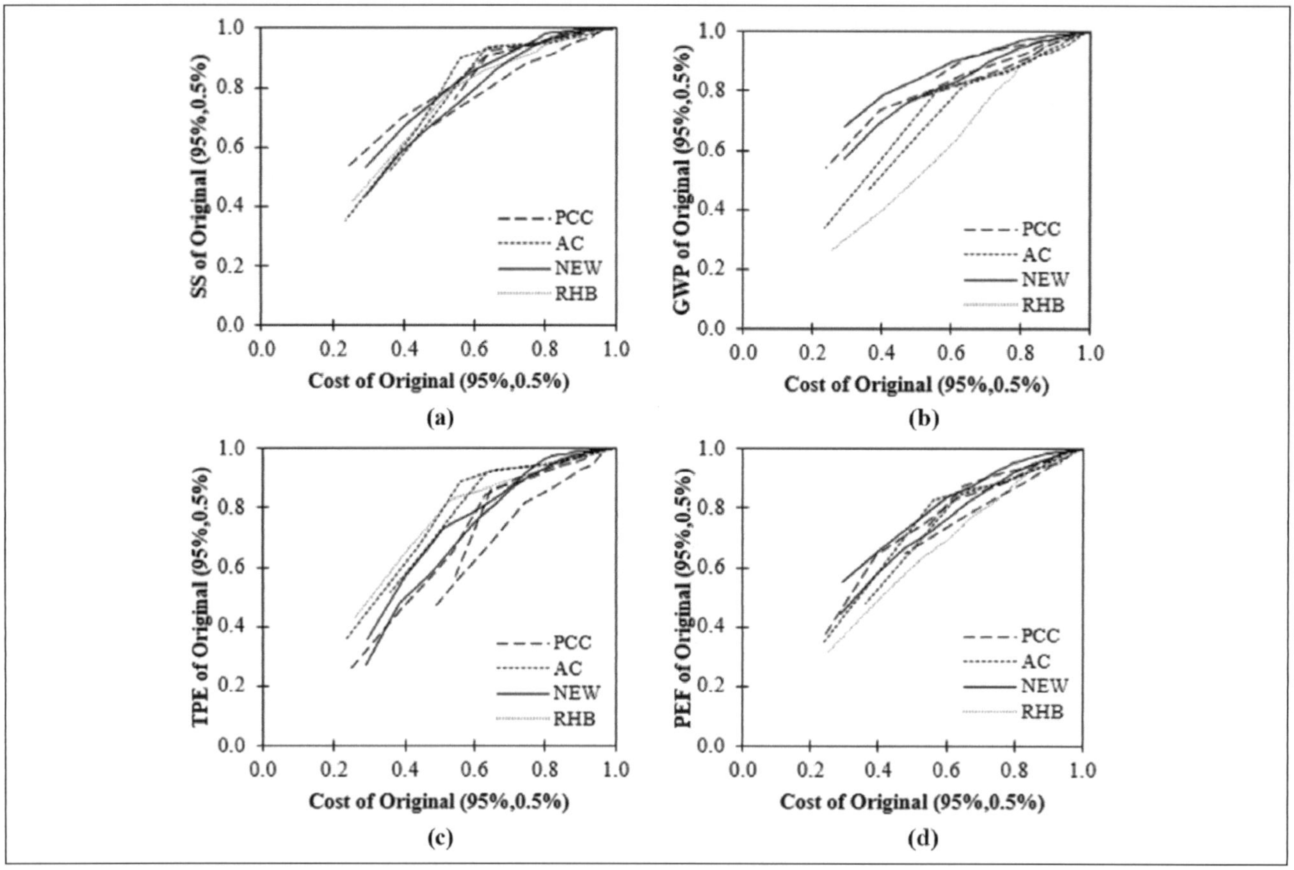

Figure 2. M&C pay item cut-off criteria sensitivity for case studies for (*a*) SS, (*b*) GWP, (*c*) TPE, and (*d*) PEF.

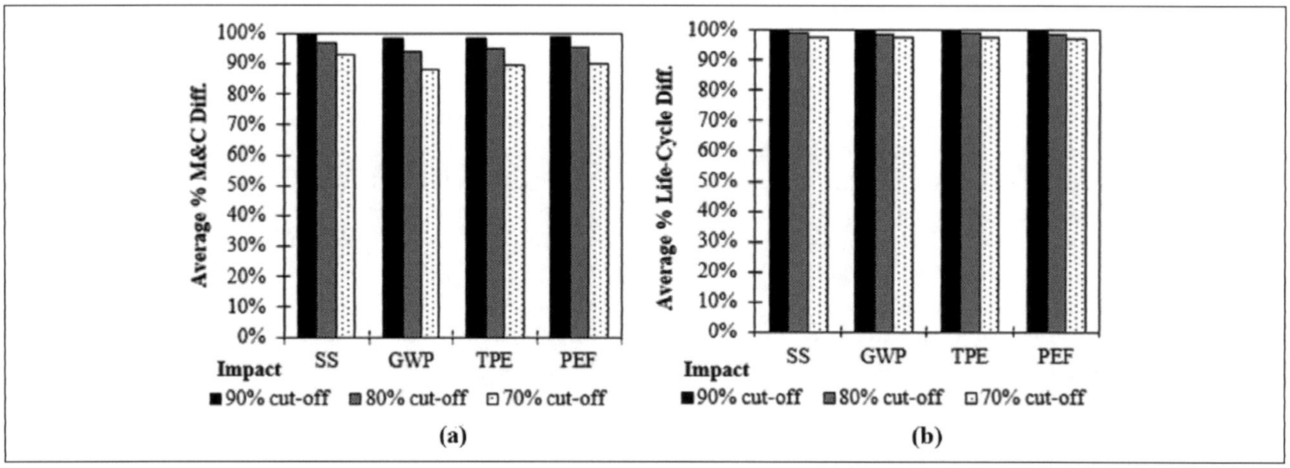

Figure 3. Average percent difference in impacts for various cut-off criteria for (*a*) M&C stage and (*b*) the entire life-cycle.

pertain to AC and PCC at EOL. EOL allocation for other recyclable materials in the pavement project system such steel and aggregate are not varied from the original cut-off method in this analysis.

The cut-off method, often used in pavement LCA studies, does not assume anything about the recyclability of the original system at EOL and thus assigns environmental impacts to the system directly responsible for producing or processing

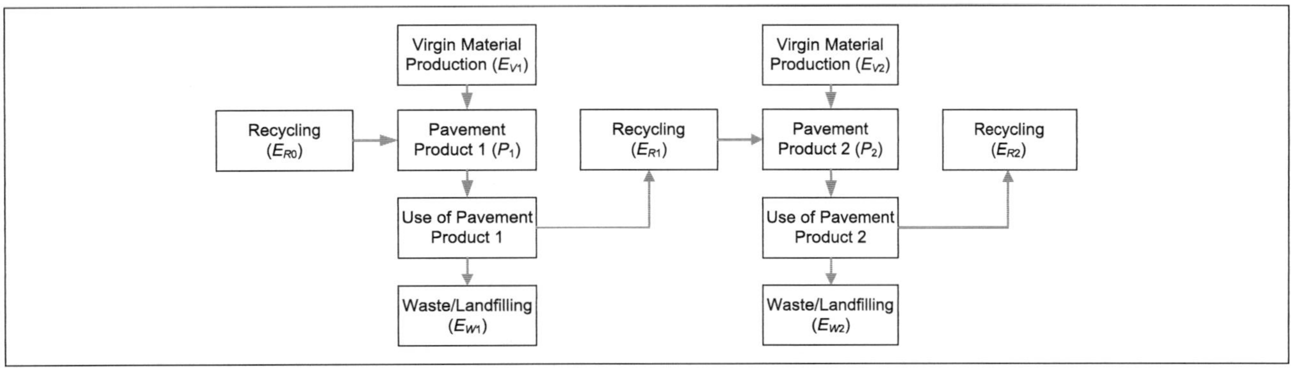

Figure 4. Chained product systems: Pavement Product 1 and Pavement Product 2, based on (*22*).

the material (*22*). The substitution method, treats the recycled material used in the future system as a substitute for virgin material and thus assigns burdens proportional to any "lost" material not continuing in the chain of product systems (*22*).

The two methods are defined explicitly in the following equations, styled after Huang et al. (*9*) and Nicholson et al. (*20*):

Cut-off Method:

$$E_{P1} = X_1 E_{V1} + (1\text{-}X_1) E_{R0} + L_1 E_{W1} \quad (1)$$

$$E_{P2} = X_2 E_{V2} + (1\text{-}X_2) E_{R1} + L_2 E_{W2} \quad (2)$$

Substitution Method:

$$E_{P1} = (1\text{-}L_1) E_{R1} + L_1 E_{V1} + L_1 E_{W1} \quad (3)$$

$$E_{P2} = (1\text{-}L_2) E_{R2} + L_2 E_{V2} + L_2 E_{W2} \quad (4)$$

where X_1, X_2 are the percentages of virgin material used in Products 1 and 2, L_1, L_2 are the percentages landfilled for Product 1 and 2, E_{P1}, E_{P2} are the environmental impacts directly attributed to Products 1 and 2, E_{V1}, E_{V2} are the environmental impacts of producing virgin materials for Product 1 and 2, E_{R0}, E_{R1}, E_{R2} are the environmental impacts of processing recycled material from Products 0, 1, and 2 for the next system, and E_{W1}, E_{W2} are the environmental impacts of landfilling materials for Products 1 and 2, respectively. The chosen allocation affects not only the EOL stage, but also the M&C as well as M&R stages.

To investigate the effect of using a cut-off versus substitution EOL allocation, four scenarios for AC1 and PCC3 were considered by evaluating impacts for each allocation method with a 5% landfilling and then a 50% landfilling assumption at EOL. From the analysis in Figure 5, *a* and *b* without the use stage impacts, it is clear that employing the substitution rather than cut-off method reduces the total impacts of the original pavement system producing the material significantly for all impacts. This is reasonable because the substitution method shares the credit of the recycled material among the producer (i.e., current system) and user of the recycled material (i.e., future system), whereas the cut-off method gives the credit to the user of the recycled material. The reductions for the original 5% landfilling varies from 39%–76%, which implies that the choice of allocation for EOL is a methodological choice with significant influence. If the landfill rate is increased to 50%, the reductions decrease to 20%–40%. In this scenario, the pavement system incurs the environmental burdens for producing the 50% of pavement that is landfilled. The differences from the 50% landfilling scenario are also notable, and for a product system with a high recycling rate in the U.S., the choice of allocation for EOL will be important. The reductions with respect to the entire life-cycle in Figure 5, *c* and *d* are notably reduced to 4%–18% for 5% landfilling and 2%–9% for 50% landfilling.

Asphalt Binder Allocation

In the original case studies, an economic allocation based on the relative market value of asphalt binder in Illinois was used to distribute a corresponding proportion of refinery processes to asphalt binder production. This allocation method was obtained from a related 2014 thesis by Yang (*21*) that calculated various allocation factors using energy content (88.9%), mass (100%), volume (75.1%), and economic value (41.9%). Each of these allocation factors were applied to the eight case study projects to determine the sensitivity of the LCA to the factor chosen.

In Figure 6, *a–d* the difference in life-cycle impacts due to different asphalt binder allocation is small and on average less than 2%. Using an allocation by mass has the largest impacts followed by that by energy and then by volume. In

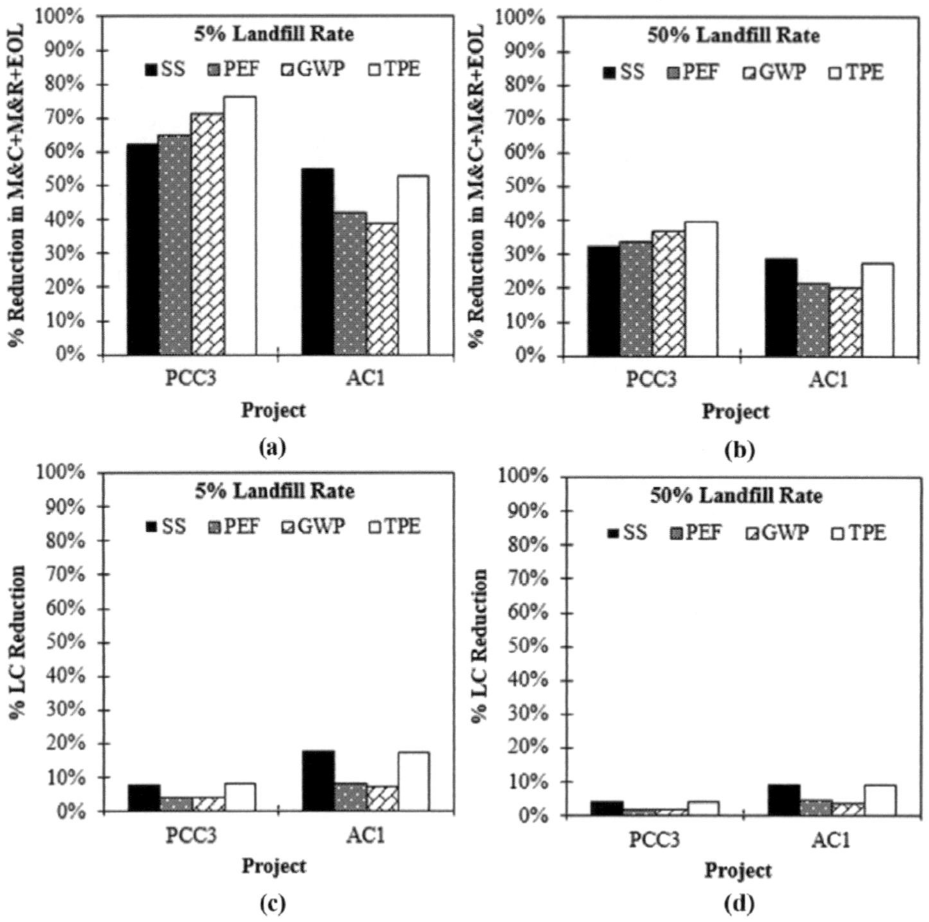

Figure 5. Effect of using a substitution versus cut-off EOL allocation scenario on impacts on (*a, b*) the M&C, M&R, and EOL stages and (*b, d*) all life-cycle stages for (*a, c*) 5% landfill rate and (*b, d*) 50% landfill rate for select projects.

Figure 6, *e–h* the difference on the results including only the M&C and M&R stages are higher and on average around 6%, which is notable. Interestingly, the difference in results is only apparent for GWP, TPE, and PEF; SS is not affected. This implies that SS is influenced largely by other processes in the production of asphalt binder (i.e., crude extraction), so that changes in impacts from refining are less significant.

Constant Traffic Growth

Another choice made in the original case studies was to assume a constant exponential traffic growth over the pavement's life-cycle. This constant rate was determined using averaged ISTHA traffic data for the appropriate toll road; however, in reality, traffic growth rate undulates from year to year. A sensitivity analysis was run to determine the effect of considering instead non-constant growth rates using project AC1, a full-depth asphalt pavement reconstruction evaluated over a 75-year analysis period.

As shown in Figure 7*a*, traffic growth rates were generated for five samples using a random number generator restricted between −2.5% and 5.5%. The original growth rate for this project was 1.47% and the mean growth rates for the five randomly generated traffic progressions were between 1.25% and 1.93%. For each of the random traffic progressions in Figure 7*a*, new vehicle miles traveled (VMT) were considered and the corresponding life-cycle impacts were calculated and analyzed. The resulting differences from the original case study are shown in Figure 7(b), where the changes in impacts are as high as ±20%. Owing to the exponential nature of the growth rate, a small perturbation in the growth percentage can have a large effect on the overall life-cycle VMT (i.e., up to 15% from Figure 7*b*). This change in VMT has a dual effect on environmental impacts as VMT is directly related to vehicle fuel consumption as well as indirectly related to the progression of roughness of the road; the latter of which in turn affects the former. Even with a small change in growth rate, the effect on life-cycle impacts can be large, as seen in these random growth scenarios.

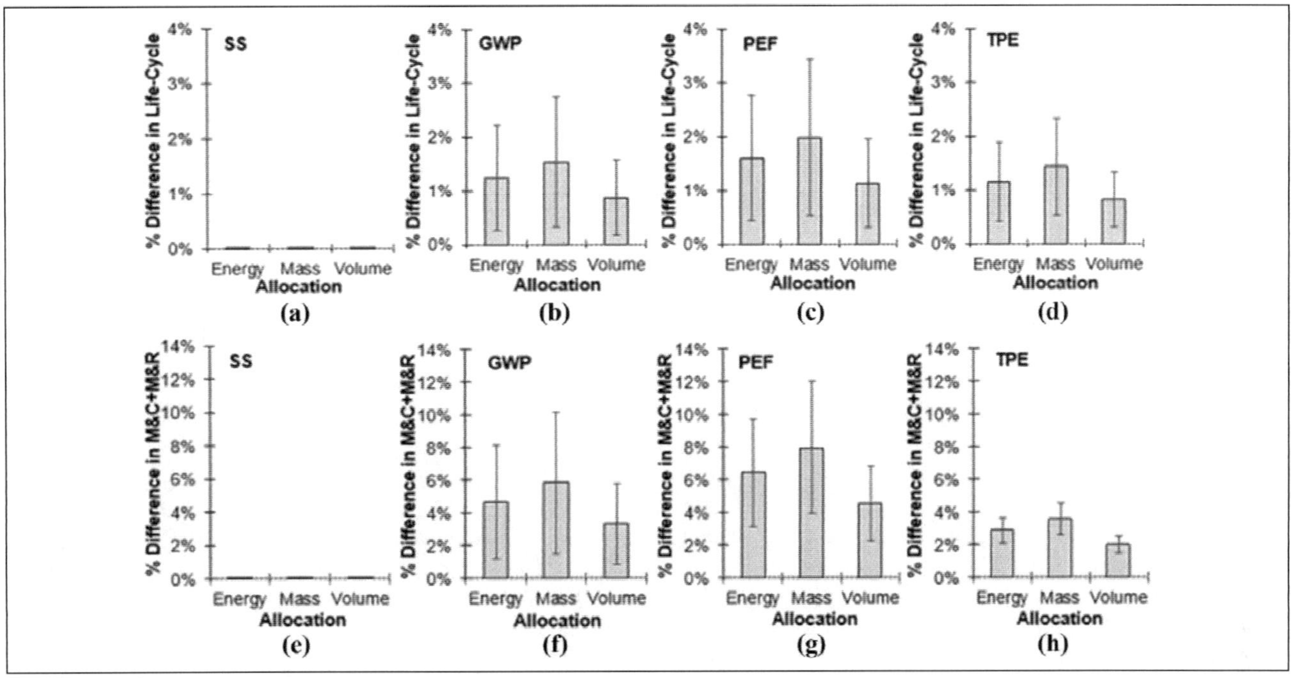

Figure 6. Differences in asphalt binder allocation for each project with respect to (*a–d*) all life-cycle stages and (*e–h*) only the M&C and M&R stages for (*a, e*) SS, (*b, f*) GWP, (*c, g*) PEF, and (*d, h*) TPE.

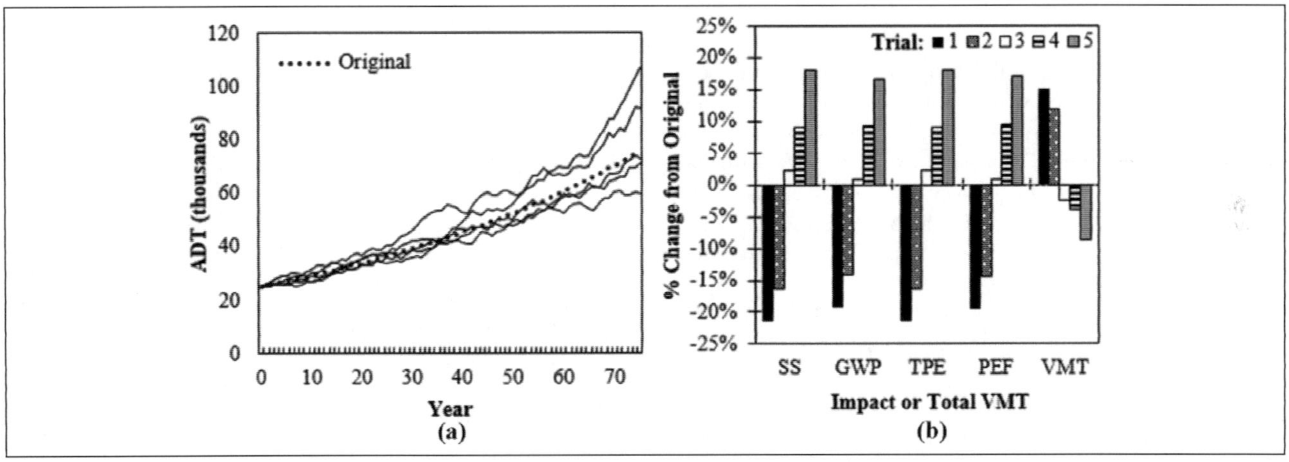

Figure 7. (*a*) Random traffic growth progressions for five trials for AC1 over 75 years and (*b*) percent variation in impacts for AC1 using five randomly generated traffic growth progressions.

Type of Energy Reported

The FHWA's Pavement LCA Framework suggests reporting a breakdown of energy types, including energy as fuel and energy as material (*3*). This issue is especially relevant for pavement because energy as material (i.e., TPE) includes feedstock energy from asphalt binder. Arguments have been presented in support of as well as against the exclusion of feedstock energy due to its influence on the energy attributed to pavement systems (*23*). A breakdown of the life-cycle primary energy consumption from the eight pavement projects are plotted in Figure 8.

The percentages of TPE from material range from 5.1% to 35.2%, with an average of 20.2% and a standard deviation of 11.8%. The projects with the highest primary energy as material (PEM) percentages are AC1 and AC2, which are full-depth asphalt reconstruction projects and have considerable asphalt binder in both the mainline and shoulders as well as in subsequent overlay rehabilitations. NEW1 and NEW3 also have high PEM; these are new PCC

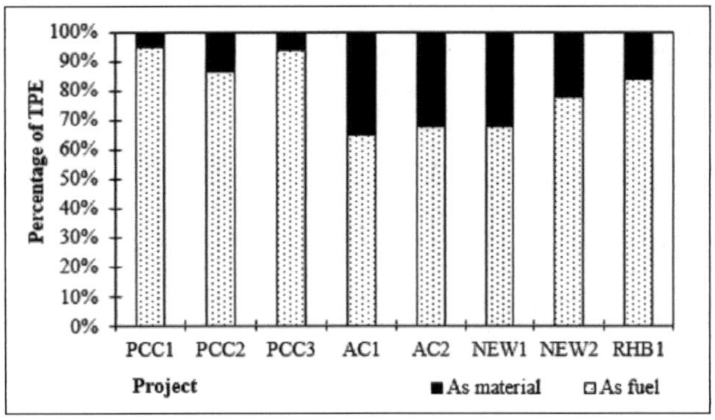

Figure 8. Percentage of TPE as fuel and material for each project.

construction projects with AC shoulders and have lower traffic volumes and thus a higher overall life-cycle impact from materials. RHB1 has a relatively high PEM percentage because it is an AC resurfacing and patching project. With the significant and varying contribution of PEM to TPE as shown in this sensitivity analysis, it is important to report both PEM and PEF.

Conclusions

The main objective of the work presented in this paper was to determine the significance of various methodological choices made in the goal and scope of a pavement LCA study. Sensitivity analyses were conducted for each of the five parameters using eight actual pavement projects, resulting in a number of key findings. A cut-off criterion using economic value was appropriate and a cut-off value below 95% would cause significant loss (up to 12% when using a 70% cut-off) of environmental impacts for the M&C stage but less so for the overall life-cycle (up to 3%). EOL allocation using a substitution method has a significant impact (up to 76%) on the M&C, M&R, and EOL stages and up to 18% on the entire life-cycle when assuming a landfill rate of 5%. The physical parameter used to determine asphalt binder allocation has a negligible impact on the overall life-cycle (less than 2%) but a notable impact (up to 6%) on the M&C and M&R stages. Traffic using a nonlinear growth rate has a significant (up to ±20%) effect on the overall life-cycle impacts. PEM contributes a notable impact (up to 35%) on the total life-cycle impacts and the gravity of this impact depends on the pavement type. Overall, it was found that the parameters examined have notable impacts on their affect stages (i.e., M&C, M&R, and EOL); however, when the entire life-cycle is considered (i.e., the use stage), the effects often diminish but may still be significant. The sensitivity of the results depends largely also on the system boundaries of the individual LCA study. Even so, the significance of the common methodological choices evaluated point toward a need to acknowledge the effect that these choices have on pavement LCA and possibly a need to standardize these choices to avoid inconsistent pavement LCA results. Further work that can be considered beyond this paper include sensitivity analyses on inventory sources used, including data from different time periods or geographical regions.

Acknowledgments

Part of this work is funded by the Illinois State Toll Highway Authority through the Illinois Center for Transportation. The authors acknowledge the input of Steve Gillen, ISTHA, and other partners of the project: Applied Research Associates, Inc., and theRightEnvironment.

References

1. *International Organization for Standardization (ISO). Environmental Management — Life-cycle Assessment - Principles and Framework.* ISO 14040:2006. Geneva, Switzerland.
2. *International Organization for Standardization (ISO). Environmental Management – Life-cycle Assessment – Requirements and Guidelines.* ISO 14044:2006(E). Geneva, Switzerland.
3. Harvey, J. T., J. Meijer, H. Ozer, I. L. Al-Qadi, A. Saboori, and A. Kendall. *Pavement Life-Cycle Assessment Framework* (Tech. Rep. No. FHWA-HIF-16-014). Federal Highway Administration, 2016.
4. Athena Institute. *A Life Cycle Perspective on Concrete and Asphalt Roadways: Embodied Primary Energy and Global Warming Potential* (Tech. Rep.). Ottawa, ON, Canada: Athena Institute, 2006.
5. Yang, R., S. Kang, H. Ozer, and I. L. Al-Qadi. Environmental and Economic Analyses of Recycled Asphalt Concrete Mixtures based on Material Production and Potential Performance. *Resources, Conservation and Recycling*, Vol. 104, 2015, pp. 141–151.
6. Chen, C., G. Habert, Y. Bouzidi, A. Jullien, and A. Ventura. LCA Allocation Procedure Used as an Incitative Method for Waste Recycling: An Application to Mineral Additions in Concrete. *Resources, Conservation and Recycling*, Vol. 54, No. 12, 2010, pp. 1231–1240.
7. Sayagh, S., A. Ventura, T. Hoang, D. Francois, and A. Jullien. Sensitivity of the LCA Allocation Procedure for BFS Recycled into Pavement Structures. *Resources, Conservation and Recycling*, Vol. 54, No. 6, 2010, pp. 348–358.
8. Blomberg, T., J. Barnes, F. Bernard, P. Dewez, S. Le Clerc, M. Pfitzmann, L. Porot, M. Southern, and R. Taylor. *Life Cycle Inventory: Bitumen* (Tech. Rep.). Brussels, Belgium: Eurobitune, 2011.
9. Huang, Y., A. Spray, and T. Parry. Sensitivity analysis of methodological choices in road pavement LCA. *The International Journal of Life Cycle Assessment*, Vol. 18, No. 1, 2013, pp. 93–101.
10. Yang, R., H. Ozer, and I. L. Al-Qadi. Regional Upstream Life-Cycle Impacts of Petroleum Products in the United States. *Journal of Cleaner Production*, Vol. 139, 2016, pp. 1138–1149.

11. Huang, Y. *Life Cycle Assessment of Use of Recycled Materials in Asphalt Pavements* (Unpublished Doctoral Dissertation). Newcastle University, Newcastle, Australia, 2007.
12. Wang, T., I.-S. Lee, A. Kendall, J. Harvey, E.-B. Lee, and C. Kim. Life Cycle Energy Consumption and GHG Emission from Pavement Rehabilitation with Different Rolling Resistance. *Journal of Cleaner Production*, Vol. 33, 2012, pp. 86–96.
13. Santero, N. J., and A. Horvath. Global Warming Potential of Pavements. *Environmental Research Letters*, Vol. 4, No. 3, 2009, 034011.
14. Loijos, A., M. Akbarian, S. Sahni, and J. Ochsendorf. Sensitivity Analysis of the Life Cycle Environmental Performance of Asphalt and Concrete Pavements. In *Proceedings of the Concrete Sustainability Conference*, 2010.
15. Santero, N., A. Loijos, and J. Ochsendorf. Greenhouse Gas Emissions Reduction Opportunities for Concrete Pavements. *Journal of Industrial Ecology*, Vol. 17, No. 6, 2013, pp. 859–868.
16. Al-Qadi, I., R. Yang, S. Kang, H. Ozer, E. Ferrebee, J. Roesler, A. Salinas, J. Meijer, W. Vavrik, and S. Gillen. Scenarios developed for improved sustainability of Illinois Tollway: Life-cycle assessment approach. Transportation Research Record: Journal of the Transportation Research Board, No. 2523, 2015, pp. 11–18.
17. Ghosh, L., L. Lu, H. Ozer, Y. Ouyang, and I. Al-Qadi. (2015). Effects of pavement surface roughness and congestion on expected freeway traffic energy consumption. Transportation Research Record: Journal of the Transportation Research Board, No. 2503, 10-19.
18. Ziyadi, M., H. Ozer, and I.L. Al-Qadi. Functional Unit Choice for Comparative Pavement LCA Involving Use-Stage with Pavement Roughness Uncertainty Quantification (UQ). *Proceedings of the Symposium on Life-Cycle Assessment of Pavements*, April 12-13, 2017, Champaign, Illinois, USA. CRC Press.
19. Bare, J., T. Gloria, and G. Norris. Development of the Method and U.S. Normalization Database for Life Cycle Impact Assessment and Sustainability Metrics. *Environmental Science and Technology*, Vol. 40, No. 16, 2006, pp. 5108–5115.
20. Lautier, A., R. K. Rosenbaum, M. Margini, J. C. Bare, R. Pierre-Olivier, and L. Deschenes. Development of Normalization Factors for Canada and the United States and Comparison with European Factors. *Science of the Total Environment*, Vol. 409, 2010, pp. 33–42.
21. Yang, R. *Development of a Pavement Life Cycle Assessment Tool utilizing Regional Data and Introducing an Asphalt Binder Model* (Unpublished master's thesis). University of Illinois at Urbana-Champaign, Champaign, IL, 2014.
22. Nicholson, A. L., E. A. Olivetti, J. R. Gregory, F. R. Field, and R. E. Kirchain. End-of-life LCA Allocation Methods: Open Loop Recycling Impacts on Robustness of Material Selection Decisions. In *ISSST'09: IEEE International Symposium on Sustainable Systems and Technology*. IEEE, 2009, pp. 1–6.
23. Santero, N. J. *Pavements and the Environment: A Life-Cycle Assessment Approach* (Unpublished doctoral dissertation). University of California, Berkeley, Berkeley, CA, 2009.

The Standing Committee on General and Emerging Pavement Design peer-reviewed this paper.

The contents of this report reflect the views of the authors, who are responsible for the facts and the accuracy of the data presented herein. The contents do not necessarily reflect the official view or policies of ISTHA or ICT. This paper does not constitute a standard, specification, or regulation.

Article

TRR
JOURNAL OF THE TRANSPORTATION RESEARCH BOARD

Transportation Research Record
2018, Vol. 2672(40) 88–96

DOI: 10.1177/0361198118794996
journals.sagepub.com/home/trr
SAGE

The Impact of Pavement Albedo on Radiative Forcing and Building Energy Demand: Comparative Analysis of Urban Neighborhoods

Xin Xu[1], Jeremy Gregory[1], and Randolph Kirchain[2]

Abstract

Albedo is the measure of the ratio of solar radiation reflected by the Earth's surface. High-albedo reflective surfaces absorb less energy and reflect more shortwave radiation. The change in radiative energy balance at the top-of-atmosphere (TOA), which is called radiative forcing (RF), reduces nearby air temperatures and influences the surrounding building energy demand (BED). The impact of reflective surfaces on RF and BED has been investigated separately by researchers through modeling and observational studies, however, no one has compared RF and BED impacts under the same context and the net effect of these two phenomena remains unclear. This paper presents a comprehensive approach to assess the net impacts of pavement albedo modification strategies in selected urban neighborhoods. We apply an adapted analytical model for RF and a hybrid model framework combining two different models for BED to estimate the impacts of increasing pavement albedo from 0.1 to 0.3 for different urban neighborhoods in Boston and Phoenix. The impact of several context-specific factors, including location, urban morphology, shadings etc., are taken into account in the models. Comparative analysis reveals that the net impact of changing pavement albedo can vary from one neighborhood to another. In Phoenix downtown, reflective pavements create net global warming potential burdens, while increasing pavement albedo results in potential savings in the Boston downtown area. This work provides insights into pavement albedo impacts at urban scale and supports more informed decisions on pavement designs that save energy and counteract some of the effects of global warming.

Urbanization has changed the landscape of the Earth and influenced the global climate significantly. In the U.S., it is estimated that in urban areas, the area of roofs varies from 20% to 25% and pavement constitutes 29% to 44% (*1*). These man-made surfaces usually absorb more incoming solar radiation than natural groundcover, raise the ambient temperature, increase buildings' energy loads, and potentially exacerbate outdoor air quality and comfort issues. Among many technological strategies, urban surface modifications, that is, increasing the reflectivity of artificial urban surfaces (mainly roofs and pavements), have been proposed as an approach to mitigate the impacts of the urban heat island (UHI) and greenhouse gas (GHG) effects.

Albedo is the ratio of solar radiation reflected by a surface to the amount it receives, ranging from 0 (complete absorption) to 1 (complete reflection). Urban surface modifications through albedo changes affect the climate through different mechanisms. One direct impact is through radiative forcing (RF), which measures the imbalance of the Earth's energy budget. Increased surface albedo results in more solar radiation reflected back into space and less remaining on the Earth. As shown in Figure 1, RF is the change in the outgoing radiation at the top-of-atmosphere (TOA), which can be translated into global warming potential (GWP). The other effect of modifying surface albedo is on building energy demand (BED) as a result of temperature changes. Increase in urban surface albedo increases the reflected *incident radiation* on nearby buildings and reduces the *ambient temperature* around buildings, which may result in changes in heating and cooling energy demands. Energy

[1]Department of Civil & Environmental Engineering, Massachusetts Institute of Technology, Cambridge, MA
[2]Institute for Data, Systems and Society, Massachusetts Institute of Technology, Cambridge, MA

Corresponding Author:
Address correspondence to Xin Xu: xux@mit.edu

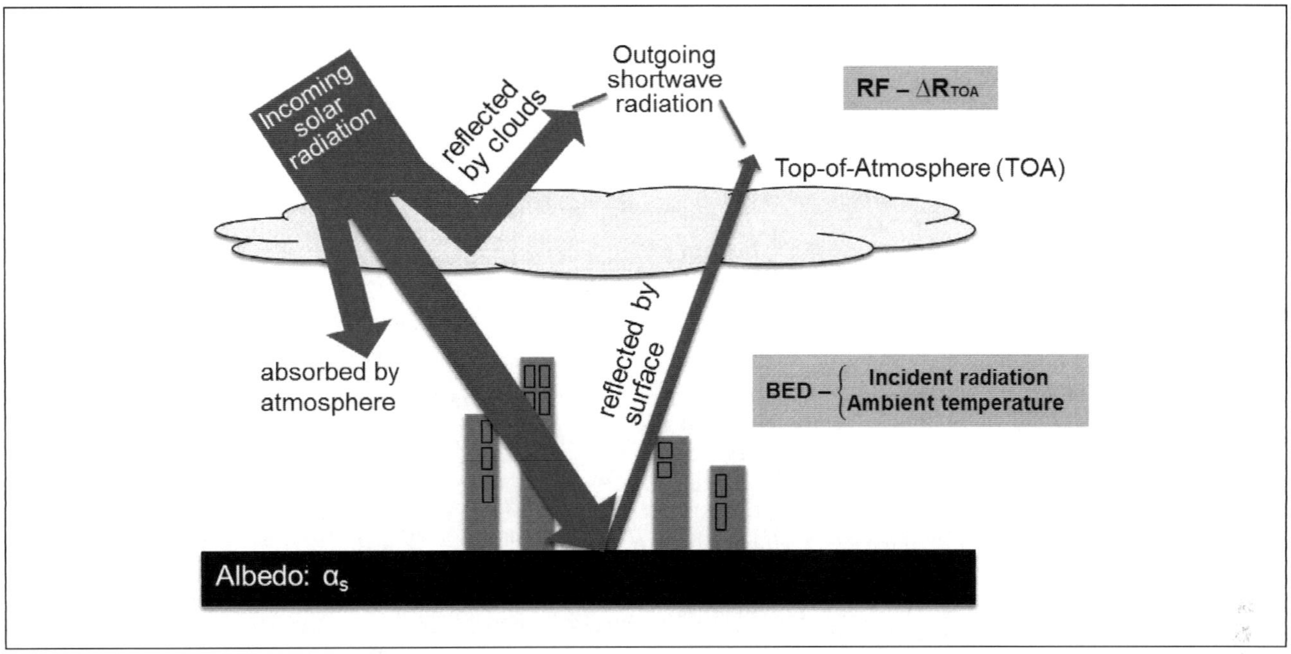

Figure 1. Schematic diagram showing the transmittance of solar radiation, and the impacts of pavement albedo through radiative forcing (RF) and building energy demand (BED).

savings or burdens due to the albedo changes can then be translated into GWP.

Many existing studies have focused on quantifying either of the two effects: RF (*2–7*) and BED (*8, 9*). However, few studies look at these two effects together and compare the magnitudes of each in the same context. There are gaps in the literature for quantifying each of these two impacts. For RF, most of the current works focus on calculating the RF due to large-scale changes in land cover such as forestation/deforestation or greenhouse agriculture (*3, 5, 6*). In most analytical models, the impact of the atmosphere on the radiative balance is modeled as global average conditions for all locations (*2, 7*) and there is no context-specific expression that is suitable for estimating the RF specifically associated with urban surface albedo modifications. The relationships between various contextual factors and albedo-induced RF have not been well characterized. The benefits of cool roofs on BED have been well demonstrated through analytical models or numerical tools (*10–12*). However, only a few studies found in the literature quantify the potential impacts of cool pavements (*8, 13*). A recent study conducted at the Lawrence Berkeley National Laboratory (*14*) compared the GWP offsets induced by RF and BED due to cool pavements in a case study of Los Angeles and Fresno, California. The results show greater GWP savings from RF over BED, but a one-time GWP offset rate provided by Akbari et al. (*15*) is used to calculate RF effect, ignoring the time dependency of albedo change and the location-specific cloudiness effects.

To fill the gaps in the literature, this paper presents a case study quantifying the net impacts of modifying pavement albedo for various neighborhoods in two selected urban areas representing very different climate zones. Location-specific models for both RF and BED have been developed and applied in this analysis accounting for the effects of contextual factors. The relationships between pavement albedo changes and the energy balance inside and outside a building are investigated. Finally, the results of the comparative analysis shed light on the effectiveness of albedo modifications as a UHI and GHG mitigation strategy.

Model Description

Analytical Model for Calculating Location-Specific RF Impacts Due to Changes in Pavement Albedo

In the literature there are well-established analytical models for calculating RF and its associated GWP for clear-sky conditions (*3, 4, 16*). These models usually oversimplify the location-dependent atmospheric conditions by either assuming a single global averaged transmittance factor or neglecting the effect of clouds. Such simplification is acceptable when the models are used for evaluating the impacts of large-scale changes in land cover, such as deforestation and urbanization. However, when it comes to pavement albedo modifications, a context-specific quantification of the RF impacts is required to support more informed decisions on their effectiveness at a local scale. In previous work, we have adapted an empirical model from the literature, and developed an analytical model for RF for cloudy-sky conditions. Details about the model development can be found in Xu et al. (*17*). A simplified description is presented here. The key of the adapted model is to develop a methodology to incorporate cloudiness into the empirical model for clear-sky RF calculation.

Clear-Sky Transmittance Factor f_a and RF at TOA. As defined by Ramaswamy et al. (*18*), RF is the change in net (down minus up) irradiance (solar plus longwave; in watts per square meter) at the TOA. In this study the RF impact associated with pavement albedo change refers to the impact on the Earth's radiative balance and global climate change. It should be differentiated from the impact of local incident radiation on building energy balance, which will be introduced in a later section. For shortwave forcing agents like surface albedo changes, the instantaneous RF at the TOA can be calculated as the change of shortwave radiation at TOA (*19*):

$$RF_{TOA} = -R_{TOA}\Delta\alpha_p \tag{1}$$

where the R_{TOA} is downward solar radiation at the TOA and $\Delta\alpha_p$ is a variation in planetary albedo. Following a procedure widely reported in the literature (*3, 4, 7, 16*), changes in planetary albedo $\Delta\alpha_p$ can be linearly related to changes in surface albedo$\Delta\alpha_s$:

$$\Delta\alpha_p = f_a\Delta\alpha_s \tag{2}$$

where f_a is a two-way transmittance factor accounting for absorption and reflection of solar radiation throughout the atmosphere. Convolving Equations 1 and 2 we have

$$RF_{TOA} = -R_{TOA}\Delta\alpha_p = -R_{TOA}f_a\Delta\alpha_s \tag{3}$$

In some early studies, f_a was estimated via radiative transfer models, but most of them applied only to clear-sky conditions (*16, 20–22*). Bright et al. (2012) estimate f_a assuming symmetric downward and upward transmittance of solar radiation (*4*):

$$f_a = K_T^2 \text{ or } T_a^2 \tag{4}$$

where K_T is the ratio of downward solar radiation at the surface to that received at the TOA and T_a is the fraction of radiation reflected from the surface that arrives back at the TOA (*16*). K_T based on satellite measurements of solar insolation at the Earth's surface can be retrieved freely online for any given location and time period ranging from a single month to 22 years (*23*). However, values for T_a are not widely reported in the literature for specific locations, which led us to develop an adapted model for cloudy-sky condition.

Estimating Cloudy-Sky Transmittance Factor f_a. Since the presence of clouds reduces the transmittance of radiation, we postulate that it could be incorporated into f_a as a multiplier g $(0<g\leq 1)$ to the clear-sky f_a and that this multiplier would be equivalent to the reduced transmittance caused by cloudiness within K_T:

$$f_{a\_cloud} = f_{a\_clear} \cdot g(cloud) \tag{5}$$

$$K_{T\_cloud} = K_{T\_clear} \cdot g(cloud) \tag{6}$$

Li and Garand established an empirical model for estimating f_a under clear-sky conditions, and found a good relationship between f_a and precipitable water and solar zenith angle. We consider $g(cloud)$ as a function of cloudiness primarily, but also want to include precipitable water and solar zenith angle as explanatory variables. Based on some empirical data, we observed a reasonable relationship among clearness index K_T, precipitable water p, and cosine of solar zenith angle c for clear-sky. Inspired by the existing model for clear-sky f_a, we obtain a simple linear regression model for K_{T_clear} with an R-square of 0.9:

$$K_{T\_clear} = f\left(\sqrt{p}, c\right) \tag{7}$$

To separate out the cloudiness effect, we compute the residuals as the ratio of all-sky clearness index K_T obtained from actual satellite measurements, to the clear-sky K_{T_clear} calculated from Equation 7:

$$Residuals = K_T/K_{T\_clear} \tag{8}$$

We ran a number of linear regression models and found that the two statistically significant explanatory variables were cloud fraction and cosine of solar zenith angle. We then fit a regression model for $g(cloud)$:

$$g(cloud) = f(cld, c) \tag{9}$$

Combining the empirical model and cloudiness regression model, we have an adapted model for cloudy-sky.

GWP Calculation Accounting for the Shading Effect in Urban Areas. To compare the GWP impacts specifically related to changes in pavement albedo, we follow the approach described above to come up with f_a values for a given location, and then apply the f_a to GWP calculation following IPCC:

$$GWP_{alb} = \frac{-\int_0^{TH} \frac{A}{A_{earth}} \cdot R_{TOA} \cdot f_a \cdot \Delta\alpha_s \, dt}{\int_0^{TH} RF_{CO_2} dt} \quad (10)$$

In this study, the functional unit is defined as a square meter (m^2) of pavement. A_{earth} is the area of the Earth, which is 5.1×10^{14} m^2. For a time horizon of 50 years ($TH = 50$), the RF for carbon dioxide ($\int_0^{TH} RF_{CO_2}\, dt$, *or Absolute GWP*) is 5.02×10^{-14}, interpolating from the 20-year and 100-year AGWP for CO_2 reported by the IPCC (*19*).

In a realistic urban setting, pavements are usually surrounded by buildings or trees, which block the sunlight and limit the amount of solar radiation that reaches the pavements. To account for the effect of shadings in an urban canyon, we further introduce an urban canyon transmittance, τ, which is defined as the ratio of the solar radiation reflected from the pavement surface at the bottom of a urban canyon, to the radiation reflected from the same road surface not in a canyon. It can also be interpreted as the transmittance or reflectance of an urban canyon. The value of τ should approach one as canyon height approaches zero, and should never exceed one. We rewrite Equation 2 to include the total transmittance of the atmosphere and transmittance of the urban canyon:

$$\Delta\alpha_p = \tau \cdot f_a \Delta\alpha_s \quad (11)$$

The value of the urban canyon transmittance τ is estimated at 0.4 based on a model called Urban Weather Generator (UWG) (*24*), where the reflected solar radiation from a shaded road surface in an urban canyon and that from an unshaded road can be simulated.

Hybrid Model for Calculating Location-Specific BED Impacts Due to Changes in Pavement Albedo

There are many existing models or simulation tools for estimating building energy consumption, such as EnergyPlus (*10*) and TRNSYS (*25*), but unfortunately none of these models were developed to capture the interactions between BED and changes in pavement around building envelopes, especially. There are two mechanisms of how pavement albedo changes can affect building energy performance. First, pavements with higher albedo reflect more *incident radiation* onto the nearby buildings, resulting in changes in building energy consumption. Secondly, increasing pavement albedo also reduces the *ambient temperature* within the canyon, altering BED in a way contradictory to the effects induced by increased *incident radiation*. Therefore, in order to estimate the net effect of changing pavement albedo has on BED, we utilize two different tools – DIVA and UWG – for the purpose of this study.

DIVA (Design, Iterate, Validate and Adapt) is a highly-optimized daylighting and energy modeling plug-in for Rhino and Grasshopper (*26*). It has the capability of effectively calculating daylight metrics (using the Radiance/Daysim engine) and energy consumption (using the EnergyPlus engine). The thermal component collects geometry, material, and schedule information associated with a building, and performs a full yearly thermal dynamic simulation. As a result, final energy consumption figures for heating and cooling are provided.

UWG is an urban design simulation tool that provides climate-specific temperatures for cityscape geometry and land use change. It estimates hourly urban canopy air temperature and humidity based on weather data from a suburban weather station. The tool has been validated in several urban settings and can satisfactorily estimate urban temperatures in different climates, weather conditions, and urban configurations. The performance of UWG is comparable to a more computationally expensive mesoscale atmospheric model.

In this analysis, we combine the above two simulation tools – DIVA and UWG – together to estimate the change in BED due to an increase in pavement albedo. The urban geometric characteristics are extracted from the 3D model using Rhinoceros or Rhino, a CAD-based modeling environment used by urban designers and architects to facilitate the integration of microclimatic considerations with the massing design. The building height, site area, building density and canyon aspect ratio (average building height/road width) are extracted from the model to calculate the required geometric parameters for UWG. The output from UWG is a modified weather file that can be used by the thermal component in DIVA. The workflow of the simulation is shown in Figure 2.

The output from the thermal component of DIVA is an estimate of annual energy demand, further broken down into heating and cooling loads in the unit of kWh per square meter of floor area. We then multiply this number by the ratio of floor area to pavement area in this neighborhood, which can be obtained from the 3D geometry in Rhino. The energy load can be translated into GWP by multiplying a CO_2 emission factor for electricity reduction provided by U.S. Environmental Protection Agency, which is 0.703 kg CO_2/kWh (*27*). To compute the change in energy demand due to albedo increase, we need to run the model twice with different pavement albedos and then take the difference of the two.

Comparative Case Study on Urban Neighborhoods

The goal of this case study is to compare the impacts of pavement albedo on RF and BED under the same

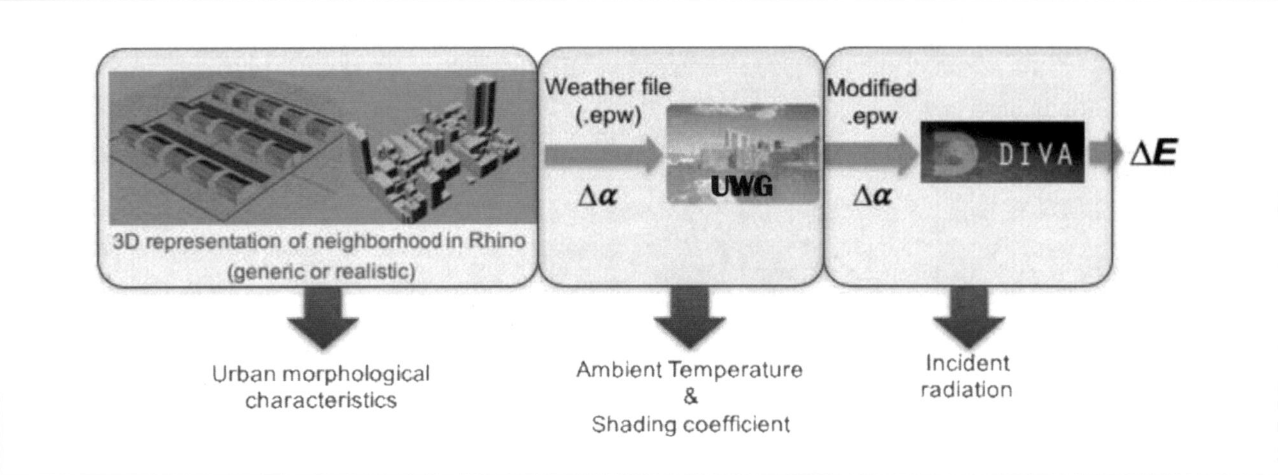

Figure 2. Simulation workflow for the hybrid model ($\Delta\alpha$ is pavement albedo change and ΔE is the associated change in BED).

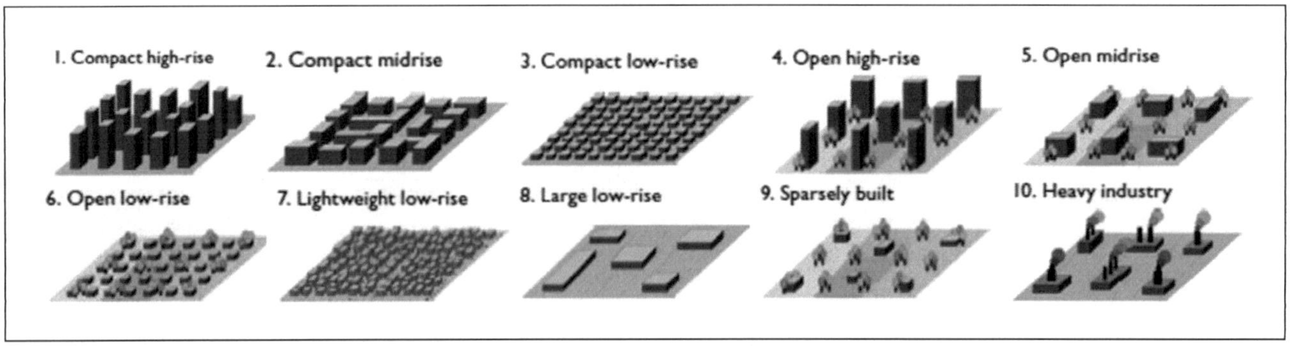

Figure 3. Illustration of 10 LCZs, from Stewart and Oke (*30*).

context, and how the relative magnitudes of these impacts vary by contextual factors. Phoenix and Boston are selected for the case study, representing two different climate zones. The latitude of the urban area determines the intensity of the incoming solar radiation, and climate conditions can affect the amount of radiation transmitted through the atmosphere. In this analysis, we calculated the GWP savings due to a 0.2 increase in pavement albedo ($\Delta\alpha_s = 0.2$) for a unit area of 1 m^2 over an analysis period of 50 years, which is commonly used for pavement life-cycle assessment in the literature (*28–30*). We further assumed that the albedo change would decay to 0.1 as pavement surfaces wear off over time.

Local Climate Zones

Urban landscapes exhibit great variations depending on morphological structure and land cover characteristics. Even within a city, there can be very distinguishable microclimate zones from one neighborhood to another. To evaluate BED at a city level more accurately, we employ a scheme to group urban neighborhoods into different zones and run building simulations within each zone.

Local climate zones (LCZs) is a classification scheme developed by Steward and Oke (*31*) for heat island studies. In this framework, LCZs are regions of uniform surface cover, structure, material, and human activity that span from hundreds of meters to several kilometers in horizontal scale. Figure 3 shows the 10 LCZs with distinct urban morphologies. More details regarding the specification of the geometries can be found in Stewart and Oke (*31*).

Data Sources

For this analysis, we obtain morphological characteristics of urban neighborhoods at zip code level from GIS shapefiles, which include parameters such as average building height, building density, aspect ratio (average

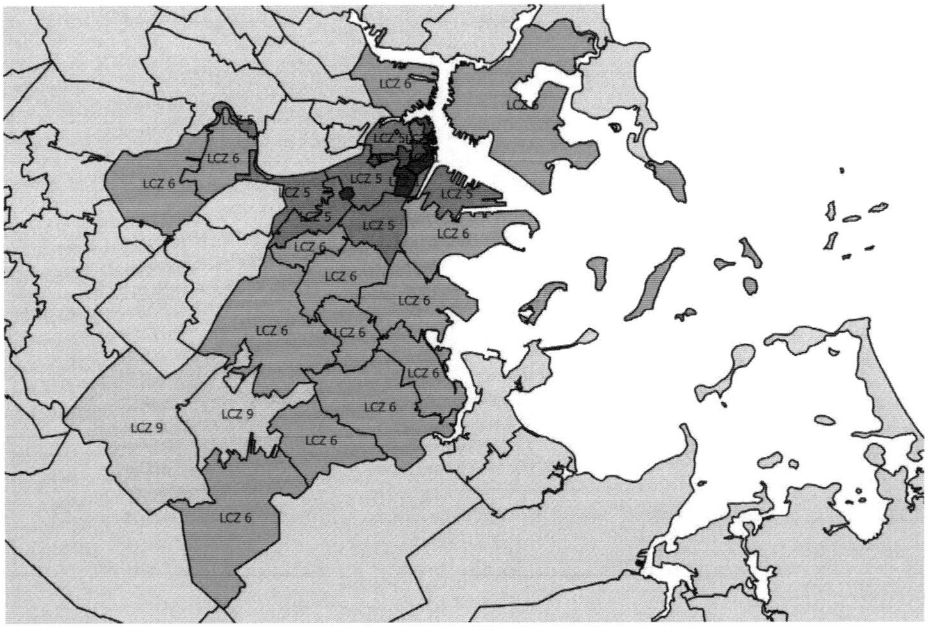

Figure 4. LCZs assigned to zip code in the city of Boston.

building height/average road width, i.e., H/W), etc. Based on this information, we assign an LCZ that falls in the ranges specified in the Supplementary Materials of Stewart and Oke (*31*). Figure 4 shows the LCZ assignments at zip code level in Boston. The darkness of the color reflects roughly the density of buildings. For example, the Financial District in downtown Boston in darkest blue is categorized as LCZ 1 (compact high-rise), while Hyde Park in south Boston is a typical residential neighborhood and is assigned LCZ 9 (sparsely built). A building prototype has been associated with each LCZ and imported into Rhino with DIVA, which is then used to simulate the average BED of the LCZ.

WRF (the Weather Research and Forecasting model) is used in this study to populate location-specific databases for solar radiation and other atmospheric parameters required for calculating the RF impact of albedo change. The modeling domain spans the continental U.S.A. with a spatial resolution of 50 km, and the simulation is run continuously for nine years from 2000 to 2008. Variables extracted from the WRF simulation include downward solar radiation at the surface R_s (or SWDOWN in WRF), downward solar radiation at the TOA R_{TOA} (or SWDNT in WRF), precipitable water p (or pw in WRF), cosine of solar zenith angle c (or COSZEN) and cloud fraction cld (or CLDFRA in WRF). For each variable, daily average values are aggregated from hourly data in WRF.

Results

The impacts of RF and BED in terms of GWP savings (in the unit of kg CO_2 equivalent per square meter of pavement modified) are calculated for all zip codes in Boston and Phoenix. We select three LCZs in each city that are most representative of the urban morphology within each city. The results are shown in Figure 5, where positive numbers indicate GWP savings while negatives suggest burdens on global warming. BED is further broken down into cooling loads and heating loads.

From Figure 5, it is clear that increasing pavement albedo results in CO_2 savings from RF in all neighborhoods in both Boston and Phoenix, but the impacts of changing pavement albedo on BED vary by urban morphology. The increase in pavement albedo often causes cooling energy to go up in summer while leading to savings in heating load during winter. The magnitudes of burdens and savings depend on the density of the neighborhood as well as the climate zone where the city is located. In a densely-built high-rise neighborhood, reflective pavements can create net GWP burdens as in the case of LCZ1 in Phoenix (Figure 5*b*). This occurs because tall buildings tend to trap multiple reflections of radiation between building walls, thereby significantly increasing cooling loads in the summer. Because winters in Phoenix are usually warm and do not require much heating energy, the effect of changing pavement albedo on heating energy demand is very minimal. The same trend

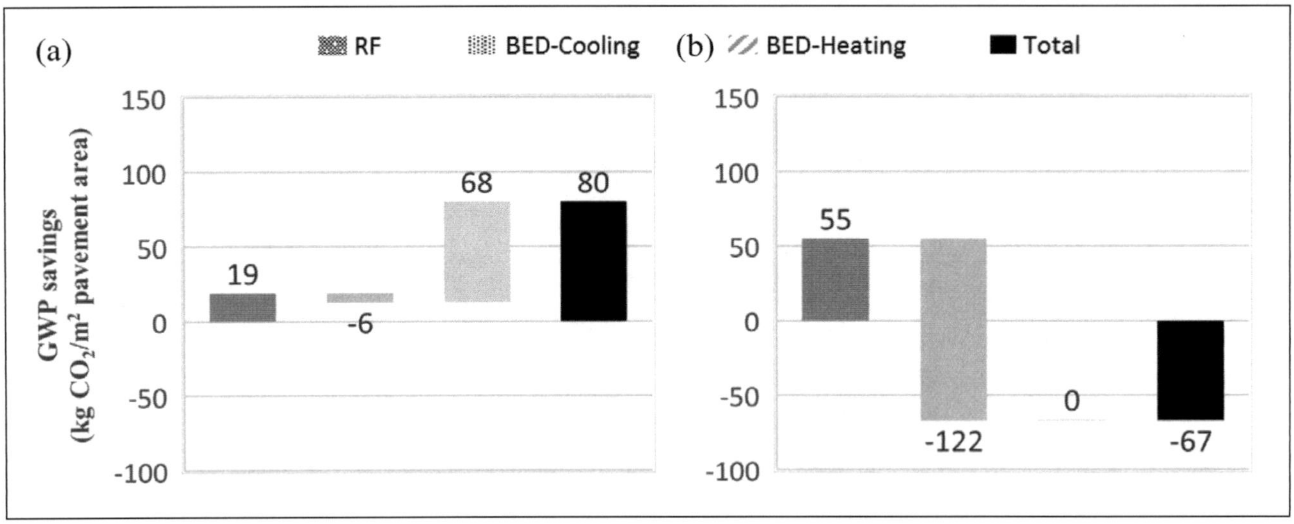

Figure 5. RF and BED savings over a 50-year period for a 0.2 increase in albedo in selected LCZs: (a) Boston LCZ1 and (b) Phoenix LCZ1.

Note: the charts in this figure are waterfall charts. RF is the first component and the others are plotted up or down relative to the one on the left. The net impacts of RF and BED are shown in black.

is not observed in the Boston downtown area (LCZ1) (Figure 5*a*) because Boston is cold in winter and demand for heating is extremely high. The potential savings from increasing pavement albedo outweigh the cooling burdens in summer. An explanation for this is that there is more solar radiation reflected from the pavement and absorbed by buildings, which leads to an increase in both ambient temperature and indoor temperature, reducing the demand for heating as a result. Similarly, for less built up areas in both Boston and Phoenix, the impact of increasing pavement albedo on BED is marginal as compared with the RF impact. This is due to the fact that roads and buildings are usually far apart and that the UHI effect is not present in these sub-urban areas. Since high-rise and densely-built districts make up only a small fraction of urban areas, the expected total savings from increasing pavement albedo at the urban scale is expected to be positive for both the greater Boston and Phoenix regions.

Conclusions

The models for quantifying RF and BED impacts due to pavement albedo changes are presented in this manuscript. To estimate the RF impact, we adapt an empirical analytical model from the literature and incorporate the effects of cloudiness and shadings in the urban area. For BED, two existing simulation tools are combined to account for the interactions between buildings and surrounding environment. A case study of the RF and BED impacts due to pavement albedo modifications in two urban areas is conducted, with location-specific data extracted from climate simulations and GIS datasets. This study is the first attempt to compare RF and BED impacts due to pavement albedo modifications under the same context, and it is a first estimate of the net benefit of reflective pavements at an urban scale while most studies in the literature only quantify the impact of RF or BED individually. Comparison of the results across the different LCZs reveals the importance of developing context-specific models for albedo impact estimation. Urban-scale impact quantification is also necessary as albedo modifications can have different and even contradicting impact on BED depending on urban morphology. Comprehensive assessment taking into account both RF and BED is recommended when evaluating reflective pavements as a global warming mitigation strategy.

There are, however, some limitations with the analytical model as well as the simulation tools, DIVA and UWG, which limit their capabilities of dealing with complex geometries, large areas, and/or multiple zones. First of all, the analytical model takes advantage of simulated data to derive statistical relationships among atmospheric transmittance and other contextual factors. The robustness of the model needs to be tested using statistical methods or validated against measured data. Second, DIVA's thermal component allows the calculation of just one thermal zone per time. Thus, each thermal zone needs to be analyzed independently, and the interaction of different zones within a building cannot be assessed. Third, UWG assumes a simple homogenous urban canyon morphology and employs a heat balance method to simulate canyon temperature. Variation and dynamics in temperature cannot be resolved, which may introduce uncertainty in evaluating the cooling effect of reflective pavements. Finally, a more comprehensive analysis of urban neighborhoods with various morphologies will be needed to gain a better understanding of the relative value of higher albedo pavements in different contexts. As a next step, the authors will extend the analysis to a higher resolution of the urban area, categorizing neighborhoods by census tract instead of zip code. The results of this study will help urban planners or decision makers make informed decisions on whether pavement albedo enhancement is an option worth pursuing in a given urban area/neighborhood.

Acknowledgments

This research was conducted as part of the Concrete Sustainability Hub at the Massachusetts Institute of Technology, which is supported by the Portland Cement Association and the Ready Mixed Concrete Research and Education Foundation. The authors acknowledge Professor C. Adam Schlosser and Dr. Liyi Xu Monier for their contributions to running the WRF climate simulations for this study. The authors also acknowledge Professor Les Norford for assistance with running the tool UWG.

Author Contributions

The authors confirm contribution to the paper as follows: study conception and design: Xin Xu, Jeremy Gregory, Randolph Kirchain; data collection: Xin Xu; analysis and interpretation of results: Xin Xu, Jeremy Gregory, Randolph Kirchain; draft manuscript preparation: Xin Xu. All authors reviewed the results and approved the final version of the manuscript.

References

1. Rose, L. S., H. Akbari, and H. Taha. *Characterizing the Fabric of the Urban Environment: A Case Study of Greater Houston, Texas*. Lawrence Berkeley National Laboratory, Berkeley, Calif., 2003.
2. Lenton, T. M., and N. E. Vaughan. The Radiative Forcing Potential of Different Climate Geoengineering Options. *Atmospheric Chemistry and Physics*, Vol. 9, No. 15, 2009, pp. 5539–5561.
3. Muñoz, I., P. Campra, and A. R. Fernández-Alba. Including CO_2-Emission Equivalence of Changes in Land Surface Albedo in Life Cycle Assessment. Methodology and Case Study on Greenhouse Agriculture. *The International Journal of Life Cycle Assessment*, Vol.15, No. 7, 2010, pp. 672–681.
4. Bright, R. M., F. Cherubini, and A. H. Strømman. Climate Impacts of Bioenergy: Inclusion of Carbon Cycle and Albedo Dynamics in Life Cycle Impact Assessment. *Environmental Impact Assessment Review*, Vol. 37, 2012, pp. 2–11.
5. Betts, R. A. Offset of the Potential Carbon Sink from Boreal Forestation by Decreases in Surface Albedo. *Nature*, Vol. 408, No. 6809, 2000, pp. 187–190.
6. López-Saldaña, G., I. Bistinas, and J. M. C. Pereira. Global Analysis of Radiative Forcing from Fire-Induced Shortwave Albedo Change. *Biogeosciences Discussions*, Vol. 11, No. 5, 2014, pp. 7775–7796.
7. Cherubini, F., R. M. Bright, and A. H. Strømman. Site-Specific Global Warming Potentials of Biogenic CO_2 for Bioenergy: Contributions from Carbon Fluxes and Albedo Dynamics. *Environmental Research Letters*, Vol. 7, No. 4, 2012, p. 45902.
8. Yaghoobian, N., and J. Kleissl. Effect of Reflective Pavements on Building Energy Use. *Urban Climate*, Vol. 2, 2012, pp. 25–42.
9. Synnefa, A., M. Santamouris, and H. Akbari. Estimating the Effect of Using Cool Coatings on Energy Loads and Thermal Comfort in Residential Buildings in Various Climatic Conditions. *Energy and Buildings*, Vol. 39, No. 11, 2007, pp. 1167–1174.
10. Jo, J. H., J. D. Carlson, J. S. Golden, and H. Bryan. An Integrated Empirical and Modeling Methodology for Analyzing Solar Reflective Roof Technologies on Commercial Buildings. *Building and Environment*, Vol. 45, No. 2, 2010, pp. 453–460.
11. Jacobson, M. Z., and J. E. Hoeve. Effects of Urban Surfaces and White Roofs on Global and Regional Climate. *Journal of Climate*, Vol. 25, No. 3, 2012, pp. 1028–1044.
12. Li, D., E. Bou-Zeid, and M. Oppenheimer. The Effectiveness of Cool and Green Roofs as Urban Heat Island Mitigation Strategies. *Environmental Research Letters*, Vol. 9, No. 5, 2014, p. 55002.
13. Santamouris, M., N. Gaitani, A. Spanou, M. Saliari, K. Giannopoulou, K. Vasilakopoulou, and T. Kardomateas. Using Cool Paving Materials to Improve Microclimate of Urban Areas – Design Realization and Results of the Flisvos Project. *Building and Environment*, Vol. 53, 2012, pp. 128–136.
14. Gilbert, H. E., P. J. Rosado, G. Ban-Weiss, J. T. Harvey, H. Li, B. H. Mandel, D. Millstein, A. Mohegh, and A. Saboori. "Energy and Environmental Consequences of a Cool Pavement Campaign." *Energy and Buildings* (2017).
15. Akbari, H., H. Damon Matthews, and D. Seto. The Long-Term Effect of Increasing the Albedo of Urban Areas. *Environmental Research Letters*, Vol. 7, No. 2, 2012, p. 24004.
16. Lenton, T. M., and N. E. Vaughan. The Radiative Forcing Potential of Different Climate Geoengineering Options. *Atmospheric Chemistry and Physics Discussions*, Vol. 9, No. 1, 2009, pp. 2559–2608.
17. Xu, X., J. Gregory, and R. Kirchain. Climate Impacts of Surface Albedo: Review and Comparative Analysis. Presented at 95th Annual Meeting of the Transportation Research Board, Washington, D.C., 2016.
18. Ramaswamy, V., O. Boucher, J. Haigh, D. Hauglustine, J. Haywood, G. Myhre, T. Nakajima, G. Y. Shi, and S. Solomon. Radiative Forcing of Climate. *Climate Change*, 2001, pp. 349–416.
19. IPCC. *Climate Change 2007: The Physical Science Basis, Summary for Policymakers, Intergovernmental Panel on*

Climate Change. Intergovernmental Panel on Climate Change (IPCC), Geneva, Switzerland, 2007.
20. Lacis, A. A., and J. Hansen. A Parameterization for the Absorption of Solar Radiation in the Earth's Atmosphere. *Journal of the Atmospheric Sciences*, Vol. 31, No. 1, 1974, pp. 118–133.
21. Chen, T. S., and G. Ohring. Clear-Sky Planetary and Surface Albedos: A Parameterization for Simple Models. *Advances in Space Research*, Vol. 5, No. 6, 1985, pp. 141–144.
22. Li, Z., and L. Garand. Estimation of Surface Albedo from Space: A Parameterization for Global Application. *Journal of Geophysical Research*, Vol. 99, No. D4, 1994, p. 8335.
23. NASA. *NASA Atmospheric Science Data Center: Surface Meteorology and Solar Energy* (Release 6.0). 2015. https://eosweb.larc.nasa.gov/cgi-bin/sse/sse.cgi?. Accessed Jul 31, 2017.
24. Bueno, B., L. Norford, J. Hidalgo, and G. Pigeon. The Urban Weather Generator. *Journal of Building Performance Simulation*, Vol. (February 2015), 2012, pp. 1–13.
25. Synnefa, A., M. Saliari, and M. Santamouris. Experimental and Numerical Assessment of the Impact of Increased Roof Reflectance on a School Building in Athens. *Energy and Buildings*, Vol. 55, 2012, pp. 7–15.
26. Jakubiec, J. A., and C. F. Reinhart. Diva 2.0: Integrating Daylight and Thermal Simulations Using Rhinoceros 3d, Daysim and Energyplus. *In Proceedings of Building Simulation*. Vol. 20, No. 11, 2011, pp. 2202–2209.
27. Environmental Protection Agency. Greenhouse Gases Equivalencies Calculator - Calculations and References. 2017. https://www.epa.gov/energy/greenhouse-gases-equivalencies-calculator-calculations-and-references. Accessed May 14, 2017
28. Ahammed, M. A., P. Eng, S. Sullivan, et al. Life Cycle Environmental Assessment Using Athena Pavement LCA Tool: A Manitoba Case Study. *In Proceedings of the Transportation Association of Canada*, Toronto, ON. 2016.
29. Xu, X., A. Noshadravan, J. Gregory, and R. Kirchain. Scenario Analysis of Comparative Pavement Life Cycle Assessment. *In Proceedings of the International Symposium on Pavement LCA*, 2014, pp. 13–26.
30. Wang, H., and R. Gangaram. *Life Cycle Assessment of Asphalt Pavement Maintenance*. Report No. CAIT-UTC-013. Center for Advanced Infrastructure and Transportation, Rutgers, The State University of New Jersey, 2014.
31. Stewart, I. D., and T. R. Oke. Local Climate Zones for Urban Temperature Studies. *Bulletin of the American Meteorological Society*, Vol. 93, No. 12, 2012, pp. 1879–1900.

The Pavements Section (AFD00) peer-reviewed this paper (18-03772).

Article

TRR
JOURNAL OF THE TRANSPORTATION RESEARCH BOARD

Transportation Research Record
2018, Vol. 2672(40) 97–107

DOI: 10.1177/0361198118783163
journals.sagepub.com/home/trr
SAGE

Towards a Reliability-Based Pavement Design using Response Surface Methods

Yared Dinegdae[1], Ibrahim Onifade[1], Bjorn Birgisson[2], Robert Lytton[3], and Dallas Little[3]

Abstract

Reliability has been incorporated in many pavement design procedures to account for the effects of inputs variabilities and uncertainties on predicted performance. The American Association of State Highway and Transportation Officials (AASHTO) mechanistic empirical pavement design guide (MEPDG) computes the reliability of pavement sections with the assumption that the variability of predicted distresses follows normal distribution. This approach does not account for the systematic contribution of each design input variability on the overall output variance. This paper evaluates a two-component reliability analysis methodology for pavement application. The two-component reliability analysis methodology uses a response surface method (RSM) for a surrogate model generation and the first order reliability method (FORM) for reliability computation. Three different response surface methods (central composite, Box–Behnken and Doehlert designs) were implemented and statistically verified for their suitability for surrogate model generation. The two-component reliability analysis methodology was further utilized for the generation of partial safety factors for the development of a load and resistance factor design (LRFD) procedure for pavement applications. Field pavement sections with a wide range in design inputs and target reliabilities were used to evaluate the proposed reliability analysis methodology. The results have shown that the three RSM can be used effectively for pavement reliability problems.

The pavement design process can be considered to be probabilistic as pavement input parameters exhibit inherent variability, the models used to determine the field variables (i.e., stress, strain and deformations) and distresses (i.e., cracking, rutting and roughness) are an approximation of the true phenomena, and there is statistical uncertainty because of inadequate data size (*1*). The combined effect of the variances associated with these variabilities and uncertainties will reflect upon the computed distress, which if considered to be a deterministic parameter will have a significant consequence on expected performance (*2–4*). Both the American Association of State Highway and Transportation Officials (AASHTO) flexible pavement design procedure and the mechanistic empirical pavement design guide (MEPDG) have recognized this fact and incorporated reliability into their respective design procedures to account for the involved variabilities and uncertainties effect (*5*, *6*).

The AASHTO MEPDG considers the individual distress quantities (e.g., rutting, fatigue cracking and thermal cracking for flexible pavements) to be key outputs of interest and model these distresses accordingly as random variables. The design reliability for each individual pavement distress mode is estimated by assuming the variability of the distress to be approximately normally distributed and using the standard error which includes the combined input variability, variability in the construction process and the model or pure error (*6*). The assumption of a normal distribution for the variability of the major distress modes without accounting directly for the influence of design inputs variabilities on the predicted distresses might not achieve the intended purpose of delivering pavement sections which perform consistently at the desired level of reliability (*7–9*).

[1]Texas A&M Engineering Experiment Station (TEES), College Station, TX
[2]Center for Infrastructure Renewal, Civil Engineering Department, Texas A&M University, College Station, TX
[3]Texas A&M Transportation Institute, Texas A&M University, College Station, TX

Corresponding Author:
Address correspondence to Yared Dinegdae: dinegdae@tamu.edu

A reliability analysis which is based on a probabilistic method of uncertainty quantification or propagation allows the effective incorporation of variabilities and uncertainties of the design inputs on the predicted performance. The first order reliability method (FORM), which is a typical example of an analytical reliability method, allows the propagation of variabilities in the design process and has been implemented for pavement reliability applications (*10–12*). Furthermore, FORM can be used to develop a reliability-based codified design procedure in the format of the load and resistance factor design (LRFD) method, which delivers pavement sections of uniform reliability level, thus providing the proper platform for life cycle cost analysis (LCCA) (*12–14*). The implementation of analytical based reliability analysis techniques for pavement application is complicated by these techniques requiring the performance function to be an explicit function of the random variables, which is unavailable in most M-E design procedures.

To overcome this complication, response surface methods can be developed and incorporated into a reliability analysis methodology for pavement reliability computation. These response surface methods are used to generate surrogate models which are an explicit function of the input variables and that effectively replace the damage and distress parameters bypassing computational analysis tools such as linear elastic analysis or finite element analysis (*15*). Response surface methodologies (RSMs) such as central composite design (CCD), Box–Behnken Design (BBD) and Doehlert designs (DOD) are appropriate for generating higher order surrogate models which can readily be implemented for pavement reliability applications (*15*). The use of a surrogate model for pavement reliability applications and pavement performance optimization is at its early stage and there are limited studies on its development and implementation.

Objective

The main objective of this paper is to evaluate the potential for the use of response surface methods for pavements reliability analysis and for the development of a LRFD procedure for pavement application. Response surface methods such as CCD, BBD, and DOD are studied and investigated for their applicability. A two-component reliability analysis methodology that uses a response surface approach for surrogate model generation and a FORM for reliability analysis is used to evaluate the reliability of fatigue cracking failure in actual field pavements using the mechanics-based design framework. Several field pavement sections with a wide range of design inputs and target reliabilities were evaluated in this study.

Methodology

Reliability Concept

Several efforts have been made in the past to address the issue of reliability for pavement design application, and to develop a sound approach for its incorporation into pavement design process (*3*, *16*, *17*). Pavement design reliability can be defined as the probability that a given pavement performance measuring parameter would remain below some critical value during the intended design period. It can be defined as considering traffic as in the AASHTO-1993 procedure or as considering distress as in the AASHTO's MEPDG (*5*, *6*). The probability of failure (P_f) and reliability (R) in the case when the variability of the independent parameters is defined by a full probability density function (pdf) can be computed using Equations 1 and 2. Equation 1 integrates the joint probability density function $f_{\mathbf{x}}(\mathbf{x})$ of the random variables $\mathbf{x}$ over the failure regions $g(\mathbf{x}) < 0$.

$$P_f = p\{g(\mathbf{x})<0\} = \int\limits_{g(\mathbf{x})<0} f_{\mathbf{x}}(\mathbf{x})d\mathbf{x} \tag{1}$$

$$R = p\{g(\mathbf{x}) \geq 0\} = 1 - P_f \tag{2}$$

The integration problem in Equation 1 has an exact solution only for special cases and numerical integration will become impractical when the number of variables exceeds two or three. Nevertheless, approximate solution to the integral can be obtained using methods such as the FORM (*18*, *19*). FORM solves the reliability problem by linearizing the limit state equation at the most likely failure point in the standard normal space and by using the minimum distance from the origin to this point which is termed the reliability index (β) to calculate the probability of failure (P_f) as follows:

$$P_f = \phi(-\beta) \tag{3}$$

In which $\varphi(\cdot)$ is the standard normal distribution function. There are a variety of FORM methods which can be used depending on the complexity of the reliability problem. In a case when the performance function involves non-normal variables and high non-linearity the algorithm suggested by Rackwitz and Fiessler is well adapted for this type of problems (*19*).

Reliability Analysis Methodology

A two-component analytical reliability analysis methodology has been suggested by Dinegdae and Birgisson (*12*, *20*) for pavement reliability computation. The first component uses a RSM for generating a surrogate model for the performance function, whereas the second component computes the reliability using the

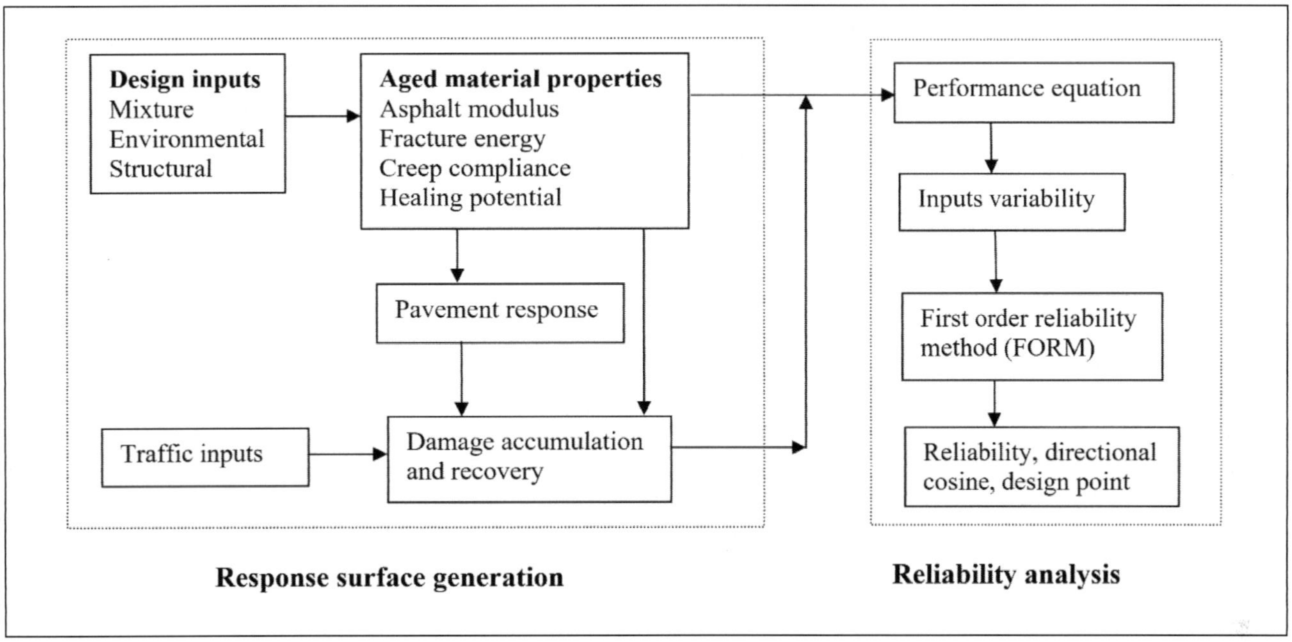

Figure 1. A two-component reliability analysis methodology.

Rackwitz–Fiessler (R–F) algorithm, which is one variety of the FORM. Figure 1 presents the flowchart for the proposed reliability analysis methodology adapted for pavement fatigue performance evaluation.

Response Surface Method. Response surface method, a collection of mathematical and statistical techniques, is used in modelling or analysis problems to describe or optimize the response caused by several input variables (*15*). It substitutes the implicit performance function based on the fit of a polynomial to the experimental data, which are located in the neighborhood of the most likely failure point or experimental region being studied (*21*). For independent input variables $x_1, x_2,..,x_k$, the RSM generates the following quadratic performance equation.

$$y = \beta_0 + \sum_{i=1}^{k} \beta_i x_i + \sum_{1 \leq i \leq j}^{k} \beta_{ij} x_i x_j + \sum_{1 \leq i \leq j}^{k} \beta_{ii} x_i^2 + \varepsilon \quad (4)$$

In which β_0 is the constant term, β_i, β_{ij} and β_{ii} are respectively the coefficients of the linear, interaction and quadratic parameters and ε is the residual associated to the experiments.

The application of RSM as an optimization technique involves many different stages. The first stage establishes those parameters which have a major effect on the response, which is followed by the choice of a RSM and the evaluation of the experiments. The third stage involves the mathematical-statistical treatment of the obtained experimental data through the fit of a polynomial function, which is followed by the evaluation of the model's fitness to make sure that the model captures the actual conditions.

For pavement reliability applications, RSMs which generate a continuous differentiable performance function in the region of the most probable failure point are required. RSMs such as BBD, CCD and DOD are appropriate for modeling higher order surrogate models which capture the interactions between parameters and critical points of the function. These symmetrical experimental designs differ from one another with respect to the way the experimental points are selected, the number of levels for variables and the number of runs and blocks performed. A brief description of the three RSMs (i.e., CCD, BBD and DOD) used in this study is presented in the subsequent sections.

Central Composite Design (CCD): The CCD generates a surrogate model for a response function by using experimental designs that consist of a full factorial design and an additional design in which experimental points are at a distance α from its center and a central point (*22*). The CCD studies all factors at five different levels (e.g., $-\alpha, -1, 0, +1, +\alpha$) and determines the required number of experimental designs (N) for a given number of random variables (k) as follows:

$$N = k^2 + 2k + c_p \quad (5)$$

In which c_p is the central point and usually mean values. The alpha (α) values determine the location of axial points in the experimental domain thus defining the characteristic of the design which can be either spherical, orthogonal, rotatable or face centered. A unity alpha value ensures a three level factorial design which is called face centered design. For a given experimental design, the alpha value is calculated as follows:

$$\alpha = 2^{(k)/4} \quad (6)$$

Box–Behnken Design (BBD): The BBD selects experimental points from the three-level factorial arrangement in a manner that allows the efficient estimation of the first and second order coefficients of the surrogate model (*23*). The BBDs are more efficient and economical than the corresponding three level factorial designs, especially for many variables. The design experimental points in BBD are located on a hypersphere, equidistant from the central point. This method studies all factors only at three levels ($-1, 0, 1$) with equally spaced intervals between these levels and the required number of experimental designs for a given number of random variables (k) being calculated as follows:

$$N = 2k(k-1) + c_p \quad (7)$$

Doehlert Design (DOD): The DOD is a practical and economical alternative in relation to other second order experimental designs as it requires few experimental points for its application. The experimental points, which describe circular, spherical and hyper-spherical domains for two, three and more than three variables respectively, are selected in a manner that emphasize the uniformity of the studied variables in the experimental domain (*24*). This method studies each variable at a different number of levels, and the required number of experimental designs for a given number of random variables (k) can be calculated as follows:

$$N = k^2 + k + c_p \quad (8)$$

Distress Model and Design Input Characterization

Mechanics-Based Analysis Framework

The mechanics-based analysis and design framework is used for the performance evaluation of field pavement sections when considering their resistance to fatigue crack initiation. It was developed on the basis of hot mix asphalt fracture mechanics and incorporates mixture morphology influence on key damage and fracture properties (*25–27*). This framework, unlike the AASHTO MEPDG, appropriately takes into account the effects of aging and healing on accumulated damage and captures the influence of strain rate through energy based damage and resistance parameters. The framework allows readily coupling of tools such as LCCA and life cycle assessment (LCA), which optimize, respectively, the different viable design alternatives for a given target reliability for economic and environmental conditions. Figure 2 presents the mechanics-based analysis and design framework flow chart.

In the mechanics-based analysis framework, performance evaluation (PE) at a given time in the analysis period is performed by comparing two energy based parameters as shown in Equation 9. The first parameter, the accumulated dissipated creep strain energy ($DCSE_{accum}$) represents the induced damage caused by applied traffic loading, whereas the dissipated creep strain energy limit ($DCSE_{lim}$) governs the mixture resistance to fracture (*25*).

$$PE(t) = DCSE_{accum}(t) - DCSE_{\lim}(t) \quad (9)$$

The evolution in the $DCSE_{lim}$ is predicted by a mixture morphology-based material model that accounts for the morphology influence through a parameter called primary structure coating thickness (t_{PS}). The primary structure coating thickness (t_{PS}) represents the mastic thickness that coats the load bearing or primary structures (PS) of the aggregate and has been observed to influence mixtures performance (*28*, *29*). Equation 10 presents the $DCSE_{lim}$ model used in the analysis framework

$$DCSE_{\lim} = k_1(t_{ps})^{-k_2}(t)^{-(k_3 + k_4 \cdot \log(t_{ps}))} \quad (10)$$

in which $k_1 = 2.38$, $k_2 = 0.79$, $k_3 = 0.33$ and $k_4 = 0.12$ are model constants.

The $DCSE_{accum}$ is a function of tensile stress (σ_{av}), creep compliance rate ($\varepsilon_{p,\max}$), hourly traffic volume (ESAL) and healing potential (h_y). The framework computes the hourly creep compliance rate and healing potential using material models and determines hourly tensile stress using an axisymmetric linear elastic analysis tool (*25*). The analysis framework was calibrated by optimizing the healing potential function accordingly with the incorporation of a single calibration parameter. Equations 11 and 12 present the accumulated damage and healing potential models respectively.

$$DCSE_{accum}(t) = 0.05 \cdot \sum_{i=1}^{nhrs} ESAL_i \cdot \sigma^2_{av,i} \cdot \dot{\varepsilon}_{p\max,i} \cdot (1 - h_{y,i}) \quad (11)$$

$$h_{ym}(t) = 1 - \left(\left[\exp\left(\frac{t_{ps}}{t}\right)^{-DCSE_i} \right]_{norm} \right)^{15.5 \cdot t_{ps} + 3.35} \quad (12)$$

The use of a single shift factor to account for the variation in the predicted and observed fatigue cracking

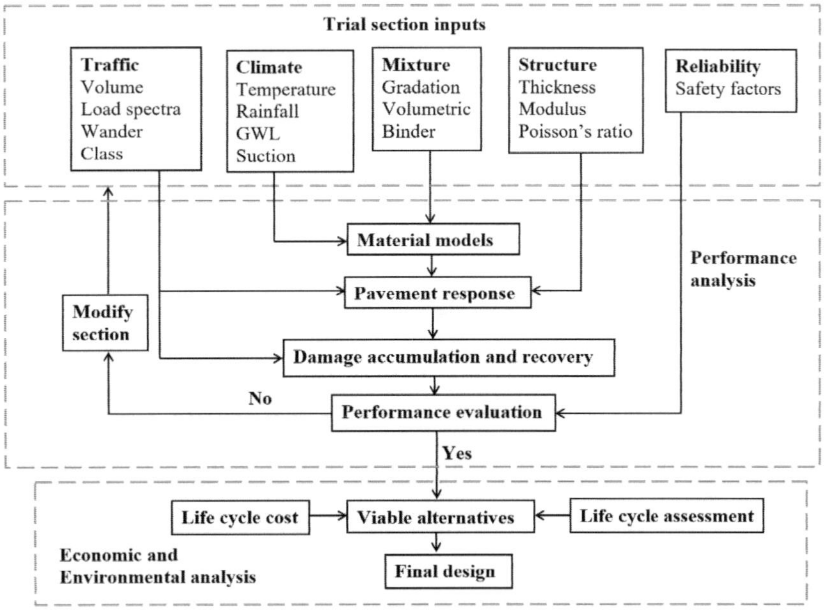

Figure 2. Flow-chart for the mechanics-based analysis and design framework.

Table 1. Summary of Information of Evaluated Pavement Sections

Section	County	Traffic/year (ESALs·10^3)	Target reliability (%)
I75-1A	Charlotte	573	95
I75-1B	Charlotte	558	95
I75-2	Lee	576	95
I-75SB	Hamilton	1,040	95
I-75SB2	Hamilton	1,190	95
US-301SB	Marion	510	95
A1A	Dade	411	90
SR60	Hillsborough	312	90
SR80-1	Lee	221	90
TPK-2	St. Lucie	166	90
US41	Hillsborough	233	90
SR563	Polk	126	85
SR16-6	Bradford	21	80
SR222	Alachua	68	80
SR18	Bradford	6	75

performances simplifies the model implementation effort and significantly reduces the systematic error associated with the use of multiple coefficients for model calibration which is the standard practice in the AASHTO MEPDG.

Pavement Sections

For the evaluation of the reliability analysis methodology, pavement sections with high quality field and laboratory data in addition to well documented field performance histories are required. For this purpose, 15 pavement sections from the state of Florida with a wide range in target reliabilities and design inputs were identified and used in this study (*14*, *25*, *30*). The target reliabilities of these pavement sections were established by following the recommendations specified in the Florida Department of Transportation pavement design guide (*31*). A list of the pavement sections with their yearly traffic volume and target reliability is presented in Table 1.

Design Inputs Variabilities

Statistical characterization of the inputs and parameters which have a significant effect on the predicted performance is a prerequisite for any reliability analysis. Based on a parametric study, design inputs such as asphalt concrete thickness (H_{AC}), base modulus (E_B), traffic volume ($ESAL_{hr}$) and dissipated creep strain energy limit ($DCSE_{lim}$) are selected as random variables. A literature review has been performed to establish the variability of these inputs in terms of coefficient of variation (COV) and distribution types. Table 2 presents the summary of findings from the literature survey and the level of variabilities which are used for this analysis. In this study, representative COV values and distribution functions which are expected under average conditions were used.

Table 2. Design Inputs Variabilities

Parameter	Variability survey (reference)	This study
Asphalt thickness (H_{AC})	Normal, COV [3–12%] (*8*) Normal, COV [3–12%] (*4*) Normal, COV [3–25%] (*9*) Normal, COV [10%] (*3*)	Normal, COV [10%]
Base modulus (E_B)	Lognormal, COV [15–50%] (*8*) Normal, COV [10–30%] (*4*) Lognormal, COV [5–60%] (*9*)	Lognormal, COV [30%]
Traffic ($ESAL_{hr}$)	Lognormal, COV [30–42%] (*9*) Lognormal, COV [42%] (*32*)	Lognormal, COV [40%]
Dissipated creep strain energy limit ($DCSE_{lim}$)	Normal, COV [35%] (*12*) Lognormal, COV [35%] (*12*)	Lognormal, COV [35%]

Note: COV = coefficient of variance.

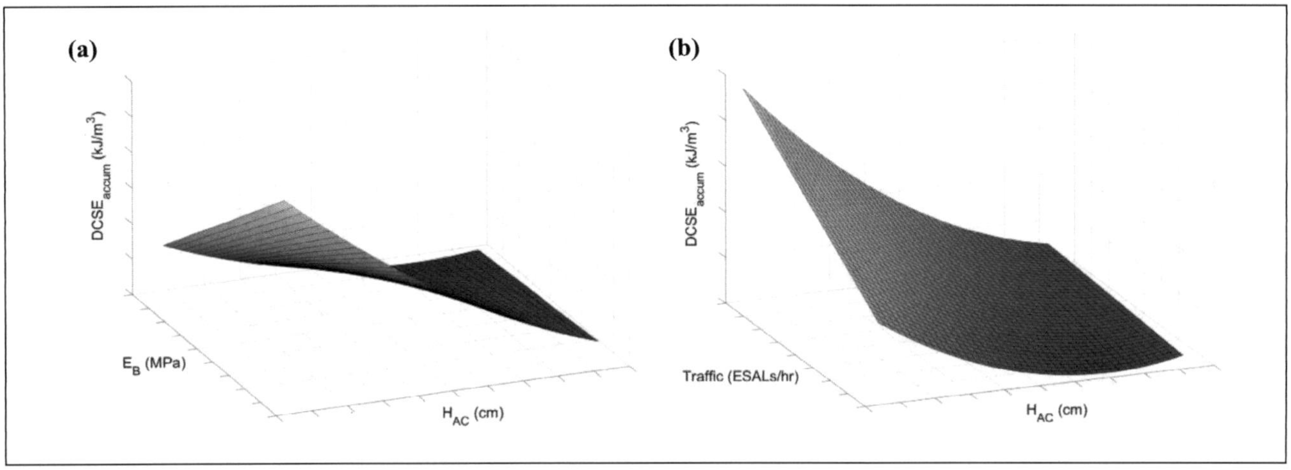

Figure 3. $DCSE_{accum}$ response surface for inputs: (*a*) H_{AC} and E_B; (*b*) H_{AC} and $ESAL_{hr}$.

Results and Discussion

The evaluation of the proposed reliability analysis methodology has been performed in two stages. The first stage evaluates the prediction capability of the three response surface methods (i.e., CCD, BBD and DOD) by comparing the predicted damage accumulation ($DCSE_{accum}$) generated from the surrogate models with the predictions from the actual model. The second stage evaluates the level of accuracy and effectiveness of the partial safety factors formulated for different target reliabilities using the three different RSMs.

Response Surface Methods Prediction Results

The inbuilt functions in MATLAB were used to generate the different level matrices for the three RSMs, which in combination with the mean (μ) and standard deviation (σ) of the three random variables (i.e., H_{AC}, E_B and $ESAL_{hr}$) were used to generate the experimental points. The least square linear regression tool was used to generate a surrogate model for the $DCSE_{accum}$ for the respective response surface methods. The generated surrogate models account for the interaction between variables and contain quadratic terms to determine the critical point. Figure 3 shows the generated response surfaces for $DCSE_{accum}$ for the case when two inputs are considered as random variables. Figure 3*a* presents the case for asphalt thickness and base modulus, whereas Figure 3*b* presents the case for asphalt thickness and hourly traffic. As can be seen in the figures, the surrogate models have captured the influence these parameters have on the damage for all the expected possible outcomes.

The effectiveness of the three RSMs (CCD, BBD and DOD) were evaluated using pavement section Charlotte I75-1A, which was used to generate the fitting coefficients for each RSM. A one way analysis of variance (ANOVA) statistical analysis was performed on the fitted

Table 3. One-Way ANOVA Result

Source	SS	DF	MS	F	P
RSM	0.03	3	0.01	0.05	.98
Error	41.87	196	0.21		
Total	41.90	199			

Note: SS = total sum of the square; DF = degree of freedom; MS = media of the square; F = statistics; *P* = probability; RSM = response surface method.

functions using randomly generated 50 experimental points. The result shows that the predictions from the three RSMs surrogate models are in a very good agreement with the actual $DCSE_{accum}$ model. Table 3 presents the results of the ANOVA analysis.

A linear regression comparison was also performed between the actual and surrogate model predicted $DCSE_{accum}$ values for the CCD, BBD and DOD RSMs respectively. Figure 4, *a–c* present the result for these comparisons while Figure 4*d* compares the three RSMs with respect to the actual model. As can be seen in Figure 4, there is a good agreement between the actual and surrogate model prediction for the range considered. The R^2 values of the CCD, BBD and DOD RSMs are 0.982, 0,977 and 0.977 respectively. In general, the results in Figure 4 show that the RSMs are able to account for the variations in the design input variables.

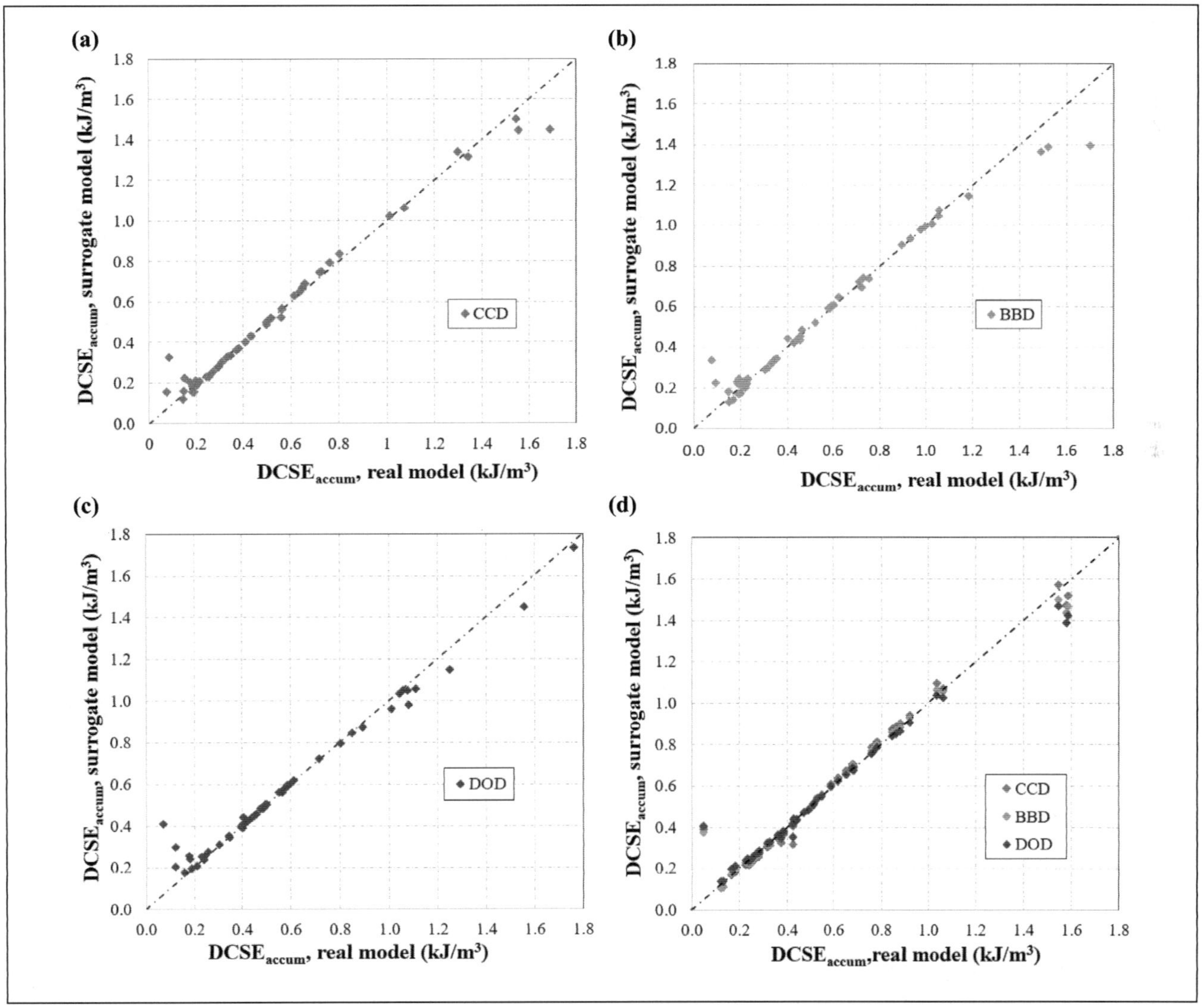

Figure 4. Comparison between actual and surrogate models generated $DCSE_{accum}$ values for (*a*) CCD, (*b*) BBD, (*c*) DOD, and (*d*) CCD, BBD, and DOD RSMs.

Reliability-Based Deterministic Design

A reliability-based deterministic design in the format of the LRFD procedure was developed to evaluate the effectiveness of the reliability analysis methodology for the respective response surface methods (i.e., CCD, BBD and DOD). LRFD is based on the limit state design concept and utilizes probabilistically derived partial safety factors to represent the variabilities associated with the design inputs (*33*). A set of partial safety factors which represent the risk level implied in the respective target reliabilities are used to optimize pavement sections for the required design conditions. The LRFD procedure recommends a single set of partial safety factors for a given target reliability which are formulated using pavement sections of various design features. The performance equation for the mechanics-based design framework can be expressed in the LRFD format as follows:

$$\gamma_{DCSE_{accum}} DCSE_{accum} \leq \phi_{DCSE_{\lim}} DCSE_{\lim} \quad (13)$$

In which $\phi_{DCSElim}$ and $\gamma_{DCSE_{accum}}$ are partial safety factors for the limit and accumulated dissipated creep strain energy values. As $DCSE_{accum}$ is the final outcome of many independent variables (i.e., H_{AC}, E_B and *Traffic*), its partial safety factor value for a given target reliability can be obtained by dividing its factored and mean values as shown in Equation 14. The partial safety factors of asphalt thickness (ϕ_{HAC}), base modulus (ϕ_{EB}), and traffic ($\gamma_{Traffic}$) are used to determine the factored $DCSE_{accum}$.

$$\gamma_{DCSE_{accum}} = \frac{DCSE_{accum}f(\phi_{H_{AC}}H_{AC}, \phi_{E_B}E_B, \gamma_{Traffic}Traffic)}{DCSE_{accum}f(H_{AC}, E_B, Traffic)} \quad (14)$$

Partial Safety Factors Formulations. The limit state equation governing the performance of the pavement structure at the designated design period needs to be established as a precursor for the reliability analysis. For the mechanics-based design framework, after a careful study of the performance histories of field pavement sections in Florida, crack initiation as a failure criterion and 15 years as a design period were recommended (*25*). The limit state equation for the mechanics-based analysis framework for any given time *t* can be expressed as follows:

$$PE(t) = DCSE_{accum}(t) - DCSE_{\lim}(t) \quad (15)$$

The second component of the reliability analysis methodology, which is a MATLAB-based R–F algorithm, was used in conjunction with each response surface methods to compute the respective reliability of the pavement sections listed in Table 1. In addition to the design reliability, analytical-based reliability analysis techniques such as the FORM provides the reliability index (β), the design or failure points and the directional cosines (α) of each random variable. The directional cosines (α) are sensitivity coefficients which provide the level of contribution of each design parameter to the overall variance and can be used in the formulation of partial safety factors for the development of a reliability-based deterministic design approach in the format of

Table 4. Partial Safety Factors for Each RSM and Target Reliabilities

		Target reliability [%]				
Parameter	RSM	75	80	85	90	95
ϕ_{HAC}	CCD	0.969	0.961	0.953	0.939	0.925
	BBD	0.969	0.961	0.952	0.938	0.928
	DOD	0.970	0.962	0.953	0.940	0.931
ϕ_{EB}	CCD	0.903	0.877	0.850	0.809	0.762
	BBD	0.905	0.879	0.853	0.810	0.778
	DOD	0.905	0.880	0.854	0.811	0.782
$\gamma_{Traffic}$	CCD	1.168	1.205	1.257	1.338	1.418
	BBD	1.166	1.203	1.255	1.329	1.388
	DOD	1.166	1.203	1.255	1.330	1.380
$\phi_{DCSElim}$	CCD	0.910	0.887	0.861	0.820	0.783
	BBD	0.908	0.885	0.858	0.815	0.791
	DOD	0.906	0.882	0.855	0.811	0.790
$\gamma_{DCSEaccum}$	CCD	1.623	1.802	2.044	2.458	2.998
	BBD	1.587	1.756	1.980	2.325	2.758
	DOD	1.548	1.707	1.917	2.244	2.586

Note: CCD = central composite design; BBD = Box–Behnken Design; DOD = Doehlert design; ϕ_{HAC} = asphalt thickness; ϕ_{EB} = base modulus; $\gamma_{Traffic}$ = traffic; $DCSE_{accum}$ = accumulated dissipated creep strain energy; $DCSE_{lim}$ = dissipated creep strain energy limit.

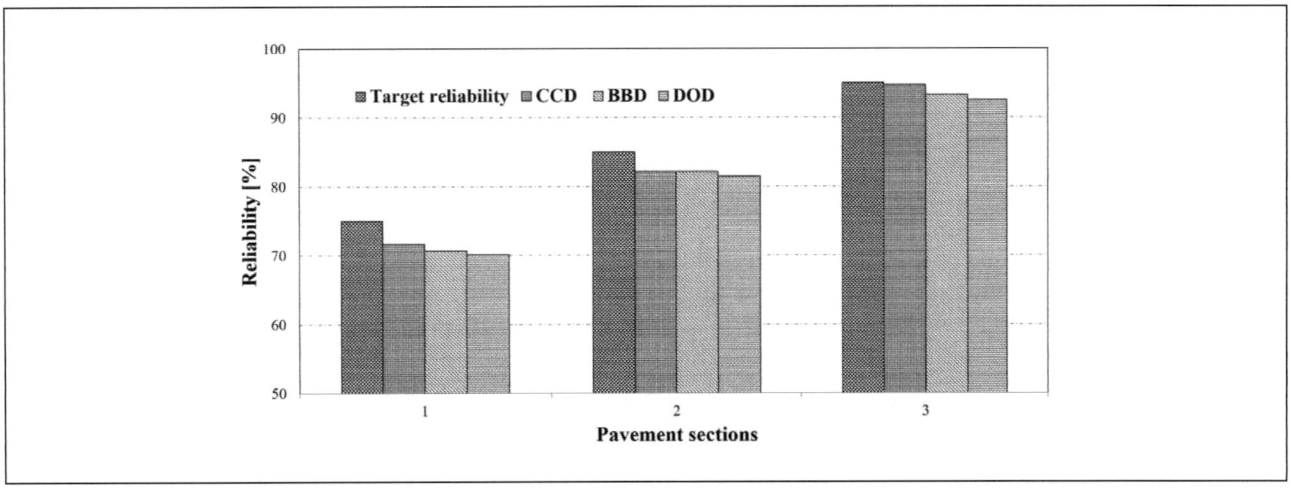

Figure 5. Comparisons between actual and target reliabilities.

LRFD. Equations 16 and 17 were used to generate a given target reliability which is represented by target reliability index (β_{tgt}) the partial safety factors for the parameters which contribute to the resistance and load effects respectively.

$$\phi_i = \frac{(\mu_i^N + \alpha_i \cdot \beta_{tgt} \cdot \sigma_i^N)}{\mu_i^N} \tag{16}$$

$$\gamma_j = \frac{(\mu_j^N + \alpha_j \cdot \beta_{tgt} \cdot \sigma_j^N)}{\mu_j^N} \tag{17}$$

In which ϕ_i and γ_j are the partial safety factors for the parameters which contribute to the resistance and load effects respectively, and μ^N, σ^N, and α are the equivalent mean, the equivalent standard deviation and the directional cosine of each design input. Table 4 presents the formulated partial safety factors for various target reliabilities and RSMs, which are representative values as they are obtained by averaging values from different pavement sections of the same reliability.

Evaluations of Partial Safety Factors. The capability of each set of partial safety factors for the respective target reliabilities and response surface methods were assessed and evaluated using pavement design examples. For this purpose, a typical four layered pavement section (Charlotte I75-1A) was designed using the respective partial safety factors for each response surface methods and for target reliabilities of 75%, 85% and 95%. The optimum pavement cross-section was obtained by optimizing the asphalt layer thickness while keeping the rest of the inputs constant. The actual reliability of each optimized section was obtained using a Monte Carlo simulation of 10,000 cycles, which is considered to be enough for generating the variability of the performance measuring parameters.

As can be seen in Figure 5, there is a very good agreement between the target reliabilities and the actual reliabilities for the three methods, especially for the 95% target reliability. Furthermore, the CCD method delivered target reliability values which are closer to the actual value than the BBD and DOD methods. This could be because CCD uses more experimental points for generating the surrogate model in comparison with both BBD and DOD. The actual reliabilities of the DOD sections were shown to be less accurate in comparison with the other two surrogate models. The result also shows that the lower the reliability the more deviation between the actual and target reliabilities. This could be because a relatively small number of pavement sections were available to formulate the partial safety factors for the lower target reliabilities (75–85%) in comparison with the higher reliability group (90–95%).

Conclusions

The AASHTO MEPDG incorporates a reliability concept which does not consider directly the effects of the involved variabilities and uncertainties on predicted performance. A reliability which is not based on the probabilistic method of uncertainty propagation might not delivered the intended benefits as it does not provide designs of uniform reliability. A reliability analysis methodology which incorporates a response surface method and first order reliability method can be used to develop a reliability based design approach in the format of the LRFD. The LRFD provides the proper platform for the appropriate appraisal of viable design options for economic and environmental conditions with tools such as LCCA and LCA.

The implemented reliability analysis methodology was shown to be effective in handling and quantifying the effects of the input variabilities on the predicted performance. The implemented response surface methods such as CCD, BBD and DOD have been used to effectively generate surrogate models for the damage accumulation in actual pavement sections with a very high degree of accuracy when compared to the actual pavement performance predictions. In addition, these methods in conjunction with the FORM based R–F algorithm are capable of estimating the reliability of pavement sections, and as such can be used in the development of LRFD for pavement application.

A good agreement between the predicted and actual target reliabilities was obtained for the three RSMs considered in this paper. The CCD provides the most accurate in comparison with the BBD and DOD RSMs. However, this can be because the CCD requires more experimental data in the determination of fitting parameters for the surrogate model. Furthermore, the result of this research shows how the influence of the different design input variables can be captured within a reliability analysis framework, and how this can be translated into a LRFD procedure for pavement application.

Acknowledgments

The authors would like to thank the Center for Infrastructure Renewal (CIR) for providing the financial support for the project under which this work was performed.

Author Contributions

The authors confirm contribution to the paper as follows: study conception and design: Dinegdae; data collection: Dinegdae; analysis and interpretation of results: Dinegdae and Onifade; draft manuscript preparation: Dinegdae. All authors reviewed the results and approved the final version of the manuscript.

References

1. Huang, Y. *Pavement Analysis and Design*, 2nd ed. Prentice Hall, Upper Saddle River, 2004, p. 505
2. Timm, D. H., B. Birgisson, and D. E. Newcomb. Variability of Mechanistic–Empirical Flexible Pavement Design Parameters. *Proc., 5th International Conference on the Bearing Capacity of Roads and Airfields*, Norway, Vol. 1, 1998.
3. Darter, M. I., B. F. McCullough, and J. L. Brown. Reliability Concepts Applied to the Texas Flexible Pavement System. *Highway Research Record*, 1972. 407: 146–161.
4. Noureldin, S. A., E. Sharaf, A. Arafah, and F. Al-Sugair. Estimation of Standard Deviation of Predicted Performance of Flexible Pavements Using AASHTO Model. *Transportation Research Record, Journal of the Transportation Research Board*, 1994. 1449: 46–56.
5. AASHTO. *Guide for Design of Pavement Structures*. American Association of State Highway and Transportation Officials, Washington, D.C., 1993.
6. ARA Inc. *Guide for Mechanistic-Empirical Design of New and Rehabilitated Pavement Structures*. Technical Report. ARA Inc., Champaign, 2004.
7. Dinegdae, Y., and B. Birgisson. Design Inputs Variabilities Influence on Pavement Performance Reliability. *Proc., 4th Chinese European Workshop Functional Pavement Design*, 2016, pp. 741–750.
8. Timm, D. H., D. E. Newcomb, B. Birgisson, and T. V. Galambos. *Incorporation of Reliability into the Minnesota Mechanistic-Empirical Pavement Design Method*. Final Report Prepared to Minnesota Department of Transportation. Minnesota Univ., Department of Civil Engineering, Minneapolis, 1999.
9. Maji, A., and A. Das. Reliability Considerations of Bituminous Pavement Design by Mechanistic-Empirical Approach. *International Journal of Pavement Engineering*, Vol. 9, No. 1, 2008, pp. 19–31.
10. Kim, H. B., and N. Buch. Reliability-Based Pavement Design Model Accounting for Inherent Variability of Design Parameters. Presented at 82nd Annual Meeting of the Transportation Research Board, Washington, D.C., 2003.
11. Retherford, J. Q., and M. McDonald. Reliability Methods Applicable to Mechanistic-Empirical Pavement Design Method. *Transportation Research Record: Journal of the Transportation Research Board*, 2010. 2154: 130–137.
12. Dinegdae, Y., and B. Birgisson. Reliability-Based Calibration for a Mechanics-Based Fatigue Cracking Design Procedure. *Road Materials and Pavement Design*, Vol. 17, No. 3, 2016, pp. 529–546.
13. Kim, H. B., R. S. Harichandran, and N. Buch. Development of Load and Resistance Factor Design Format for Flexible Pavements. *Canadian Journal of Civil Engineering*, Vol. 25, 1998, pp. 880–885.
14. Dinegdae, Y. H. *Reliability-Based Design Procedure for Flexible Pavements*. Licentiate thesis. KTH Royal Institute of Technology, Stockholm, Sweden, 2015.
15. Montgomery, D. C. *Design and Analysis of Experiments: Response Surface Method and Designs*. John Wiley and Sons, New Jersey, 2005.
16. Lamer, A. C., and F. Moavenzadeh. Reliability of Highway Pavements. *Highway Research Record*, 1971. 362: 1–8.
17. Darter, M. I., W. R. Hudson, and J. L. Brown. Statistical Variation of Flexible Pavement Properties and Their Consideration in Design. *Proceedings of the Association of Asphalt Paving Technologists*, Vol. 42, 1973, pp. 589–615.
18. Hasofer, A. M., and N. C. Lind. An Exact and Invariant Second-Moment Code Format. *Journal of Engineering Mechanics Division*, Vol. 100, No. 1, 1974, pp. 111–121.
19. Rackwitz, R., and B. Fiessler. Structural Reliability under Combined Random Load Sequences. *Computers and Structures*, Vol. 9, No. 5, 1978, pp. 484–494.
20. Dinegdae, Y., and B. Birgisson. Reliability-Based Design Procedure for Fatigue Cracking in Asphalt Pavements. *Transportation Research Record: Journal of the Transportation Research Board*, 2016. 2583: 127–133.

21. Wu, Y. T. *Efficient Methods for Mechanical and Structural Reliability Analysis and Design*. PhD thesis. University of Arizona, Tucson, Arizona, 1984.
22. Box, G. E. P., and K. B. Wilson. Series B (Methodological). *Journal of the Royal Statistical Society*, Vol. 13, No. 1, 1951, pp. 1–45.
23. Box, G. E. P., and D. W. Behnken. Some New Three Level Designs for the Study of Quantitative Variables. *Technometrics*, Vol. 2, No. 4, 1960, pp. 455–475.
24. Doehlert, D. H. Uniform Shell Designs. *Applied Statistics*, Vol. 19, 1970, pp. 231–239.
25. Dinegdae, Y. H., I. Onifade, D. Jelagin, and B. Birgisson. Mechanics Based Top Down Fatigue Cracking Initiation Prediction Design Framework for Asphalt Pavements. *Journal of Road Materials and Pavement Design*, Vol. 16, No. 4, 2015, pp. 907–927.
26. Onifade, I., Y. Dinegdae, and B. Birgisson. Hierarchical Approach for Fatigue Cracking Performance Evaluation in Asphalt Pavements. *Frontiers of Structural and Civil Engineering*, Vol. 11, No. 3, 2017, pp. 1–13.
27. Dinegdae, Y. H, and B. Birgisson. Effects of Truck Traffic on Top-Down Fatigue Cracking Performance of Flexible Pavements Using a New Mechanics-Based Analysis Framework. *Road Materials and Pavement Design*, Vol. 19, No. 1, 2018, pp. 182–200.
28. Kumar, D. P., B. Birgisson, D. Jelagin, and N. Kringos. Investigation of the Asphalt Mixture Morphology Influence on Its Ageing Susceptibility. *Materials and Structures*, Vol. 48, No. 4, 2015, pp. 987–1000.
29. Lira, B., D. Jelagin, and B. Birgisson. Gradation-Based Framework for Asphalt Mixture. *Materials and Structure*, Vol. 46, No. 8, 2013, pp. 1401–1414.
30. *NCHRP Project 1-37A: Mechanistic-Empirical Design of New and Rehabilitated Pavement Structures*. Transportation Research Board of the National Academies, Washington, D.C., 2004.
31. Florida Department of Transportation (FDOT). *Roadway Designs/Pavement Managements/Reports*, 2013. www.dot.state.fl.us. Accessed April 15, 2015.
32. AASHTO. *Proposed American Association of State Highway and Transportation Officials Guide for Design of Pavement Structures*. NCHRP Project 20-7/24, Vol. 2, Washington, D.C., 1985.
33. AASHTO. *LRFD Highway Bridge Design Specifications, SI Units*. American Association of State Highway and Transportation Officials, Washington, D.C., First Edition (1994 with 1996 and 1997 Interims).

The Standing Committee on General and Emerging Pavement Design (AFD30) peer-reviewed this paper (18-05731).

Article

TRR
JOURNAL OF THE TRANSPORTATION RESEARCH BOARD

Transportation Research Record
2018, Vol. 2672(40) 108–121

DOI: 10.1177/0361198118797781
journals.sagepub.com/home/trr
SAGE

Incorporation of Endurance Limit in the Mechanistic-Empirical Flexible Perpetual Pavement Design

Sheng Hu[1], Sang-Ick Lee[1], Lubinda F. Walubita[1], Fujie Zhou[1], and Tom Scullion[1]

Abstract

In recent years, there has been a push toward designing long-lasting thick hot mix asphalt (HMA) pavements, commonly referred to as a perpetual pavements (PP). For these pavements, it is expected that bottom-up fatigue cracking does not occur if the strain level is below a certain limit that is called the HMA fatigue endurance limit (EL). This paper proposed a mechanistic-empirical PP design method based on this EL concept. The ELs of 12 HMA mixtures were determined using simplified viscoelastic continuum damage testing and the influential factors were comparatively investigated. It was found that HMA mixtures seem to have different EL values based on mix type and test temperatures. There is not just a single EL value that can be used for all mixtures. Thus, default EL criteria for different mixtures under different climatic conditions were developed and incorporated into the Texas Mechanistic-Empirical Flexible Pavement Design System (TxME). As a demonstration and case study, one Texas PP test section with weigh-in-motion traffic data was simulated by TxME. The corresponding TxME inputs/outputs in terms of the PP structure, material properties, traffic loading, environmental conditions, and ELs were demonstrated. The corresponding TxME modeling results were consistent with the actual observed field performance of the in-service PP section.

In recent years, there has been a push toward designing long-lasting thick hot mix asphalt (HMA) pavement, commonly referred to as perpetual pavement (PP). The Asphalt Pavement Alliance (APA) has defined a PP as "an asphalt pavement designed and built to last longer than 50 years without requiring major structural rehabilitation or reconstruction, and needing only periodic surface renewal in response to distresses confined to the top of the pavement" (*1*). For PPs, it is expected that bottom-up fatigue cracking does not occur if the strain level is below the HMA fatigue endurance limit (EL). Therefore, additional pavement thickness, greater than that required for keeping strains below the EL, would not provide additional life. This concept has significant design and economic implications.

Since 2001, the State of Texas has been designing and constructing PPs on some of its heavily trafficked highways where the expected 20-year truck-traffic estimate of 18-kip equivalent single-axle loads (ESALs) is in excess of 30 million. To date, there are more than 10 in-service PP sections that are currently being monitored and evaluated for data population in the Texas flexible pavements and overlays database (*2*). The Texas Department of Transportation (TxDOT) also routinely collects traffic data using permanent weigh-in-motion (WIM) stations on some of the PP sections. Thus, it is possible to validate and calibrate the EL concept and PP design method using the actual measured performance and traffic data along with representative material properties and climatic conditions.

This paper first provides brief background information about the HMA EL and field-measured strain distribution, and proposed a mechanistic-empirical (ME) PP design method. The ELs of 12 HMA mixtures were determined using the simplified viscoelastic continuum damage (S-VECD) test (*3*). Based on the S-VECD test results, the default EL criteria for different mixtures were developed and incorporated into the Texas Mechanistic-Empirical Flexible Pavement Design System (TxME) (*4*). Finally, one Texas PP test section with WIM traffic data

[1]Texas A&M Transportation Institute, College Station, TX

Corresponding Author:
Address correspondence to Sheng Hu: s-hu@tti.tamu.edu

was simulated and the corresponding TxME inputs/outputs were demonstrated.

Background

Endurance Limit for HMA Mixtures

The concept of an EL is widely recognized in many areas of materials science, especially that of ferrous metals. Barret et al. describe the EL for metals as being a stress below which for un-cracked materials, the plot of stress versus cycles to failure becomes essentially horizontal and fatigue does not occur (*5*). Although this "limit" has been extensively studied and defined in metal and other material areas, relatively fewer works have been done for HMA, a typical viscoelastic material. Monismith and McLean first demonstrated the existence of a fatigue EL below which asphalt mixtures tend to have an extraordinarily long fatigue life and proposed an EL of 70 micro-strains for asphalt pavements (*6*). The log-log relationship between strain and bending cycles converged below 70 micro-strains at approximately 5 million load cycles. Maupin and Freeman noted a similar convergence (*7*). Nunn and Nishizawa et al. proposed concepts for long-life pavements for which classical bottom-up fatigue cracking would not occur (*8*, *9*). Nishizawa et al. reported an EL of 200 micro-strains based on the analysis of in-service pavements in Japan (*9*). Similarly, strain levels at the bottom of the asphalt layer of between 96 and 158 micro-strains were calculated based on backcalculated stiffness data from the falling weight deflectometer for a long-life pavement in Kansas (*10*). Other researchers propose that one should limit the strain anywhere from 60 to 100 micro-strains based on laboratory testing (*11*). Another experimental pavement project in China allowed PP design to reach the less conservative value of 125 micro-strains (*12*).

The National Center for Asphalt Technology (NCAT) has led a research effort for National Cooperative Highway Research Program (NCHRP) Project 9-38 to investigate the EL for HMA (*13*). This study involved conducting fatigue tests for a number of HMA mixtures over a wide range of strain levels. The tests have been conducted up to 50 million load cycles to failure. The Asphalt Institute has also been involved in the portion of the work to test samples having fatigue lives up to 50 million load cycles. The primary objectives of that study were to determine if HMA mixtures do have an EL and to provide guidance on determining this limit for various mixture types. The results indicated that HMA mixtures appeared to have an EL that varies with mix types, that is, there is not just one limit that can be used for all mixtures.

Most recently, NCHRP Project 09-44A "Validating an Endurance Limit for HMA Pavements: Laboratory Experiment and Algorithm Development" (*14*) further identified that the EL varies with temperature, pavement design conditions, binder grade, binder content, air voids, and the rest period between the load applications.

Field-Measured Strain Distribution from NCAT Test Track

Although laboratory testing showed that HMA mixtures may have an EL, further verification and validation was still needed from field strain measurements. In this respect, a study at the NCAT Pavement Test Track provided very interesting insights. In 2000, NCAT built 46 test sections with a minimum 23 inches of HMA pavements and then rebuilt some of them with thinner thickness in 2003. The experiments at these sections started from 2000, 2003, and 2006, respectively, and around 10 million ESALs of traffic loading was applied on each experimental test section (*15*). Figure 1*a* shows the pavement structure of eight PP test sections of the 2003 NCAT Test Track and Figure 1*b* shows the cumulative distributions of the estimated strain. Test section N8 was excluded because of other investigating needs. After around 19 million ESALs of trafficking, five of the test sections including N1, N2, N5, N6, and N7 showed fatigue cracking. By contrast, Sections N3 and N4 did not exhibit any distresses. As shown in Figure 1*b*, N1, N2, N5, N6, and N7 have larger strain than N3 and N4 for any given percentile. So the strain distribution of N3 and N4 can be used as references of criteria.

Based on the measured strains from 2003 and 2006 NCAT Test Track sections, Willis (*15*) proposed a cumulative frequency distribution of allowable strains for PP design. Table 1 lists the strain limit values for each percentile.

Methodology of ME PP Design

According to the preceding discussions, several considerations should be given when incorporating EL into ME PP design procedures:

- EL should be mixture dependent. Different asphalt mixtures should have different ELs because the binder grade, binder content, and gradation type have influence on the EL value.
- The temperature effect on the EL should be considered. A pavement section in a cold climatic area should have a lower EL value than that in a hot climatic area even if the structural thickness, layer materials, and traffic loading are the same.
- Consideration should be given to whether the EL is really best represented by a single value or a distributed strain profile.

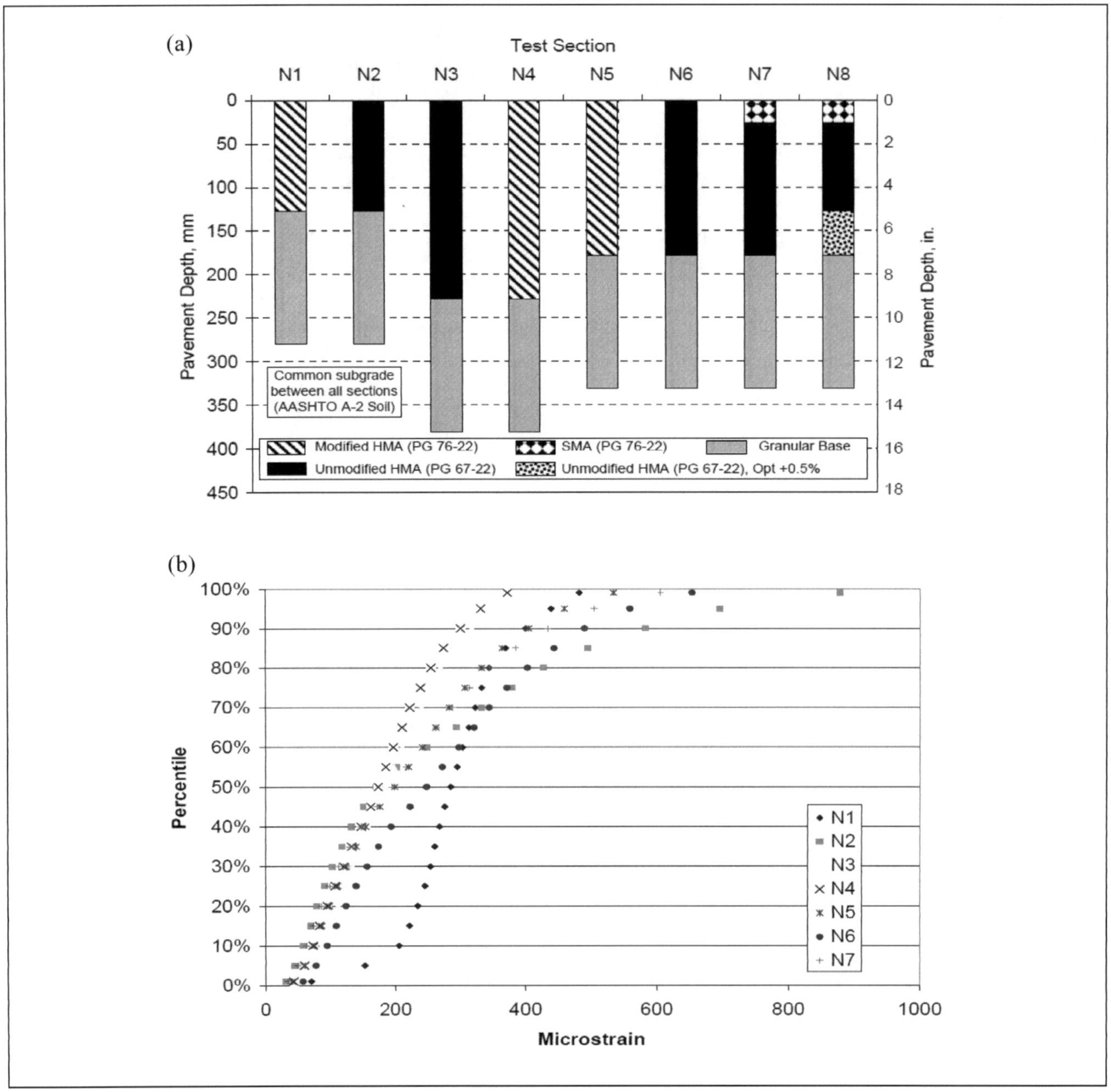

Figure 1. NCAT 2003 test sections: (*a*) pavement thickness; (*b*) cumulative strain distribution (*15*).

Based on these considerations, a two-level method of ME PP design was proposed as follows: one level is to implement the EL criteria and the other level is to implement strain distribution criteria. In this study, both levels of the method were incorporated into the TxME design software. In the TxME, users can select either load spectra (Level 1) or ESALs as the traffic load input (Level 2). The maximum tensile strain at the bottom of asphalt layer can be calculated based on multi-layer linear elastic theory solutions for each specific axle load. The corresponding two-level ME PP design method is described below:

- When the traffic input is ESALs, the strain calculation is actually based on one single standard 18-kip axle load. The tensile strains at the bottom of the asphalt layer will be determined at the average annual temperature and compared with the

Table 1. Strain Distribution Criteria for Perpetual Pavements (*15*)

Percentile	Fatigue limit
99%	394
95%	346
90%	310
85%	282
80%	263
75%	247
70%	232
65%	218
60%	205
55%	193
50%	181
45%	168
40%	155
35%	143
30%	132
25%	122
20%	112
15%	101
10%	90
5%	72
1%	49

single-value EL criterion. This single EL criterion should be dependent on mixture type and climatic conditions too.

- When the traffic input is load spectra, the traffic is distributed into different axle types such as single-axle, tandem-axle, tridem-axle, and quad-axle according to the vehicle volume information. The corresponding repetitions of each type of axle can be determined based on the annual average daily traffic (AADT) information. These repetitions are further distributed into numbers for each axle type based on different load levels. Thus, the tensile strains at the bottom of asphalt layer under different axle load levels and temperature conditions can be determined and the corresponding strain distribution can be evaluated. This strain distribution will be compared with some user-defined strain distribution criteria.

EL Determination

Test Methods for EL Measurement

Traditionally, the four-point beam fatigue test was conducted to determine the EL (*16*). Recently, two other uniaxial fatigue tests including the S-VECD test and the repeated direct tension test based on energy-based mechanics approach (*17*, *18*) have been used to determine the EL. Figure 2 summarizes and compares the features of each test method. Note that in this figure AMPT means asphalt mixture performance tester, and MTS means material testing system. Based on this comparison, the S-VECD method was selected as the EL test in this study, because while the test should be performed at one temperature, the ELs at different temperatures can easily be determined (*19*, *20*).

Analysis of the S-VECD Test Data

The key function of the S-VECD model is the damage characteristic curve that relates the overall amount of damage in a specimen to the pseudo secant modulus, or material integrity (*3*). A detailed description of the pseudo modulus and pseudo strain concepts, as well as a detailed derivation of the S-VECD model can be found elsewhere (*19*, *20*).

The linear visco-elastic (LVE) characterization procedure found in both standard test procedures AASHTO TP 62 and AASHTO TP 79/PP 61 can be used to characterize the dynamic modulus, which needs to be incorporated into the S-VECD model. It is important that the complete S-VECD protocol is compatible with the AMPT capabilities because this device is likely to become the standard asphalt mixture test equipment.

The ALPHA-Fatigue software (v 3.1.5) developed by Underwood and Kim was used to analyze the S-VECD test data and determine the fatigue parameters and ELs. With the dynamic modulus and fatigue test results, the ALPHA-Fatigue software produces two outputs, namely the damage characteristic curve and the energy-based failure criterion. Figure 3 shows an example of analysis results from the ALHPA-Fatigue software. It can be seen from Figure 3*a* that the load cycle to failure for the first replicate is 26,476 cycles at a strain level of 300 $\mu\varepsilon$, 5,815 cycles for the second replicate at a strain level of 350 $\mu\varepsilon$, and 895 cycles for the third replicate at a strain level of 400 $\mu\varepsilon$. The determined ELs at different temperatures are 51, 54, 61, 66, and 78 $\mu\varepsilon$ at 5, 10, 15, 20, and 25°C, respectively, as shown in Figure 3*b*.

Endurance Limits of Different HMA Mixtures

To determine ELs for different HMA mixtures and to provide default EL criteria for the suggested ME PP design, five plant mixtures and seven laboratory-designed mixtures were selected for the S-VECD test in this study. For testing each mixture, at least three replicates were made for S-VECD testing. Also, three replicates were made for dynamic modulus testing because the S-VECD model needs to incorporate the dynamic modulus for the LVE characterization. In addition, five overlay tester (OT) replicates were fabricated and tested according to procedure Tex-248-F (*21*). For the laboratory-designed

Item	Beam Fatigue	S-VECD	RDT
Test Machine	Beam Fatigue Apparatus	AMPT	MTS
Sample Size	380mm × 50mm × 63mm	100mm Dia. × 130mm Tall	100mm Dia. × 150mm Tall
No. of Samples	>4	3	4
Analysis Program	Excel template	Alpha-F Software	Excel template
Standard Test Procedure	-Yes, AASHTO T 321	- Yes, AASHTO TP107	- No
Advantage	- Result is straightforward	- Simple to run - Easy calc. of EL at diff. temp.	- Simple to run
Disadvantage	- Need long testing time - Sample is more difficult to fabricate	- Need dynamic modulus result	- Need to run different sets of test for certain temp.

Figure 2. Comparison of the EL test methods.

Table 2. EL Values for Different HMA Mixtures at Different Temperatures

					EL ($\mu\varepsilon$) at different temperatures				
No.	Mix type	Binder type	Gradation	Other data	5°C	10°C	15°C	20°C	25°C
1	Plant mixture	PG 64-22	Dense graded type B	Highway US 82	34	35	38	40	46
2		PG 64-22	Dense graded type C	HighwaySH 7	54	58	67	74	90
3		PG 64-22	Dense graded type C	HighwaySH 304	33	37	45	52	67
4		PG 64-22	SP-C	20% RAP	39	40	43	47	58
5		PG 76-22	SP-D	14% RAP	29	28	29	30	36
6	Lab-designed mixture	PG 64-22	SP-C	AC 4.8%	30	33	38	45	62
7				AC 5.3%	44	49	56	66	89
8		PG 64-22	Dense graded type C	AC 4.7%	39	42	47	56	78
9				AC 5.2%	37	42	50	61	85
10				AC 5.7%	37	44	53	69	102
11		PG 64-22	SP-D	AC 5.9%	36	40	46	54	73
12				AC 6.4%	57	64	75	90	123

Note: PG = performance grade; SP = superpave; RAP = reclaimed asphalt pavement.

mixtures, the asphalt content (AC) varied for each mix type to evaluate the effect of the AC on the EL. Table 2 summarizes the ELs for the mixtures selected.

For the convenience of comparison, Figure 4*a* plots the ELs of all plant mixtures together. Similarly, Figure 4*b* compares the ELs at different AC levels for a Type C laboratory-designed mixture. It can be seen in the plots that temperature, gradation, binder grade, AC, and RAP contents have significant influence on the EL values.

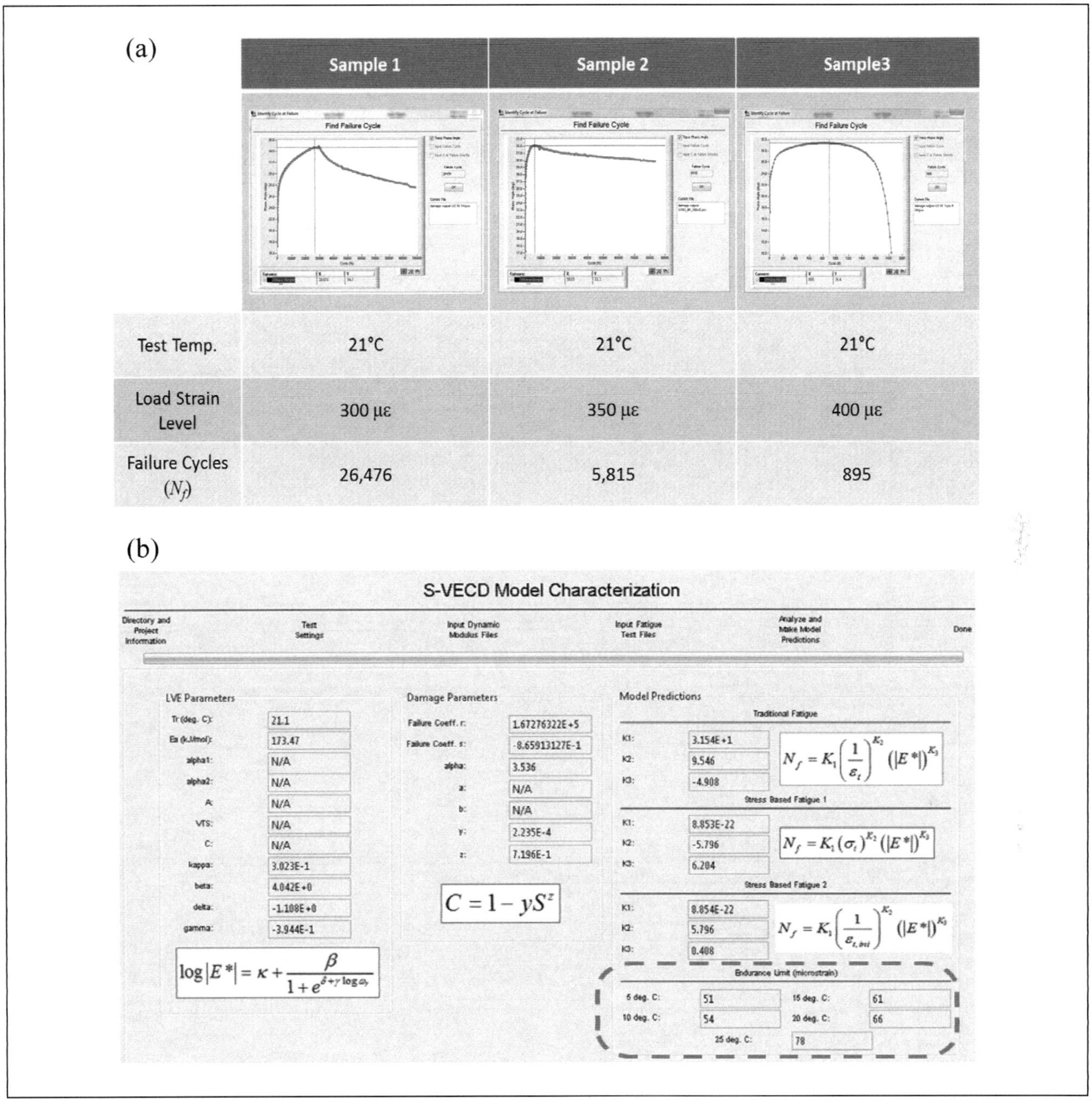

Figure 3. (*a*) S-VECD test results; (*b*) EL determination using the ALPHA-Fatigue software.

Default of EL Criteria for Different HMA Mixtures

Because the EL value is related to both the mixture itself and temperature, it is not appropriate to assume one EL value for one typical mixture. In the TxME, default OT cycles were incorporated in the system for most typical Texas mixtures depending on their mixture type, aggregate type/gradation, binder grade, RAP/RAS content, and so forth. Thus, to develop the default EL criteria for different mixtures at different temperature conditions, the relationship between the EL value and corresponding OT cycles were developed, as seen in Figure 5.

In Figure 5, it can be seen that the ELs have a good relationship with the OT cycles, particularly at 25°C, 20°C, and 15°C, with relatively high coefficient of determination (R^2). For the ELs at lower temperatures, the

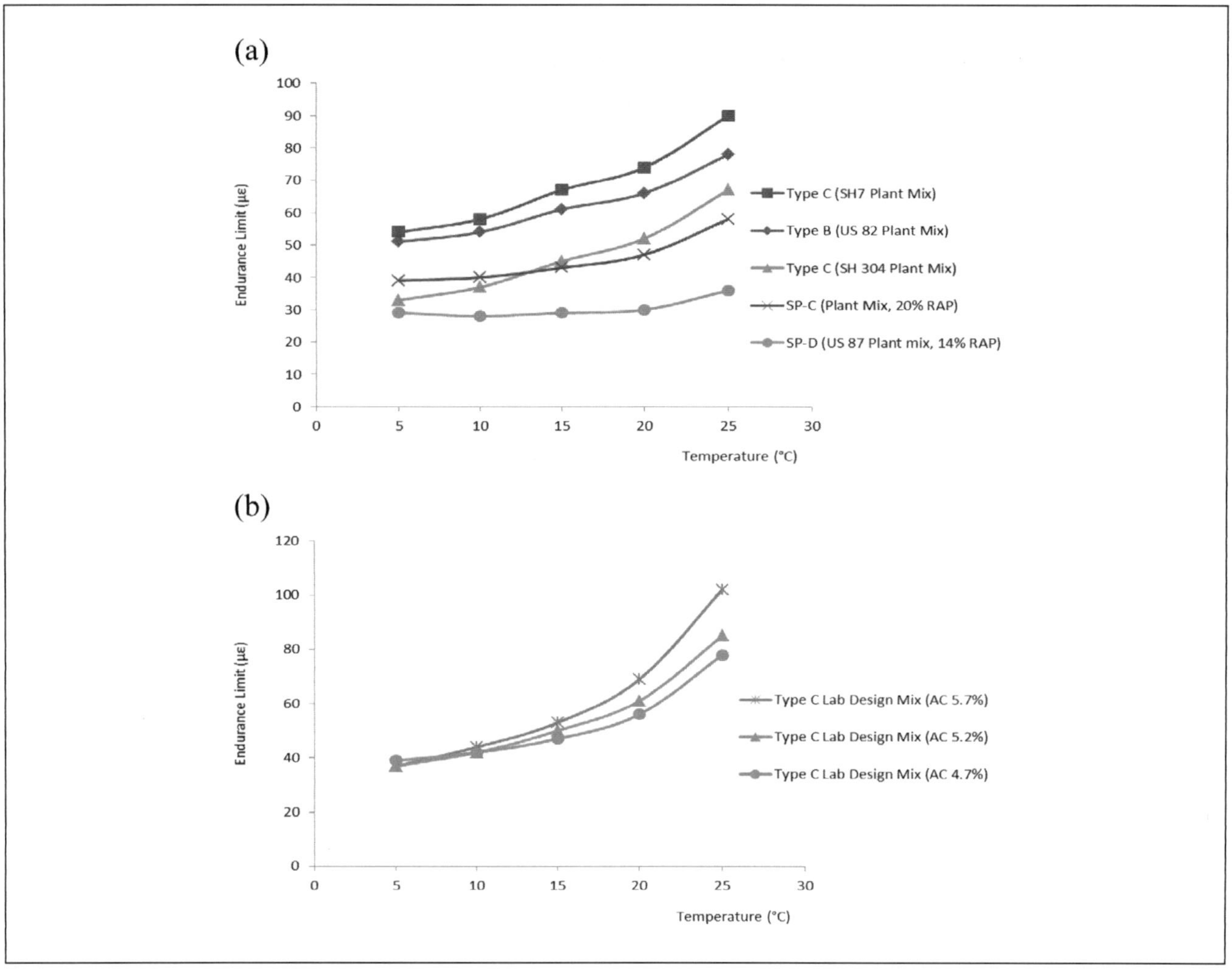

Figure 4. EL comparisons among (*a*) plant mixtures and (*b*) lab-designed mixtures.

Table 3. Suggested ELs at Different Temperatures Based on OT Cycles

	Endurance limits ($\mu\varepsilon$) at different temperatures				
OT cycles	25°C	20°C	15°C	10°C	5°C
3	34	31	31	29	29
5	40	34	33	31	31
15	52	42	39	35	34
20	56	44	40	36	34
50	66	51	45	39	37
100	74	56	48	42	39
200	82	61	51	45	40
300	86	64	53	46	42
400	89	66	55	47	42
500	92	67	56	48	43
600	94	68	57	49	43
700	96	70	58	49	44
800	97	71	58	50	44
900	99	71	59	50	44
1000	100	72	59	51	45
1500	104	75	62	52	46
3000	112	80	65	55	48

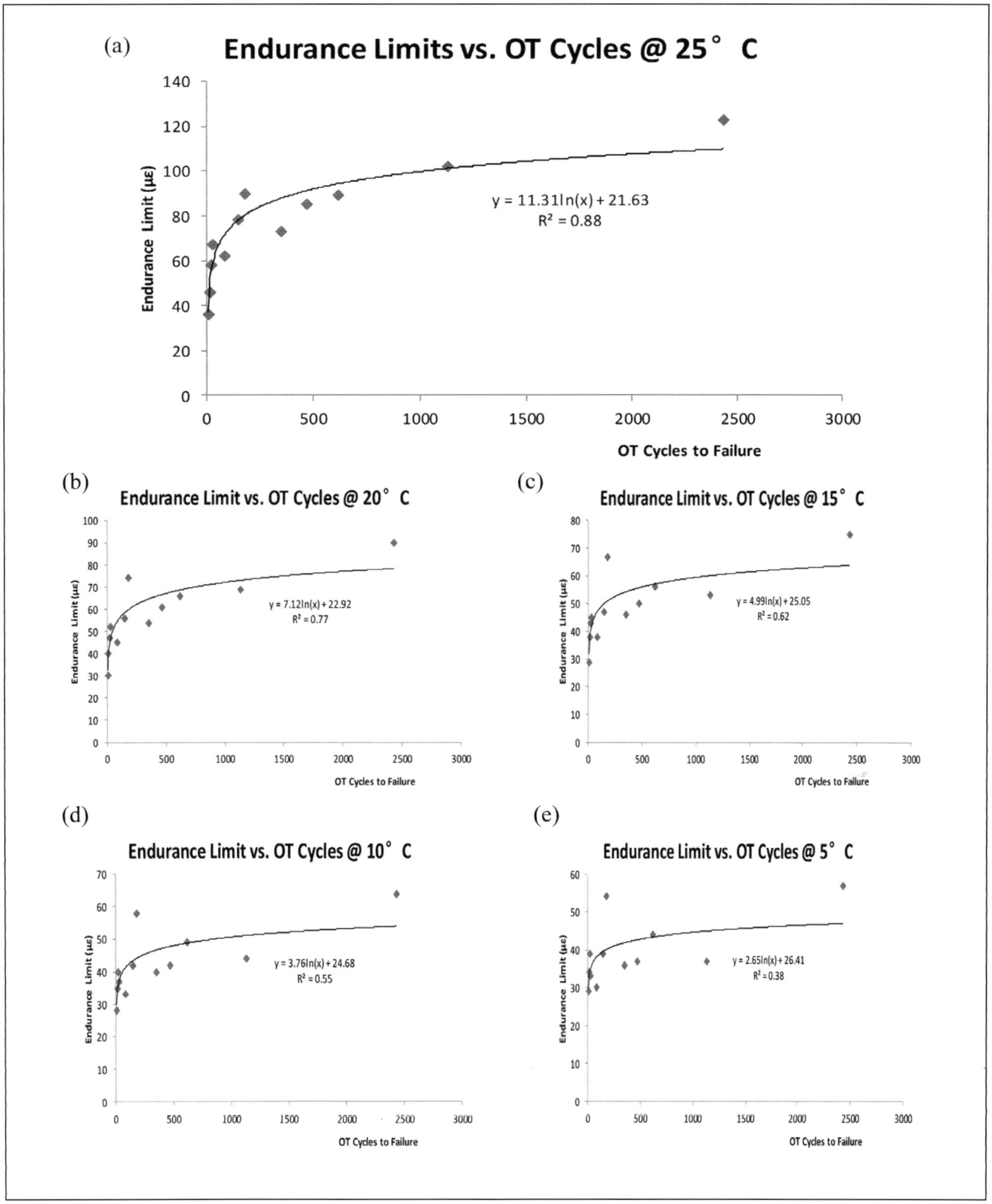

Figure 5. Endurance limits at (*a*) 25°C, (*b*) 20°C, (*c*) 15°C, (*d*) 10°C, and (e) 5°C vs. OT cycles.

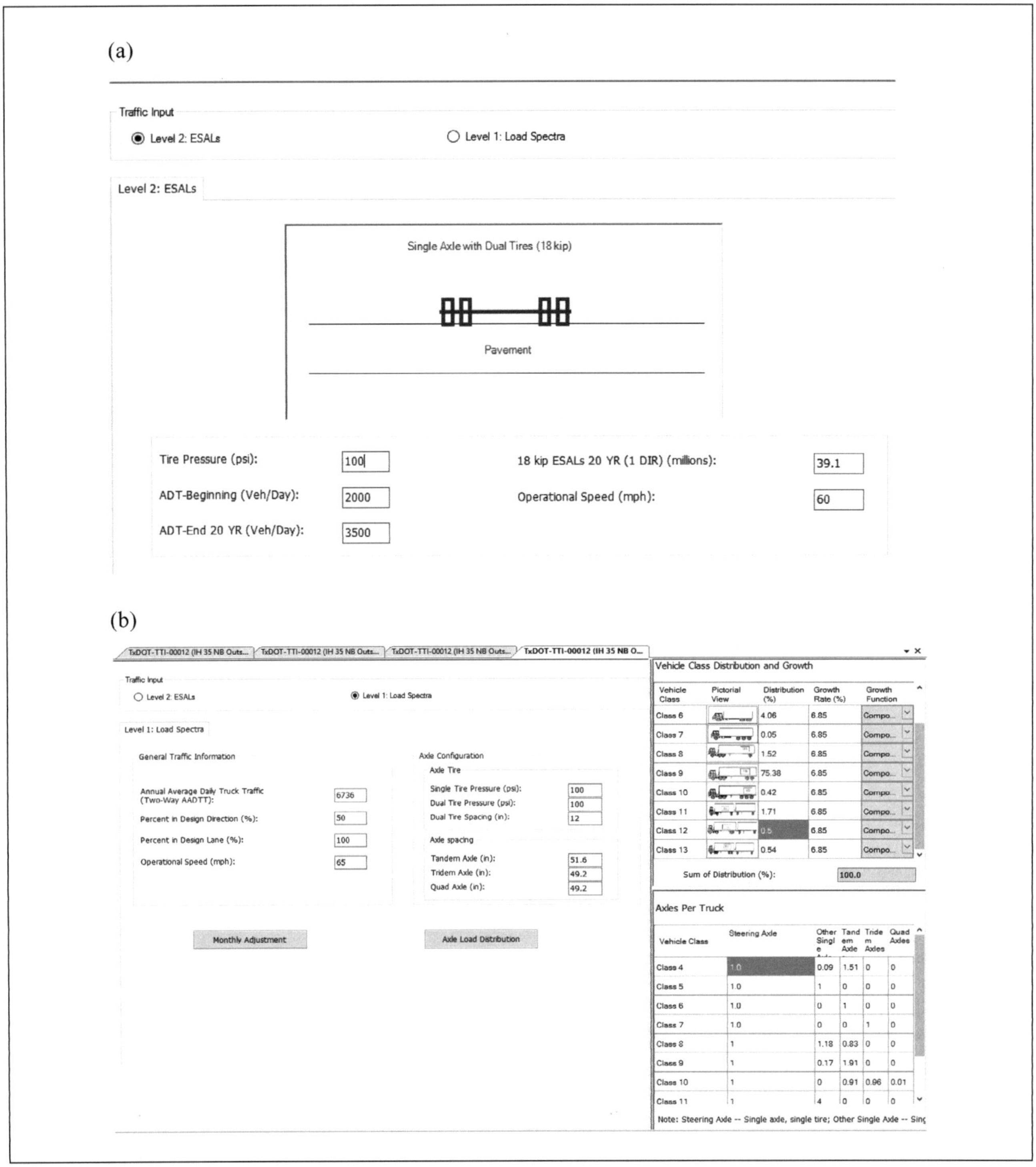

Figure 6. TxME traffic input screen for (*a*) Level 2, ESALs and (*b*) Level 1, load spectra.

relationship shows lower R^2 values. Based on the above-established relationships and regression equations, the preliminarily default ELs were determined in accordance with the OT cycles, as listed in Table 3. With these tabulated results and given the OT cycles, the EL at any temperature can be easily interpolated.

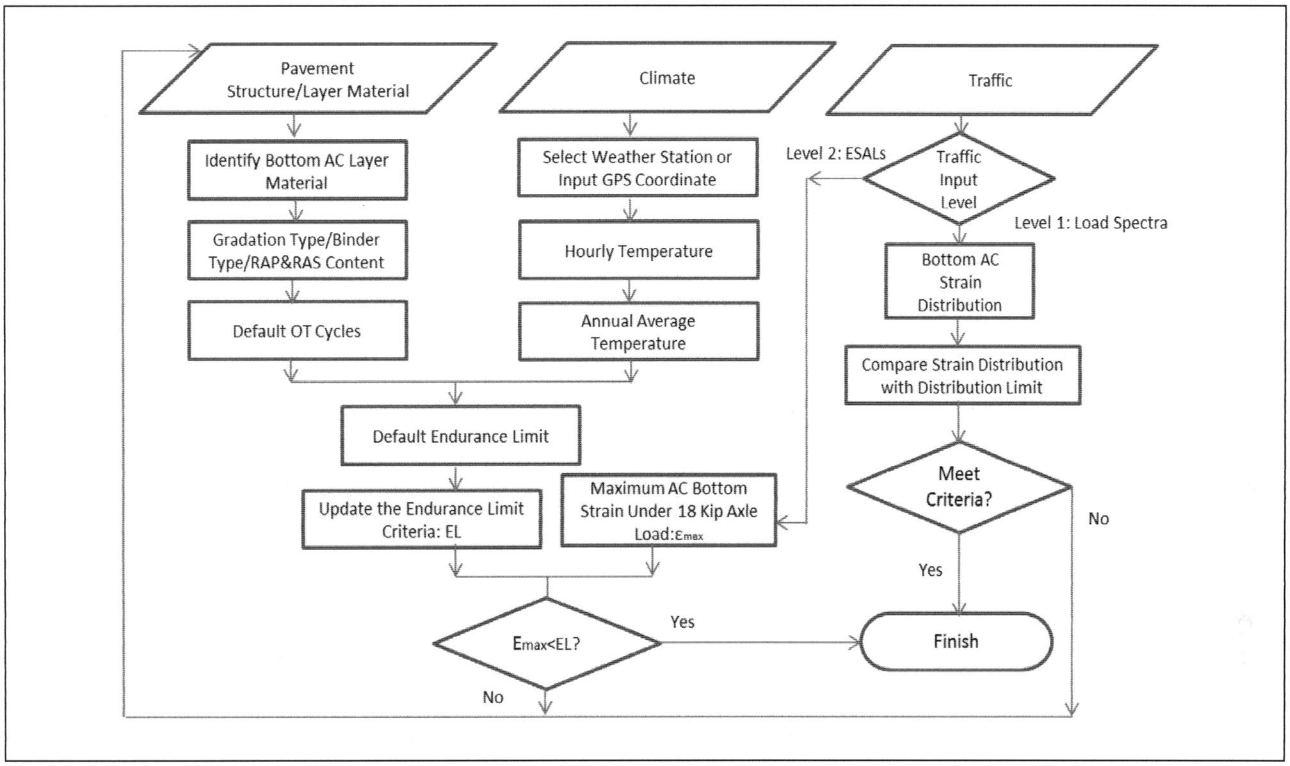

Figure 7. Flow chart of the ME PP design approach.

Incorporation of the ME PP Design Method into TxME

As discussed previously, a two-level of ME PP design method was incorporated into the TxME. Depending on the traffic input level, the TxME will computationally decide whether to calculate a single maximum strain value or strain distribution and compare the result with the corresponding criteria. Figure 6, *a* and *b* show the TxME traffic input screen for Level 1 and Level 2, respectively. Note that for Level 2, users only need to input some simple information such as ESALs, but for Level 1 much more detailed information is required, such as AADT, vehicle class distribution, axles per truck, axle load distribution, monthly adjustment, and so forth. All this detailed information can be obtained from traffic WIM data.

Figure 7 shows the flow chart that illustrates the PP design approach. Note that when traffic input is Level 2 (ESALs), the design criterion is a single EL value; however, when the traffic input is Level 1 (load spectra), the design criteria is a pre-defined strain distribution. In this study, the single EL criterion default is tied to the default OT cycles and the strain distribution criteria adopts the suggested data in Table 1. It is noted that these criteria can be changed by the user.

Demonstration Case Study

Below is an example of using the enhanced TxME to analyze a Texas PP section on IH 35 in La Salle County, Laredo District. There is a permanent WIM station in this location, and thus the detailed load spectra data are available. Figure 8, *a* and *b* show the traffic summary and pavement structure (northbound outside lane), respectively. The dynamic modulus and other material properties determined in the laboratory using plant mixture from each HMA layer were entered into the TxME.

If users select Level 2 (ESALs) for the traffic input, TxME calculates the hourly temperature, annual average temperature, and other parameters based on the section location or weather station information. Next, the TxME calculates the EL value (52 $\mu\varepsilon$ in this case) based on the bottom HMA layer (Superpave C, PG 64-22) properties and average annual temperature (71.1°F), as shown in Figure 9. Note that at any time when users change the pavement structure or climatic information, the EL value will be automatically re-calculated. In addition, users can manually change the EL criterion input if they have conducted the EL test.

If users select Level 1 (load spectra) for the traffic input, TxME calculates the strain distribution according to the load spectra and compares with the strain

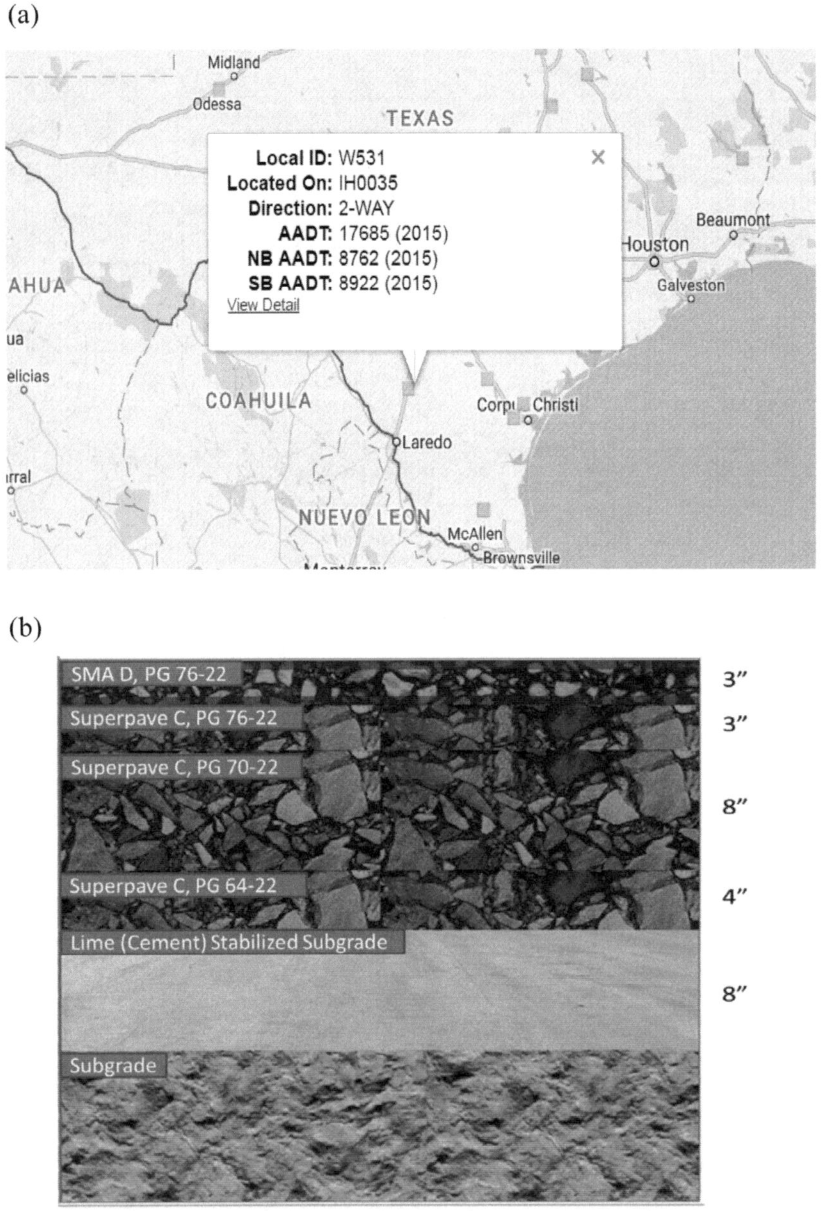

Figure 8. IH 35 PP test section: (*a*) location and traffic; (*b*) structure.

distribution limit. Figure 10 shows the analysis results for both Level 1 and Level 2 traffic inputs. Figure 10*a* shows that with the 18-kip ESALs input, the maximum strain at the AC bottom was determined to be 33.7 $\mu\varepsilon$, which is less than the EL criterion of 52 $\mu\varepsilon$. Figure 10*b* shows that with load spectra input, the determined strain distribution curve is on the left side of the strain limit curve, which means that for a given percentile, the determined strain is less than the corresponding strain limit. Thus, both cases indicate that the IH 35 PP section meets the "perpetual" criteria, which is consistent with the actual measured performance on the in-service IH 35 as of summer 2017 that no fatigue cracking was observed on this section since construction in 2004.

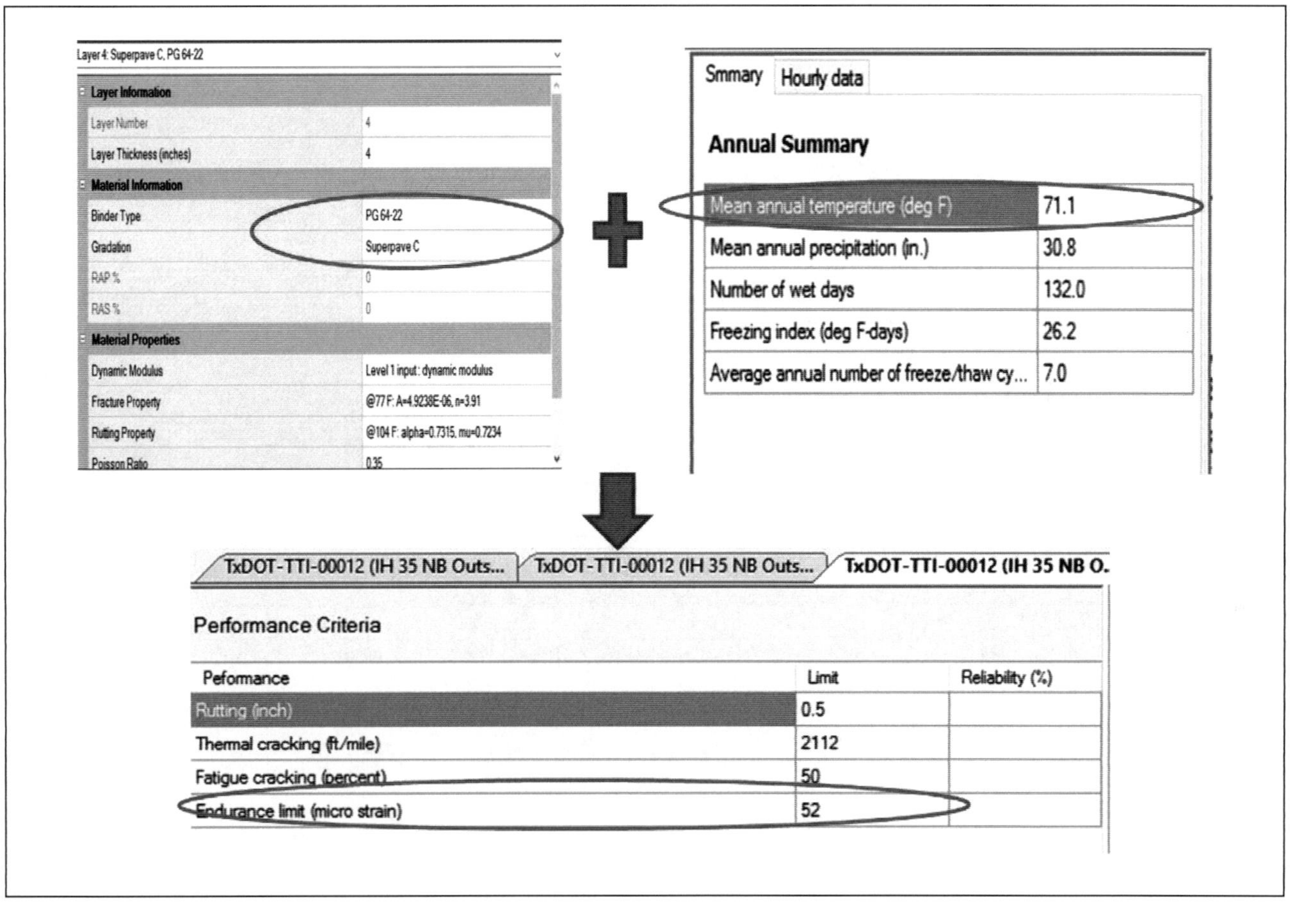

Figure 9. TxME EL criteria determined from user input data.

Summary and Conclusion

This paper proposed and implemented a methodology for incorporating EL into the ME PP design. The ELs of 12 HMA mixtures were determined in the laboratory using the S-VECD test. Based on these test results, default EL criteria were developed and incorporated into the TxME system. A typical Texas PP section with WIM-measured traffic data was utilized as a demonstration case study. Based on the work presented in this paper, the following conclusions and recommendations are made:

- HMA mixture appears to have EL and varies with mixture type—therefore, there is not just one EL that can be used for all mixtures.
- The S-VECD test was identified as an appropriate test method for determining the EL. The corresponding standard test procedure and data-analyzing software are available as well.
- Based on laboratory testing of 12 typical Texas mixtures, the following mix-design variables were found to have a significant influence on the EL: mixture type, aggregate gradation, binder grade/content, and RAP/RAS content.
- Correlative relationships were established between the ELs and OT cycles and permit the arithmetical determination of EL value at any temperature for given OT cycles.
- Temperature was found to have an increasing effect on the EL, with higher values recorded at high temperatures, and should be considered in ME PP design. Thus, a PP section in a cold climatic area should have a correspondingly lower EL criterion even if the structural thicknesses, layer materials, and traffic loading are the same.
- Two levels of ME PP design methods were proposed when incorporating the EL into TxME design system, namely:
    - When the traffic input is ESALs (Level 2), the tensile strain at the bottom of asphalt layer is determined based on the 18-kip ESALs at the

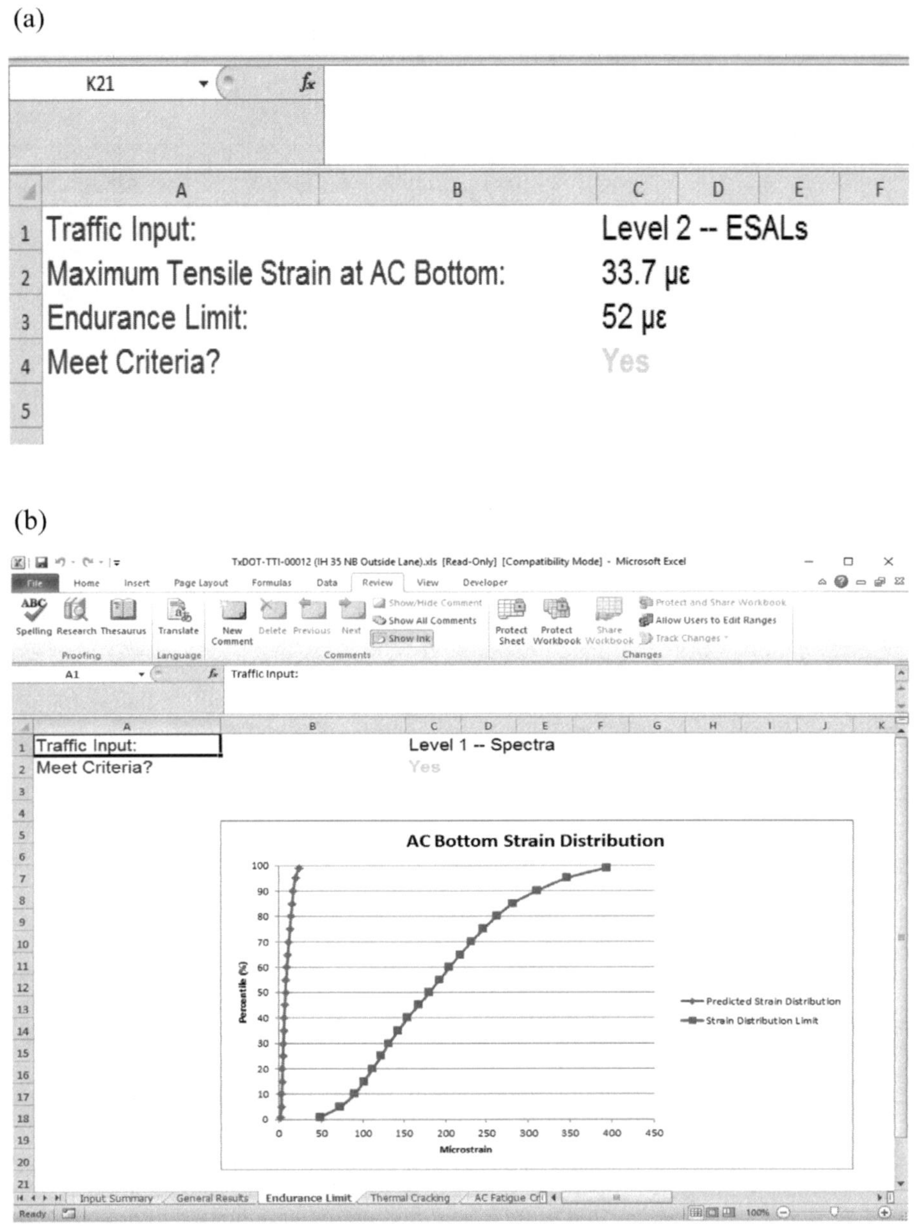

Figure 10. TxME PP design output when traffic input is (*a*) ESALs and (*b*) load spectra.

average annual temperature and compared with a single-value EL criterion. This single-value EL criterion should be either a default value based on mixture type, binder type, and climatic condition, or directly determined from laboratory testing.

- When the traffic input is load spectra (Level 1), the tensile strain distribution under different load levels and temperature conditions is determined and compared with the user-defined strain distribution criteria.

One Texas PP test section (IH 35, La Salle County, Laredo) with WIM-measured traffic data was simulated using the enhanced TxME design system. The corresponding TxME inputs/outputs in terms of the PP structure, material properties, traffic loading, environmental conditions, and EL were successfully demonstrated, with the modeling results matching the actual in-service field performance of the PP structure. However, additional validation and calibration of the enhanced TxME PP design system are continuing with other in-service PP test sections as field performance and traffic data are progressively collected.

Acknowledgments

This study was made possible by the Texas Department of Transportation in cooperation with FHWA. The authors thank the project manager Wade Odell and other members of the project monitoring committee for their valuable guidance and insights on this study.

Author Contributions

The authors confirm contribution to the paper as follows: study conception and design: SH and S-IL; data collection: SH, S-IL, LFW, and FZ; Analysis and interpretation of results: SH and S-IL; draft manuscript preparation: SH. All authors reviewed the results and approved the final version of the manuscript.

References

1. Newcomb, D. *Perpetual Pavements: A Synthesis*. APA 101. Asphalt Pavement Alliance, Orange Park, FL, 2002.
2. Walubita, L. F., S. Hu, and T. Scullion. *The Texas Perpetual Pavement Database – Users' Manual*. Report FHWA/TX-09/0-4822-4. Texas A&M Transportation Institute, College Station, TX, 2009.
3. Underwood, B. S., Y. R. Kim, and M. N. Guddati. Improved Calculation Method of Damage Parameter in Viscoelastic Continuum Damage Model. *International Journal of Pavement Engineering*, Vol. 11, No. 6, 2010, pp. 459–476.
4. Hu, S., F. Zhou, and T. Scullion. *Development of Texas Mechanistic-Empirical Flexible Pavement Design System (TxME)*. Report FHWA/TX-14/0-6622-2. Texas A&M Transportation Institute, College Station, TX, 2014.
5. Barret, C. R., W. D. Nix, and A. S. Tetelman. *The Principles of Engineering Materials*. Prentice Hall, Englewood Cliffs, NJ, 1973.
6. Monismith, C. L., and D. B. McLean. Structural Design Considerations. *Proceedings of the Association of Asphalt Paving Technologists*, Vol. 41, 1972, pp. 258–314.
7. Maupin, G. W. Jr., and J. R. Freeman, Jr. *Simple Procedure for Fatigue Characterization of Bituminous Concrete*. Final Report No. FHWA-RD-76-102. Federal Highway Administration, Washington, D.C., 1976.
8. Nunn, M. Long-Life Flexible Pavement. *Proc., 8th International Conference on Asphalt Pavements*, Seattle, WA, 1997.
9. Nishizawa, T., S. Shimeno, and M. Sekiguchi. Fatigue Analysis of Asphalt Pavements with Thick Asphalt Mixture Layer. *Proc., 8th International Conference on Asphalt Pavements*, Vol. 2. University of Washington, Seattle, WA, 1997, pp. 969–976.
10. Wu, Z., Z. Q. Siddique, and A. J. Gisi. Kansas Turnpike. *An Example of Long Lasting Asphalt Pavement*. International Symposium on Design and Construction of Long Lasting Asphalt Pavements, National Center for Asphalt Technology, Auburn, AL, 2004, pp. 857–876.
11. Romanoschi, S. A., A. J. Gisi, M. Portillo, and C. Dumitru. The First Findings from the Kansas Perpetual Pavements Experiment. *Transportation Research Record: Journal of the Transportation Research Board*, 2008. 2068: 41–48.
12. Yang, Y., X. Gao, W. Lin, D. H. Timm, A. L. Priest, G. A. Huber, and D. A. Andrewski. Perpetual Pavement Design in China. *Proc., International Conference on Perpetual Pavement*, The Ohio Research Institute for Transportation and the Environment, Athens, OH, 2005.
13. Prowell, B. D., E. R. Brown, R. M. Anderson, J. S. Daniel, H. Von Quintus, S. Shen, S. H. Carpenter, S. Bhattacharjee, and S. Maghsoodloo. *NCHRP Report 646: Endurance Limit of Hot Mix Asphalt Mixtures to Prevent Fatigue Cracking in Flexible Pavements*. Transportation Research Board of the National Academies, Washington, D.C., 2010.
14. Witczak, M., M. Mamlouk, M. Souliman, and W. Zeiada. *NCHRP Report 762: Laboratory Validation of an Endurance Limit for Asphalt Pavements*. Transportation Research Board of the National Academies, Washington, D.C., 2013.
15. Willis, J. R. *Field-Based Strain Thresholds for Flexible Perpetual Pavement Design*. Doctoral dissertation. Auburn University, Auburn, AL, 2008.
16. ASTM International. *Standard Test Method for Determining Fatigue Failure of Compacted Asphalt Concrete Subjected to Repeated Flexural Bending*. ASTM Designation: D 7460 – 10. ASTM International, West Conshohocken, PA, 2010.
17. Luo, X., R. Luo, and R. L. Lytton. Energy-Based Mechanistic Approach to Characterize Crack Growth of Asphalt Mixtures. *Journal of Materials in Civil Engineering*, Vol. 25, No. 9, 2013, pp. 1198–1208.
18. Luo, X., R. Luo, and R. L. Lytton. Energy-Based Mechanistic Approach for Damage Characterization of Pre-Flawed Visco-Elasto-Plastic Materials. *Mechanics of Materials*, Vol. 70, 2014, pp. 18–32.
19. Kim, Y. R., and D. N. Little. One-Dimensional Constitutive Modeling of Asphalt Concrete. *Journal of Engineering Mechanics*, Vol. 116, No. 4, 1990, pp. 751–772.
20. Kim, Y. R., M. N. Guddati, B. S. Underwood, T. Y. Yun, V. Subramanian, S. Savadatti, and S. Thirunavukkarasu. *Development of a Multiaxial VEPCD-FEP + +*. Publication FHWA-HRT-08-073. FHWA, U.S. Department of Transportation, Washington, D.C., 2008.
21. Texas Department of Transportation. *Test Procedure for Overlay Test*. TxDOT Designation Tex-248-F. https://ftp.dot.state.tx.us/pub/txdot-info/cst/TMS/200-F_series/pdfs/bit248.pdf. Accessed June, 2017.

The Standing Committee on General and Emerging Pavement Design (AFD30) peer-reviewed this paper (18-05196).

Article

TRR
JOURNAL OF THE TRANSPORTATION RESEARCH BOARD

Transportation Research Record
2018, Vol. 2672(40) 122–131

DOI: 10.1177/0361198118781657
journals.sagepub.com/home/trr
SAGE

Pavement Risk Assessment for Future Extreme Precipitation Events under Climate Change

Donghui Lu[1], Susan L. Tighe[1], and Wei-Chau Xie[2]

Abstract

Pavement infrastructure is experiencing unanticipated climate conditions caused by global warming. Extreme weather events, such as extreme precipitations, are increasing in intensity and frequency, creating rising concern in pavement vulnerability and resilience analysis. Previous design approaches based on historical climate data may no longer be adequate for addressing future conditions. To promote pavement resilience under climate change, assessing pavement risk for extreme events is essential for prioritizing vulnerable infrastructure and developing adaptation strategies. The objective of this study is to develop a quantitative evaluation methodology for assessing pavement risk from extreme precipitations under climate change. Hazard analysis, fragility modeling, and cost estimation are the three major components for risk evaluation. An ensemble of 24 global climate models is used for predicting future extreme precipitations under various climate-forcing scenarios. The Mechanistic-Empirical Pavement Design Guide is employed to simulate performance change for performing fragility modeling. Risk assessment models considering a full range of hazards were used to quantify risk of asset value loss over specified analysis periods. Results indicate that future extreme precipitation events are expected to cause an increased medium risk of asset value loss. However, high uncertainties are involved in the estimation owing to variations in predicted climates. Major pavement damages do not necessarily equate with highest risk because the probability of occurrence of major damage is relatively lower. The proposed approach provides a practical tool for analyzing the interaction among extreme precipitation levels, pavement designs, damage states, occurrence probability, and asset value at risk.

Road infrastructures are valuable assets. They are important to Canada's economy given that huge amounts of goods are transported by truck, and most people travel by car. The deterioration of roads is affected by the interaction of climate, pavement structure, and traffic. The impact of climate change on pavement infrastructure has become a rising concern, as current pavements are designed based on historical climate data, which do not adequately reflect current and future climate. During the life cycle, pavements may be subjected to certain unanticipated risk as a result of extreme weather events, such as flooding, drought, and freezing rain, and can be vulnerable when exposed to them. Specifically, as predicted by the Intergovernmental Panel on Climate Change (IPCC), it is very likely that the frequency and intensity of extreme precipitation will increase in the mid-latitude region (*1*). Accordingly, the effects of potential flooding on pavement infrastructure will have an influence on pavement planning, designing, constructing, and management.

Previous research has investigated the short-term impact of extreme hydrologic events on pavement structural damages (*2–5*). Floodwater weakens pavements and accelerates the deterioration rapidly rather than gradually, especially when a certain inundation situation occurs. Climate stressors under climate change, including temperature, precipitation, and others, are used to predict pavement performance (*6–9*). By incorporating climate change impacts into an infrastructure preservation and asset management process, the most cost-effective approach for the life cycle can be achieved (*10*). Risk analysis of the impact of climate change provides a

[1]Centre for Pavement and Transportation Technology, Department of Civil and Environmental Engineering, University of Waterloo, Waterloo, ON, Canada
[2]Department of Civil and Environmental Engineering, University of Waterloo, Waterloo, ON, Canada

Corresponding Author:
Address correspondence to Donghui Lu: donghui.lu@uwaterloo.ca

framework for identifying, assessing, and prioritizing risks, which assists the decision making in generating adaptation strategies to the changing climate (*11*). Since most research on the long-term climate change impacts focuses mainly on performance predictions instead of quantifying the risk, a comprehensive understanding of climate change risk could help promote infrastructure system resiliency.

Climate change hazards and infrastructure vulnerability are two basic elements in risk assessment. The standard approach of climate change impact assessment described in IPCC is a scenario-driven approach, where future climate is estimated under a given future climate-forcing scenario. Local climate is downscaled from global climate models (GCM) by analytical methods. The Fifth Assessment Report (AR5) of IPCC provides four climate-forcing scenarios that are representative concentration pathways (RCPs): RCP 2.6, RCP 4.5, RCP 6.0, and RCP 8.5 (*1*). This paper will incorporate various climate-forcing scenarios into the risk assessment for pavement infrastructure under future extreme precipitations events.

Vulnerability is a consequence of adverse events. It requires fragility analysis and cost estimation for infrastructure assets. Fragility analysis has caught the most attention in seismic risk assessment and has been increasingly employed in other natural hazard risk analyses recently. In this paper, the Mechanistic-Empirical Pavement Design Guide (MEPDG) is used for pavement performance prediction for fragility analysis given different climate conditions. This study specifically focuses on risk of flexible pavements under climate change, induced by extreme precipitation events. The main contributions are to provide a pavement risk quantification framework and methodology, and to present a case study on the risk assessment considering the changing climate. Ultimately, the risk assessment would offer a foundation for prioritizing risk and developing climate adaptation strategies for achieving more resilient pavement infrastructure.

Scope and Objective

The objective of this study is to develop a quantitative evaluation methodology and framework for assessing pavement infrastructure risks from extreme precipitation events in changing climate conditions. The risk concerning the occurrence of different damages and loss of pavement asset value are estimated.

Methodology

Risk is the possibility that undesired outcomes will happen. Climate change risks for infrastructure result from the interaction of climate-related hazards with the exposure and vulnerability of infrastructure. It is often represented as the probability of occurrence of hazardous events multiplied by the consequences if these events occur. Risk can be quantified by numerical integration of probability of hazard, pavement fragility, and cost:

$$\text{Risk} = \sum_{All} P\{\text{EP}\} \times P\{\text{Damage}|\text{EP}\} \times P\{\text{Cost}|(\text{Damage}|\text{EP})\} \qquad (1)$$

where $P\{\text{EP}\}$ is the occurrence of extreme precipitation hazards; $P\{\text{Damage}|\text{EP}\}$ is the pavement fragility representing the probability of exceeding certain damage when the level of extreme precipitation hazard occurs; $P\{\text{Cost}|(\text{Damage}|\text{EP})\}$ is the cost of possible damage given certain extreme precipitation hazard. It can be seen from Equation 1 that high risk can result not only from high occurrence rate but also from low occurrence rate of very severe consequences. Therefore, the assessment should cover the full range of hazards.

Hazard characteristics can be obtained from future climate prediction, and described by hazard curves, which provide the relationship between level of extreme events and annual probability of exceedance. By simulating or observing pavement performance changes with and without extreme events, fragility functions and curves can be generated indicating the probability of pavement damage given a certain hazard. The numerical integration of hazard, fragility, and cost is the risk (Figure 1).

Future Extreme Precipitation under Climate Change

Extreme precipitation depth is used for estimating the level of extreme precipitations events and their consequences, although flooding-related indicators, such as duration and flow velocity, can also illustrate the hazard characteristics. Neverthelesss, it is common to use depth to estimate damage nowadays, and it is assumed that extreme precipitation depth can reflect flood potential.

For predicting the future climate, climate forcings are input in GCM and regional climate models (RCM), and the downscaling method can be used to link GCM/RCM to local study areas under climate change (*1*). The local future climate can then be predicted. Reliable rainfall intensity estimates are critical for infrastructure planning and drainage design. Intensity-Duration-Frequency (IDF) curves are often used to describe the frequencies of extreme rainfall events. Based on future climate, the IDF can be updated (*12*). Then, future hazard curves indicating the relationship between extreme precipitation level and annual exceedance probability can be obtained from the updated IDF curves in the analysis return period

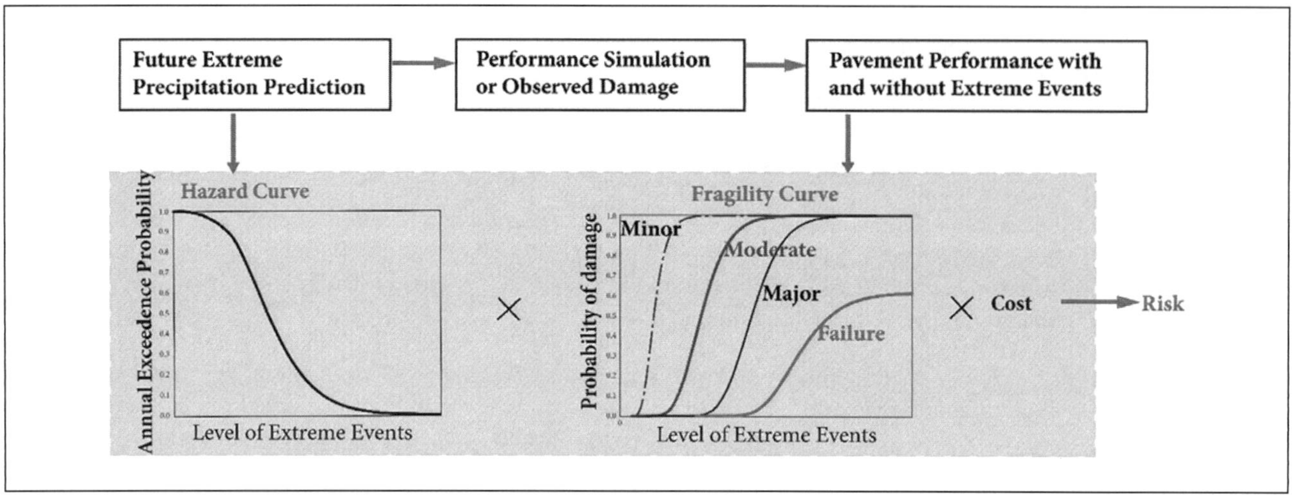

Figure 1. Pavement climate change risk quantification methodology.

Table 1. Global Climate Models Used for Predicting Future Extreme Precipitation

Modeling center	Model	Modeling center	Model	Modeling center	Model
CCCma	CanESM2	LASGCESS	FGOALS_g2	IPSL	IPSL-CM5A-MR
CCSM	CCSM4	EC-EARTH	EC-EARTH	MIROC	MIROC5
CNRM	CNRM-CM5	NOAA GFDL	GFDL-CM3	MIROC	MIROC-ESM
CSIRO	CSIRO-Mk3-6-0	NOAA GFDL	GFDL-ESM2G	MIROC	MIROC-ESM-CHEM
CESM	CESM1-CAM5	NOAA GFDL	GFDL-ESM2M	MPI-M	MPI-ESM-LR
MRI	MRI-CGCM3	MOHC	HadGEM2-AO	MPI-M	MPI-ESM-MR
BNU	BNU-ESM	MOHC	HadGEM2-ES	BCC	bcc_csm1_1 m
BCC	bcc_csm1_1	IPSL	IPSL-CM5A-LR	NOR	NorESM1-M

range. The procedures are described in the following sections.

Climate-Forcing Scenarios

The AR5 of the IPCC provides four climate-forcing scenarios. The scenarios RCP 2.6 (lower bound), RCP 4.5 (intermediate level), and RCP 8.5 (higher bound) are used for predicting future extreme precipitations events in this study. RCP 2.6 represents a pathway where radiative forcing peaks at approximately 3 W/m^2 before 2100 and then declines. RCP 4.5 refers to an intermediate stabilization pathway in which radiative forcing is stabilized at approximately 4.5 W/m^2 after 2100. RCP 8.5 is a high pathway for which radiative forcing reaches over 8.5 W/m^2 by 2100 and continues to rise for some time.

Local Climate Prediction Method

Selected Coupled Model Intercomparison Project Phase 5 (CMIP5) projection models are used as GCM models for future climate prediction. In this study, an ensemble of 24 GCM is employed for predicting future precipitations events. The models (*13*) are listed in Table 1.

The method of downscaling is for estimating high spatial resolution climate data from low-resolution GCM output. The inverse square distance weighting method is applied to achieve the spatial interpolation. The equidistance quantile matching method (*12*) is used for generating future IDF curves under climate change. The process is implemented by using IDF-CC tool (*13*). The hazard curves are then converted from IDF curves.

Future Hazard Curves and Uncertainty

Toronto Lester B. Pearson International airport was selected as the local area for this case study. The historical data are based on the precipitation data from 1950 to 2013. Future extreme precipitations (2017–2100) are estimated based on projected climate from RCP 2.6, RCP 4.5, and RCP 8.5 scenarios. Figure 2 shows the annual exceedance probability and extreme precipitation depths (24-hour) for 2, 5, 10, 25, 50, and 100-year return periods under RCP 2.6, RCP 4.5, and RCP 8.5 with historical

Table 2. Uncertainty of Extreme Precipitation for 100-Year Return Period with Upper Bound and Lower Bound

Duration (hour)	Historical (mm)	RCP 2.6 (mm)	RCP 4.5 (mm)	RCP 8.5 (mm)
1	59.81	67.91 (−22.38%, + 23.06%)	72.79 (−23.16%, + 17.34%)	79.05 (−36.66%, + 32.90%)
6	96.54	110.00 (−22.26%, + 22.52%)	118.12 (−23.73%, + 17.76%)	128.53 (−37.49%, + 33.64%)
12	107.31	121.83 (−22.37%, + 23.07%)	130.59 (−23.16%, + 17.34%)	141.83 (−36.66%, + 32.90%)
24	119.50	134.90 (−22.11%, + 22.81%)	142.77 (−22.92%, + 17.15%)	155.06 (−36.30%, + 32.58%)

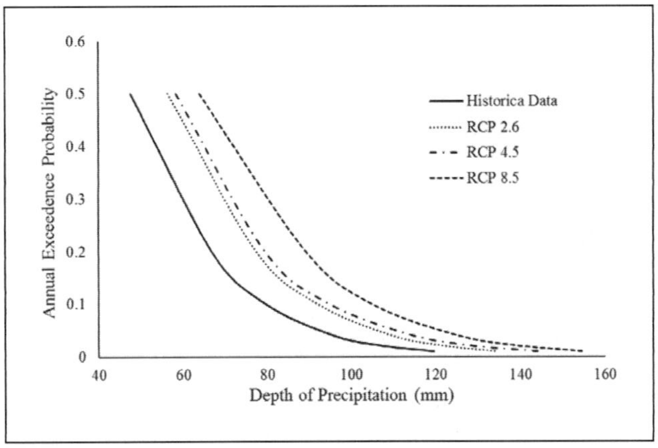

Figure 2. Hazard curves for future and historical extreme precipitation events at Toronto Lester B. Pearson Airport.

precipitation for Toronto Lester B. Pearson Airport. The y axis is annual exceedance probability which is calculated using 1 divided by the return period values. There is a variation in the predictions from the 24 GCMs; the uncertainty of future precipitation magnitude for the 100-year return period in percentage is presented in Table 2.

Pavement Fragility Analysis Method

Pavement fragility is the conditional probability of a damage given a level of extreme events, and it can be described by fragility functions and fragility curves (*14*, *15*). The methods for conducting fragility analysis are described in the following sections.

Fragility Function

This section describes a method to generate empirical fragility functions. Historical observations are generally preferred. However, if there are no adequate real-world data, pavement performance simulation from MEPDG can also be used to estimate the fragility functions.

The common assumption is that the damages of a structure given extreme precipitation level are lognormally distributed. Hence, the fragility functions can be estimated using lognormal cumulative distribution functions:

$$P_{\text{Fragility}} = P(\text{Damage}|\text{EP} = x) = \Phi\left(\frac{\ln x - \alpha}{\beta}\right) \quad (2)$$

where

$P(\text{Damage} \mid \text{EP} = x)$ is the probability that a flood level with EP = x will result in a certain damage of the pavement structure;

EP is extreme precipitation level;

$\Phi()$ is the cumulative distribution function of standard normal distribution;

α and β are logarithmic mean and logarithmic standard deviation of EP.

Estimating α and β from pavement structure damage analysis is required for calibrating Equation 2. There are several statistical approaches for estimating parameters in the fragility function. Maximum likelihood method has been shown to be a reasonable method in a variety of cases (*16*, *17*). Assuming that observation of damage or no damage from each flood level is independent of the observations from other flood levels, the probability of observing k_i collapses out of n_i with flood level $EP = x_i$ is given by the binomial distribution:

$$P_{k_i} = C_{n_i}^{k_i} p_i^{k_i} (1 - p_i)^{n_i - k_i} \quad (3)$$

where

p_i is the probability that $EP = x_i$ causes certain pavement damage;

$C_{n_i}^{k_i}$ is the binomial coefficient.

The maximum likelihood approach identifies the parameters for fragility functions, which provides the highest probability of having the observed pavement damage data. Hence, the likelihood function is the product of the binomial probabilities at each flood level for the entire data set:

$$L = \prod_{i=1}^{m} C_{n_i}^{k_i} p_i^{k_i} (1 - p_i)^{n_i - k_i} \quad (4)$$

where

L is the likelihood function;

m is the number of flood levels.

Table 3. Damage States for Pavement

Damage state	Damage level	Description
PDS0	Insignificant	Insignificant pavement performance change
PDS1	Minor	Very slight pavement performance change
PDS2	Moderate	Medium pavement performance change
PDS3	Major	Severe pavement performance change, and may not be accessible
PDS4	Collapse	Totally damaged and needs to be replaced by new pavement

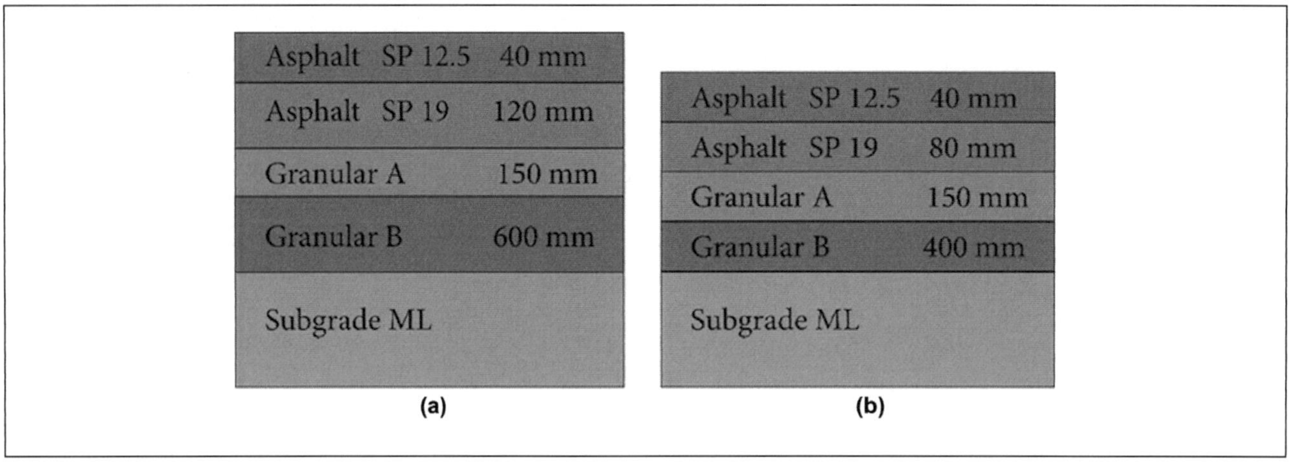

Figure 3. Pavement cross-sections: (*a*) arterial pavement and (*b*) collector pavement.

Substituting Equation 2 for p_i, the parameters of the fragility function are explicit in the likelihood function. In order to maximize the likelihood, it is easier to maximize the logarithm of the function:

$$L = \prod_{i=1}^{m} C_{n_i}^{k_i} \left\{ \Phi\left(\frac{\ln x_i - \alpha}{\beta} \right) \right\}^{k_i} \left\{ 1 - \Phi\left(\frac{\ln x_i - \alpha}{\beta} \right) \right\}^{n_i - k_i} \tag{5}$$

Different types of damage can be defined by damage states (Table 3). The damage state descriptions can be related to the pavement structure damage or performance change. HAZUS multi-hazard methodology (*18*) is a tool for estimating road vulnerability and fragility for earthquakes. The damage states in HAZUS are defined in terms of ground settlement; however, in this research, they are defined according to pavement performance change with and without extreme events. The serviceability is emphasized when analyzing pavement fragility.

The fragility curves for each damage state can be generated based on pavement damage data for each level. The threshold value of each damage should be defined from the perspective of serviceability, functional capability, economical losses, or other methods, such as expert opinions for identifying severity of damage.

Pavement Damage Simulation

This research utilizes MEPDG and performance prediction methods to investigate pavement performance change caused by extreme precipitation events. The climate module of MEPDG provides the state-of-the-art tool to estimate the impact of variation of precipitation on pavement life cycle performance change. AASHTOWare Pavement ME Design version 2.3 is used to conduct the MEPDG analysis.

Typical design of pavement cross-sections (*19*) are used for fragility analysis (Figure 3). Regarding the critical traffic information, the two-way average annual daily truck traffic is 2800 for cross-section (a), and 400 for (b); the truck traffic in the design lane is 80% and 100% for sections (a) and (b), respectively. In Ontario, SP 12.5 is an asphalt layer containing PG 64-28 with unit weight of 2,460 kg/m^3, and SP 19 is an asphalt layer containing PG 58-28 with unit weight of 2,460 kg/m^3. Granular A and Granular B indicate quarried bedrock or boulders, cobbles, gravel, sand, and fines from naturally formed deposits, while ML indicates silts, sandy silts, or diatomaceous soils.

As mentioned earlier, the climate data are from Toronto Lester B. Pearson Airport. The climate file is prepared by changing the precipitation data. This study

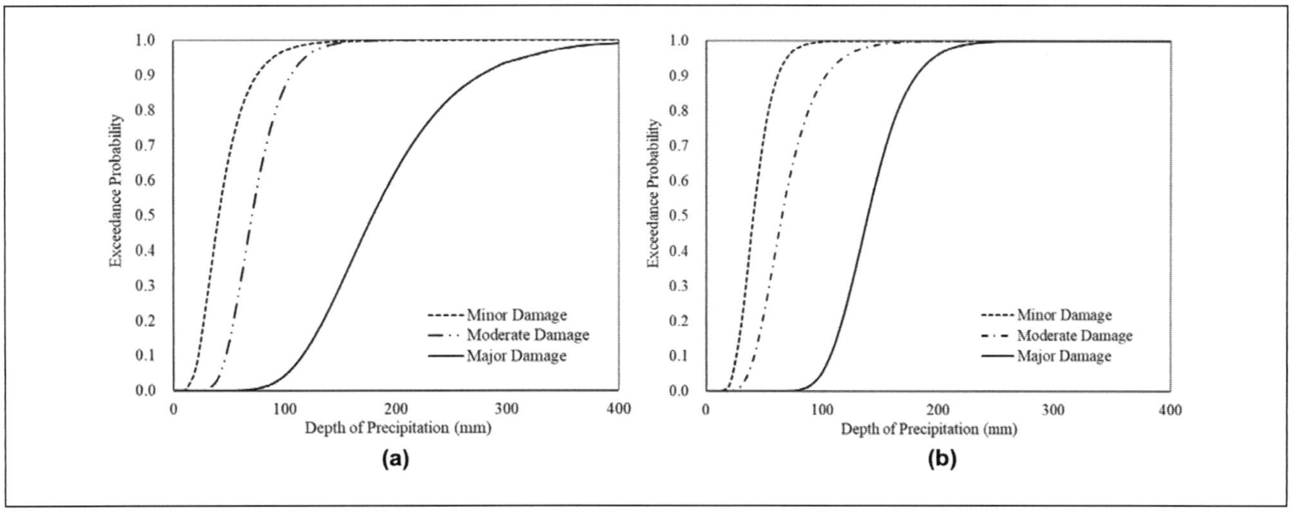

Figure 4. Fragility curves: (*a*) arterial pavement and (*b*) collector pavement.

Table 4. Damage Ratio Results from Pavement Performance Simulation

Precipitation scenarios (mm)	Arterial pavement damage	Collector pavement damage
20	0.80%	0.47%
40	0.80%	0.93%
60	1.20%	1.40%
80	1.60%	1.86%
100	2.00%	1.86%
120	2.00%	1.86%
150	2.40%	2.79%

assumes that the 7-day defined precipitation events would occur each year in May based on the situation in Ontario. The input 24-hour precipitation values are 0, 20, 40, 60, 80, 100, 120, 150 mm. The performance of a new asphalt pavement with 20-year design life is simulated, with the initial international roughness index (IRI) value of 0.8 m/km. Reliability values of 80% and 90% for collector and arterial pavement, respectively, are also assigned to account for uncertainties in predicting pavement deterioration. Pavement damages caused by extreme precipitation are described by damage ratio. The equation for calculating damage ratio is:

$$\kappa = \frac{(\mathrm{IRI}_t - \mathrm{IRI}_i)}{\mathrm{IRI}_i} \tag{6}$$

where κ is damage ratio, IRI_t is terminal international roughness value with extreme precipitation, and IRI_i is terminal international roughness value without extreme precipitation.

The damage ratios are calculated based on the performance simulations (Table 4). The results for arterial and collector pavement are different. As precipitation depth increases, collector pavements experience a larger jump in damage compared with arterial pavements. These results are used for fragility modeling. In the future, observed data should be collected to generate more realistic fragility functions.

Generation of Pavement Fragility Curves

The damage ratios are taken as the mean values for Monte Carlo simulation of lognormal distribution. The standard deviation is assumed to be 0.2% for each level. Monte Carlo simulation is used to generate 1000 events following the lognormal distribution. Using Equations 2 to 5, the fragility curves are generated (Figure 4). In this case, the damage state thresholds for minor, moderate, and major damage are assumed to be exceeding a damage ratio of 1%, 1.5%, and 2.5%, respectively for both arterial and collector pavements. Alternatively, for pavements with different functionality, the damage states can be defined differently, which could be justified by expert opinion.

It can be seen from the fragility curves that the probability of pavement damage increases with increases of extreme precipitation level, and it is easier for minor damage to occur than major damage for each pavement type. In addition, every structure has its fragility function, and different pavement designs follow different trends of fragility curves. In the future, more accurate fragility models could be calibrated for different pavement structures using observed performance change, and more meaningful damage state threshold values should be proposed.

Pavement Risk Assessment

High risk can result not only from high occurrence but also from low probability with very severe consequences. As a result, risk assessment covering the full range (2, 5, 10, 25, 50, 100-year) of hazards should be evaluated. Annual exceedance probability of certain pavement damage is calculated by Equation 7,

$$\gamma = -\int_0^{\infty} P_{Fragility}(\mathrm{DP})\frac{dH(\mathrm{DP})}{d(\mathrm{DP})}d(\mathrm{DP}) \tag{7}$$

where

γ is the annual exceedance probability of certain pavement damage;

DP is depth of precipitation;

H is extreme precipitation hazard.

A numerical method is usually applied to quantify the integral in Equation 7,

$$\gamma = -\sum_{i=1}^{N} P_{\mathrm{Fragility}}\left(\mathrm{DP}^{(i)}\right)\Delta H\left(\mathrm{DP}^{(i)}\right) \tag{8}$$

where N is the number of intervals for DP.

Value at risk is defined as the potential loss of asset value as a result of certain damage from extreme events in analysis periods. The pavement asset value estimation is based on the written down replacement cost asset valuation method wherein the asset value is adjusted with respect to the change of pavement conditions. Providing the straight-line method is the easiest depreciation method; depreciation expense is recognized evenly over the analysis period. Then, the asset value for each year in the analysis period can be calculated using Equation 9,

$$V_n = ICC - n(\mathrm{ICC} - \mathrm{Sal})\zeta \tag{9}$$

where

V_n is the asset value at year n;

ICC is the initial construction cost;

Sal is the salvage value at the end of the 20-year design life;

n is year number;

ζ is the depreciation rate.

Equations 10 to 13 are used to calculate the risk of asset value loss in the analysis period:

$$R_{\mathrm{total}} = \sum_{j=1}^{\infty}\sum_{m}^{q} \gamma_j{}^m R_j{}^m \tag{10}$$

$$R_j{}^m = \sum_{n=1}^{20} N_n^m V_n \kappa \tag{11}$$

$$R_j{}^m = \sum_{n=1}^{20} N_n^m V_n\, \kappa \tag{12}$$

$$\gamma_j{}^m = \prod_{n=1}^{20} \gamma^{N_n^m}\left(\frac{1}{1+\gamma}\right)^x \tag{13}$$

where

R_{total} is the total risk of asset value loss for any number of occurrences of extreme events in specified analysis periods;

$R_j{}^m$ is the risk of asset value loss for the mth combination for j times of occurrences of extreme events in the analysis period;

q is the total number of combinations to satisfy Equation 12;

$\gamma_j{}^m$ is the probability of certain pavement damage for the mth combination for j times of occurrence of extreme events in the analysis period;

N_n^m is the number of occurrences of extreme events in the nth year;

κ is the damage ratio from Equation 6;

x is the number of $N_n^m = 0$.

Results of the Case Study

The risks of occurrence indicated by medium annual exceedance probability of arterial and collector pavements for different pavement damage states under various climate change scenarios were estimated (Table 5). The upper bound and lower bound values are also presented to show the uncertainties. Compared with the historical condition, climate change may cause an increased occurrence of various damages from the median estimation. It can also be seen from the median estimation that with increased climate forcing, the values of annual exceedance probability for all damage states and pavement designs are increasing. However, high uncertainty is evident given the upper and lower bound. In the estimation of probability of occurrence, uncertainty exists as a result of the variation of future climate predictions.

For value at risk, typical initial construction costs per kilometer for the arterial and collector pavement are CAD \$1,065,744 and CAD \$430,236, respectively, and the salvage values at the end of the 20th year are CAD \$646,135.6 and CAD \$260,721.2, respectively (*19*). The depreciation rate ζ is 0.05. It is noted that using Equations 10 to 13, the risk for any times occurrence of extreme events can be calculated. For this case study, the risks of asset value loss are calculated based on the assumption that there is a maximum one-time extreme event per year, for simplifying the comparison of the results.

The medium value at risk per kilometer for arterial and collector pavements with upper and lower bounds

Table 5. Annual Exceedance Probability for Different Pavement Damage States under Various Climate Change Scenarios with Upper Bound and Lower Bound

	Arterial pavement		
Scenarios	Major	Moderate	Minor
Historical data	0.0036	0.1978	0.3989
RCP 2.6	0.0086 (−0.0066, + 0.0144)	0.2757 (−0.1057, + 0.0889)	0.4331 (−0.0484, + 0.0273)
RCP 4.5	0.0109 (−0.0082, + 0.0166)	0.2917 (−0.0924, + 0.0852)	0.4385 (−0.0372, + 0.0248)
RCP 8.5	0.0184 (−0.0167, + 0.0431)	0.3432 (−0.1492, + 0.0895)	0.4543 (−0.0517, + 0.0222)
	Collector pavement		
Scenarios	Major	Moderate	Minor
Historical data	0.0041	0.2481	0.4403
RCP 2.6	0.0125 (−0.0108, + 0.0278)	0.3182 (−0.0960, + 0.0724)	0.4682 (−0.0411, + 0.0152)
RCP 4.5	0.0173 (−0.0149, + 0.0320)	0.3317 (−0.0810, + 0.0684)	0.4718 (−0.0285, + 0.0125)
RCP 8.5	0.0313 (−0.0304, + 0.0821)	0.3735 (−0.1254, + 0.0696)	0.4804 (−0.0337, + 0.0078)

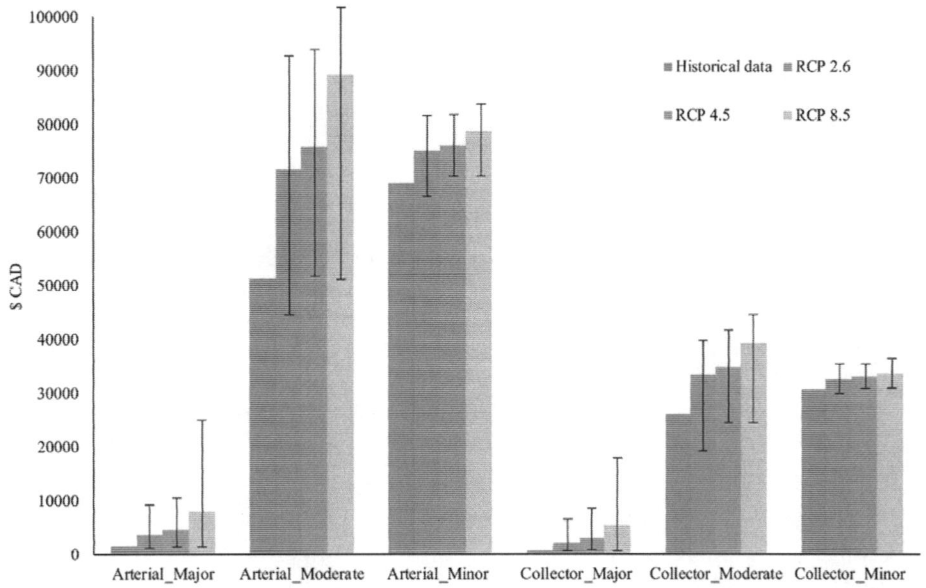

Figure 5. Value at risk per kilometer for arterial and collector pavements. Range of risk varies by damage state and RCP scenario considered. The bars represent the upper bound and lower bound risk.

are presented in Figure 5. For all climate scenarios, the medium risk for arterial pavements is higher than for collector pavements. As the climate forcing increases, the medium risk increases as well. Decision makers should not neglect the consequences of minor hazards. Although minor damage is relatively small, the high occurrence of minor damage leads to a high potential of asset value loss. On the other hand, it is noticeable that the occurrence of major pavement damage does not necessarily mean the highest risk in terms of asset value loss because the probability of occurrence of major damage is relatively lower. As Figure 5 shows, there is no large gap between moderate and minor damage risk within each pavement type. This indicates that the difference between moderate and minor damage is not significant regarding asset values loss potential using the damage state threshold values in this case study. However, various threshold values can lead to different risks of asset value loss. Comparing with the minor damage risk, the variation of asset value at risk is higher for major and moderate damage compared with minor damage. This result indicates that the uncertainty of the value at risk is lower when estimating the minor damage. It is estimated that the risk of increased intensity of extreme precipitation will add approximately as high as CAD $112,471.28, and CAD $46,487.81 per kilometer for moderate damage of arterial and collector pavements, respectively, with respect to the predicted worst climate scenario over the 2017–2100 period.

This risk assessment methodology reveals the impact of a full range of hazards on pavement asset value at risk over the analysis period given different damage states. In the future, more meaningful threshold values should be determined when more observations of damage data can be collected. Nevertheless, the proposed risk assessment framework and methods can quantify the risk of asset value loss, and indicate the interaction between value at risk, probability of occurrence, damage states, and climate change forcing (extreme precipitation) for different types of pavements.

Summary and Conclusions

This paper presents a method to assess pavement infrastructure risk under various climate change scenarios considering extreme precipitation events. This risk assessment method that takes the full range of hazards into account is able to quantify risk of asset value loss over analysis periods. The key contribution of this work is to use the integration of future extreme precipitation hazards, infrastructure fragility, and cost to quantify the pavement value at risk. An ensemble of 24 GCM is utilized for predicting future extreme precipitations (2017–2100) under climate-forcing scenarios of RCP 2.6, RCP 4.5, and RCP 8.5. In order to simulate pavement performance changes for fragility modeling, the MEPDG method is employed. The probability of occurrence and asset value risk per kilometer are discussed for different climate-forcing scenarios, damage states, and functionality of pavements. In summary, the following conclusions can be drawn from this study:

1. The proposed risk quantification method provides a practical way of assessing the implication of various climate change scenarios on the risk of asset value loss for different pavement structures with a variety of damage states for extreme precipitation events. Besides, the framework and methods allow decision makers get a better understanding of risk assessment procedures.
2. The risk assessment framework can be used to analyze the interactions among extreme precipitation levels, pavement structures designs, damage states, damage occurrence probability, and asset value at risk.
3. Fragility analysis shows that the fragility of different pavement structures is different. The fragility modeling method is able to generate fragility functions and curves, which clearly show the trend for the probability of exceeding certain pavement damage given an extreme precipitation level.
4. Under the changing climate conditions, future extreme precipitation events are expected to result in an increased medium risk of pavement asset value losses. However, there are high uncertainties involved in the risk estimation as a result of the variations of predicted future climates.
5. In the case study, considering the climate from 2017 to 2100, the extreme precipitations from RCP 8.5 climate scenario will add about as high as CAD $112,471.28 and CAD $46,487.81 per kilometer for arterial and collector pavements, respectively. These risks of asset value loss are approximately 10% of the initial construction cost for both types of pavements.
6. The risk of major damage is not the highest when compared with the risks for minor and moderate damage, because of the lower occurrence of major damage. From this aspect, it can be seen that the proposed risk assessment methodology is able to provide a comprehensive risk indicator to address the combination of hazard magnitude, infrastructure fragility and cost.

The results of the work are based on many assumptions. Additional work is required; the work presented in this study points out the following recommendations:

1. Observations of the pavement performance change should be collected to calibrate fragility functions in the future. In addition, more realistic threshold values of damage states should be determined by investigating real pavement performance changes and expert judgment.
2. Large uncertainty exists in the climate projection models and fragility models. The fragility functions in this study are actually the estimated mean exceedance probability. Risk quantification models should be improved by considering the uncertainty in fragility models.
3. Value at risk is determined based on the potential asset value losses in this study; risk from the aspect of life cycle costs, including maintenance, preservation, and rehabilitation costs, should also be evaluated.

Acknowledgments

The research for this paper was supported by the Natural Sciences and Engineering Research Council of Canada (NSERC), Norman W. McLeod Chair Partners, and Centre for Pavement and Transportation Technology (CPATT) Partners.

Author Contributions

The authors confirm contribution to the paper as follows: study conception and design: Donghui Lu, Susan Tighe, Wei-Chau Xie; data collection: Donghui Lu; analysis and interpretation of results: Donghui Lu, Susan Tighe, Wei-Chau Xie; draft manuscript preparation: Donghui Lu. All authors reviewed the results and approved the final version of the manuscript.

References

1. Pachauri, R. K., M. R. Allen, V. R. Barros, J. Broome, W. Cramer, R. Christ, and N. K. Dubash. *Climate Change 2014 Synthesis Report*. Contribution of Working Groups I, II and III to the Fifth Assessment Report of the Intergovernmental Panel on Climate Change. Cambridge University Press, Cambridge, 2014.
2. Zhang, Z., Z. Wu, M. Martinez, and K. Gaspard. Pavement Structures Damage Caused by Hurricane Katrina Flooding. *Journal of Geotechnical and Geoenvironmental Engineering*, Vol. 134, No. 5, 2008, pp. 633–643. https://dx.doi.org/10.1061/(ASCE)1090-0241(2008)134:5(633)
3. Helali, K., M. Robson, R. Nicholson, and W. Bekheet. Importance of a Pavement Management System in Assessing Pavement Damage from Natural Disasters: A Case Study to Assess the Damage from Hurricanes Katrina and Rita in Jefferson Parish, Louisiana. Presented at 7th International Conference on Managing Pavement Assets, Calgary, AB, 2008.
4. Sultana, M., G. Chai, T. Martin, and S. Chowdhury. A Study on the Flood Affected Flexible Pavements in Australia. Presented at 9th International Conference on Road and Airfield Pavement Technology, Dalian, China, 2015.
5. Tari, Y. H., S. S. Shamsabadi, R. Birken, and M. Wang. Deterioration Modeling for Condition Assessment of Flexible Pavements Considering Extreme Weather Events. Presented at SPIE 9437, Structural Health Monitoring and Inspection of Advanced Materials, Aerospace, and Civil Infrastructure, San Diego, CA, 2015.
6. Mallick, B. R., J. M. Radzicki, S. J. Daniel, and J. Jacobs. Use of System Dynamics to Understand the Long Term Impact of Climate Change on Pavement Performance and Maintenance Cost. *Transportation Research Record: Journal of the Transportation Research Board*, 2014. 2455: 1–9.
7. Mills, B. N., S. L. Tighe, J. Andrey, J. T. Smith, S. Parm, and K. Huen. *The Road Well-Travelled: Implications of Climate Change for Pavement Infrastructure in Southern Canada*. Final Technical Report. Environment Canada, 2007.
8. Mills, B. N., S. L. Tighe, J. Andrey, J. T. Smith, and K. Huen. Climate Change Implications for Flexible Pavement Design and Performance in Southern Canada. *Journal of Transportation Engineering*, Vol. 135, No. 10, 2009, pp. 773–782. https://dx.doi.org/10.1061/(ASCE)0733-947X(2009)135:10(773)
9. Meagher, W., J. S. Daniel, J. Jacobs, and E. Linder. Method for Evaluating Implications of Climate Change for Design and Performance of Flexible Pavements. *Transportation Research Record: Journal of the Transportation Research Board*, 2012. 2305: 111–120.
10. Meyer, M. D., A. Amekudzi, and J. P. O'Har. Transportation Asset Management Systems and Climate Change: Adaptive Systems Management Approach. *Transportation Research Record: Journal of the Transportation Research Board*, 2010. 2160: 12–20.
11. Noble, M., D. Bruce, and J. Egener. *Overview of the Risk Management Approach to Adaptation to Climate Change in Canada*. Natural Resources Canada, Ottawa, ON, Canada, 2005.
12. Srivastav, R. K., A. Schardong, and S. P. Simonovic. Equidistance Quantile Matching Method for Updating IDF Curves Under Climate Change. *Water Resources Management*, Vol. 28, No. 9, 2014, pp. 2539–2562. https://dx.doi.org/10.1007/s11269-014-0626-y
13. Simonovic, S. P., A. Schardong, D. Sandink, and R. Srivastav. A Web-Based Tool for the Development of Intensity Duration Frequency Curves Under Changing Climate. *Environmental Modelling and Software*, Vol. 81, 2016, pp. 136–153. https://dx.doi.org/10.1016/j.envsoft.2016.03.016
14. Lu, D., S. L. Tighe, and W. C. Xie. Pavement Fragility Modeling Framework and Build-In Resilience Strategies for Flood Hazard. Presented at 96th Annual Meeting of the Transportation Research Board, Washington, D.C., 2017.
15. Deshpande, V. P., I. D. Damnjanovic, and P. Gardoni. Modeling Pavement Fragility. *Journal of Transportation Engineering*, Vol. 136, No. 6, 2010, pp. 592–596. https://dx.doi.org/10.1061/(ASCE)TE.1943-5436.0000111
16. Straub, D., and A. Der Kiureghian. Improved Seismic Fragility Modeling from Empirical Data. *Structural Safety*, Vol. 30, No. 4, 2008, pp. 320–336. https://dx.doi.org/10.1016/j.strusafe.2007.05.004
17. Shinozuka, M., M. Q. Feng, J. Lee, and T. Naganuma. Statistical Analysis of Fragility Curves. *Journal of Engineering Mechanics*, Vol. 126, No. 12, 2000, pp. 1224–1231. https://dx.doi.org/10.1061/(ASCE)0733-9399(2000)126:12(1224)
18. Kircher, C. A., R. V. Whitman, and W. T. Holmes. HAZUS Earthquake Loss Estimation Methods. *Natural Hazards Review*, Vol. 7, No. 2, 2006, pp. 45–59. https://dx.doi.org/10.1061/(ASCE)1527-6988(2006)7:2(45)
19. Applied Research Associate (personal communication). Methodology for the Development of Equivalent Pavement Structural Design Matrix for Municipal Roadways: Including Maintenance and Rehabilitation Schedules and Life Cycle Cost Analysis. Toronto, Canada, 2015.

The Standing Committee on General and Emerging Pavement Design (AFD30) peer-reviewed this paper (18-00643).

Article

TRR
JOURNAL OF THE TRANSPORTATION RESEARCH BOARD

Transportation Research Record
2018, Vol. 2672(40) 132–142

DOI: 10.1177/0361198118781682
journals.sagepub.com/home/trr
SAGE

Assessment of Equivalent Thickness Design Principles for Geosynthetic Reinforced Pavements by Way of Accelerated Testing

W. Jeremy Robinson[1], Jeb. S. Tingle[2], Gregory J. Norwood[2], and Isaac L. Howard[3]

Abstract
The Engineer Research and Development Center (ERDC) of the U.S. Army Corps of Engineers has performed multiple laboratory and full-scale evaluations of geosynthetic reinforced pavements. One result from early geosynthetic reinforced pavement evaluations was a pavement design methodology implemented in ETL 1110-1-189: *Use of Geogrids in Pavement Construction*. Since that time, the evaluations have been primarily focused on comparing performance between varying types of geosynthetic products. While the studies have independently compared the discrete performance of single geosynthetic reinforced sections to unreinforced sections, a comprehensive analysis of available data has not been performed to validate or refine the implemented design methodology. The objective of this effort was to assemble available data from laboratory and full-scale testing conducted at ERDC for the primary purpose of assessing the flexible pavement design methodology presented in ETL 1110-1-189. Simplifying assumptions were made to allow comparison of varying loading and pavement structure conditions. This assessment found that the combined dataset supports the original design curve produced with the equivalent thickness methodology described in ETL 1110-1-189. The updated dataset would reduce the equivalent reinforced thickness by approximately 1.0-inches (25.4 mm) at unreinforced thicknesses less than 14 inches (356 mm), providing a slightly more conservative result. The adjusted data converged with the original equivalency chart at an unreinforced thickness of approximately 16 inches (406 mm).

The Engineer Research and Development Center (ERDC) of the U.S. Army Corps of Engineers (USACE) has performed multiple laboratory and full-scale evaluations of geosynthetic reinforced pavements. One result from early geosynthetic reinforced pavement evaluations was a pavement design methodology based on two subgrade California bearing ratios (CBR) values. This method is implemented in ETL 1110-1-189: *Use of Geogrids in Pavement Construction* (*1*) in the form of an equivalent pavement thickness chart (Figure 1) that is considered valid for subgrade CBR values ranging from 0.5 to 8.0. Simply, an unreinforced flexible pavement consisting of an asphalt (AC) and base layer (denoted on the y-axis in Figure 1) is designed for the given subgrade conditions. A reinforced pavement thickness is determined by entering the chart with the unreinforced pavement thickness, intersecting the equivalency curve, and drawing a vertical line to the equivalent reinforced pavement thickness (denoted on the x-axis in Figure 1). The reinforced aggregate thickness is determined by subtracting the unreinforced pavement thickness and equivalent reinforced pavement thickness, while maintaining the same asphalt layer thickness, thereby reducing the aggregate layer thickness only. After development of the equivalent pavement thickness chart, evaluations have been primarily focused on comparing performance between varying types and manufacturer's geosynthetics. While the studies have independently compared the performance of discrete geosynthetic reinforced sections to unreinforced sections, a comprehensive analysis of available data has not been performed for one purpose, in particular to evaluate the current state of practice within USACE that was developed with considerably less data than is currently available.

[1]Research Civil Engineer & Doctoral Student, U.S. Army Engineer Research and Development Center & Mississippi State University, Vicksburg, MS
[2]Research Civil Engineer, U.S. Army Engineer Research and Development Center, Vicksburg, MS
[3]Materials and Construction Industries Chair, Civil and Environmental Engineering Department, Mississippi State University, Mississippi State, MS

Corresponding Author:
Address correspondence to W. Jeremy Robinson: Jeremy.Robinson@usace.army.mil

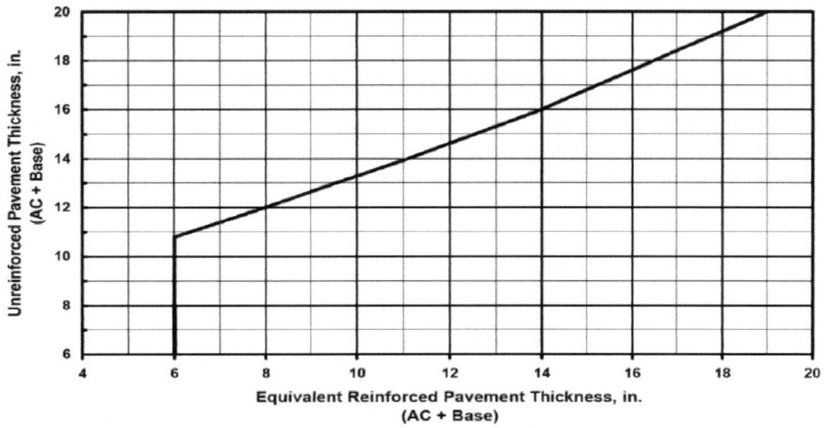

Figure 1. Equivalent pavement thickness chart from ETL 1110-1-189 (*1*).

Objective and Scope

The objectives of this effort were to assemble available data from laboratory and full-scale testing conducted at ERDC, document the various material parameters that have been investigated, and draw appropriate conclusions from the combined dataset. The data utilized here span a time frame beginning in the early 1990s when USACE began investigating inclusion of geosynthetics into pavement structures and incorporates data collected as recently as 2016. The additional data collected since implementation of ETL 1110-1-189 are useful to assess the current USACE design methodology or determine if adjustments to the design methodology are required. A total of nine different studies were assessed consisting of a total of 31 different test sections.

Pavement and Material Properties for Lab- and Full-Scale Testing

Inclusion of geosynthetics in pavement structures by ERDC has been investigated from the early 1960s to as recently as 2016. However, this summary only includes ERDC geosynthetic experiments conducted since the 1990s. As new geosynthetic products became available, laboratory-scale evaluations used a 6-ft (1.8-m) by 6-ft (1.8-m) box with a cyclic plate load setup as shown in Figure 2*a* and *b*. Full-scale test sections were trafficked using a single-wheel load cart as shown in Figure 2*c* and *d* and/or a heavy vehicle simulator (HVS) as shown in Figure 2*e* and *f*.

In summarizing the materials used in past experiments, it is difficult to provide comprehensive descriptions of each material used. However, this section provides a general summary of material properties while detailed descriptions can be found in references (*2–8*). A typical cross-section showing the various layers and construction quality control (QC) parameters is presented in Figure 3.

For the experiments included in this paper, the design subgrade consisted of a locally available high-plasticity clay having liquid limits ranging from 65 to 90, plastic limits ranging from 22 to 29, and plasticity indices ranging from 38 to 62. According to the ASTM D2487 Unified Soil Classification System, the soil was classified as a high-plasticity clay (CH) and an A-7-6 according to the AASHTO M 145 classification system. The percent fines (P200) ranged from 95% to 98%. The review of modified proctor data for the CH subgrade soil showed that the maximum dry density ranged from 92 to 104 lbs/ft^3 (1,474–1,666 kg/m^3) at optimum moisture contents ranging from 19% to 23%. Design CBR of the CH subgrade for the different experiments ranged from 1% to 8% as determined by ASTM D4429.

A subbase material was used in only one study (Norwood, 2016, personal communication), and consisted of a locally available granular material comprised of approximately 98% sand. The subbase material was classified as a poorly-graded sand according to D2487 and an A-1-b according to M 145. Additionally, the material selected was found to meet the subbase requirements of the FAA P-154 specification.

Multiple base course materials have been investigated consisting of crushed limestone, crushed chert gravel, and rounded clay gravel. Modified proctor maximum dry densities for the base materials ranged from 149 lbs/ft^3 (2,387 kg/m^3) for the crushed limestone base to 137 lbs/ft^3 (2,195 kg/m^3) for the rounded clay gravel base. Optimum moisture contents ranged from 3.8% to 5.9%.

Pavement surfacing has consisted of no surfacing (aggregate-surfaced roadways), rubber mat surfacing (simulated flexible pavement surfaces), double bituminous surface treatments, and hot-mix asphalt (HMA) surfacing. HMA surfacing thickness ranged from 2 to 5 inches (50–127 mm) and were typical of local 3/8-inch (9.5-mm) nominal maximum aggregate size (NMAS) mixtures used by the Mississippi Department of Transportation (MDOT). Voids in mineral aggregate (VMA) ranged from 13% to 15% and asphalt

Figure 2. Photographs of accelerated testing equipment: (*a*) cyclic plate load laboratory equipment; (*b*) close-up of cyclic plate test; (*c*) single-wheel load cart; (*d*) inside single-wheel load cart; (e) HVS; (*f*) dual-tandem truck gear on HVS.

binder content (P_b) ranged from 5.2% to 5.7%. Recycled asphalt pavement (RAP) content was up to 20% by weight of the aggregate blend for cases where data were available.

Description of Previous Research Efforts

The first research effort included in this summary was published in 1993 (*2*). The intent of the study was to evaluate flexible pavements for light aircraft. The study consisted of 12 individual test items at subgrade CBR values of 3 and 8. The pavement structure consisted of 6-in. (152-mm); 10-in. (254-mm); 12-in. (305-mm); 14-in. (356-mm); and 18-in. (457-mm) crushed limestone base layers. All sections were surfaced with a 2-in. (50-mm) thick HMA layer. The geosynthetics were punched and drawn as well as woven biaxial geogrids located at the base/subgrade interface in all sections with one product located in the middle of a 14-in. (356-mm) base section. Control (unreinforced) items were constructed at base thicknesses of 6 in. (152 mm) and 10 in. (254-mm)

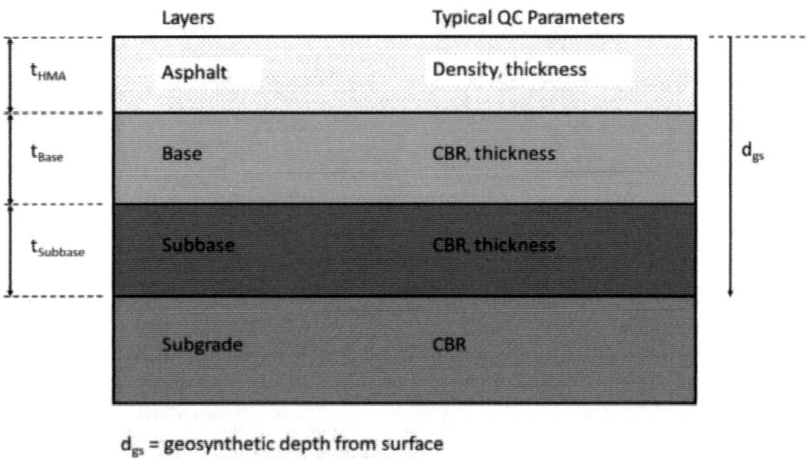

Figure 3. Typical cross-section.

over an 8 CBR CH subgrade and 12-in. (305-mm), 14-in. (356-mm), and 18-in. (457-mm) base thicknesses over a 3 CBR CH subgrade. Each test item was trafficked with a single-wheel C-130 load cart at a total load of 30,000 lbs (133.5 kN) and a tire contact pressure of 68 psi (468 kPa).

The study concluded that the inclusion of geogrid reinforcement in base courses for flexible pavements could provide structural improvement. It was noted that geogrid performance is a function of depth of placement and it was recommended that the minimum placement depth should be 6 in. (152 mm) (as measured from the pavement surface). Importantly, a thickness reduction chart was developed presenting a relationship between unreinforced total pavement thickness and equivalent reinforced total pavement thickness. This relationship is presented in USACE ETL-1110-1-189 (*1*) as the means of determining pavement thickness for reinforced flexible pavement sections surfaced with asphalt concrete. It is noted that the procedure is valid for subgrade CBR values equal to or less than 8. The ETL recommends that for subgrade strengths greater than 8 CBR a full-scale test section be constructed to determine the effect of geogrid on base course reduction as the benefit of geogrid inclusion was expected to diminish with increasing subgrade strength.

Tingle and Jersey (*3*) performed laboratory-scale testing of geosynthetic reinforced aggregate-surfaced sections. A total of six sections were evaluated consisting of five sections with 14 in. (356 mm) of crushed limestone and one section with 20 in. (508 mm) of crushed limestone. One of the five 14-in. (356-mm) aggregate sections was unreinforced and the 20-in. (508-mm) aggregate section was unreinforced. Geosynthetic reinforcement consisted of a punched and drawn biaxial geogrid and a needle-punched nonwoven polypropylene geotextile. A design subgrade CBR of 1 was targeted for all sections. Cyclic plate load testing was accomplished through a 12-in. (305-mm) diameter steel plate with a total load of 9,000 lbs (40 kN) and an 80 psi (551 kPa) contact pressure.

It was concluded that all reinforced pavement sections showed an improvement in rutting performance when compared with the unreinforced pavement sections as determined by traffic benefit ratios (TBR) (ratio of load cycles of a reinforced pavement structure to a defined failure state to load cycles of the same unreinforced pavement structure at the same failure state) in excess of 1.0. It was hypothesized that at very low subgrade strengths, the separation function provided by geotextiles was the primary means of reinforcement and base reinforcement was a secondary function. It was further (*3*) hypothesized that there may be a maximum depth of placement based on a Boussinesq analysis indicating that the horizontal stress approaches zero at a depth of 16 in. (406 mm) below the pavement surface.

Tingle and Jersey (*4*) evaluated the performance of eight full-scale aggregate road sections that included three different aggregate materials (crushed limestone, crushed chert gravel, and rounded clay gravel). Three of the eight 6-in. (152-mm) aggregate road sections were unreinforced (control) sections for each aggregate type. Geosynthetics included in the study consisted of a punched and drawn biaxial geogrid and a needle-punched nonwoven polypropylene geotextile. All sections were constructed over a 4 CBR clay subgrade. The test sections were trafficked with a dual-wheel tandem axle truck having a total gross weight of 43,650 lbs (194 kN) and a tire contact pressure of 50 psi (344 kPa).

The results of (*4*) indicated that reinforced pavement sections displayed improved rutting resistance when compared with unreinforced sections for all aggregate types tested. The clay gravel was found to be the best performer, followed by the crushed limestone, and finally the crushed chert gravel. The authors attribute the clay gravel sections' improved performance to natural cementation stemming from drying of the clay gravel base, noting that moisture susceptibility was not a test variable. It was found that the geogrid-reinforced crushed limestone section outperformed the geotextile-reinforced crushed limestone section, whereas the geotextile- and geogrid-reinforced crushed limestone section performed the best overall.

Jersey and Tingle (*5*) performed cyclic plate load tests on three geogrid-reinforced highway pavement sections over a 3 CBR subgrade. The pavement sections consisted of 6-in. (152-mm) crushed limestone base over the 3 CBR CH subgrade that was surfaced with a 0.5-in. (12.5-mm) thick rubber mat. The rubber mat was used to simulate a flexible asphalt pavement surface layer. Each item was subjected to a series of load levels at the following magnitudes: 3,000 lb (13 kN), 6,000 lb (27 kN), 9,000 lb (40 kN), 11,000 lb (49 kN), 13,000 lb (58 kN), and 15,000 lb (67 kN) for a duration of 5,000 cycles or until a permanent deformation of 0.5 in (12.5 mm). Three different geogrids were evaluated consisting of two triaxial and one biaxial geogrid. An unreinforced control section was not included in this study; therefore conclusions about performance improvement could not be reached. It was noted that relative improvements in performance between the three geogrids were observed.

Performance of geosynthetic reinforced pavement sections surfaced with a double bituminous surface treatment was evaluated by Norwood and Tingle (*6*). The test section contained one item with a 6-in. (152-mm) crushed limestone base reinforced with a triaxial geogrid and one item with an 8-in. (203-mm) crushed limestone unreinforced base both placed over a 6 CBR subgrade. The 1-in.-thick (25.4-mm) double bituminous surface treatment was considered typical for low-volume road placement. Loading was accomplished by a single axle load cart outfitted with a single axle dual-wheel truck gear loaded to a nominal load of 10,000 lb (44.5 kN) and 120 psi (827.4 kPa) tire pressure.

It was concluded that both items performed equally well through the application of 60,400 equivalent single axle loads (ESALs) despite the unreinforced pavement section having an additional 2 in. (50 mm) of base. Falling-weight deflectometer tests showed that the stiffness of the unreinforced item decreased more rapidly when compared to the stiffness of the reinforced item.

Norwood (2017, personal communication) constructed nine full-scale test items with structural cross-sections of 4-in (101-mm) HMA over 14-in. (356-mm) crushed limestone base on an 8 CBR subgrade to evaluate the inclusion of geosynthetics in relatively thick airfield pavement sections. Biaxial and multiaxial geogrids were installed at the base/subgrade interface, with the exception of an unreinforced item and one item where the biaxial geogrid was installed mid-depth in the base layer. Trafficking of the test section was accomplished using an HVS outfitted with a single C-17 wheel. The nominal wheel load was 45,000 lbs (200 kN) at a tire pressure of 142 psi (978 kPa).

Test results were mixed with the unreinforced control section performing both better and worse than the reinforced sections. An approximate 30% improvement in rutting resistance was observed in the best performing reinforced section when compared with the control. Conversely, the worst performing reinforced item experienced 20% more rutting when compared with the unreinforced control item. The researchers suggested that the overall stiffness of the sections may have contributed to the minimal improvement observed by geogrid inclusion. It was noted that all geogrids may not provide the same reinforcement benefit.

Norwood and Tingle (*7*) constructed two test items with structural cross-sections of 4-in. (101-mm) HMA over 8-in. (203-mm) crushed limestone base and 3-in. (76.2-mm) HMA over 6-in. (152-mm) crushed limestone base reinforced with a triaxial geogrid. The design subgrade CBR was 6% for both sections. Trafficking of the section was accomplished using an HVS outfitted with a dual-tandem truck gear. The unreinforced section was subjected to a total of 811,200 ESALs; however, a malfunction in the HVS environmental chamber caused the pavement temperature of the reinforced section to rise well beyond the specified range. It is noted that the rate of surface rutting dramatically increased at the time of the environmental control malfunction. It was stated that the reinforced section performed equally as well as the unreinforced section until the significant temperature spike.

Norwood (2016, personal communication) performed cyclic plate load tests to investigate the effect of geosynthetic reinforcement airfield pavements. The laboratory-scale sections consisted of 12-in. (305-mm) subbase, 7-in. (178-mm) base, and 5-in. (127-mm) HMA surfacing. A subgrade CBR of 3% was targeted for all sections. Four test items were evaluated; three items reinforced with geogrid at the base/subbase interface and an unreinforced control item. Plate loading consisted of a total load of 28,800 lbs (128 kN) and a contact pressure of 254 psi (1,750 kPa).

It was found that rutting resistance improved with geogrid inclusion when compared with the unreinforced control item evidenced by TBR values ranging from 20 to 30. Post-test forensic investigation found that subgrade rutting was observed in the unreinforced control item, whereas no measurable rutting was observed in any of the geogrid-reinforced items.

Robinson et al. (*8*) evaluated two full-scale test sections trafficked using a dual-tandem truck gear on the HVS. Both sections consisted of 6-in. (152-mm) crushed limestone base and 3-in. (76.2-mm) HMA surfacing constructed over a 3 CBR CH subgrade. Pavements were reinforced with multiaxial geogrids, one of which was previously evaluated by Norwood and Tingle (*7*) and was retested because of a temperature malfunction. The dual-wheel tandem axle truck gear was used to apply 811,200 ESALs to each test item. Comparisons were made to the unreinforced section previously trafficked by Norwood and Tingle (*7*).

Geogrid inclusion improved rutting resistance when compared with previously tested unreinforced pavements. Additionally, it was suggested that the inclusion of the multiaxial geogrids provided a benefit at least equivalent to the benefit provided by 1 in. (25.4 mm) of additional HMA and an additional 2 in. (50.8 mm) of base material.

Normalization of Data for Analysis

The studies described in the previous section were performed to evaluate, at the time, emerging geosynthetic products or to address specific sponsor questions. Therefore, each study test plan was designed as a stand-alone project; at the time, consideration was not given to future compilation of individual datasets. Loading conditions and structural thicknesses were selected for each specific study, and as such, significantly differing loading conditions and structural thicknesses were observed.

In order to compile previous studies into an overall data set, varying loading conditions and varying structural thicknesses were normalized to common parameters. Specifically, tire and/or plate loading configurations were converted to ESALs and HMA thicknesses were converted to equivalent base course thicknesses (t_{equiv}). Note the term t_{equiv} is considered the same as the pavement thickness term shown in Figure 1 due to maintaining the same asphalt thickness in the ETL 1110-1-189 procedure. One can observe a similar approach in the results of the AASHO Road Test and subsequently the 1993 AASHTO Guide for Design of Pavement Structures (*9*), which has been used extensively by state departments of transportation. The authors acknowledge that this approach is a substantial simplification based on the load-dependency nature of the paving materials utilized. The procedure used for each conversion is described below.

Traffic Conversion

Various loading configurations and conditions were applied in the studies compiled for this effort. A majority of the test items were trafficked or loaded with loads and tire pressures generally in the range of that anticipated in a highway condition. It was determined that in order to provide a reasonable comparison of the various studies the applied loading would be converted to ESALs. Previous studies requiring traffic conversion were Webster (30,000 lbs [133.5 kN] at 68 psi [468 kPa]) (*2*) and Norwood (45,000 lbs [200 kN] at 142 psi [978 kPa]) (2017, personal communication). The Asphalt Institute's equivalent axle load factors (*10*) were plotted and a best-fit polynomial trend line was fitted through the data. The trend line equation was then used to extrapolate the aircraft loading to an equivalent single axle load factor (ESALF). The ESALF calculated for Webster's loading condition was 114 and for Norwood's loading condition was 634. The authors acknowledge that the ESALFs obtained are well beyond published values and represent a very large and approximate extrapolation beyond the AASHO Road Test.

Asphalt Thickness Conversion

Asphalt thicknesses were converted to equivalent base thicknesses using the AASHTO structural coefficients of 0.44 for asphalt and 0.14 for crushed stone base, which are typically specified by MDOT for the local materials used in each study. Each asphalt thickness was multiplied by 0.44 and then divided by 0.14 to obtain an equivalent base thickness. The equivalent thicknesses were then added to the underlying base thickness to calculate an overall equivalent aggregate thickness ($t_{equiv.}$). It is noted that variations in the properties of asphalt placed may have occurred when the MDOT transitioned from the Marshall Mix design method to the Superpave mix design method (approximately 1998); however, for simplicity a standard HMA structural coefficient was used. Use of a universal asphalt coefficient of 0.44 is approximate, and it should be noted that modern asphalt pavements have values used above 0.44 in some cases (e.g. 0.50–0.54).

Limited asphalt mixture properties were available, as the asphalt surfacing was not typically considered a test variable for each individual study. Key properties documented from each study are presented in Table 1.

Selection of Analysis Data

The relationship between equivalent aggregate thickness ($t_{equiv.}$) and loading (ESALs) for all reinforced test items is presented in Figure 4*a*. Variations in base course type, geosynthetic placement location, and geosynthetic type were investigated by removing each group from the overall dataset and observing changes in the regression.

It was found that a majority of the pavement structures consisted of crushed limestone base course material. Pavement structures that did not include crushed limestone were excluded from the analysis (four test items) due to significantly different documented strength characteristics. Figure 4*b* presents the relationship for reinforced sections containing crushed limestone base course material. An upward shift was observed in the regression (intercept of +0.03 to +1.7) indicating that base course type could influence performance.

Al-Qadi et al. (*11*) and Abu-Farsakh and Chen (*12*) concluded that pavement performance was improved when the geosynthetics were installed in the upper one-third of thick base layers. Haas et al. (*13*) suggested that the optimum placement location was the mid-point of the base layer. It was found that for a majority of the test items, the geosynthetic was placed at the subgrade/base interface. Two items were found to have a geosynthetic installed mid-depth in the base course and four items had a geosynthetic installed at the base/subbase interface and were excluded from the analysis

Table 1. Asphalt Mixture Properties

Reference	Design method	NMAS (mm)	RAP (%)	P_b (%)	VMA (%)	P200 (%)
2	Marshall	9.5	NR	NR	NR	NR
Norwood, 2017, personal communication	Superpave	9.5	NR	NR	15.8	NR
7	Superpave	9.5	15	5.7	15.4	3.8
8	Superpave	9.5	20	5.7	15.1	5.1
Norwood, 2016, personal communication	Superpave	12.5	NR	5.2	13.0	5.0

Note: P_b = percent binder by mass of the mixture; P200 = percent passing #200 sieve size; NR = not reported.

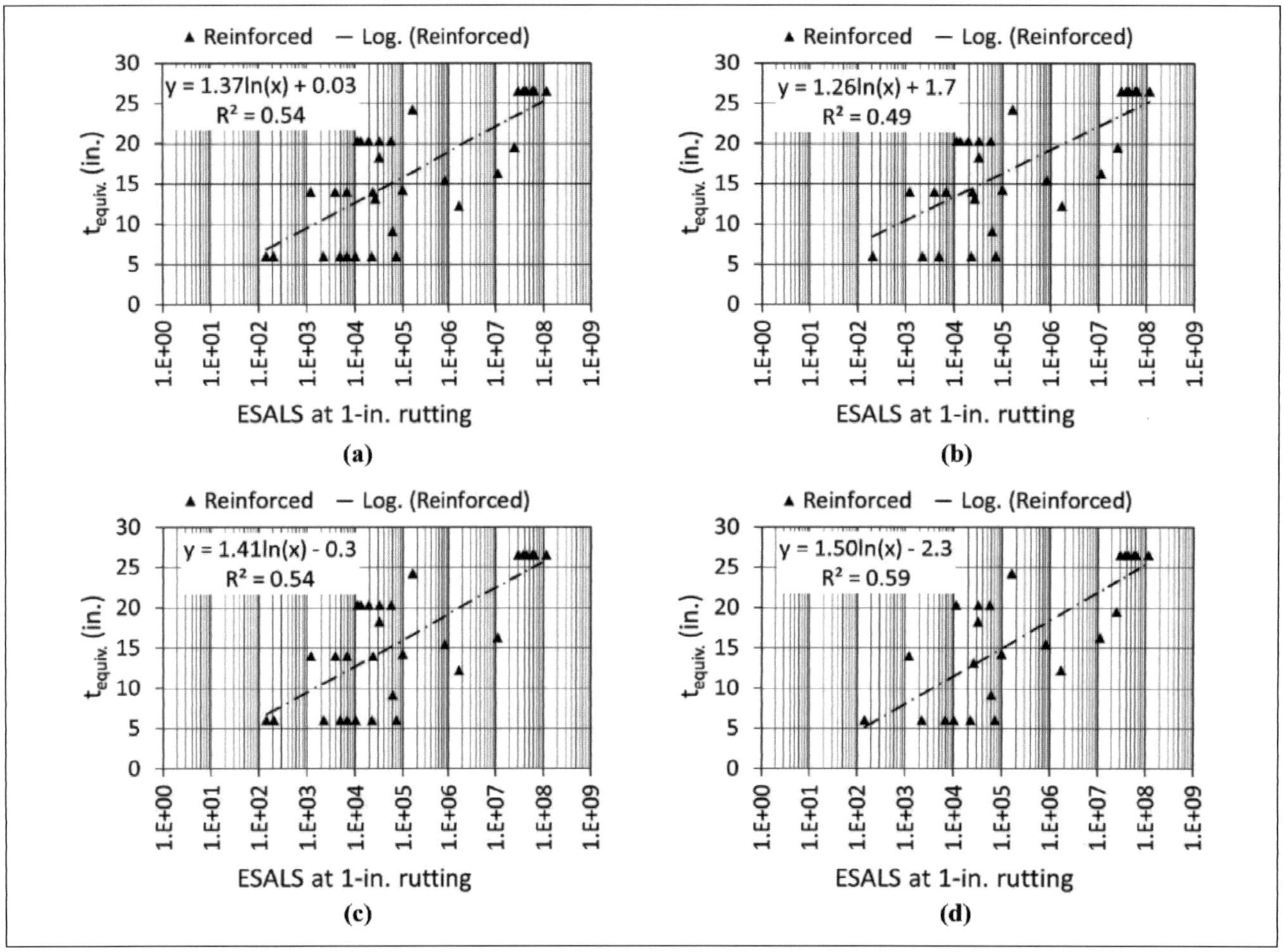

Figure 4. Effect of analysis variable on reinforced pavement regression: (*a*) all reinforced test items; (*b*) all limestone base test items; (*c*) all items reinforced at base/subgrade; (*d*) all geogrid test items.

as shown in Figure 4*c*. Of the two items where a geosynthetic was installed mid-depth in the base course, one item (constructed on a 3 CBR subgrade and $t_{equiv.}$ of 20 in. [508 mm]) showed improved performance when compared with the unreinforced item. The second item (constructed on an 8 CBR subgrade and $t_{equiv.}$ of 27 in. [686 mm]) showed poor performance when compared with the unreinforced item. It was found that including only items with a geosynthetic installed at the subgrade/base interface did not significantly change the observed trend.

It has been suggested that the reinforcement mechanisms of geotextile and geogrids are different, and that different performance improvement could be observed (*14*, *15*). Test items (seven total) containing a geotextile or geogrid/geotextile combination were excluded because of potentially differing performance enhancement characteristics (reinforcement

Table 2. Plate Load Testing

Reference	HMA (in.)	Base (in.)	Subbase (in.)	Subgrade CBR	Grid type	dgs (in.)	$t_{equiv.}$ (in.)	ESALs at 1-in. rutting
5	0	6	0	3	Triaxial	6	6	22,737
	0	6	0	3	Biaxial	6	6	10,229
	0	6	0	3	Triaxial	6	6	72,197
3	0	14	0	1	Biaxial	14	14	1,212

Note: dgs = depth of geosynthetic below surface; 1 in. = 25.4 mm.

Table 3. Full-Scale Testing

Reference	HMA (in.)	Base (in.)	Subbase (in.)	Subgrade CBR	Grid type	dgs (in.)	$t_{equiv.}$ (in.)	ESALs at 1-in. rutting
8	3	6	0	6	Triaxial 1	9	15	811,200*
	3	6	0	6	Triaxial 2	9	15	811,200*
6	1	8	0	6	None	–	11	60,400*
	1	6	0	6	Triaxial	7	9	60,400*
7	4	8	0	6	None	–	21	1,996,800
Jersey, 2009, personal communication	2	8	0	3	None	–	14	13,000
	3	8	0	3	None	–	17	27,870
	2	8	0	3	Triaxial	10	14	100,000
Norwood, 2017, personal communication	4	14	0	8	None	–	27	44,976,650
	4	14	0	8	Biaxial	18	27	63,737,000
	4	14	0	8	Triaxial	18	27	37,335,860
	4	14	0	8	Biaxial	18	27	58,126,869
	4	14	0	8	Biaxial	18	27	43,774,572
	4	14	0	8	Biaxial	18	27	113,451,860
	4	14	0	8	Biaxial	18	27	29,445,219
	4	14	0	8	Biaxial	18	27	114,726,600
4	0	6	0	4	None	–	6	23
	0	6	0	4	Geogrid	6	6	2,205
2	2	10	0	8	None	–	16	1,702,650
	2	10	0	8	Biaxial	12	16	11,351,000
	2	6	0	8	Biaxial	8	12	1,702,650
	2	6	0	8	None	–	12	76,052
	2	18	0	3	None	–	24	128,380
	2	18	0	3	Biaxial	20	24	162,546
	2	12	0	3	Biaxial	14	18	32,010
	2	14	0	3	Biaxial	16	20	56,755
	2	14	0	3	Biaxial	16	20	32,350

Note: dgs = depth of geosynthetic below surface; 1 in. = 25.4 mm.
*Test terminated before 1-in. rutting.

vs. separation). Figure 4*d* presents the relationship for reinforced items containing only geogrid, and a downward shift was observed in the regression (intercept of +0.03 to –2.3) indicating that geosynthetic type could influence performance.

Data selected for analysis is shown in Table 2 for laboratory-scale test sections and Table 3 for full-scale test sections. It is noted that the loading condition in Norwood (2016, personal communication) (four items) was not considered in this analysis due to contact pressure of 254 psi (1,750 kPa) being well outside that anticipated in a highway scenario. A total of 31 different test items were included in the analysis (4 from plate load testing and 27 from full-scale testing).

Results

The relationship between equivalent aggregate thickness (t_{equiv}) and loading (ESALs) was plotted for both reinforced and unreinforced test items using the approach described in

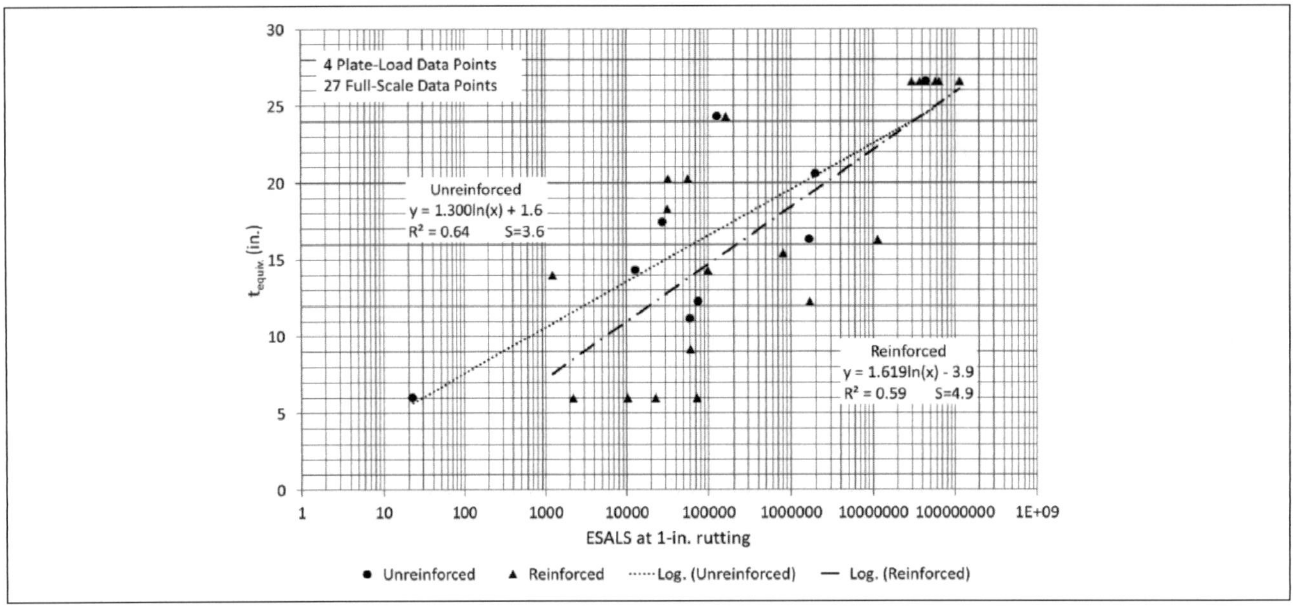

Figure 5. Relationship between equivalent aggregate thickness and ESALs for 1-in. rutting.

Webster, 1993, and is shown in Figure 5. It is noted that the aggregate equivalency chart derived by Webster and implemented in ETL 1110-1-189 was derived from a total of eight data points (two points per line) and as such overall variability could not be determined. When the overall dataset is combined, it is observed that the fitted equations have R^2 values of approximately 0.64 and 0.59 and standard error (S) of 3.6 and 4.9 for unreinforced and reinforced items, respectively.

Data were reviewed to determine if an increase in subgrade CBR provided for an increase in rutting performance within the range of CBR values investigated (1 to 8). It would be expected that items with higher subgrade CBR would fall above the regression line and lower subgrade CBR would fall below the regression line. It was found that no clear trend existed between subgrade CBR and the derived regression line, with CBR values falling both above and below each regression line.

Using the derived trends for unreinforced and reinforced test items, equivalent aggregate thickness values were calculated at levels ranging from 100 to 2,000,000 ESALs in approximately logarithmic intervals. The calculated unreinforced thicknesses were then plotted on the y-axis and the calculated reinforced thicknesses were plotted on the x-axis as shown in Figure 6. The combined data show relatively good agreement with the original equivalency chart. It is noted that the updated dataset reduces the equivalent thickness by approximately 1 inch (25.4 mm) at unreinforced thickness less than 14 inches (356 mm) and converges with the original equivalency chart at an unreinforced thickness of approximately 16 inches (406 mm). Additionally, it was found that as the geogrid depth of placement increases, the performance benefit provided by the geogrid decreases. Review of the relationship in Figure 6 indicates that after a depth of approximately 22 inches (559 mm) from the surface little to no performance improvement is realized. A 16-in.-depth (406-mm) was suggested by Tingle and Jersey (*3*) and Kinney et al. (*16*). Tingle and Jersey (*3*) hypothesized that the performance benefit decrease could be attributed to reduced horizontal stress and/or deflection at increasing depth resulting in failure to mobilize tensile behavior in the geosynthetics. Kinney et al. (*16*) found that the traffic benefit ratio (ratio of cycles for a reinforced section to an identical unreinforced section) decreased from a value in excess of 10 for base thicknesses of 10 inches (254 mm) or less to a value of 1 at base thicknesses of about 14 inches (356 mm). Al-Qadi et al. (*11*) found that for aggregate layer thicknesses ranging from 8 inches (203 mm) to 18 inches (457 mm) geogrid was effective in reducing horizontal shear deformation. It was further (*11*) concluded that for thicker base layers, optimal geogrid placement location was at the upper third of the layer, suggesting that performance benefit decreases with placement depth.

Conclusion

Multiple test sections have been constructed and trafficked at ERDC over the course of approximately 25 years to assess geosynthetic inclusion in pavement structures. The pavement test sections were constructed in a uniform manner with minor variability utilizing similar materials and construction techniques. The uniformity of construction and materials allow for meaningful comparisons between test sections to be made. Varying geogrid products in varying

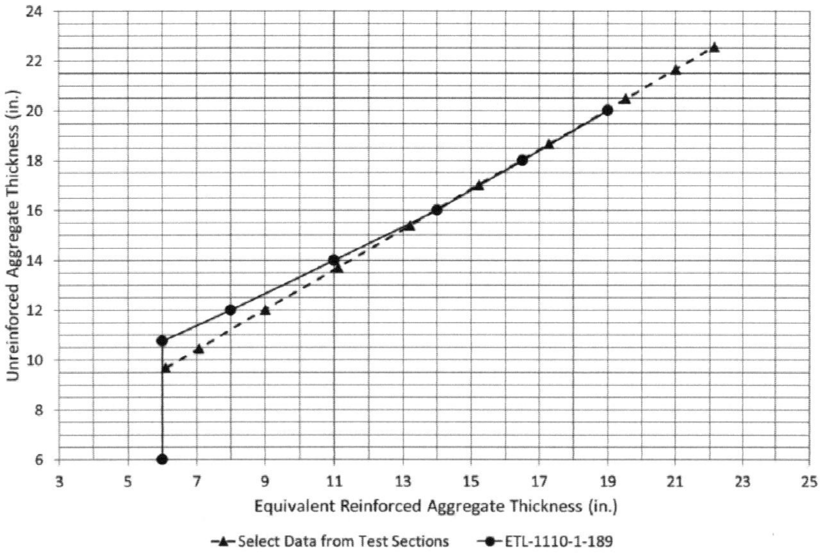

Figure 6. Relationship between unreinforced and reinforced aggregate thickness.

pavement structures have been evaluated using a range of loading conditions. A majority of the evaluations performed at ERDC have consisted of a pavement structure comprised of HMA surface, crushed limestone aggregate base, and a high-plasticity clay subgrade. Combination of the test data yielded the following:

1. Review of studies indicates that in terms of rutting performance, geogrid stabilized sections performed equal to or better than companion unstabilized sections.
2. Improvement appears to diminish with increasing depth of geogrid placement and approaches no distinguishable improvement around approximately 22 inches.
3. The combined data support the equivalent thickness chart recommended by Webster (*2*) and implemented in ETL 1110-1-189 (*1*). The adjusted data reduces the equivalent reinforced thickness by approximately 1.0 inches (25.4 mm) at unreinforced thickness less than 14 inches (356 mm), providing a slightly more conservative result. The adjusted data converged with the original equivalency chart at an unreinforced thickness of approximately 16 inches (406 mm).

Acknowledgments

The tests described and the resulting data presented here, unless otherwise noted, were obtained from research sponsored by Tensar International, the Federal Aviation Administration, and the Air Force Civil Engineering Center and performed by the U.S. Army Engineer Research and Development Center. Permission was granted by the Director, Geotechnical and Structures Laboratory, to publish this information.

Author Contributions

The authors confirm contributing to the paper as follows: study conception and design: W. Jeremy Robinson, Jeb. S. Tingle, Gregory J. Norwood, Isaac L. Howard; data collection: W. Jeremy Robinson; analysis and interpretation of results: W. Jeremy Robinson, Jeb. S. Tingle, Isaac L. Howard; draft manuscript preparation: W. Jeremy Robinson, Jeb. S. Tingle, Gregory J. Norwood, Isaac L. Howard. All authors reviewed the results and approved the final version of the manuscript.

References

1. U.S. Army Corps of Engineers. *Use of Geogrids in Pavement Construction*. Technical Letter ETL 1110-1-189. U.S. Army Corps of Engineers, Washington, D.C., 2003.
2. Webster, S. L. *Geogrid Reinforced Base Courses for Flexible Pavements for Light Aircraft: Test Section Construction, Behavior Under Traffic, Laboratory Tests, and Design Criteria*. Technical Report WES/TR/GL-93-6. Army Engineer Waterways Experiment Station, Vicksburg, Miss., 1993.
3. Tingle, J., and S. Jersey. Cyclic Plate Load Testing of Geosynthetic-Reinforced Unbound Aggregate Roads. *Transportation Research Record: Journal of the Transportation Research Board,* 2005. 1936: 60–69.
4. Tingle, J., and S. Jersey. Full-Scale Evaluation of Geosynthetic-Reinforced Aggregate Roads. *Transportation Research Record: Journal of the Transportation Research Board*, 2009. 2116: 96–107.
5. Jersey, S. R., and J. S. Tingle. *Cyclic Plate Testing of Geogrid-Reinforced Highway Pavements*. No. ERDC/GSL-TR-09-36. Engineer Research and Development Center Geotechnical and Structures Laboratory, Vicksburg, Miss., 2009.
6. Norwood, G. J., and J. S. Tingle. *Performance of Geogrid-Stabilized Gravel Flexible Base with Bituminous Surface*

Treatment. No. ERDC/GSL-TR-14-15. Engineer Research and Development Center Geotechnical and Structures Laboratory, Vicksburg, Miss., 2014.
7. Norwood, G. J., and J. S. Tingle. *Performance of Geogrid-Stabilized Flexible Pavements*. No. ERDC/GSL-TR-14-28. Engineer Research and Development Center Geotechnical and Structures Laboratory, Vicksburg, Miss., 2014.
8. Robinson, W. J., J. S. Tingle, and G. J. Norwood. *Full-Scale Accelerated Testing of Multi-axial Geogrid Stabilized Flexible Pavements*. No. ERDC/GSL TR-17-9. Engineer Research and Development Center Geotechnical and Structures Laboratory, Vicksburg, Miss., 2017.
9. Association of American Highway and Transportation Officials. *AASHTO Guide for Design of Pavement Structures*. Association of American Highway and Transportation Officials, Washington, D.C., 1993.
10. Huang, Y. H. *Pavement Analysis and Design*. Pearson Prentice Hall, Upper Saddle River, N.J., 2004.
11. Al-Qadi, I. L., S. H. Dessouky, J. Kwon, and E. Tutumluer. Geogrid-Reinforced Low-Volume Flexible Pavements: Pavement Response and Geogrid Optimal Location. *Journal of Transportation Engineering*, Vol. 138, No. 9, 2012, pp. 1083–1090.
12. Abu-Farsakh, M. Y., and Q. Chen. Evaluation of Geogrid Base Reinforcement in Flexible Pavement using Cyclic Plate Load Testing. *International Journal of Pavement Engineering*, Vol. 12, No. 3, 2011, pp. 275–288.
13. Haas, R., J. Walls, and R. G. Carroll. Geogrid Reinforcement of Granular Bases in Flexible Pavements. *Transportation Research Record: Journal of the Transportation Research Board,* 1988. 1188: 19–27.
14. Al-Qadi, I. L., T. L. Brandon, R. J. Valentine, B. A. Lacina, and T. E. Smith. Laboratory Evaluation of Geosynthetic-Reinforced Pavement Sections. *Transportation Research Record: Journal of the Transportation Research Board,* 1994. 1439: 25–31.
15. Barksdale, R. D., S. F. Brown, and F. Chan. *NCHRP Report 315: Potential Benefits of Geosynthetics in Flexible Pavement Systems*. TRB, National Research Council, Washington, D.C, 1989.
16. Kinney, T., J. Abbott, and J. Schuler. Benefits of Using Geogrids for Base Reinforcement with Regard to Rutting. *Transportation Research Record: Journal of the Transportation Research Board,* 1998. 1611: 86–96.

The Standing Committee on General and Emerging Pavement Design (AFD30) peer-reviewed this paper (18-01497).

Article

TRR
JOURNAL OF THE TRANSPORTATION RESEARCH BOARD

Transportation Research Record
2018, Vol. 2672(40) 143–154

DOI: 10.1177/0361198118776479
journals.sagepub.com/home/trr
SAGE

Mechanistic-Empirical Simulations and Life-Cycle Cost Analysis to Determine the Cost and Performance Effectiveness of Asphalt Mixtures Containing Recycled Materials

Erdem Coleri[1], Yuqi Zhang[1], and Blaine M. Wruck[1]

Abstract

Use of reclaimed asphalt pavements (RAP) and recycled asphalt shingles (RAS) in asphalt paving, although considered as sustainable, is a practice that agencies are reluctant to employ because of the unpredictability of asphalt mixes containing recycled materials. The asphalt binder in RAP/RAS is aged and stiffened, which reduces ductility of the pavement. Consequentially, a pavement can exhibit unsatisfactory fatigue performance and have the potential for early cracking failure. Although methods exist to counteract the brittle behavior of pavements containing RAP/RAS (namely binder-grade bumping, binder-grade dumping and high binder content), they are not accounted for in mechanistic-empirical (ME) pavement design. Additionally, the cost benefits of using RAP/RAS in pavements are not easily calculated. For these reasons, characterization of fatigue performance for asphalt pavements containing RAP/RAS in ME design software needs to be accomplished and a life-cycle cost analysis (LCCA) framework for pavements containing RAP/RAS needs to be developed so that agencies can make informed decisions about RAP/RAS use in asphalt mixtures. In this study, laboratory test results for asphalt mixtures with different combinations of RAP/RAS contents, binder contents, and binder types were used to calculate ME pavement model coefficients to perform forward calculations to determine pavement performance. Using predicted performance from ME models, LCCAs were conducted to determine the cost benefits of using binder-grade bumping/dumping and high binder content in Oregon asphalt mixtures. These strategies are expected to increase RAP/RAS use in asphalt mixtures, reduce life-cycle costs, improve the cracking performance and encourage widespread use of RAP/RAS asphalt mixtures.

Asphalt is an advantageous material to use in pavement construction because of its recyclability. Using reclaimed asphalt pavement (RAP) and recycled asphalt shingles (RAS) in asphalt paving provides direct cost savings to agencies by displacing raw materials and ultimately yields a more sustainable pavement. Supplementing 50% RAP/RAS into hot mix asphalt (HMA) mixes can save agencies up to 35% on the cost of raw materials in asphalt paving projects (*1*). In 2014, the use of RAP/RAS on roads in the United States displaced 20 million barrels of oil and 68 million tons of aggregate (*2*). This resulted in a saving of \$2.8 billion based on binder cost of \$550/ton and aggregate cost of \$9.50/ton, according to a study by National Asphalt Pavement Association (NAPA) (*2*).

The main caveat to using high proportions of recycled materials in asphalt pavements is the reduced durability brought on by aged and stiffened asphalt binder contained in RAP/RAS. In particular, fatigue cracking is a failure mechanism that is common in asphalt pavements with high recycled materials content (*3*). The aged binder is less ductile and facilitates fast crack propagation once a crack is initiated, which can result in early fatigue failure of the pavement structure. RAS particularly contributes to reduced durability because RAS binder is much stiffer than conventional asphalt binder and is also subjected to a greater degree of aging than RAP.

Binder-grade bumping/dumping and high binder content are the strategies evaluated in this study to combat premature fatigue cracking of asphalt pavements containing recycled materials. Binder-grade bumping refers

[1]School of Civil and Construction Engineering, Oregon State University, Corvallis, OR

Corresponding Author:
Address correspondence to Blaine M. Wruck: wruckb@oregonstate.edu

to increasing the high temperature performance grade (PG) by one increment. Binder-grade dumping refers to reducing the high temperature grade by one increment. In general, the use of a softer binder grade is expected to yield HMA mixtures with better cracking performance than mixes with stiffer binders (*3*). The problem of premature fatigue cracking can also be potentially overcome by increasing the binder content. Increasing virgin binder content can help to counteract the brittle nature of high RAP/RAS mixes by increasing the ductility of the mix (*4*).

These strategies can preserve the fatigue resistance of the pavement without causing rutting failure problems by achieving a balance between the excess stiffness of RAP/RAS and the high ductility of softer binders and higher binder contents. Binder-grade bumping/dumping and high binder content are expected to increase RAP/RAS use in asphalt mixtures, reduce life-cycle costs, improve the cracking performance and encourage widespread use of RAP/RAS asphalt mixtures. These two strategies are explored in detail in (*4*).

This study focused on use of Mechanistic-Empirical Pavement Design Guide (MEPDG) models (*5*) to quantify the impact of RAP content, RAS content, binder content, and binder type on in-situ rutting and bottom-up fatigue cracking (alligator cracking) performance. Laboratory test results from (*4*) were used to develop MEPDG models for different RAP/RAS mixtures. For this purpose, Level 1 MEPDG simulations (the level with highest accuracy) were conducted by using the dynamic modulus (DM) and binder dynamic shear rheometer test results from (*4*). Measured asphalt mixture properties (effective binder content) for all tested asphalt mixtures were also used to improve model predictions. Effective binder content refers to the amount of virgin binder plus the amount of binder in RAP/RAS material that is actively blending with the virgin binder in the mix (*4*).

The use of RAP and RAS in HMA paving is often considered a cost-saving measure (*2*). Although it can make the pavement more susceptible to cracking failure, reductions in cost and environmental impacts brought on by its use mean that it is considered a sustainable alternative to asphalt mixtures with all-virgin materials. However, contractors and agencies who are unable to accurately quantify savings brought on by using RAP or RAS in HMA mixes may be discouraged from using these materials because of their reduction in HMA cracking resistance. Cost analysis becomes even more difficult when different binder types and binder contents are incorporated into the mix, which are practices discussed in this study for combating cracking susceptibility. The culmination of these factors yields a necessity for a simple way to analyze different mix design options.

As part of this study, a material cost calculation tool was developed to quantify asphalt mixture costs for different binder contents, binder types and RAP/RAS contents. Although the use of binder-grade bumping/dumping and high virgin binder content strategies generally increases the cost of virgin binder used in the asphalt mixture, increased RAP/RAS content and improved RAP/RAS performance can reduce the overall life-cycle cost of pavements utilizing recycled asphalt materials. Using the predicted performance from ME models and cost calculations for different combinations of RAP content, binder content, and binder type, life-cycle cost analyses (LCCAs) were conducted to investigate the performance and cost benefits of using binder-grade bumping/dumping and high binder content in Oregon RAP/RAS mixes.

The key objectives of this study are to:

- Predict the impact of RAP/RAS content, virgin binder grade and virgin binder content on in-situ cracking and rutting performance using ME modeling;
- Develop a cost calculation tool to compare the unit cost of asphalt mixtures with different RAP/RAS contents, binder contents, and binder types; and
- Quantify the cost benefits of increasing RAP/RAS content in asphalt mixtures using ME modeling and LCCA.

MEPDG Rutting and Fatigue Cracking Models

Experimental results from (*4*) were used as inputs for MEPDG simulations in this study. In particular, semi-circular bending (SCB), DM, and flow number (FN) tests were conducted. A full experimental plan for testing and modeling is shown in Table 1.

Fatigue Cracking Models

Fatigue cracking models used in this study were adopted from MEPDG (*5*). However, it should be noted that MEPDG longitudinal (top-down cracking) cracking models were not used for simulations because the accuracy of these models was determined to be low (*6*). Findings of NCHRP 9-30 (*7*) also suggested excluding current longitudinal cracking models in the local calibration guide. The model coefficients used in the MEPDG bottom-up fatigue cracking model encompassed the effect of excess stiffness from RAS binder as these coefficients were determined from test results of mixtures containing RAP/RAS in (*4*).

Rutting Models

Rutting in the asphalt and unbound layers are separately predicted in MEPDG. Total surface rutting is calculated by summing the predicted rutting in all layers. Because rutting in aggregate base and subgrade layers is not expected to be significant in Oregon (*6*), calibration factors for aggregate base and subgrade layers are set to zero in this study.

Details of the models and equations used in the fatigue cracking and rutting simulations can be found in (*4*).

Level 1 MEPDG Simulations using Experimental Results

Three hierarchical levels of inputs are available in MEPDG (*5*). Level 1 input represents the highest accuracy and requires laboratory or field test results to characterize material properties. Level 2 analysis requires binder, aggregate and general mixture properties and uses these input variables to estimate HMA stiffness by using correlation functions embedded in MEPDG. Level 3 analysis will provide the lowest accuracy. Average values for material properties for the U.S. are selected by the user as input parameters. Experiments are not conducted to measure any material properties to use as input variables for modeling.

In this study, MEPDG model coefficients were developed to quantify the impact of RAP content, RAS content, binder content, and binder type on in-situ rutting and bottom-up fatigue cracking performance. For this purpose, Level 1 MEPDG simulations (highest accuracy) were conducted by using the DM and binder dynamic shear rheometer test results and field calibration constants from (*4*). Measured asphalt mixture properties (effective binder content) for all tested asphalt mixtures were also used to improve model predictions. Table 1 shows all cases modeled with MEPDG to determine the impact of climate, RAP content, RAS content, binder content, and binder type on in-situ rutting and bottom-up fatigue cracking performance.

Mastersolver V2.2, a spreadsheet developed by Dr. Ramon Bonaquist of Advanced Asphalt Technologies, was used to develop master curves and parameters required to perform Level 1 MEPDG analysis. Calculated parameters under the MEPDGINPUT tab (*4*) were used as input parameters to conduct Level 1 MEPDG analysis.

A local calibration was performed by using the pavement management system data of Oregon Department of Transportation (ODOT) (*6*). In this study, bottom-up fatigue cracking and rutting model calibration factors from (*6*) were used to perform Level 1 MEPDG simulations. Calibration coefficients from local calibration and Level 1 inputs are expected to provide realistic performance predictions for the cases analyzed in this study.

MEPDG model coefficients for identical structures were developed for all cases and the material properties for new asphalt overlays used the DM test results obtained for different mixture types (*4*). The typical structure used for model development for the initial rehabilitation is shown in Figure 1*a*. A full friction interface was assumed for all layers. Although the assumption of full friction between asphalt layers is made in this study, the friction level might decrease in field conditions because of damage accumulated at the bond interface between asphalt layers. MEPDG rehabilitation analysis was conducted. The top 25.4 mm of the existing 76.2 mm layer was milled for the analysis (a 50.8 mm-thick layer was left for modeling). After milling, a 76.2 mm-thick asphalt overlay was constructed as the new layer. Level 1 parameters were entered for the top 76.2 mm overlay and the existing asphalt layer was assumed to be in "Fair" condition with a PG64-22 binder. After running the

Table 1. Cases Modeled with MEPDG

Category	Tests	Binder content	RAP content	Binder grade	Climate	Traffic (AADTT[1])	Number of models
High RAP	SCB	6.0%	30%	PG 58-34	Ontario	6,000	36
	DM	6.4%	40%	PG 64-22	Portland		
	FN	6.8%		PG 76-22			
RAP&RAS	SCB	6.0%	ABR[2] equal	PG 64-22	Ontario	6,000	8
	DM	6.8%	to 30% RAP	PG 76-22	Portland		
	FN						
Low RAP	SCB	6.0%	0%	PG 64-22	Ontario	6,000	16
	DM	6.8%	15%	PG 76-22	Portland		
	FN						

Note: [1]AADTT = average annual daily truck traffic. [2]ABR = Asphalt Binder Replacement.

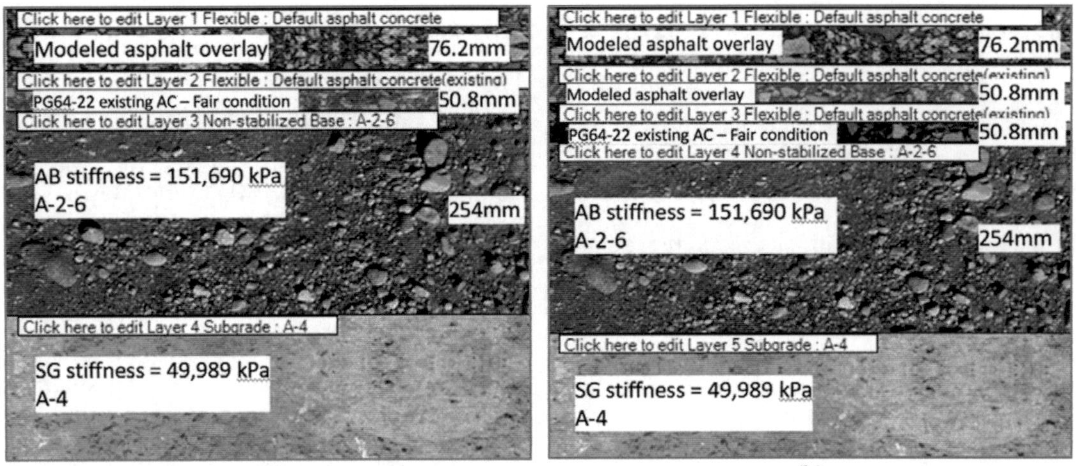

Figure 1. Cross sections of structures used for MEPDG modeling: (*a*) initial rehabilitation (overlay) and (*b*) structure after the second rehabilitation.

initial rehabilitation model and finding the service life, the top layer from the model was milled and a second 76.2 mm overlay was constructed (Figure 1*b*). These simulations were continued until the total service life of modeled structures exceeded 50 years, which was the analysis period for the LCCA. If a structure did not fail within a 50-year period, a 50-year design life was assumed and used for the LCCA. In this study, bottom-up fatigue cracking is considered to be the only distress for LCCA because almost none of the sections were failing from rutting.

Analyses were conducted with an initial traffic level of 6,000 average annual daily truck traffic (AADTT) and an annual traffic growth rate of 3%. For the second and third rehabilitation models, the previous year's traffic from the last rehabilitation was used as the starting traffic level for the following rehabilitation to maintain a continuous traffic growth rate for the section. Climate stations from Portland, OR and Ontario, OR were used for simulations. Because Ontario is colder in winter and warmer in summer when compared with Portland, cracking and rutting accumulation rates are expected to be higher.

Cost Calculation Tool

In this study, a cost calculation tool was created that allows the users to compare mix design strategies against one another to calculate the potential savings they can realize by choosing mix designs with different RAP contents, RAS contents, binder contents, and binder types. This simple tool, created using a Microsoft Excel spreadsheet, is meant to increase incentive for agencies to use recycled materials in their HMA mixes, thereby increasing the sustainability and cost effectiveness of asphalt pavement construction. Given the geometry of a pavement section and pertinent material cost data, the contractor or the agency (or both) can evaluate the estimated material cost of implementing a particular mix design strategy for their project.

To use the tool, the user must input data about their HMA mix design, such as target density, binder content, and recycled materials content. Input data about pavement section geometry, such as length, lane width, number of lanes, and compacted layer thickness, must also be entered. The user also inputs their unit costs for binder, aggregate and recycled materials (RAP and RAS). The tool will automatically calculate the volume, weight, and cost of HMA material that is anticipated for the target density and pavement section geometry.

The tool can compare up to four different mix strategies at once. This means the user can evaluate differences in costs for up to four different binder types, binder contents, and RAP/RAS contents. A summary spreadsheet compares the various mix design options. This sheet shows the cost differences between each individual mix design, as well as an indication of minimum and maximum cost options. A simple bar chart shows a side-by-side comparison of each mix design strategy to visualize the costs of each option.

In this study, sample material costs were used for cost calculations. These costs were obtained from local providers in Oregon. The following costs were used to calculate the total material cost of asphalt mixtures:

- RAP: \$20/ton,
- RAS: \$40/ton,
- Aggregate: \$13/ton,
- PG58-34 binder: \$425/ton,
- PG64-22 binder: \$375/ton, and
- PG76-22 binder: \$450/ton.

Life-Cycle Cost Analysis

In this study, LCCAs were performed by using the service lives for pavement structures obtained from the MEPDG simulations described in Table 1. The material costs of each strategy were calculated by using the cost calculation tool developed in this study. Material production costs, plant operation costs, and profit were not considered in the analysis.

First, analyses were performed by only considering material costs to be able to compare the impact of RAP content, RAS content, binder content, binder grade, and climate on life-cycle costs. Then, analyses were performed by considering agency costs that were composed of material, equipment, labor, traffic handling, and lane closure costs.

In this study, each section was assumed to be a single-lane having a width of 3.7 m and a length of 1.61 km. Material costs were calculated for all the sections based on their thicknesses. The cost calculation tool developed in this study was used to estimate material costs.

Costs corresponding to lane closure and traffic handling were considered to be \$1,000 per day per lane (*8*). To estimate operation days, lane closure, and traffic handling costs, a rate of application of 500 tons per day was assumed for HMA overlays (*9*). Because each 1.61 km section requires about 1,200 tons, lane closure and traffic handling costs for a 1.61 km 1-lane section were assumed to be \$2,400 [(1,200 tons/500 ton/day)*\$1,000/day].

Based on the data obtained from (*10*), equipment and total labor costs for an asphalt overlay were estimated using

$$\text{Equipment and total labor costs for HMA overlay} = 0.19^{*}(\text{total of construction and material costs}) \quad (1)$$

Total agency costs were calculated by summing up the equipment, total labor, material, traffic handling, and lane closure costs. After finding the service lives for each strategy, agency costs were estimated for each year at which the treatment was applied. Net present value (NPV) of agency costs were determined afterwards using a 4% interest rate (*11*) for a 50-year analysis period by using Equation 2. At the end of the analysis period, the NPV of the salvage value of each section was computed as the agency benefits and added to the NPV of agency costs.

$$\text{NPV} = \sum_{t=0}^{T} \frac{C_t}{(1+r)^t} \quad (2)$$

where
C_t = estimated agency costs at year t,
r = interest rate, and
T = number of time periods.

Results and Discussion

In this section, results of the MEPDG simulations are presented. MEPDG predictions, material costs, and LCCA are used to evaluate the cost effectiveness of asphalt mixtures with different RAP contents, RAS contents, binder contents, and binder types. LCCAs were performed by only considering material costs (raw material costs only; production costs are not considered) and by considering total agency costs including material, equipment, labor, traffic handling, and lane closure costs.

MEPDG Performance Predictions

Rutting. Binder type was observed to be the most significant factor controlling asphalt concrete rutting. This result agrees with the FN test results given in (*4*). It was also observed that increasing binder content and decreasing RAP content would increase rutting. Results also showed that none of the sections fail from rutting within the first 15 years for failure criteria of 12.7 mm rut depth and the majority of the sections do not fail within the first 25 years. Because asphalt aging is going to significantly increase asphalt stiffness during this long time period, it is highly likely that none of the sections will fail from rutting. For this reason, bottom-up fatigue cracking was used as the only failure criterion for the LCCA.

Bottom-Up Fatigue Cracking. Predicted bottom-up cracking for all the 30% and 40% RAP cases (Table 1) for Portland and Ontario climates is shown in Figure 2. Bottom-up cracking for all sections and RAP/RAS contents for Portland and Ontario climates are given in Figures 3 and 4, respectively.

It can be observed that the cracking accumulation rate for Ontario is higher than the Portland climate because of lower winter temperatures for Ontario. Binder content was observed to be the most significant factor controlling bottom-up cracking. This result agrees with the SCB test results given in (*4*). Increasing the binder content while decreasing the RAP content was also observed to decrease cracking accumulation. For the sections with PG76-22 binder, cracking performance was observed to be significantly lower. Increasing binder content and decreasing RAP content for this binder type did

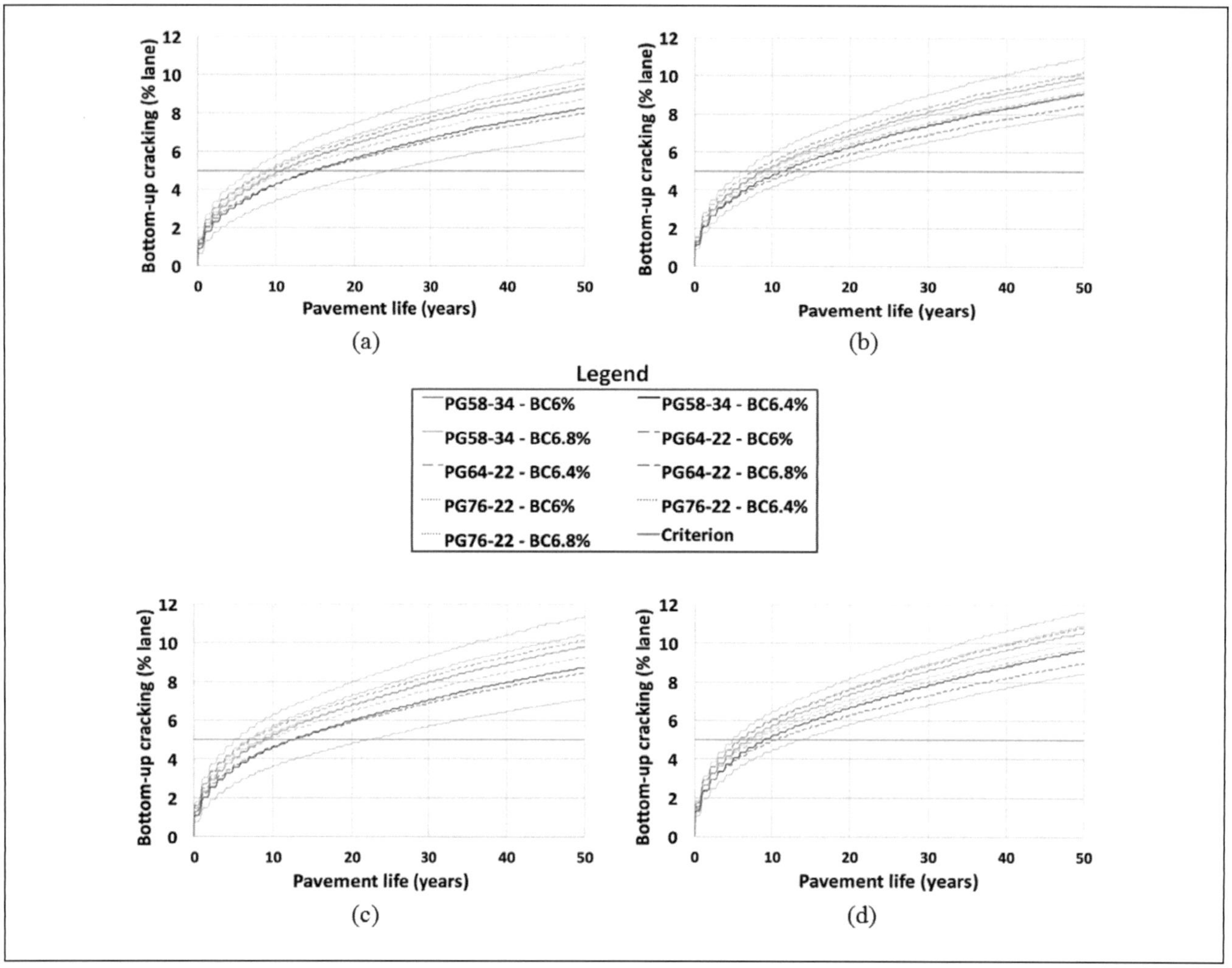

Figure 2. Predicted bottom-up cracking (alligator) for mixes with 30% and 40% RAP: (*a*) Portland—30% RAP, (*b*) Portland—40% RAP, (*c*) Ontario—30% RAP, and (*d*) Ontario—40% RAP.

not appear to create significant changes in performance. However, it should be noted that according to (*4*), asphalt mixtures with PG76-22 binder and higher binder contents (7.2%–7.8%) can create mixtures highly resistant to both rutting and cracking. DM tests should be conducted with the suggested high binder content mixtures to be able to develop MEPDG models and simulate their rutting and cracking performance. Increased performance for PG76-22 mixtures with higher binder contents may result in reduced life-cycle costs.

The SCB/Flexibility Index (FI) test results from (*4*) are correlated with MEPDG simulation outputs and MEPDG-predicted cracking in this study. Figure 5 shows the results of SCB tests from (*4*) for all cases in Table 1, as measured by FI.

As is seen in Figure 5, FI increases as the binder content increases for mixtures with the same RAP contents. Asphalt mixtures with lower binder contents are more brittle and more susceptible to cracking. Results also show that increasing RAP content results in more brittle mixtures and reduces the cracking resistance. Moreover, using softer binder increases FI. These results reaffirm that the binder content is the controlling element in the bottom-up cracking resistance of the pavement.

All rutting and cracking curves presented in this section are for the structure shown in Figure 1*a* (a 76.2 mm asphalt overlay on a 50.8 mm milled existing asphalt layer). To be able to determine service lives after initial failure, simulations were also performed after the first failure (pavement structure shown in Figure 1*b*) by milling 25.4 mm from the existing overlay surface and placing a new 76.2 mm asphalt overlay. Traffic levels were also adjusted for all simulations to have a continuous increase in AADTT of 3% throughout the analysis

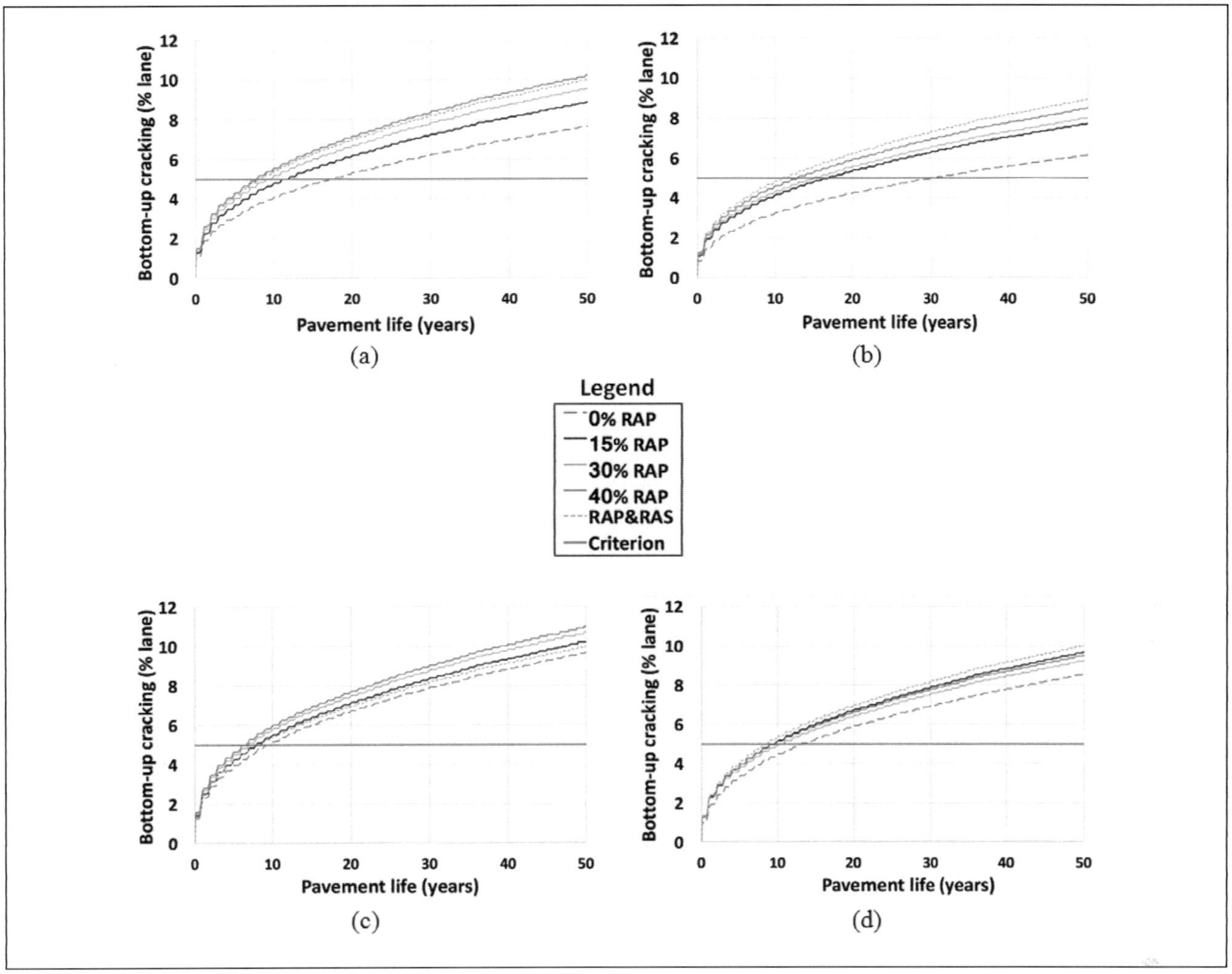

Figure 3. Predicted bottom-up cracking (alligator) for Portland for mixes with PG64-22 and PG76-22 binders: (*a*) PG64-22—6% binder content, (*b*) PG64-22—6.8% binder content, (*c*) PG76-22—6% binder content, and (*d*) PG76-22—6.8% binder content.

period. This process was repeated until the total life of all constructed pavements reached or exceeded the 50-year analysis period.

Figure 6 shows the bottom-up cracking performance curves for the section with PG76-22 binder, RAP&RAS, 6% binder content and the Portland climate. In this study, the asphalt overlay was constructed when the bottom-up cracking reached 5% of the lane area (considered as the rehabilitation trigger value or failure criterion). Because total service life did not reach the 50-year analysis period after two rehabilitations, the structure in Figure 1*b* was milled 25.4 mm and a 76.2 mm-thick asphalt overlay was constructed for a third time. The section thickness, which increased by 50.8 mm with every asphalt overlay construction, was observed to increase the service life. Using the service lives for each construction (6.83, 30, and 48.33 years for the case in Figure 6) and calculated costs, LCCA was performed and is explained in the next section.

Life-Cycle Cost Analysis

Results of LCCA by Only Considering Material Costs. LCCA was performed by following the process described in (*4*). Material costs of each strategy (for the cases outlined in Table 1) were calculated by using the cost calculation tool developed in this study. Material production costs, plant operation costs, and profit were not considered in the analysis. In this section, analyses were performed by only considering material costs to be able to compare the impact of RAP content, RAS content, binder content, binder grade, and climate on life-cycle costs. Figure 7 compares the NPVs for all cases. An interest rate of 4% was used for NPV calculations (*11*).

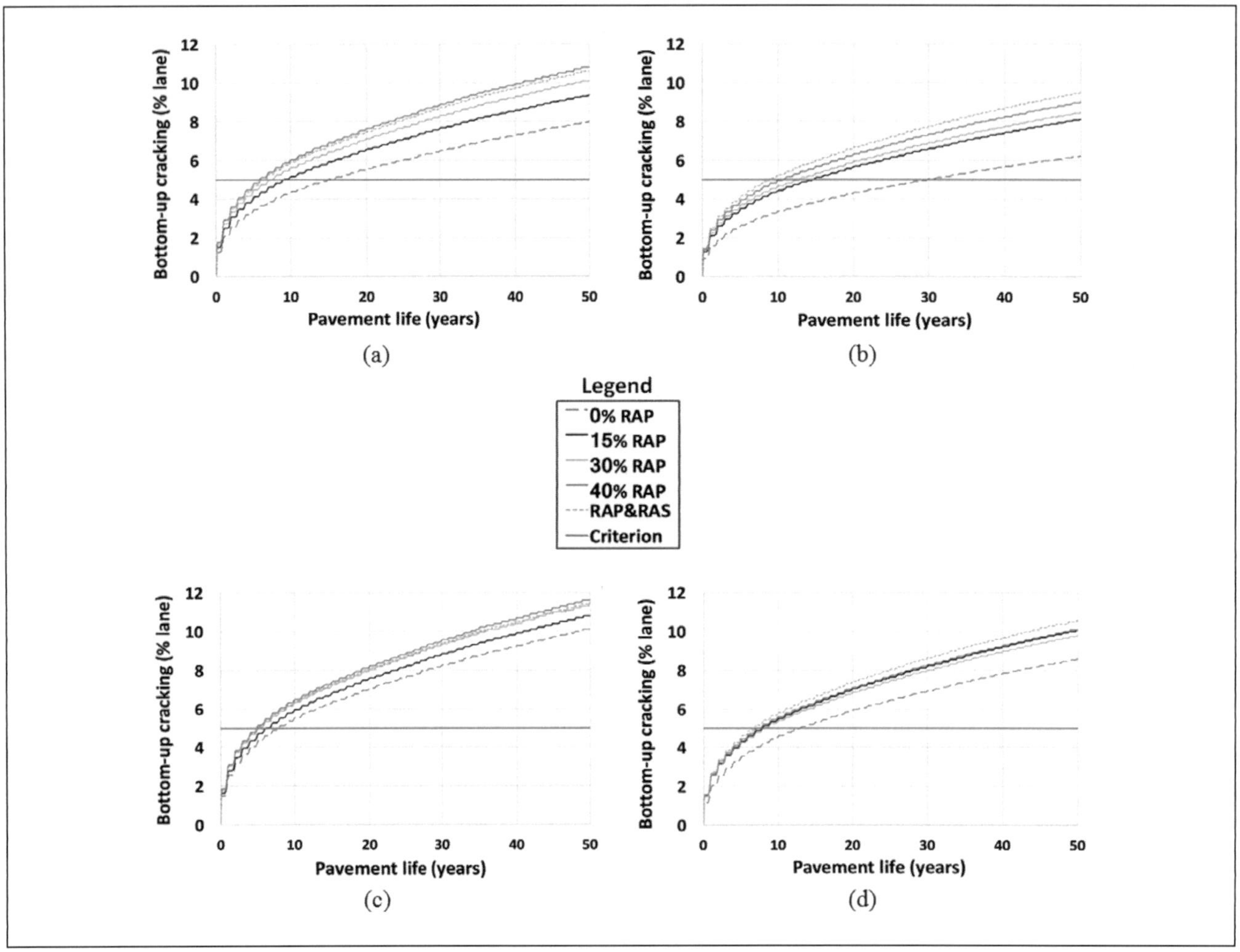

Figure 4. Predicted bottom-up cracking (alligator) for Ontario for mixes with PG64-22 and PG76-22 binders: (*a*) PG64-22—6% binder content, (*b*) PG64-22—6.8% binder content, (*c*) PG76-22—6% binder content, and (*d*) PG76-22—6.8% binder content.

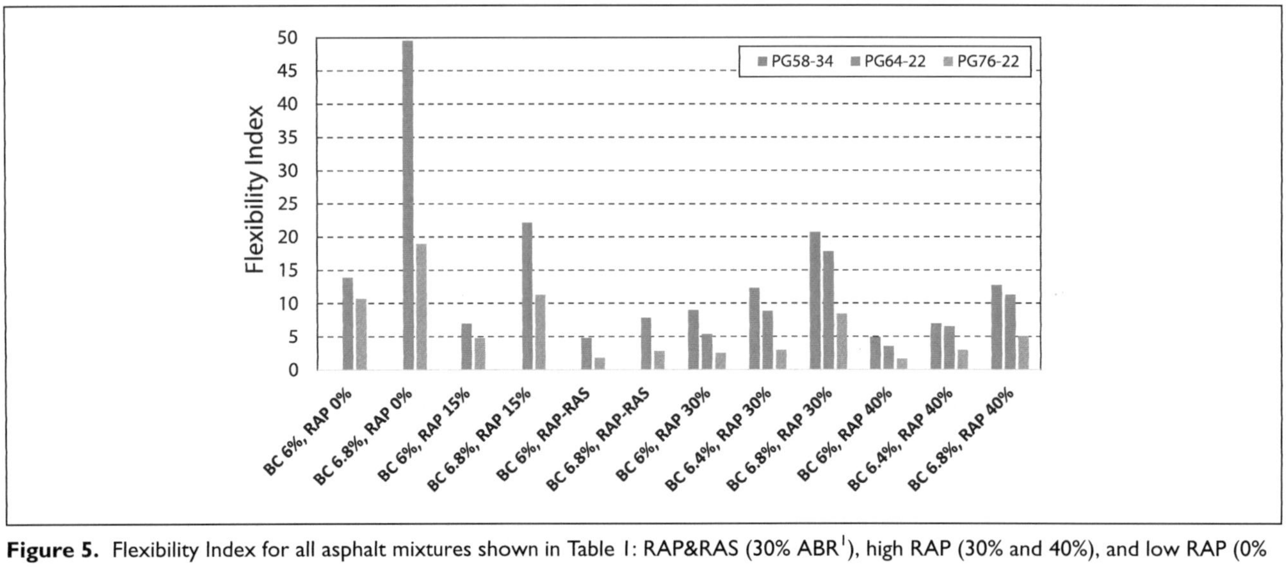

Figure 5. Flexibility Index for all asphalt mixtures shown in Table 1: RAP&RAS (30% ABR[1]), high RAP (30% and 40%), and low RAP (0% and 15%) (*4*).

Note: [1]ABR = Asphalt Binder Replacement.

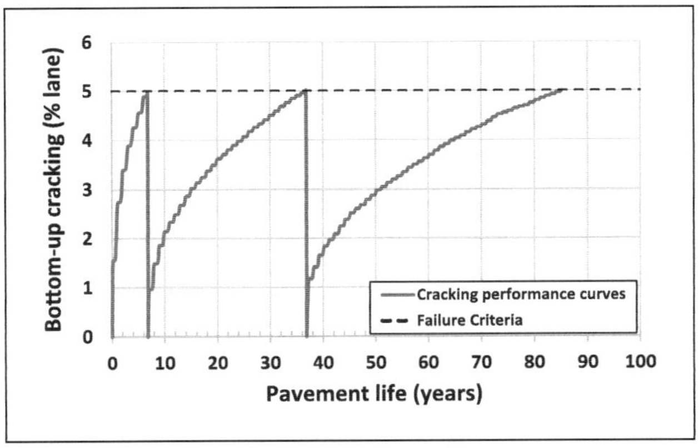

Figure 6. Cracking performance curves for the section with PG76-22 binder, RAP&RAS, 6% binder content, and Portland climate.

It can be observed from Figure 7 that although increasing binder content increases the mixture costs, increased performance generally results in higher cracking resistance and lower life-cycle costs. This result suggests that increasing the binder content of an asphalt mixture is an effective strategy to improve the condition of the highway network and reduce long-term costs. In addition, higher RAP contents generally result in lower life-cycle costs. However, the case with PG64-22 binder, 0% RAP and 6.8% binder content has a significantly higher performance which results in lower life-cycle costs for both climates. As the cost of PG64-22 binder ($375/ton) is lower than PG58-34 ($425/ton) and PG76-22 ($450/ton) binders, life-cycle costs for the sections constructed with PG64-22 were also determined to be lower. Because of lower binder cost for PG64-22, changes in binder content and RAP content were observed to have a less significant effect on life-cycle costs of sections constructed with PG64-22 binder. However, it is worth noting that this result may change as the price of asphalt binder fluctuates. Because of colder temperatures, simulations with the Ontario climate resulted in lower cracking resistance. Lower cracking resistance increased the life-cycle costs for these sections.

For the sections with PG76-22 binder, cracking performance was observed to be significantly lower. Calculated life-cycle costs were also higher for the section with PG76-22 binder. However, it should be noted that according to (*4*), asphalt mixtures with PG76-22 binder and higher binder contents (7.2%–7.8%) can create mixtures highly resistant to both rutting and cracking. DM tests should be conducted with the suggested high binder content mixtures to be able to develop MEPDG models and simulate their rutting and cracking performance. Increased performance for PG76-22 mixtures with higher binder contents may result in reduced life-cycle costs.

Results of LCCA by Considering Total Agency Costs. The procedure described in (*4*) was followed to calculate total agency costs. After finding the service lives for each strategy, agency costs were estimated for each year at which the treatment was applied. The NPV of agency costs for each strategy was determined afterwards using Equation 2 with a 4% interest rate (*11*) and a 50-year analysis period.

Figure 8 shows the correlation between calculated NPVs obtained using material costs and total agency costs. It can be observed that using the total agency costs did not change the ranking of the life-cycle costs for mixtures with PG58-34 and PG64-22 binders. However, because mixtures with PG76-22 binder have shorter service lives and more than two rehabilitations are generally required within the 50-year analysis period, using total agency costs rather than only material costs changed the life-cycle cost rankings for a very few cases. For instance, when only material costs were used for LCCA, the asphalt mixture with 15% RAP and 6.8% binder content had the highest NPV. The mixture with RAP&RAS and 6% binder content had the highest NPV when the total agency costs were used for LCCA. However, in general, using total agency costs rather than only material costs did not create any major changes in conclusions. Although using total agency costs increases the calculated NPVs (as expected), a strong linear correlation between NPVs for material costs and total agency costs suggests that using total agency costs rather than only material costs did not change the rankings and conclusions except for the three cases shown in Figure 8 (PG76-22-BC6.8%-RAP15%-Ontario, PG76-22-BC6.4%-RAP30%-Portland, and PG76-22-BC6.8%-RAP30%-Portland).

Summary and Conclusions

This study focused on development of MEPDG models to quantify the impact of RAP content, RAS content, binder content, and binder type on in-situ rutting and bottom-up fatigue cracking performance. A material cost calculation tool was also developed to calculate the asphalt mixture costs for different combinations of binder content, binder type, and RAP/RAS content. Using the MEPDG-predicted performance curves and calculated material and total agency costs, LCCAs were performed to determine the most cost effective strategies.

The analyses presented have yielded the following conclusions:

- Binder type was observed to be the most significant factor controlling asphalt concrete rutting. Increasing binder content and decreasing RAP content was also observed to increase rutting.

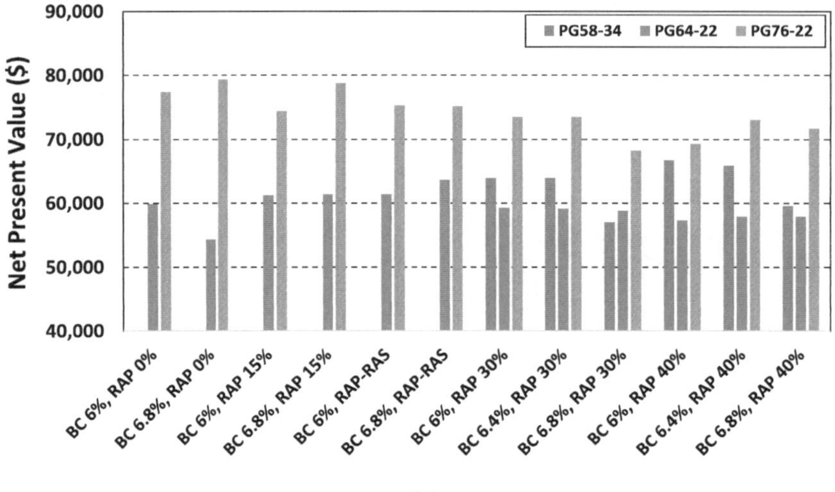

(a)

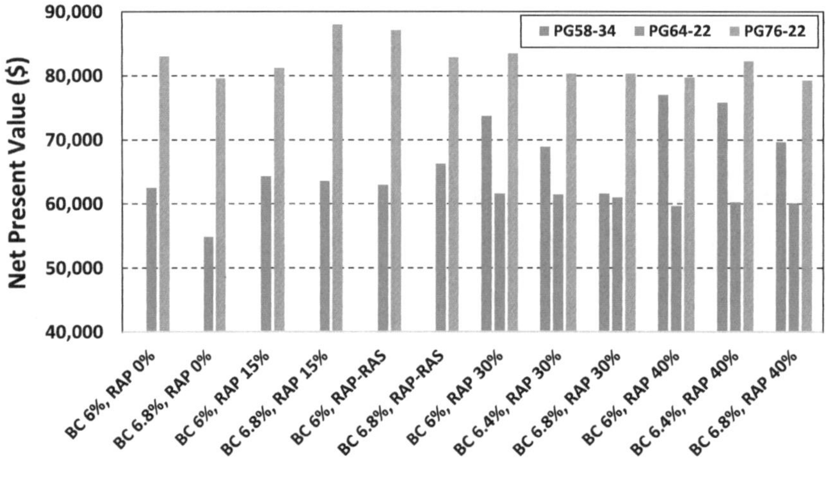

(b)

Figure 7. Calculated material cost net present values for all cases, considering only material costs for (*a*) Portland and (*b*) Ontario.

- None of the sections considered failed from rutting within the first 15 years of analysis for a failure criterion of 12.7 mm rut depth, and the majority of the sections did not fail within the first 25 years of analysis. Because asphalt aging significantly increases pavement stiffness during this long time period, it is highly likely that none of the sections will fail from rutting. For this reason, bottom-up fatigue cracking was used as the only failure criterion for LCCA.
- The cracking accumulation rate for Ontario is higher than that of the Portland climate because of lower winter temperatures for Ontario.
- Binder content was observed to be the most significant factor controlling bottom-up cracking. Increasing binder content and decreasing RAP content was observed to decrease cracking accumulation.
- For the sections with PG76-22 binder, cracking performance was observed to be significantly lower. Increasing binder content and decreasing RAP content did not appear to create significant changes in performance. However, it should be noted that according to (*4*), asphalt mixtures with PG76-22 binder and higher binder contents (7.2%–7.8%) can create mixtures highly resistant

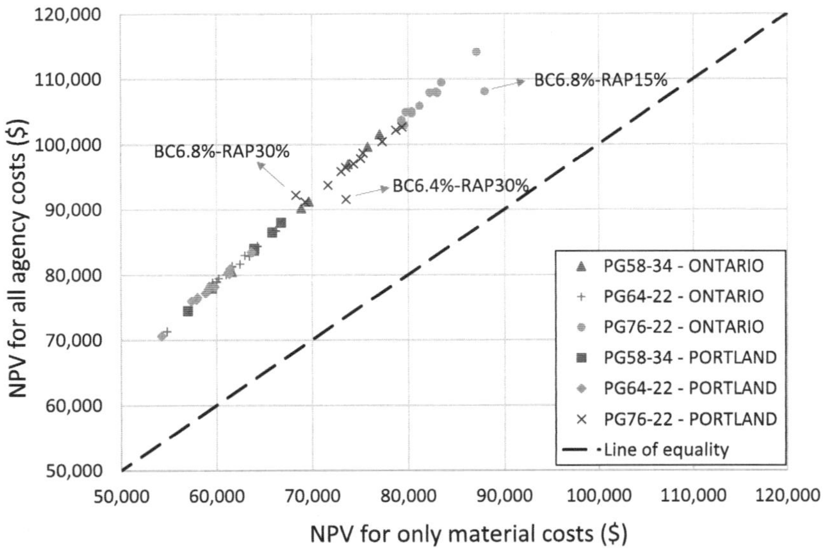

Figure 8. Correlation between NPVs calculated by using material and agency costs.
Note: NPV = net present value.

to both rutting and cracking. DM tests should be conducted with the suggested high binder content mixtures to be able to develop MEPDG models and simulate their rutting and cracking performance. Increased performance for PG76-22 mixtures with higher binder contents may result in reduced life-cycle costs.

- Increasing the binder content of an asphalt mixture is an effective strategy to improve the condition of the highway network and reduce long-term costs.
- Higher RAP contents generally result in lower life-cycle costs.
- Because of colder temperatures, simulations with the Ontario climate resulted in lower cracking resistance. Lower cracking resistance increased the life-cycle costs for these sections.
- Although using total agency costs increased the calculated NPVs (as expected), a strong linear correlation between NPVs for material costs and total agency costs suggests that using total agency costs rather than only material costs did not change the rankings and conclusions except for three cases shown in Figure 8 (PG76-22-BC6.8%-RAP15%-Ontario, PG76-22-BC6.4%-RAP30%-Portland, and PG76-22-BC6.8%-RAP30%-Portland).

Suggestions for Future Work

In this study, MEPDG simulations were conducted to determine the rutting and bottom-up cracking performance of different asphalt mixtures. MEPDG longitudinal cracking (top-down cracking) models were not used for simulations because the accuracy of these models was determined to be low (*6*, *7*). Because top-down cracking is the major distress mode in Oregon, using top-down cracking models for performance prediction and LCCA will provide more realistic results. More effective models that can explain the mechanism behind top-down cracking are currently being developed in research project NCHRP 01-52. The analysis performed in this research study should also be performed with top-down fatigue cracking as the main failure mode to further evaluate the cost and performance effectiveness of different high RAP/RAS strategies in Oregon.

It would behoove this research topic to have some specific data on the properties of RAP/RAS binder. Because of financial and temporal constraints on this study, binder testing on these materials could not be performed. It is suggested that binder extraction and testing be performed on these materials to gain insight on the properties of the recycled binder and their effect on the performance of asphalt mixtures containing RAP/RAS.

For the sections with PG76-22 binder, cracking performance was observed to be significantly lower. Increasing binder content and decreasing RAP content did not appear to create significant changes in performance for these cases. However, it should be noted that according to (*4*), asphalt mixtures with PG76-22 binder and higher binder contents (7.2%–7.8%) can create mixtures highly resistant to both rutting and cracking. DM tests should be conducted with the suggested high binder content mixtures to be able to develop MEPDG models and simulate their rutting and cracking performance. Increased performance for PG76-22 mixtures with higher binder contents may result in reduced life-cycle costs.

Acknowledgments

This paper describes research activities that were requested and sponsored by the Oregon Department of Transportation (ODOT). Federal and State sponsorship and interest are gratefully acknowledged. The authors thank the members of the ODOT Project Technical Advisory Committee and ODOT research for their advice and assistance in the preparation of this report. The authors would like to thank Shashwath Sreedhar and Sogol Haddadi for their help with laboratory testing.

Author Contributions

The authors confirm contribution to the paper as follows: study conception and design: EC = 70%, YZ = 20%, BW = 10%; data collection: EC = 50%, YZ = 40%, BW = 10%; analysis and interpretation of results: EC = 50%, YZ = 20%, BW = 30%; draft manuscript preparation: EC = 60%, YZ = 5%, BW = 35%. All authors reviewed the results and approved the final version of the manuscript.

References

1. Willis, J. R., P. Turner, G. Julian, A. J. Taylor, N. Tran, and F. Padula. *Effects of Changing Virgin Binder Grade and Content on RAP Mixture Properties.* NCAT Report No. 12-03. National Center for Asphalt Technology, Auburn, Ala., 2012.
2. Hansen, K. R., and A. Copeland. *Asphalt Pavement Industry Survey on Recycled Materials and Warm Mix Asphalt Usage.* Information Series 138, 5th ed. National Asphalt Pavement Association, Lanham, Md., 2014.
3. Al-Qadi, I. L., Q. Aurangzeb, S. H. Carpenter, W. J. Pine, and J. Trepanier. *Impact of High RAP Content on Structural and Performance Properties of Asphalt Mixtures.* FHWA Report FHWA-ICT-12-002. Illinois Center for Transportation, Urbana, 2012.
4. Coleri, E., S. Haddadi, S. Sreedhar, S. Lewis, Y. Zhang, and B. Wruck. *Binder-Grade Bumping and High Binder Content to Improve Performance of RAP-RAS Mixtures.* Final Report SPR 797. Draft report (under review by TAC), Oregon Department of Transportation, Salem, 2017.
5. AASHTO. *AASHTOWare-Pavement ME Design 1.0^{TM}.* American Association of State Highway and Transportation Officials, Washington, D.C., 2008.
6. Williams, R. S., and R. Shaidur. *Mechanistic Design Guide Calibration for Pavement Rehabilitation.* Final Report SPR 718. Oregon Department of Transportation, 2013.
7. Von Quintus, H., J. Mallela, R. Bonaquist, C. Schwartz, and R. L. Carvalho. *Calibration of Rutting Models for HMA Structural and Mixture Design.* NCHRP Report 9-30A, Final Report No. 719. NCHRP, Transportation Research Board, Washington, D.C., 2009.
8. Herbsman, Z. J., and C. R. Glagola. Lane Rental—Innovative Way to Reduce Road Construction Time. *Journal of Construction Engineering and Management*, Vol. 124, No. 5, 1998, pp. 411–417. https://doi.org/10.1061/(ASCE)0733-9364(1998)124:5(411).
9. North Carolina Department of Transportation. *Guidance for Determining Contract Times.* North Carolina Department of Transportation, 2014.
10. Berkeley Lab, Energy End-Use Forecasting Group. *New Pavement Pricing.* http://enduse.lbl.gov/Info/pavements_LCCA_web.xls. Accessed February 14, 2017.
11. Zhang, H., G. A. Keoleian, M. D. Lepech, and A. Kendall. Life-Cycle Optimization of Pavement Overlay Systems. *Journal of Infrastructure Systems*, Vol. 16, No. 4, 2010, pp. 310–322. https://doi.org/10.1061/(ASCE)IS.1943-555X.0000042.

The Standing Committee on General and Emerging Pavement Design (AFD30) peer-reviewed this paper (18-01079).

Article

TRR
JOURNAL OF THE TRANSPORTATION RESEARCH BOARD

Transportation Research Record
2018, Vol. 2672(40) 155–165

DOI: 10.1177/0361198118790097
journals.sagepub.com/home/trr
SSAGE

Quality Management Program for Pavement Condition Data Collected on the National Highway System

Pedro A. Serigos[1], Beth A. Visintine[1], Amy L. Simpson[1], Gonzalo R. Rada[1], and Jonathan L. Groeger[1]

Abstract

This paper proposes a framework for a pavement condition data Quality Management Program (QMP). The framework complies with the final rule for national performance management measure regulations published by the Federal Highway Administration (FHWA) to implement requirements from the Moving Ahead for Progress in the 21st Century Act and the Fixing America's Surface Transportation Act. The quality check procedures and criteria recommended for the different elements of the framework are based on the experience and recommendations that resulted from developing a QMP for collecting approximately 10,000 miles of pavement condition data on Interstate Highways using an automated measurement system. The findings from implementation of the study QMP and the intervention needed for automated data collection condition metrics were documented for future improvements.

The two latest surface transportation authorization bills – the Moving Ahead for Progress in the 21st Century Act (MAP-21) (*1*) and the Fixing America's Surface Transportation (FAST) Act (*2*) – have required that performance measures be established for the National Highway System (NHS). As a result of their enactment, the Federal Highway Administration (FHWA) issued a Notice of Proposed Rulemaking (NPRM) in January 2015 (*3*) to establish performance measures to assess the condition of the pavements on the NHS. In January 2017, the final rule (*4*) was published and became effective on May 20th, 2017.

The final rule on pavement condition measures requires State Departments of Transportation (DOTs) to develop and implement a Quality Management Program (QMP) for the condition data collected to report pavement metrics for the Highway Performance Monitoring System (HPMS). State DOTs must submit their QMP to FHWA for approval not later than one year after the effective date of the rule and posterior changes to the QMP must be documented and resubmitted to FHWA for approval. The main goal of the QMP is to ensure acceptable quality data and enhanced consistency through documenting and formalizing the quality control (QC) and quality assurance (QA) processes and methods already implemented by the agencies and incorporating missing QC/QA processes for the collected data within a systematic and traceable framework. Ensuring adequate quality data and consistency will allow for more reliable estimates for the performance measures used to support and monitor strategies toward achieving the performance targets established in asset management plans for the NHS.

This paper proposes a QMP framework following the requirements of the final rule. The QC/QA procedures and criteria recommended for the different elements of the framework are based on the experience and recommendations that resulted from the recent FHWA study "Interstate Pavement Condition Sampling" (*5*) (referred to in this paper as "IS Condition study"), and expanded based on the final rules published after the study was completed. Some of the findings from the IS Condition study were used to support requirements subsequently defined in the final rule, such as the pavement section length and lane required to collect and report data to HPMS, among other aspects. The data management and quality evaluations developed for the IS Condition study for the collection of pavement condition data on approximately 10,000 miles of Interstate Highways (IH) using

[1]Amec Foster Wheeler, Beltsville, MD

Corresponding Author:
Address correspondence to Pedro A. Serigos: Pedro.Serigos@amecfw.com

an automated measurement system were discussed and approved by FHWA. These data quality processes and methods cover most of the QMP components required by the rule for State DOTs and allowed for detecting quality issues to be considered when developing a pavement data QMP.

The next section of the paper provides a brief description of the main components and data elements specified in the rule regarding the pavement condition data QMP. The following three sections present: (1) the proposed data QMP framework containing a list of recommended elements within each required component; (2) a discussion of implementation aspects with a description of QC/QA processes and criteria for each of the recommended QMP elements; and (3) lessons learned from selected cases of data quality issues from the IS Condition study. The last section of the paper provides the main conclusions with recommendations for agencies to develop pavement condition data QMP to comply with the rules promulgated by the MAP-21 and FAST acts.

Regulation Related to Pavement Condition Data QMP

The pavement condition data QMP framework developed in this paper was prepared in response to item §490.319(c) of the final rule for national performance management measure regulations published by the FHWA on January 18, 2017 (*4*) to implement requirements from the MAP–21 (*1*) and FAST acts (*2*). Item §490.319(c) requires that each State DOT "... *shall develop and utilize a Data Quality Management Program, approved by FHWA that addresses the quality of all data collected, regardless of the method of acquisition, to report the pavement condition metrics, discussed in §490.311, and data elements discussed in §490.309(c)*" (*4*).

Metrics and Inventory Data Elements

The pavement condition metrics included in §490.311 consist of: (i) the International Roughness Index (IRI); (ii) the percentage of cracking (cracking percent) for all pavement types; (iii) the average depth of rutting for asphalt pavements; and (iv) the average height of faulting for jointed concrete pavements. Item §490.311 also specifies that the present serviceability rating (PSR) may be used as an alternative to the IRI, cracking percent, rutting, and faulting pavement condition metrics for the mainline highways on the non-Interstate NHS with posted speed limits of less than 40 mph. Item §490.311 provides the definition and specifications for calculating each of these condition metrics for each pavement surface type, as well as other requirements for reporting these data to the HPMS, such as the section length, and time and location reference information. In addition, §490.309(c) provides the requirements for reporting three additional inventory data elements: through lanes, surface type, and structure type.

Based on the condition metrics included in §490.311, each DOT must report the following four pavement performance measures, as indicated in §490.307: (1) percentage of pavements of the Interstate System in good condition; (2) percentage of pavements of the Interstate System in poor condition; (3) percentage of pavements of the non-Interstate NHS in good condition; and (4) percentage of pavements of the non-Interstate NHS in poor condition.

Required QMP Components

Item §490.319(c) requires that the data QMP submitted by DOTs include, as a minimum, methods and processes for five components: (1) data collection equipment calibration and certification; (2) certification process for persons performing manual data collection; (3) data quality control measures to be conducted before data collection begins and periodically during the data collection program; (4) data sampling, review and checking processes; and (5) error resolution procedures and data acceptance criteria.

Framework for Pavement Condition Data QMP

This section describes a general framework recommended for DOTs to develop their data QMP to comply with item §490.319(c) of the third final rule for FHWA's national performance management measure regulations (*4*). The section is organized in accordance with the five required data QMP components. The FHWA rules do not specify what elements are to be included within each component. The elements proposed in the framework for inclusion in each component are based on the experience and recommendations resulting from the data QMP developed as part of the FHWA's IS Condition study (*5*) and expanded based on the final rules published afterwards.

QMP Component 1: Data Collection Equipment Calibration and Certification

This component involves the procedures for ensuring that the equipment used by the DOT for collecting the data to be reported to HPMS have the acceptable quality. The following two main actions are recommended for each type of pavement condition metric collected:

- Specify data collection guidelines and standards for automated measurement methods.
- Establish and document procedures for calibration and certification of equipment before data collection.

QMP Component 2: Certification Process for Persons Performing Manual Data Collection

This component has the goal of ensuring that the aptitude of personnel involved in manual data collection methods is appropriate. The following two main actions are recommended for each type of pavement condition data manually collected:

- Specify data collection guidelines and standards for manual data collection methods.
- Establish and document certification processes for manual data collection raters.

QMP Component 3: Data Quality Control Measures Conducted Both before Data Collection Begins and Periodically during the Data Collection Program

This component comprises the methods and criteria for controlling the quality of measured data before and during data collection. The following three main actions are recommended for each type of pavement condition data:

- Establish and document procedures for periodic quality checks of data collection equipment and manual raters.
- Establish and document procedures to continuously monitor data being collected and checks at the end of the day.
- Monitor ambient conditions and specify how to address issues.

QMP Component 4: Data Sampling, Review, and Checking Processes

This component comprises the methods and criteria for controlling the quality of measurements after data collection. The following two main actions are recommended for each type of pavement condition data:

- Establish and document a method to sample the collected data for quality checks.
- Develop and document processes to perform data quality checks on collected data and to review flagged data.

QMP Component 5: Error Resolution Procedures and Data Acceptance Criteria

This component comprises the acceptance criteria and error resolution procedures for the quality checks involved in the data QMP. The following actions are recommended:

- Establish reporting requirements for tracking procedures within the QMP including logs for tracking QC procedures and acceptance of deliverables.
- Establish data acceptance criteria and error resolution procedure for issues identified through the QC program.

Recommended Data QMP Elements Based on "IS Condition" Study

This section discusses implementation aspects and provides recommendations for each of the elements proposed within the required data QMP components. The recommended quality checks and processes for each QMP element were developed for the IS Condition study, except for the ones corresponding to the second component as no manual data were collected for the study.

QMP Component 1: Data Collection Equipment Calibration and Certification

Data Collection Guidelines and Standards for Automated Measurement Methods. The standard and guidelines for all data types are specified in item §490.309(3) of the rule. State DOTs must meet the required standards and their associated reporting format for each piece of measuring equipment used to collect the data must be reported to HPMS. For instance, the IRI data collection device must comply with American Association of State Highway and Transportation Officials (AASHTO) Standard M328-14 (*6*), and the five-point collection of rutting data method must comply with AASHTO Standard R48-10 (*7*). The specified standards and guidelines for some of the data types are to be collected in accordance with the HPMS Field Manual (*8*). The HPMS Field Manual describes methodological aspects for measuring each data type and indicates the standards to be followed.

Several State DOTs use guidelines and standards to specify their data collection methods and distress definitions that are not compatible with the ones required by the item §490.309(3) of the rule. A possible approach used in the past by some DOTs to overcome this discrepancy is to produce two different pavement condition datasets from the raw data collected by the measuring equipment: one based on the guidelines and standards defined by the DOTs to be used in their Pavement

Management System (PMS) and the second one based on the guidelines and standards required by the FHWA rule to be reported to HPMS.

Methods and Procedures for Calibration and Certification of Equipment. State DOTs are required to define and document calibration and certification methods and procedures for each type of equipment (e.g., the ones recommended by the equipment manufacturer or developed in-house) involved in the collection of the data types reported to HPMS as part of the QMP. As the FHWA rule does not specify standard methods for certifying the required data types—it only mentions that DOTs are expected to specify the AASHTO Standard R56 (*9*), or an equivalent standard, for certifying IRI measuring equipment—State DOTs must design certification procedures and acceptance criteria for FHWA approval. The AASHTO PP-67 (*10*) and PP-70 (*11*) provide guidance and recommendations for validating automated measurements of surface distresses and transverse profile data. These protocols recommend conducting a field experiment to compare measurements taken by the automated systems to reference measurements collected by certified manual raters on a set of pavement sections selected by the DOT to be representative of the pavement system being measured. In addition, there are efforts currently underway for developing certification standards for these data types; however, these standards are not yet ready for implementation.

The certification procedures carried out in the IS Condition study for inertial profiling devices followed the AASHTO Standard R56-14 (*9*), and the one for the automated measurement of transverse profile to calculate rutting data followed the guidelines provided in the AASHTO PP 70-14 (*11*) standard. The allowed measurement error for the inertial profiler devices in the project was: (a) "10 runs of Road Surface Profilometer (RSP) data having a minimum repeatability of 92 percent compared with profile measured with a reference device," and (b) "10 runs of RSP data having a minimum accuracy of 90 percent compared with profile measured with a reference device." The allowed measurement error for the transverse profile measuring device in the project was "10 runs of data have a standard deviation of less than 1 mm rut depth compared with rutting hand measurements taken in the field." Refer to the IS Condition Addendum Report (*12*) for more details on the allowed repeatability levels during QC checks for each data type.

QMP Component 2: Certification Process for Persons Performing Manual Data Collection

Data Collection Guidelines and Standards for Manual Data Collection Methods. State DOTs must meet the required standards and their associated reporting format for all data collection methods involving manual data collection personnel. As part of this, DOTs should evaluate potential discrepancies between the pavement condition definitions currently adopted by the DOT and the ones defined in the HPMS. If discrepancies are identified, the DOT should evaluate remedial actions to meet the required standards, such as developing equations to translate the collected condition metrics or re-training personnel to collect data using the required definition when translation is not technically possible.

Certification Processes for Manual Data Collection Raters. State DOTs must document the procedures for certifying and training personnel involved in manual data collection of the data types reported to HPMS. Several DOTs have experience developing and conducting certification procedures for manual data collection personnel. These procedures should be reviewed and, if necessary, modified for compliance with the current rules; e.g., ensuring that surface distresses are rated in accordance with the definitions and methods from the HPMS manual (*8*).

A reference procedure for certifying manual raters of pavement surface distress is the one used by the FHWA's Long-Term Pavement Performance (LTPP) program (*13*). This procedure consists of evaluating the accuracy of manually collected distress surveys on a set of pavement sections using a consensus survey performed by a committee of two to four "master" manual raters as a reference. The consensus survey is most frequently conducted by three raters with vast experience in the manual data collection of surface distresses. The consensus survey involves the master raters completing the distress survey together discussing their interpretation of each detected distress. The resulting distress maps and statistics, produced as the agreement among the master raters, are used as reference for certifying manual raters (refer to Rada et al. (*13*) for more information regarding the accreditation process for LTPP manual raters). In addition, it is recommended that DOTs include field experiments to control for the manual raters' repeatability and reproducibility when designing their certification processes. These two properties can be controlled by limiting the variability of survey results produced by the same rater, and among different raters, on the same test sections and under the same experimental conditions.

Manual raters may also be involved with the condition assessment of pavement surfaces through digital images. This includes the assessment of pavement conditions from unrated images collected by a survey vehicle as well as the review of distress maps generated by automated distress measurement systems. Certification procedures for personnel involved in these tasks should evaluate the accuracy, repeatability and reproducibility of the manual raters using a set of consensus distress maps as a reference. The consensus distress maps can be produced by a crew of experienced manual raters completing the distress survey together discussing their interpretation of each detected distress, similarly to the procedures for collecting ground truth data for certifying LTPP manual distress raters in the field.

QMP Component 3: Data QC Measures Conducted Both before Data Collection Begins and Periodically during the Data Collection Program

Periodic Quality Checks of Data Collection Equipment and Manual Raters. Certified equipment and personnel involved in the collection of each reported data type should be periodically checked for accuracy and precision to ensure their capacity is maintained over time. These periodic checks are a supplement to the certification processes defined in the first two QMP components. The frequency and procedure for these periodic checks will be defined by the DOT for each data type and measurement method according to its needs and the intended use of the data. For reference, AASHTO Standard PP-70 (*11*) suggests performing routine checks monthly where the measurements of transverse profiles taken by the equipment are compared with reference profile data manually collected on a set of test sections. A similar field test can be designed by the DOT to conduct routine checks of accuracy and repeatability for each piece of equipment and manual rater. Equipment or personnel that does not pass the acceptance criteria for the periodic checks should not be allowed to collect data until the issue is resolved and the equipment is re-certified.

State DOTs should select a certain portion of their networks (5% to 10% of the network mileage is recommended based on analysis performed for the IS study and on practical constraints) to be collected for QC purposes. The amount of mileage to be collected should be broken up into groups of segments selected randomly across the network. DOTs should specify the allowable difference between repeated runs of collected QC data as well as the allowable percentage of data that exceed this allowable difference before data collection is ceased to investigate the issue. In addition, DOTs that use multiple data collection vehicles for data collection should consider implementing a similar requirement for data consistency between the multiple vehicles.

The IS Condition study's QMP required that monthly repeatability checks be performed on the equipment. These checks required five repeat runs on a short (but no less than 1,000 Ft) pavement section. Data from the monthly repeatability checks were required to be within 5% of each other for each data collection item—IRI, cracking, rutting and/or faulting. Figure 1 illustrates the monthly IRI repeatability checks conducted during the study, where roughness was within the 5% limit. If either questionable data and/or improperly functioning equipment were suspected, the data collection team was

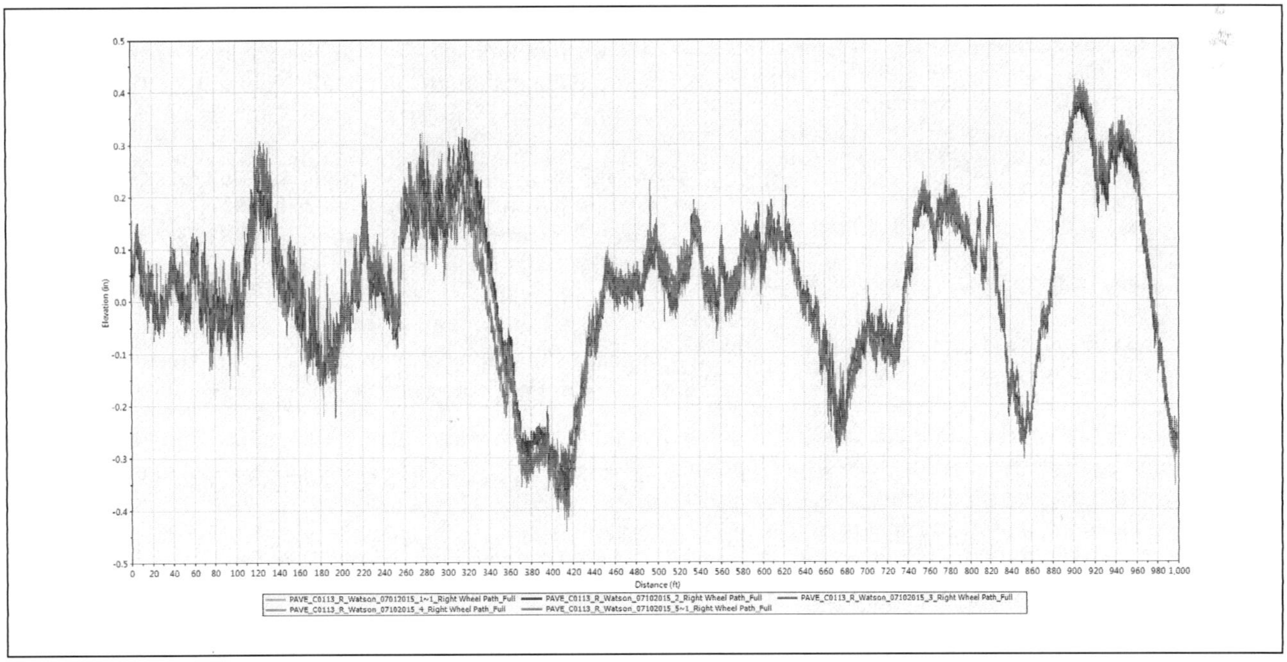

Figure 1. Monthly repeatability testing for IRI.

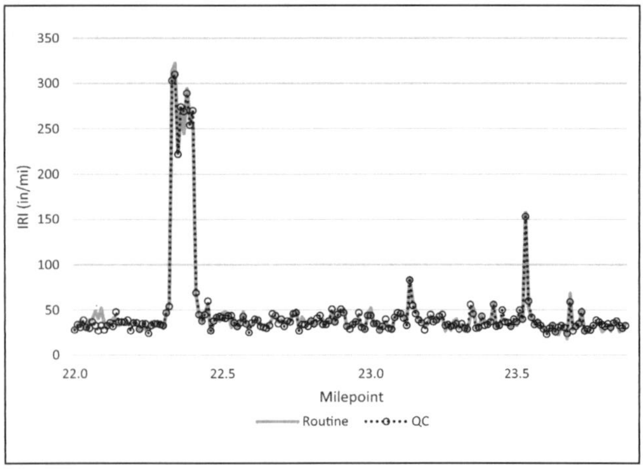

Figure 2. Comparison of routine versus QC data collection.

required to cease data collection activities immediately until the issue(s) in question was resolved; however, data collection cessation was not required during the study because the quality of the data was considered very good.

In addition to the verification sites, repeat data were collected throughout the study route for QC purposes to provide supplementary data to check the repeatability of the data collection. About 5% or 500 miles of the study's approximately 10,000-mile route was specifically collected for QC purposes. These QC data were compared with the previously collected data, as illustrated in Figure 2 for 21 500-Ft segments of IRI data (refer to (*12*) for more details). Segments that had measurements that differed by more than 5% were identified. If more than 10% of the QC data were identified as differing by more than 5%, the data collection team was required to stop data collection and investigate the potential cause(s) of the difference; however, this was not required during the study.

Monitor Data Being Collected and Checked at the End of the Day. State DOTs should document quality-check procedures performed during data collection and at the end of the data-collection day to prevent requiring recollection of data at a later date and so that equipment or other data collection issues can be addressed in a timely manner to prevent requiring recollection of multiple days of data. These checks may be conducted by contracted staff when data collection is outsourced to a service provider, in which case the DOT's role should consist of—as a minimum—providing input on acceptance criteria and data elements to check, revising QC checks documentation upon receipt and, based on the QC results, making the decision of whether or not to recollect data.

For the IS Condition study, the data collection contractor was to follow established guidelines and to continuously monitor the data being collected to check that data appeared reasonable and that equipment was functioning properly. In addition to the automated data checks performed by the system during data collection, data were checked at the end of the day to make sure major flaws (e.g., out-of-range values, blurry images, profile spikes) were not visually detectable. It was also considered important that the data collection technicians know, precisely, the correct pavements to be surveyed within each State route along with the correct direction, lane and milepost.

To minimize subjective acceptance criteria for the visual checks conducted during data collection, it is recommended that an automated QC program be developed and incorporated into the measurement system software. The development of an automated program for checking QC aspects in real time, such as out-of-range measurements, would allow for more consistent, traceable and timely detection of equipment malfunctions.

Monitor Ambient Conditions. It is important to verify that the equipment is operated within the manufacturer's specifications when planning for, or performing data collection, as extreme ambient conditions can significantly affect the quality of the data. It is recommended that DOTs include ambient condition considerations applicable to their local conditions in their data QMP and specify how to address issues; e.g., stop data collection once temperatures reach the specified limit for sensor capabilities.

In developing QC plans for the IS Condition study, the data collection contractor was required to address controls and mitigation efforts for impacts caused by ambient conditions. It was required that the contractor stop data collection immediately once ambient conditions beyond those under which the equipment sensors could properly operate occurred. As reported in the IS Condition final report, extreme temperatures were encountered during the data collection for the project and the sensors and collected data were monitored to check that the quality of data was not affected and that the equipment was functioning properly (*5*).

QMP Component 4: Data Sampling, Review, and Checking Processes

Method to Sample the Collected Data for Quality Checks. State DOTs should document the methods and criteria used to sample the collected pavement network to perform quality checks for each data type. The resulting sample should be representative of the pavement network being measured and the methods used for obtaining the sample should promote the use of unbiased data.

For the selected sample to be an unbiased representation of the collected pavement network, it is recommended that the DOTs apply a stratified sampling technique covering a series of factors such as annual average daily traffic levels, the functional system and the speed limit of the measured network, among others. DOTs should report the factors and levels selected to achieve representativeness of the pavement network conditions, as well as the sample size determined for each stratum. The precision level adopted for determining the sample size for each combination of factors should reflect their relative importance. As a reference, the HPMS Field Manual (see Table 6.2 (*8*)) tabulates the precision level to use for sampling the pavement network as a function of the highway urban code and facility type; e.g., the HPMS Field Manual requires the use of a 90% confidence level and 10% precision rate when calculating the sample size for the strata corresponding to minor arterials in rural areas whereas the precision rate is 5% for the case of rural IH.

Processes for Quality Checks and Review of Flagged Data. DOTs should develop and document processes and criteria to assess the quality of data after collection and to review the flagged data. Quality checks should control to ensure that the collected data are not corrupt, and that they complete (i.e., no missing required data elements), within their valid range, and consistent. In addition, it is recommended that DOTs include temporal analysis as part of the data quality checks to detect unexpected changes in condition over time indicating potential data collection issues or possibly an issue with the location referencing of the collected data. Data not passing the quality checks should be flagged and subjected to further review to identify any systemic problems with the collection.

The IS Condition project QC review included the development of an automated QC program to check for data completeness, range of values, and consistency. The complete list of checks included in the QC program can be found in the Data Storage and Quality Review Plan appendix to the report (*5*)—examples include roughness values between 40 and 250 in/mi, faulting less than 1 in., and a difference in rut depth between wheel-path values at the same location not greater than 0.25 in. Flagged data not passing the stipulated checks were reviewed further, including comparing the flagged data values to the pavement images to assess whether the data were reasonable. If the flagged data were deemed reasonable, they were added to the project database. Data not passing the reasonableness check from the image review were discussed with the data collection team for further assessment and error resolution.

The review of flagged data was conducted by a QA auditor who was separate from the day-to-day data collection effort. The QA auditor was a principal engineer with over 25 years' experience in the pavement field. The flagged data were reviewed to identify causes for the flags, such as missing data values or values failing the range checks, as well as to identify if there were systemic problems with the collection based on the type and frequency of the flags. The data flags were reviewed using both the data and the images. The images provided some perspective on the data that could not be seen by just looking at the data; e.g., large ruts could be observed in some forward images. In addition, faulting data were reviewed in combination with ride quality data to confirm that they—faulting and ride quality—were both consistent (refer to (*12*) for detailed information on the QA audits).

The next section of this paper describes representative examples of flagged data and the findings from the review performed on them. An important component of the IS Condition study's QMP that involved a significant level of effort consisted of the quality check and review of pavement surface images collected by the automated cracking detection system. Simpson et al. (*14*) recommended that images should be reviewed for systematic errors in data collection and processing for a minimum of 5% of the images. Simpson et al. also recommended that if a systematic error or an error rate greater than 15% was identified during the review, then an increase in the percentage of images to be reviewed was warranted and necessary. The review of images for the IS Condition study included the following elements:

- For each day of data collection, images were reviewed by visually checking that they were of sufficient quality to enable distress takeoffs—e.g., not too bright or blurry—and they captured expected road features—e.g., painted stripes—without excessive distortion or abnormalities. The review of images was performed for five miles at the first mileage collected of the day, at the quarter mileage collected, at the half mileage collected, at the three-quarter mileage collected and at the last mileage collected. This allowed for data collected to be reviewed at different times of the day under different lighting conditions. In addition, 5% of the event markers, such as bridges, were compared with the Right-of-Way images each day to help check the location referencing.
- Areas of cracking identified in downward images were compared with forward images to assess the numerical quantity of cracking reported. About

15% of locations were selected to compare the areas of cracking (numerical) to the images and to compare the downward and forward images to assess the quantity of cracking reported. AASHTO PP 67-16 provides more detail concerning an acceptable result of the automated crack detection process including (*10*):
  - 33% of the cracks less than 0.12 in. and 60% of cracks from 0.12 in. to less than 0.2 in. wide are mapped.
  - 85% of the cracks 0.2 in. or wider are mapped.
  - A crack is considered mapped if at least 85% of its length is detected.

Data collection for the IS Condition study was only performed in 2015, which did not allow for the conduct temporal analyses of pavement conditions. However, the earlier referenced FHWA Infrastructure Health study (*14*) provides recommended criteria for the minimum and maximum allowable annual change in pavement condition, which can be applied to those data passing those data-quality checks previously addressed in this paper. The recommended criteria for flagging data based on annual pavement condition changes are as follows:

- IRI increase > 10 in./mile per year or decrease > 5 in./mile
- Rut depth increase > 0.1 in./year or decrease > 0.05 in.
- Faulting increase > 0.08 in./year or decrease > 0.04 in.
- Percent cracking > 10% / year or decrease > 5%

QMP Component 5: Error Resolution Procedures and Data Acceptance Criteria

Each State DOT must develop acceptance criteria and error resolution procedures for the different quality checks involved in the data QMP. Setting the acceptance criteria includes defining the target level of data accuracy, precision, and reproducibility to accept the collected data. If the collected data do not pass the acceptance criteria, there should be error-resolution procedures in place outlining the steps to be taken to address the issues. Such procedures should include the basis for recommendations such as reprocessing or recollecting the data failing a specified check, or the rejection of the deliverable.

The IS Condition study's QMP defined acceptance criteria for all its quality checks. As an example, the acceptance criteria after review of flagged data from the automated quality check of collected data were the following:

- If data appears to be an outlying record, data were maintained and stored in database.
- If data appear to be part of a persistent issue (more than four records flagged for a day for the same reason), the contractor was notified.
- If more than 10 records are flagged for same issue, a stop-work order was issued to the contractor to sort out the problem and correct any equipment issues.

The error resolution procedure defines the steps to be taken to address issues with data not passing the acceptance criteria. The error resolution procedure should also include a feedback or documentation aspect that logs all issues identified, the procedure used to address the issues, and any additional outcomes. This should be used to improve the data QMP or kept for troubleshooting similar issues at a future time.

The flowchart from Figure 3 shows the simplified error resolution approach developed for the IS Condition study (*5*). The proposed methodology sorts the data issues into the following four types:

- Procedural: An error such as an incorrect method of calculating percent cracking requires an adjustment to the method or procedure. With procedural errors, there is no issue with the raw data, but with the post-processing procedures used to report the data; e.g., slabs with both longitudinal cracks and transverse cracks being counted as a cracked slab for each type of cracking instead of being counted as a single cracked slab.
- Processing: An example of these errors includes not selecting the proper filters during processing, such as including sealed cracks in the calculation. Processing errors require reprocessing of the raw data.
- Data quality and omission: These errors include poor image quality, poor accuracy due to equipment malfunction or out of calibration. They require re-collection of the data.
- Data correctness: These errors include collecting the wrong route, data elements or using an incorrect standard for data collection. These errors result in rejection of the data.

The feedback mechanism loops back to both the data issue sorting and the recommended procedures to address the errors. This enables the approach to be improved and updated. Each DOT should develop a more detailed approach to address errors encountered within their system.

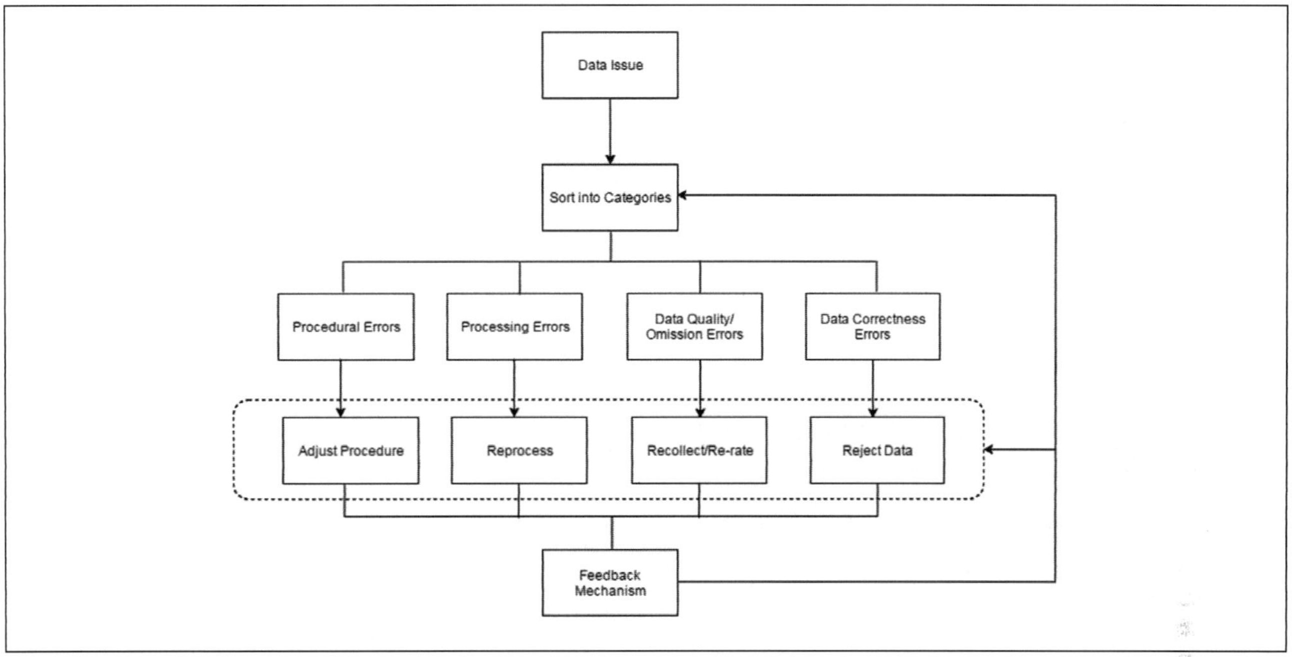

Figure 3. Error resolution methodology (*5*).

Cases of Flagged Data from Study QC Checks

This section reports selected examples of flagged data elements from the IS Condition study QC checks. Documenting the findings from the implementation of a data QMP is a critical activity, as it can highlight the importance of such a program as well as suggest other areas that should be included in updated versions of the QMP to address lessons learned. The findings from the implementation can also be used in the feedback loop of the error resolution procedures.

Throughout the IS Condition study, several items were flagged by elements of the data QMP where corrections were required. Selected examples include:

- Range and completeness checks that were part of the automated QC program identified issues with the cracking percent calculation for Jointed Plain Concrete Pavement (JPCP), where 0.1% of the reported values for the data submittal were greater than 100%, as well as issues with surface type and cracking percent being null. This error falls into the category of "Procedural Errors" (Figure 3). The procedure for calculating the percent cracking for JPCP was corrected by the data collection contractor, which required re-computation of cracking percent for the entire JPCP data set. The values greater than 100% were the result of an issue with the calculation algorithm that resulted in double counting cracked slabs in those cases where the slabs had both transverse and longitudinal cracking.
- Data consistency checks showed that summary values were reported for incorrect intervals. This error falls into the category of "Processing Errors" (Figure 3). The reporting interval for the data was corrected and values were re-processed for the 507 miles of IH collected.
- Review of forward-facing images for clarity revealed that there was a shadow created by the van when data collection was performed late in the afternoon and the survey van travelled in an easterly direction. This error falls into the category of "Data Quality" (Figure 3). Data collection was stopped until the problem was addressed, and the data for the affected six miles were recollected.
- Review of pavement images collected by a 3-D system revealed that cracks filled with sand (most likely due to wind) were not being properly identified by the automated crack detection software. Figure 4 shows the intensity image (i.e., color information) and the range image (i.e., depth information) of a stretch of pavement surface where some visible cracks filled with sand were not reported because no depth was detected in the range image. This error falls into the category of "Data Quality" (Figure 3). Efforts to automate the detection of these cracks by the data collection contractor were unsuccessful. As a result,

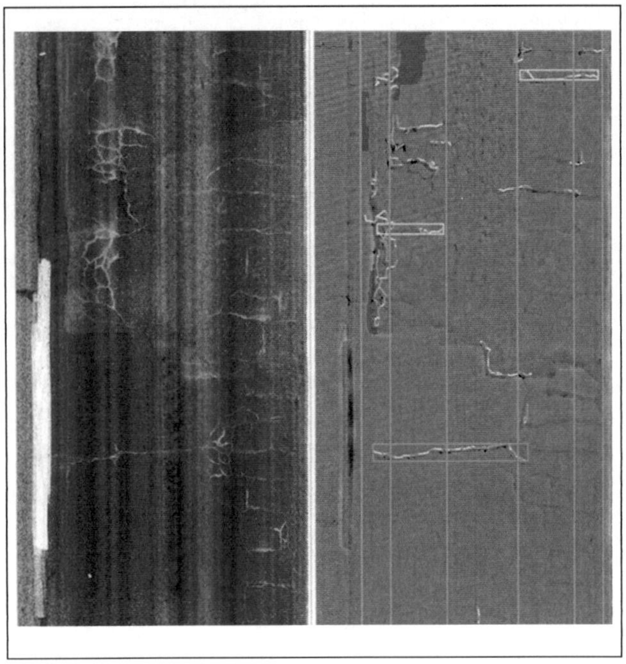

Figure 4. Sand-filled crack flagged data example.

significant effort was spent in: (1) identifying where those cracks occurred; (2) performing manual take-offs of those cracks from the images; and (3) integrating the resulting take-offs with the rest of the cracking data. Fortunately, the occurrence of the sand-filled cracks on the data collection route was rare, with most of it occurring along I-10 in California and Arizona as well as along I-15 in California. The larger number of cracks detected by the manual intervention of images resulted in an average increase in cracking of 1.5% for the affected segments.

Summary and Conclusions

This paper proposes a framework for a pavement condition data QMP for State DOTs to comply with the final rule for national performance management measure regulations published by the FHWA to implement requirements from the MAP–21 (*1*) and the FAST acts (*2*). Most of the QC/QA procedures and criteria recommended within the proposed framework were developed as part of the IS Condition study conducted for the FHWA for the collection of approximately 10,000 miles of pavement condition data on the IH. These QC/QA processes and criteria were discussed and approved by the FHWA within the context of the study. As a result, the recommended QMP framework covers the five required components of the rule and includes QA/QC processes that were developed and tested under actual conditions in the field, and with the support and approval of the FHWA. In addition, the paper presents the findings and lessons for selected cases where data were flagged as a result of the QC/QA processes implemented as part of the IS Condition study.

Author Contributions

The authors confirm contributions to the paper as follows: study conception and design: Rada, Groeger, and Serigos; data collection: Visintine and Simpson; analysis and interpretation of results: Simpson, Visintine, Rada, and Serigos; draft manuscript preparation: Serigos, Visintine, Simpson, Rada, and Groeger. All authors reviewed the results and approved the final version of the manuscript.

References

1. *Moving Ahead for Progress in the 21st Century Act*, H.R. 4348, 2012.
2. *Fixing America's Surface Transportation Act*, H. R. 22, 2015.
3. Federal Register. *National Performance Management Measures; Assessing Pavement Condition for the National Highway Performance Program and Bridge Condition for the National Highway Performance Program*. 80 FR 325. Vol. 80, No. 2, 2015.
4. Federal Register. *National Performance Management Measures; Assessing Pavement Condition for the National Highway Performance Program and Bridge Condition for the National Highway Performance Program*. 80 FR 8250. Vol. 82, No. 11, 2017.
5. Simpson, A. L., G. R. Rada, B. A. Visintine, and J. L. Groeger. *Interstate Pavement Condition Sampling*. FHWA-HIF-17-022. Washington, D.C., 2016.
6. American Association of State and Highway Transportation Officials. *Standard Equipment Specification for Inertial Profiler*. AASHTO M 328-14. American Association of State and Highway Transportation Officials, 2014.
7. American Association of State and Highway Transportation Officials. *Standard Practice for Determining Rut Depth in Pavements* — 2003. AASHTO R48-10. American Association of State Highway and Transportation Officials, 2013.
8. Federal Highway Administration. *Highway Performance Monitoring System – Field Manual*, 2016.
9. American Association of State and Highway Transportation Officials. *Standard Practice for Certification of Inertial Profiling Systems*. AASHTO R 56-14. American Association of State and Highway Transportation Officials, 2014.
10. American Association of State and Highway Transportation Officials. *Standard Practice for Quantifying Cracks in Asphalt Pavement Surfaces from Collected Images Utilizing Automated Methods*. AASHTO PP 67-10. American Association of State Highway and Transportation Officials, 2010.
11. American Association of State and Highway Transportation Officials. *Standard Practice for Collecting the Transverse Pavement Profile*. AASHTO PP 70-10. American Association of State Highway and Transportation Officials, 2010.

12. Simpson, A. L., J. M. Bryce, B. A. Visintine, G. R. Rada, and J.L. Groeger. *Interstate Pavement Condition Sampling: Report Addendum*. FHWA Publication No. (Under review by FHWA for possible publication). Federal Highway Administration, U.S. Department of Transportation, Washington, D.C., 2017.
13. Rada, G. R., J. S. Miller, W. Y. Bellinger, and R. B. Rogers. Accreditation of Strategic Highway Research Program Long-term Pavement Performance Pavement Distress Raters. *Transportation Research Record: Journal of the Transportation Research Board*, 1993. 1410: 1–10.
14. Simpson, A. L., G. R. Rada, B. Visintine, J. Groeger, and J. Guerre. *Improving FHWA's Ability to Assess Highway Infrastructure Health: Development of Next Generation Pavement Performance Measures*. Report No. FHWA-HIF-13-042. Federal Highway Administration, Washington, D.C., 2013.

The Standing Committee on Pavement Management Systems (AFD10) peer-reviewed this paper (18-00420).

Article

TRR
JOURNAL OF THE TRANSPORTATION RESEARCH BOARD

Transportation Research Record
2018, Vol. 2672(40) 166–175

DOI: 10.1177/0361198118788430
journals.sagepub.com/home/trr
SAGE

Creating a Predictive Model for Pavement Deterioration using Geographic Weighted Regression

Maddalena Romano[1], Maxwell S. Siegel[1], and Hui Yan (Terri) Chan[1]

Abstract

This paper represents an effort by the New York City Department of Transportation (NYCDOT) to develop, by means of geographically weighted regression (GWR), an empirical pavement deterioration model for the five boroughs of New York City. Evaluation of probabilistic and deterministic models from existing literature, as well as statistical modeling techniques and applications, were examined to inform model selection. A model using GWR was selected as the preferred methodology. Validation tests demonstrated an adjusted R^2 of .808 and a standard error (σ) representing 4.8% of the mean. Emphasis is placed on the novelty and practicality of GWR as a means for pavement deterioration prediction. Future applications of this model will aid decision makers in strategically allocating limited funds to priority projects.

New York City Department of Transportation (NYCDOT) is tasked with maintaining a safe, resilient, and sustainable transportation network of about 20,000 lane miles that serves the mobility needs of New Yorkers. With limited funding, however, projects need to be prioritized according to the greatest need and/or most efficient use of resources in order to minimize life cycle costs. Maintenance costs increase as roadways deteriorate, which creates a need for decision makers to take action before costs dramatically rise. A reliable pavement forecasting model can act as a powerful tool for strategic planning of capital projects, providing ways for decision makers to visualize different "what if" scenarios for treatment and maintenance, and judge the cost/benefit and risk involved with each choice.

This paper proposes a statistical model for forecasting future pavement conditions of the roadway network using available pavement condition data and variables known to contribute to deterioration. Model construction considered the spatial variability of pavement deterioration as well as the agency's intended use of the forecasts. The condition rating is the dependent variable for this study. Independent variables analyzed are the previous year's observed rating, years since last resurfacing, number of travel lanes, street cut permits, traffic volumes, truck routes and bus routes. After a review of contemporary pavement modeling techniques, an evaluation of the intended uses of model output, and an evaluation of the structure of available data, this study decided to use a deterministic regression modeling technique. Baseline model diagnostics proved significant, and quality assurance exercises ensured that the data was cleaned appropriately. This study also performed exploratory regression analysis to ensure correctly assumed correlation of explanatory variables and prediction validity were scrutinized at the segment and network level to ensure breadth of predictive power.

The organization of this paper is as follows: First, a review is presented of all relevant literature, summarizing the common statistical techniques, their limitations, and how this study will contribute to the literature and address some of these limitations. Then the model's conceptual design is detailed, including variable and regression method selection. Reported next is the final model form, diagnostics and predictive power of the model, defending certain use cases. Lastly, the paper concludes with a discussion of how the results have or have not met the objective of this study, and suggests areas of future research.

[1]Asset Management Unit, Office of the Executive Deputy Commissioner for Strategic & Agency Services, New York City Department of Transportation, New York, NY

Corresponding Author:
Address correspondence to Maddalena Romano: mromano1@dot.nyc.gov

Background

This section reviews existing research surrounding the algebraic nature of, and known contributors to, pavement deterioration, common practice contemporary modeling techniques, and nuanced statistical rules of such techniques.

Age of pavement, measured as time since the last major maintenance event (i.e., resurfacing or reconstruction), is the basis for the vast majority of deterioration models, though the methodology and form of the relationship differs. A familiar model form is put forth by FHWA (*1*), where given an assumed 50 year useful life, pavement deterioration reaches a tipping point at about 37.5 years, and then accelerates more rapidly over the next 12 years. Khraibani et al. (*2*) used a model with polynomial variables to explain the relationship between their cracking index and age. George et al. (*3*) used a power function, where rating is a function of age to the power of a coefficient, and Katicha et al. (*4*) featured a natural log transformation which amplifies their time variable. Xu et al. (*5*) analyzed several polynomial forms to identify their pavement condition indices' relationship with age.

Traffic loads (volume and weight) also contribute to pavement deterioration (*6*). The use and structure of variables relating to the effect of traffic loads differs based on available data sources. Equivalent Single Axel Load (ESAL) is the standard metric developed by AASHTO (*6*), however it requires a myriad of data over an extended period of time to extrapolate the load each axel bears on varying pavement surfaces. An acceptable alternative to ESAL is data on Average Annual Daily Traffic (AADT) (*7*, *5*), which is more easily attained. AADT captures the number of vehicles but does not specify the weight or class of vehicles. A proxy to ESAL can be developed by controlling for heavy duty vehicles (bus and truck) (*8*). The last recorded condition index value is often used to "account for unexplained variability by the model" (*4*). While age, environmental, and structural variables account for a portion of variability and street rating, the last observed condition narrows error associated with the stochastic nature of rating indices.

Modeling spatial phenomena can take many forms based on the data structure and nature of the desired output or prediction. There are generally three methodologies utilized for modeling relationships over space: Bayesian Spatial Modeling, Spatial Markov Chains and Regression (*9–12*). Bayesian Spatial Modeling and Spatial Markov Chains present restrictions to variable type and prediction ability. Bayesian Spatial Modeling relies on conditional probabilities of multiple events (*9*). The events serve as variables for the explanation of a certain phenomenon. In the case of pavement deterioration, some variables can be measured in terms of their conditional probabilities (i.e., street cuts). Most explanatory variables, however, are not probabilistic (i.e., years since maintenance). Markov Chains are often considered because of their ability to predict categorical values in variables (*13*), however the majority of data associated with this study is continuous and the desired output is more granular than integer level. Regression is featured in the majority of deterioration modeling research publications, and presents the fewest restrictions to variable structures, output forms, and use cases. Desire for analyses related to specific impact of individual variables was considered when choosing a method.

Many applications of statistical models rely on regression analysis to evaluate relationships between variables, most commonly using the Ordinary Least Squares (OLS) technique (*14*, *15*). This method fits a global model to the variables and creates a single equation to represent that model—relying on the assumption that data is randomly distributed over space (*16*). Spatial statistics practitioners rely on tests for spatial autocorrelation (Global and Local Moran's I) to evaluate whether randomness or clustering is exhibited by residuals of an OLS model. If the results of the Moran's I tests demonstrate clustering in the data, a model with a single equation will not be appropriate, since that global model would overgeneralize in unique localities (*17*, *18*). In such cases, a local modeling technique, like Geographically Weighted Regression (GWR), would more accurately represent the data. GWR can account for local variability in data by fitting individual regression equations to all features in the data (*19*). Recently, GWR has been applied to forecasting pavement deterioration (*9*) to interpolate distress levels of pavement data within New Mexico, where the state lacks resources to assess all roadways. Their results, with a high level of confidence ($R^2 > .9$), highlight the success possible with GWR and touts the notion that GWR can actually "do a surprisingly good job of compensating for unknown variables missing from a model specification" (*16*).

The following section outlines a methodology similar to that of Zhang et al. (*9*) to estimate the condition rating of street segments using available explanatory data. Modifications have been made to both the method and the variables applied to reflect the unique characteristics of New York City's pavement.

Methods

The methods are organized into five sub-sections: data, dependent variable, explanatory variables, model testing, and prediction validation. Data was collected, cleaned, and merged into a composite dataset from which all variables were derived and tested. Methods were examined and validated to find the best fitting model and regression technique. Equation 1 represents the model layout from which the methodology is based:

$$C = \beta_0 + \beta_1 + \beta_{2-7} \quad (1)$$

where
C = pavement condition assessment rating
β_0 = y-intercept
β_1 = third order polynomial transformation ($Y_i + Y_i^2 + Y_i^3$) of time/age variable
$\beta_{2\text{-}7}$ = linear explanatory variables ($\beta_{2\text{-}7}$)

Data

Street geometry is based on New York City Department of City Planning (DCP) street centerline feature class "LION". Data representing street condition, permitting, maintenance events, and truck routes are collected and maintained by NYCDOT. Bus route was obtained from the Metropolitan Transit Authority (MTA). AADT data was obtained from New York State Department of Transportation (NYSDOT). Data were compiled into a single spatial dataset representing street segments and their associated attributes.

Dependent Variable

Similar to the use of Pavement Condition Index (PCI) by Nunez and Shahin (*20*) and the use of Critical Condition Index by Katicha et al. (*4*), a metric describing physical integrity of the roadway is used to establish the deterioration modeling. The dependent variable chosen for this model is the pavement condition assessment rating (*C*). NYCDOT collects condition assessments using a 1–10 scale. Condition is determined by a windshield survey using a trained inspector's judgment of number, extent, and severity of distresses, and is recorded at the integer level. Deterioration is measured as the drop in rating over a period of time, requiring at least two ratings for comparison.

Explanatory Variables

Seven explanatory variables were selected based on industry knowledge of pavement deterioration causes. Pavement condition is most associated with age, or time since last major resurfacing/reconstruction event. Model performance depends greatly upon accurately fitting this relationship with the correct algebraic form. The relationship between age and condition rating is explored to best fit the observed pattern of decay. The scatterplot of current rating plotted against years since last resurfacing indicates a non-linear relationship (Figure 1). Second and third degree polynomials were tested in addition to a logarithmic transformation. The third degree polynomial featured the highest level of correlation, accounting for 17.2% of the variance in rating compared to 16% and 14.8% for that of the second order polynomial and the linear term, respectively. The resulting form of the age variable is defined in Equation 2 as:

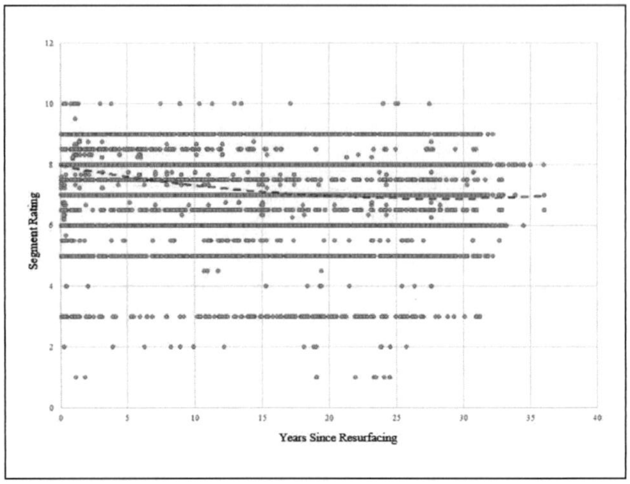

Figure 1. Non-linear relationship between rating and age.

$$\beta_1 = Y_i + Y_i^2 + Y_i^3 \quad (2)$$

where
β_1 = third order polynomial transformation of time/age variable
Y = number of years since last resurfacing for some segment i

In order to obtain meaningful data regarding the impact of traffic loads on roadway condition, it is critical to control for the size of the road (β_2). For the purposes of this study, it is assumed that wider roadways are built to withstand greater load, represented by AADT (*21*). Street width is represented by the number of travel lanes, and is denoted in Equation 3 as:

$$\beta_2 = T_i \quad (3)$$

where
β_2 = street width
T = number of travel lanes for each segment i

Street opening permits are used as a proxy measure of the influence of street cuts on deterioration rates (β_3), representing the number of permits taken out on each segment since the last resurfacing. This data was cleaned to ensure only permits that were acted upon were included. While the number of permits does not equate to the actual number of street cuts performed on a segment, it is assumed that a similar ratio of street cuts to permits exists across New York City. The street cut variable is derived from Equation 4 as:

$$\beta_3 = \Sigma(P_i)_{D2-D1} \quad (4)$$

where

β_3 = number of street cut permits since last resurfacing

P = number of permits along some segment i

D_1 = date of last resurfacing

D_2 = date of current rating

The last condition rating (β_4) is used to mitigate the error associated with unexplained variance in condition rating. Using the last observed rating, the degree of error is drastically limited, and model confidence increased (*15*). Equation 5 describes the last pavement condition assessment rating as:

$$\beta_4 = (R_i)_L \quad (5)$$

where

β_4 = last pavement condition assessment rating

R = condition rating of some segment i

L = previous pavement condition rating event

The remaining three explanatory variables all relate to the traffic classification and volume along the roadway. Each play vital roles in understanding the relationship between large, heavy vehicles along varying sized roadways. Truck routes (β_5) and bus routes (β_6) are treated identically in the model, as binary explanatory variables representing the presence or lack of such routes. Binary variables can be tricky to use within regression, particularly with GWR, because of local multicollinearity. To ensure there was no multicollinearity occurring, a detailed exploratory regression analysis was performed. The variance inflation factor (VIF) of less than 1.5 confirmed that these two binary variables were not producing multicollinearity in the model.

The third traffic related variable is AADT (β_7), which quantifies the traffic volume for a stretch of roadway for an average day. To extrapolate the AADT values from the available AADT data for 20,000 street segments (the entire LION dataset consists of over 100,000 street segments), a Python script was developed to match physically and characteristically congruent segments. AADT values were transferred to the segments which possessed no data, to expand the variable to all features. The segment match was based on the proximity and the congruency of six attributes: street name (within ¼ mile), number of lanes, directionality (within 45 degrees), presence/absence of truck route and bus route, and functional class.

OLS Model and Diagnostics

The three traffic load related variables and the number of travel lanes variable required greater scrutiny as they were found to influence each other within the model and featured extremely high p-values (>.4). Each of the four variables (β_2, β_5, β_6, β_7) created greater variance in coefficient direction and value, and were thus removed from the model.

Model building, exploration, and testing was systematically performed in R-Studio. Within R-Studio, Cran, Lattice and Mosaic packages were used to evaluate and plot individual relationships between variables; residual distributions of model iterations; validity of variables; and varying algebraic forms related to deterioration with age. All OLS-related testing was performed within this R environment.

The model diagnostics for the resulting model were significant, with a correlation coefficient (R^2) of .66 and a standard error of .55 (7.8% of mean). The explanatory variables were analyzed using OLS regression and the model featured in Equation 6:

$$C_i = \beta_{i0} + \beta_{i1}(V_1) + \beta_{i1}^2(V_2) + \beta_{i1}^3(V_3) + \beta_{i3}(V_4) + \beta_{i4}(V_5) \quad (6)$$

where

C = pavement condition assessment rating for all street segments i

β_0 = y-intercept for all street segments i

β_1 = third order polynomial transformation of time/age variable for all street segments i

β_3 = number of street cut permits since last resurfacing for all street segments i

β_4 = last pavement condition assessment rating for all street segments i

$V_{1\text{-}5}$ = coefficients for each variable for each street segment, including coefficients for β_1

Table 1 demonstrates the influence of each variable on the current rating, in order of significance. Each coefficient represents the industry acknowledged expected direction of influence. P-values indicate more than 99% significance in each variable. The three variables associated with the polynomial term for years since last resurfacing feature a high VIF value. While typically concerning in linear regression as a sign of multicollinearity, it is proven to be exaggerated in the case of polynomial regression, where redundancy is enacted intentionally (i.e., $X^1 + X^2 + X^3$) (*22, 23*). Therefore, the high VIF values can safely be ignored.

To evaluate the validity of the OLS method, this study performed an analysis of the randomness of model residual distribution. Exploratory analysis indicated a clustered spatial distribution of the residuals, which was confirmed by the results of the Global Moran's I test for autocorrelation and the Anselin Local Moran's I test. Global Moran's I is regarded as the statistical test of appropriateness for OLS vs GWR in spatial regression modeling (*17, 18*). The method calculates how alike

Table 1. OLS Variable Summary

Variable	Coefficient	*P*-value	VIF
β_0 intercept	0.826	<.000001	NA
β_4 previous rating	0.838	<.000001	1.155353
β_1 years since resurfacing	0.023	<.000001	49.9
β_1^2 years since resurfacing^2	–0.001449	<.000001	263.918
β_1^3 years since resurfacing^3	0.0000198	<.000001	103.589
β_3 street cuts	–0.0011	<.000001	1.12

values near in space are, compared to those further away. Anselin Local Moran's I utilizes a threshold of similarity among residuals to determine whether a feature is spatially clustered. There is no metric for significance like the Global Moran's I, but amount and distribution of clustered features indicates support of the Global Moran's I result.

For the OLS model, the z-score of 37.96 indicated statistical significance (a less than 1% chance that the clustering is the result of a random chance), and the positive Global Moran's I score of .19 indicated clustering (a negative score indicates dispersion). This clustering suggests that the relationships being modeled behave differently in different parts of the study area. A local model, such as GWR, accommodates regional variation by creating an equation for every single feature in our dataset using nearby features in a specified bandwidth (i.e. radius) rather than all of the features in the dataset, allowing the relationships to change across the study area. Based on these results, the authors chose to use GWR as the methodology.

This study used the GWR Tool provided by the ESRI ArcGIS Spatial Analyst extension, and specified a fixed kernel bandwidth of 2500 feet. A fixed bandwidth was chosen instead of an adaptive bandwidth because of the relatively uniform density of street segments across New York City. At 2500 feet, both the number of neighbors (i.e. condition number) and steadiness of coefficients are optimized. The GWR tool produced an output feature class containing the dynamic coefficient for each variable for each street segment, as well as the variable inputs, the intercept, and the predicted future rating.

Prediction Validation

In addition to statistical diagnostics, accurate prediction is critical in attributing confidence to a model. The GWR model's predictive power was tested using a hindcast method to predict all non-resurfaced segments between 2009 and 2017 to evaluate the accuracy and variance of yearly predictions against the observed ratings. The observed mean ratings for all segments in the population sample were plotted along with the model prediction for the respective year. Additionally, the difference in means were analyzed using a paired T-test to evaluate the null hypothesis (i.e., "there is no difference between the means of actual and predicted ratings").

Results and Discussion

This section covers the GWR Model and Diagnostics, Residual Analysis and Local and Global Prediction Power.

GWR Model and Diagnostics

The final OLS model featured in the previous section was deemed suitable for GWR in the form of Equation 7:

$$C_i = \boldsymbol{\beta}_{i0}^{1-80,808} + \left(\boldsymbol{\beta}_{i1}{}^{1-80,808}\left(V_1{}^{1-80,808}\right)\right) + \left(\boldsymbol{\beta}_{i1}{}^{21-80,808}\left(V_2{}^{1-80,808}\right)\right) + \left(\boldsymbol{\beta}_{i1}{}^{31-80,808}\left(V_3{}^{1-80,808}\right)\right) + \left(\boldsymbol{\beta}_{i3}{}^{1-80,808}\left(V_4{}^{1-80,808}\right)\right) + \left(\boldsymbol{\beta}_{i4}{}^{1-80,808}\left(V_5{}^{1-80,808}\right)\right) \tag{7}$$

where

C = pavement condition assessment rating for all records i *(1-80,808)*

β_0 = y-intercept for all street segments i

β_1 = third order polynomial transformation of time/age variable for all street segments i

β_3 = number of street cut permits since last resurfacing for all street segments i

β_4 = last pavement condition assessment rating for all street segments i

V_1-V_5 = dynamic coefficients for each variable for each record, including coefficients β_1

GWR model diagnostics include R^2 (confidence level), σ (standard error) and condition number. The R^2 (measure of the goodness of fit) of the model is .805 and the adjusted R^2 is .808. The R^2 of the GWR is .11 higher than that of OLS, indicating better prediction confidence with a localized method. The standard error (standard deviation of the residuals) for this model is .348—or 4.8% of the mean predicted condition rating, compared to a standard error of .55 for OLS—7.8% of the mean. This means that for an average condition rating of 7.18,

66.7% of predictions would fall between 7.53 and 6.84. Lastly, the sum of all residuals is 261.12, representing .0037 per segment, a negligible amount. This statistic indicates strength in predicting an aggregate of segments.

The condition number indicates the stability of coefficients for each feature modeled. There is a sacrifice between the bandwidth and the condition number. As the bandwidth expands, the coefficients are more closely aligned with the global OLS values, signaling stability. As the bandwidth tightens, the pool of features from which the regression predicts is smaller, which can lead to bias in coefficient values. Condition numbers under 30 are accepted as stable. Polynomial variables, however, can introduce overstated redundancy similar to the observed VIF inflation. The polynomial GWR model produces condition numbers over 30. However, a linear GWR model (i.e., X^1 instead of $X^1 + X^2 + X^3$) was tested, and 90% of those values fell below 30. The stable condition numbers of the linear model provide evidence that the condition numbers from the polynomials GWR model are inflated, and thus can safely be ignored (*24*).

Residual Analysis

The R^2 and standard error present a holistic evaluation of the model performance. It is also important to evaluate the local performance of the model to ensure that predictions along short stretches of pavement are accurate. In order to evaluate local performance, this study also inspected the distribution of residuals.

Figure 2 below features a histogram of the frequency of classified residual values. The GWR model produces a much smoother distribution of residuals than OLS. This local model accounts for regional variations by adjusting

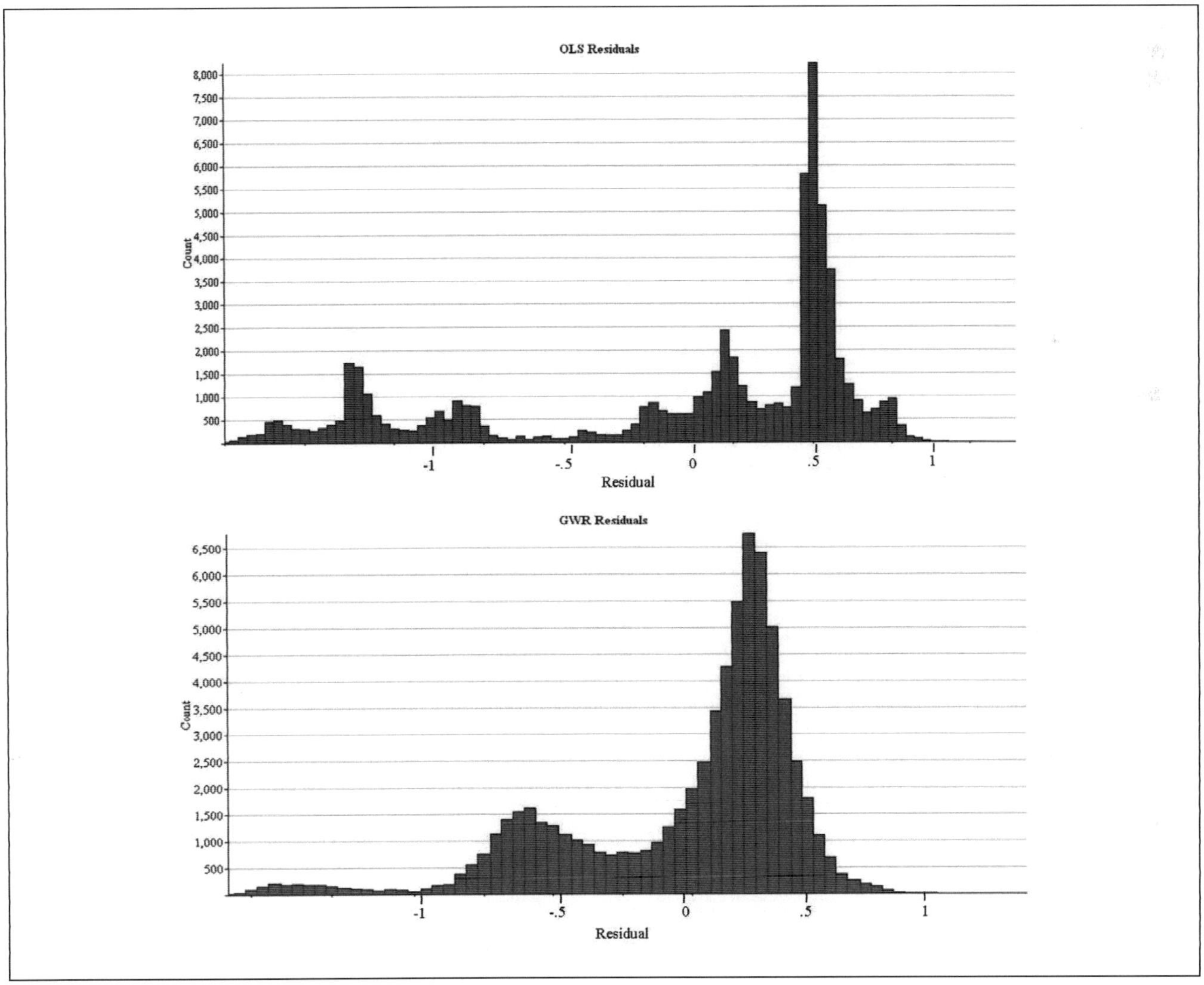

Figure 2. Comparison of residual distribution.

the coefficients accordingly, reducing spikes in the residuals and producing a better model fit. However, a skewed binomial distribution is visible in each. The binomial distribution is caused by the continuous nature of the predictions (two decimal places) versus the integer-level nature of the observed values. The model predicts an average condition rating decrease of .3 points per segment. The residual is the difference between the predicted and the observed value, thus there will be two bumps in frequency: one around + .3 and one around −.7. The segment (most frequently) either stays the same rating, or drops one point, creating these two common residual values on either side of zero.

Local Prediction Power

The local prediction power of this model hinges upon the integer-level ratings that are currently collected at NYCDOT. These integer ratings mean that any model prediction which is + /−.5 of the observed value should be considered correct. Additionally, the model begins to show its reliability after multiple years of ratings, where several decimal level drops in predicted rating can more closely be evaluated to the integer-level observed ratings. The easiest example here is a segment which is forecasted to decrease from an 8 to a 7.7 to a 7.4 and to a 7.1 over a three-year period. If the rating decreases from 8 to 7 anytime within that three year period, the model is proven to have only been off by .1. This is drastically different to evaluating the model year by year, where the residuals are inflated because of these independent integer ratings.

Global Prediction Power

For the purpose of this paper, emphasis is placed on the prediction power of the entirety of the population of segments. Therefore, the individual GWR predictions were aggregated and averaged into a global metric of prediction power. The global prediction power of this model is significant overall. Furthermore, the under- and over-predictions on each side of the binomial distribution neutralize their impact when predicting along a stretch of roadway as well as the city as a whole. A greater number of over predictions of lesser value, and a lesser number of under predictions of greater value create an aggregate error of .003 per segment. A hindcast was performed to test the model's prediction power for the entire city over a period of several years. The final GWR model was used to predict and compare all segments which were not resurfaced between 2009 and 2017. Figure 3 below plots the observed mean ratings between 2009 and 2017, as well as the GWR-predicted mean ratings, respectively. Figure 3 also features a 95% confidence band derived from the predicted mean values. This band represents the range within which the model predicts the observed mean value will lie, plus or minus two standard deviations of the difference between the predicted and observed means (.24).

A paired two-tailed t-test was performed to test the significance of the difference between means. The t-statistic was generated using Equation 8 below.

$$\boldsymbol{t} = \boldsymbol{\mu}/(\boldsymbol{s}/\sqrt{\boldsymbol{h}}) \tag{8}$$

where

μ = mean of difference (.09)

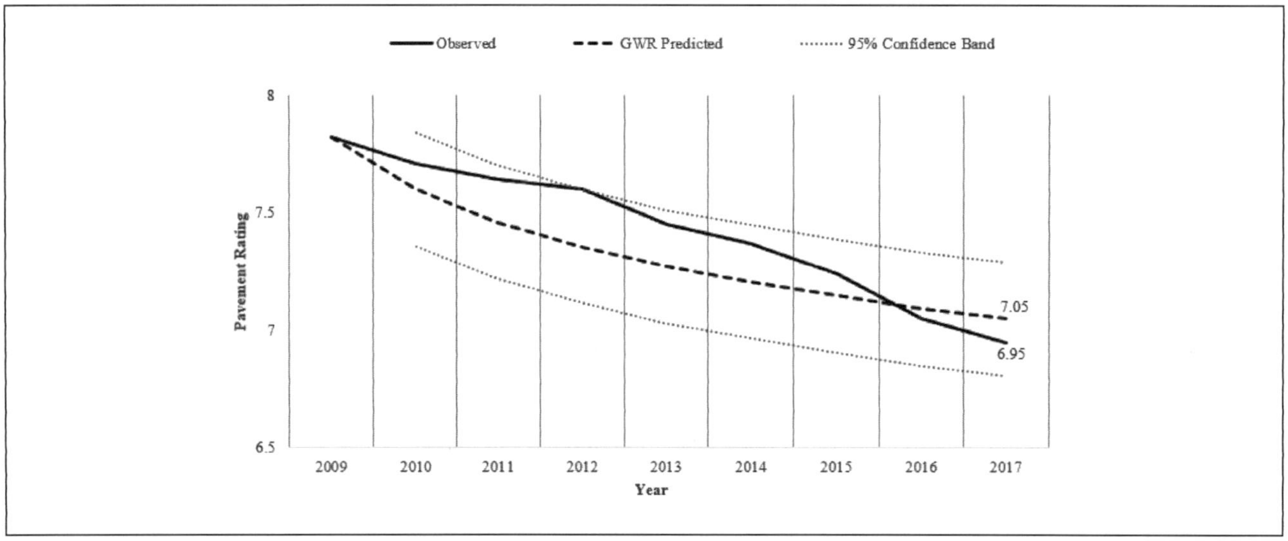

Figure 3. Hindcast results: actual drop vs predicted drop (citywide).

s = standard deviation of difference of means (.12)
h = number of paired means (8)

The results of this test produced a t-statistic of 2.251 and a p-value of .0521. This means that, with 95% confidence level, the predicted rating mean is not statistically different from the observed values between 2009 and 2017. The results of this t-test are enlightening. It demonstrates that the model is over predicting deterioration by .09 points on average each year. However, the degree to which the model is over-predicting deterioration is not statistically significant (p-value of .0521). This means that statistically speaking, it cannot be concluded that the two means (observed and predicted) are different, deeming the predictions "similar" to the observed means. Predicted values lower than observed values is a sign that the time period 2016–2017 featured a higher rate of deterioration than did the years featured in the hindcast. This is in keeping with the idea that as pavement continues to age, deterioration hastens. Bearing this in mind, the results of the hindcast and associated t-test are encouraging with regards to model prediction accuracy.

Conclusion

This effort endeavors to add GWR to the list of supported methodologies and techniques for pavement deterioration modeling. It stresses the importance of testing for spatial autocorrelation and the known natures of pavement deterioration patterns with time. A large roadway network poses difficulties for statisticians, as do subjective data, and data stored in many different formats. After data gathering and cleaning, two methods were tested for their validity. Through an understanding of the desired model output, the nature of pavement deterioration and the spatial relationships of our phenomena, a model was chosen incorporating three of the seven independent variables examined, and GWR was chosen as the methodology that produced the best fitting model. The confidence level of the model increased by 17% with GWR ($R^2 \approx .808$) when compared with OLS ($R^2 \approx .68$). Exploratory analysis indicates that the remaining issues with model confidence lie with the datasets themselves. It is the understanding of this paper that, with better data, the confidence in the model's forecasting accuracy will increase.

Limitations

Missing from this analysis are the age of pavement and pavement profiles (structure of the roadway), as this data is currently not available for all street segments. As such, their effects on the dependent variable are not able to be considered at this time. Elevation and climate attributes were all tested in early OLS versions of the model, but none proved significant to the prediction of deterioration. Information on the quantity, type, and weight of truck traffic is also unavailable. Data quality is also a concern. Current pavement condition assessment practices do not provide the desired level of objectivity or granularity. Windshield inspections, though performed by experienced personnel, exhibit inconsistencies in the data due to subjectivity. In addition, street cut data are only available as a proxy variable using street permit data. While internal confirmation has assured us that the majority of permits are acted upon, there is still an assumption that the ratio of permits to cuts is stable across space. Lack of knowledge pertaining to the size or depth of street cuts also limits the power of the variable in the model.

Further Research

Currently, data on pavement age is only available for a fraction of streets. Pavement cores, from which pavement profiles can be gleaned, are also available for only a fraction of streets. Future efforts will investigate the best ways to obtain and incorporate data on pavement age and profiles, as well as data (or a reliable proxy) for ESAL and climate variables, into future versions of this model. In addition, NYCDOT is currently exploring new methods for collecting pavement data and calculating PCI (*25–27*) to increase sophistication of model inputs. Future applications of this model include developing tools for trade-off analysis ("what if" scenarios) and assessing the possible effects of climate change on the condition of assets.

To ensure the model predicts accurately, future research will also include more prediction validation analyses. Planned testing includes a time series analysis using an autoregressive integrated moving average (ARIMA); analysis of residuals across several years to ensure randomness; hindcast modelling for a greater number of prior years; and analysis of random groups of individual street segments to assess the global and local predictive power of the model.

Acknowledgments

The authors would like to thank NYCDOT Commissioner Polly Trottenberg, and Executive Deputy Commissioner for Strategic and Agency Services Joseph H. Jarrin, as well as the many individuals at NYCDOT that provided input to the project report.

Author Contributions

The authors confirm contribution to the paper as follows: study conception and design: Maddalena Romano; data collection: Maxwell Siegel, Terri Chan; analysis and interpretation of results: Maddalena Romano, Maxwell Siegel, Terri Chan; draft manuscript preparation: Maddalena Romano, Maxwell Siegel, Terri Chan. All authors reviewed the results and approved the final version of the manuscript.

References

1. Elkins, G. E., T. M. Thompson, J. L. Groeger, B. Visintine, and G. R. Rada. *Reformulated Pavement Remaining Service Life Framework*. FHWA-HRT-13-038. U.S. Department of Transportation, Federal Highway Administration, Washington, D.C., 2013. http://www.fhwa.dot.gov/publications/research/infrastructure/pavements/13038/13038.pdf. Accessed July 1, 2017.
2. Khraibani, H., T. Lorino, P. Lepert, and J. M. Marion. Nonlinear Mixed-Effects Model for The Evaluation and Prediction of Pavement Deterioration. *Journal of Transportation Engineering*, Vol. 138, No. 2, 2010, pp. 149–156. https://doi.org/10.1061/(ASCE)TE.1943-5436.0000257.
3. George, K. P., A. S. Rajagopal, and K. L. Lim. Models for Predicting Pavement Deterioration. *Transportation Research Record: Journal of the Transportation Research Board*, 1989. 1215: 1–7.
4. Katicha, S. W., S. Ercisli, G. W. Flintsch, J. M. Bryce, and B. K. Diefenderfer. Diefenderfer. *Development of Enhanced Pavement Deterioration Curves*. Virginia Department of Transportation, University of Virginia, Virginia Transportation Research Council, 2016. http://www.virginiadot.org/vtrc/main/online_reports/pdf/17-r7.pdf.
5. Xu, G., L. Bai, and Z. Sun. *Pavement Deterioration Prediction Model and Project Selection for Kentucky Highways*. Operation and Pavement Management Branch, Kentucky Transportation Cabinet, 2014. https://vtechworks.lib.vt.edu/bitstream/handle/10919/56457/ICMPA9-000071.PDF?sequence = 2&isAllowed = y.
6. FHWA. Reformulated Pavement Remaining Service Life Framework. Publication No. FHWA-HRT-13-038. US Department of Transportation, Federal Highway Administraion. Washington, D.C., 2013. https://www.fhwa.dot.-gov/publications/research/infrastructure/pavements/13038/13038.pdf
7. Chen, D., T. L. Cavalline, D. S. Thompson, and V. O. Ogunro. *Development and Validation of Pavement Deterioration Models and Analysis Weight Factors for the NCDOT Pavement Management System*. Department of Engineering Technology and Construction Management, University of North Carolina at Charlotte, 2014. https://connect.ncdot.gov/projects/planning/RNAProjDocs/2011-01-Phase%20I_Final%20Report.pdf.
8. Boile, M., P. Narayanan, and K. Ozbay. Impact of Buses on Highway Infrastructure. Case Study for New Jersey State. *Transportation Research Record: Journal of the Transportation Research Board*, 2003. 1841: 32–40.
9. Zhang, S., C. D. Lippitt, and S. M. Bogus. Pavement Surface Condition Estimation Based on Geospatial Modelling. *Annals of GIS*, Vol. 23, No. 3, 2017, pp. 1–15. http://dx.doi.org/10.1080/19475683.2017.1325404.
10. Gupta, A., P. Kumar, and R. Rastogi. Critical Review of Flexible Pavement Performance Models. *KSCE Journal of Civil Engineering*, Vol. 18, No. 1, 2014, pp. 142–148. https://doi.org/10.1007/s12205-014-0255-2.
11. Li, W. *Markov Chains Random Field Theory*. University of Connecticut Department of Geography. http://gis.geog.uconn.edu/weidong/MCG/MCRF.htm. Accessed July 1, 2017.
12. Katicha, S. W., G. W. Flintsch, J. Bryce, and B. Diefenderfer. Modeling Pavement Condition and Deterioration: Linear Empirical Bayes Approach. Presented at 96th Annual Meeting of the Transportation Research Board, Washington, D.C., 2017.
13. LeSage, J. P., and R. K. Pace. *Introduction to Spatial Econometrics (Statistics, Textbooks and Monographs)*. CRC Press, Boca Raton, FL, 2009.
14. Wooldridge, J. M. *Introductory Econometrics: A Modern Approach* (6th ed.). Cengage Learning, Boston, MA, 2016.
15. Hamilton, L. C. *Regression with Graphics: A Second Course in Applied Statistics*. Duxbury Press, Belmont, Calif., 1992.
16. Griffith, D. *Spatial Autocorrelation*. University of Texas at Dallas, 2005. http://www.utdallas.edu/~dagriffith/Taiwan_lectures/background%20readings/ESM-2005.pdf. Accessed July 1, 2017.
17. ESRI. Spatial Autocorrelation (Global Moran's I). *ArcGIS Pro Tool Reference*, 2017. http://pro.arcgis.com/en/pro-app/tool-reference/spatial-statistics/spatial-autocorrelation.htm. Accessed July 1, 2017.
18. ESRI. Geographically Weighted Regression (GWR). *ArcMap Tools*, 2016. http://desktop.arcgis.com/en/arcmap/10.4/tools/spatial-statistics-toolbox/geographically-weighted-regression.htm. Accessed July 1, 2017.
19. Fotheringham, A. S., C. Brunsdon, and M. Charlton. *Geographically Weighted Regression: The Analysis of Spatially Varying Relationships*. John Wiley & Sons, Chichester, 2002.
20. Nunez, M. M., and M. Shahin. Pavement Condition Data Analysis and Modeling. *Transportation Research Record: Journal of the Transportation Research Board*, 1986. 1070: 125–132.
21. FHWA. Flexible Pavement (Appendix C: 1993 AASHTO Design Method, section C.1). *Geotechnical Aspects of Pavements Reference Manual*. US Department of Transportation, Federal Highway Administration, Washington, D.C., 2017. https://www.fhwa.dot.gov/engineering/geotech/pubs/05037/ac.cfm.
22. Baldigara, T., and M. Koic. *Modelling the International Tourism Demand in Croatia Using a Polynomial Regression Analysis*. University of Rijeka, Faculty of Tourism and Hospitality Management in Opatija, 2015. http://www.visokaturisticka.edu.rs/docs/tupos/tupos15/rad3.pdf.
23. Allison, P. When Can You Safely Ignore Multicollinearity? *Statistical Horizons*, 2012. https://statisticalhorizons.com/multicollinearity. Accessed July 1, 2017.
24. Brunsdon, C., M. Charlton, and P. Harris. Living With Collinearity in Local Regression Models. Presented at the 10th International Symposium on Spatial Accuracy Assessment in Natural Resources and Environmental Sciences, Florianópolis, Santa Catarina, Brazil, 2012.
25. McGhee, K. H. *NCHRP Synthesis 334: Automated Pavement Distress Collection Techniques*. Transportation

Research Board of the National Academies, Washington, D.C., 2004.
26. Tighe, S., L. Ningyuan, and T. Kazmierowski. Evaluation of Semiautomated and Automated Pavement Distress Collection for Network-Level Pavement Management. *Transportation Research Record: Journal of the Transportation Research Board*, 2008. 2084: 11–17.
27. Vavrik, W. R., L. D. Evans, J. A. Stefanski, and S. Sargand. PCR Evaluation-Considering Transition from Manual to Semi-Automated Pavement Distress Collection and Analysis. Ohio Department of Transportation, Office of Statewide Planning and Research, State Job Number 134668, Columbus, OH, 2013. http://worldcat.org/arcviewer/7/OHI/2014/02/25/H1393345070680/viewer/file1.pdf. Accessed July 1, 2017.

The Standing Committee on Pavement Management Systems (AFD10) peer-reviewed this paper (18-03144).

The contents of this article reflect the views of the authors and do not necessarily reflect the official views of NYCDOT.

Article

TRR
JOURNAL OF THE TRANSPORTATION RESEARCH BOARD

Transportation Research Record
2018, Vol. 2672(40) 176–185

DOI: 10.1177/0361198118792120
journals.sagepub.com/home/trr
SAGE

Determining the Age and Smoothness of Asphalt and Concrete Pavements at the Time of First Rehabilitation using Long-Term Pavement Performance Program Data

Mary Robbins[1*], Nam Tran[2], and Audrey Copeland[3]

Abstract

Initial performance period is an important input in life-cycle cost analysis (LCCA). An objective of this study was thus to determine actual initial performance periods, as the pavement age at first rehabilitation, for asphalt and concrete pavements using Long-Term Pavement Performance (LTPP) program data. In addition, most agencies use International Roughness Index (IRI), a measure of pavement roughness applicable to both asphalt and concrete pavements, in their decision-making and performance-evaluation process. A secondary objective was, therefore, to determine the pavement roughness condition at the time of first rehabilitation using the same dataset. Based on surveys of highway agencies, initial performance periods frequently used in LCCA for asphalt pavements are between 10 and 15 years, while the average asphalt pavement age at time of first rehabilitation in the LTPP program was found to be approximately 18 years. For concrete pavements, most initial performance periods used in LCCA are between 20 and 25 years, whereas the average concrete pavement age at the time of first rehabilitation in the LTPP program is about 24 years. This suggests initial performance period values used for LCCA do not adequately represent the actual age of asphalt pavements at the time of first rehabilitation, while they are generally representative of actual concrete pavement age at the time of first rehabilitation. Also, it was found that asphalt pavements are typically rehabilitated when they are in good or fair condition according to Federal Highway Administration (FHWA) IRI criteria whereas concrete pavements are typically not rehabilitated until the pavement is in fair or poor condition.

State highway agencies (SHAs) use life-cycle cost analysis (LCCA) to choose the most cost-effective project alternatives when planning new construction or reconstruction of their roadways. LCCA takes into account all anticipated costs over the life of each pavement alternative.

Figure 1 shows the ideal life cycles of two competing pavement alternatives under consideration within the analysis period in an LCCA. It is assumed that the two alternatives provide the same level of performance or benefits to the project's users; thus, the alternatives can be compared solely on the basis of cost (*1*).

Figure 1 also shows two types of performance periods that are typically considered in LCCA: initial performance period and rehabilitation performance period. Initial performance period represents the average time in years for a newly constructed or reconstructed pavement to reach an agency's threshold for first rehabilitation. Rehabilitation performance period is the length of time for a rehabilitated pavement to reach an agency's rehabilitation threshold.

The initial performance period can be significantly different for competing alternatives, and it affects the timing of future maintenance and rehabilitation (M&R) activities, in turn, affecting the life cycle cost of each pavement alternative (*3*). Since initial performance period plays such a critical role in LCCA, two questions arise: first, what is the actual initial performance period for each pavement type; and, second, is the accurate initial performance period being used in LCCA?

[1]Ohio Research Institute for Transportation and the Environment, Ohio University, Lancaster, OH
[2]National Center for Asphalt Technology, Auburn, AL
[3]National Asphalt Pavement Association, Lanham, MD
*Work completed while Assistant Research Professor at National Center for Asphalt Technology, Auburn, Alabama

Corresponding Author:
Address correspondence to Mary Robbins: robbinm1@ohio.edu

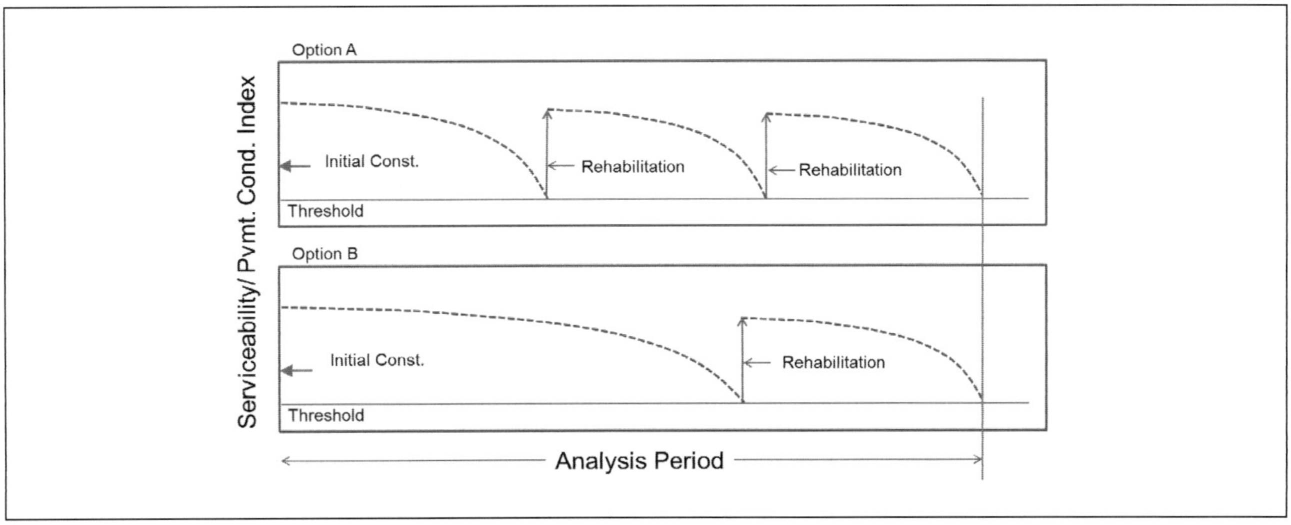

Figure 1. Ideal life cycle diagrams of two hypothetical pavement alternates (2).

Objective and Scope

The first objective of this study was to determine actual initial performance periods, as the age of the pavement at first rehabilitation, for both asphalt concrete (AC) and Portland cement concrete (PCC) pavements based on historical data. Data utilized for analyses in this study were from the Long-Term Pavement Performance (LTPP) program standard data release (SDR) 28. In addition, since all potential pavement alternatives are assumed to provide the same level of performance (in terms of function and ride quality) in LCCA, another objective was to determine the pavement roughness condition at the time of first rehabilitation using the same dataset.

In the following sections, current practices for determining the initial performance period are discussed. This is followed by analyses for determining the time to first rehabilitation and pavement roughness at the time of first rehabilitation. Finally, key findings from the analyses are summarized.

Current Practices

A literature search and a survey of SHAs were conducted in this study to gather information about pavement service life and rehabilitation activities considered in LCCA for both AC and PCC pavements. Key findings from these efforts are summarized in the following subsections.

Rehabilitation Techniques and Initial Performance Periods

Four surveys of SHAs about their LCCA procedures were conducted in the last ten years. One was conducted by the Mississippi Department of Transportation (DOT) in 2007 (*4*), followed by two surveys conducted for the South Carolina DOT in 2008 (*5*). The most recent survey was conducted for this study in 2015 (*6*).

In these surveys, participants were asked to list the rehabilitation treatments for AC and PCC pavements considered in their LCCA procedures. The first rehabilitation activities most commonly listed by SHAs included slab replacement, diamond grinding, and joint repair for rigid pavements, and milling with structural overlay for flexible pavements.

Agencies were also asked to provide information about the year each treatment is accounted for in their LCCA (*4*, *5*). Based on their responses, the reported initial performance periods corresponding to the first rehabilitation activities ranged between 10 and 26 years for flexible pavements, with the majority of agencies reporting values falling between 10 and 15 years. For rigid pavements, the reported initial performance periods ranged from 10 to 35 years, with the majority of agencies reporting a value between 20 and 25 years. These wide ranges in initial performance period reported among the participating SHAs could be in part due to the different M&R strategies that vary from one agency to another.

Methods for Determining Initial Performance Period and Timing of First Rehabilitation

While the initial performance periods used in LCCA for each pavement type are important, the method by which agencies arrive at these values is of most concern. To gain a better understanding of the current methods agencies are using, the questionnaire issued in this study asked agencies to provide information on how the initial performance period is determined for use in LCCA and how SHAs determine the timing of first rehabilitation (*6*).

Based on the survey responses from 30 SHAs, methods used to determine initial performance period for use in LCCA are varied and multifaceted, as summarized below (*6*):

- Two agencies indicated that initial performance period is determined on a project-by-project basis.
- Fifteen SHAs indicated that they determine initial performance period based on historical performance or PMS records. For two of those states, initial performance period determined using this method applied to only one pavement type or only for Interstates and parkways.
- Ten SHAs reported that they determine initial performance period for LCCA based on engineering judgment or experience. This includes committee consensus and/or basing the decision on state, regional, or national practice.
- Six agencies considered various distresses or condition indexes to establish initial performance period while two other states considered pavement design life when establishing initial performance period for LCCA.

When agencies were asked to provide information on how they have determined the actual timing of rehabilitation, half of the participating SHAs indicated some types of performance index are typically used as part of the decision-making process (*6*). Various performance measures are considered in SHAs' indexes for determining the timing of rehabilitation for AC and PCC pavements, but not all performance measures apply to both pavement types. Unlike other distresses such as cracking or faulting, International Roughness Index (IRI) is determined in the same manner regardless of the pavement type. IRI is a calculated value and serves as an objective measurement of the longitudinal profile of a pavement. Responses to the questionnaire indicate that most agencies utilize IRI in the decision-making process, and it is generally utilized as part of an index or in combination with other factors.

Determining Initial Performance Period using LTPP Data

An analysis was conducted in this study to determine the actual time to first rehabilitation utilizing LTPP program standard data release (SDR) 28. Data gathered under the LTPP program are considered comprehensive and consistent. LTPP data were suggested in the 1998 FHWA publication on LCCA as a potential source for determining performance periods and activity timing for LCCA (*3*). A literature search conducted in this study did not find any previous research to determine the actual time to first rehabilitation for asphalt and concrete pavements using the LTPP program database.

For this analysis, pavement sections included in LTPP experiments across the United States and Canada were utilized. LTPP experiments include both flexible and rigid pavement sections, each of which is 500 feet long and one lane wide. They are in four climatic zones: "wet, freeze," "wet, non-freeze," "dry, freeze," and "wet, non-freeze."

There are two classifications of experiments: General Pavement Studies (GPS) and Specific Pavement Studies (SPS). GPS sections were constructed and in service before being accepted into the LTPP program, whereas SPS sections were constructed on existing routes based on experiments designed by the LTPP program (*7*).

The data available in SDR 28 were first evaluated to select the following experiments for this analysis:

- Pavement sections included in Experiments GPS-1 through GPS-5 were from existing in-service pavements constructed before acceptance into the LTPP program. The remaining GPS experiments (GPS-6 through GPS-9) included pavement sections which were either overlaid before entering the LTPP program, or rehabilitated after being in the program to be purposely reclassified to a new GPS experiment, making it difficult to determine accurate initial performance period. Thus, only GPS-1 through GPS-5 experiments were selected for further analysis.
- Experiments SPS-5 through SPS-7 and SPS-9 were included in this analysis. As part of the SPS-5, 6 and 7 experiments focusing on rehabilitation, the intention was to include existing pavements in their first performance period, having received no previous rehabilitation and were "part of an agency's planned rehabilitation program." It is believed pavements for which rehabilitation was already planned were also sought for the SPS-9A experiments, which aimed to study AC overlays on existing CRCP, JPCC, or AC pavements. Another condition for nomination to these six SPS studies was that the existing pavements exhibit pavement distress which is typical of the distresses experienced in that state (*7–11*).

Table 1 shows the LTPP experiments that were included in this analysis. These pavement sections represented pavement structures in common use and representative of good engineering practices in terms of pavement design and pavement materials, and the timing of

Table 1. LTPP Experiments Considered for Determining Initial Performance Period

Experiment	Pavement type	Experiment title
GPS-1	AC on granular base	Asphalt Concrete (AC) Pavement on Granular Base
GPS-2	AC on bound base	AC Pavement on Bound Base
GPS-3	JPCP	Jointed Plain Concrete Pavement (JPCP)
GPS-4	JRCP	Jointed Reinforced Concrete Pavement (JRCP)
GPS-5	CRCP	Continuously Reinforced Concrete Pavement (CRCP)
SPS-5	AC	Rehabilitation of AC Pavements
SPS-6	JPCC	Rehabilitation of Jointed Portland Cement Concrete (JPCC)
SPS-7	PCC	Bonded PCC Overlays of Concrete Pavements
SPS-9C	CRCP	AC Overlay on CRCP
SPS-9J	JPCC	AC Overlay on JPCC
SPS-9O	AC	AC Overlay on AC Pavement

Table 2. LTPP Rehabilitation Types Considered for Determining Initial Performance Period

Code	Rehabilitation type	Pavement type
7	PCC slab replacement (sq. yards)	PCC
12	Grinding surface (sq. yards)	PCC
14	Pressure grout subsealing (no. of holes)	PCC
16	Asphalt subsealing (no. of holes)	AC
19	Asphalt concrete overlay (sq. yards)	AC/PCC
20	Portland cement concrete overlay (sq. yards)	AC/PCC
43	Hot-mix recycled asphalt concrete (overlay) (sq. yards)	AC/PCC
44	Cold-mix recycled asphalt concrete (overlay) (sq. yards)	AC/PCC
45	Heater scarification, surface recycled asphalt concrete (sq. yards)	AC
46	Fracture treatment (crack-and-seat) of PCC pavement as base for new AC surface (sq. yards)	PCC
47	Fracture treatment (crack-and-seat) of PCC pavement as base for new PCC surface (sq. yards)	PCC
48	Recycled portland cement concrete (overlay) (sq. yards)	PCC
50	Joint load transfer restoration in PCC pavements (linear feet)	PCC
51	Mill off AC and overlay with AC (sq. yards)	AC
52	Mill off AC and overlay with PCC (sq. yards)	AC
55	Mill existing pavement and overlay with hot-mix recycled AC.	AC/PCC
56	Mill existing pavement and overlay with cold-mix recycled AC.	AC/PCC

rehabilitation was decided by the agency itself (*7*). Therefore, the time to first rehabilitation determined in this analysis should be representative of actual practices, and, by including pavements from across the United States and Canada, bias to one practice is minimized.

A list of rehabilitation activities was developed for use in this analysis, as shown in Table 2. This list was determined based on information regarding types of rehabilitation activities considered by SHAs in their LCCA procedures obtained through previous surveys (*4*, *5*), the questionnaire issued for this study (*6*), and the rehabilitation activities listed in the LTPP Information Management System User Guide (*7*). Code 12 can be applied to both AC and PCC pavements, so care was taken to identify grinding on PCC pavements only, to be consistent with the practices reported by SHAs.

Using the rehabilitation activities listed in Table 2, the date associated with the first occurrence of one or more of these rehabilitation activities was identified. Any pavement sections that had received rehabilitation before being accepted into the LTPP program were removed from the dataset. Using construction date and date of first rehabilitation, the actual initial performance periods were determined and summarized.

While the LTPP Program database includes comprehensive, quality information reported by SHAs and Canadian provinces to the LTPP regional centers and every effort was made in this analysis to extract unbiased, accurate data, there may still be biases and errors in the extracted dataset. Thus, the dataset for each LTPP experiment listed in Table 1 was first evaluated. The data were then combined based on the original surface type and experiments with like surface type, as follows. In some cases, when all or most of the test sections at a SPS project site were rehabilitated at once, the average time to first rehabilitation was taken for the site:

- For AC pavements, the times to first rehabilitation for the pavement sections from GPS-1 (on granular base) and GPS-2 (on bound base) experiments were combined with the times to first

(a)

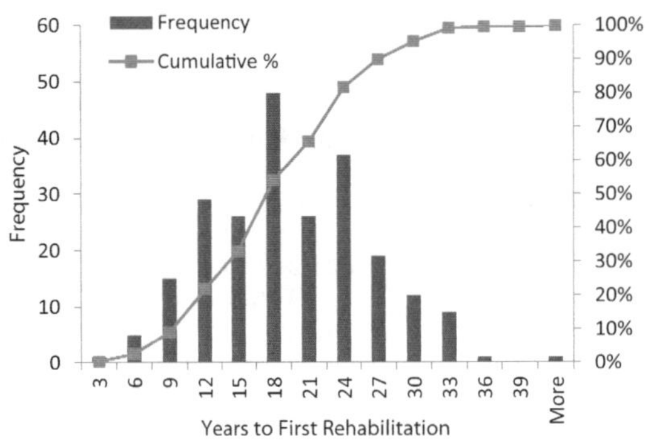

Statistics	AC
Mean	17.93
Median	17.14
Mode	21.22
Std Deviation	6.80
Range	41.04
Minimum	0.07
Maximum	41.11
Confidence Level (95%)	0.89
Count	229

(b)

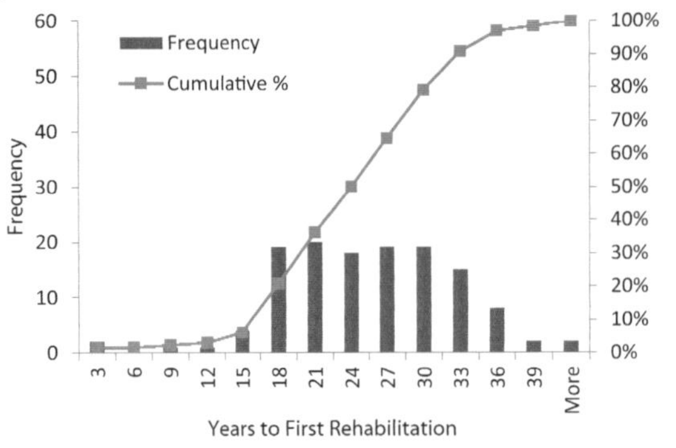

Statistics	PCC
Mean	23.84
Median	23.90
Mode	21.91
Std Deviation	7.16
Range	41.32
Minimum	1.12
Maximum	42.44
Confidence Level (95%)	1.24
Count	130

Figure 2. Distribution of time to first rehabilitation for AC and PCC pavements.

rehabilitation for AC pavement sites from the SPS-5 (rehabilitation of AC pavements) and SPS-9O (AC overlay on AC pavement) experiments.

- Similarly, for PCC pavements the times to first rehabilitation for the pavement sections from GPS-3 (JPCP), GPS-4 (JRCP), and GPS-5 (CRCP) experiments were combined with the times to the first rehabilitation for pavement sites from the SPS-6 (rehabilitation of JPCC), SPS-7 (bonded PCC overlay of concrete pavements), SPS-9C (AC overlay on CRCP), and SPS-9J (AC overlay on JPCC) experiments.

Figure 2, *a* and *b* show the distribution of the time to first rehabilitation for AC and PCC pavements, respectively. The data for both AC and PCC pavements appear to be normally distributed. The largest number of AC pavements (48 of 229) was rehabilitated between 15 and 18 years. The distribution of the PCC pavements was fairly uniform between 15 and 30 years, and does not display a definitive peak. The 50th percentile of the cumulative distribution is between 21 and 24 years which is the center of this uniform period.

Figure 2 also summarizes the statistics computed for the two combined datasets. The average time to first rehabilitation for AC pavements was found to be 17.93 years, with 95% confidence that the mean falls between 17.05 and 18.82 years. On average, the time to first rehabilitation for PCC pavements was found to be 23.84 years. The time to first rehabilitation for PCC pavements was slightly more variable resulting in a wider 95% confidence interval of 22.60 to 25.08 years. The minimum and maximum values were very similar among the two pavement types; however, the minimum values appeared to be unrealistic values and likely outliers.

The wide ranges in the time to first rehabilitation and the very short initial performance periods (i.e., the minimum values) shown in Figure 2 for both pavement types suggest there are likely some errors in reporting data and/or missing information about the dates for initial construction and/or first rehabilitation activities. However, there are not a significant number of pavement sections on either end of the distribution, as evident in Figure 2. Thus, to eliminate some possibly erroneous values, the data could be limited to the middle 90% of the distribution for further analysis. This results in lower

Table 3. Summary of Middle 90% of Pavement Ages at Time of First Rehabilitation

Pavement type	No. of sections	Average	Minimum	Maximum	Standard deviation
AC	206	17.68	7.09	28.93	5.51
PCC	121	23.84	12.88	35.44	5.79

Table 4. Statistics for Time to First Rehabilitation by AC and PCC Pavement Types

			Age at first rehabilitation, years			
Experiment	Pavement type	No. of sections (GPS)/sites (SPS)	Average	Min	Max	Standard deviation
GPS-1	AC on granular base	111	17.92	7.09	28.18	5.36
GPS-2	AC on bound base	76	16.82	7.75	28.93	5.70
GPS-3	JPCP	41	23.40	14.68	34.68	5.98
GPS-4	JRCP	27	23.88	15.07	35.44	5.68
GPS-5	CRCP	28	23.80	14.76	33.69	6.11
SPS-5	AC	16	18.90	9.01	25.75	4.74
SPS-9O	AC	3	23.99	16.87	28.16	6.20
SPS-6	PCC	14	24.79	17.19	31.98	5.55
SPS-7	PCC	4	23.23	12.88	34.83	9.20
SPS-9C	CRCP	2	22.29	21.75	22.84	N/A
SPS-9J	JPCC	5	25.88	22.97	30.27	3.51

and upper boundaries of 6.71 and 29.15 years for AC pavements, and 12.03 years and 35.64 years for the PCC pavements. The data outside these boundaries were removed and the average, minimum, and maximum pavement ages at the time of first rehabilitation were determined for the data in the middle 90% of each pavement type. These values are summarized in Table 3.

In the surveys (*4–6*), many agencies have or continue to use historical values to estimate initial performance period for LCCA. Based on these surveys, the majority of SHAs reported a value for initial performance period in LCCA between 10 and 15 years for AC pavements. Those values are 3–8 years sooner than the national average (17.68 years) for first rehabilitation of AC pavements shown in Table 3. For PCC pavements, the majority of SHAs was using an initial performance period between 20 and 25 years, which align well with the national average age of 23.84 years shown in Table 3.

Further analysis was conducted on the middle 90% of the data distribution to determine if any trends in the time to first rehabilitation were associated with pavement type (or experiment), climatic zone, region, or functional classification. Results of this analysis related to pavement type are discussed in the subsections below. More information about this analysis can be found in a report (*6*).

Time to First Rehabilitation by Pavement Type and Experiment

The average, minimum, maximum, and standard deviation for the time to first rehabilitation by pavement type (or experiment) were summarized in Table 4. AC pavements on granular base had an average pavement age of 17.9 years, which was slightly more than one year longer than AC pavements on bound base. The times to first rehabilitation for the three GPS experiments for JPCP, JRCP, and CRCP were between 23.40 and 23.88 years, on average. The times to first rehabilitation for AC pavements in the SPS experiments were longer than those in the GPS experiment, on average. This was also true for two of the SPS experiments for PCC pavements: SPS-6 and SPS-9J.

Pavement Roughness at Time of First Rehabilitation

The times to first rehabilitation were determined for pavement sections in the LTPP experiments in the previous section. Although historical times to first rehabilitation can be useful in LCCA, the performance level at first rehabilitation is critical in determining the initial performance period in LCCA. One such performance parameter, pavement roughness, in terms of IRI, was found to be used frequently, either directly or indirectly in determining timing of actual rehabilitation based on the questionnaire conducted in this study (*6*). IRI is a consistent and repeatable measure of pavement roughness. Furthermore, it has been reported that the public perceives pavement roughness as an important indicator of performance (*12*). Therefore, this section sought to document the levels of pavement roughness, in terms of

Table 5. Summary of Last MRI Value before First Rehabilitation by Experiment

Experiment	Pavement type	No.	Avg MRI (in./mi)	Min MRI (in./mi)	Max MRI (in./mi)	Std. Dev.	COV
AC pavements							
GPS-1	AC on Granular base	101	119.10	30.18	359.01	58.86	49.4%
GPS-2	AC on bound base	65	102.01	49.90	260.35	43.75	42.9%
AC - combined		166	112.41	30.18	359.01	53.97	48.0%
PCC pavements							
GPS-3	JPCP	41	138.85	81.32	260.66	50.26	36.2%
GPS-4	JRCP	22	130.91	86.02	229.16	32.86	25.1%
GPS-5	CRCP	27	112.39	48.29	257.88	45.68	40.7%
PCC - combined		90	128.97	48.29	260.66	46.14	35.8%

IRI, at the time of first rehabilitation as well as the rate of progression of pavement roughness with pavement age leading up to the first rehabilitation.

Pavement Roughness Measured before First Rehabilitation

An evaluation of IRI data collected for the LTPP experiments listed in Table 1 showed limited IRI data available for the SPS experiments measured before first rehabilitation. Thus, IRI data analyzed were limited to AC pavement sections in the GPS-1 and GPS-2 experiments and PCC pavement sections in the GPS-3, 4, and 5 experiments. Additionally, the upper and lower bounds for the middle 90% of the initial performance period distribution were applied to each dataset to guard against potential errors in the LTPP database. This resulted in a combined 166 AC pavement sections from the GPS-1 and GPS-2 experiments, and 90 PCC pavement sections from the GPS-3, 4 and 5 experiments.

Further evaluation of the IRI dataset showed some limitations with the LTPP data that should be noted. First, the last IRI measurement was not always made immediately before rehabilitation and, in some cases, was taken as much as four years before first rehabilitation. Second, detailed IRI measurements on pavement sections in the GPS experiment are not available before acceptance into the LTPP program. Therefore, the available IRI measurements in the LTPP database often do not span the entire life of the pavement.

For each run, the average IRI of the two wheelpaths was calculated and stored in the LTPP database as the mean IRI (MRI) of the two wheelpaths. An average of the replicate MRI values for each date were calculated and utilized in the analyses presented in this section.

Table 5 presents statistics on the last MRI measured before the first rehabilitation for each experiment and pavement type. Pavement roughness measured before first rehabilitation varied widely for both AC and PCC pavements. Among AC pavements, pavement roughness ranged from as smooth as 30 in./mi to as rough as 359 in./mi with an average MRI of 112.41 in./mi. The 95% confidence interval about the mean for AC pavements was 104.14 in./mi to 120.68 in./mi. The range for PCC pavements was a bit narrower with MRI ranging from 48 in./mi to 260 in./mi, but PCC pavements had a higher average of 128.97 in./mi for the last MRI measured before the first rehabilitation. The 95% confidence interval about the mean for PCC pavements was 119.34 in./mi to 138.64 in./mi.

Also, as shown in Table 5, AC pavements on granular base (GPS-1 experiment) on average were rougher than AC pavements on bound base (GPS-2 experiment) before rehabilitation. AC pavements on granular base were the most variable amongst AC and PCC pavements investigated, while AC pavements in the GPS-2 experiment were the smoothest pavements. JPCPs (GPS-3 experiment) were the roughest pavements among the five experiments before first rehabilitation with an average MRI of 139 in./mi. PCC pavements in the GPS-5 experiment (CRCP) represented the smoothest PCC pavements with an average for the last MRI measured before the first rehabilitation less than that found for JPCPs by nearly 26 in./mi.

When categorizing the IRI data based on FHWA categories (*13*) for ride quality (very good, good, fair, poor, and very poor) associated with IRI measurements as shown in Table 6, it was found that PCC pavements were more likely to reach poor ride quality before the first rehabilitation. MRI values for AC pavements most frequently fell into the good ride quality category, indicating AC pavements were commonly rehabilitated while in better ride quality condition and smoother than PCC pavements at the time of first rehabilitation. More than 85% of AC or PCC pavement sections were rehabilitated before reaching FHWA's (*14*) unacceptable threshold level for ride quality of 170 in./mi.

IRI at the time of rehabilitation is of interest to better understand the performance of each pavement type at the time first rehabilitation was performed. Therefore, further analysis is presented in the following subsection to examine deterioration of the pavements with respect to roughness over time.

Table 6. Ride Quality (IRI) before Rehabilitation

Pavement type	Percent of total pavement sections				
	Very good[b] <60 in/mi	Good 60–94 in./mi	Fair 95–119 in./mi	Poor 120–170 in./mi	Very poor >170 in./mi
AC pavements	9.6%	34.3%	24.1%	17.5%	14.5%
PCC pavements[a]	1.1%	23.3%	26.7%	34.4%	14.4%

[a]Sum is not 100% due to rounding.
[b]Categories after (*13*).

Table 7. Rate of Pavement Roughness Progression by Experiment (Rate > 0, $R^2 \geq 50\%$)

Experiment	Pavement type	No.	Percent of total	Mean rate (in./mi/yr)	Min rate (in./mi/yr)	Max rate (in./mi/yr)
AC pavements						
GPS-1	AC on granular base	59	55.7%	6.76	0.64	24.52
GPS-2	AC on bound base	47	44.3%	4.67	0.72	14.64
AC – combined		106	100.0%	5.83	0.64	24.52
PCC pavements						
GPS-3	JPCP	24	54.5%	3.74	0.59	16.71
GPS-4	JRCP	11	25.0%	3.39	0.74	11.37
GPS-5	CRCP	9	20.5%	3.73	0.47	7.29
PCC – combined		44	100.0%	3.65	0.47	16.71

Deterioration Rate of Ride Quality

Further analysis was conducted to determine the deterioration rate for IRI over time by fitting a linear regression to MRI and pavement age for each pavement section which had at least three MRI measurements before the first rehabilitation. It should be noted that the *y*-intercept was not fixed at the origin. The percent distribution of the rate of pavement roughness progression revealed roughness on AC and PCC pavements most frequently increased at a rate between 0 and 3 in./mi/yr. A small percentage of AC and PCC pavement sections revealed negative rates of pavement roughness which could be due to either inconsistent data or maintenance activities. While more than 30% of the AC pavement sections had R^2 values between 75% and 90% for the individual linear relationships, coefficients of determination were more evenly distributed for PCC pavement sections. However, percentages of the total number of pavement sections with R^2 values greater than 90% were similar among the two pavement types: 21.8% for AC pavement sections and 17.8% for PCC pavement sections.

The data were reduced again to include only pavement sections with positive rates of roughness progression and moderate or better linear relationships (R^2 greater than or equal to 50%). Approximately 75% of the AC pavement sections and 60% of the PCC pavement sections met these criteria.

Rates for pavement roughness progression were summarized for each pavement type and experiment in Table 7. The average rate among individual AC pavement sections was 5.83 in./mi/yr. The average rate among individual PCC pavements was smaller at 3.65 in./mi/year. While differences among the PCC experiments (GPS-3, 4, and 5) were negligible, the average rate of roughness progression was greater by more than 2 in./mi/year for AC pavement sections on granular base than AC pavement sections on bound base.

Conclusions and Recommendations

The objective of this study was to determine actual initial performance periods, as the age of the pavement at first rehabilitation, and pavement roughness at the time of first rehabilitation for both asphalt and concrete pavements using the LTPP program standard data release 28. The study included (1) a literature search and a survey of SHAs to synthesize current practices for determining initial performance period and the timing of first rehabilitation; and (2) analyses to determine the time to first rehabilitation and pavement roughness at the time of first rehabilitation for both pavement types using the LTPP data. Based on the results obtained in this study, the following conclusions can be drawn:

- Procedures used to determine the initial performance period for use in LCCA tend to be multilayered, and complex, especially those that are based on one or more condition or performance

indexes. Practices vary by SHA; therefore, direct comparisons could not be made. Commonalities among practices for determining the initial performance period for LCCA were noted:
    - Agencies utilize PMS data to estimate the initial performance period. It is unclear if PMS data refers to historical pavement performance measures at the time of rehabilitation, or historical timing of rehabilitation, or a combination of both.
    - Expert opinion or engineering judgment were reported as methods used either as alone, or in combination with other methods.
    - Various indexes were included in the responses provided.
- A review of SHAs decision-making process and performance measures used to determine the actual timing of rehabilitation revealed that practices are also varied among agencies. Various types of condition indexes as rehabilitation triggers in addition to other factors were often reported. The condition indexes, generally, included one or more distresses unique to a pavement type. Common distresses (not including pavement roughness) for AC and PCC pavements are not equivalent, and, as such, the condition indexes and associated thresholds are not comparable between unlike pavement types. Thus, actual practices and criteria for determining the timing of rehabilitation do not appear to be based on achieving equal levels of performance.
- Initial performance periods commonly used for LCCA do not adequately represent the actual age of AC pavements at the time of first rehabilitation. In previous research, the initial performance period used in LCCA for AC pavements was most frequently reported as a value between 10 and 15 years. The results of the analyses conducted here revealed the average pavement age at the time of the first rehabilitation for existing AC pavements in the LTPP program was found to be 17.7 years.
- The initial performance period commonly used in LCCA for PCC pavements is generally representative of actual PCC pavement age at the time of first rehabilitation. The initial performance period used in LCCA for PCC pavements reported in previous literature was most frequently between 20 and 25 years. This analysis showed the average pavement age of existing PCC pavements in the LTPP program of 23.8 years.
- Given that the majority (85%) of AC and PCC pavement sections were rehabilitated before reaching 170 in./mi, it can be concluded that an IRI of 170 in./mi is too high as a rehabilitation trigger.
- Generally, AC pavements are smoother than PCC pavements at the time of first rehabilitation even though AC pavements develop pavement roughness at a faster rate in the years before the first rehabilitation than PCC pavements. AC pavements were first rehabilitated while in good or fair condition whereas PCC pavements were rehabilitated in fair or poor condition.

Recommendations are made for determining the initial performance period of AC and PCC pavements for use in LCCA. The recommendations are based on the analyses conducted on the initial performance periods for AC and PCC pavements in the LTPP program as well as the review of SHAs' LCCA practices for determining actual timing of the first major rehabilitation of AC and PCC pavements:

- It is recommended that agencies consider actual measured performance for determining initial performance period for LCCA to ensure consistent levels of performance are being compared among unlike pavement types. Practices for determining initial performance period vary by agency; efforts should, therefore, be made to develop a standard practice of determining initial performance period for LCCA.
- If used, historical timing should be combined with a performance measure and threshold common to all pavement types.
- Performance indicators and associated thresholds used to determine initial performance period in LCCA should be indicative of when rehabilitation is truly necessitated and should strike a balance between structural and functional performance. Due to the absence of a common performance parameter which can equate both structural and functional performance among different pavement types, IRI may be considered for use in determining initial performance period for LCCA.
- An equivalent IRI threshold for use in determining initial performance period for LCCA should be balanced with structural performance for each pavement type, but should not be compromised to accommodate budget constraints, user delay or constructability issues which may be associated with historical PMS data.
- Given on-going advances in material characterization, pavement design, and construction, the threshold value for use in determining initial performance period in LCCA should be re-examined periodically to capture the changes in pavement performance due to changes in design and materials.

- Based on the finding of this report, it is recommended that an IRI value lower than 170 in./mile be used as a threshold for first rehabilitation, as this was much greater than the IRI at first rehabilitation for the majority of pavements studied.

Acknowledgments

The authors wish to acknowledge funding provided by the National Asphalt Pavement Association and the State Asphalt Pavement Associations. The authors would like to thank the LTPP program staff who helped extract pavement data needed for this study.

Author Contributions

The authors confirm contribution to the paper as follows: study conception and design: Mary Robbins, Nam Tran, Audrey Copeland; data collection: Mary Robbins; analysis and interpretation of results: Mary Robbins, Nam Tran; draft manuscript preparation: Mary Robbins, Nam Tran, Audrey Copeland. All authors reviewed the results and approved the final version of the manuscript.

References

1. U.S. Department of Transportation. *Life-Cycle Cost Analysis Primer.* Federal Highway Administration, U.S. Department of Transportation, Washington, D.C., 2002.
2. West, R., N. Tran, M. Musselman, J. Skolnik, and M. Brooks. *A review of the Alabama Department of Transportation's Policies and Procedures for Life-Cycle Cost Analysis for Pavement Type Selection.* NCAT Report 13-06. Auburn University, Auburn, AL, 2013.
3. Walls, J. III, and M. Smith. *Life-Cycle Cost Analysis in Pavement Design – Interim Technical Bulletin.* FHWA-SA-98-079. Federal Highway Administration, U.S. Department of Transportation, Washington, D.C., 1998.
4. Mississippi Department of Transportation. *Life Cycle Cost Analysis for Pavement Type Selection.* AASHTO RAC Member Survey Results. AASHTO, Washington, D.C., 2007. http://research.transportation.org/_layouts/15/AASHTORAC/RACSurveyResultDetail.aspx?SurveyID=119. Accessed July 2017.
5. Rangaraju, P. R., S. Amirkhanian, and Z. Guven. *Life Cycle Cost Analysis for Pavement Type Selection.* FHWA-SC-08-01. South Carolina Department of Transportation, Columbia, SC, 2008.
6. Robbins, M., and N. Tran. *Review of Initial Service Life Determination in Life Cycle Cost Analysis (LCCA) Procedures and In Practice.* NCAT Report 18-02. National Center for Asphalt Technology, Auburn, AL, 2018.
7. Elkins, G. E., T. Thompson, B. Ostrom, A. Simpson, and B. Visintine. *Long-Term Pavement Performance Information Management System User Guide.* FHWA-RD-03-088 (revision). Federal Highway Administration, McLean, VA, 2013.
8. *Specific Pavement Studies Guidelines for Nomination and Evaluation of Candidate Projects for Experiment SPS-5 Rehabilitation of Asphalt Concrete Pavements.* Operational Memorandum No. SHRP-LTPP-OM-006. SHRP, National Research Council, Washington, D.C., 1989.
9. *Specific Pavement Studies Guidelines for Nomination and Evaluation of Candidate Projects for Experiment SPS-6 Rehabilitation of Jointed Portland Cement Concrete Pavements.* Operational Memorandum No. SHRP-LTPP-OM-007. SHRP, National Research Council, Washington, D.C., 1989.
10. *Specific Pavement Studies Guidelines for Nomination and Evaluation of Candidate Projects for Experiment SPS-7 Bonded Portland Cement Concrete Overlays.* Operational Memorandum No. SHRP-LTPP-OM-011. SHRP, National Research Council, Washington, D.C., 1990.
11. U.S. Department of Transportation. *Specific Pavement Studies Guidelines for Nomination and Evaluation of Candidate Projects for Experiment SPS-9A Superpave™ Asphalt Binder Study.* Federal Highway Administration, McLean, VA, 1995.
12. Robbins, M., and N. Tran. *A Synthesis Report: Value of Pavement Smoothness and Ride Quality to Roadway Users and the Impact of Pavement Roughness on Vehicle Operating Costs.* NCAT Report 16-03. National Center for Asphalt Technology, Auburn, AL, 2016.
13. U.S. Department of Transportation. *1999 Status of the Nation's Highways, Bridges and Transit: Conditions and Performance, Report to Congress.* U.S. Department of Transportation, Washington, D.C., 2000.
14. U.S. Department of Transportation. *1998 National Strategic Plan.* Federal Highway Administration, U.S. Department of Transportation, Washington, D.C., 1998.

The Standing Committee on Pavement Management Systems (AFD10) peer-reviewed this paper (18-03354).

Article

TRR
JOURNAL OF THE TRANSPORTATION RESEARCH BOARD

Transportation Research Record
2018, Vol. 2672(40) 186–196

DOI: 10.1177/0361198118757969
journals.sagepub.com/home/trr
SAGE

Impact of Wide-Base Tires on Pavements: A National Study

Imad L. Al-Qadi[1], Jaime A. Hernandez[1], Angeli Gamez[1], Mojtaba Ziyadi[1], Osman Erman Gungor[1], and Seunggu Kang[1]

Abstract

This paper summarizes a multi-year effort comparing the new-generation wide-base tires (NG-WBT) and dual-tire assembly from a holistic point of view. The tires were compared considering not only pavement damage but also environmental impact. Numerical modeling, prediction methods, experimental measurements, and life-cycle assessment were combined to provide recommendations about the use of NG-WBT. A finite element (FE) approach considering variables that are usually omitted in the conventional analysis of flexible pavement was used for modeling pavement structures combining layer thickness, material properties, tire load, tire-inflation pressure, and pavement type (interstate and low volume). A prediction tool, ICT-Wide, was developed based on an artificial neural network to obtain critical pavement responses in cases excluded from the FE analysis matrix. Based on the bottom-up fatigue cracking, permanent deformation, and international roughness index, the life-cycle energy consumption, cost, and green-house gas emissions were estimated. To make this research useful for state departments of transportation and practitioners, a modification to AASHTOware is proposed to account for NG-WBT. The revision is based on two adjustment factors, one accounting for the discrepancy between the AASHTOware approach and the FE model of this study, and the other addressing the impact of NG-WBT. Although greater pavement damage may result from NG-WBT, for the analyzed cases, the extra pavement damage may be outweighed by the environmental benefits when NG-WBT market penetration is considered.

Wide-base tires (WBT) provide numerous advantages compared with conventional dual-tire assembly (DTA). Regarding fuel economy, some studies indicate that fuel efficiency of WBT is 10% higher than that of DTA (*1,2*), or between 3.5 and 12% (*3*), and WBT can improve fuel economy up to 18% when combined with aerodynamic devices (*4*). In addition, WBT increases the hauling capacity of trucks by 2% because it is lighter than DTA (*5*); easier to inspect, repair, and maintain (*3*); uses less rubber material, so the amount of disposable materials at the end-of-life is diminished; has similar or slightly better performance than DTA with regard to handling and comfort (*5*); produces slightly less noise (*5*); and reduces emissions (*6*).

Despite its benefits, the impact of WBT on road infrastructure has been a concern for almost four decades since the inception of first-generation wide-base tires (FG-WBT) in the early 1980s. After FG-WBT were proved to be considerably more damaging to roads than conventional DTA (*7–12*), new-generation wide-base tires (NG-WBT) were introduced in an attempt to preserve the advantages of WBT while reducing pavement damage. The impact of NG-WBT on road infrastructure through accelerated pavement testing and experimental measurements (*13–20*) and analytical and numerical methods (*16,21*) has been studied; however, no evidence has been provided about their net benefit.

This study reflects a relevant step forward, where a comprehensive approach is used to compare NG-WBT with DTA. Numerical modeling, prediction methods, experimental measurements, and environmental impact assessment were combined to provide recommendations about the use of NG-WBT. Numerical modeling allowed the calculation of pavement responses based on an ample combination of variables, which incorporated accurate variables such as measured tire–pavement contact loads. A prediction tool based on an artificial neural network (ANN) was developed to obtain critical pavement responses in cases excluded from the finite element (FE) analysis matrix. The tool can promptly predict critical pavement responses without performing FE calculations. The experimental program aimed to validate the results of the numerical model and verify the trends inferred from the FE model for the scenarios of the testing program. The collected information was compiled in a database that users could access to retrieve experimental results. The environmental impacts were determined using life-cycle assessment (LCA)

[1]University of Illinois at Urbana-Champaign, Illinois Center for Transportation, Rantoul, IL

Corresponding Author:
Address correspondence to Jaime A. Hernandez: hrnndzr2@illinois.edu

(a) (b)

(c)

Figure 1. Tested tires and stress-in-motion system: (*a*) stress-in-motion systems; (*b*) new-generation wide-base tire 445/R22.4; (*c*) dual-tire assembly 275/80R22.5

based on bottom-up fatigue cracking, permanent deformation, and international roughness index (IRI). LCA estimated life-cycle energy consumption, cost, and green-house gas (GHG) emissions. Finally, to make the outcome of this research effort readily implementable by state departments of transportation and practitioners, a modification to AASHTOware is proposed to account for NG-WBT. The revision is based on two adjustment factors, one accounting for the discrepancy between the AASHTOware approach and the FE model of this study, and the other addressing the impact of NG-WBT.

Tire–Pavement Contact Loads

The dual stress-in-motion (SIM) Mk IV system was used for contact load measurement (see Figure 1*a*). The system consists of two SIM pad assemblies; each pad consists of an array of 21 instrumented pins installed across the center portion of the assembly (*22*). The pins are instrumented with strain gauges, which are calibrated to convert strain to load. Static tire ink imprints were made using black roof paint on white paper under the HVS Mk III accelerated pavement testing system to obtain the contact area for each loading case.

Three-dimensional (3D) tire–pavement contact loads were measured for the NG-WBT 445/50 R22.5 and DTA 275/80 R22.5; the tires are presented in Figure 1, *b* and *c*. Four values of tire-inflation pressure *S* at 552, 690, 758, and 862 kPa, and five tire loadings *P* at 26.6, 35.6, 44.4, 62.3, and 80 kN, were applied to the tires to measure the 3D contact loads. In addition, differential tire-inflation pressure was considered for DTA; one tire of the assembly was kept at 758 kPa while two values, 413 and 552 kPa, were applied to the other.

Based on the measurements, peak forces per unit length in the vertical, longitudinal, and transverse directions are relatively higher for NG-WBT than DTA. In addition, regardless of the tire type, the magnitude of the tangential contact stresses is considerably high, which highly influences pavement responses. Although both tires have the same applied load and tire-inflation pressure, the ratios between the peak forces are different. Therefore, the effect of NG-WBT and DTA on pavement responses and performance cannot be assumed to be the same.

The contact area, obtained from static imprints, is also different for both tires. The contact area increased as the applied load reached 80 kN. Furthermore, the contact area was greater for DTA than NG-WBT for all loading cases, and the contact area ratio between DTA and NG-WBT was as high as 1.3. The average contact length for DTA was approximately 8% shorter than that for NG-WBT. Additional details regarding contact loads measurement and analysis can be found elsewhere (*23,24*).

Pavement Response and Damage

FE modeling was used to calculate critical pavement responses. The 3D model considers appropriate material characterization for each layer in the pavement structure and accurate representation of truck loading regarding direction, magnitude, and amplitude (*25*). The numerical predictions of critical pavement responses were contrasted with experimental measurements to ensure model validity (*26*). Various layer thicknesses for the asphalt concrete (AC) and base layers were combined to simulate low-volume and interstate roadways, termed as thin and thick pavements, respectively. Each pavement structure included two extreme values of thicknesses, in which the thin pavement simulated AC layer thicknesses of 75 and 125 mm, while the thick pavement considered three AC layers with total AC thicknesses of 125 and 412.5 mm. On the other hand, the granular base considered 150 and 600 mm layers. Regarding material property inputs, AC layers were assumed to be linear viscoelastic, while the base layer was assumed to be orthotropic stress-dependent (for thin pavement only) and linear elastic otherwise. Details of the material property selection can be found elsewhere (*25*).

Twelve loading cases were simulated to compare the impact of NG-WBT and DTA. As seen in Table 1, the load P was varied from 26.6 to 80.0 kN and the tire-inflation pressure S ranged from 552 to 862 kPa. Two DTA cases with differential tire-inflation pressures were included to represent realistic conditions in which the inner tire typically has a lower inflation pressure because of the difficulty of reaching the inner tire during maintenance.

Trends were fairly similar when comparing pavement responses from NG-WBT and DTA with respect to various pavement layer thicknesses, material properties, and loading conditions. It was noticeable that NG-WBT produced higher critical pavement responses than DTA in most scenarios. In addition, the applied load governed the pavement responses more significantly than the tire-inflation pressure. Detailed comparison of the critical strains can be found elsewhere (*25*).

Transfer functions used by the mechanistic-empirical pavement design guide (MEPDG) were used to relate mechanistic strain responses to pavement damage by calculating the number of repetitions to failure. A ratio of the number of repetitions between the DTA and NG-WBT, DW, was calculated to compare the number of repetitions to failure of DTA, N_{DTA}, and the allowable number of repetition of NG-WBT, $N_{\text{NG-WBT}}$. It is worth noting that $DW < 1$ indicates that the DTA has a lower number of allowable repetitions than NG-WBT.

$$DW = \frac{N_{\text{DTA}}}{N_{\text{NG-WBT}}} \tag{1}$$

Table 1. Simulated Loading Cases in Finite Element Analysis

Loading case	Tire	P (kN)	S (kPa)
L1	NG-WBT	26.6	552
L2	NG-WBT	26.6	862
L3	NG-WBT	80.0	552
L4	NG-WBT	80.0	862
L5	DTA	26.6	552
L6	DTA	26.6	862
L7	DTA	26.6	552/758
L8	DTA	80.0	552
L9	DTA	80.0	862
L10	DTA	80.0	552/758
L11	NG-WBT	44.4	758
L12	DTA	44.4	758

Considering five pavement distresses, the cumulative impact of DTA relative to NG-WBT is compared using a combined ratio, CDW. All failure mechanisms considered are combined using a logarithmic weighing factor, a_i, to account for varying orders of magnitude among the number of repetitions to failure of each distress.

$$CDW = a_1 DW_{\text{BU}} + a_2 DW_{\text{TDS}} + a_3 DW_{\text{TDT}} + a_4 DW_{\text{RS}} + a_5 DW_{\text{RH}} \tag{2}$$

$$a_i = \frac{\frac{1}{\log(N_i)}}{\sum_{j=1}^{5} \frac{1}{\log(N_j)}} \tag{3}$$

where DW_{BU}, DW_{TDS}, DW_{TDT}, DW_{RS}, and DW_{RH} are the DW for bottom-up fatigue cracking, near-surface cracking caused by shear strain, near-surface cracking caused by tensile strain, subgrade rutting, and AC rutting, respectively. In the following plots, each group of bars corresponds to the same load and tire-inflation pressure. For instance, L6:L2 compares DTA and NG-WBT subjected to $P = 26.6$ kN and $S = 862$ kPa.

Thin Pavement Structures

The average damage ratios from the various material properties for thin pavements are shown in Figure 2. Lower damage for DW_{TDT} than other fatigue cases was noted; the average damage ratio for near-surface fatigue cracking caused by surface tensile strain was 1.28, while for bottom-up and near-surface caused by shear, the average damage ratios were 2.13 and 2.34, respectively (see Figure 2, *a–c*). For bottom-up fatigue cracking, DW_{BU} decreased as the magnitude of the applied load increased, and it was less sensitive to layer thickness when the load was high. On the other hand, DW_{TDS}

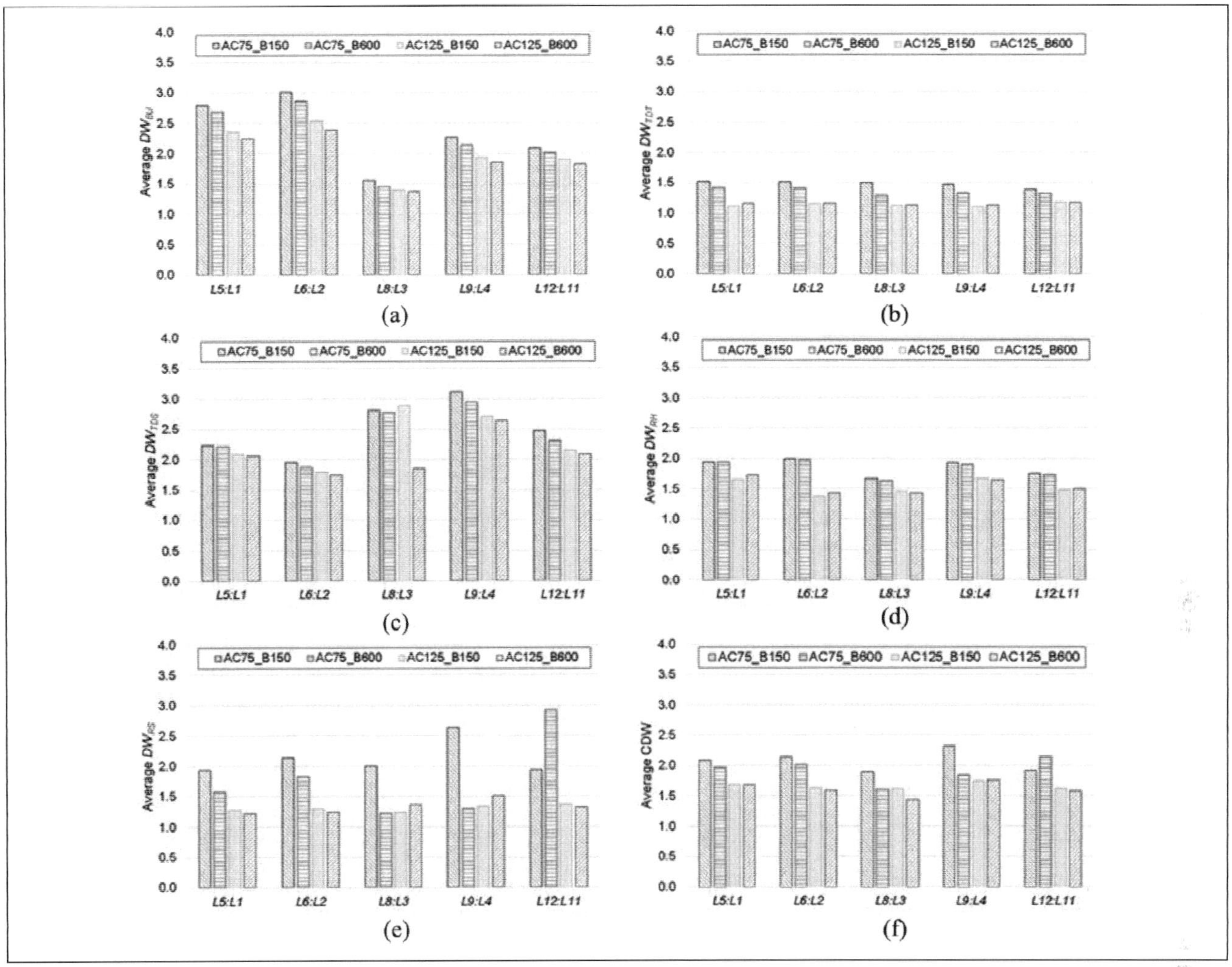

Figure 2. Ratio of the allowable number of repetitions to failure comparing DTA to NG-WBT for thin pavement structures: (a) *DW* for bottom-up cracking; (b) *DW* for top-down cracking due to tensile strain; (c) *DW* for top-down cracking due to shear strain; (d) *DW* for AC rutting; (e) *DW* for subgrade rutting; (f) CDW considering all five distresses.

increased as the load increased. For AC = 125 mm, average DW_{TDT} was not affected by base thickness, applied load, or tire-inflation pressure.

Base thickness minimally affected the average difference between DW_{RH} of NG-WBT and DTA, as seen in Figure 2*d*. A more relevant role of tire-inflation pressure was observed for AC rutting when compared with fatigue cracking if the applied load was low and the AC thickness was 125 mm. On the other hand, tire-inflation pressure increased DW_{RH} when the applied load was high regardless of the layer thicknesses.

From Figure 2*e*, the average damage ratio for subgrade rutting indicated that if AC = 125 mm, the effect of base thickness was less significant than for AC=75 mm because the magnitude of vertical strain in the base layer is smaller for the thick AC layer. Also, the effect of tire-inflation pressure is more relevant in the case of AC = 75 mm because of the proximity of the subgrade to the loaded area. It should be noted that the average damage ratios for subgrade rutting are low compared with the other distresses. This agrees with the previous observation about the reduction in difference between NG-WBT and DTA as pavement depth increases. Combining all five pavement distresses, the highest average ratio was 2.4 for a load of 26.6 kN, tire-inflation pressure of 862 kPa, AC thickness of 75 mm, and base thickness of 150 mm. The average value in Figure 2*f* is 1.8, with the highest magnitude generally corresponding to pavement structures with thinnest AC. This suggests that the impact of NG-WBT could be significant on weak thin pavements.

Thick Pavement Structures

Based on Figure 3, DW_{BU} was highest at 3.2 when AC = 125 mm and B = 600 mm, although one could observe that the relative difference between the two tire types diminishes as AC thickness increases to 412.5 mm, regardless of

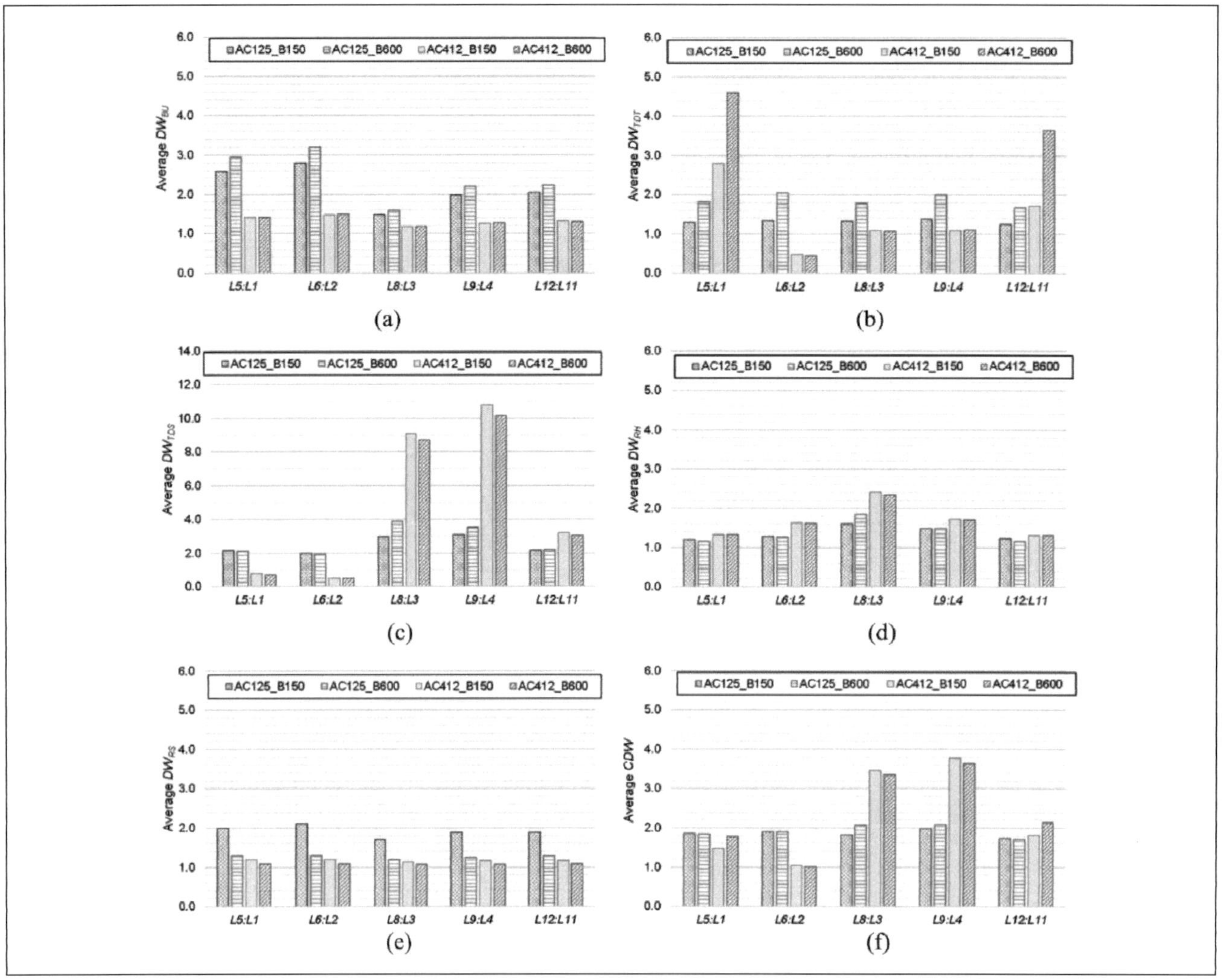

Figure 3. Ratio of the allowable number of repetitions to failure comparing DTA with NG-WBT for thick pavement structures: (a) *DW* for bottom-up cracking; (b) *DW* for top-down cracking due to tensile strain; (c) *DW* for top-down cracking due to shear strain; (d) *DW* for AC rutting; (e) *DW* for subgrade rutting; (f) CDW considering all five distresses.

the loading condition. Conversely, when considering the near-surface cracking caused by the surface tensile strain, DTA can impose between 0.4 to 4.6 times more loading repetitions than that of NG-WBT cases. Increasing the AC thickness to 412.5 mm (with B = 600 mm) generated the maximum DW_{TDT} of 4.6 when P = 26.6 kN. Although the range of average ratios significantly increased when P = 26.6 kN, the surface tensile strain difference between NG-WBT and DTA cases was less than 4 $\mu\varepsilon$. The resulting number of repetitions to failure were a magnitude different from the ones induced by near-surface shear strain and were much greater than 1.0^{10} repetitions when AC = 412.5 mm, which verifies that top-down cracking due to surface tensile strain is not the governing distress.

On the other hand, DW_{TDS} significantly increased as the applied load became greater than 26.6 kN, as observed in Figure 3*c*. This increase was predominant when the AC layer was 412.5 mm thick and P = 80 kN, in which DW_{TDS} ranged from 8.7 to 10.8. One could infer that for thick pavements, the effect of the critical shear strain within the AC had a relatively high impact on near-surface damage, most especially under NG-WBT loading.

Based on Figure 3*d*, DW_{RH} were relatively low and resulted in values equal to or less than 1.5 for most cases, which indicates that the difference between DTA and NG-WBT is not as significant. However, under P = 80 kN and S = 862 kPa, DTA resulted in higher allowable repetitions than NG-WBT as DW_{RH} increased up to 2.4. Similarly, DW_{RS} ranged from 1.1 to 2.1, as seen in Figure 3*e*. Comparing the four pavement structures, DW_{RS} was the highest when AC = 125 mm and B = 150 mm, whereas the minimal difference between DTA and NG-WBT was observed for the remaining

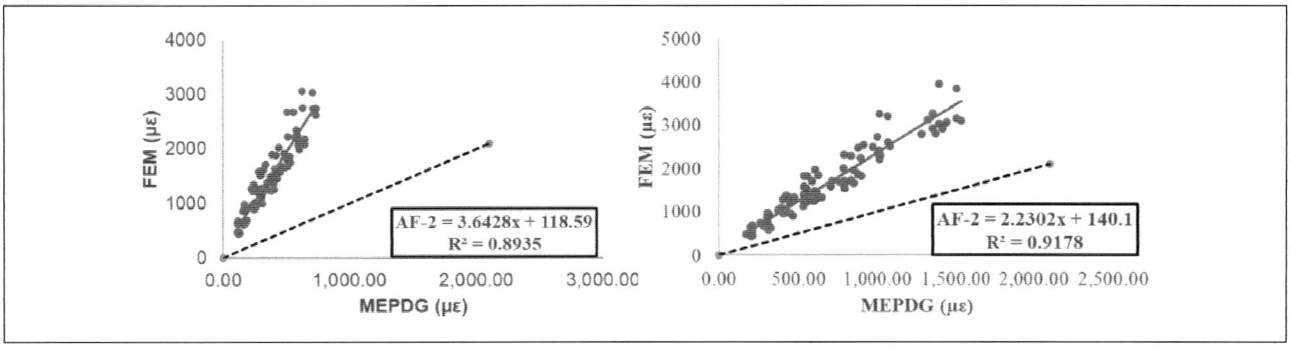

Figure 4. Maximum compressive strain within base for strong and weak base layers, respectively.

three thick pavement structures. Combining all the five distresses, the resulting *CDW* varied from 1.0 to 3.8 (Figure 3*f*). Consequently, for an applied load of 44.4 kN, the *CDW* remained relatively similar despite the change in pavement layer thickness, in which the ratios range from 1.7 to 2.1.

The damage induced by DTA with differential tire-inflation pressure (S = 552/758 kPa) and an applied load of 26.6 kN was higher than the DTA loading case subjected to 44.4 kN and uniform tire-inflation pressure of 758 kPa. The difference was higher for low-volume roads than for interstate highways. In general, the impact of NG-WBT is less on interstate highway structure than on low-volume roads.

Adjustment Factors

Although the MEPDG has a more theoretically grounded methodology for pavement analysis, as compared with traditional pavement design guides, it has several limitations and unrealistic simplifications that may result in inaccurate response predictions. These limitations and simplifications are mostly caused by the assumptions resulting from the use of multi-layered elastic theory (*27*). The FE model presented in this study can overcome these limitations by simulating tire–pavement interaction more realistically. However, it is computationally too expensive to adopt an FE model into the MEPDG framework. Therefore, two factors were created to adjust the pavement responses obtained from MEPDG in accordance with the developed FE model.

Adjustment Factor 2, AF2, adjusts the pavement responses obtained from MEPDG under DTA loading. The mechanistic part of the MEPDG procedure was implemented by exploiting the capabilities of Matlab and AutoHotkey. The main steps of implementing the MEPDG procedure are:

- Subdivision of pavement structure in sublayers
- Calculation of dynamic modulus at mid-depth of each sublayer
- Creation of input file
- Running JULEA (Linear Elastic Computer Program used by MEPDG)
- Post-processing to obtain pavement responses

Differences in loading conditions between FE and MEPDG (three-dimensionality and non-uniformity of the contact stresses), material characterization, and layer interaction introduce serious challenges that complicate the development of AF2. To obtain statistically good correlation for AF2, the cases were divided into three categories: (i) thick pavement, (ii) thin pavement with strong base material, and (iii) thin pavement with weak base material. A sample of AF2 is given for only two responses in Figure 4. Details about AF2 development and the results for all responses can be found in Gungor et al. (*27*).

Adjustment Factor 1, AF1, was developed to convert the response resulting from DTA into NG-WBT because MEPDG cannot simulate NG-WBT. AF1 was created considering the same material properties and pavement structures. The only difference was the contact stresses and contact areas, which were measured under the same tire load for NG-WBT and DTA. AF1 is presented as a linear function of the DTA response, and a sample for two responses is provided in Figure 5. The full list of all the responses can be found in Gungor et al. (*28*).

Artificial Neural Network

ANN modeling was used as a prediction tool to estimate pavement responses because of the technical and computational challenges of calculating pavement responses for every possible combination of input variables using the FE method. In this study, multilayer perceptron (MLP) with Levenberg–Marquardt, a fast adoptive training algorithm, was used for training all models. The sigmoidal function was also utilized for the hidden layer. Details about the ANN methodology are well established in the literature and can be found elsewhere (*29,30*).

Given the cases simulated via FE method, ANN models interpolate the responses for any arbitrary case within the analysis matrix, for a fraction of a second relative to FE simulation time. Eleven responses, including longitudinal and

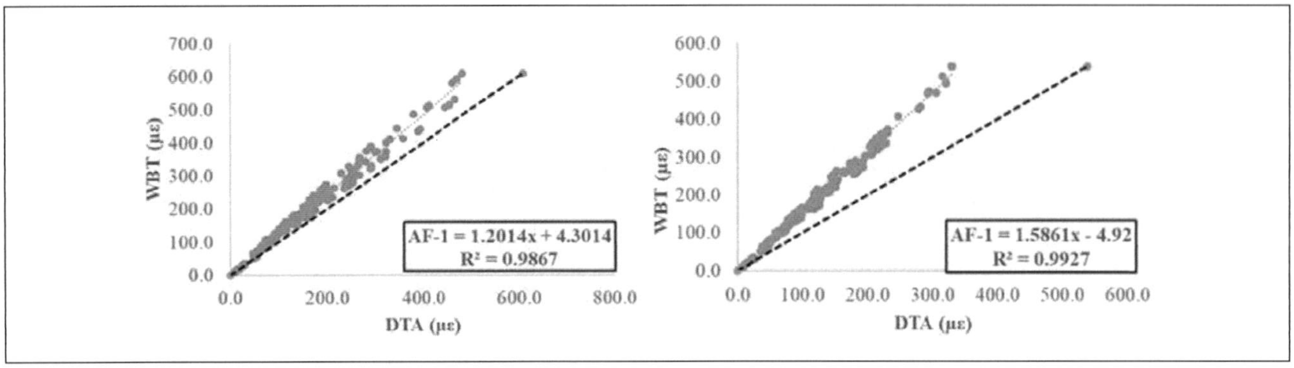

Figure 5. Maximum tensile strain at the bottom of the AC along the longitudinal and transverse direction, respectively.

transverse strains at surface and bottom of AC and vertical and shear strains in base and top of subgrade, were calculated. Critical variables affecting the responses, categorized as either loading or structure, were used as input. Detailed information regarding model development process and training results can be found elsewhere (*30*).

The material properties, especially complex modulus, are not readily available. Therefore, two levels of input properties were defined to account for data availability issues: Level 1 for detailed material properties and Level 2 for modulus of each layer at room temperature only. According to the results, the average error for all models was 4.0% with an R-squared value of 0.99, which indicates the high level of accuracy of ANN predictions. In addition, a k-fold cross-validation technique was used to test and obtain the optimum models. The lowest coefficient of correlation from all critical strains was 0.959 for the vertical strain in the AC of the thick pavement. In addition, the highest normalized root-mean-square error for thick pavement was 3.35% for the vertical shear in the subgrade.

A sensitivity analysis using two methods, the missing data problem and incremental sensitivity (also called one-at-a-time analysis), was implemented to analyze the ANN models (*29*). According to the sensitivity analysis, the pavement structure had the highest influence in both scenarios, and the tire type played an important role in pavement response calculation; in which a 10–30% influence on the response was found. This range of influence was nearly equal to the effect of tire-inflation pressure.

ICT-Wide Tool

To assist agencies, designers, and practitioners in the evaluation of the effect of NG-WBT on pavements, all trained models were incorporated into a tool called the Illinois Center for Transportation wide-base tool (ICT-Wide tool). The tool is a Matlab-based, stand-alone program that can be installed on any Windows operating system. The tool is capable of predicting critical responses of pavements as well as calculating damage to the pavement using both Mechanistic-Empirical and Asphalt Institute methods. Figure 6 shows the interface of the tool. For a 10% NG-WBT market penetration, the pavement impact increase could be outweighed by the environmental benefits.

Experimental Pavement Sections

Three new test sites were built and instrumented in Florida, California, and Ohio (*31*). Strain, stress, and temperature were measured at various locations within the pavement structures. The analysis of the experimental measurements showed 30% higher strain at the bottom of AC for NG-WBT. The impact of the applied load was different for each tire. Longitudinal strain at the bottom of the AC increased by 50% and 40% after changing the load from 26.7 to 80.0 kN for DTA and NG-WBT, respectively. Similarly, an increase in tire-inflation pressure from 552 to 827 kPa resulted in approximately 10% increase in strain at the bottom of AC for DTA and 20% for NG-WBT. For the same change in tire-inflation pressure, surface transverse strains augmented 30% and 40% for DTA and NG-WBT were measured, respectively.

The research team also gathered pavement responses from previous projects that used NG-WBT. Existing and new data were compiled and organized in an online database that could be made available to the public. The online database was developed to organize data and utilize both new and existing data, considering NG-WBT loading on pavements. The database management procedure included pre-processing data, filtering and smoothing, max/min response extraction, post-processing and summary, and online application design.

Life-Cycle Assessment and Life-Cycle Costing

The use of NG-WBT can improve the fuel efficiency of trucks by 1–3.5% (*3*), but at the same time it could cause a greater impact on pavement. Therefore, a holistic study

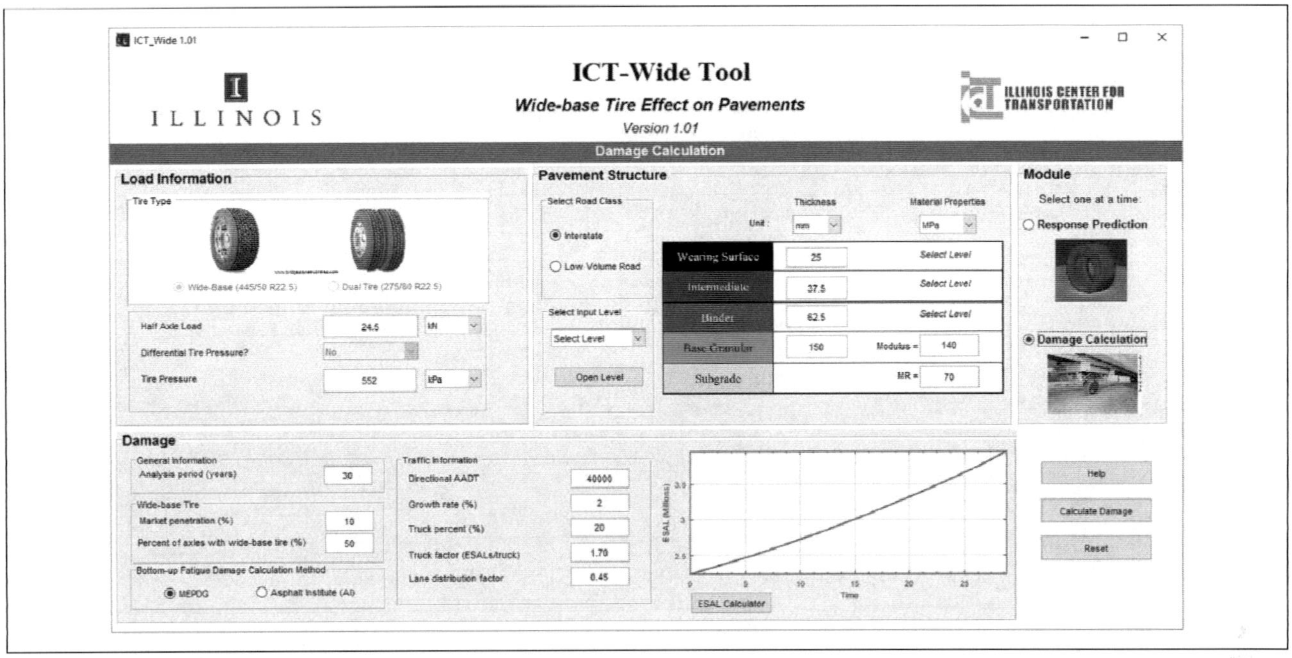

Figure 6. Main page of ICT-Wide tool.

considering the entire life-cycle of a pavement is needed to properly assess the net environmental benefit of NG-WBT compared with DTA.

This study used LCA, a technique that compiles and evaluates inputs and outputs (i.e., environmental impacts) of a product system during its life-cycle (*32*). Goal and scope definition, methodology, and key assumptions regarding the LCA study are described in detail elsewhere (*33*). Relevant pavement life-cycle stages considered are material, construction, and use-stage (rolling resistance [RR] due to roughness and fuel savings from NG-WBT) of AC overlay. The result of the LCA study is annualized and the functional unit used is two-mile-two-lane-year. The study also considered different levels of NG-WBT market penetration (5, 10, 50, and 100%); 0% market penetration implies 100% DTA and is considered as the baseline. Two pavement sections with different AC overlay thicknesses (75 mm and 125 mm) were also considered; the pavement sections share the same base and subgrade layers, consisting of 250 mm milled and recompacted base, 320 mm old aggregate base, 200 mm tipped and compacted subgrade on top of clayey subgrade.

Based on the scenario-based case study, the net energy saving for each life-cycle stage by using different levels of NG-WBT market penetration is calculated and shown in Figure 7. The three scenarios considered are as follows: (i) Scenario I: standard DTA and NG-WBT; (ii) Scenario II: DTA and NG-WBT have the same IRI performance but different fatigue cracking potential; and (iii) Scenario III: DTA and NG-WBT have the same fatigue cracking potential but different IRI performance. Details of the models used to conduct the case study are described in Kang et al. (*33*). It is evident in most cases of this particular case study that the use of NG-WBT brings positive life-cycle environmental savings compared with DTA, even if additional pavement damage is considered. The only exception is with the thin AC overlay section of Scenario III, in which increased burdens in material and construction stages, and RR surpassed the savings from NG-WBT at high market penetrations.

For Scenario I, the net life-cycle costing benefits were positive; those for Scenarios II and III were slightly negative at certain market penetrations (50 and 100%) for the thick section (*33*). However, environmental and economic benefits are sensitive to the method used to determine the pavement performance.

Summary and Conclusions

The main objective of this study was to present a multi-year effort that compared the impact of NG-WBT and DTA on pavements and environment. Contact loads of the NG-WBT and DTA were measured and incorporated in a numerical model to realistically simulate pavement structures, materials, and loading conditions. The model was validated with various field test sections across the United States. Good agreement was noted between FE analysis results and field-instrumented pavement responses when using proper material characterization parameters, especially for vertical pressure and tensile strain at the bottom of AC in transverse direction.

It was evident that NG-WBT responses were generally greater than DTA for pavements. The difference decreased

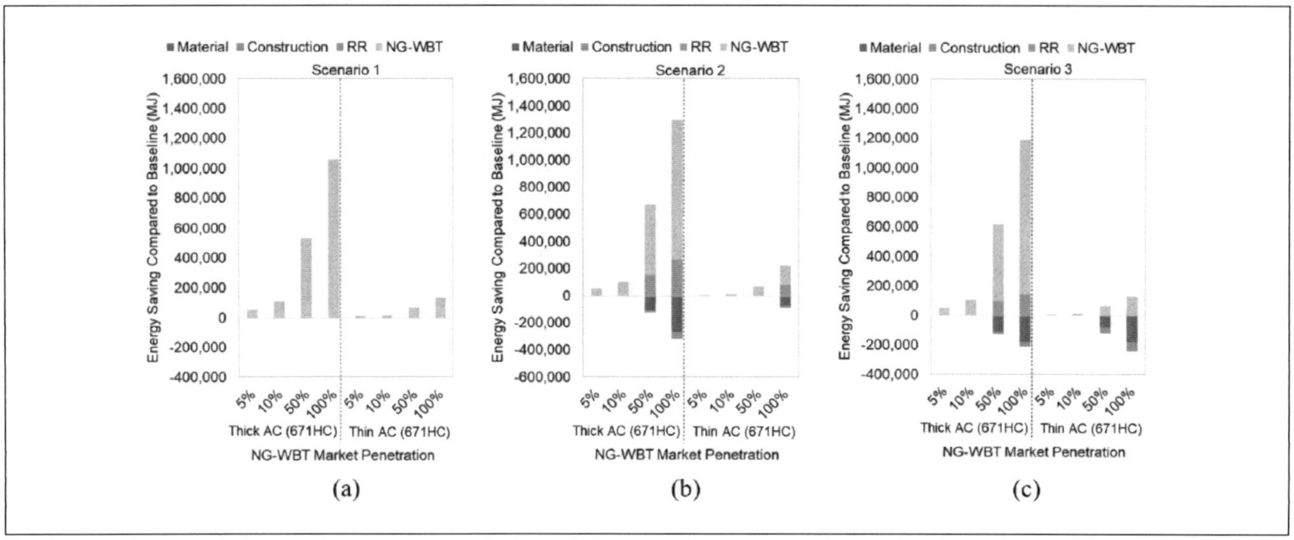

Figure 7. Net energy saving of using NG-WBT for (*a*) Scenario I, (*b*) Scenario II, and (*c*) Scenario III (*33*).

with depth and a few cases showed higher responses for DTA (e.g., shear strain in the subgrade in thin pavement). In addition, DTA with differential tire-inflation pressure produced higher damage than DTA with the same tire-inflation pressure when compared with the tire with the highest tire-inflation pressure. As expected, different behavior was observed for low-volume (thin) and interstate roads (thick). On one hand, for low-volume roads, subgrade rutting was the controlling distress of the pavement life; the difference between the impacts of the two tires on subgrade failure was minimal. For interstate highways, the influence of NG-WBT was greater than that of DTA, in which the controlling distress in AC was top-down or near-surface cracking and possible AC rutting.

Since MEPDG may not be used directly for predicting responses of NG-WBT, two adjustment factors were developed to modify the pavement responses obtained from MEPDG to account for model complexity and different contact mechanisms of DTA and NG-WBT. Furthermore, to simplify the process of using NG-WBT and allow for implementation by agencies, an ANN tool was developed to predict the pavement response without running FE. The tool results rendered by the ANN surrogate models were highly accurate, with an average prediction error less than 5% and R-square values higher than 0.95. It was also observed that although NG-WBT could cause greater pavement response than DTA for some roads, NG-WBT demonstrated significant improvement compared with the FG-WBT.

From the environmental point of view and based on LCA models that considered NG-WBT market penetration percentages, NG-WBT could save energy and reduce GHG and emissions, depending on corresponding pavement performance. Based on the results of the study and using the developed tools, for a NG-WBT market penetration of 10%, the pavement impact increase could be outweighed by the environmental benefits as well as cost savings in energy.

Acknowledgments

This project was conducted in cooperation with the Illinois Center for Transportation; the U.S. Department of Transportation, Federal Highway Administration; Rubber Manufacturers Association, and the following state departments of transportation: Illinois, Minnesota, Montana, New York, Ohio, Oklahoma, Texas, and Virginia. The involvement of the University of California at Davis, Ohio University, and the state departments of transportation of Florida, Ohio, and South Dakota is appreciated.

The input and feedback by Eric Weaver is greatly appreciated. The contribution of other partners, including John Harvey, Rongzong Wu, James Greene, Bouzid Choubane, Morris De Beer, and Tom Scarpas, Shad Sargan, and Issam Khoury are also acknowledged.

References

1. Muster, T. *Fuel Savings Potential and Costs Considerations for US Class 8 Heavy Duty Trucks through Resistance Reductions and Improved Propulsion Technologies Until 2020.* MIT EL 00-001. Massachusetts Institute of Technology, Cambridge, Massachusetts, 2000.
2. Franzese, O., H. Knee, and L. Slezak. Effect of Wide-Based Single Tires on Fuel Efficiency of Class 8 Combination Trucks. *Transportation Research Record: Journal of the Transportation Research Board*, 2010. 2191: 1–7.
3. GENIVAR Consulting Group. *Economic Study Use of Supersingle Tires by Heavy Vehicles Operating in Quebec.* Report published by GENIVAR Consulting Group, Montreal, QC, Canada, 2005.
4. Bachman, L. J., A. Erb, and C. L. Bynum. *Effect of Single Wide Tires and Trailer Aerodynamics on Fuel Economy and NOx Emissions of Class 8 Line-Haul Tractor-Trailers.* Society of

Automotive Engineers (SAE) Technical Paper 2005-01-3551. SAE International, Warrendale, 2005.

5. Markstaller, M., A. Pearson, and I. Janajreh. *On Vehicle Testing of Michelin New Wide Base Tire*. Society of Automotive Engineers (SAE) Technical Paper 2000-01-3432. SAE International, Warrendale, 2000.
6. Ang-Olson, J., and W. Schroeer. Energy Efficiency Strategies for Freight Trucking: Potential Impact on Fuel Use and Greenhouse Gas Emissions. *Transportation Research Record: Journal of the Transportation Research Board*, 2002. 1815: 11–18.
7. Huhtala, M., J. Pihlajamaki, and M. Pienimaki. Effects of Tires and Tire Pressures on Road Pavements. *Transportation Research Record: Journal of the Transportation Research Board*, 1989. 1227: 107–114.
8. Sebaaly, P. E., and N. Tabatabaee. Effect of Tire Parameters on Pavement damage and load- equivalency factors. *Journal of Transportation Engineering*, Vol. 118, No. 6, 1992, pp. 805–819.
9. Harvey, J., and L. Popescu. *Rutting of Caltrans Asphalt Concrete and Asphalt-Rubber Hot Mix under Different Wheels, Tires and Temperatures–Accelerated Pavement Testing Evaluation*. Report prepared for California Department of Transportation. University of California, Berkeley, 2000.
10. Gillespie, T. D. *NCHRP Report 353: Effects of Heavy-vehicle Characteristics on Pavement Response and Performance*. HRB, National Research Council, Washington, D.C., 1993.
11. Siddharthan, R. V., J. Yao, and P. E. Sebaaly. Pavement Strain from Moving Dynamic 3D Load Distribution. *Journal of Transportation Engineering*, Vol. 124, No. 6, 1998, pp. 557–566.
12. Myers, L., R. Roque, B. Ruth, and C. Drakos. Measurement of Contact Stresses for Different Truck Tire Types to Evaluate their Influence on Near-Surface Cracking and Rutting. *Transportation Research Record: Journal of the Transportation Research Board*, 1999. 1655: 175–184.
13. COST 334. *Effects of Wide Single Tires and Dual Tires*. Final Report of the Action. European Cooperation in the Field of Scientific and Technical Research, Brussels, Belgium, 2004.
14. Pierre, P., G. Doré, and L. Vagile. *Characterization and Evaluation of Tire-Roadway Interface Stresses. Ministry of Transport*. University of Laval, Quebec, Canada, 2003.
15. Al-Qadi, I. L., P. J. Yoo, M. A. Elseifi, and I. Janajreh. Effects of Tire Configurations on Pavement Damage. *Journal of the Association of Asphalt Paving Technologists*, Vol. 74, No. 1, 2005, pp. 921–961.
16. Priest, A., and D. Timm. Mechanistic Comparison of Wide-Base Single versus Standard Dual Tire Configurations. *Transportation Research Record: Journal of the Transportation Research Board*, 2006. 1949: 155–163.
17. Al-Qadi, I., S. Dessouky, J. Kwon, and E. Tutumluer. Geogrid in Flexible Pavements: Vali- Dated Mechanism. *Transportation Research Record: Journal of the Transportation Research Board*, 2008. 2045: 102–109.
18. Greene, J., U. Toros, S. Kim, T. Byron, and B. Choubane. Impact of Wide-Base Single Tires on Pavement Damage. *Transportation Research Record: Journal of the Transportation Research Board*, 2010. 2155: 82–90.
19. Xue, W., and E. Weaver. Pavement Shear Strain Response to Dual and Wide-Base Tires. *Transportation Research Record: Journal of the Transportation Research Board*, 2011. 2225: 155–164.
20. Grellet, D., G. Doré, J.-P. Bilodeau, and T. Gauliard. Wide-base Single-Tire and Dual-Tire Assemblies: Comparison Based on Experimental Pavement Response and Predicted Damage. *Transportation Research Record: Journal of the Transportation Research Board*, 2013. 2369: 47–56.
21. Al-Qadi, I., H. Wang, P. Yoo, and S. Dessouky. Dynamic Analysis and In Situ Validation of Perpetual Pavement Response to Vehicular Loading. *Transportation Research Record: Journal of the Transportation Research Board*, 2008. 2087: 29–39.
22. De Beer, M., and C. Fisher. Stress-In-Motion (SIM) System for Capturing Tri-Axial Tyre–Road Interaction in the Contact Patch. *Measurement*, Vol. 46, No. 7, 2013, pp. 2155–2173.
23. Hernandez, J., A. Gamez, I. Al-Qadi, and M. De Beer. Analytical Approach for Predicting Three-Dimensional Tire-Pavement Contact Load. *Transportation Research Record: Journal of the Transportation Research Board*, 2014. 2456: 75–84.
24. Hernandez, J. A., I. Al-Qadi, and M. De Beer. Impact of Tire Loading and Tire Pressure on Measured 3D Contact Stresses. In *Airfield and Highway Pavement 2013: Sustainable and Efficient Pavements* (S. Murrell and I. L. Al-Qadi, eds.), ASCE, Los Angeles, 2013, pp. 551–560.
25. Hernandez, J. A., A. Gamez, and I. L. Al-Qadi. Effect of Wide-Base Tires on Nationwide Flexible Pavement Systems: Numerical Modeling. *Transportation Research Record: Journal of the Transportation Research Board*, 2016. 2590: 104–112.
26. Gungor, O. E., I. L. Al-Qadi, A. Gamez, and J. A. Hernandez. In-Situ Validation of Three- Dimensional Pavement Finite Element Models. In *The Roles of Accelerated Pavement Testing in Pavement Sustainability* (J. Aguiar-Moya, A. Vargas-Nordcbeck, F. Leiva-Villacorta, and L. Loría-Salazar, eds.), Springer, Cham, 2016, pp. 145–159.
27. Gungor, O. E., I. L. Al-Qadi, A. Gamez, and J. A. Hernandez. Development of Adjustment Factors for MEPDG Pavement Responses Utilizing Finite-Element Analysis. *Journal of Transportation Engineering, Part A: Systems*, Vol. 143, No. 7, 2017, p. 04017022.
28. Gungor, O. E., J. A. Hernandez, A. Gamez, and I. L. Al-Qadi. Quantitative Assessment of the Effect of Wide-Base Tires on Pavement Response by Finite Element Analysis. *Transportation Research Record: Journal of the Transportation Research Board*, 2016. 2590: 37–43.
29. Ziyadi, M., and I. L. Al-Qadi. Efficient Surrogate Method for Predicting Pavement Response to Various Tire Configurations. *Neural Computing and Applications*, Vol. 28, No. 6, 2016, pp. 1355–1367.
30. Moghaddam, F. R., S. Afandizadeh, and M. Ziyadi. Prediction of Accident Severity Using Artificial Neural Networks. *International Journal of Civil Engineering*, Vol. 9, No. 1, 2011, p. 41.
31. Ziyadi, M., and I. L. Al-Qadi. Tire Type Effect on Pavement Responses; Accelerated Pavement Testing Results. In *The Roles*

of Accelerated Pavement Testing in Pavement Sustainability (J. Aguiar-Moya, A. Vargas-Nordcbeck, F. Leiva-Villacorta, and L. Loría-Salazar, eds.), Springer, Cham, 2016, pp. 277–288.

32. Standard, I. Environmental Management-Life Cycle Assessment Requirements and Guidelines, Vol. 14044. ISO, 2006.
33. Kang, S., I. Al-Qadi, H. Ozer, M. Ziyadi, and J. Harvey. Environmental and Economic Impact of Using New Generation Wide-Base Tires. *Transportation Research Part D: Transport and Environment*, Vol. In Press, 2018.

The Standing Committee on Full-Scale Accelerated Pavement Testing peer-reviewed this paper (18-03063).

The contents of this report reflect the view of the authors, who are responsible for the facts and the accuracy of the data presented here. The contents do not necessarily reflect the official views or policies of the Illinois Center for Transportation, the Federal Highway Administration, or the participating partners. This report does not constitute a standard, specification, or regulation.

Article

TRR
JOURNAL OF THE TRANSPORTATION RESEARCH BOARD

Development of Structural Design Guidelines for Porous Asphalt Pavement

Transportation Research Record
2018, Vol. 2672(40) 197–206

DOI: 10.1177/0361198118758335
journals.sagepub.com/home/trr
SAGE

Kevin D. Hall[1] and Charles W. Schwartz[2]

Abstract

Porous asphalt pavements allow designers to introduce more sustainability into projects and lessen their environmental impact. Current design procedures are based primarily on hydrologic considerations; comparatively little attention has been paid to their structural design aspects. As their use grows, a design procedure and representative material structural properties are needed to ensure that porous pavements do not deteriorate excessively under traffic loads. The objective of this project was to develop a simple, easy to apply design procedure for the structural design of porous asphalt pavements. Two methodologies were considered for such a structural design procedure: (*a*) the 1993 AASHTO Pavement Design Guide empirical approach, and (*b*) the mechanistic–empirical approach employed by the AASHTOWare Pavement ME Design software. A multifactor evaluation indicated the empirical 1993 AASHTO design procedure to be the most appropriate platform at this time. It is noted, however, that both design procedures lack validation of porous asphalt pavements against field performance. AASHTO design parameters and associated material characteristics are recommended, based on an extensive literature review. For "thin" open-graded base structures (12 in. or less), the AASHTO procedure is performed as published in the 1993 Guide. For "thick" base structures (>12 in.), the base/subgrade combination is considered a *composite* system which supports the porous asphalt layer; an equivalent deflection-based approach is described to estimate the composite resilient modulus of the foundation system, prior to applying the 1993 AASHTO design procedure.

Porous asphalt pavements are an increasingly popular approach for introducing more sustainability into projects and lessening their environmental impact. Currently, the design of porous asphalt pavements is based primarily on hydrologic considerations—that is, providing sufficient storage capacity for the gradual release of captured storm water into the environment. Porous asphalt is typically used for parking lot and other low volume road applications; consequently, comparatively little attention has been paid to their structural design aspects. However, even parking lot and low volume residential pavements see periodic heavy truck and bus traffic. A design procedure and representative material structural properties are required to ensure that porous pavements do not deteriorate excessively under traffic loads.

The objective of this project was to develop a simple, easily applied design procedure for the structural design of porous asphalt pavements. The focus here is on the structural design procedure and its required inputs. Site selection, hydrologic design, mixture design for porous asphalt and asphalt-treated permeable base (ATPB), construction, and maintenance are included only insofar as they affect structural design and performance. In addition, secondary features such as noise reduction, spray and splash reduction, water quality, pollutant removal, etc. are not addressed.

Background

Porous asphalt pavements have historically been designed and constructed to serve as stormwater storage and infiltration systems. The structure of the system is designed to maximize infiltration and water flow, as illustrated in Figure 1. The porous surface allows stormwater to pass into a "stone recharge bed," which stores water as it slowly infiltrates into the subgrade.

Hydrologically, the key design elements of the system include the infiltration rate and capacity of the subgrade soil and the expected runoff volume, which is related to climate. The runoff volume and the ability of the subgrade soil to "accept" water governs the thickness and properties of the stone recharge bed. In general, porous pavements are feasible for subgrade infiltration rates ranging from 0.1 to 10 in./hour. In addition, it is desirable that the stone recharge bed be able to drain within 12 to 72 hours. These criteria result in recharge beds ranging from 12 to 36 in. in thickness.

[1]University of Arkansas, Fayetteville, AR
[2]University of Maryland, College Park, MD

Corresponding Author:
Address correspondence to Kevin D. Hall: kdhall@uark.edu

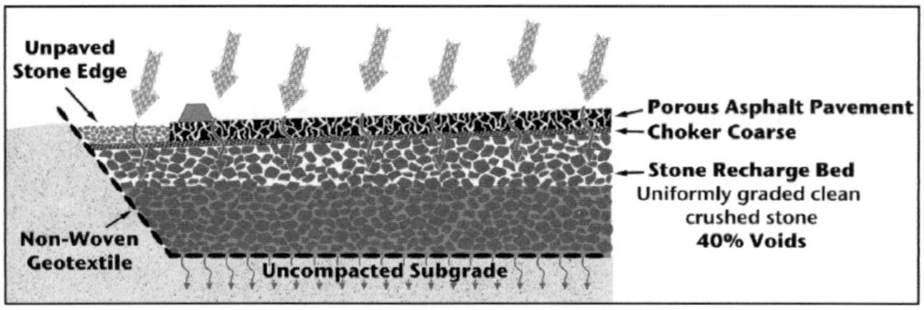

Figure 1. Typical cross section of porous pavement (*1*).

The hydrologic design aspects of porous pavement systems not only represent a completely different basis for design compared with conventional flexible asphalt pavements but are also antithetical to common structural pavement design principles. Flexible pavements designed for structural load carrying capacity are designed to keep water out of the unbound layers and often include drainage systems for the rapid removal of any water that does infiltrate. Porous asphalt pavements, conversely, are designed to hold the water and then release it slowly into the subgrade. This is detrimental from a structural design standpoint, particularly if the subgrade soil has high moisture susceptibility. Schwartz and Hall (*2*) present a comparison of typical layers and their associated design purposes for conventional flexible and porous asphalt pavements. Note, however, when considering typical porous pavement structures within the context of the AASHTO (1993) empirical pavement design methodology, that even a somewhat conservative estimation of layer properties implies a structure capable of carrying traffic loads corresponding to low- to medium-volume roadways (*2*).

Case Studies

Nearly all of the case studies in the literature are for porous asphalt pavements designed for hydrologic purposes under light traffic loading. These are not relevant to the objectives of this project, so only a sampling of case studies will be presented.

Early international experience with porous asphalt pavements is summarized in TRR 1265 *Porous Asphalt Pavements: An International Perspective* (*3*). McAdams (*4*) cites several porous asphalt parking lot projects constructed in porous asphalt in the 1980s that were still functioning well at the time of writing. The NAPA Information Series 131 *Porous Asphalt Pavements for Stormwater Management* also provides descriptions of early porous asphalt pavement successes (*1*).

As examples of more recent case studies on porous asphalt pavements, Roseen et al. (*5*) evaluated a project in New Hampshire in which cold climate and significant frost penetration were thought to be potential problems. Surface infiltration capacity and water quality were the key parameters tracked in the study. The authors concluded that "The porous asphalt pavement system performed impressively as a stormwater management strategy despite cold-climate challenges." Al-Rubaei et al. (*6*) evaluated the effect of clogging on the long-term infiltration capacity and porosity of two 18- and 24-year-old porous asphalt pavements in Sweden. Although the long-term infiltration capacity was far below initial values and at best only partially restored after washing and vacuum cleaning, the pavement still had the capacity to infiltrate an intense design rainfall (100 year average return interval, 15 min duration), underlining that porous asphalt can be a resilient feature even under unfavorable conditions.

Peabody (*7*) describes the construction and initial performance of a heavily trafficked (16750 AADT) porous asphalt pavement at the entrance to a shopping mall in Maine. After one year of service, the pavement was performing well from a rutting and cracking perspective, although there were some areas of raveling in the open-graded friction course (OGFC) surface. The Maine Department of Transportation (DOT) has continued to monitor the performance of this pavement, but to date no formal updates have been published. However, Peabody states, "[F]rom a pavement performance perspective, this road is still holding up very well." (*8*)

As stated, nearly all of the case studies in the literature deal with the hydrologic aspects of porous asphalt pavement performance. Key issues include: clogging of the porous asphalt surface; restoration of the surface through vacuuming or other cleaning; quality of the water exfiltrating from the system, including concentrations/removal of heavy metals and other contaminants; porous asphalt mixture durability (e.g., raveling and cracking because of increased photooxidative degradation); freeze–thaw resistance; and other related functional performance issues. Rutting and distortion under heavy loads are noted, but these are not design considerations. Other functional aspects of porous asphalt pavements mentioned include reduced noise, increased wet weather friction and visibility, and mitigation of urban heat island effects.

Hydrologic Design Methods

Hydrologic design of porous asphalt pavements is beyond the scope of this paper. An excellent summary of hydrologic design considerations and procedures is given in (*1*). This manual has become an industry standard for porous asphalt design.

Other guidance for the hydrologic design of porous asphalt pavements can be found in state design manuals from Pennsylvania (*9*), New Jersey (*10*), New Hampshire (*11*), California (*12*), Wisconsin (*13, 14*), and others. The Federal Highway Administration also provides guidance at the national level (*15*).

Structural Design Methods

The current state of the art provides two approaches for the structural design of flexible pavements: empirical and mechanistic–empirical.

Empirical Design. The widely-used AASHTO empirical design method is based upon the AASHO Road Test conducted at the end of the 1950s. This test measured the relationship between pavement performance and traffic for pavements representing a range of structural capacities. Pavement performance was characterized in terms of a subjective composite measure, the pavement serviceability index (PSI; originally the present serviceability rating or PSR) on a scale of 0 (very poor) to 5 (excellent); traffic was quantified by 18 kip equivalent standard axle loads (ESALs); and structural capacity for flexible pavements was characterized in terms of an empirical structural number (SN). The SN value is the sum of the structural contributions of all of the *n* finite-thickness layers in the pavement structure; the structural contribution of a given layer is the product of the layer thickness and a structural layer coefficient (a_i) representing the material comprising the layer (*16*).

Mechanistic–Empirical Design. The relatively new mechanistic–empirical design procedure as implemented in the AASHTOSare Pavement ME Design software was developed for conventional pavements under relatively high traffic loadings (*17*). It is neither intended nor calibrated for lower volume roads, and thus it may not be appropriate for many porous asphalt pavement design scenarios. The required inputs are either difficult to define (e.g., traffic load spectra) or largely unknown (e.g., dynamic modulus, permanent deformation properties for porous asphalt, resilient moduli for coarse-stone base courses) at present. The national calibration data set of field sites contains no porous asphalt pavements. It is the opinion of the authors that the Pavement ME Design approach is excessive for porous asphalt pavements and would not find wide acceptance.

Researchers at the University of California – Davis Pavement Research Center (UCPRC) have formulated a conceptual outline for a mechanistic–empirical design procedure for porous asphalt pavements (*18*). It should be stressed that this is a conceptual framework and is not intended as a final design procedure. The basic steps in this design procedure include: (*a*) Choose a base thickness based on hydrologic requirements; (*b*) Choose a porous asphalt thickness based on predicted fatigue for a given traffic level based on mechanistic–empirical damage; and (*c*) Check that shear stress/strength ratio at top of subgrade is less than 0.3 to prevent excessive subgrade rutting.

Evaluation Of Existing Design Methods

Evaluation Criteria

Two design methodologies were considered for a structural design procedure for porous asphalt pavements: (*a*) the empirical approach based on the 1993 AASHTO Pavement Design Guide; and (*b*) the mechanistic–empirical approach implemented in the AASHTOWare Pavement ME Design software.

Criteria for evaluating the design methodologies included the following issues:

- Characterization of poorly defined traffic;
- Characterization of porous asphalt pavement materials;
- Adaptability to the special hydrologic characteristics of porous asphalt pavements (e.g., storage versus drainage);
- Validation against field performance;
- Availability of design tools;
- Appropriateness for agency technology level.

When evaluating these criteria, it should be kept in mind that most porous asphalt pavements are designed by county, municipal, and other local agencies. The technical capabilities of these local agencies are usually limited as compared to state DOTs. Table 1 summarizes the evaluation of these criteria against the empirical 1993 AASHTO and mechanistic–empirical Pavement ME Design approaches.

Recommendation

Based on the evaluations in Table 1, the empirical 1993 AASHTO design procedure is the clear choice for the structural design of porous asphalt pavements. Its traffic and material characterization and adaptability to the special hydrologic conditions of porous asphalt pavements are much more suited to agency practice, particularly at the local level, than is Pavement ME Design. The availability of design tools

Table 1. Evaluation of Structural Design Methods

Criterion	1993 AASHTO	Pavement ME Design
Characterization of traffic	ESALs. Design input is in $\log_{10}$(ESALs), which reduces sensitivity to traffic level. Agencies have long experience in estimating ESALs for higher volume roads; less experience for estimating ESALs for low-volume pavements (e.g., parking lots, residential streets, industrial and commercial areas. Traffic characterization for commercial, industrial, and residential areas can be adapted to porous asphalt pavements.	Load spectra: frequency distributions of vehicle types and axle loads. Load spectra largely unknown and difficult to estimate for low-volume pavements (e.g., parking lots, residential streets, industrial and commercial areas). Pavement ME Design was developed for high-traffic-volume pavements and its applicability to low-volume pavements is unclear.
Characterization of materials	Layer coefficients. Layer coefficients are related to resilient moduli (M_R). Agencies have confident values for layer coefficients for conventional flexible pavement materials; values for porous asphalt pavement materials are less clear.	Dynamic modulus (E*) and repeated load permanent deformation (RLPD) properties of asphalt materials; resilient moduli (M_R) for unbound materials. Many agencies (and nearly all local agencies) have little experience with E* and RLPD properties, more experience with M_R (although not at the local agency level).
Adaptability to hydrologic characteristics	Drainage ability of materials incorporated through layer moisture coefficients; can be adapted to porous asphalt pavement characteristics.	Drainage is not explicitly considered; must be incorporated indirectly through material property values.
Validation against field performance	Largely based on the original AASHO road test data. No porous asphalt pavements were included in this test.	National calibration/validation against LTPP field sites. No porous asphalt pavements are included in this dataset.
Availability of design tools	Widely available (DARwin, PaveXpress, Pavement Interactive). Design equation can also be solved using Excel, other simple computational tools.	Requires Pavement ME Design software. Pavement ME Design software is a sophisticated and expensive software package that is generally available only at the state DOT level and not at the local agency level.
Appropriateness for agency technology level	Design approach has been long used in practice. Most agency pavement personnel have been exposed to this at some point, even at the local level.	Many state DOTs possess appropriate technical capabilities (either in house or through consultants) for using Pavement ME Design. Local agencies generally have had little if any exposure to Pavement ME Design software and methodology.

and the required level of agency technological capability are also clear advantages for the 1993 AASHTO procedure.

Both design procedures are lacking in validation of porous asphalt pavements against field performance. This in part is because there is very little field performance data available at present for porous asphalt pavements under heavier traffic loadings. The collection of these field performance data is an avenue of future research.

Structural Design Methodology

The structural design of porous asphalt pavements is best performed using the widely-used AASHTO empirical design method. Structural capacity for flexible pavements was characterized in terms of an empirical structural number (SN). The SN value is the sum of the structural contributions of all of the n finite-thickness layers in the pavement structure

$$SN = \sum_{i-1}^{n} a_i D_i m_i \quad (1)$$

in which a_i is an empirical structural layer coefficient that is a function of material type and location in the pavement, D_i is the layer thickness, and m_i is a drainage coefficient (relevant only for unbound pavement layers). The culmination of the design system was the 1993 version of the AASHTO Pavement Design Guide (*16*):

$$\log_{10}(W_{18}) = Z_R S_o + 9.36 \log_{10}(SN+1) - 0.20 + \frac{\log_{10}\left[\frac{\Delta PSI}{4.2-1.5}\right]}{0.40 + \frac{1094}{(SN+1)^{5.19}}} + 2.32 \log_{10}(M_R) - 8.07 \quad (2)$$

where

W_{18} = number of 18-kip equivalent single axle loads (ESAL) expected over the design life of the pavement;

Z_R = standard normal deviate, corresponding to the level of design reliability;

S_o = standard deviation associated with the design equation;

SN = design structural number of the pavement;

ΔPSI = allowable change in the Present Serviceability Index (PSI) over the design life of the pavement;

M_R = subgrade resilient modulus (psi).

There is no closed-form solution for the AASHTO design Equation 2. Typically, all inputs are fixed and the *SN* is varied until the calculated value of W_{18} converges with the design value.

Once the required SN is determined from Equation 2, design of the pavement structure is completed by providing appropriate paving layers and thicknesses. Each layer is assigned a structural coefficient (a_i), which accounts for the layer's contribution to the overall structural capacity of the pavement. The a_i value for each layer is related to the material type. The AASHTO Guide provides guidance for selecting appropriate a_i values, and most agencies using the AASHTO methodology have developed representative values for materials used in their pavements. The total structural capacity of the pavement (*SN*) is calculated as the sum of the products of the a_i value and thickness d_i of each layer. Once the a_i values for each layer are established, layer thicknesses are adjusted until the overall *SN* of the pavement meets or exceeds the required *SN* from the AASHTO design equation.

Structural Design Inputs

Design Traffic (W_{18}). The 1993 AASHTO design procedure quantifies traffic loading in terms of 18-kip equivalent single axle loads (ESALs). Most roadway agencies have existing systems in place for estimating design ESALs for a given project; initially, such systems will suffice for the purpose of providing the traffic input for porous pavement design. However, current AASHTO methods for determining ESALs, which are based on the effect of traffic loads on dense-graded hot-mix asphalt, may not be entirely appropriate for porous asphalt pavements. Other organizations such as NAPA (*19*) provide traffic estimates for various vehicle and facility types.

Reliability, Standard Deviation, and Allowable Change in PSI. Most roadway agencies have little or no experience related to the structural design of porous asphalt pavements. It is appropriate, therefore, to use existing policies regarding the selection of design values for reliability (Z_R and S_o) and allowable performance (ΔPSI).

AASHTO Drainage Coefficient. To address concerns related to climatic effects, the 1986 version of the AASHTO Guide introduced a layer drainage coefficient (m_i). A primary purpose of the drainage coefficient is to modify the estimation of structural capacity (a_i) for unbound material layers as a function of drainage and moisture. There are two inputs for estimating the m_i value for a given layer: (*a*) the quality of drainage, related to the time required to remove water from the layer; and (*b*) the percentage of time the pavement structure is exposed to near-saturation conditions.

Given that the purpose of a porous pavement is to provide relatively slow (12–72 hour) drainage of the stored water into the underlying subgrade, it is reasonable to assume that the quality of drainage is roughly in the "good" (water removed within one day) category. It is also reasonable to assume that the percent of time the pavement is exposed to moisture conditions approaching saturation is likely no less than 5% for a porous asphalt pavement. For these assumptions, the AASHTO Guide recommends values for *m* ranging from 1.15 to 1.0. For purposes of porous pavement design, it is recommended to (conservatively) assign a drainage coefficient $m = 1.0$ for all situations.

Material Characterization

Definitions of the structural material types most commonly used in porous asphalt pavement systems are as follows:

- *Porous asphalt.* An open-graded asphalt mixture similar to open-graded friction courses (OGFC). Porous asphalts typically have a 9.5 mm nominal maximum aggregate size (NMAS), high air-void volume (16% minimum, typically), 5 to 6% asphalt content (less for larger NMAS mixes), and polymer-modified asphalts, often with fibers to limit draindown. Mixture design considerations are given in the NAPA Information Series 115 *Design, Construction, and Maintenance of Open-Graded Asphalt Friction Courses* and ASTM D7064 *Standard Practice for Open-Graded Friction Course (OGFC)*.
- *ATPB.* ATPB is similar to porous asphalt/OGFC except that it has a lower asphalt content, typically 3 to 3.5%.
- *Open-graded base.* Typically, a uniformly graded crushed stone with large aggregates, minimal fines, and a high void ratio (typically around 40%).

Early Studies. Several early porous asphalt pavement projects and studies for heavier traffic conditions are summarized in (*1*), including studies conducted in Arizona (*20, 21*), Oregon (*22*), and Vermont (*23*). Table 2 summarizes AASHTO structural layer coefficients (a_i) reported in these early studies. There is little variation reported in values representing the porous asphalt surface and the open-graded stone base (aggregate recharge bed). On the other hand, the a_i values reported for ATPB vary significantly.

Later Studies. Schwartz and Hall (*2*) detail more recent studies related to material characterization for materials used, or

Table 2. Recommended AASHTO Layer Coefficients (ai) for Porous Pavements from Early Studies

Material	NAPA (2008)	Arizona (1991)	Oregon (1992)	Oregon (1991)	Vermont (2006)
Porous Asphalt	0.40–0.42	0.40	0.42	0.42	
Asphalt-Treated Permeable Base	0.30–0.35	0.20	0.14–0.19	0.24	0.33
Open-Graded Stone Base	0.10–0.14	0.11			

Table 3. Reported Stiffness of High-Void/Open-Graded Asphalt Mixes

Source *(ref)*	Test Type	Stiffness[a] (ksi)
Jones et al. (*24*)	Flexural Dynamic Modulus	100–300
Goh & You (25)	Dynamic Modulus	350–800
Seo et al. (26)	Dynamic Modulus	580 (at 15% air voids)

[a]Stiffness values are expressed for 70F and 10Hz, corresponding approximately to the resilient modulus of the material.

are similar to those used, in typical porous asphalt pavements. A summary is provided here.

Porous Asphalt. Porous asphalt is very similar to OGFC, a material with which the paving industry has wide experience. However, OFGC is generally not treated as a structural layer, and consequently the structural properties are usually not quantified.

The 1993 AASHTO Guide relates a_l to stiffness in terms of resilient modulus. The AASHTO layer coefficient versus stiffness relationship reaches a limit of $a_1 = 0.44$ at a resilient modulus of 440 ksi. The historical data given earlier in Table 2 indicates that structural coefficients assigned for porous asphalt surfaces deviate only slightly from this value.

The AASHTO relationship represents typical dense-graded asphalt mixtures having in-place air voids of around 7%. Porous asphalt surfaces are typically constructed at much higher in-place void levels, for example, 15 to 20%. Table 3 summarizes stress measurements performed on high-void or open-graded mixes; Schwartz and Hall (*2*) provide additional details of the studies cited. The stiffness levels reported in Table 3 roughly correspond to AASHTO layer coefficient values ranging from 0.18 to 0.55.

Davis and Timm (*27*) estimated asphalt a_1 values using data from the NCAT test track. They recommended 0.54 as the average a_1 value for modern-day asphalt concrete. The results of both Seo (*26*) and Davis and Timm (*27*) suggest the traditional a_1 value of 0.44 may be conservative—and, by extension, the corresponding values typically used for other asphalt mixture types may also be conservative. This suggests the use of higher a_1 values for modern asphalt mixtures compared with mixes used at the AASHO road test may be appropriate. However, a total revision of a_i values for the AASHTO pavement design methodology is well beyond the scope of the present study.

For the purposes of characterizing the porous asphalt surface for use in AASHTO-based pavement design, continuing the current practice based on an a_1 value of 0.44 for dense-graded asphalt surfaces is recommended. To estimate the a_1 value for porous asphalt, three concepts are pertinent and applicable here:

1. It is generally accepted that low density in an asphalt surface mix translates to comparatively poorer performance. This is why there is so much focus on density in field quality assurance testing.
2. Porous asphalt mixtures typically feature a relatively open gradation compared with dense-graded mixes and are compacted to not less than 15 to 20% air voids (over twice the level of dense-graded mixes).
3. Limited field observations of porous asphalt pavements have shown generally good structural performance.

Given the above, it is reasonable to characterize the structural capacity of porous asphalt as "less" than a traditional dense-graded mix. In the absence of published research specifically addressing the effect of high air voids and other porous asphalt characteristics on pavement performance, an a_1 value of 0.40 is recommended for porous asphalt surfaces.

ATPB. Zhang and colleagues (*28*) reported ATPB stiffness values ranging from 400 to 2000 MPa (58–290 ksi), and concluded, "The stiffness of conventional asphalt concrete can be 5.5–7 times larger than the stiffness of ATPB even in a dry state." As shown in Table 2, agencies have reported a wide variation in the structural coefficient (a_2) used to represent ATPB. The AASHTO Guide presents a relationship between the structural layer coefficient a_2 and the stiffness of the bituminous-treated base (in terms of resilient modulus). While the AASHTO relationship represents typical dense-graded bituminous-treated base mixes, which may differ from ATPB used in a porous pavement structure, the few reported stiffness values for ATPB would correspond to AASHTO a_2 values similar to those shown in Table 2. As discussed earlier, it is likely that layer coefficients presented in the AASHTO Guide are conservative for modern asphalt mixes. A structural layer coefficient a_2 of 0.30 to 0.33 is therefore recommended for ATPB.

Open Graded Granular Base. Jones (*24*) tested four coarse uniform base materials, including granite, basalt, and alluvium. The measured resilient modulus values at low confinement ranged between 100 and 200 MPa (14.5 and 29 ksi). Compared with other results from the literature (*29, 30*), the resilient moduli of the materials in this study generally showed similar magnitudes and trends regarding the effects of grading and particle size.

Zhang (*28*) tested a typical Wisconsin open-graded aggregate commonly used for drainage layers. Resilient modulus tests were performed at 15% and 20% air voids. Laboratory samples could not be prepared at higher air voids, so discrete element modeling was used to simulate the resilient response at higher air voids up to 45%. The resilient modulus values ranged from 100 to 300 MPa (~15–45 ksi) at 15% air voids and from 70 to 300 MPa (~10–45 ksi) at 20% air voids. The results also indicated an approximate 85% reduction in resilient modulus as the air voids increase from 20% to 40%. Zhang notes that "[T]he resilient moduli obtained from the DEM simulation . . . are conservative (smaller) compared with the laboratory-determined values. . . . [Nonetheless,] the variation caused by air void content is still appreciable and cannot be neglected."

Table 2 indicates little variation in the a_2 structural coefficient used to characterize the open-graded stone base (recharge reservoir) featured in a porous pavement. The 1993 AASHTO Guide relationship between the a_2 structural layer coefficient and the stiffness of granular base suggests typical a_2 values between 0.07 and 0.10 and is thus recommended for the design of porous pavement structures.

Subgrade. Jones (*24*) tested silt and clay subgrades at different compacted densities and moisture conditions. The lowest measured resilient modulus values (at low confining stress, lowest compaction, and highest moisture content) were about 40 MPa (5.8 ksi) for the silt and between 50 and 90 MPa (7.2–13 ksi) for the clay. The soils could not be tested at saturated conditions because the samples failed during preparation. The default MEDPG Level 3 resilient modulus values for these subgrade materials are 24.5 and 14.5 ksi, respectively, for the silt and clay. These Level 3 values are intended to correspond to optimum moisture and density compacted conditions. The lowest values measured by Jones were approximately 25% and 50% of the Mechanistic–Empirical Pavement Design Guide (MEPDG) Level 3 values for the silt and clay, respectively. These reduced stiffnesses are very broadly compatible with the moisture corrections in the MEPDG for conditions other than optimum moisture and density.

Resilient modulus values for flexible pavement subgrade typically used in design by agencies generally correspond to subgrade compaction levels achieved in conventional flexible pavements. However, subgrades under porous asphalt pavements are often uncompacted in order to increase infiltration rates. Given the hydrologic purposes of porous pavements, subgrades under these will often be at much higher moisture contents than subgrades under conventional pavements (i.e., under pavements with a relatively impermeable surface and perhaps containing internal drainage systems). Following the philosophy of the moisture and density adjustment factors employed in the MEPDG, it is recommended that the typical flexible pavement subgrade modulus values used by agencies be reduced by 25 to 50% for porous asphalt pavements.

Limit of Effective Structural Thickness

Porous asphalt pavements often have very thick coarse stone bases for hydrologic purposes—that is, to provide adequate storage capacity after a major rainfall event. However, not all of this thickness may be engaged in providing structural load carrying capacity. There is likely a limiting thickness beyond which the base does not provide any additional structural capacity—for example, any additional contribution to *SN*. The AASHO Road Test data unfortunately do not provide any insight into this, as the maximum granular base layer thickness was only 9 in.

To illustrate the concept of a limiting structural thickness, consider the scenario depicted in Figure 2. Six inches of a full-depth porous asphalt layer is placed directly on top of a semi-infinite subgrade consisting of material similar to the coarse stone reservoir material having a resilient modulus of 20,000 psi (roughly equivalent to a layer coefficient of a_2 = 0.10). This gives *SN* of 0.40*6 in. or 2.40—which, using the pavement design data shown in Figure 2, is sufficient to carry 2.3 million ESALs. Next, consider the more usual scenario of porous asphalt over a hydrology-dictated thick coarse stone base over a soft subgrade. For the purposes of this example, let us assume 6 in. of porous asphalt over 36 in. of coarse stone reservoir over a silty clay subgrade having a resilient modulus of 4,000 psi. Assuming that the entire thickness of the coarse stone base contributes to structural capacity, this gives a total *SN* of 0.40*6 in. plus 0.1*36 in. or 6.00, which corresponds to a design traffic of approximately 41.5 million ESALs.

Comparing these two scenarios, the key question is this: *For 6 inches of porous asphalt, how can a finite 36 in. of coarse stone over a soft subgrade provide 18 times the traffic capacity as the same 6 in. of porous asphalt over an infinitely thick coarse stone layer?* The answer, of course, is that it cannot. There must be a limitation to how much of the coarse stone layer thickness contributes to structural capacity. Comparisons of 93 AASHTO designs against multilayer elastic analyses suggest that the granular base layer adds very little additional structural capacity beyond a thickness of about 12 to 18 inches. Unfortunately, there is no simple *a priori* method for accurately determining the maximum effective structural thickness for the base.

Instead of treating the coarse stone base as a finite layer with its own structural layer coefficient and SN, it can be viewed alternatively as enhancing the subgrade of the pavement. This is precisely the concept behind undercut-and-backfill remediation for soft subgrade areas encountered during

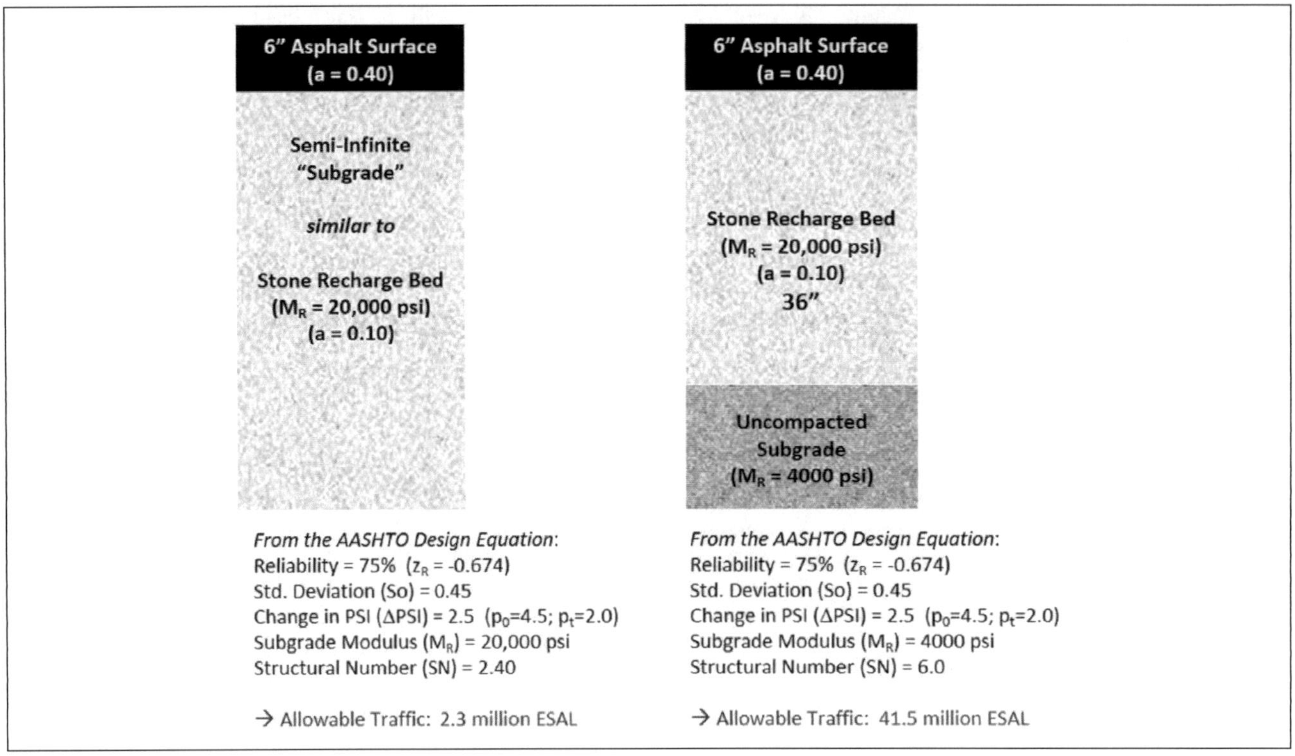

Figure 2. Composite subgrade concept for stone recharge bed.

construction. Some soft subgrade is excavated and replaced by coarse granular material to increase the overall effective stiffness of the subgrade. This is usually done to provide a stable work platform during construction. But, if excess coarse stone over subgrade is planned beneath the entire pavement, this can then be incorporated into the structural design.

The recommended process for incorporating the coarse stone base into a composite subgrade is to use an equivalent-deflection approach based in the theory of elasticity. The two-layer, coarse-stone-over-subgrade system is converted into an equivalent homogeneous (composite) subgrade. This process is described in the following steps:

Step 1. Select Analysis Parameters

Establish values for: Stone base modulus (E_{base}), existing subgrade resilient modulus (M_R), base thickness (D_2), applied load (q), load diameter (a).

Step 2. Estimate Deflection for 2-Layer System

Use Burmister's procedure for estimating deflection under load. To illustrate the procedure, the finite base thickness example from Figure 2 will be continued here. Equation 3 shows Burmister's surface deflection equation:

$$w_0 = \frac{1.5qa}{E_2} F_2 \tag{3}$$

where w_0 = deflection under load (in)
q = applied load (psi)
a = diameter of circular load (in)
$h_1 = D_2$ = base thickness (in)
$E_1 = E_{base}$ = base resilient modulus (psi)
$E_2 = M_R$ = subgrade resilient modulus (psi)
F_2 = Burmister Deflection Factor.

The Burmister Deflection Factor (F_2) for (h_1 / a = 6) and (E_1 / E_2 = 5.0) is 0.28; Schwartz and Hall (*2*) provide a simple graphical method for estimating this factor.

Calculate Two-Layer Surface Deflection (w_0) from Equation 3:

$$w_0 = \frac{1.5*(100\,psi)*(6\,in)}{4000\,psi} * 0.28 = 0.063\,inches$$

Step 3: Calculate Resilient Modulus of an Equivalent 1-Layer Composite System

Use Burmister's procedure for estimating deflection under load. Equation 4 represents Burmister's surface deflection equation for a single-layer system:

$$E = \frac{1.5qa}{w_0} \tag{4}$$

where: $E = M_R$ = subgrade resilient modulus (psi)

Table 4. Recommended Structural Properties for Porous Asphalt Pavement Layers

Layer	Recommended Property Value	Comments
Porous asphalt	a_1 = 0.40	Minimum thickness set to protect coarse stone base.
ATPB	a_2 = 0.30–0.33	
Coarse stone base (reservoir)	Thick base (>12 in.)	Incorporate coarse stone base thickness into effective subgrade modulus.
	Thin base (12 in. or less): a_2 = 0.07–0.10	Apply the 93 AASHTO methodology and take the larger porous asphalt thickness.
Uncompacted subgrade	Reduce typical design values by 25–50% for lack of compaction and high moisture content; Use geosynthetic separator layer between subgrade and coarse stone base.	

w_0 = deflection under load (in) – *from Step 2*
q = applied load (psi)
a = diameter of circular load (in)
Continuing the example calculation:

$$E = \frac{1.5*(100\,psi)*(6\,in)}{0.063\,in} = 14,285\,psi$$

Thus, the 36 in. of coarse stone over a 4,000 psi subgrade provides a composite subgrade resilient modulus of 14,285 psi to support the pavement system.

The composite subgrade modulus calculation relies upon the assumption that there are two distinct layers. In order to achieve this in practice, a geosynthetic separation is required between the coarse stone base and the fine grained subgrade. Otherwise, coarse aggregate penetration into the subgrade and fines migration and contamination of the coarse granular base would blur the boundary between the two distinct layers.

For coarse stone thicknesses of 12 in. or less (not usual for porous asphalt systems), the pavement can alternatively be designed as a conventional flexible pavement using the standard 93 AASHTO approach with appropriate layer coefficients (e.g., a_2= 0.1, m_2 = 1.0 for coarse stone base). The more conservative porous asphalt thickness value (composite subgrade versus conventional 93 AASHTO) should be selected for the design. The minimum porous asphalt thickness required to protect the coarse stone base can be determined from the 93 AASHTO equation by treating the stone base as the foundation layer.

Summary Of Recommendations

The recommended structural properties for the design of porous asphalt pavements using the empirical 1993 AASHTO design procedure are summarized in Table 4. Values for materials characterization should be re-examined when more field data are available for the performance of porous asphalt pavements under heavier traffic conditions.

Acknowledgments

This paper was prepared under research sponsored by the National Asphalt Pavement Association (NAPA), the State Asphalt Pavement Associations (SAPA), and the Federal Highway Administration (FHWA) through a joint cooperative agreement with NAPA.

References

1. National Asphalt Pavement Association (NAPA). *Porous Asphalt Pavements for Stormwater Management: Design, Construction, and Maintenance Guide*. Information Series 131. NAPA, Lanham, Md., 2008.
2. Schwartz, C. W., and K. D. Hall. *Development of Structural Design Guidelines for Porous Asphalt Pavements*. Interim Report. National Asphalt Pavement Association, Lanham, Md., 2017.
3. Transportation Research Board. Porous Asphalt Pavements: An International Perspective. In *Transportation Research Record 1265*, Transportation Research Board, Washington, D.C., 1990.
4. Adams, M. C. Porous Asphalt Pavement with Recharge Beds: 20 Years and Still Working. *Stormwater Magazine*, May–June, Vol. 4, 2003, 24–32.
5. Roseen, R., T. Ballestero, J. Houle, J. Briggs, and K. Houle. Water Quality and Hydrologic Performance of a Porous Asphalt Pavement as a Storm-Water Treatment Strategy in a Cold Climate. *Journal of Environmental Engineering*, Vol. 138, No. 2, 2012, pp. 81–89.
6. Al-Rubaei, A., A. Stenglein, M. Viklander, and G. Blecken. Long-Term Hydraulic Performance of Porous Asphalt Pavements in Northern Sweden. *Journal of Irrigation and Drainage Engineering*, Vol. 139, No. 6, 2013, pp. 499–505.
7. Peabody, D. *Performance of a Porous Pavement System on the Maine Mall Road in South Portland: Construction and Interim Report*. Technical Report 10-05. Maine Department of Transportation, Transportation Research Division, Augusta, Maine, 2010. ntl.bts.gov/lib/52000/52400/52486/report1005ci1.pdf. Accessed November 14, 2016.
8. Peabody, D. Personal Communication, December 2016.
9. Pennsylvania Department of Environmental Protection. *Pennsylvania Stormwater Best Management Practices Manual*. Pennsylvania Department of Environmental Protection, Pennsylvania, 2009. www.elibrary.dep.state.pa.us/dsweb/View/Collection- 8305. Accessed November 10, 2016.
10. *New Jersey Stormwater Best Practices Manual. Revised 2009*. State of New Jersey Department of Environmental Protection, Trenton, N.J., 2004. http://www.njstormwater.org/ bmp_manual2.htm. Accessed November 15, 2016.
11. University of New Hampshire Stormwater Center (UNHSC). *Regular Inspection and Maintenance Guidance for Porous*

Pavements. UNHSC, Durham, N.H., 2011. www.unh.edu/unhsc/sites/unh.edu.unhsc/files/UNHSC%. Accessed December 13, 2016.
12. Caltrans. *Pervious Pavement Design Guidance*. California Department of Transportation, Sacramento, Calif., 2014. http://www.dot.ca.gov/hq/oppd/stormwtr/bmp/DG-Pervious-Pvm_082114.pdf. Accessed September 8, 2015.
13. Wisconsin Asphalt Pavement Association. *Porous Asphalt Pavements*. Wisconsin Asphalt Pavement Association, Madison, Wis., 2015. http://www.wispave.org/wp-content/uploads/dlm_uploads/WAPA_Tech_Bulletin_Porous_Asphalt_Pavements_2015-09.pdf. Accessed December 18, 2018.
14. Wisconsin Department of Natural Resources. *Conservation Practice Standard 1008: Permeable Pavement*. Wisconsin Department of Natural Resources, Madison, Wis., 2014. http://dnr.wi.gov/topic/stormwater/documents/PermeablePavement1008.pdf. Accessed October 31, 2016.
15. Federal Highway Administration (FHWA). *TechBrief: Porous Asphalt Pavements with Stone Reservoirs*. FHWA-HIF-15-009, FHWA, Washington, D.C., 2015.
16. American Association of State Highway and Transportation Officials (AASHTO). *Guide for Design of Pavement Structures*. AASHTO, Washington, D.C., 1993.
17. American Association of State Highway and Transportation Officials (AASHTO). *Mechanistic-Empirical Pavement Design Guide: A Manual of Practice*, 2nd ed. AASHTO, Washington, D.C., 2015.
18. Li, H., D. Jones, and J. Harvey. Development of Mechanistic–Empirical Design Procedure for Fully Permeable Pavement under Heavy Traffic. In *Transportation Research Record 2305, Journal of the Transportation Research Board*. Transportation Research Board of the National Academies, Washington, D.C., 2012, pp. 83–94.
19. National Asphalt Pavement Association. *Design of Hot Mix Asphalt Pavements for Commercial, Industrial, and Residential Areas*. Information Series 209, National Asphalt Pavement Association, Lanham, Md., 2002.
20. Hossain, M., L. A. Scofield, and W. R. Meier, Jr., Porous Pavement for Control of Highway Runoff in Arizona: Performance to Date. In *Transportation Research Record 1354*, Transportation Research Board, Washington, D.C., 1992, pp. 45–54.
21. Hossain, M., and L. A. Scofield. *Porous Pavement for the Control of Highway Run-Off*. Final Report. Arizona Department of Transportation, Phoenix, Ariz., 1991. https://apps.azdot.gov/ADOTLibrary/publications/project_reports/PDF/AZ352.pdf. Accessed December 5, 2016.
22. Zhou, H., L. Moore, J. Huddleston, and J. Gower. *Free Draining Base Material Properties*. Final Report. Oregon Department of Transportation, Salem, Ore., 1992. http://ntl.bts.gov/lib/42000/42000/42023/Free_Drainage_Base.pdf. Accessed December 5, 2016.
23. Pologruto, M. Study of In Situ Pavement Material Properties Determined from FWD Testing. *Journal of Transportation Engineering*, Vol. 132 No. 9, 2006, pp. 742–750.
24Jones, D., J. Harvey, H. Li, T. Wang, R. Wu, and B. Campbell. *Laboratory Testing and Modeling for Structural Performance of Fully Permeable Pavements: Final Report*. UCPRC-RR-2010-01, University of California Pavement Research Center, University of California–Davis, 2010.
25. Goh, S., and Z. You. Mechanical Properties of Porous Asphalt Pavement Materials with Warm Mix Asphalt and RAP. *Journal of Transportation Engineering*, Vol. 138, No. 1, 2012, pp. 90–97.
26. Seo, Y., O. El-Haggan, M. King, S. Lee, and Y. R. Kim. Air Void Models for the Dynamic Modulus, Fatigue Cracking, and Rutting of Asphalt Concrete. *Journal of Materials in Civil Engineering*, Vol, 19, No. 10, 2007, pp. 874–883.
27. Davis, K., and D. Timm. Recalibration of the Asphalt Layer Coefficient. *Journal of Transportation Engineering*, Vol. 137, No. 1, 2011, pp. 22–27.
28. Zhang, Y., X. Wang, Y. Li, L. Wang, and R. Boudreau. Determination of Mechanical Properties for Unstabilized Drainage Layer and Implementation in Mechanistic–Empirical Pavement Design Guide. *Journal of Materials in Civil Engineering*, Vol. 29, No. 2, 2016, p. 04016213. http://ascelibrary.org/doi/10.1061/%28ASCE%29MT.1943-5533.0001716.
29. Tutumluer, E., and U. Seyhan. Laboratory Determination of Anisotropic Aggregate Resilient Moduli Using an Innovative Test Device. In *Transportation Research Record 1687*, Transportation Research Board, National Research Council, Washington, D.C., 1999, pp. 13–21.
30Applied Research Associates. *Resilient Modulus Testing Open-Graded Drainage Layer Aggregates*. Applied Research Associates, Toronto, Ontario, Canada, 2005.

The Standing Committee on Design and Rehabilitation of Asphalt Pavements (AFD60) peer-reviewed this paper (18-01088).

Article

TRR
JOURNAL OF THE TRANSPORTATION RESEARCH BOARD

Transportation Research Record
2018, Vol. 2672(40) 207–216
© National Academy of Sciences:
Transportation Research Board 2018
Article reuse guidelines:
sagepub.com/journals-permissions
DOI: 10.1177/0361198118759013
journals.sagepub.com/home/trr
SAGE

Local Calibration of the MEPDG Distress and Performance Models for Ontario's Flexible Roads: Overview, Impacts, and Reflection

Xian-Xun Yuan[1] and Iliya Nemtsov[1]

Abstract

Built upon a seven-year local calibration study of Ontario's flexible pavements, this paper provides a summary of the calibration results and design impact and, more importantly, shares the experience and lessons learned from the process. The best results have been achieved on the local calibration of the rutting, bottom-up fatigue cracking, and international roughness index (IRI) distress models minimizing the residual sum of squares (RSS) while maintaining the average bias at zero. Significant efforts have been made to calibrate the other distress models with limited success. A design impact study found that local calibration of the rutting models was very important, whereas the alligator fatigue cracking did not usually govern the design in Ontario, although the global model was found to under-predict the cracking damage. The performance of the calibrated IRI model in the design of heavy traffic freeways for both reconstructed and rehabilitated sections was unsatisfactory and needs further study. The paper also presents several open questions for future research. These include the handling of section-length effects of observed cracking data, the determination of initial IRI, the updating of standard deviation functions and the overall reliability models, and the prioritization of pavement research under the new paradigm of the Mechanistic–Empirical Pavement Design Guide (MEPDG).

The AASHTO Mechanistic–Empirical Pavement Design Guide (MEPDG) represents a paradigm shift in pavement design method. Unlike its predecessors, which were largely empirically based, the MEPDG establishes a direct tie between pavement distresses and various design inputs through mechanistic analyses and empirical models. Multilayer elastic theory, cumulative fatigue damage theory, and fracture mechanics-based crack propagation theory are used to provide the theoretical bases for the prediction of various distresses including permanent deformation (or rutting), fatigue cracking, thermal cracking, and reflection cracking. Sophisticated material (e.g., dynamic modulus and creep compliance), climate and traffic models are included to improve the correlation between distresses and structural responses calculated by the mechanistic modules. Due to the complexity of the distresses (e.g., nonlinear viscoelastoplasticity in permanent deformation, and singularity and randomness in cracking), empirical transfer models are developed to translate the mechanistic calculation into the prediction of field observable pavement distresses. Finally, the overall pavement performance is evaluated by the international roughness index (IRI), a composite performance indicator that is an empirical function of rut depth, fatigue cracking, thermal cracking, and other site factors. Adequate design of pavement structure requires the abovementioned distresses and IRI to be within their respective target threshold values with a given reliability. The design method has been packaged in a user-friendly working platform now called the AASHTOWare Pavement ME software (originally DARWin-ME).

Although the empirical distress and performance models (or "global calibration") have been extensively studied under multiple NCHRP projects including 1-37A (*1*), 1-40 (*2*) and 9-30A (*3*) over the past 15 years, local calibration is recommended for transportation agencies before they implement a new design method. The local calibration study serves two main purposes: first, it assesses the correctness and accuracy of the global models in the prediction of pavement distresses and performance and, second, the correctness and accuracy can be further improved by comparing the predicted and observed distress and IRI data. Over the past decade, several local calibration studies have been carried out and almost every State Department of Transportation (DOT) that did the study found the necessity of local calibration [e.g., (*4–18*)]. As one of the leading transportation agencies in Canada, the

[1]Department of Civil Engineering, Ryerson University, Toronto, Ontario, Canada

Corresponding Author:
Address correspondence to Xian-Xun Yuan: arnold.yuan@ryerson.ca

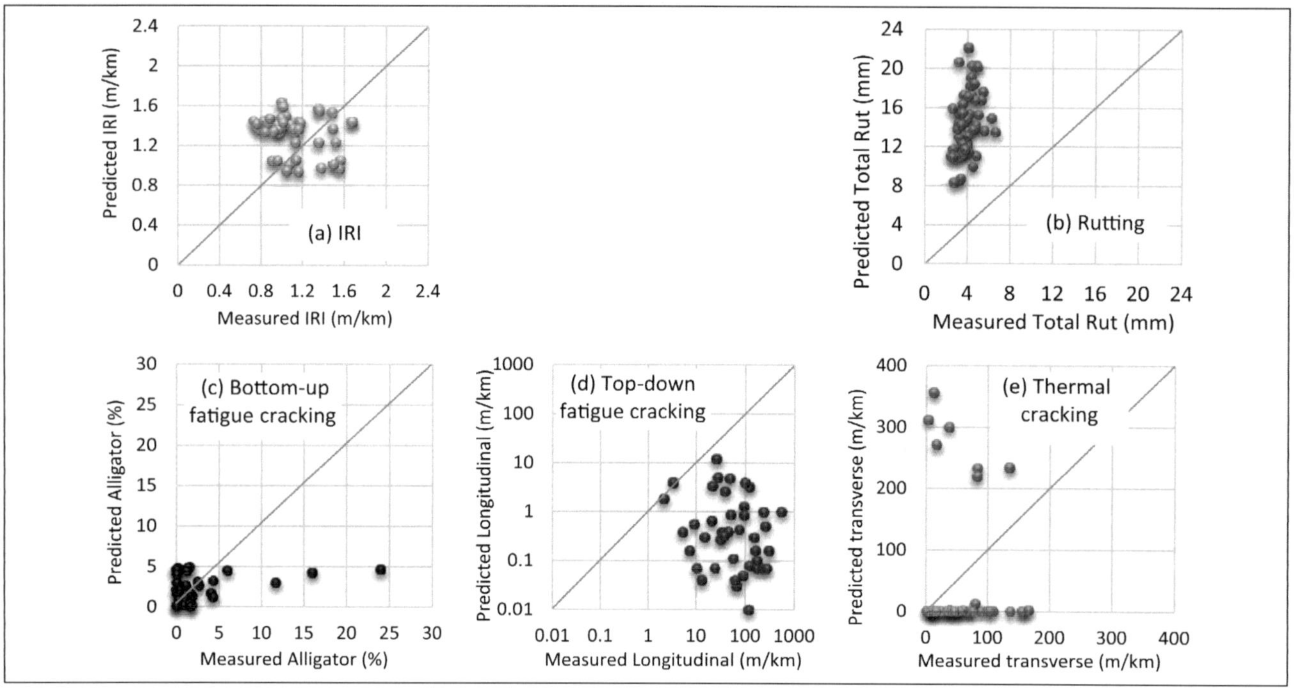

Figure 1. Comparison of the observed and predicted distresses from the global models.

Ministry of Transportation of Ontario (MTO) has been mandated to implement the MEPDG for future pavement design.

In Ontario there are more than 16,500 centerline kilometers of highway pavements under MTO's jurisdiction. These pavements are divided, under the pavement management system (PMS), into about 1,800 pavement sections. The majority of these sections are flexible pavements. Early pavements were mainly built with Marshall mixes. Several hot mix asphalts of Marshall type, including DFC, HL 1, HL 2, HDBC, etc. [refer to OPSS 1150 (*19*)] were used in different functional layers. In 2001 MTO started to introduce Superpave mix including stone mastic asphalt into pavement construction as the top asphalt layers. The most frequently used Superpave mixes include SP 12.5 (and its varieties), 19.0 and 25.0 [refer to OPSS 1151 (*20*)]. By 2006 all new and reconstructed asphalt pavements and overlaid pavements in provincial roads had been Superpave specified. On the other hand, MTO is well known for its practice of PMS. The second-generation PMS (or PMS-2), started in the mid-1990s, contains more than 30 years of pavement performance data.

To perform the local calibration, since 2010 three research projects have been commissioned to Ryerson University under the support of the MTO Highway Infrastructure Innovation Funding Program (HIIFP). Due to the limited test road sections in Ontario, pavement performance data from PMS-2 was used for the local calibration. The first project focused on the development of a local calibration database, need analysis, and preliminary calibration. The second project expanded the calibration database to include Superpave sections and focused mainly on calibration of the rutting models. The third and last project continued on the local calibration for the cracking models and IRI model by using more accurate performance data collected by the new ARAN 9000 system (*21*). While the first two projects exclusively studied flexible pavements, the last project was also extended to include rigid pavements (conducted by the University of Waterloo), though rigid roads represent only a small portion of the provincial road system.

A detailed summary of the major efforts and findings of the study was presented recently at the Transportation Association of Canada conference (*22*). Major challenges faced in the database development were discussed in (*23*) and (*24*). Some intermediate local calibration results for the rutting, fatigue cracking and IRI models were reported in (*25–30*). This paper presents the impact of local calibration for Ontario's flexible pavement design and shares some reflections from the seven-year study. For completeness, a summary of the local calibration results is presented first.

Summary of Local Calibration Results

Need for Local Calibration

Figure 1 compares the measured distresses with the predictions by the default global models for (a) IRI, (b) total rut depth, (c) bottom-up (alligator) fatigue cracking, (d) top-down (longitudinal) fatigue cracking, and (e) thermal cracking. While the IRI model provides unbiased predictions for the sections in the Superpave section, the trend with the measured IRI is not clear. As reported by many other local calibration studies, the total rutting is greatly over-predicted. The two fatigue cracking models generally under-predict the cracking damage. The thermal cracking prediction exhibits the greatest variation.

As mentioned above, the observed distress and IRI data were collected by the ARAN 9000 system, which was deployed in 2013. After one-year trial data collection, mainly data collected in 2015 were used for the local calibration study. This data represents the most accurate measurements available in Ontario. It should be noted, however, that the data were collected for PMS purposes and not for local calibration. Therefore, the cracking data were pre-processed before being used in this study. Basically, the total (i.e., sum of light, medium, and severe) alligator cracks on the two wheel paths were considered to be bottom-up fatigue cracking. Similarly, the total longitudinal cracks on the wheel paths were considered to be top-down fatigue cracking. To be consistent with the global calibration, a weighted average of all transverse cracks on the lane was used as the thermal cracking data.

The pavements sections of the local calibration database developed for the study include both new or reconstruction sections and rehabilitation sections. For rehabilitation sections, it is impossible to separate the new fatigue and thermal cracking from the reflection cracking. This partly explains the large variations observed in the cracking predictions.

Results

The results of the local calibration study are summarized in Table 1. As shown, except the IRI model, each type of distress involves several empirical functions to be locally calibrated. From 46 carefully selected road segments [for more details, refer to (*22*)] the local calibration coefficients are highlighted with bold face. Among the six groups of empirical models that are subject to local calibration, only the rutting model, bottom-up fatigue cracking model, and the IRI model have been calibrated. Figure 2 shows the comparison of the measured distresses and the predicted distresses from the three calibrated models. These calibrated models have also been validated by a separate validation dataset; for more details, refer to (*22*).

Trial local calibration for the top-down and thermal cracking models was conducted. The reflection cracking model involve many local calibration coefficients which are unique for different overlay types. Therefore, proper local calibration requires knowledge of existing conditions, and the current Level 3 inputs were considered inadequate to provide any meaningful and significant result. To identify the most sensitive coefficients, sensitivity analyses were performed; details are presented in the subsequent subsection. These three cracking models cannot be properly calibrated at the current stage. There are several reasons for this, but the major one is that these models are still subject to substantial improvement or global calibration from the original model developers. For the reflection cracking model for AC over AC sections, the difficulty of separating the observed surface cracks from new and reflection cracks adds another technical challenge to the calibration. Because of the three uncalibrated cracking models, the IRI model was only partially calibrated, with C_3 and C_4 remaining their default global values.

Sensitivity Analysis of the Reflection Cracking Model

Recently the reflection cracking model in the MEPDG underwent a thorough modification from its first version. While the original model is largely empirical and based solely on the bottom-up fatigue cracking of the underlying layers, the recent version (2.2.0 and later) of the AASHTOWare considers both fatigue and thermal cracks (*31*). Moreover, it takes the Paris–Erdogan crack propagation theory as the basic modeling skeleton, rather than Miner's cumulative damage theory as used in the fatigue cracking model. As a result, the new reflection cracking model contains five C s and three k s in the fatigue cracking module and a similar eight coefficients in the thermal cracking module. In total there are 16 coefficients that can be changed. Since the reflection cracking model is relatively new, it is interesting to see how sensitive the coefficients are relative to the prediction of the total reflection fatigue and thermal cracking. Note that the three k s functions exactly in the same way as C_1, C_2 and C_3. Therefore, the relatively independent coefficients are only the five C s for either fatigue or thermal reflection cracking.

To perform the sensitivity analysis, five AC-over-AC sections were selected from the local calibration database, each representing a typical overlay design of typical environmental and traffic characteristics in Ontario. By changing the five Cs within the range between −100% and +100%, the percentage changes in the reflection fatigue and thermal cracking were calculated. The averaged percentage changes in the reflection cracking against the percentage change in the coefficients are plotted in Figure 3. The letter "F" in front of the C represents the coefficients for the fatigue model while "T" represents the thermal model.

For the reflection fatigue cracking model, FC2 and FC4 are the most sensitive coefficients with opposite direction. The relative change rates are nonlinear and asymmetric. FC3 represents the weighting factor for thermal cracking in the fatigue model, and thus the sensitivity is zero, as expected. Meanwhile, four out of the five sections exhibit very similar sensitivity. The reason for the one different section was not clear.

The sensitivity of the reflection thermal cracking exhibits a pattern very different than that of the fatigue cracking

Table 1. Summary of Local Calibration Results (46 Sections)

Empirical models	Expression	Calibrated coefficients	Summary statistics
Rutting	$\frac{\varepsilon_{p,AC}}{\varepsilon_{r,AC}} = k_z\beta_{AC}10^{-3.3541}T^{1.5606\beta_T}N^{0.4791\beta_N}$ $\frac{\varepsilon_{p,GB}}{\varepsilon_{r,GB}} = 2.03\beta_{BG}\phi(N,\alpha)$ $\frac{\varepsilon_{p,SG}}{\varepsilon_{r,SG}} = 1.35\beta_{SG}\phi(N,\alpha)$	$\beta_{AC} = 1.7692$ $\beta_T = 1.0$ $\beta_N = 0.6262$ $\beta_{GB} = 0.0968$ $\beta_{SG} = 0.2787$	Bias = 0 SD = 1.0 mm
Fatigue Cracking	$N_f = 0.007566 C_V C_H \beta_f \varepsilon_t^{-3.9492\beta_t} E^{-1.281\beta_E}$ BT: $FC_{bt}(t) = \frac{100\%}{1+\exp(C_1C_1' - C_2C_2' \log_{10}100D)}$	All βs remain 1.0 $C_1 = 0.5236$ $C_2 = 0.1404$	Bias = 0; SD = 6.14%
	TD: $FC_{top}(t) = \frac{10560}{1+\exp(C_3C_3' - C_4C_4' \log_{10}100D)}$	Tried, but not calibrated	NA
Thermal Cracking	$A = k_t\beta_{t1}10^{4.389-2.52\log_{10}(E_{AC}\sigma_m n)}$ $TC = \beta_{t2}\Phi\left[\frac{1}{\sigma_d}\log\left(\frac{C_d}{h_{AC}}\right)\right]$	Tried, but not calibrated	NA
Reflection Cracking	$\Delta C_{bend} = k_1\sum AK_b^n dN$ $\Delta C_{shear} = k_2\sum AK_s^n dN$ $\Delta C_{thermal} = k_3\sum AK_t^n dN$ $D_T = C_1\Delta C_{bend} + C_2\Delta C_{shear} + C_3\Delta C_{thermal}$ $RC = \frac{\alpha}{C_4 + \exp(C_5 \log_{10} D_T)} \times 100\%$	Sensitivity analysis done, not calibrated	NA
IRI	$IRI = IRI_0 + C_1RD + C_2FC + C_3TC + C_4SF$	$C_1 = 55.096$ (c.f. 40.0) $C_2 = 1.088$ (c.f. 0.400) $C_3 = 0.008$ (global) $C_4 = 0.015$ (global)	Bias = 0 SD = 0.30 m/km

model in that TC4 is the only sensitive coefficient. However, with greater change in TC5, changes in thermal cracking were observed. The variation of the sensitivity of TC4 among the five pavement sections was negligible.

Impact of Local Calibration on Pavement Design

While significant changes have been determined in the local calibration coefficients of the rutting, bottom-up fatigue cracking, and IRI models, the practical significance of these changes in terms of pavement structural design remains an interesting question to explore. This section reports the findings from the before- and after-calibration trial designs.

To evaluate the significance, three new or reconstruction sections and two rehabilitated (AC-over-AC) sections were selected for the trial design. The function, construction type, layer structure, target reliability, and design life of each section are listed in Table 2. The threshold values of IRI, total rut depth, AC rut depth and alligator cracking are taken from the Ontario Default Parameter for MEPDG

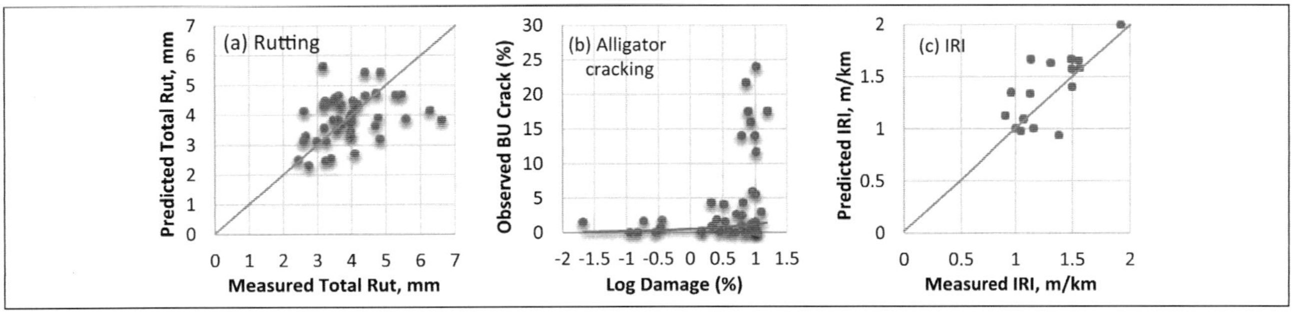

Figure 2. Comparison of the observed and predicted distresses from the locally calibrated models.

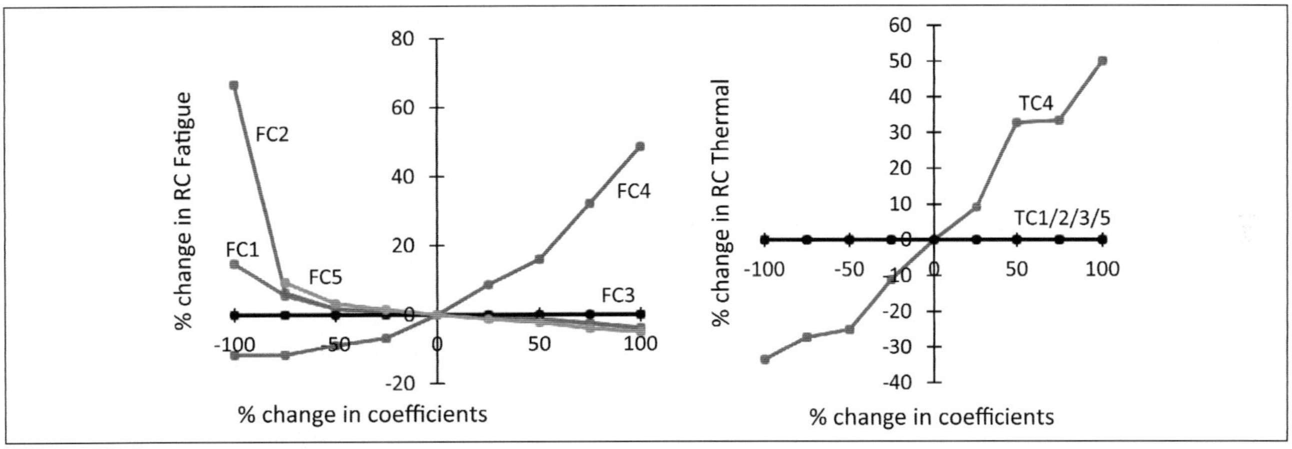

Figure 3. Sensitivity of the local calibration coefficients of the reflection cracking (RC) models.

Table 2. Trial Design Sections and Impacts of the MEPDG before and after Local Calibration

Section	114		698		835		1217		1260	
Function	Arterial		Arterial		Collector		Freeway		Freeway	
Construction	Overlay		Recon		New		New		Overlay	
Structure	40 SP12.5FC1 50 SP19 50 old SP19 225 Gran A SM Soil		40 SP12.5 50 SP19 200 Pulverized 190 Gran A 150 Gran B SM soil		40 SP12.5 FC1 100 SP19 150 Gran A 450 Gran B1 CL Soil		40 SP12.5FC2 120 SP19 180 SP25 550 Gran A CL-ML soil		40 SP12.5FC2 40 SP19 200 old HDB 300 old Gran A 300 old Gran B Cl soil	
Target reliability	85%		85%		75%		95%		95%	
Design life (years)	10		18		18		18		10	
Global (G)/Calibrated (C)	G	C	G	C	G	C	G	C	G	C
	Increase in AC Layer Thickness (mm) for different distress modes									
IRI	50	0	40	0	0	0	>500	>500	>500	>500
AC Bottom-up Cracking	0	0	0	0	0	0	0	0	0	0
Deformation – AC only	0	0	0	0	0	0	>500	0	>500	0
Deformation – Total	50	0	20	0	100	0	100	0	300	0

(*32*). It is important to note that all five sections were designed in accordance with Ontario's Pavement Design and Rehabilitation Manual (*33*) and therefore they are deemed to satisfy the design requirements. The design life for new and reconstructed sections is 15 to 18 years whereas that of overlay sections is 8 to 10 years. In this study, the upper limits are used.

For each section, two design iterations were carried out, one based on the global calibration models and the other on the locally calibrated models. Only IRI, total rut depth, AC rut depth, and alligator fatigue cracking were used as the design criteria. Because the other cracking distresses have not been calibrated, they were not considered in the design. For either iteration, when a certain design criterion is not met, the thickness of the binding course (usually the second AC layer from the top) is increased until the criterion is satisfied. For Section 1217, the thickness of the second and third layers may be increased. The increases of the layer thickness are tabulated in the bottom half of Table 2.

The results show that before local calibration the default MEPDG models would disqualify the existing pavement design. In particular, Section 698 would require the least amount of design increase due to IRI and total rutting. Section 835 would be disqualified only for the total rutting. For the two freeway sections with heavy traffic, however, IRI, AC rutting, and total rutting would all demand an enhanced design. More importantly, although the requirements for total rut depth could be met by increasing the AC layer thickness, the IRI and AC rutting requirements could not be honored even after the AC layer thickness has artificially been increased to 500 mm, the upper limit of the AC layer suggested by the software.

Using the locally calibrated distress and performance models, the MEPDG would qualify the first three sections. Interestingly, the achieved reliability of IRI of the three sections is very close to the target reliability. Meanwhile, the two freeway sections (1217 and 1260) would still be disqualified by the calibrated MEPDG. More importantly, regardless how the AC layer thickness and granular layers are changed, the IRI value at the end of design life would still be above the design threshold, while the other distress models would have achieved 100% reliability. The reason for this might be in the site factor of the IRI equation. However, this hypothesis is subject to further study. It is also interesting to observe that the fatigue cracking does not control the design for all of the five sections, either before or after the local calibration.

Experience and Lessons Learned

Experience has been gained and many lessons learned from the local calibration study. We have developed our own local calibration methods for rutting models, using both longitudinal and cross-sectional calibration; the work was reported in (*27*). In addition, the method of determining the traffic and temperature exponents for the AC rutting model is unique; the details can be found in (*25*). We have also tried different validation methods with results to be reported separately. In the following, four important open issues are presented and discussed.

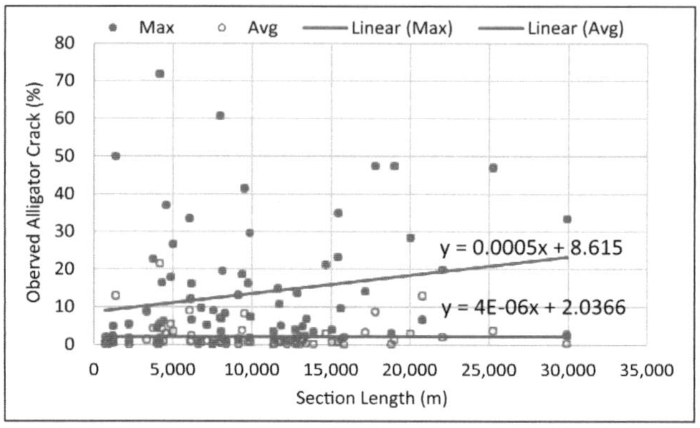

Figure 4. Comparison of the maximum and averaged alligator cracking.

Section-Length Effects on Cracking Models

An issue that has not been widely discussed in local calibration literature is the effect of section length on the observed cracking data. The majority of local calibration studies, including this one, use field pavement sections and field observed pavement performance data. This means that the pavement sections are not the standard 500 ft (~150 m) long. For example, the sections used in this study vary in length from 0.7 km to 30 km. Within each section, a vast number of standard segments of 500 ft could be extracted. Associated with each segment is one observed crack data. As a result, a question arising naturally is: which statistic (e.g., algebraic average, geometric mean, or maximum) of the segment data should be used as the observed crack data for the whole pavement section?

The algebraic average seems to be widely used in practice. However, this practice would underestimate the variability of the cracking data in local calibration. Since the standard deviation of a mean is roughly inversely proportional to the square root of the number of sample points, the standard deviation of a standard long term pavement performance (LTPP) section would be 10 times that of a 15 km-long field pavement section, if all other things remain the same.

Alternatively, the maximum value can be used. Obviously, the longer sections tend to report greater cracking damage data. Blindly used, this can cause unnecessary bias in local calibration. For the purpose of illustration, Figure 4 depicts the significant difference between the two aggregation methods for alligator cracking. The mean and standard deviation of the maxima are 13.3% and 15.7%, respectively, whereas those of the averages of the same sections are just 2.07% and 3.49%. More interestingly, the maximum data shows an increasing trend with the section length, whereas the average data remains almost constant with a negligible slope in the trend line. In addition, while the variation of the maximum data tends to increase with the length, it is obvious that the variation of the average data is decreasing. How to handle these length effects is an open question for further study. On the other hand, it also brings up the issue whether the PMS data are reliable enough to support local calibration.

The Initial IRI Value

A debate arose during the local calibration what IRI_0 value should be used in the local calibration and future design. IRI_0 refers to the roughness of pavement within six months after construction completion. According to the MEPDG software manual (*34*), "[T]he initial IRI value provided must be what is typically attained in the field." By default it is 1 m/km (*32*). The MTO's current practice in payment adjustment (*35*) expects the initial IRI of a readily accepted pavement to be in the range between 0.65 and 1 m/km. Two questions arose from the debate: What IRI_0 value should be used in the local calibration? Can we use the IRI_0 used in or determined by the local calibration in future design, and if not, how do we determine it?

As to the first question, three schemes have been suggested. The first one is to use the actual initial IRI values as a Level 1 input for the calibration. This usually results in higher R^2 value. However, some argue that this method should not be used because IRI_0 is a part of pavement design specifications, which should not be the value that is achieved, but one that is expected. Others argue that the actual IRI_0 was used only in local calibration to determine C_1 to C_4 of the IRI model. In future design, the IRI_0 value should be a unified, rather than section-specific, value that reflects the owner or authorization's expectation of the initial IRI. However, there was concern that the IRI model calibrated by this means would underestimate the variability of the IRI residuals and thus over-predict the pavement reliability.

The second scheme was to treat the IRI_0 as a local calibration coefficient and let the optimization (a simple linear regression) determine the optimal one. However, it was soon realized that sometimes the calculated IRI_0 will be a negative value, particularly when only one year of IRI data are used for the local calibration. Those supporting this scheme argue that this method at least captures the growth trend of IRI over time, which is the main purpose of the local calibration. They further suggested that in future design the IRI_0 value should be determined separately.

The third scheme was to use a default value that is determined by the agency in both local calibration and future design. This suggestion treats the initial IRI as an exogenous variable of local calibration. However, this method often provides a local calibration result that is even poorer than the global calibration model.

As to the second question, it is generally agreed that the IRI_0 value for future design should be determined separately. To determine this value, the post-construction IRI measurements can be used to perform the statistical analysis. The mean, standard deviation and probability distribution of the initial IRI can be determined. If necessary, regression models that relate it to other explanatory variables such as road functional class, AC type, geographical zone, season of construction, etc. can be developed. After the statistical analyses, the determination of the initial IRI as a policy number still requires several considerations. First, one needs to consider the current smoothness-related payment adjustment practice. This does not necessarily suggest that the current 100% pay line be used. Rather, impact analysis can be conducted if a change is initiated. Second, the impact of the design reliability needs to be assessed. The MEPDG uses exclusively a residual-based method for the reliability analysis. For the IRI model, only the residual standard deviation of the global calibration is considered. Now that the IRI_0 is treated as an exogenous variable, which is also a random variable, the variability should be included in the standard deviation function. In addition, if the suggested IRI_0 value is not the mean value of the current practice, then the calculation of the reliability needs to be adjusted.

Reliability Models

Another important set of empirical models in the MEPDG suite that has not been much discussed across the community is the standard deviation models. Despite its importance in pavement reliability design, some of the standard deviation functions (e.g. for rutting and IRI) used in the software are improperly documented. Local calibration will change the amount and/or profile of the residuals. However, there is little guidance on what to do with the standard deviations after local calibration. Currently, the AASHOTware does not allow users to change the standard deviation models. This has affected the efficiency of local calibration. Bayesian statistics-based methods are available for the updating of standard deviation models, and further research must be done to re-examine the residual-based reliability method.

The hierarchical levels of accuracy of input data represent an important innovation to the MEPDG. From the perspective of reliability design, however, this apparent strength has also added sophisticated difficulty in the treatment of uncertainties and reliability in a coherent manner. Note that, unlike the AASHTO 1993 design method, the MEPDG does not consider the variability of design inputs in evaluating the overall standard deviation of the distresses or damages. Except for IRI, the current estimation of the overall standard deviation for the non-IRI distresses includes only the residual standard deviation. This means that the prediction uncertainty is underestimated, and the extent of the underestimation has yet to be determined. A few researchers [e.g., (*36, 37*)] have proposed using the Latin hypercube technique and response surface to quantify the propagation of the input uncertainties to prediction of the pavement distresses. In our view, simulation is not a suitable candidate for this; rather the traditional load resistance factored design should be further investigated (*38*).

Paradigm Shifts in Pavement Design and Pavement Research

The MEPDG is a very complicated pavement analysis system. The several mechanistic models adopted by the MEPDG are at different stages of maturity. For example, the rutting and bottom-up fatigue cracking models are widely accepted in the community, whereas the top-down fatigue cracking and reflection cracking models are still subject to further development and verification. On the other hand, new materials and new construction and rehabilitation technologies are emerging every year, if not every day. All of these will drive local calibration as a dynamic process. This is both good and bad news. It is good because it provides a common platform for continuous improvements in pavement design theory and practice. However, it can be also unwelcome for transportation agencies as it requires continual investment in local calibration. Without a proper strategic research plan, local calibration can turn into an endless investment trap. With shrinking research funding for applied research, almost all transportation agencies in North America are facing difficult choices of prioritizing research investment. The MEPDG does represent a paradigm shift in design method indeed, but does it also invoke a paradigm shift in pavement research? Shall all future pavement research be based on the MEPDG platform? Within the platform, should focus be on the enhancement of mechanistic modules or on the refinement of local calibration? These issues are worth serious consideration.

Conclusions

Ontario's local calibration has come a long way. Extensive experience has accumulated during this study. The best results have been achieved on the calibration of the rutting models and the bottom-up fatigue cracking model. Earnest efforts have been made to try to calibrate the other models, with limited success. It was originally hoped that with the new ARAN data the cracking models could be reasonably calibrated to the local conditions and practice. However, due to the pending development of the top-down and thermal cracking global models, the local calibration work for these models has been postponed.

The trial design study finds that local calibration of the rutting models is very important. Without it, the global model would provide overdesigned pavement structures. In contrast, the alligator fatigue cracking does not often govern the design in Ontario, although the global model was found to underpredict the cracking damage. Local calibration of the IRI model is also important. The calibrated IRI model qualifies the current pavement design. However, the IRI model needs further study mainly to finish calibrating its rutting and fatigue cracking components, but also to improve its performance in heavy traffic freeways for both reconstructed and rehabilitated sections.

This is just the beginning and there are many outstanding areas for improvement to calibrate the MEPDG distress and performance models fully. These include, but are not limited to, the determination of initial IRI, sensitivity analysis of different protocols to preprocess the observed cracking data, calibration of the other three cracking models, collecting and including longer distress time series into the calibration database, verification of the time trend of the empirical models through longitudinal calibration, calibration of the standard deviation functions, and finally exploration and development of enhanced mechanistic models and theories. It is important to prioritize investment in pavement research.

Acknowledgments

The financial support of the MTO HIIFP in 2010–11, 2013–15, and 2015–17 is gratefully acknowledged. Warren Lee and Ningyuan Li of the Pavement Design and Foundation Section, MTO, have provided tremendous assistance throughout the last two projects. Joseph Ponniah, a former senior pavement research engineer of MTO, provided technical assistance on the database development during the early stage of the research program. Former graduate students Gulfam Jannat, Afzal Waseem, Gyan Gautam, Sifat Ahmed, and Maryam Amir provided important inputs for the work presented in this paper. Thanks are due also to Dr. Medhat Shehata of Ryerson University for his generous discussions and suggestions.

References

1. NCHRP. *Guide for Mechanistic Empirical Design of New and Rehabilitated Pavement Structures*. ARA, Inc., ERES Division 505 West University Avenue Champaign, Ill., 2004.
2. Brown, S. F., M. R. Thompson, and E. J. Barenberg. *Research Results Digest 307*. Research Results Digest, National Cooperative Highway Research Program. 2006.
3. Von Quintus, H. L., J. Mallela, R. Bonaquist, C. W. Schwartz, and R. L. Carvalho. *Calibration of Rutting Models for Structural and Mix Design*. NCHRP Report 719. AASHTO, Washington, D.C., 2012.
4. Donahue, J. Local Calibration of the MEPDG for HMA Pavements in Missouri. *Journal of the Association of Asphalt Paving Technologists*, Vol. 77, 2008, pp. 975–984.
5. Hudson, W. R., W. Visser, C. L. Monismith, and C. Dougan. Using State PMS Data to Validate the New Mechanistic/Empirical Pavement Design Guide (MEPDG). *Proc., 7th International Conference on Managing Pavement Assets*. Calgary, Alberta, Canada, 2008.
6. Khazanovich, L., I. Yut, S. Husein, C. Turgeon, and T. Burnham. Adaptation of Mechanistic-Empirical Pavement Design Guide for Design of Minnesota Low-Volume Portland Cement Concrete Pavements. *Transportation Research Record: Journal of the Transportation Research Board*, 2008. 2087: 57–67.
7. Muthadi, N. R., and Y. R. Kim. Local Calibration of Mechanistic-Empirical Pavement Design Guide for Flexible Pavement Design. *Transportation Research Record: Journal of the Transportation Research Board*, 2008. 2087: 131–141.

8. Von Quintus, H. L. *Local Calibration of MEPDG - An Overview of Selected Studies*. Presented at 2008 Annual Meeting of the Association of Asphalt Paving Technologists, AAPT, Philadelphia, Pa., 2008.
9. Banerjee, A., J. P. Aguiar-Moya, and J. A. Prozzi. Calibration of Mechanistic-Empirical Pavement Design Guide Permanent Deformation Models. *Transportation Research Record: Journal of the Transportation Research Board*, 2009. 2094: 12–20.
10. Darter, M. I., L. Titus-Glover, and H. Von Quintus. *Implementation of the Mechanistic–Empirical Pavement Design Guide in Utah: Validation, Calibration, and Development of the UDOT MEPDG User's Guide*. No. UT-09.11. Utah Department of Transportation, 2009.
11. Glover, L. T., and J. Mallela. *Guidelines for Implementing NCHRP 1-37A M-E Design Procedures in Ohio: Volume 4—MEPDG Models Validation & Recalibration*. Final Report. Office of Research and Development, Ohio Department of Transportation, 2009.
12. Li, J., L. M. Pierce, and J. Uhlmeyer. Calibration of Flexible Pavement in Mechanistic-Empirical Pavement Design Guide for Washington State. *Transportation Research Record: Journal of the Transportation Research Board*, 2009. 2095: 73–83.
13. Kim, S., H. Ceylan, K. Gopalakrishnan, and O. Smadi. Use of Pavement Management Information System for Verification of Mechanistic-Empirical Pavement Design Guide Performance Predictions. *Transportation Research Record: Journal of the Transportation Research Board*, 2010. 2153: 30–39.
14. Souliman, M. I., M. Mamlouk, M. M. El-Basyouny, and M. Mamlouk. Calibration of the AASHTO MEPDG for Flexible Pavement for Arizona Conditions. Presented at 89th Annual Meeting of Transportation Research Board, Washington, D.C., 2010.
15. Xiao, D. X., K. C. P. Wang, and K. D. Hall. *Matching Distress Definitions in Field and in MEPDG for Local Calibration*. Presentation to Pavement Evaluation 2010. University of Arkansas, Roanoke, Virginia, 2010.
16. Williams, R. C., and R. Shaidur. *Mechanistic-Empirical Pavement Design Guide Calibration for Pavement Rehabilitation*. Final Report SPR 718. Oregon Department of Transportation, 2013.
17. Mallela, J., L. Titus-Glover, S. Sadasivam, B. Bhattacharya, M. Darter, and H. Von Quintus. *Implementation of the Mechanistic–Empirical Pavement Design Guide for Colorado*. Report No. CDOT-2013-4. Colorado Department of Transportation, 2013.
18. Tarefder, R., and J. I. Rodriguez-Ruiz. Local Calibration of MEPDG for Flexible Pavements in New Mexico. *Journal of Transportation Engineering*, Vol. 139, No. 10, 2013, pp. 981–991.
19. Ontario Provincial Standard Specification. Material Specification for Hot Mix Asphalt. *OPSS 1150*. Ontario Provincial Standard Specification, 2008. www.raqsb.mto.gov.on.ca.
20. Ontario Provincial Standard Specification. Material Specification for Superpave and Stone Mastic Asphalt Mixtures. *OPSS 1151*. Ontario Provincial Standard Specification, 2006. www.raqsb.mto.gov.on.ca.
21. Fugro Roadware. *Laser XVP*. 2013. http://www.roadware.com/related/Laser-XVP_2014.pdf.
22. Yuan, X.-X., W. Lee, and N. Li. Ontario's Local Calibration of the MEPDG Distress and Performance Models for Flexible Roads: A Summary. *The 2017 Conference of the Transportation Association of Canada*. St. John's, NL, 2017.
23. Jannat, G. E. *Database Development for Ontario's Local Calibration of Mechanistic-Empirical Pavement Design Guide (MEPDG) Distress Models*. Department of Civil Engineering, Ryerson University, 2012.
24. Jannat, G.E., et al., Database Development for Ontario's Local Calibration of the MEPDG Distress Models for Flexible Pavements. *Proc., 9th International Transportation Specialty Conference, CSCE*, Edmonton, AB, 2012.
25. Gautam, G. P., et al., Local Calibration of the MEPDG Rutting Models for Ontario's Flexible Roads: Recent Findings. Presented at 95th Annual Meeting of the Transportation Research Board, Washington, D.C., 2016.
26. Jannat, G. E., X.-X. Yuan, and M. Shehata. Development of Regression Equations for Local Calibration of Rutting and IRI as Predicted by the MEPDG Models for Flexible Pavements Using Ontario's Long-Term PMS Data. *International Journal of Pavement Engineering*, Vol. 17, No. 2, 2015, pp. 166–175.
27. Waseem, A., and X.-X. Yuan. Longitudinal Local Calibration of MEPDG Permanent Deformation Models for Reconstructed Flexible Pavements Using PMS Data. *International Journal of Pavement Research and Technology*, Vol. 6, No. 4, 2013, pp. 304–312.
28. Waseem, A. *Methodology Development and Local Calibration of MEPDG Permanent Deformation Models for Ontarios Flexible Pavements*. MS thesis. Department of Civil Engineering, Ryerson University, 2013.
29. Ahmed, C. J. S. E. *Local Calibration of Cracking Models of MEPDG for Ontario's Flexible Pavements*. Department of Civil Engineering, Ryerson University, Toronto, Ontario, Canada, 2017.
30. Amir, M. *Local Calibration of MEPDG for Superpave Pavements in Ontaro: Enhanced Rutting Calibration and Preliminary Study on IRI Model*. Department of Civil Engineering, Ryerson University, Toronto, Ontario, Canada, 2017.
31. Lytton, R. L., et al., Models for Predicting Reflection Cracking of Hot-Mix Asphalt Overlays. NCHRP Report 669. National Cooperative Highway Research Program, Washington, D.C., 2010.
32. MTO. *Ontario's Default Parameters for AASHTOWare Pavement ME Design*. Interim Report. Ontario Ministry of Transportation, Pavement and Foundations Section, Materials Engineering and Research Office, Downsview, Ontario, Canada, 2014.
33. MTO. *Pavement Deisgn and Rehabilitation Manual* (2nd ed). Ontario Ministry of Transportation, Downsview, Ontario, Canada, 2013.
34. AASHTO. *AASHTOware Pavement M-E Design v2.0 Help Manual*. AASHTOWare, 2014.
35. Ontario Provincial Standard Specification. *Asphaltic Concrete Payment Adjustment for Surface Smoothness Based on Quality Assurance Measurements Taken by an Inertial Profiler*. Special Provision No. 103F31. 2014.
36. Aguiar-Moya, J. P., J. A. Prozzi, and L. Manuel. Framework for Reliability Analysis Using the Mechanistic-Empirical Pavement Design Guide. *Proc., 8th International Transportation Specialty*

Conference, CSCE. Canadian Society of Civil Engineering, Winnipeg, Manitoba, 2010, pp. 1274–1283.

37. Thyagarajan, S., B. Muhunthan, N. Sivaneswaran, and K. Petros. Efficient simulation techniques for reliability analysis of flexible pavements using the mechanistic-empirical pavement design guide. *Journal of Transportation Engineering*, Vol. 137, No. 11, 2011, pp. 796–804.

38. Retherford, J., and M. McDonald. Reliability Methods Applicable to Mechanistic-Empirical Pavement Design Method. *Transportation Research Record: Journal of the Transportation Research Board*, 2010. 2154: 130–137.

The Standing Committee on Design and Rehabilitation of Asphalt Pavements (AFD60) peer-reviewed this paper (18-04022).

TRR
JOURNAL OF THE TRANSPORTATION RESEARCH BOARD

Article

Transportation Research Record
2018, Vol. 2672(40) 217–227

DOI: 10.1177/0361198118756873
journals.sagepub.com/home/trr
SAGE

Fatigue Performance Prediction of Asphalt Pavements with FlexPAVE™, the S-VECD Model, and DR Failure Criterion

Yizhuang David Wang[1], **Behrooz Keshavarzi**[1], **and Y. Richard Kim**[1]

Reliable predictions of asphalt materials and pavement performance are important elements in mixture design, mechanistic-empirical pavement design, and performance-related specifications. This paper presents FlexPAVE™, a pavement performance prediction program. FlexPAVE™ is a three-dimensional finite element program that is capable of moving load analysis under realistic climatic conditions. It utilizes the simplified viscoelastic continuum damage (S-VECD) model to predict asphalt pavement fatigue life. This S-VECD model currently incorporates the so-called G^R failure criterion to define the failure of asphalt mixtures. In this study, a new failure criterion for the S-VECD model, designated as the D^R criterion, has been developed to remedy some of the shortcomings of the G^R failure criterion. This D^R criterion has been implemented successfully in FlexPAVE™. In this paper, FlexPAVE™ is used to simulate the fatigue performance of field test sections. These test sections include various pavement structures, such as perpetual pavements and accelerated load facility test pavements in the United States, South Korea, and China, as well as various materials, such as warm-mix asphalt, reclaimed asphalt pavement, and mixtures with modified binders. The D^R-based FlexPAVE™ predictions have yielded good agreement with the field measurements and show more reasonable trends compared to predictions obtained using the G^R failure criterion.

The prediction of asphalt pavement performance has received much attention because it is central to the success of mechanistic–empirical pavement design and performance-related specifications. Reliable and accurate performance predictions allow state agencies to optimize pavement structures in pavement designs, and can help provide incentives and disincentives for paving projects based on the predicted service life. The North Carolina State University research team has developed a three-dimensional finite element program, FlexPAVE™ (formerly known as the LVECD program), under the auspices of the Federal Highway Administration (FHWA), with the goal to provide state highway agencies and contractors with an efficient and mechanistically sound tool for asphalt pavement performance predictions.

FlexPAVE™ utilizes Fourier transform-based finite element analysis with moving loads. FlexPAVE™ simulations can account for the effects of temperature using the temperature–time history data obtained from the Enhanced Integrated Climatic Model (*1,2*). The simplified viscoelastic continuum damage (S-VECD) model is embedded in FlexPAVE™ for the fatigue damage computations. The S-VECD model is a material model that describes changes in the constitutive relationship as fatigue damage grows (*3–7*) and characterizes the fundamental material properties that are independent of temperature, mode of loading (i.e., control strain, control stress, and monotonic loading), and loading amplitude. To characterize failure in a material, the S-VECD model needs a failure criterion. For this purpose, Sabouri and Kim developed the so-called G^R failure criterion (*8*), which is based on pseudo energy and is independent of temperature, mode of loading, and loading amplitude. The G^R failure criterion has been implemented in FlexPAVE™, and the resultant software package has yielded good predictions of observed cracking in various pavements (*9–11*). Wang et al. evaluated and compared the prediction accuracy of Pavement ME Design (formerly known as the MEPDG) and FlexPAVE™ using material and field data and reported that, although both programs can provide reasonable results, FlexPAVE™ yields better agreement with field measurements due to the fundamental models it applies (*12*).

[1]Department of Civil, Construction, and Environmental Engineering, North Carolina State University, Raleigh, NC

Corresponding Author:
Address correspondence to Y. Richard Kim: kim@ncsu.edu

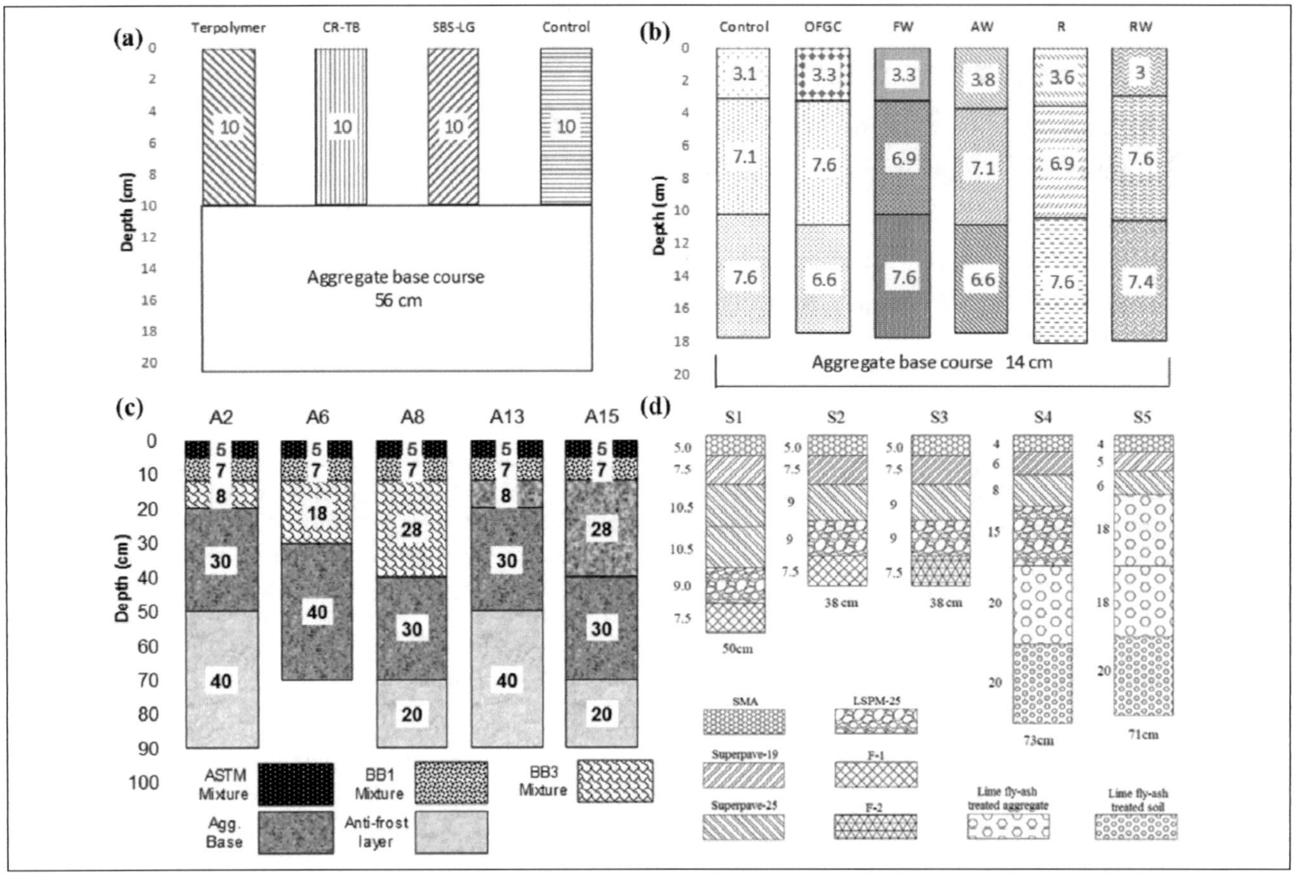

Figure 1. Materials and pavement test structures used in this study: (*a*) FHWA ALF, (*b*) NCAT, (*c*) KEC, and (*d*) Binzhou pavements.

Recently, Wang and Kim developed a new failure criterion, designated as the D^R criterion, to overcome some extrapolation issues the G^R failure criterion causes in pavement performance predictions (*13*). After a brief introduction to the S-VECD model, the G^R and D^R failure criteria, and the implementation of the D^R criterion into the FlexPAVE™ program, this paper then discusses the advantages of using the new D^R failure criterion in field simulations and presents comparisons between predicted performance and field measurements.

Study Materials and Pavements

To evaluate the prediction accuracy of the FlexPAVE™ program together with the S-VECD model and one of the two failure criteria (G^R and D^R were each tested for comparative purposes), different asphalt mixtures and pavement structures were tested and modeled in this study. The study asphalt mixtures included conventional Superpave mixes, polymer-modified mixes, warm-mix asphalt mixtures, mixtures with different reclaimed asphalt pavement (RAP) contents, a stone–mastic asphalt mix, and fatigue-resistant mixtures. The field projects included the FHWA's Accelerated Load Facility (ALF) test sections, the National Center for Asphalt Technology (NCAT) test track, the Korea Expressway Corporation (KEC) test roads, and perpetual pavements in Binzhou, China. Figure 1 presents schematic illustrations of the study mixtures and pavement structures. Basic information about these materials and field sections can be found in previous publications (*12,14*).

Models and Failure Criteria

The S-VECD Model

The S-VECD model is a pseudo energy-based fatigue model that reflects changes in the material constitutive relationship due to the development of damage. A damage evolution law is based on the elastic–viscoelastic correspondence principle, which allows the entire loading history to be taken into account for the damage calculation, as well as on the time–temperature superposition principle. The S-VECD model applies the pseudo stiffness concept to indicate the integrity of a material under loading. The pseudo stiffness value is 1 when the material is intact; this

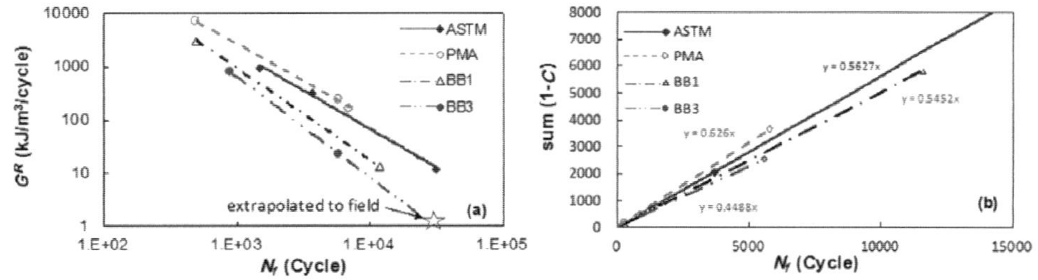

Figure 2. Failure criteria for the asphalt mixtures used in the KEC test road: (*a*) G^R failure criterion and (*b*) D^R failure criterion. Note: each data point represents one cyclic fatigue test sample at one strain level.

value decreases as damage accumulates. The model defines a unique relationship (the so-called *damage characteristic curve*) between pseudo stiffness (C) and the amount of cumulative damage (S) for a given mixture. This relationship is independent of mode of loading, temperature, and loading amplitude; thus, it is considered a fundamental material property. The damage characteristic curve represents how damage grows in the material; however, a failure criterion is needed to predict the failure of the material. For this study, the fatigue properties of the mixtures were obtained from cyclic direct tension tests following AASHTO TP 107 (*8,12,13*). The following sections present two failure criteria based on the S-VECD model and the dissipated pseudo strain energy concept.

The G^R Failure Criterion

The G^R failure criterion is based on the experimental observation that, when damage accumulates in fatigue test specimens, the number of load cycles has a power relationship with the accumulated dissipated pseudo strain energy (*8*). This observation results in the definition of G^R shown in Equation 1 and a linear relationship between G^R and the number of load cycles to failure (N_f) in log–log scale:

$$G^R = \frac{\int_0^{N_f} \frac{1}{2}(1-C)\left(\varepsilon_{0,ta}^R\right)^2 dN}{\left(N_f\right)^2} \tag{1}$$

where $\varepsilon_{0,ta}^R$ is the pseudo strain amplitude for the duration of the stress in tension, and N is the number of loading cycles. Figure 2*a* presents the linear relationships between G^R and N_f in log–log scale for asphalt mixtures obtained from the KEC test road. The ASTM and PMA mixtures are unmodified and polymer-modified surface mixtures, respectively, and the BB1 and BB3 mixtures are base mixtures with different aggregate gradations. A higher position of the G^R versus N_f line indicates a better crack-resistant material. Figure 2*a* shows that the PMA mixture is the most crack-resistant material and that the base mixtures are less crack-resistant than the surface mixtures.

The DR Failure Criterion

The D^R failure criterion is based on experimental observations by Wang and Kim (*13*) that the average loss of integrity per cycle throughout an asphalt mixture's fatigue life, that is, D^R in the following equation, is constant regardless of temperature, mode of loading, and load amplitude:

$$D^R = \frac{\int_0^{N_f}(1-C)dN}{N_f} = \frac{\text{sum}(1-C)}{N_f} \tag{2}$$

According to the definition of D^R given in Equation 2, D^R is the slope of the linear relationship between the sum of $(1 - C)$ to failure and N_f, as shown in Figure 2*b*. Wang and Kim demonstrated that the slope of D^R is a ductility parameter because it indicates the brittleness or ductility of a mixture (*13*). For example, D^R values typically would decrease as RAP content increases, and high D^R values have been observed for mixtures with polymer-modified binder (*13,14*). These trends can be seen in Figure 2*b*; that is, the PMA mixture has a higher slope than the ASTM mixture, and the base mixtures have lower slopes than the surface mixtures. It is also noted that the ranking based on the G^R criterion, as shown in Figure 2*a*, is the same as the ranking based on the D^R criterion, as shown in Figure 2*b*.

However, there are two important differences between the G^R criterion and the D^R criterion. The first difference is that the D^R failure criterion is based on the linear relationship between the sum of $(1 - C)$ to failure and N_f in arithmetic scale, whereas the G^R criterion is in log–log scale. The second difference is that the linear relationship between the sum of $(1 - C)$ to failure and N_f passes through the origin, as shown in Figure 2*b*. Therefore, for the D^R criterion, the only property that needs to be characterized is the slope of the linear relationship between the sum of $(1 - C)$ to failure and N_f, that is, D^R; whereas for the G^R criterion, both the slope and intercept of the linear relationship between G^R and N_f in log–log scale need to be determined. Because there are fewer parameters in the D^R criterion than in the G^R criterion, fewer tests are required to characterize the D^R criterion than the G^R criterion or, if the same number of tests is used for characterization, the D^R criterion offers more reliability than the G^R criterion.

These two primary differences between the two failure criteria become important when pavement cracking performance is predicted using a structural model such as FlexPAVE™. Because of the differences between the strain level in the pavement under actual traffic loading and the strain level applied in laboratory tests, extrapolation is necessary for pavement performance predictions. For the G^R criterion, the extrapolation is performed in log–log scale, which can cause a significant impact on the fatigue life that is predicted for actual pavements due to variability in the material characterization tests. However, for the D^R criterion, this sensitivity to data variability can be reduced because the extrapolation is done in arithmetic scale and only one parameter, that is, D^R, needs to be characterized instead of the two parameters required for the G^R criterion. The impact of these differences is demonstrated further using pavement performance predictions discussed in the following sections.

FlexPAVE™

FlexPAVE™ can be used to predict critical pavement distresses such as cracking and rutting. The program employs a finite element method in a three-dimensional layered system to obtain mechanical responses (i.e., stresses and strains) under moving traffic loads. The computational time during analysis is reduced dramatically by applying the fast Fourier transform in both the time and spatial domains. FlexPAVE™ predicts the fatigue damage throughout the pavement design life using the S-VECD model. Damage contours of cross-sections of the asphalt layers are generated from the analysis and plotted to indicate the location of damage (e.g., top-down versus bottom-up cracking) as well as the level of damage.

FlexPAVE™ uses two overlapping triangles to form the reference cross-section area within which the level of damage is calculated (*15*). Figure 3 illustrates these overlapping triangles. The top inverted triangle has a 170-cm wide base that is located at the top of the surface layer and a vertex that is located at the bottom of the bottom asphalt layer. The 12-cm wide base of the second triangle is located at the bottom of the bottom asphalt layer and its vertex is positioned at the surface layer.

The percentage of damage (referred to as "percent damage") is defined as the ratio of the sum of the damage factors within the reference cross-section area to the reference cross-section area,

$$\text{Percent Damage} = \frac{\sum_{i=1}^{M} (\text{Damage Factor})_i \times A_i}{\sum_{i=1}^{M} A_i} \tag{3}$$

where:

i = nodal point number in finite element mesh;
M = total number of nodal points in finite element mesh;
A_i = area represented by nodal point i in finite element mesh; and
$\sum A_i$ = reference area.

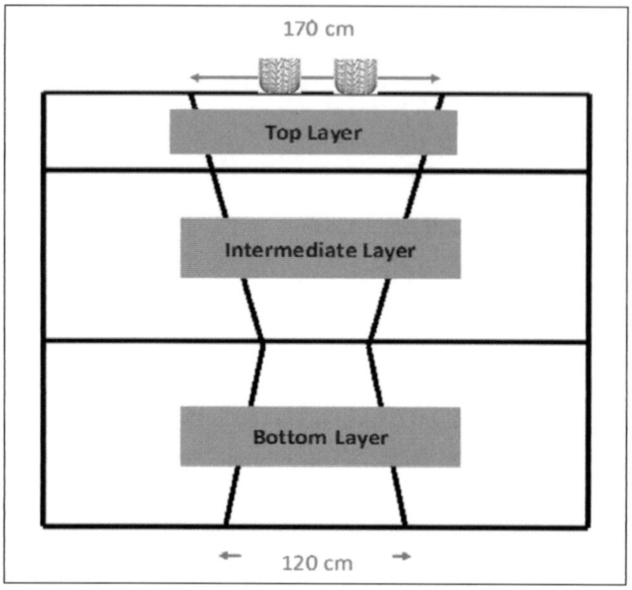

Figure 3. Reference area for "percent damage" definition in FlexPAVE™.

When the G^R criterion is used as the failure criterion, the damage factor is defined using Minor's law:

$$\text{Damage Factor} = \frac{N}{N_f} \tag{4}$$

When the D^R criterion is used in FlexPAVE™, the damage factor in Equation (3) is defined as

$$\text{Damage Factor} = \frac{1 - C_{avg}}{D^R} \tag{5}$$

where C_{avg} is the average pseudo stiffness per cycle up to the current number of loading cycles. Therefore, the damage factor defined by the D^R criterion is the ratio of $(1 - C_{avg})$, which represents the average reduction in pseudo stiffness per cycle up to the current number of load cycles, to D^R, which represents the average reduction in pseudo stiffness per cycle up to failure.

Comparison of Predicted Fatigue Damage Results Using the Two Failure Criteria

This section presents a comparison of the fatigue damage predictions obtained using the G^R and D^R criteria, and the results demonstrate the advantages of using the D^R criterion over the G^R criterion. Two pavement structures, Sections A2 and A13 from the KEC test road (see Figure 1), are used as comparative examples. Section A2 is a full-depth asphalt pavement with three asphalt layers, and Section A13 is an asphalt pavement with an aggregate base course (ABC). The total asphalt layer thickness for Section

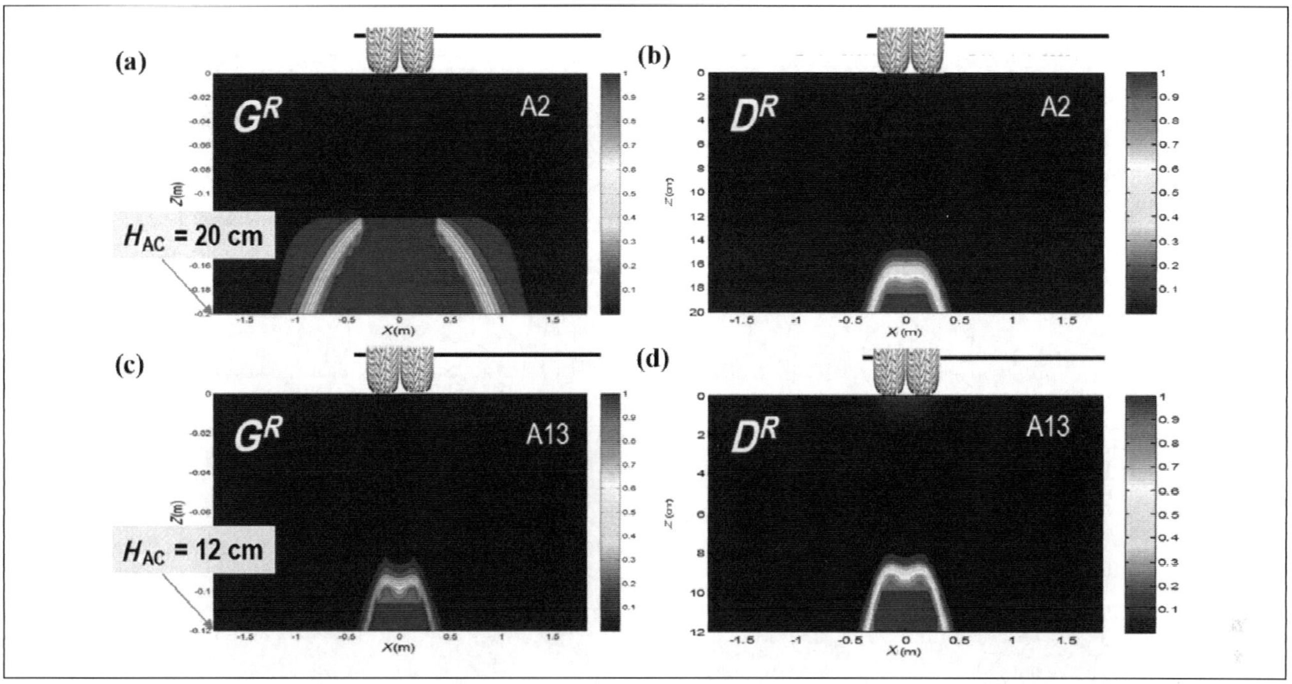

Figure 4. Damage contours: (*a*) Section A2 based on the G^R criterion; (*b*) Section A2 based on the D^R criterion; (*c*) Section A13 based on the G^R criterion; and (*d*) Section A13 based on the D^R criterion.

A2 is 20 cm, and the total asphalt layer thickness for Section A13 is 12 cm.

The fatigue performance of these pavements was simulated in FlexPAVE™ using material properties obtained from laboratory tests. Figure 4, *a–d* present the damage contours for Section A2 and Section A13 after 6 years of loading, as predicted by FlexPAVE™. It is noted that the scale shown on the right side of each contour plot in Figure 4 and all the subsequent contour plots in the paper indicates the damage factor calculated from Equation 5. In these contour plots, damage factors of zero and one in red and blue colors represent a completely failed element and an intact element, respectively.

Figure 4, *a* and *c* present the damage contours of the two sections using the G^R failure criterion and Figure 4, *b* and *d* present the damage contours of the two sections using the D^R failure criterion. Figure 4, *a* and *c* show that Section A2, the full-depth pavement with the 20-cm asphalt layer thickness, experienced much more fatigue damage than Section A13, the ABC pavement with the 12-cm asphalt layer thickness, when the G^R failure criterion was used for the predictions. However, these prediction results differ from the field measurements; this discrepancy is due to the extrapolation that was carried out in log–log scale using the G^R criterion. That is, in full-depth pavements, the strain at the bottom of the asphalt layer is less than in the thinner ABC pavements, and less strain in full-depth pavements requires greater extrapolation using the G^R criterion. The greater the extrapolation, the greater the impact data variability will have on the predictions.

In Figure 4, *b* and *d*, the D^R failure criterion is shown to predict more damage for the ABC pavement than for the full-depth pavement, and the predicted ranking matches the field measurements well. Moreover, it is worth pointing out that the model coefficients for the D^R criterion were obtained from the same test data that were used to characterize the G^R criterion. Therefore, as shown in the A2 and A13 examples, the fatigue performance predictions based on the D^R criterion reflect the field observations better than those based on the G^R failure criterion, because the D^R criterion predictions are less affected by the variability in material testing. Variability affects the accuracy and pavement predictions of the D^R criterion less than the G^R failure criterion because the D^R criterion requires fewer tests and extrapolation is in arithmetic scale, not logarithmic scale.

Simulation Results and Discussion

This section presents a brief introduction to the field sections followed by the FlexPAVE™ simulation results that were obtained using the D^R failure criterion.

FHWA ALF Pavements

The FHWA's ALF contains test lanes that were constructed using similar structures (10 cm single asphalt layers and aggregate bases) and different asphalt mixtures. These ALF pavements were conditioned at a median temperature (18°C) while a single-wheel heavy load was applied in cycles. For

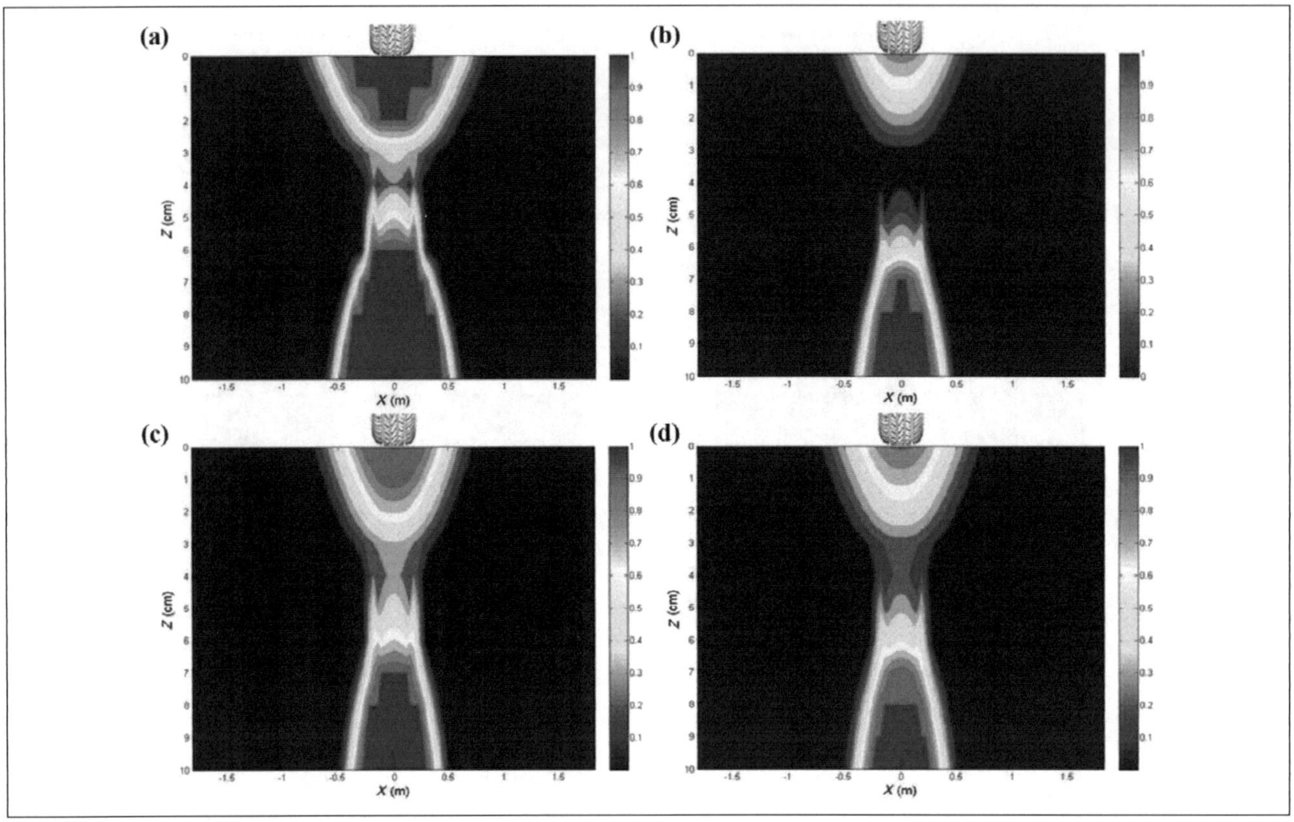

Figure 5. Contours for damage factors used in 20-year simulations of FHWA ALF test sections: (*a*) control, (*b*) CR-TB, (*c*) terpolymer, and (*d*) SBS-LG mixtures.

this study, three mixtures with modified binders (CR-TB, terpolymer, SBS-LG) and one control section were selected from the test lanes. The mixtures were characterized in the laboratory, and the simulations were performed using FlexPAVE™.

Figure 5 presents the damage contours for the FHWA ALF asphalt pavement sections and Figure 6 presents the predicted "percent damage" and the cracking area (%) measured from the field. Note that the terpolymer section is not included in Figure 6 for comparisons with the field data because of the reported "lime nugget distribution" problem found in the terpolymer section during construction (*16*); that is, the method used to add the lime led to the formation of large lime particles in the mixtures. According to Figures 5 and 6, both the predicted "percent damage" and the measured cracking area indicate that the sections with modified binders performed better than the control section. Among the test sections that contained modifed binder, the section with the SBS binder performed the best according to the field measurements, but the prediction results indicate that the SBS section is the second best test section [although the difference in predicted damage between the best section (CR-TB) and the SBS section is relatively small]. However, if only the "failed points" (dark red) in the damage contours are counted, the SBS section would be considered to perform the best of the three sections. This observation may warrant the need to investigate different ways to represent the damage level simulated by FlexPAVE™. In addition to ranking the pavement sections that contain different mixtures, the curves in semi-log scale indicate increasing rates for both the damage evolution curves and field measurements.

NCAT Test Track

FlexPAVE™ simulations of sections from the NCAT test track were performed in this study. Six sections (S8, S9, S10, S11, N10, and N11) from the 2009 research cycle were selected; see Figure 1. The asphalt mixtures in these sections were a control (C) mixture, WMA with additive (AW), foamed warm-mix asphalt (FW), open-graded friction course (OGFC) mixtures, mixtures with a high RAP content (R), and WMA with a high RAP content (RW). These mixtures were tested in the laboratory, and the structures, truck loading, and climate conditions were modeled using FlexPAVE™.

Figure 7 presents the predicted damage contours for the NCAT test sections. Figure 8, *a* and *b* illustrate the predicted "percent damage" and the field-measured cracking area (%), respectively. Figure 8*c* presents a comparison between the

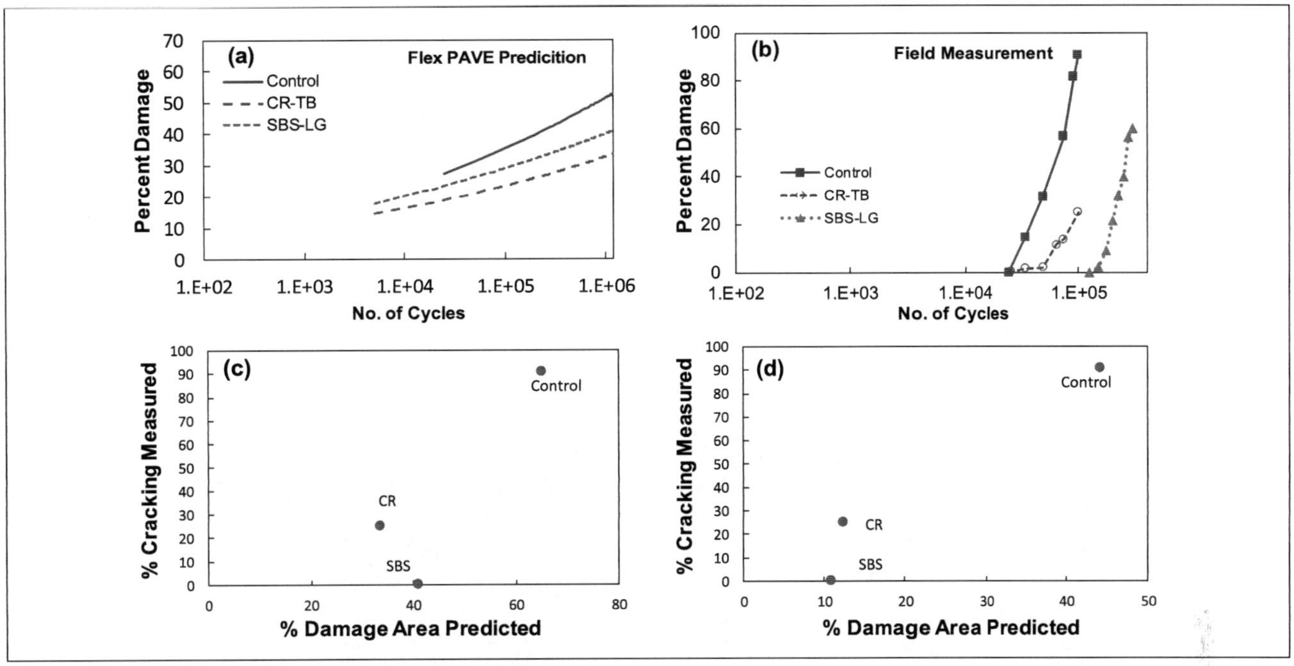

Figure 6. Cracking evolution derived from (*a*) FlexPAVE™, (*b*) FHWA ALF test section results, (*c*) predicted damage versus measured cracking, and (*d*) predicted damage versus measured cracking if only total failed nodes are counted.

predicted "percent damage" and the measured cracking area at the end of the experiment, that is, at 40 months. Figure 8*c* shows that the predicted ranking of fatigue damage for those sections is in good agreement with the field measurements.

The measured cracking data presented in Figure 8*b* that were obtained from the section that contains the OGFC mixture as the surface mix show an interesting trend, with relatively slow crack growth in the earlier life and then a dramatic increase in the cracking area. This behavior is captured somewhat by FlexPAVE™, as shown in Figure 8*a*. Moreover, top-down cracking is evident from the damage contours shown in Figure 7*f*, which also is confirmed by observation of the field cores (*17*). Note that the current version of FlexPAVE™ does not include an aging model, nor does it update the stiffness values of asphalt layers as damage accumulates in those layers in the fatigue simulations. Therefore, the amount of predicted damage will increase after those features are added in the next version of FlexPAVE™. With those planned features, the severity of top-down cracking is expected to increase more than for bottom-up cracking because of much greater aging as well as a greater stress redistribution in the top portion of the simulated pavement.

On the other hand, FlexPAVE™ was able to predict the low amount of damage for the RW and R sections in the field. It is commonly assumed that pavement sections that contain a high proportion of RAP will exhibit significant amounts of cracking because of the inherent pre-aged brittle material, but this assumption is not the case according to the field observations of the NCAT test sections. The good agreement between the results from FlexPAVE™ and the field measurements indicates that this discrepancy between the common assumption and the field observations of the RAP sections can be predicted by the mechanistic models employed in FlexPAVE™ in conjunction with the laboratory data. Based on the material test results, the mixtures that contain high percentages of RAP have high modulus values. When those materials are placed on a roadway, the mechanical responses under the traffic load, such as tensile strain and shear stress, have much lower values than for materials that are softer than RAP mixtures. Therefore, low amounts of cracking were observed even though the mixtures with high RAP contents were brittle.

KEC Test Road

This section presents the FlexPAVE™ simulation results for Sections A2, A6, A8, A13, and A15 in the KEC test road along with measured performance data (see Figure 1). Figure 9*a* presents the damage evolution simulated by FlexPAVE™. The total asphalt layer thicknesses of Sections A2, A6, and A8 are 20, 30, and 40 cm, respectively. Sections A13 and A15 both contain 12-cm thick asphalt layers, but the aggregate base in Section A15 is much thicker than in Section A13. Figure 9*a* shows that the predicted "percent damage" decreases as the asphalt layer thickness decreases. Also, the "percent damage" decreases when the aggregate base thickness increases, according to the results for Sections A13 and A15. The comparison between the predicted "percent damage" shown in

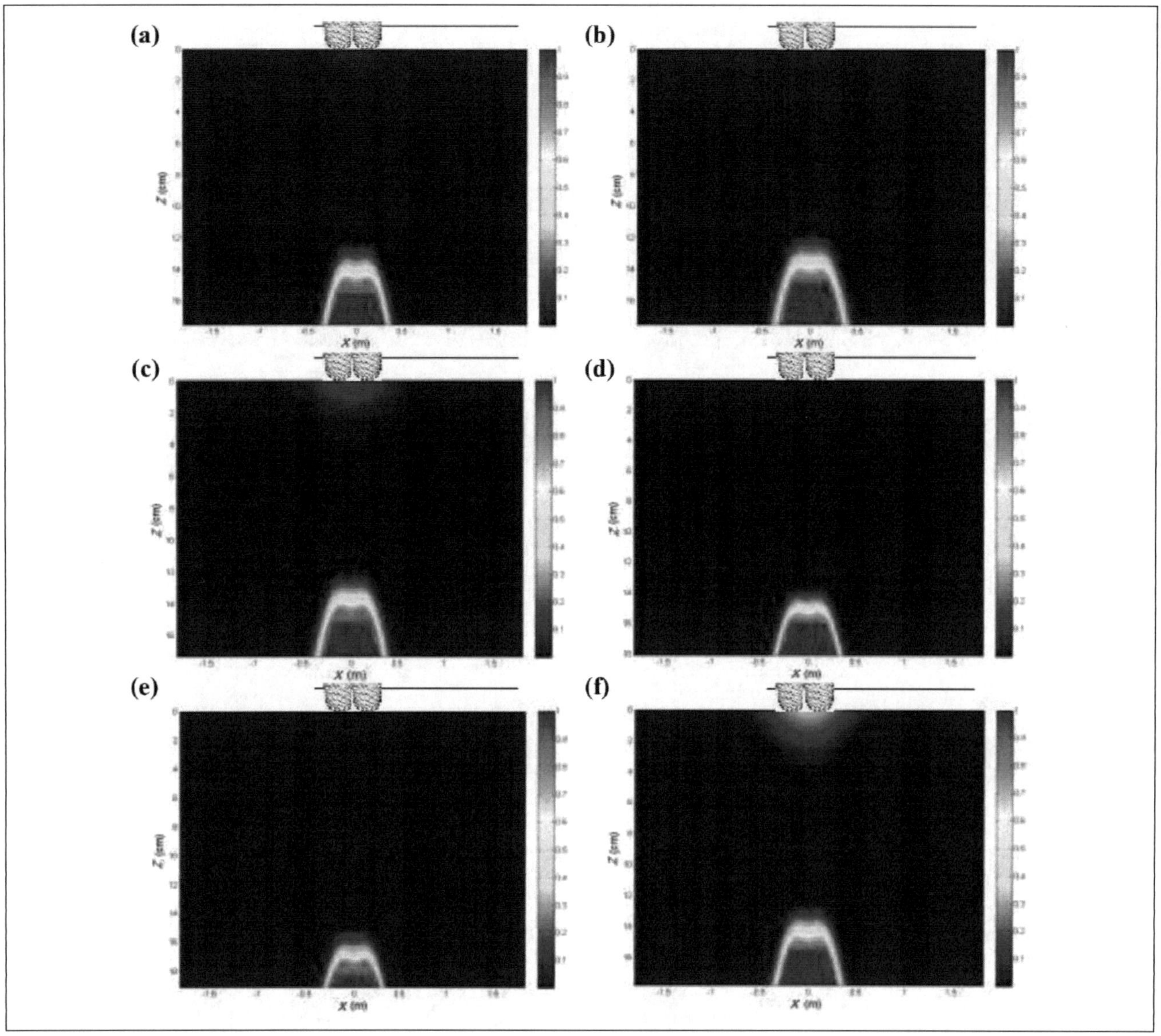

Figure 7. Contours for 40-month simulation results obtained from FlexPAVE™ for NCAT mixtures: (*a*) NCAT-Control, (*b*) NCAT-AW, (*c*) NCAT-FW, (*d*) NCAT-RW, (*e*) NCAT-R, and (*f*) NCAT-OGFC.

Figure 9*a* and the measured cracking area shown in Figure 9*b* indicates that the predicted ranking and damage evolution trend agree well with the field performance.

Binzhou Perpetual Pavements

The main goal of the Binzhou project was to evaluate different perpetual pavement designs. Several pavement sections were used for this study, three of which (Sections 1, 2, and 3) have fatigue-resistant layers as their bottom asphalt layers. The other two pavement sections are thinner and have a thick lime fly-ash-treated aggregate base and subgrade. Figure 1 presents the pavement layouts. Unfortunately, no detailed field cracking information was available for the project. The only known fatigue-related information is that Section 5 failed within 5 years and was replaced.

Owing to the limited field performance information, the purpose of the simulations was to rank the sections based on the predicted damage levels and evaluate the results using engineering judgment. Figure 10 presents the damage contours. As shown, Section 1 experienced the least fatigue damage due to the thickness (50 cm) of the asphalt layers. Relatively more damage can be observed in Sections 2 and 3 that have thinner (38 cm) asphalt layers. Of these two sections, Section 3 yielded less damage. The good performance of Section 3 is due to its superior material (F-2) in the

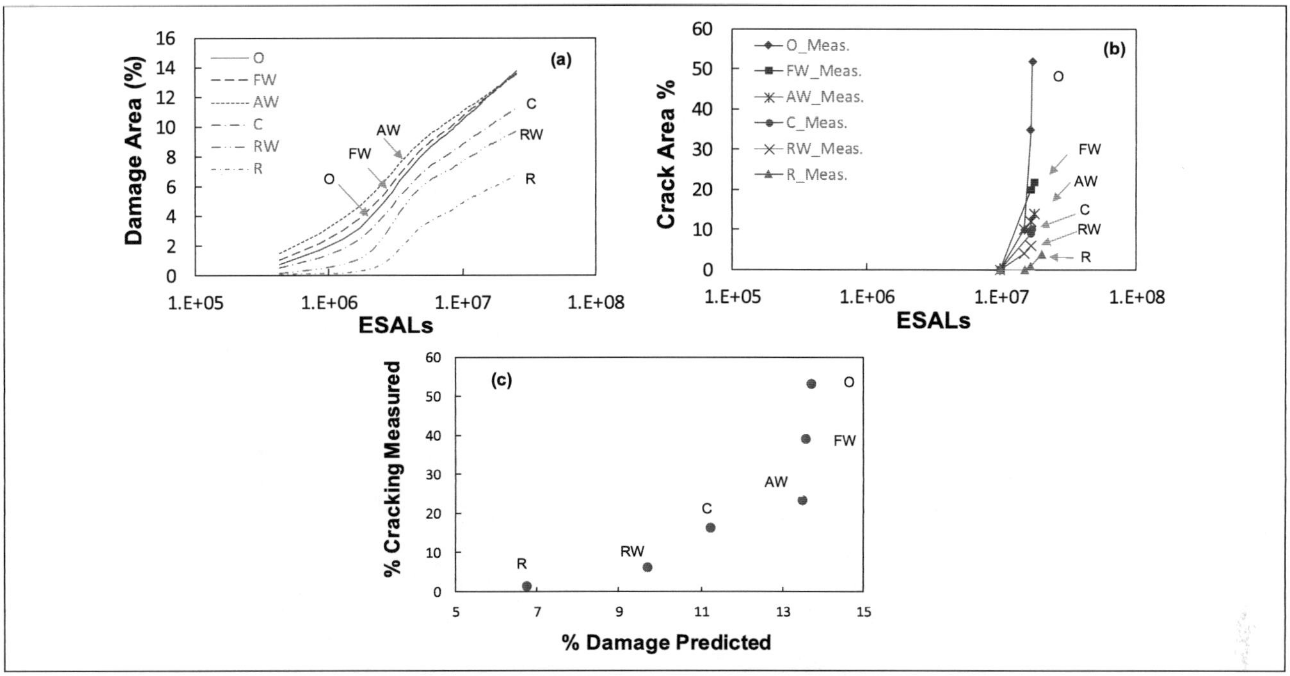

Figure 8. Cracking evolution derived from (*a*) FlexPAVE™, (*b*) NCAT test track results, and (*c*) predicted "percent damage" versus measured cracking area at the end of the experiment.

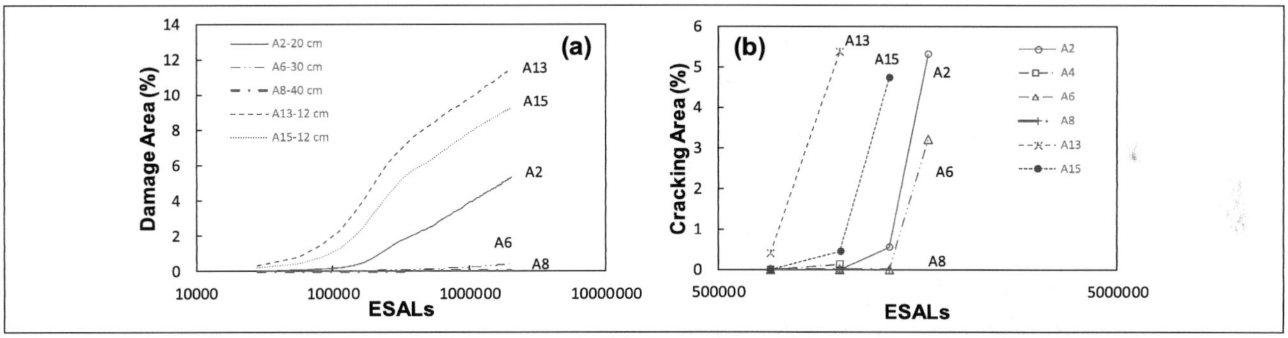

Figure 9. (*a*) "Percent damage" for KEC test road simulations and (*b*) KEC field results. Note that the asphalt layer thicknesses are indicated in the legend.

fatigue-resistant layer. In contrast with Sections 2 and 3, large amounts of damage can be observed in Section 4; the asphalt layer thickness is about the same as in Sections 2 and 3, but the fatigue-resistant layer was replaced with a 20-cm lime fly-ash-treated base. According to the damage observed, the fatigue-resistant layer functioned properly. Section 5 is a regular semi-rigid pavement with 6 in. of asphalt mixture. A significant amount of damage was found in this section from both the predicted results and the field measurements. In Sections 1, 2, 3, and 4, the fatigue damage is located mostly in the layer where a large stone porous mixture was placed. This material is known to provide drainage for a pavement system but offers poor fatigue resistance. In summary, this study shows that FlexPAVE™ and the D^R failure criterion together yield results that are in line with engineering judgment and field observations.

Conclusions

The D^R failure criterion, which is implemented in arithmetic scale, was found to yield more realistic fatigue cracking performance predictions than the G^R failure criterion, which is implemented in log–log scale. In addition, the fact that the D^R failure criterion requires the characterization of only one parameter, that is, the slope of the linear relationship between the sum of (1-*C*) to failure and N_f, yields more reliable performance predictions than the G^R failure criterion for the same number of tests, because the

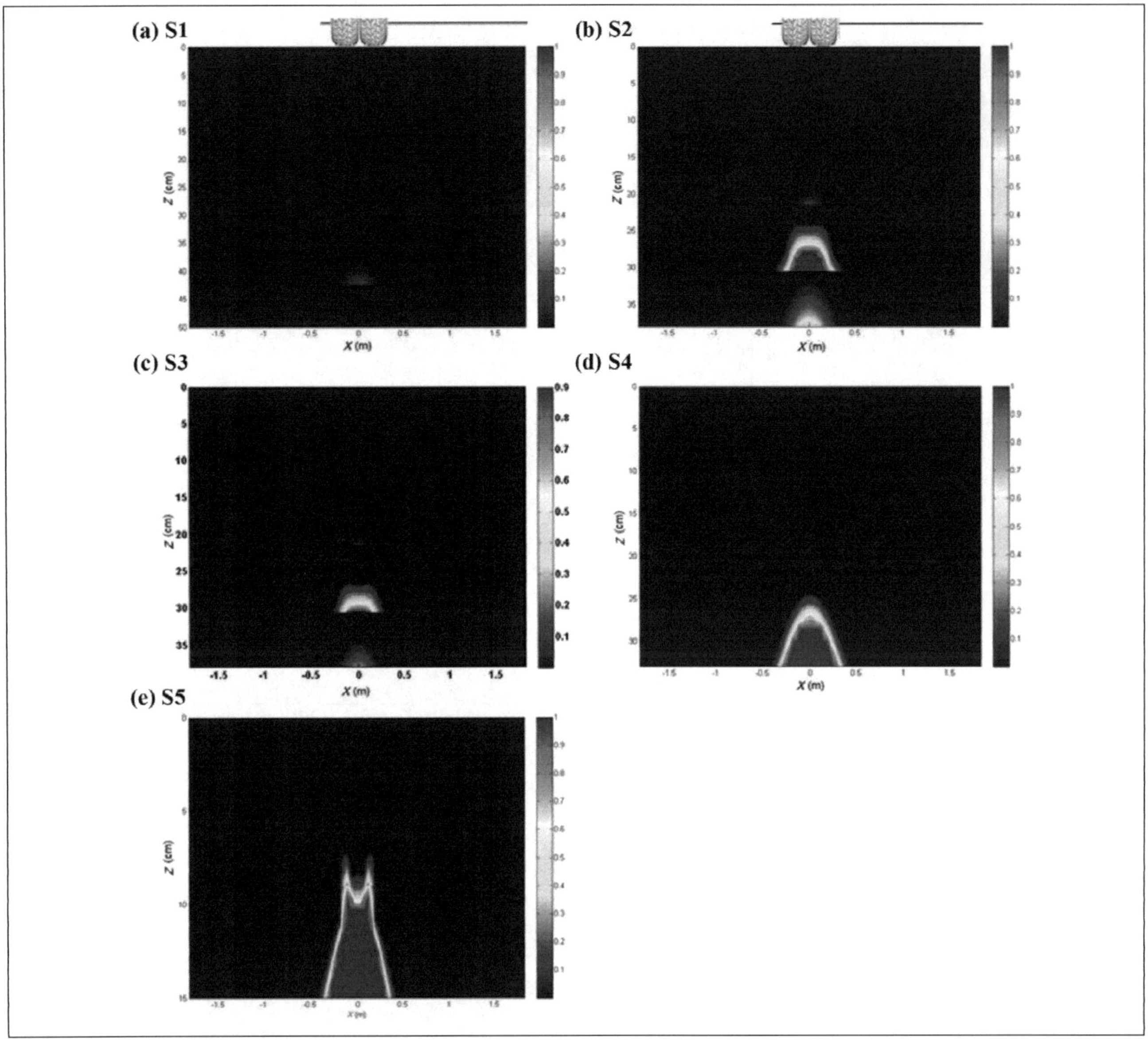

Figure 10. Binzhou test road sections: 8-year simulation results obtained from FlexPAVE™.

G^R failure criterion requires the characterization of both the slope and intercept of the linear relationship between G^R and N_f in log–log scale.

Various pavement sections with different asphalt mixtures and pavement structures were modeled using FlexPAVE™ and the D^R failure criterion. The comparisons between predicted damage and measured cracking from the field indicate that FlexPAVE™, together with the D^R failure criterion, can reasonably predict and rank pavement performance and capture the effects of different materials and pavement structures reasonably well. Note that the prediction results have not yet been calibrated due to the limited number of field sections. Development of transfer functions via calibration is a focus for future research.

Acknowledgment

The authors acknowledge the financial support from the Federal Highway Administration under the DTFH61-08-H-00005 Project.

References

1. Eslaminia, M., and M. N. Guddati. Fourier-Finite Element Analysis of Pavements under Moving Vehicular Loading. *International Journal of Pavement Engineering*, Vol. 17, 2016, pp. 602–614.
2. Eslaminia, M., S. Thirunavukkarasu, M. N. Guddati, and Y. R. Kim. Accelerated Pavement Performance Modeling Using Layered Viscoelastic Analysis. *Proceedings of the 7th International RILEM Conference on Cracking in Pavements*, Delft, Netherlands, 2012, pp. 20–22.

3. Kim, Y. R., and D. N. Little. One-dimensional Constitutive Modeling of Asphalt Concrete. *ASCE Journal of Engineering Mechanics*, Vol. 116, No. 4, 1990, pp. 751–772.
4. Daniel, J. S., and Y. R. Kim. Development of a Simplified Fatigue Test and Analysis Procedure Using a Viscoelastic Continuum Damage Model. *Journal of the Association of Asphalt Paving Technologists*, Vol. 71, 2002, pp. 619–650.
5. Chehab, G. R., Y. R. Kim, R. A. Schapery, M. Witczack, and R. Bonaquist. Characterization of Asphalt Concrete in Uniaxial Tension Using a Viscoelastoplastic Model. *Journal of the Association of Asphalt Paving Technology*, Vol. 72, 2003, pp. 315–355.
6. Underwood, B. S., C. Baek, and Y. R. Kim. Simplified Viscoelastic Continuum Damage Model as Platform for Asphalt Concrete Fatigue Analysis. *Transportation Research Record: Journal of the Transportation Research Board*, 2012. 2296: 35–45.
7. Underwood, B. S., Y. R. Kim, and M. N. Guddati. Characterization and Performance Prediction of ALF Mixtures Using a Viscoelastoplastic Continuum Damage Model. *Journal of the Association of Asphalt Paving Technologists*, Vol. 75, 2006, pp. 577–636.
8. Sabouri, M., and Y. R. Kim. Development of a Failure Criterion for Asphalt Mixtures under Different Modes of Fatigue Loading. *Transportation Research Record: Journal of the Transportation Research Board*, 2014. 2247: 117–125.
9. Norouzi, A., D. Kim, and Y. R. Kim. Numerical Evaluation of Pavement Design Parameters for the Fatigue Cracking and Rutting Performance of Asphalt Pavements. *Materials and Structures*, Vol. 49, 2015, pp. 3619–3634.
10. Norouzi, A., M. Sabouri, and Y. R. Kim. Evaluation of the Fatigue Performance of High RAP Asphalt Mixtures. *Proceedings of the 12th International Society for Asphalt Pavements*, Raleigh, NC, 2014.
11. Cao, W., A. Norouzi, and Y. R. Kim. Application of Viscoelastic Continuum Damage Approach to Predict Fatigue Performance of Binzhou Perpetual Pavements. *Journal of Traffic and Transportation Engineering* (English Edition), Vol. 3, 2016, pp. 104–115.
12. Wang, Y., A. H. Norouzi, and Y. R. Kim. Comparison of Fatigue Cracking Performance of Asphalt Pavements Predicted by Pavement ME and LVECD Programs. *Transportation Research Record: Journal of the Transportation Research Board*, 2016. 2590: 44–55.
13. Wang, Y., and Y. R. Kim. Development of a Pseudo Strain Energy-Based Fatigue Failure Criterion for Asphalt Mixtures. *International Journal of Pavement Engineering*, 2017, DOI: 10.1080/10298436.2017.1394100.
14. Wang, Y., and Y. R. Kim. Development of a Simple Fatigue Index Parameter for Asphalt Concrete Based on the Viscoelastic Continuum Damage Theory. *Journal of Association of Asphalt Paving Technologies*, 2018, submitted.
15. Kim, Y. R., M. N. Guddati, Y. T. Choi, D. Kim, A. Norouzi, Y. Wang, B. Keshavarzi, M. Ashouri, and A. D. Wargo. *Development of Asphalt Mixture Performance-Related Specifications*. Final report, Contract No. DTFH61-08-H-00005, North Carolina State University, Raleigh, N.C., 2017
16. Gibson, N., X. Qi, A. Shenoy, G. Al-Khateeb, M. Kutay, A. Andriescu, K. Stuart, J. Youtcheff, and T. Harman. *Performing Testing for Superpave and Structural Validation*. FHWA-HRT-11-045 Report. Washington DC, FHWA, 2012.
17. West, R., D. Timm, R. Willis, B. Powell, N. Tran, D. Watson, M. Sakhaeifar, R. Brown, M. Robbins, A. Vargas-Nordcbeck, F. L. Villacorta, X. Guo, and J. Nelson. *Phase IV NCAT Pavement Test Track Findings*. Report 12-10, National Center for Asphalt Technology, 2012.

The Standing Committee on Design and Rehabilitation of Asphalt Pavements peer-reviewed this paper (18-05648).

Article

TRR
JOURNAL OF THE TRANSPORTATION RESEARCH BOARD

Transportation Research Record
2018, Vol. 2672(40) 228–241

DOI: 10.1177/0361198118777090
journals.sagepub.com/home/trr
SAGE

Damage Assessment for ME Rehabilitation Design of Modified Asphalt Pavements: Challenges and Findings

Jhony Habbouche[1], Elie Y. Hajj[1], Peter E. Sebaaly[1], and Nathan E. Morian[2]

Abstract

The overall objective of this study was to assess the use of Level 1 analysis for mechanistic-empirical (ME) rehabilitation designs of deteriorated polymer-modified asphalt concrete (AC) pavements in Nevada using the AASHTOWare® Pavement ME software. This research also explored the possible implementation of a hybrid approach for AC damage characterization to overcome the challenges associated with the use of the Witczak model for estimating the undamaged dynamic modulus master curve of the existing AC layer. Two rehabilitation field projects were used as part of this study. The experimental plan involved falling weight deflectometer (FWD) testing in the right wheelpath before rehabilitation, analysis of core samples, estimation of an equivalent undamaged dynamic modulus, and estimation of equivalent damaged dynamic modulus from FWD backcalculation. The proposed hybrid approach consisted of conducting laboratory dynamic modulus testing on the collected core samples and estimating an equivalent undamaged dynamic modulus at the same FWD testing temperature and loading frequency. The pre-overlay damage, characterized based on the approach in Pavement ME Design software (i.e., using a Witczak prediction model and backcalculated modulus), showed overly high values that did not match with the collected pre-overlay distress data on either of the rehabilitation projects. Based on the findings from this study, the hybrid approach was recommended for implementation by Nevada Department of Transportation (NDOT) when designing AC overlay over polymer-modified asphalt pavements in Nevada. Recommendations for user inputs were also provided for future consideration in Pavement ME Design software.

Over the last 35 years, state highway agencies (SHAs) have changed their emphasis from the construction of new roads to the maintenance and rehabilitation of existing infrastructure (*1*). The Nevada Department of Transportation (NDOT) uses various maintenance and rehabilitation repair strategies to improve the overall state pavement network condition (*2*, *3*). Structural asphalt concrete (AC) overlays represent a wide variety of treatments used in Nevada to rehabilitate deteriorated roads. These AC overlays are at times applied on pavements having medium to high levels of load-related distresses (e.g., fatigue cracking) which make the use of preventive maintenance repair strategies ineffective or not cost effective (*4*). Many factors may influence the design of these AC overlays including traffic, climate, pavement layers structure, material properties, and performance criteria.

The AASHTOWare® Pavement ME Design™ software (hereinafter referred to as Pavement ME Design) is a state-of-the-art tool for designing and evaluating new and rehabilitated flexible pavement structures based on mechanistic-empirical (ME) principles. The software allows for the prediction of selected pavement distresses and smoothness over a specified design or analysis period based on project-specific data for traffic, climate, and materials. The adequacy of a pavement structure is then evaluated by checking the estimated distresses and smoothness at a given reliability level at the end of the design or analysis period against specified target limit values (*5*). Many SHAs have implemented or are in the process of implementing the Pavement ME Design for their new and rehabilitated pavement designs through intensive local calibrations of performance models and transfer functions used in the prediction of pavement distresses.

[1]Pavement Engineering and Science Program (PESP), Department of Civil and Environmental Engineering, University of Nevada, Reno, NV
[2]Nevada Department of Transportation, Carson City, NV

Corresponding Author:
Address correspondence to Jhony Habbouche:
johnny.habbouche.91@gmail.com

The philosophy and general approach for selecting or determining design inputs for corresponding pavement structural materials in the Mechanistic-Empirical Pavement Design Guide (MEPDG) is a hierarchical (i.e., level) system. Three levels are defined for the material characterization in the Pavement ME Design: (*a*) Level 1, which is the most current implementable procedure available, normally involving comprehensive laboratory or field tests; (*b*) Level 2, consisting of compiled inputs estimated through correlations with other material properties that are measured in the laboratory or field; and (*c*) Level 3, which requires the designer to estimate the most appropriate design input value of the material property based on experience with little or no testing. The design of a new or rehabilitated pavement system widely affects the selection of the hierarchical level while configuring input parameters for a particular material of the pavement structure itself (*5*).

Many SHAs have recognized the benefits of polymer-modified AC mixes in resisting multiple modes of climate- and load-induced distresses in flexible pavements. For the past 20 years, NDOT has mandated the use of these mixtures in the construction and rehabilitation of the state's road network throughout the entire depth of the AC layers, not just in the top lift, because of their observed benefits in resisting rutting and fatigue cracking (*6*). In 2015, NDOT began implementing the use of Pavement ME Design for new and rehabilitated flexible pavements (AC overlay over AC) using a comprehensive materials database for asphalt mixtures used in Nevada (*5*). The local calibration of the rutting and fatigue performance models was performed for all three NDOT districts (I, II, and III) using Level 1 inputs for AC overlay and Level 2 or Level 3 inputs for existing AC pavement as outlined in the NDOT *Manual for Designing Flexible Pavements in Nevada Using the AASHTOWare Pavement-ME Design* (*5*).

For rehabilitation design of AC overlay over AC pavements, the Pavement ME Design software accounts for the pre-overlay damage of the existing AC layer by first calculating an "undamaged" modulus ($E_{\text{AC-undamaged}}$) and then adjusting this modulus for damage using the pre-overlay condition. The undamaged modulus is established using the Witczak regression predictive model (*4*). The pre-overlay damage is defined using one of the following three approaches: (*a*) at rehabilitation Level 1, backcalculated AC modulus, determined using nondestructive testing (NDT) such as falling weight deflectometer (FWD) conducted in the wheelpath, is used as the basis for characterizing existing pavement condition; (*b*) at rehabilitation Level 2, the condition is defined by the extent of bottom-up fatigue cracking (FC) at the surface; and (*c*) at rehabilitation Level 3, the condition is defined by the overall pavement condition rating (*5*).

Several alternative predictive relationships including the Witczak model have been established to estimate $E_{\text{AC-undamaged}}$ using simple material properties and volumetrics as inputs. This model was developed in 1999 based on 2,750 data points from 205 asphalt mixtures, mostly including unmodified asphalt binders. The Witczak prediction model, expressed in Equation 1 (*5*), predicts $|E^*|$ of a field-aged AC layer as a function of aggregate gradation, mix air voids, effective binder content, and binder viscosity at different temperatures and under various loading frequencies. This model is used to predict the $E_{\text{AC-undamaged}}$ master curve used in the Pavement ME Design.

$$\log|E^*| = 3.750063 + 0.02932\rho_{200} - 0.001767(\rho_{200})^2 - 0.002841\rho_4 - 0.058097V_a - 0.802208\left(\frac{V_{b_{\text{eff}}}}{V_{b_{\text{eff}}} + V_a}\right) + \frac{3.871977 - 0.0021\rho_4 + 0.003958\rho_{38} - 0.000017(\rho_{38})^2 + 0.005470\rho_{34}}{1 + e^{(-0.603313 - 0.313351\log(f) - 0.393532\log(\eta))}} \quad (1)$$

where

$|E^*|$ = dynamic modulus in psi,
η = bitumen viscosity in 10^6 Poise,
f = load frequency in Hz,
V_a = air void content in %,
$V_{b_{\text{eff}}}$ = effective bitumen content in % by volume,
ρ_{34}, ρ_{38}, and ρ_4 = cumulative % retained in ¾ in. (19 mm), 3/8 in. (9 mm), and #4 (4.75 mm) sieves, respectively, and
ρ_{200} = % passing #200 (0.075 mm) sieve.

Certainly, this and numerous other evaluations of the predictive model have been conducted, such as (*7–10*), with subsequent revisions to this model being recommended as a result (*11, 12*). Currently, such modifications have not been incorporated nor is it permissible to revise the relationship within the Pavement ME Design for polymer-modified AC mixtures. Accordingly, the Witczak relationship may not necessarily be appropriate when conducting a Level 1 Pavement ME rehabilitation design of polymer-modified AC pavements. This study will provide recommendations on how to proceed with necessary rehabilitation designs utilizing the existing Pavement ME Design methodology.

Research Objective and Scope

Based on the varying success of using the Witczak predictive model implemented in the Pavement ME Design software, the challenge of this study was to assess the use of Level 1 analysis for ME rehabilitation designs of deteriorated polymer-modified AC pavements in Nevada. Additionally, this study explored the possible implementation of a hybrid approach for AC damage characterization to overcome the challenges of using the Witczak prediction model for the estimation of the $E_{AC\text{-undamaged}}$ master curve of the existing polymer-modified AC layer. The proposed hybrid approach consists of a combination of both Level 1 and Level 2 analyses that fits within the current limitations of the Pavement ME Design software.

To this end, two rehabilitation field projects were used as part of this study: (*a*) US95 Las Vegas in Southern Nevada (District I), and (*b*) US395 Carson City Freeway in Northern Nevada (District II). A PG76-22NV and a PG64-28NV polymer-modified asphalt binder were used on each of the US95 and US395 projects for the AC overlay as well as in the existing AC layer, respectively. The "NV" extension indicates that these asphalt binders have been graded with the NDOT PG-plus system which includes the Superpave performance grade (PG) binder system plus the following properties (*13*):

- For PG76-22NV—ductility at 39°F (4°C) for original binder, non-recoverable creep compliance and creep recovery for rolling thin film oven (RTFO) binder;
- For PG64-28NV—ductility at 39°F (4°C) for original and RTFO binder, toughness and tenacity at 77°F (25°C) for original binder.

Figure 1 illustrates the overall experimental program of this study. FWD testing was conducted on both rehabilitation projects and field cores from existing pavements were sampled correspondingly. The FWD data were analyzed to determine backcalculated moduli of damaged AC, base, and subgrade layers. The cores were evaluated in the laboratory to determine component measures of binder and aggregate to develop an undamaged dynamic modulus master curve using the Witczak prediction model. Additionally, dynamic modulus testing was conducted on the field cores using the Asphalt Mixture Performance Tester (AMPT) at various frequencies and temperatures. The bottom-up FC damage in the existing AC layer was then characterized using the relative difference between the damaged backcalculated moduli value at the associated FWD testing temperature and loading frequency and either (*a*) the Witczak undamaged modulus (MEPDG approach), or (*b*) the laboratory measured undamaged dynamic modulus of core samples (hybrid approach). Pavement rehabilitation designs were then conducted for the two field projects using the Pavement ME Design software and comparisons of the designed AC overlays thicknesses were then established.

Analysis of FWD Data

The ME rehabilitation design using Level 1 analysis requires an accurate characterization of the existing AC pavement. Accordingly, FWD testing was conducted on each of the respective field projects at a loading level of 12,000 lb (5,443 kg). MODULUS 6.1 software (*14*) was then used to backcalculate the moduli of the in-situ layers from measured surface deflections. The linear elastic layered theory was adopted to backcalculate the various moduli while acknowledging the following basic assumptions: (*a*) uniformity of the surface load distribution over a circular area; (*b*) continuity in vertical stresses and strains at horizontal interfaces; (*c*) homogeneity, isotropy, and linear elasticity of the evaluated materials; (*d*) infinite dimension in horizontal dimension of existing layers; and (*e*) semi-infinite consideration of the bottom layer. To ensure a realistic estimation of backcalculated moduli for the tested pavement structures, the backcalculation process took into consideration the existence of a bedrock or apparent stiff layer, the possible stress-dependency and non-linearity of unbound materials, the variation in pavement layer thicknesses, and the magnitude of the relative stiffness between successive layers.

On both field projects, the FWD load had a pulse duration around 30 milliseconds (t_{FWD}) resulting in a loading frequency of approximately 33.3 Hz ($f_{FWD} = 1/t_{FWD}$) according to the MEPDG approach (*14*).

The FWD measurements on US95 were collected from the right wheelpath before rehabilitation (i.e., before milling) between mileposts 126.539 and 127.914. Field cores were sampled respectively from between the wheelpaths. The thicknesses of the various layers were determined from the field cores as shown in Figure 2*a*. The pavement layer thicknesses are generally assumed to be constant over the entire evaluated pavement section. The pavement structure consisted of a total of approximately 8.5 in. (216 mm) of AC and 15 in. (381 mm) of aggregate base over a subgrade. The existing AC was composed of multiple thin (i.e., layer thickness less than half of the FWD plate radius) AC layers or lifts with possibly different material properties (Figure 2*a*). The backcalculation of the moduli of such thin AC layers with similar characteristics from measured deflection basins is quite challenging. Accordingly, one of the following two approaches can be implemented to overcome such a challenge: (1) attribute a fixed modulus for each of the thin AC layers (based on laboratory testing), or (2) combine a thin AC layer with similar adjacent layers. In this study, the thin AC layers were combined into one full AC layer for which equivalent damaged (backcalculated modulus for the combined AC layer) and undamaged moduli were determined.

Air temperatures were measured using an infrared sensor mounted on the FWD equipment and corresponding pavement temperatures were determined using the BELLS3 equation (*15*) at one-third of the AC depth. An average modulus of 876.2, 45.1, and 26.6 ksi (6,040, 311, and 183 MPa) was backcalculated at a pavement

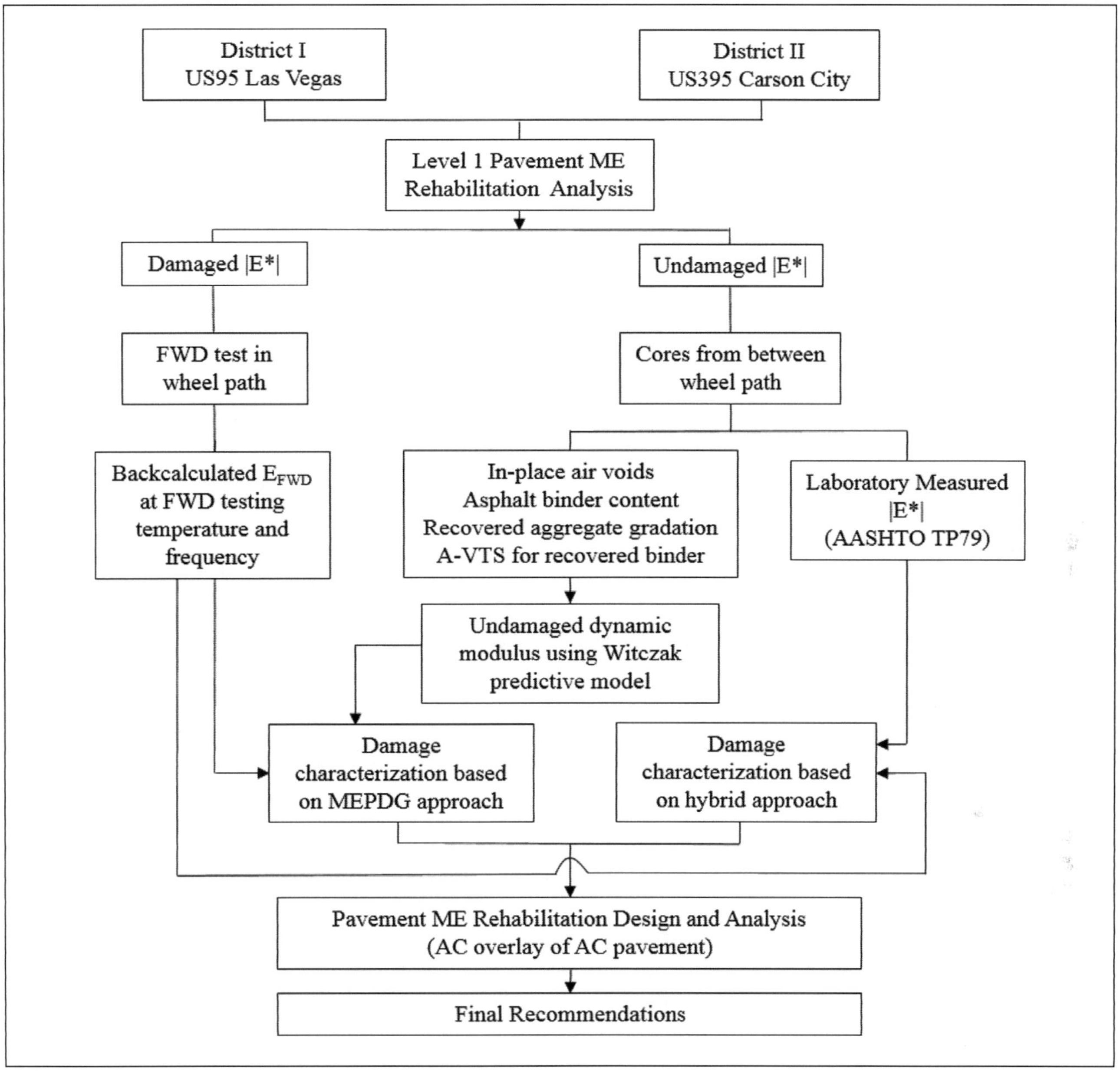

Figure 1. Flowchart summary of the proposed experimental work plan.

temperature of 97.6°F (36.4°C) for the combined AC layer, base layer, and subgrade from US95, respectively.

On the other hand, the FWD measurements on the Northbound lane of the Carson City Freeway (i.e., US395) were collected from the wheelpath on the inside lane between mileposts 7.050 and 8.200. Field cores (refer to Figure 2*b*) were also sampled from locations proximate to where FWD testing was conducted. The pavement structure for US395 consisted of approximately 6.4 in. (163 mm) of AC over a 14 in. (356 mm) aggregate base layer and a subgrade. In this case, the AC layer consisted of three thin AC layers or lifts. An average modulus of 517.9, 23.4, and 14.5 ksi (3,571, 161, and 100 MPa) was backcalculated at a pavement temperature of 90.8°F (32.7°C) for the combined AC layer, base layer, and subgrade, respectively.

Damage Characterization Based on MEPDG

The Pavement ME uses the dynamic modulus ($|E^*|$) master curve to evaluate the structural response of the

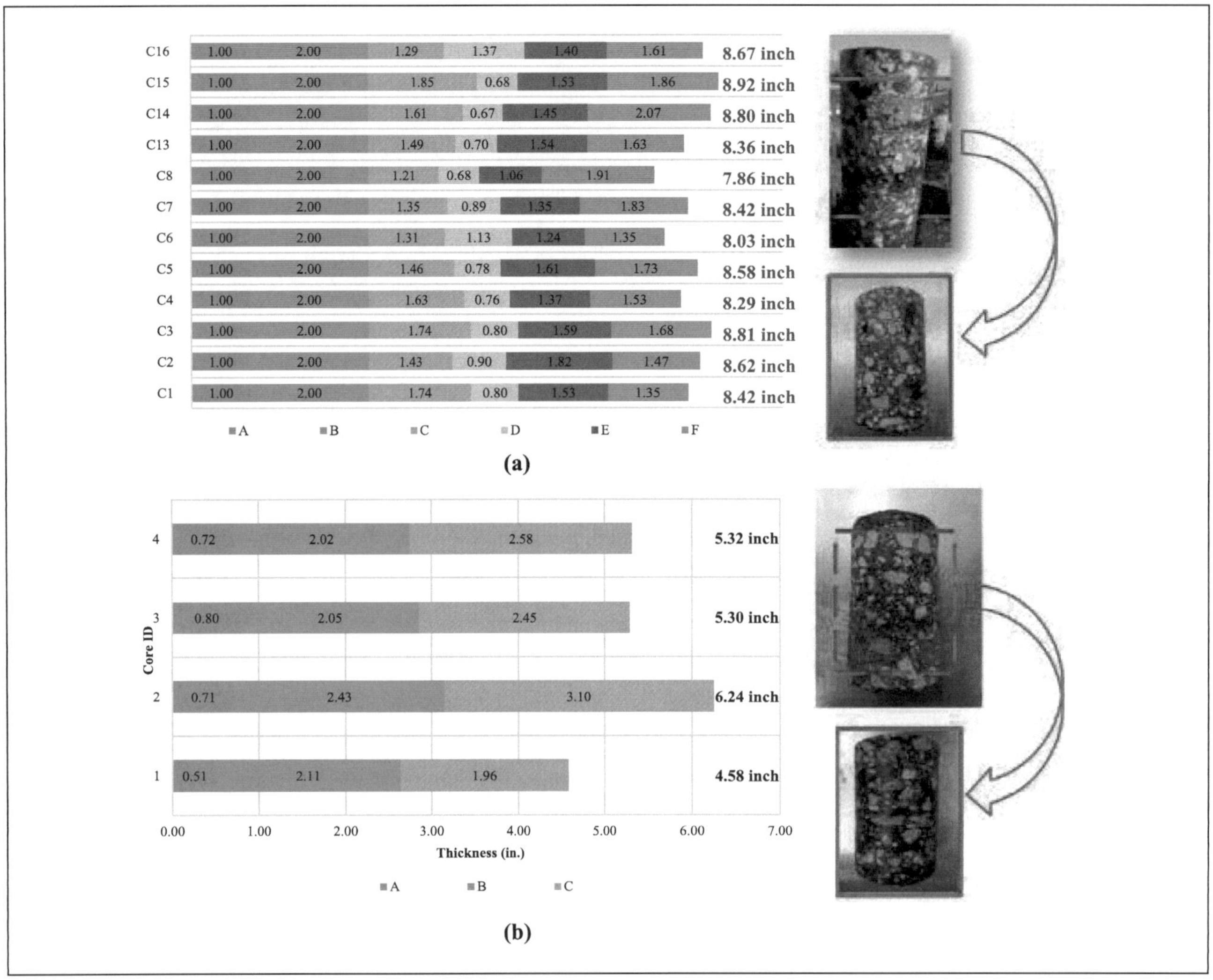

Figure 2. Field core samples and pre-existing AC sub-layers' thicknesses for (*a*) US95, and (*b*) US395.

asphalt pavement under various combinations of traffic loads, speed, and environmental conditions. For rehabilitation designs, the $|E^*|$ of the existing AC layer is adjusted to account for the load-induced bottom-up FC damage. For all three analysis levels (i.e., Level 1, 2, and 3) of AC overlay over AC pavements in Pavement ME Design software, the $|E^*|$ of the existing AC layer in its undamaged state is determined using the Witczak prediction model (Equation 1) which is mathematically modeled by a sigmoidal function (*4*) as

$$\log|E^*| = \delta + \frac{\alpha}{1 + e^{\beta + \gamma \log t_r}} \tag{2}$$

where $|E^*|$ is the dynamic modulus in ksi, δ, α, β, and γ are dynamic modulus master curve fitting parameters, and t_r is the reduced time determined function of the loading time t and the time-temperature shifting expressed in seconds.

The damage accumulation in the AC layer is then calculated or estimated over its service life depending on the selected level of analysis. For Level 1 analysis (see Figure 3*a*), the damaged dynamic modulus master curve is modeled through a vertical downward shift of the undamaged $|E^*|$ in such a way that it passes through the backcalculated modulus at the corresponding testing temperature and loading frequency. The damage d_{AC} is then quantified (*4*) by

$$E_{AC-damaged} = 10^{\delta} + \frac{E_{AC-undamaged} - 10^{\delta}}{1 + e^{-0.3 + 5\log d_{AC}}} \tag{3}$$

Meanwhile for Level 2 analysis, the damage accumulated in the AC layer is determined directly from the amount of bottom-up FC using the transfer function (*5*)

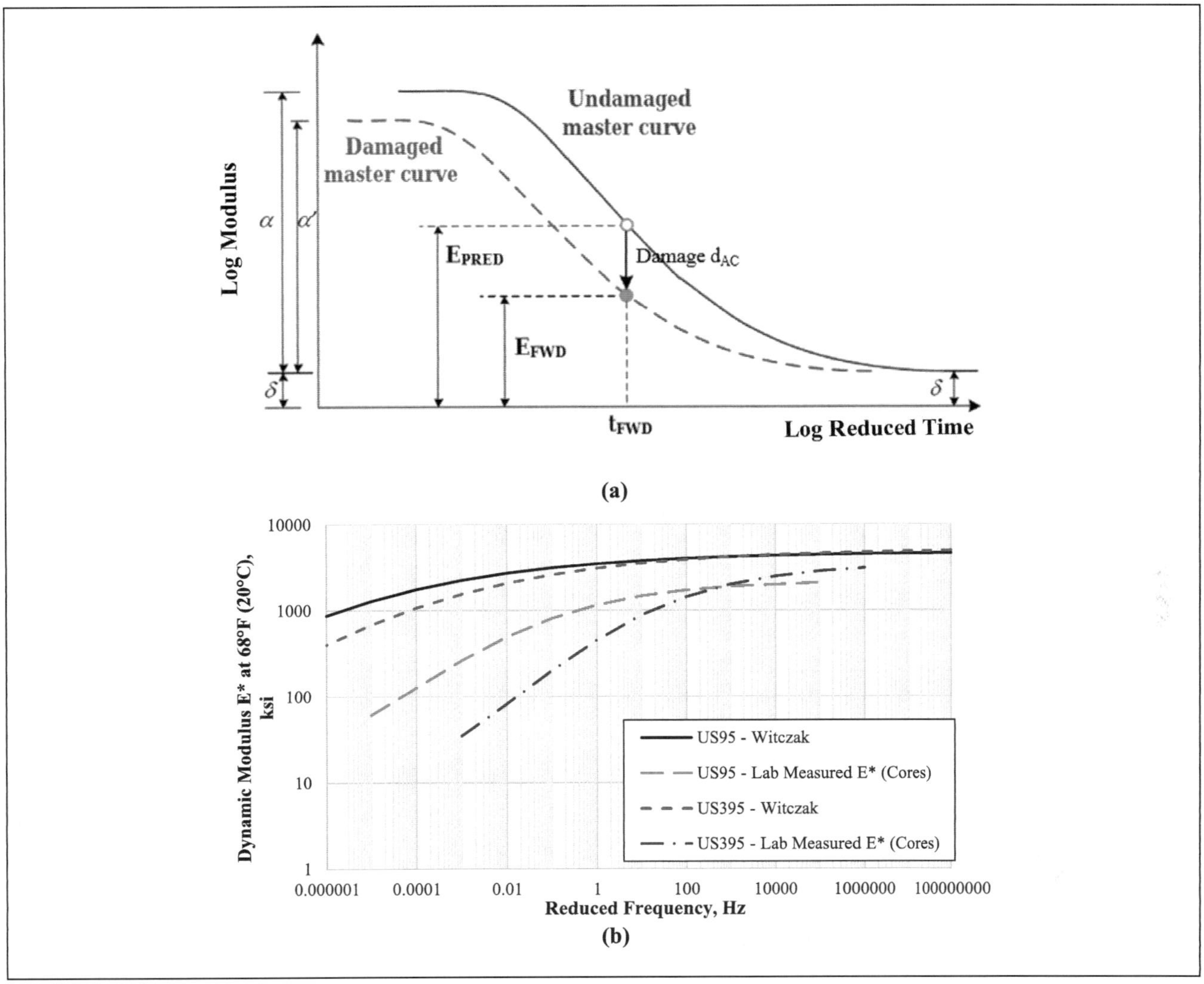

Figure 3. (*a*) MEPDG Level 1 approach for damage characterization, and (*b*) undamaged dynamic modulus master curves for existing AC layers at 68°F determined using Witczak prediction model and laboratory measured modulus on field cores.

$$FC_{\text{Bottom-up}} = \left(\frac{1}{60}\right)\frac{C_4}{1 + e^{C_1 C_1' + C_2 C_2' \log(d_{AC}*100)}} \quad (4)$$

where $FC_{\text{Bottom-up}}$ is the area of alligator cracking that initiates at the bottom of the existing AC layer in percent of total lane area, d_{AC} is the cumulative damage index at the bottom of the AC layer, and $C_{1,2,4}$ are locally calibrated regression constants ($C_1 = C_2 = 0.8$, and $C_4 = 6000$). Equation 5 and Equation 6 are used to calculate the C'_1 and C'_2 parameters, respectively.

$$C_1' = -2*C_2' \quad (5)$$

$$C_2' = -2.40874 - 39.748*(1 + H_{AC})^{-2.856} \quad (6)$$

where H_{AC} is the total thickness of the AC layer, in inches.

On the other hand, for Level 3 analysis, typical cumulative damage values are attributed to the existing AC layer based on pavement condition rating (i.e., excellent, good, fair, poor, and very poor) (*4*).

Undamaged Dynamic Modulus Master Curve of Existing AC Layer

As mentioned earlier, the Pavement ME Design software calculates an undamaged master curve of the existing AC layer using the Witczak prediction model. The basic inputs for this model are material properties and volumetrics. Figure 4 presents the experimental plan for evaluating the cores sampled from each of the rehabilitation field projects. The following properties were determined

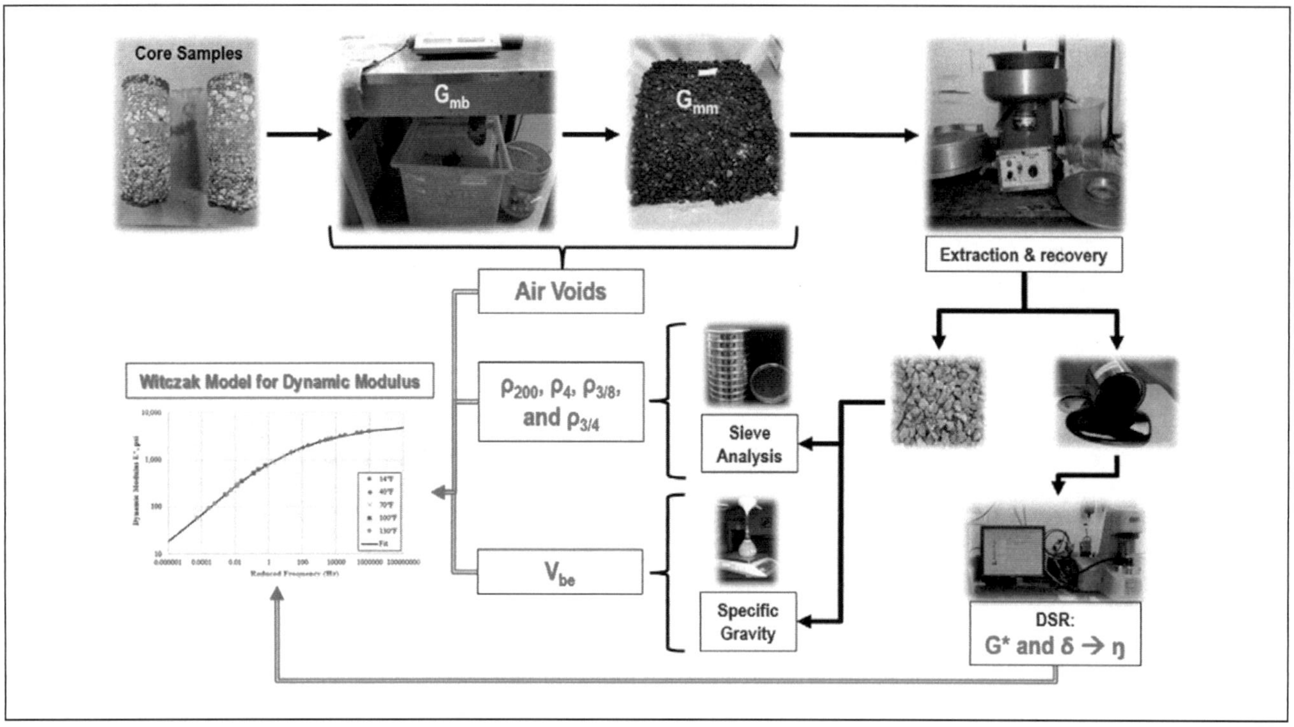

Figure 4. Flowchart summary of the proposed experimental work plan for evaluating sample cores.

for the core samples: asphalt binder content, viscosity-temperature susceptibility (A-VTS) of recovered asphalt binder, gradation of extracted aggregates, and in-place air voids (from measured bulk specific gravity G_{mb} and maximum theoretical specific gravity G_{mm} on core samples) (refer to Table 1). The calculated undamaged master curves of the existing AC layer on US95 and US395 are illustrated in Figure 3*b*. A reference temperature of 68°F (20°C) was selected representing an intermediate temperature for FC. A review of the Witczak data (i.e., $E_{AC\text{-}undamged}$) presented in Figure 3*b* reveals the following observations:

- The Witczak model for the existing polymer-modified AC layers on US95 and US395 resulted in a relatively high value for the undamaged dynamic modulus property [i.e., greater than 3,500 ksi (24,131 MPa) at 10 Hz]. Accordingly, the Witczak prediction model for predicted $|E^*|$ tends to overestimate the $E_{AC\text{-}undamaged}$ of used polymer-modified AC mixes.
- A higher $E_{AC\text{-}undamaged}$ master curve was observed for the existing AC layer from the US95 project when compared with the master curve determined for US395, as would be expected from the age and binder grade.

Estimation of FC Damage in Existing AC

The damage characterization of the existing AC layer remains a critical input for the design of an AC overlay over AC pavement. For a Level 1 analysis, this damage is calculated based on $E_{AC\text{-}undamaged}$ determined using the Witczak prediction model and $E_{AC\text{-}damaged}$ determined from FWD backcalculation at testing temperature and loading frequency (refer to Figure 5, *a* and *b*). Equation 3 is used to solve for d_{AC}. The structural condition of the existing AC pavement can be determined from the attributed damage value as follows: (*a*) Excellent for d_{AC} between 0.00 and 0.20, (*b*) Good for d_{AC} between 0.20 and 0.40, (*c*) Fair for d_{AC} between 0.40 and 0.80, (*d*) Poor for d_{AC} between 0.80 and 1.20, and (*e*) Very Poor for d_{AC} greater than 1.20. Table 2 summarizes the calculated damage values as well as their equivalent percentage of bottom-up FC using the corresponding locally calibrated transfer function. A damage value of 1.86 and 63% of equivalent FC were estimated for US95 rating the existing AC layer as very poor ($d_{AC} > 1.20$). However, the distress survey data before rehabilitation revealed low to moderate block cracking with no visible rutting greater than 0.1 in. (2.5 mm) throughout the entire project length indicating an overestimation of the pre-overlay damage using Level 1 analysis. On the other hand, a damage value of 2.41 and a 68.5% of equivalent FC were

Table 1. Summary of Input Parameters for Witczak Model and Traffic Data Inputs for Pavement ME Analysis

Parameters		US95 Las Vegas	US395 Carson City Freeway
Material properties and volumetric input parameters for Witczak prediction model			
Aggregate gradation	$\rho_{3/4}$	0.5	5.1
	$\rho_{3/8}$	20.6	31.8
	ρ_4	46.7	50.5
	ρ_{200}	7.6	5.4
Air Voids, V_a	G_{mb}	2.302	2.305
	G_{mm}	2.480	2.449
	V_a (%)	7.2	5.9
Effective Volume of Binder, V_{be}	% AC by TWM[a]	4.20	5.60
	G_{sb}[b]	2.497	2.573
	G_{se}[c]	2.647	2.673
	V_{be} (%)	11.7	13.8
Viscosity (cP)	at 40°F	3,532,921,903,310	251,658,729,260
	at 70°F	44,239,666,647	3,224,652,028
	at 100°F	1,532,528,698	90,990,809
	at 130°F	91,712,220	4,284,646
Traffic data inputs for Pavement ME analysis			
Two-directional average annual daily traffic (AADT)		9,120	8,080
Truck percent of ADT, %		15.9	19.6
Directional split, %		50	50
Truck factor		1.032	1.032
Number of lanes per direction		2	2
Lane distribution factor (80–100), %		90	90
Average annual growth rate, %		3.91	3.00
Operational speed, mph		60	60
FHWA Vehicle Classification	Class 4	4.59	16.13
	Class 5	11.73	40.63
	Class 6	1.79	5.40
	Class 7	0.01	0.01
	Class 8	8.29	12.04
	Class 9	57.53	19.51
	Class 10	1.40	0.12
	Class 11	2.55	3.85
	Class 12	1.40	0.65
	Class 13	10.71	1.66

[a]denotes total weight of mix; [b]denotes bulk specific gravity of the aggregate blend; [c]denotes bulk effective specific gravity of the aggregate blend.

observed for US395, also rating the existing AC layer as very poor ($d_{AC} > 1.20$). These observations support the overestimation of the pre-overlay damage when the Witczak prediction model for undamaged dynamic modulus along with the FWD damaged dynamic modulus are used in a Pavement ME Level 1 analysis.

Hybrid Approach for FC Damage Characterization

The characterization of FC damage for the existing AC layers using the Witczak prediction model in combination with FWD backcalculated modulus resulted in overestimated damage values indicating the need for a different approach referred to as hybrid approach. Accordingly, dynamic modulus testing using the AMPT was proposed on core samples to characterize the FC damage through a comparison with the FWD backcalculated modulus. The damage characterized using the hybrid approach is then compared with the damage estimated based on Level 1 analysis in Pavement ME Design software.

Laboratory Measured Dynamic Modulus Master Curve of Field Cores

The $|E^*|$ property of the sampled field cores was evaluated under various combinations of loading frequency and temperature in accordance with AASHTO TP 79 (*16*). Because the cores were sampled from between the wheelpaths for US95 and from the wheelpath for US395, their $|E^*|$ property will provide an indication of the undamaged and damaged modulus, respectively, as well as the overall quality of the existing AC layer. The test specimen consisted of a 4.0 in. (102 mm) diameter by 6.0 in. (153 mm) height sample cored from the center of a field core as shown in Figure 2, *a* and *b*. It should be mentioned that the cored specimens were trimmed from both sides (i.e., upper and lower part) to reach the target height, which was 6.0 in. (153 mm). Additionally, the visual inspection of the field cores and cored $|E^*|$ samples did not reveal any debonding between the AC layer lifts. Figure 3*b* shows the undamaged $|E^*|$ master curves at 68°F (20°C) for the US95 and U395 cores. A review of the $|E^*|$ data presented in Figure 3*b* reveals the following observations:

- The combination of aggregate source, asphalt binder grade and viscosity, and volumetrics had a significant impact on the magnitude of the laboratory measured $|E^*|$ property. Higher E^* values were observed for the US95 project when compared with the US395 regardless of the sampling location of the cores (between or from wheelpaths).
- The measured dynamic moduli on core samples from both rehabilitation field projects (i.e., US95 and US395) were much lower than the ones determined using the Witczak prediction model. The $E_{\text{AC-undamaged}}$ values determined using the Witczak prediction model at 68°F (20°C) and 10Hz were about 4.4 and 2.5 times higher than the measured dynamic modulus on cores from US95 and US395, respectively. This again indicates the overestimation of the Witczak prediction model for the dynamic modulus of polymer-modified AC mixes.

In summary, the noticeably higher values observed for the dynamic modulus estimated using the Witczak prediction model indicate that the ME approach for predicting undamaged dynamic modulus of an existing AC layer might not be appropriate for the rehabilitation designs of polymer-modified AC pavements in Nevada.

Laboratory Estimation of FC Damage in Existing AC

The pre-overlay damage of the existing AC layer is evaluated using the FWD backcalculated modulus and the measured dynamic modulus on core samples at the corresponding FWD testing temperature and loading frequency (Equation 3) (refer to Figure 5, *c* and *d*). Table 2 summarizes the calculated damage values as well as their equivalent percentage of bottom-up FC using the corresponding locally calibrated transfer function (*5*). A damage value of 0.585, 3.2 times lower than the one estimated following the Pavement ME Level 1 analysis, along with a respective FC of 39% were estimated for US95, thus indicating the existing AC layer as being in fair condition ($0.40 < d_{AC} < 0.80$). On the other hand, a damage value could not be calculated using the same approach for the US395 project because the laboratory measured $|E^*|$ and the FWD backcalculated modulus are similar and both represent a damaged state for the existing AC. This situation occurred because the cores were obtained from the wheelpath. However, this exercise proved the good correlations between FWD and laboratory testing in estimating the damaged dynamic modulus of the existing AC layer.

Pavement Rehabilitation Analysis

The purpose of this section is to conduct AC overlay analyses using the Pavement ME Design software (version 2.3.1) in accordance with the latest version of the NDOT design manual for new and rehabilitated flexible pavements (*5*). Three analyses were conducted for each rehabilitation project: (*a*) MEPDG approach (Level 1)—a Level 1 analysis that uses the Witczak prediction model in combination with backcalculated FWD modulus at respective pavement temperature and loading frequency; (*b*) hybrid approach—a Level 2 analysis that uses the Witczak prediction model in combination with the bottom-up FC for the damage of the existing AC layer estimated from the difference between the laboratory measured dynamic modulus on core samples and the FWD backcalculated modulus; and (*c*) MEPDG approach (Level 3)—a Level 3 approach that uses a Witczak model in combination with the pre-overlay damage estimated from the pavement condition rating.

It should be noted that the proposed hybrid approach stems from the following two limitations: (*a*) the Pavement ME Design software mandates the use of the Witczak prediction model to estimate the undamaged dynamic modulus at all three level of analyses (i.e., Level 1, 2 and 3), and (*b*) the regression constants for the Witczak prediction model cannot be modified in the current version of the Pavement ME Design software to reflect polymer-modified AC mixtures in the existing AC pavement.

Pavement ME Design Data Input

The main user data inputs needed for conducting an AC rehabilitation design using the Pavement ME Design software can be summarized as follows:

- General information: including the pavement type (in this case AC over AC), design life (20 years), existing and pavement construction, and traffic opening time.
- Performance criteria: the AC bottom-up FC (maximum limit of 15%), the permanent deformation

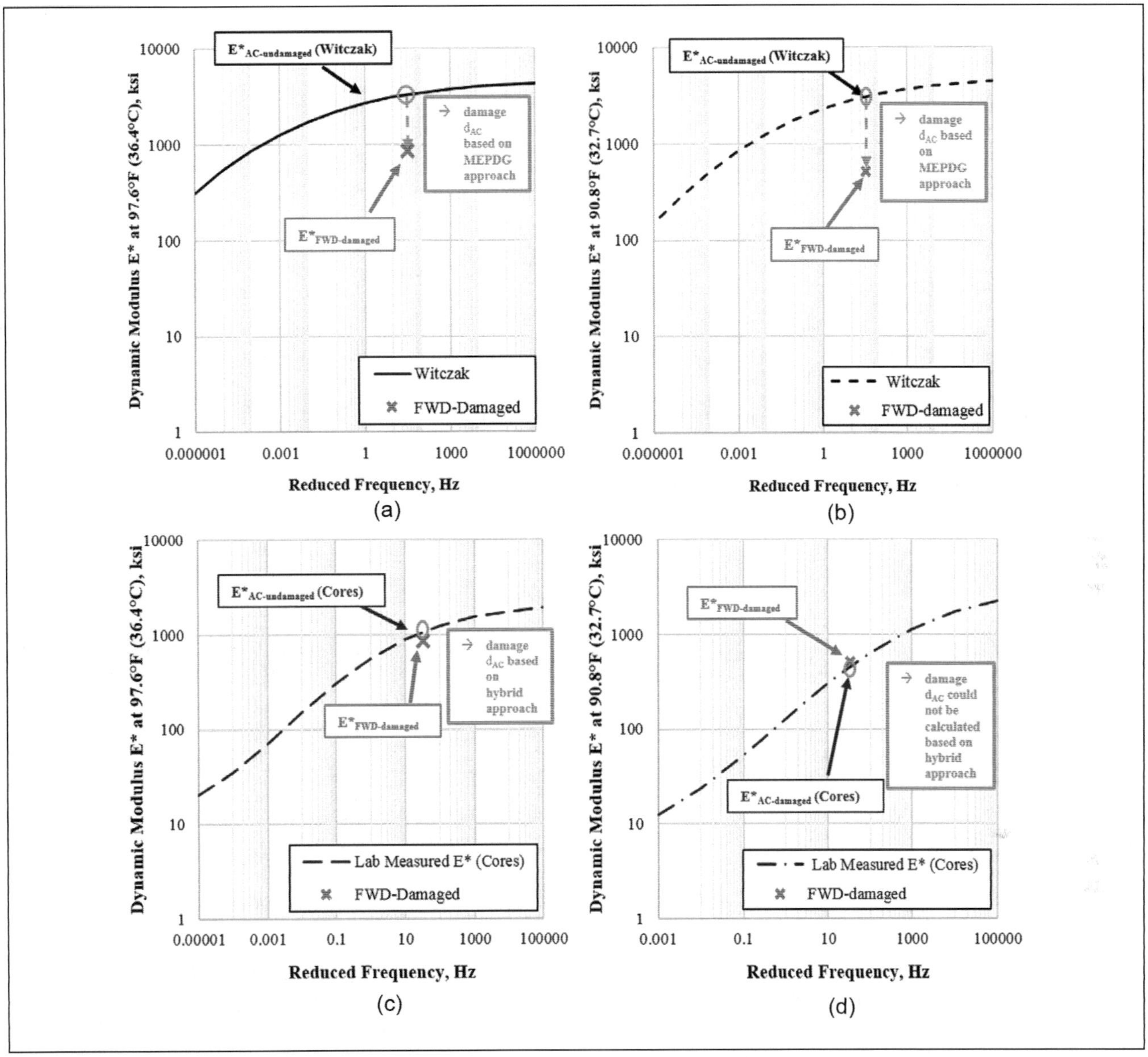

Figure 5. Pre-overlay damage characterization for AC layers based on: (*a*) MEPDG approach for US95, (*b*) MEPDG approach for US395, (*c*) hybrid approach for US95, and (*d*) hybrid approach for US395.

Table 2. Damage Characterization of Existing AC Layers on US95 and US395

Project ID	Pavement Temperature (°F)	Loading Frequency (Hz)	Approach	$E_{AC\text{-damaged}}$ (ksi)	$E_{AC\text{-undamaged}}$ (ksi)	Damage d_{AC}	Bottom-up FC (% of Total Lane Area)
US95	97.6	33.3	MEPDG (FWD & Witczak)	876.2	3,307.0	1.86	63.0
			Hybrid (FWD & measured *E** on core)	876.2	1,057.4	0.58	38.7
US395	90.8	33.3	MEPDG (FWD & Witczak)	517.9	3,100.0	2.41	68.5
			Hybrid (FWD & measured *E** on core)	517.9	458.2	NA[a]	NA

Note: [a]NA denotes not applicable.

Table 3. Summary Table of US95 and US395 Pavement ME Design Results

Project ID	Analysis Approach (Level)	Inputs	Milling (in.)	AC Overlay Thickness (in.)	AC Existing (in.)	Total Rutting (in.)	AC Rutting (in.)	AC Bottom-up FC (% Lane Area)
Target distress at specified reliability						0.50	0.15	15.00
US95	MEPDG (Level 1)	Witczak + FWD damage + FWD $E_{base\ \&\ subgrade}$	2.0	5.5	6.5	0.05	0.04	9.56
	Hybrid (Level 1 and Level 2)	Witczak + % FC + FWD $E_{base\ \&\ subgrade}$	2.0	3.0	6.5	0.06	0.05	0.45
	MEPDG (Level 3)	Witczak + pavement rating + FWD $E_{base\ \&\ subgrade}$	2.0	3.5	6.5	0.05	0.05	7.81
US395	MEPDG (Level 1)	Witczak + FWD damage + FWD $E_{base\ \&\ subgrade}$	1.0	10.0	5.4	0.08	0.00	12.40
	MEPDG (Level 3)	Witczak + pavement rating + FWD $E_{base\ \&\ subgrade}$	1.0	4.0	5.4	0.03	0.01	13.50

of the total pavement structure [maximum limit of 0.50 in. (12.5 mm)], and the permanent deformation of the AC layer [maximum limit of 0.15 in. (3.8 mm)] were the only ones considered in this study.

- Traffic: the specifics of the US95 and US395 projects were based on the 2015 traffic information summarized in Table 1.
- Climate: close proximity weather stations were selected for each rehabilitation project according to its location (i.e., longitude, latitude, elevation, depth of water table, and climate stations).
- Pavement structure definition and materials: these inputs were selected in accordance with the NDOT Design Manual for Flexible Pavements. The $E_{AC\text{-}damaged}$ determined from FWD testing and analysis in the previous section were also in the analysis.
- AC layer properties: mainly emphasizing the condition of existing AC flexible pavements. For the Level 1 analysis, the data from FWD analysis and Witczak prediction model are used to characterize the damage in the AC existing pavement. For the Level 2 analysis, the estimated percentage of alligator cracking on the existing AC pavement surface before the placement of AC overlay is entered and then internally converted into a damage value using a locally calibrated bottom-up FC transfer function.
- Calibration factors: The performance models and transfer functions from the Pavement ME Design for flexible pavement were calibrated to Nevada's local conditions for material, traffic, and climate (*5*). It should be mentioned that the calibration was limited to the AC bottom-up FC and permanent deformation (rutting).

Results and Discussions

Table 3 summarizes the results of the Pavement ME Design analyses completed for both rehabilitation projects (i.e., US95 and US395). These results lead to the following observations:

- The use of the Witczak prediction model in combination with the FWD damaged dynamic modulus in a Pavement ME Level 1 analysis resulted in a much thicker AC overlay for both rehabilitation projects to satisfy the target distresses and selected performance criteria (i.e., AC rutting, total rutting, and AC bottom-up FC).
- For US95, a thickness of 2 in. (51 mm) was milled from the existing pavement. The Pavement ME Level 1 analysis (i.e., MEPDG approach) resulted in a required AC overlay thickness of 5.5 in. (140 mm); meanwhile, the analysis following the proposed hybrid approach resulted in an AC overlay of 3.0 in. (76 mm). This is consistent with the estimated damage using the hybrid approach in comparison with the relatively high damage estimated following the MEPDG approach for the existing polymer-modified asphalt layers. Additionally, a Level 3 analysis was performed based on the pavement condition rating and resulted in an AC

overlay thickness of 3.5 in. (89 mm). It should be mentioned that similar AC rutting and total rutting performance were observed for all three analyses. However, the lowest AC percent of bottom-up FC was observed for the hybrid approach analysis.

- For US395, a thickness of 1 in. (51 mm) was milled from the existing pavement. The Pavement ME Design Level 1 analysis resulted in a required AC overlay thickness of 10.0 in. (254 mm) to satisfy the target distresses and selected performance criteria. The analysis based on the hybrid approach could not be conducted for US395 because both FWD testing and core samples were collected from the wheelpath and a corresponding damage value (d_{AC}) could not be estimated. Therefore, a Level 3 analysis was conducted based on the pavement condition rating and resulted in a significantly thinner AC overlay of 4.0 in. (102 mm) supporting again the idea of pre-overlay damage overestimation when Witczak prediction model is used.

In summary, the Pavement ME analysis using Level 1 inputs for rehabilitation design of polymer-modified AC layers resulted in a thick and unrealistic estimation for the AC overlay. This has been attributed to the high value for the characterized FC damage of the existing AC layer. On the other hand, the Pavement ME Design analysis using the hybrid approach that is a combination of Level 1 and Level 2 data inputs resulted in more realistic AC overlay thicknesses.

Summary and Recommendations

The overall objective of this study was to assess the use of Level 1 analysis for Pavement ME Design rehabilitation of deteriorated polymer-modified AC pavements in Nevada. Additionally, this research explored the possible implementation of a hybrid approach for AC damage characterization to overcome the challenges of using the Witczak prediction model for the estimation of the undamaged dynamic modulus master curve of the existing polymer-modified AC layer. Two rehabilitation field projects were used as part of this effort: (*a*) US95 Las Vegas in Southern Nevada (District I), and (*b*) US395 Carson City Freeway in Northern Nevada (District II). The work plan included the use of FWD testing in the AASHTOWare® Pavement ME Design software for AC overlays over AC pavements and involved the following major steps:

- Collection of FWD measurements in right wheelpath before rehabilitation.
- Collection of core samples from between wheelpaths on the same lane where FWD testing was conducted.
- Estimation of equivalent undamaged dynamic modulus, $E_{AC\text{-undamaged}}$, using the Witczak prediction model and core properties which resulted in an overestimation of the dynamic modulus of polymer-modified asphalt mixtures.
- Estimation of equivalent damaged dynamic modulus, $E_{AC\text{-damaged}}$, using FWD backcalculation modulus.

The proposed hybrid approach consisted of a combination of both Level 1 and Level 2 analyses that fits within the current limitations of the Pavement ME Design software. This approach consisted of the following additional steps:

- Conducting of laboratory dynamic modulus testing on the collected core samples and determination of corresponding undamaged dynamic modulus master curves.
- Estimation of an equivalent undamaged dynamic modulus at the same temperature and FWD loading frequency using the laboratory developed master curve and backcalculated modulus.

The Pavement ME Design analysis for the two rehabilitation projects conducted using the Witczak prediction model for undamaged dynamic modulus and FWD damaged dynamic modulus resulted in a higher damage characterization for the existing AC layers and led to the need for a significantly thicker AC overlay to satisfy the target distresses and selected performance criteria. On the other hand, the Pavement ME Design analysis conducted using the measured undamaged dynamic modulus on core samples combined with the damaged dynamic modulus from FWD backcalculation showed reasonable results. The implementation of the hybrid approach in the Pavement ME Design software requires the use of a combination of Level 1 and Level 2 data inputs.

In summary, the proposed hybrid approach has been proposed to NDOT when conducting rehabilitation designs using the Pavement ME Design software. This approach is recommended because: (*a*) the Witczak model is mandated for the estimation of undamaged dynamic modulus, and (*b*) the regression constants for the Witczak prediction model cannot be modified in the current version of the Pavement ME Design software. The hybrid approach consists of the following eight steps:

- Step 1—Conduct FWD testing in the right wheelpath before rehabilitation.

- Step 2—Backcalculate the damaged dynamic modulus of the existing AC layer(s).
- Step 3—Collect core samples from between the wheelpaths to reflect the undamaged existing AC layer.
- Step 4—Measure the asphalt binder content, A-VTS, aggregate gradation, and in-place air voids for collected core samples.
- Step 5—Measure the equivalent undamaged dynamic modulus on collected core samples following AASHTO TP79.
- Step 6—Characterize bottom-up FC damage by comparing the damaged (i.e., backcalculated modulus from Step 2) to the undamaged (i.e., measured on core samples in Step 5) dynamic modulus using the approach described in MEPDG.
- Step 7—Calculate the bottom-up FC for the determined damage in Step 6 using the locally calibrated transfer function.
- Step 8—Run the Pavement ME Design software to determine the AC overlay thickness according to Level 2 inputs for the existing AC layer(s) using the core properties measured in Step 4 as well as the bottom-up FC calculated in Step 6. A Level 1 input can still be used for the new AC overlay mixture.

To overcome the issues described in this study, it is also suggested that the Pavement ME Design software be modified to allow for one or both of the following two options:

- Provide the user with the option to input the laboratory measured dynamic modulus data on core samples for the estimation of undamaged dynamic modulus of an existing AC layer.
- Provide the user with the capability to input locally calibrated regression constants for the Witczak prediction model.

Acknowledgments

This study was sponsored by Nevada Department of Transportation (NDOT) and the Western Regional Superpave Center (WRSC) at the University of Nevada, Reno (UNR).

Author Contributions

The authors confirm contribution to the paper as follows: study conception and design: Jhony Habbouche (UNR) and Elie Y. Hajj (UNR); data collection: Nathan E. Morian (Nevada DOT); analysis and interpretation of results: Jhony Habbouche (UNR), Elie Y. Hajj (UNR), and Peter E. Sebaaly (UNR); draft manuscript preparation: Jhony Habbouche (UNR), Elie Y. Hajj (UNR), Peter E. Sebaaly (UNR), and Nathan E. Morian (Nevada DOT). All authors reviewed the results and approved the final version of the manuscript.

References

1. Newcomb, D. E. *Thin Asphalt Overlays for Pavement Preservation.* Information Series 135. National Asphalt Pavement Association, Lanham, MD, 2009.
2. Hajj, E. Y., P. E. Sebaaly, and J. P. C. Azofeifa. *Reflective Cracking of Flexible Pavements, Phase III: Laboratory Evaluation.* Research Report No. 13KJ-2. Nevada Department of Transportation, University of Nevada, Reno, 2011.
3. Nevada Department of Transportation. *State Highway Preservation Program.* Report to the 2015 Legislature, NDOT, 2015.
4. Applied Research Associates, Inc., *Guide for Mechanistic-Empirical Design of New & Rehabilitated Pavement Structures.* Final Report NCHRP 1-37A. National Cooperative Highway Research Program, Transportation Research Board, National Research Council, Champaign, Illinois, March, 2004.
5. Hajj, E.Y., P. E. Sebaaly, and P. Nabhan. *Manual for Designing Flexible Pavements in Nevada Using AASHTO-Ware® Pavement-ME Design.* Nevada Department of Transportation, Western Regional Superpave Center (WRSC), University of Nevada, Reno, 2015.
6. Habbouche, J., P. E. Sebaaly, and E. Y. Hajj. *Structural Coefficient for High Polymer Modified Asphalt Mixes in Florida, Task 1: Literature Review.* Project BE321. Florida Department of Transportation, University of Nevada Reno, Western Regional Superpave Center, 2017.
7. Dongré, R., L. Myers, J. D'Angelo, C. Paugh, and J. Gudimettla. Field Evaluation of Witczak and Hirsch Models for Predicting Dynamic Modulus of Hot-Mix Asphalt (with Discussion). *Journal of the Association of Asphalt Paving Technologists*, Vol. 74, 2005, pp. 381–442.
8. Ceylan, H., C. Schwartz, S. Kim, and K. Gopalakrishnan. Accuracy of Predictive Models for Dynamic Modulus of Hot-Mix Asphalt. *Journal of Materials in Civil Engineering*, Vol. 21, No. 6, 2009, pp. 286–293.
9. Singh, D., M. Zaman, and S. Commuri, Evaluation of Predictive Models for Estimating Dynamic Modulus of Hot-Mix Asphalt in Oklahoma. *Transportation Research Record: Journal of the Transportation Research Board,* 2011. 2210: 57–72.
10. Yousefdoost, S., B. Vuong, I. Rickards, P. Armstrong, and B. Sullivan. Evaluation of Dynamic Modulus Predictive Models for Typical Australian Asphalt Mixes. *Proc., 15th International Flexible Pavements Conference, Australian Asphalt Pavement Association*, Brisbane, Australia, 22–25 September 2013.
11. Al-Khateeb, G., A. Shenoy, N. Gibson, and T. Harman. *A New Simplistic Model for Dynamic Modulus Predictions of Asphalt Paving Mixtures.* Association of Asphalt Paving Technologists Annual Meeting, Savannah, GA, 2006.
12. Bari, J., and M. W. Witczak. Development of a New Revised Version of the Witczak E* Predictive Model for Hot Mix Asphalt Mixtures (with Discussion). *Journal of*

the Association of Asphalt Paving Technologists, Vol. 75, 2006, pp. 381–423.
13. Tedford, D. *Asphalt Binder*. Materials Re Section 703—Bituminous Materials. Nevada Department of Transportation, Nevada, 2015.
14. Liu, W., and T. Scullion. *Modulus 6.0 for Windows: User's Manual*. FHWA/TX-05/0-1869-2. Texas Transportation Institute, The Texas A&M University System, Federal Highway Administration, Texas, 2001.
15. Lukanen, E. O., R. Stubstad, and R. Briggs. *Temperature Predictions and Adjustment Factors for Asphalt Pavement*. FHWA-RD-98-085. Braun Intertec Corporation, Federal Highway Administration, Minneapolis, MN, 2000.
16. American Association of State Highway and Transportation Officials. *Determining the Dynamic Modulus and Flow Number for Hot Mix Asphalt (HMA) Using the Asphalt Mixture Performance Tester (AMPT)*. AASHTO TP79. American Association of State Highway and Transportation Officials, Washington, D.C., 2013.

The Standing Committee on Design and Rehabilitation of Asphalt Pavements (AFD60) peer-reviewed this paper (18-03000).

The contents of this study reflect the views of the authors, who are responsible for the facts and the accuracy of the data presented here. The contents do not necessarily reflect the official views or policies of any of the sponsors at the time of publication.

Article

TRR
JOURNAL OF THE TRANSPORTATION RESEARCH BOARD

Transportation Research Record
2018, Vol. 2672(40) 242–253

DOI: 10.1177/0361198118798724
journals.sagepub.com/home/trr
SAGE

Comparing the Bonded Concrete Overlays of Asphalt-Mechanistic Empirical Design Procedure and the Short Jointed Plain Concrete Pavement Module in the Pavement Mechanistic Empirical Design Procedure

Kevin Alland[1], Julie M. Vandenbossche[1], John W. DeSantis[1], Mark B. Snyder[2], and Lev Khazanovich[1]

Abstract

Bonded concrete overlays of asphalt pavements (BCOA) consist of a concrete overlay placed on an existing asphalt or composite pavement. This technique is intended as a cost-effective rehabilitation solution for marginally distressed in-service asphalt or composite pavements. BCOA with panel sizes between 4.5 ft and 8.5 ft have become popular as they reduce curling stresses while keeping the longitudinal joints out of the wheelpath. The BCOA-ME (mechanistic empirical) design procedure and Pavement ME short jointed plain concrete pavement (SJPCP) module can both be used to design BCOA with mid-size panels. However, these design procedures differ in the assumptions used to develop the mechanistic computational model, fatigue models used to predict failure, treatment of environmental conditions, estimate of asphalt stiffness, consideration of structural fibers, the application of traffic loading, and the calibration process. This results in the procedures producing different overlay thicknesses and predicted distresses. The strengths and limitations of each procedure are evaluated and comparisons are made between the design thicknesses obtained from them.

Bonded concrete overlays of asphalt pavements (BCOA) consist of a concrete overlay placed on an existing asphalt or composite pavement. The existing asphalt is often milled prior to the overlay to remove any surface distress and to increase the surface area to promote a greater bond between the Portland cement concrete (PCC) and asphalt. This technique is intended as a cost-effective rehabilitation solution for marginally distressed in-service asphalt or composite pavements. BCOA typically have smaller panel sizes than traditional jointed plain concrete pavements (JPCP) to reduce curling and warping stresses and bending stresses produced by applied loads (*1–3*).

Several procedures have been developed for designing BCOA structures. Pioneering work in this area include BCOA design procedures developed by the Portland Cement Association (PCA), American Concrete Pavement Association (ACPA), and the Colorado Department of Transportation (Colorado DOT) (*4, 5*). However, these procedures were developed before long-term performance data was available for these structures, and therefore were primarily mechanistically based on projected failure modes. As long-term performance data became available, the failure mechanisms for these structures were re-evaluated. The newly defined failure modes, along with the performance data available for calibration were used to develop a mechanistic-empirical (ME) design procedure, BCOA-ME, under the FWHA-sponsored Pooled Fund Study TPF-5-165 (*6*). Some of the failure modes within the BCOA-ME design procedure were then incorporated into the American Association of State Highway Transportation Officials (AASHTO) Pavement ME design procedure in the short jointed plain concrete pavement (SJPCP) module (*7*).

[1]University of Pittsburgh, Pittsburgh, PA
[2]Pavement Engineering and Research Consultants, LLC, Bridgeville, PA

Corresponding Author:
Address correspondence to Julie M. Vandenbossche: jmv7@pitt.edu

For BCOA structures, the critical failure mode, which should be considered in design, varies based on the panel size (*8*). For BCOAs with panel widths less than or equal to 4.5 ft, the primary failure mode is corner breaks. With a longitudinal joint spacing less than 4.5 ft, the wheelpath tends to fall on or near a longitudinal joint, increasing the stress in the corner of the slab. A stress prediction model for corner breaks in BCOA was developed by the PCA (*3*). This stress prediction model was incorporated in the BCOA-ME design procedure and was calibrated for overlay thicknesses between 3 and 6.5 in. These panels sizes are more commonly used on low-volume roads or intersections and were therefore not considered in the Pavement ME for the SJPCP module.

For BCOAs with panel widths between 4.5 and 8.5 ft, the primary failure mode is longitudinal cracking in the wheelpath, initiating at the transverse joint (*9*). This failure mode is considered in both the BCOA-ME design procedure and the SJPCP module in Pavement ME. The BCOA-ME design procedure considers overlay thicknesses between 3 and 6.5 in and recommends that overlays 6.5 in or greater be designed as a new concrete pavement on an asphalt stabilized base. The Pavement ME for SJPCP considers overlay thicknesses between 4 and 8 in.

For BCOA with panels the full width of the lane, the primary failure mode is transverse cracking at mid-slab, similar to the failure mode in traditional JPCPs. Longitudinal cracks can also develop at mid-lane. However, these cracks are not in the wheelpath and therefore were shown at Minnesota Road Research Facility (MnROAD) not to affect the serviceability of the road (*10*). The BCOA-ME design procedure uses the stress equation developed by Colorado DOT to design these structures for overlay thicknesses between 4 and 6.5 in. Concrete overlays with full lane width panels greater than 6.5-in. thick act as new JPCP on an asphalt base, and can be designed using traditional pavement design procedures, such as the JPCP module in Pavement ME (*6*).

BCOA with panel widths between 4.5 and 8.5 ft have become popular, as they reduce curling stresses, while keeping the longitudinal joints out of the wheelpaths. The BCOA-ME design procedure and Pavement ME SJPCP module can both be used to design BCOA with panels in this size range. These two procedures are intended to be equivalent, and were developed using similar calibration databases. However, there are some design assumption differences between the two procedures. This results in the development of different overlay thicknesses and predicted distresses when using the same inputs. The differences between the design procedures are explored below by comparing the theory used to develop each procedure, analyzing the sensitivity of each procedure to the input values, and evaluating the performance of each procedure using three design examples.

Comparison of Theory

The goal of the Pavement ME SJPCP module is to implement the theories and assumptions used in the BCOA-ME design procedure in the computational framework of the Pavement ME software. However, there are several differences in the theory used to develop the design procedures. These include differences in:

- Computational models
- Fatigue models
- Treatment of environmental conditions
- Estimation of asphalt stiffness
- Consideration of structural fibers
- Application of traffic loading
- Calibration

Computational Model

Both the BCOA-ME design procedure and the Pavement ME SJPCP design module use a computational model to estimate stresses in the pavement. For panels with slab lengths between 4.5 and 6.5 ft, both procedures consider transverse tensile stress at the bottom of the slab near the intersection of the transverse joint and the wheelpath to be the critical stress. However, the software used and boundary conditions assumed vary between design procedures. The BCOA-ME design procedure computational model was developed in the general purpose 3-D finite element method (FEM) software, Abaqus. In the BCOA-ME computational model, the asphalt layer is treated as an independent layer bonded to the overlay layer. The overlay layer is discontinuous at the joints, whereas the asphalt is continuous through the joints. The BCOA-ME computational model allows moment transfer through the joint when this load is applied, whereas the joint will act as a hinge in the model incorporated into the SJPCP module. This moment transfer will influence the deflection basin, and ultimately induce stress when a wheel load is applied. Continuity in the asphalt layer will also lead to additional restraint to slab deformation. This will lead to higher stresses due to an applied temperature gradient, but may also prevent the asphalt layer beneath the joint from losing contact with the base or subgrade layer.

The SJPCP module computational model was developed in ISLAB, a pavement-specific "2.5-D" FEM software based on plate theory (*11*). The SJPCP module computational model treats the asphalt and overlay layers as a single equivalent "plate" with a flexural stiffness equal to the composite stiffness of the two layers, assuming full bonding. All load transfer at the joints is provided by shear springs, which provide a load transfer efficiency (LTE) of 80%, with no continuity assumed in the asphalt.

The importance of the BCOA joint behavior for the overlay performance has been observed in prior field studies (*12*). DeSantis et al. have shown that depth of joint activation, through just the overlay or through both the overlay and asphalt layer, depends on the flexural stiffness ratio (FSR) and panel sizes (*13*). The FSR is the ratio of the flexural stiffness of the concrete, to the flexural stiffness of the asphalt. When the FSR is greater than 0.9, the joint likely activates completely through the asphalt. When the FSR is less than 0.4, joints likely only activate through the concrete. Structures with a FSR between these values can have both depths of activation along the same overlay. When a joint activates through the asphalt, preventing moment transfer, the LTE calculated from falling weight deflectometer (FWD) measurements is often significantly lower than the 80%, which is the default in the SJPCP module (*13*).

The SJPCP module allows lower values of LTE to be input but was calibrated assuming a LTE of 80%. The SJPCP module considers the presence of a tied shoulder, which is not considered in the BCOA-ME design procedure. A tied shoulder in the SJPCP module was found to increase longitudinal cracking, likely due to increased restraint to slab curvature. Data is not available to determine whether this trend reflects the actual performance of in-service pavements.

Fatigue Prediction

The relationship between the fatigue stress and the allowable wheel loads, also varies between the design procedures. The BCOA-ME procedure incorporates the ACPA fatigue equation (Equation 1) to calculate the allowable loads as a function of the applied stress at a given level of reliability (*14*):

$$Log(N_f) = \left[\frac{-SR^{-10.24}*log(R)}{0.0112}\right]^{0.217} \quad (1)$$

where

N_f is the allowable number of applied loads at a given set of conditions;

SR is the stress ratio, defined as the ratio between the applied stress and the modulus of rupture (MOR) of the concrete overlay; and

R is the prescribed reliability level.

In the SJPCP module, fatigue equations are used to calculate damage, which a transfer function then uses to calculate the predicted fatigue cracking. The fatigue equation used in the SJPCP module estimates the allowable load applications under a given set of design conditions using Equation 2:

$$Log(N_f) = 2.0*SR^{-1.22} - 0.4371 \quad (2)$$

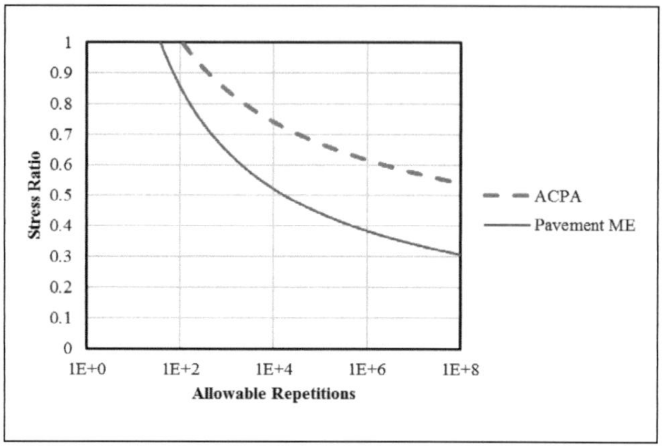

Figure 1. Comparison between ACPA and Pavement ME fatigue equations.

Reliability is not considered in the SJPCP module fatigue equation. Reliability is instead applied to the predicted distress. The allowable loads at a given stress ratio for both fatigue models can be seen in Figure 1. A reliability level of 50% was used for the ACPA fatigue equation in Figure 1. As can be seen, there is a significant difference between the fatigue equations used in the design procedures. The ACPA fatigue equation curve is considerably steeper than the one used in the SJPCP module. This can lead to a large difference in calculated damage.

Reliability in the SJPCP module is accounted for based on calculating the standard error of the predicted cracking after the transfer functions have been applied to estimate distress from damage in a manner similar to the reliability analysis for other types of rigid pavements and overlays (*15*). The BCOA-ME procedure accounts for the reliability in the fatigue damage calculation.

In spite of the seemingly substantial differences in the fatigue damage and reliability analysis, the empirical part of the design procedures may be able to reconcile the difference in the fatigue model, as the procedures are calibrated over similar calibration databases.

Environmental Conditions

The two design procedures also consider environmental conditions differently. The BCOA-ME procedure uses the concept of strain based effective equivalent linear temperature gradient (EELTG) and effective elastic modulus of the asphalt layer (Ehma) to account for environmental effects on the slab curvature and the stiffness of the asphalt respectively (*16, 17*). This concept involves determining a single value for EELTG and monthly values of Ehma, which produces equivalent damage to an hourly analysis. To estimate the values of the equivalent parameters, virtual BCOA structures were analyzed at

198 weather stations using the nonlinear temperature profile in the slab and asphalt temperature calculated using the enhanced integrated climatic model (EICM) at hourly increments. The EELTG and monthly Ehma values were calculated for each virtual BCOA structure based on equivalent damage. The dataset of calculated EELTG and Ehma for the virtual structures were used to develop predictive equations to estimate the EELTG and Ehma as a function of the section location, elevation and sunshine zone, and temperature zone. This eliminates the need to perform EICM analyses during each analysis increment in the design process. This analysis does not consider the effect of additional stresses due to the nonlinearity of the temperature gradient. However, these stresses were found to be negligible for BCOA structures (*17*).

The SJPCP module uses the EICM functionality built into Pavement ME to develop hourly estimates of the ELTG, stresses due to the nonlinearity of the temperature gradient, and mid-depth asphalt temperature based on data from one or more weather stations. This may reduce some variability in predicting the EELTG and Ehma, but increases the computational time required to run a design.

In the BCOA-ME design procedure, the stresses caused by traffic loading and the stress caused by slab curvature are calculated in two separate steps and added together. This assumes no interaction between slab curvature and the stress generated by traffic loading. The interaction between the wheel load generated stress, and the temperature gradient generated stress was found to be negligible due to the short slab lengths (*18*). A preliminary analysis also found that the coefficient of thermal expansion (CTE) of the concrete did not significantly affect the stress due to the temperature gradient within the range of CTEs typically assumed for concrete; again, this is due to the short slab lengths. Therefore, the CTE was not considered in the analysis (*18*).

The SJPCP module uses a neural network to predict the total stress due to slab curvature and wheel loads from a single analysis. This technique allows interaction between the slab curvature and the stress generated from traffic loading. The CTE was considered in this neural network as part of Korenov's non-dimensional temperature gradient (*19*). It should be noted that the structural module used in the SJPCP module does not allow separation between the overlay and asphalt, but does allow separation between the asphalt layer and subgrade.

Asphalt Stiffness

The two procedures also estimate the stiffness of the asphalt differently. In the BCOA-ME design procedure, the location of the project is used to select a typical binder grade using the LTPP-Bind software (*20*). A mastercurve for undamaged asphalt is then estimated using the Witczak equation, assuming a typical aggregate gradation and binder content (*21*). In the BCOA-ME procedure, the asphalt stiffness is decreased to account for fatigue damage in the asphalt layer based on the condition of the existing asphalt pavement. For an existing asphalt pavement in "adequate" condition (i.e., 0% to 8% fatigue cracking), a damage factor of 0.4 was assumed based on the fatigue cracking transfer function in the Pavement ME module for designing new asphalt pavements. For an existing asphalt pavement in "marginal" condition, (i.e., 8% to 15% fatigue cracking), a damage factor of 0.6 was assumed based on the fatigue cracking transfer function (*22*). The stiffness reduction formula from the Pavement ME module for designing new asphalt pavements was used to estimate the stiffness reduction based on the damage. This results in a stiffness reduction of 10% for pavements in adequate condition, and 20% for pavements in marginal condition. The effect of partial debonding between the asphalt and overlay is accounted for in the BCOA-ME design procedure using a stress adjustment factor. This stress adjustment factor is a function of the thickness of the existing asphalt and the thickness of the overlay.

The mastercurve for undamaged asphalt can be estimated using the Witczak equation in the SJPCP module. The binder grade is selected manually. The SJPCP module also allows the selection of the aggregate gradation and binder content. The asphalt stiffness reduction in the SJPCP module was not based on the condition of the existing asphalt pavement, but rather was used as a calibration factor. The SJPCP module does not allow the designer to account for the condition of the existing asphalt pavement. This is discussed further under the calibration section.

Stiffness of Supporting Layers

The stiffness of the layers below the asphalt is also treated differently in each of the design procedures. In the computational models, all layers beneath the asphalt are treated as a dense liquid foundation, with a stiffness defined by the composite modulus of subgrade reaction (k-value). The BCOA-ME design procedure requires the k-value to be provided by the user. This value can also be estimated using the ACPA k-value calculator (*23*). The k-value in the BCOA-ME design procedure is assumed to remain constant throughout the year.

The SJPCP module allows the stiffness of each layer to be entered as an elastic modulus for stabilized layers or a resilient modulus for unstabilized layers. The program uses layered elastic analysis to determine the k-value that produces the same response as the layered elastic structure when a virtual FWD load is applied

(*22*). The *k*-value in the Pavement ME design procedure changes throughout the year based on the temperature of asphalt stabilized layers, and the moisture content of unstabilized layers (*22*).

Treatment of Structural Fibers

In the BCOA-ME design procedure, structural (macro) fibers are accounted for by increasing the effective MOR, using the procedure developed by the Illinois Center for Transportation, to account for the additional energy required to propagate a fatigue crack (*24*). This procedure increases the effective MOR based on the type of fiber used and the volume fraction of fiber. The SJPCP module does not consider the potential performance benefit of structural fibers.

Traffic

The procedure for accounting for traffic loading varies between the design procedures. The mechanistic analysis in the BCOA-ME design procedure is based on equivalent single axle loads (ESALs), similar to the AASHTO 1993 design guide (*25*). A tool is provided as part of the procedure to convert a traffic load spectrum into ESALs. The SJPCP module does not convert the load spectrum into ESALs. Instead, damage is accumulated for all axle configurations and weights in the load spectrum. A typical load spectrum was used in calibration.

Calibration

The two design procedures were calibrated differently. The sections used to calibrate the BCOA-ME were primarily a subset of the sections used for calibrating the SJPCP module. The BCOA-ME calibration database included 12 sections with mid-size panels and the SJPCP module calibration included 30 sections with mid-size panels. The calibration database for both design procedures was limited in that the calibration section with the maximum number of ESALs was 11 million.

Several calibration sections were included in the SJPCP module calibration database that were not included in the BCOA-ME database. They include the following:

- The driving and passing lanes of MnROAD cells 114 to 914. These cells have very short section lengths, all less than 150 ft, and some less than 50 ft, which can lead to high variability in the observed cracking so they were not included in the calibration database for the BCOA-ME.
- MnROAD cells 160 and 162. These cells had not been constructed when BCOA-ME was calibrated. These sections were constructed in 2013, and have not developed distress as of 2017.
- MnROAD cells 61 and 63. These cells have unsealed joints, and the distress observed on these sections appears to be influenced by the lack of joint sealant, so they were not included in the calibration database for the BCOA-ME.

There are also some calibration sections that were included in the BCOA-ME calibration database but not the SJPCP module calibration database:

- MnROAD Cell 95. This overlay is less than 3-in. thick, and is therefore outside of the inference space for the SJPCP module. It also contains structural fibers.
- US-85 in Colorado.

Both design procedures utilized the passing lane of CO 119 in the calibration database. Only 0.5% of the slabs exhibited cracking in 2004 (7 years in service), the last data point used for BCOA-ME calibration. Additional data from this section was available for the calibration of the SJPCP module, with up to 17% longitudinal cracking observed in 2016 (19 years in service). However, further investigation of this distress showed the longitudinal cracking was influenced by transverse cracks that developed over the tie bars that were too close to the pavement surface.

For calibration of the BCOA-ME design procedure, the *k*-value at MnROAD was defined based on FWD deflection data. The *k*-value for the Colorado sections was defined in the research performed by Colorado DOT (*26*). The *k*-value for other sections was estimated using the ACPA *k*-value calculator. The stiffness of the layers beneath the asphalt layer used in calibrating the SJPCP module was not reported.

The asphalt mixture used in the BCOA-ME is different from the default mixture used in calibrating the SJPCP module. The typical mixture in the BCOA-ME procedure has 5.8% air voids, and an effective binder content of 12.5%. This mixture is based on laboratory testing of the BCOA sections at MnROAD, which contains 25 of the 30 SJPCP calibration sections. The mixture used in calibrating the SJPCP module was not reported. The default mixture has 7% air voids, and an effective binder content of 11.6%. The voids in the mineral aggregate for the two procedures are similar, 18.3% for BCOA-ME and 18.6% for the SJPCP module. The Witczak prediction of the δ parameter, the minimum asphalt stiffness, is the only parameter in the Witczak equation affected by the mix proportions. The other terms are solely functions of the aggregate gradation and binder viscosity. The prediction of the δ parameter considers both the air void content, and the percentage of voids filled with asphalt. Therefore, the predicted δ parameter for these two mixes might be similar if the default values in Pavement ME were used.

The parameters used for calibration vary between the design procedures. For the BCOA-ME, calibration is performed by modifying the stress adjustment factor. The stress adjustment factor is given by Equation 3:

$$log_{10}F_{adj} = C_0 + C_1{*}Log(H_{pcc}) + C_2{*}\log(H_{HMA}) + C_3{*}H_{HMA}^2 \quad (3)$$

where H_{pcc} is the thickness of the overlay and H_{HMA} is the thickness of the existing asphalt layer. The regression parameters C_{0-3} are used to perform the calibration. There is no inherent distress transfer function associated with the BCOA-ME design procedure. Instead, 100% fatigue damage is considered to correspond to 25% of the slabs having longitudinal cracks. For calibration purposes, slab cracking is assumed to be linearly related to damage and a linear transfer function was used such that a damage level of 100% corresponds to a 25% level of percent slab cracking. The values of the calibration parameters $C_{0-3,}$ are established in a two-step process. First, a value of F_{adj} that results in a good fit between the predicted and the measured distress at each data point is estimated. Next, a transformed linear regression is performed to estimate the regression coefficients.

The calibration of the SJPCP module is based on two separate relationships. The first parameter, considered a tuning parameter, is the stiffness reduction in the asphalt layer. A damage factor of 1.1 was found to fit the data best, this corresponds to 65% fatigue cracking, and an asphalt stiffness reduction of 48%. The assumed fatigue damage in the existing pavement of 1.1 is not based on the existing condition of the pavement, but is used as a regression coefficient to account for the following: 1. fatigue in the asphalt pavement, 2. material deterioration of the asphalt layer, and 3. partial debonding between the asphalt and overlay layer. The calibration sections at MnROAD, which account for 25 of the 30 SJPCP calibration sections, exhibited no fatigue cracking prior to the placement of the overlay. Therefore, this calibration factor may not adequately consider the effect of distress in the existing asphalt pavement.

The second relationship used for calibrating the SJPCP module is the transfer function. Cracking is estimated as a function of fatigue damage using Equation 4:

$$CRK = \frac{1}{1 + C_1{*}D^{C_2}} \quad (4)$$

where

CRK is the estimated fraction of slabs with longitudinal cracks;

D is the accumulated fatigue damage; and

C_1 and C_2 are regression coefficients.

This provides 3 df for calibration of the SJPCP design procedure, compared with 4 df for calibrating the BCOA-ME design procedure.

Sensitivity Analysis

The previous section has demonstrated that there are significant differences in the approaches used to develop the BCOA-ME and the SJPCP module. Although, the sections used for calibrating the BCOA-ME is a subset of the calibration sections used for calibrating the SJPCP module, so it is possible that the end results still might be comparable, even though the approaches differ.

A sensitivity analysis was conducted to evaluate the effects of different variables on the overlay thickness required by these design procedures at various traffic levels. The five variables and design levels considered in the sensitivity analysis are shown in Table 1. The analysis was performed at the 85% reliability level, which is the recommended reliability level for the BCOA-ME design procedure, with a failure criterion of 25% slabs cracked, which is the default in both design procedures. To develop these plots, the BCOA-ME analyses were performed at seven traffic levels, and the thickness required for each traffic level was calculated. The Pavement ME analyses were conducted at various thicknesses, starting at the minimum thickness of 4 in. and increasing in 0.2-in. increments. The cumulative traffic that caused failure

Table 1. Input Variables Considered in Sensitivity Analysis

Input	Baseline value	High value	Low value
Location (climate)	Minneapolis, MN	Minneapolis, MN	Sacramento, CA
Slab size	6 ft × 6 ft	Not varied	
CTE	5.5 $\mu\varepsilon$/F	6.5 $\mu\varepsilon$/F	4.5 $\mu\varepsilon$/F
HMA thickness (Hhma)	6 in.	8 in.	4 in.
k-value (static)	150 psi/in.	200 psi/in.	100 psi/in.
MOR	650 psi	750 psi	550 psi

Note: CTE = coefficient of thermal expansion; HMA = hot-mix asphalt; MOR = modulus of rupture.

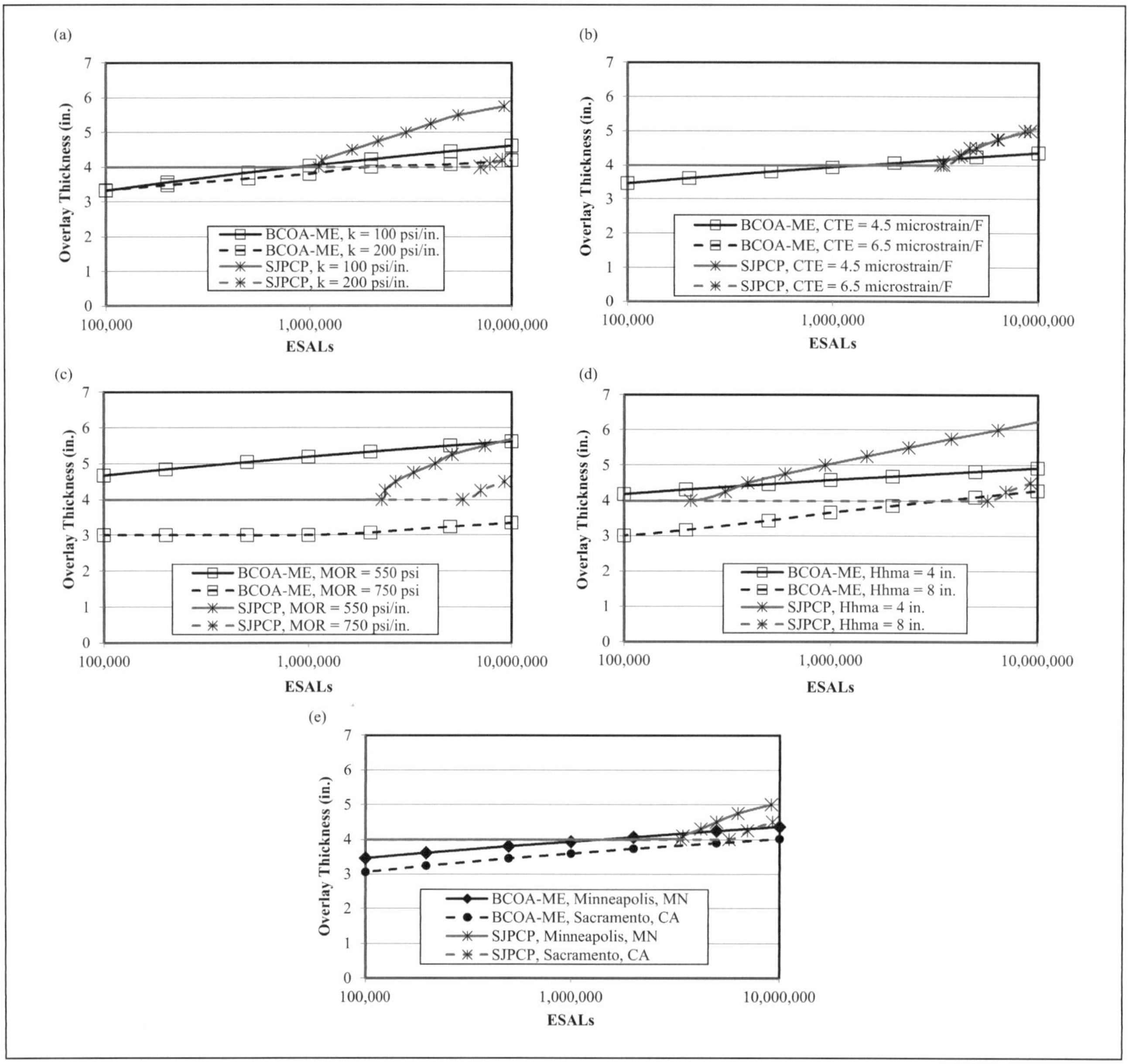

Figure 2. Sensitivity analysis results: (*a*) to the *k*-value; (*b*) to the CTE of the concrete overlay; (*c*) to the MOR of the concrete overlay; (*d*) to the thickness of the asphalt layer; (e) to the climate.

was calculated at each thickness, and converted to ESALs using the estimate within the Pavement ME design procedure, until the estimated traffic exceeded 10 million ESALs.

The resilient modulus of an 8-in. thick AASHTO A-1-A granular base layer and an infinite AASHTO A-7-5 subgrade layer in the SJPCP module was adjusted until the dynamic *k*-value was equal to double the targeted static *k*-value. The dynamic *k*-value during the summer months was used, because nearly all the predicted damage in the SJPCP module occurs during these months.

The predicted overlay thickness at various traffic levels for each of the values considered in the sensitivity analysis can be seen in Figure 2. Figure 2*a* shows the results of the sensitivity analysis of each design procedure to *k*-value. It is interesting to note that the predicted design thickness is comparable at a *k*-value = 200 psi/in. (Note: the minimum design thickness for the SJPCP module is 4 in.) The required thickness from the BCOA-ME is substantially thinner than for the SJPCP module at a *k*-value of 100 psi/in. It is evident that the SJPCP module is much more sensitive to *k*-value than the

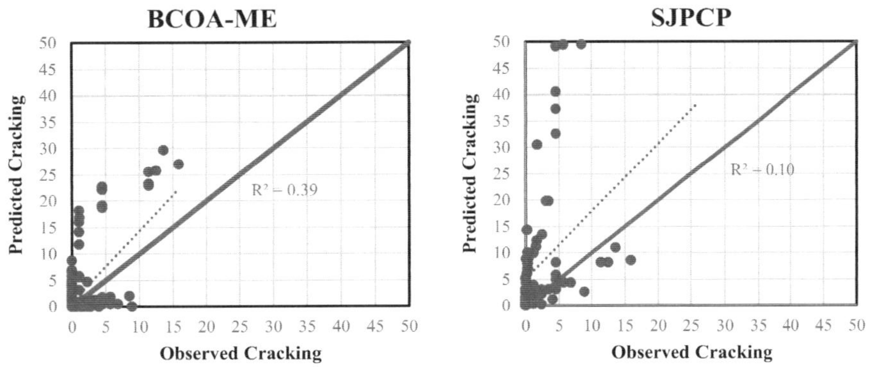

Figure 3. Measured vs. predicted cracking.

BCOA-ME. This could be due to the treatment of joints in the computational models. In the SJPCP module computational model, the asphalt is not continuous through the joint, unlike the BCOA-ME computational model, which most likely contributes to the increased sensitivity to the *k*-value in the SJPCP module.

Figure 2*b* shows the results of the sensitivity analysis of each design procedure to the CTE of the concrete overlay. As previously stated, BCOA-ME does not consider CTE in the design of mid-size slabs. The SJPCP module considers this parameter but is not sensitive to it. The small panel size minimizes the effects of curling, making the overlay thickness insensitive to the concrete CTE in both design procedures.

Figure 2*c* shows the results of the sensitivity analysis of each design procedure to MOR of the concrete overlay. The BCOA-ME design procedure is much more sensitive to MOR than the SJPCP module. This is likely due to the different fatigue models used. As can be seen in Figure 1, the relationship between damage and the stress ratio is much steeper for the ACPA fatigue model used in the BCOA-ME design procedure than it is for the Pavement ME fatigue model. Therefore, changes in the MOR, which directly affect the stress ratio, has more of an effect for BCOA-ME, which utilizes the ACPA model, than for SJPCP module.

Figure 2*d* shows the results of the sensitivity analysis of each design procedure to thickness of the asphalt layer. The SJPCP module requires significantly thicker overlays for thinner asphalt layers and comparable results with a thick asphalt layer. It should be noted, however, that the Pavement ME design thickness for thick asphalt layer are masked by the 4-in. minimum overlay thickness.

For the BCOA-ME procedure, the stress adjustment factor, shown in Equation 3, is dependent on the thickness of the asphalt. A thicker asphalt layer will lead to a higher stress adjustment factor. This considers the fact that thicker asphalt structures provide diminishing returns as the thickness increases. The additional increase in asphalt thickness does not provide diminishing returns for the SJPCP module. The thinner asphalt sections tend to result in thicker overlays when using the SJPCP module.

The final predicted overlay thickness would likely still be the same if the size of the calibration database was sufficiently large enough to account for the thickness of the asphalt in the calibration process. Unfortunately, only two sections in either of the calibration databases used for these design procedures included asphalt layers with thicknesses less than 5 in. The SH-119 test sections in Colorado and Highway 4 in Piatt County, Illinois, both have asphalt layers with thicknesses less than 5 in. In addition, the passing lane of SH-119 Section 3 in Colorado appears to be suffering distress due to cracking above the tie bars along the longitudinal joint at midlane and is exhibiting signs of alkali-silica reaction. This makes the passing lane of this section inappropriate for use in calibration.

Figure 2*e* shows the results of the sensitivity analysis of each design procedure to climate. Minneapolis, MN, receives 58% sunshine and has an average temperature of 46°F, whereas Sacramento receives 78% sunshine and has an annual average temperature of 70°F. Both design procedures appear to be moderately sensitive to geographic location, with overlays designed for Minneapolis being approximately 0.5-in. thicker than overlays designed for Sacramento. This may be due to the wider range of temperatures experienced in Minneapolis. The asphalt binder used in the original asphalt pavement needs to be sufficiently soft to prevent thermal cracking in the winter, yet sufficiently stiff enough to carry load in the hot summer months, when most BCOA damage occurs in the SJPCP module. Regions that experience

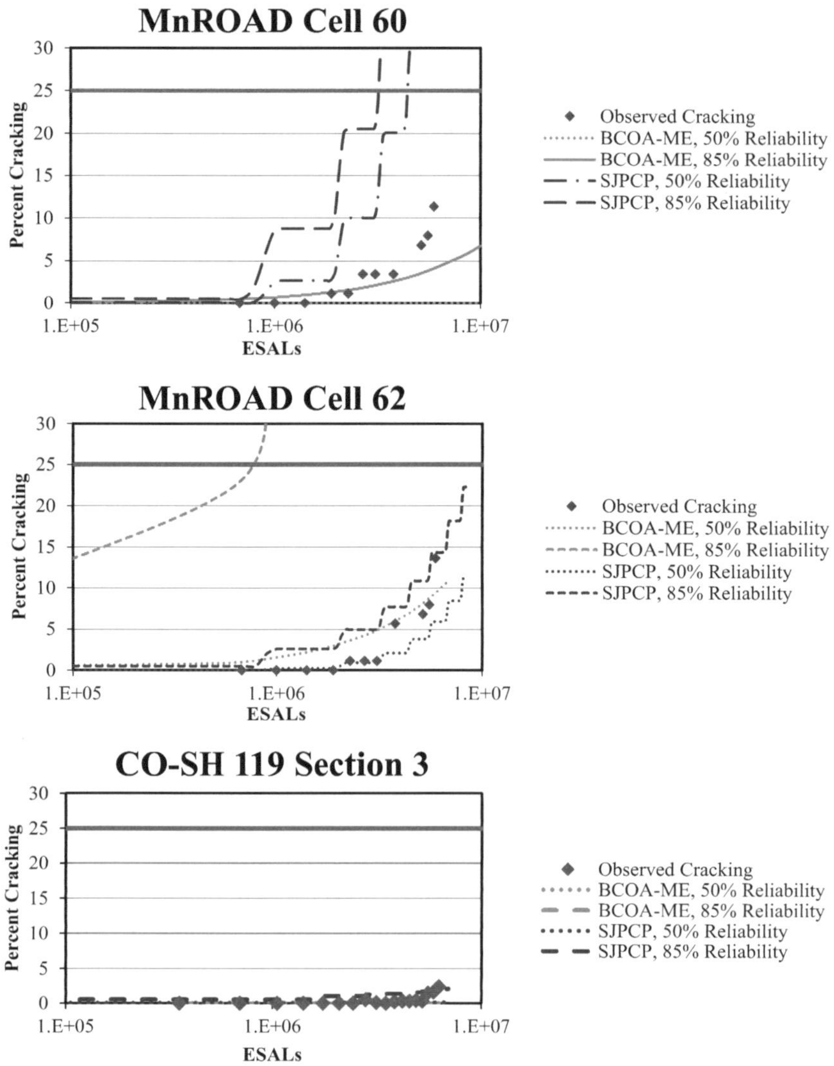

Figure 4. Design examples.

narrow ranges of temperature can use binders that are stiffer in the summer, as thermal cracking in the winter is less of a concern. The BCOA-ME design procedure estimates an Ehma of 380,000 psi in Minneapolis, compared with 480,000 psi in Sacramento. The similar sensitivity to geographic location between the two design procedures indicates that the use of an EELTG and an Ehma based on equivalent damage in the BCOA-ME design procedure does not significantly affect the predicted design.

Calibration Fit

It can be seen that while in some instances BCOA-ME and the SJPCP module produce similar results, the results can also vary greatly. Therefore, it is important to observe how well each design procedure fits the calibration data. The measured versus predicted values for each design procedure are shown in Figure 3, for most sections for which performance data is available. The following sections were not included in the fit:

- MnROAD Cell 95, the thickness of the concrete is less than the minimum design thickness for the SJPCP module and it contains structural fibers.
- MnROAD cells 114 to 914, these panels have short section lengths, increasing variability in measured distress making it challenging to define the true average performance.

- MnROAD cells 160 and 162, these panels were only constructed in 2013, have accumulated only 5 million ESALs and are not currently exhibiting any distress.
- Colorado SH-119 research Section 3 (passing lanes), analysis of the cracking revealed that the longitudinal cracking was influenced by transverse cracks that occurred above the tie bars.

It should be noted that a single data point is not shown on the fit of the SJPCP module, Highway 4 in Piatt County, Illinois with an observed cracking of 1%, and a predicted cracking of 70%, in order to keep scales sufficiently small to observe trends. It should also be noted that a stronger fit ($R^2 = 0.87$) has been reported between the observed cracking and predicted cracking for the SJPCP module (*7*). However, that fit could not be replicated based on the inputs provided by the respective state agencies for these sections or the values included in the database provided as part of the final report produced on the calibration of the SJPCP module (*7*). The database in the final report included all input values used for calibration accept for the values used to characterize the stiffness of the subgrade and base layers. Using the best available information for the subgrade stiffness, the predicted performance obtained through this study was substantially different than what was reported in the final report on the calibration of the SJPCP module (*7*). Furthermore, within the final report for the SJPCP module, there is a discrepancy between the predicted cracking for MnROAD Cell 63 reported in the design example and that reported under the sensitivity analysis. The sensitivity analysis reports a predicted cracking of 15% after 20 years for Cell 63. This study showed that these results could be replicated. In contrast, in the design examples section of the report, the predicted cracking for Cell 63 is reported to be around 50% at 7.5 years, which closely matches the observed cracking. This result could not be replicated using any reasonable value for subgrade stiffness. The predicted cracking values presented with the design examples were used to evaluate the fit of the calibration in the final report. Cell 63 is the most distressed section in the SJPCP calibration database by far, and the observations are strong influence points on the calibration fit. If the predicted distress presented in the sensitivity analysis section were used instead of the predicted distress in the design examples of the final report, the R^2 value would have been reduced. The artificially high R^2 does not solely affect the apparent strength of the tool. Pavement ME uses the fit of the calibration to calculate the standard error of the prediction, which is used to account for reliability. This then results in the design procedure being less sensitive to increased reliability.

Although many BCOAs have recently been constructed, the number of sections with adequate distress and design information for model development is still relatively small. As additional performance data becomes available, the predictions of both design procedures should improve. It would be desirable to have a better fit between the predicted and observed performance for both the BCOA-ME and the SJPCP module. But, it is also important to note, that most calibrations of currently used mechanistic-empirical rigid pavement design procedures include pavement sections with much more distress than allowable for typical thresholds set for in-service pavements. These high distress sections are influence points that control the fit of the calibrations. If only pavements with distress levels below or near the defined failure level are used to evaluate these design procedures, the fits would be less optimistic. The practical power of a design procedure is based on the ability of the design procedure to predict cracking below failure and the time to failure, at a given reliability level. Therefore, evaluating the fit of the procedure based on these "high distress" influence points is overly optimistic. The distress level of the available BCOA calibration data is all below the allowable level, resulting in a less optimistic fit than traditionally seen with mechanistic-empirical design procedures.

Design Examples

In addition to examining the sensitivity of each of the procedures to design inputs, it is also important to look at the prediction of distress at several design examples. Three design examples are considered:

- MnROAD Cell 60, 5-in. thick overlay, with MOR of 595 psi/in., 7-in. thick asphalt layer and a *k*-value of 315 psi/in.
- MnROAD Cell 62, 4-in. thick overlay, with MOR of 575 psi/in., 8-in. thick asphalt layer and a *k*-value of 315 psi/in.
- Colorado SH-119 (driving lane), 5.1-in. thick overlay, with MOR of 612 psi/in., 3.3-in. thick asphalt layer and a *k*-value of 150 psi/in.

The measured and predicted distress levels for the sections can be seen in Figure 4. For all sections, there is a significant difference between the predicted distress using the SJPCP module, and the predicted distress using BCOA-ME. At MnROAD Cell 60, the SJPCP module overestimates distress, whereas BCOA-ME underestimates distress. At MnROAD Cell 62, the BCOA-ME and the SJPCP module both accurately estimate distress. Both design procedures also predict no distress at Colorado SH-119, where only 2% of slabs are cracked after an estimated 7 million ESALs. The SJPCP module is less sensitive to the desired reliability level than BCOA-ME for all three design sections. This is not surprising since a high *r*-squared was reported (*7*) between the predicted and observed distress.

Recommendations

The BCOA-ME design procedure, and the SJPCP module in Pavement ME are unique design procedures, each with strengths and weaknesses. Some limitations to both procedures that should be considered when designing BCOAs include:

- The number of sections available for calibration and the range of design parameters represented are limited.
- The section with the maximum number of ESALs included in the calibration database for both procedures had only 11 million ESALs. Care should be taken when extrapolating beyond this traffic level.
- Both procedures only address fatigue cracking with neither procedure considering faulting as a failure criterion.

The Pavement ME SJPCP design procedure is preferred when:

- The design traffic consists of nonstandard load configurations, vehicle distributions in the traffic stream, or both elements.

BCOA-ME design procedure is preferred when:

- The designer wants to consider the existing condition of the asphalt pavement in the design process.
- A lower-volume road is being rehabilitated and the thickness of the overlay might be less than 4 in. or panel sizes smaller than 4.5 ft are desired. The SJPCP module cannot be used for these pavements.
- Uncertainty exists regarding the stiffness of the layers beneath the asphalt as the SJPCP module seems to be overly sensitive to the stiffness of these layers.
- Structural fibers will be used in the paving mixture. The SJPCP module does not account for the additional fatigue life provided by structural fibers.

Another potential concern is that it appears the SJPCP module is not adequately considering design reliability. The reported calibration fit of the SJPCP module, which is used to account for reliability, was not replicated in the current analysis. Therefore, this procedure might not properly account for reliability.

Author Contributions

The authors confirm contribution to the paper as follows: study conception and design: JV, MS, LK; data collection: KA, JD; analysis and interpretation of results: KA, JV, JD; draft manuscript preparation: KA, JV, MS, JD. All authors reviewed the results and approved the final version of the manuscript.

References

1. Mack, J. W., L. W. Cole, and J. P. Mohsen. Analytical Considerations for Thin Concrete Overlays of Asphalt. *Transportation Research Record: Journal of the Transportation Research Board*, 1993. 1388: 167–173.
2. Mack, J. W., L. D. Hawbaker, and L. W. Cole. Ultra-Thin Whitetopping - State-of-the-practice for Thin Concrete Overlays of Asphalt. *Transportation Research Record: Journal of the Transportation Research Board*, 1998. 1610: 39–43.
3. Wu, C.-L., S. M. Tarr, T. M. Refai, M. A. Nagi, and M. J. Sheehan. Development of Ultra-Thin Whitetopping Design Procedure. *PCA R&D Serial*, No. 2124, 1998.
4. Wu, C., S. Tarr, T. Refai, M. Nagi, and M. Sheehan. Development of Ultra-Thin Whitetopping Design Procedure. *PCA R&D Serial*, 1998.
5. Tarr, S. M., M. J. Sheehan, and P. A. Okamoto. *Guidelines for the Thickness Design of Bonded Whitetopping Pavement in the State of Colorado*. Report No. CDOT-DTD-R-98-10. Colorado Department of Transportation, Denver, CO, 1998.
6. Vandenbossche, J. M., N. Dufalla, and Z. Li. Bonded Concrete Overlay of Asphalt Mechanical-Empirical Design Procedure. *International Journal of Pavement Engineering*, Vol. 8436, 2016, pp. 1–12.
7. Bhattacharya, B., A. Gotlif, and M. I. Darter. Implementation of the Thin Bonded Concrete Overlay of Existing Asphalt Pavement Design Procedure in the AASHTOWare Pavement ME Design. *Transportation Research Record: Journal of the Transportation Research Board*, 2017. 2641: 12–20.
8. Barman, M., N. Dufalla, Z. Li, F. Mu, and J. M. Vandenbossche. Development of a Rational Mechanistic Empirical Design Guide for Thin and Ultra-Thin. *Task 4 Report: Climatic Considerations*, University of Pittsburgh, Pittsburgh, PA, 2011.
9. Li, Z., and J. M. Vandenbossche. Redefining the Failure Mode for Thin and Ultra-Thin Whitetopping with a 1.8- × 1.8-m (6- × 6-ft) Joint Spacing. *Transportation Research Record: Journal of the Transportation Research Board*, 2013. 2368: 133–144.
10. Snyder, M. B. Lessons Learned from MnROAD (1992–2007): Whitetopping Design, Construction, Performance and Rehabilitation. Presented at 88th Annual Meeting of the Transportation Research Board, Washington, D.C., 2009.

11. Khazanovich, L., H. T. Yu, S. Rao, K. Galasova, E. Shats, and R. Jones. ISLAB2000-Finite Element Analysis Program for Rigid and Composite Pavements. *User's Guide*. ERES Consultants, Champaign, IL, 2000.
12. Pereira, D. D. S., J. T. Balbo, and L. Khazanovich. Theoretical and Field Evaluation of Interaction between Ultra-Thin Whitetopping and Existing Asphalt Pavement. *International Journal of Pavement Engineering*, Vol. 7, No. 4, 2006, pp. 251–260.
13. DeSantis, J. W., J. M. Vandenbossche, K. Alland, S. Sachs, T. Burnham, and A. Montenegro. Joint Performance in Bonded Concrete Overlays of Asphalt. *Proc., 11th International Conference on Concrete Pavement Design and Rehabilitation*. San Antonio, TX, 2016.
14. Riley, R. C., L. Titus-Glover, J. Mallela, S. Waalkes, and M. Darter. Incorporation of Probabilistic Concepts into Fatigue Analysis of Ultra-Thin Whitetopping as Developed for the American Concrete Pavement Association. *Proc., International Conference on Best Practices for Ultra-Thin and Thin Whitetoppings*. Denver, CO, 2005.
15. Darter, M., L. Khazanovich, T. Yu, and J. Mallela. Reliability Analysis of Cracking and Faulting Prediction in the New Mechanistic-Empirical Pavement Design Procedure. *Transportation Research Record: Journal of the Transportation Research Board*, 2005. 1936: 150–160.
16. Sachs, S., J. M. Vandenbossche, Z. Li, and M. Barman. Accounting for Temperature Susceptibility of Asphalt Stiffness When Designing Bonded Concrete Overlays of Asphalt Pavements. *Journal of Transportation Engineering*, Vol. 142, No. 10, 2016, pp. 1–7.
17. Mu, F., and J. Vandenbossche. Establishing Effective Linear Temperature Gradients for Ultra-Thin Bonded Concrete Overlays on Asphalt Pavements. *Transportation Research Record: Journal of the Transportation Research Board*, 2012. 2305: 24–31.
18. Li, Z., J. M. Vandenbossche, and N. Dufalla. Structural Model for Longitudinal Cracking in Bonded Whitetopping with a 1.83 m × 1.83 m Joint Spacing. *Journal of Transportation Engineering, Part B: Pavements*, Vol. 143, No. 4, 2017, p. 4017015.
19. Korenev, B. G., and E. I. Chernigovskaya. *Analysis of Plates on Elastic Foundation*. Gosstroiizdat, Moscow, 1962. (in Russian)
20. Mohseni, A. LTPPBind Software 2.1. Federal Highway Administration, US Department of Transportation, McLean, VA, 2003.
21. Andrei, D., M. W. Witczak, and M. W. Mirza. *Development of Revised Predictive Model for the Dynamic Complex Modulus of Asphalt Mixtures*. Washington, D.C., 2005.
22. ARA Inc. *Guide for Mechanistic-Empirical Design of New and Rehabilitated Pavement Structures*. ARA Inc., Washington, D.C., 2004.
23. ACPA Static *k*-Value Calculator.
24. Roesler, J., A. Bordelon, and M. Beyer. *Design and Concrete Material Requirements for Ultra-Thin Whitetopping*. Illinois Center for Transportation, Champaign, IL, 2007.
25. American Association of State Highway and Transportation Officials. *AASHTO Guide for Design of Pavement Structures, 1993*. American Association of State Highway and Transportation Officials, Washington, D.C., 1993.
26. Sheehan, M.J., S.M. Tarr, and S. Tayabji. Instrumentation and Field Testing of Thin Whitetopping Pavement in Colorado and Revision of the Existing Colorado Thin Whitetopping Procedure. Colorado Department of Transportation Research Branch, Denver, CO, 2004.

The Standing Committee on Design and Rehabilitation of Concrete Pavements (AFD50) peer-reviewed this paper (18-06116).

Article

TRR
JOURNAL OF THE TRANSPORTATION RESEARCH BOARD

Transportation Research Record
2018, Vol. 2672(40) 254–263
© National Academy of Sciences:
Transportation Research Board 2018
Article reuse guidelines:
sagepub.com/journals-permissions
DOI: 10.1177/0361198118792325
journals.sagepub.com/home/trr
SAGE

Establishing the Interlayer Structural Response for Unbonded Concrete Overlays of Existing Concrete Pavements

Steven G. Sachs[1], Julie M. Vandenbossche[1], Derek Tompkins[2], and Lev Khazanovich[1]

Abstract

An improved mechanistic empirical design procedure for unbonded concrete overlays of existing concrete pavements (UBOLs) should account for the effect of the interlayer on the structural response of the pavement. One approach is to use the Totsky model to characterize the interlayer. The Totsky model treats the interlayer as a bed of springs between two plates and is currently incorporated into the rigid pavement finite element software ISLAB. A difficulty encountered in implementing this model is that there are currently no guidelines as to what the interlayer *k*-value should be for different types of interlayers. The interlayer can be constructed of new or aged asphalt (open or dense graded) or a nonwoven geotextile fabric. To establish the *k*-values that accurately characterize each of these materials, an ISLAB model of a laboratory test was created so the *k*-values could be established by matching the measured and calculated difference between the deflections in the overlay and existing pavement. To supplement the use of the laboratory data in establishing the Totsky interlayer *k*-value, an analysis was carried out using falling weight deflectometer (FWD) data from UBOLs at the Minnesota Road Research Facility (MnROAD). Analyses were then performed to determine if the difference between *k*-values for different interlayer materials are statistically significant, and if the results from the laboratory analysis match those obtained from the MnROAD field data. The Totsky *k*-value recommended for use when modeling the response of an UBOL with an asphalt interlayer is 3500 psi/in and 425 psi/in for a fabric interlayer.

Unbonded concrete overlays of existing concrete pavements (UBOLs) are a rehabilitation technique that places a new concrete pavement over an existing aged concrete pavement separating the two layers with an interlayer. The interlayer acts as a shear plane to prevent distress in the existing concrete pavement from reflecting up into the overlay. Common interlayer types include hot mix asphalt (either open or dense graded) or a nonwoven geotextile fabric. Previous overlay design procedures do not directly account for the presence of the interlayer (*1–6*). For example, in Pavement ME, the structural model used for UBOLs is identical to that used for new rigid pavement design where the interlayer is treated as the base layer and is combined with the overlay to create an effective slab (*2*). As a result, it is not possible to directly account for the effect of the interlayer on the stresses and deflections in the overlay. Therefore, future design procedures should account for the effect of the interlayer on the response of the structure. This can be achieved by using a Totsky model, which treats the interlayer as a bed of springs separating two plates representing the concrete overlay and the underlying existing, distressed concrete layer, the existing pavement then rests on a Winkler foundation. Figure 1 shows an unbonded concrete overlay of an existing concrete pavement (UBOL) modeled with a Totsky interlayer. The only parameter to define for the Totsky interlayer is the spring stiffness, k.

The advantage of the Totsky model is that it can model the "cushioning" property of the interlayer in a straightforward, computationally efficient manner. ISLAB is a finite element method analysis software for pavements that incorporates the Totsky model to specifically model unbonded concrete overlays (*7–9*). The Totsky model consists of eight node elements with 24 degrees of freedom. The first four nodes of the overlay plate are placed at the neutral axis of this upper plate,

[1]University of Pittsburgh, Pittsburgh, PA
[2]American Engineering Testing, Inc., MN

Corresponding Author:
Address correspondence to Julie M. Vandenbossche: jmv7@pitt.edu

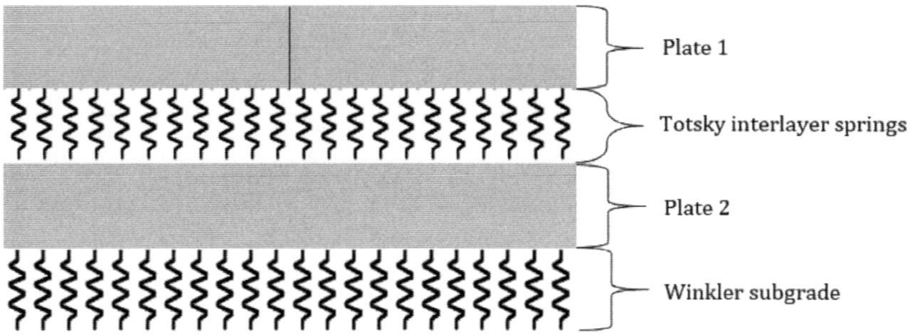

Figure 1. Totsky model for layer interface in an UBOL system adopted in ISLAB.

while the other four nodes of the existing plate are placed at the neutral axis of the lower plate. The plate elements model the bending, while the springs accommodate the direct compression occurring in the interlayer. Within ISLAB, it is assumed that the stiffness matrices of the spring interlayer and Winkler subgrade are dependent only on nodal displacements and not on the nodal rotations. For larger general-purpose finite element method packages, a Totsky interlayer spring approach can be taken, or the interlayer can be modeled as an elastic or viscoelastic solid. The implementation of this approach is time consuming given the extent of features available and the relative difficulty in their implementation.

The results from laboratory beam testing are used to create a model in ISLAB to define the Totsky *k*-value representative of different asphalt and fabric interlayers. To supplement this analysis, falling weight deflectometer (FWD) data from UBOL test cells at the Minnesota Road Research Facility (MnROAD) are used to establish interlayer *k*-values using ISLAB. Statistical tests are then performed on the *k*-values obtained from the laboratory and from the FWD testing at MnROAD.

Laboratory Investigation

The UBOL beam specimens tested consist of an existing Portland Cement Concrete (PCC) layer on the bottom, an interlayer, and an overlay PCC layer on the top, as shown in Figure 2. The dimensions of the UBOL beams consist of a 6 in height, 6 in width, and 30 in length. Approximately a 1/8 in discontinuity is sawed in the lower concrete beam (representing the existing pavement) of the UBOL specimen to simulate a joint or crack, as seen in the schematic in Figure 2*b*. The saw cut is full depth in the existing (lower) concrete beam. In the test configuration, there are no restraints on either the top or the bottom layer. A gap was created under the central portion of the beam by removing part of the artificial foundation so that there was no support for a length spanning 10 in at the central portion of the beam. Thus, the middle third of the beam, centered under the load, is unsupported. A neoprene pad with effective *k*-value of 200 psi/in is placed under the supported region. The specimen is then loaded with a quasi-static load (under load control at 30 pounds per second directly over the sawn discontinuity) until a reflective crack is generated through the interlayer into the overlay concrete beam. To characterize the strength of the overlay, a 6 in × 6 in × 24 in modulus of rupture beam was cast from the same batch of concrete as each overlay beam. The flexural strength of the modulus of rupture beams was tested according to ASTM C78 (*10*). Therefore, the flexural strength of the overlay PCC is known for each specimen where a reflective crack is generated.

Interlayers

The nonwoven geotextile fabrics used for this study were manufactured by Propex and consisted of a dense (thicker) and a less dense (thinner) fabric. The thicker fabric (F15) is sold under the brand name Propex and weighs 15 oz/yd^2. A thinner fabric (F10) was made specifically for this study and weighs 10 oz/yd^2.

Beams simulating the existing concrete pavement were cast using the mixture design designated as Mixture 1 in Table 1. The concrete mix for the lower beam of the specimens with the fabric interlayer has a water to cementitious material ratio (w/cm) of 0.36 with a target flexural strength of approximately 850 psi. The fabric is attached to the existing concrete beam with a geotextile glue manufactured by 3M and sold under the brand name Scotch-Weld HoldFast 70 Adhesive. Typically, the geotextile fabric is pinned with metal fasteners to the existing concrete but a geotextile adhesive has also been used during the construction of an UBOL in Minnesota. The adhesive was used instead of the metal fasteners to avoid damage to the existing beam. In addition, the use of the adhesive eliminates any effects in the performance owing to the location and spacing of the fasteners. Once the fabric was attached, concrete beams simulating the

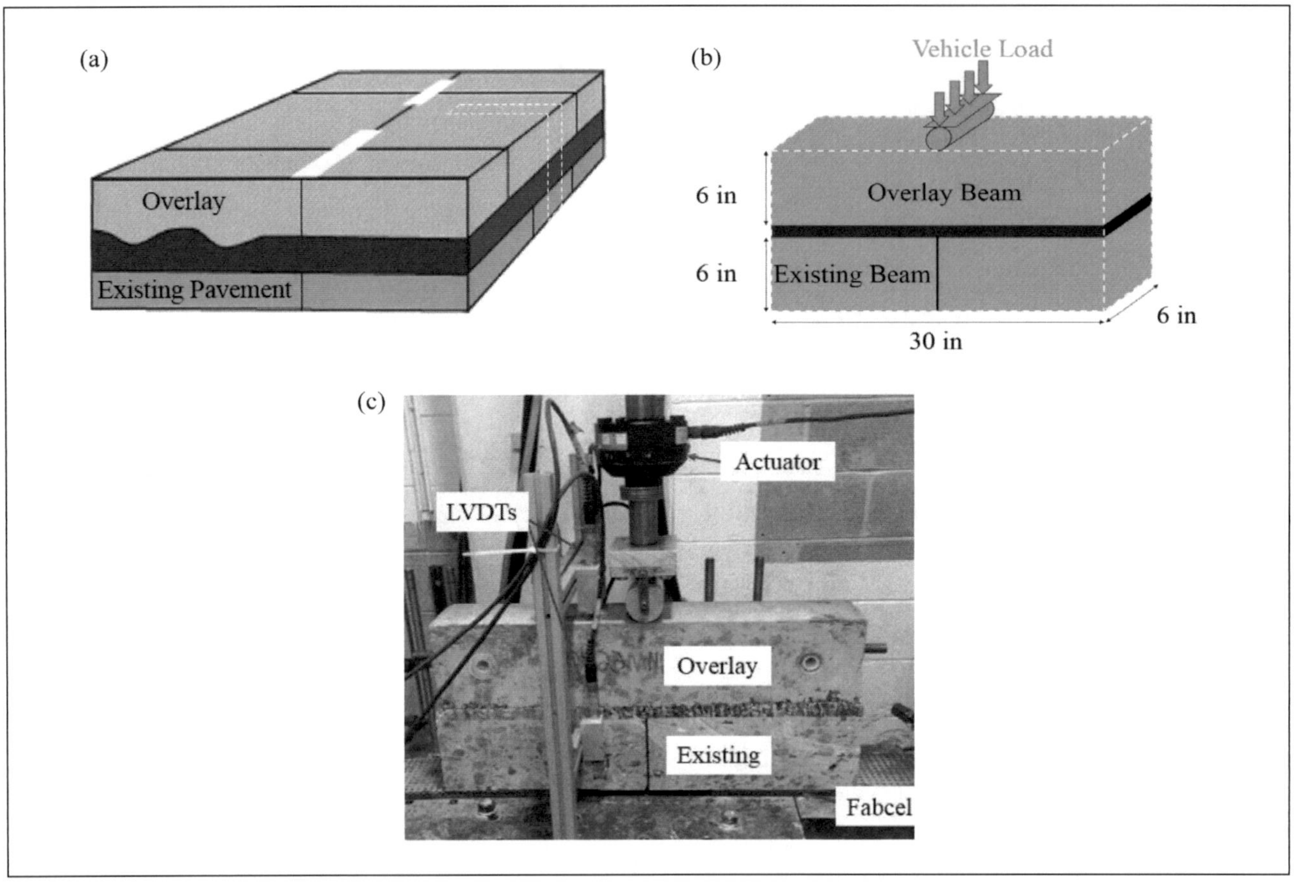

Figure 2. Replicating the UBOL in the laboratory: (*a*) in service pavement; (*b*) UBOL specimen; and (*c*) reflective cracking specimen in loading frame.

Table 1. Target Mixture Designs for Casting the UBOL Specimens

Material	Weight (lb/cy)	Volume fraction
Mixture 1 for casting beams representative of the existing slab		
Coarse aggregate	1918	0.42
Fine aggregate	1163	0.26
Cement	650	0.12
Water	234	0.14
Air content (6%)	-	0.06
Superplasticizer, Sikament SPMN	17 oz per 100 lbs of cement	
Air entrainer, Sika AIR-360	3 oz per 100 lbs of cement	
Mixture 2 for casting beams representative of the overlay		
Coarse aggregate	2053	0.45
Fine aggregate	1023	0.23
Cement	600	0.11
Water	252	0.15
Air content (6%)	-	0.06
Air entrainer, Sika AIR-360	2 oz per 100 lbs of cement	

Table 2. Description of the UBOL Specimens Tested

UBOL specimen	Overlay beam	Existing beam	Interlayer	Average asphalt thickness (in)	Roadway
F15	Mixture 2	Mixture 1	15 oz/yd^2 fabric	-	-
F10	Mixture 2	Mixture 1	10 oz/yd^2 fabric	-	-
MIDAU	Mixture 2	JPCP (15 yrs old)	Dense graded, aged, unmilled (15 yrs old)	1	US-131 Rockford, MI
MIOAU	Mixture 2	JPCP (11 yrs old)	Open graded, aged, unmilled (11 yrs old)	2	US-131 Kalamazoo, MI
MNDAM	Mixture 2	JPCP (5 yrs old)	Dense graded, aged, milled (5 yrs old)	0.875	I-94, Albertville, MN (MnROAD)
MNDAU	Mixture 2	JPCP (5 yrs old)	Dense graded, aged, unmilled (5 yrs old)	2.75	I-94, Albertville, MN (MnROAD)
MNONU	Mixture 2	JRCP (≈ 26 yrs old)	Open graded, new, unmilled (≈ 1 week)	1.75	US-169, MN
PADNU	Mixture 2	JRCP (40 yrs old)	Dense graded, new, unmilled (≈ 1 week)	1	SR 50 Bridgeville, PA

Note: JPCP = Jointed Plain Concrete Pavement; JRCP = Jointed Reinforced Concrete Pavement; UBOL = unbonded concrete overlay of existing concrete pavements.

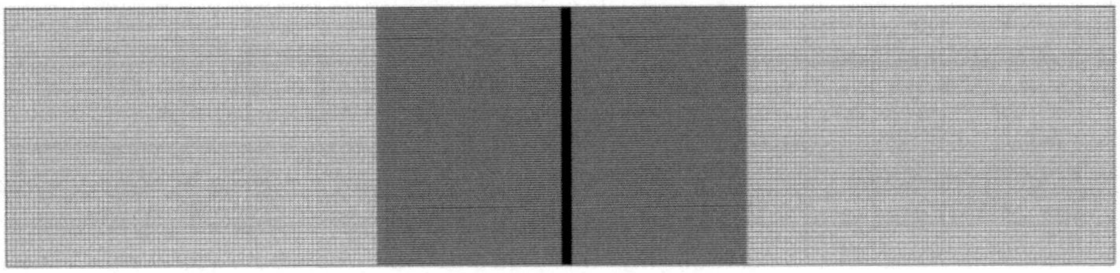

Figure 3. ISLAB two-dimensional model of reflective cracking test with top-view showing the mesh and load area (unsupported area highlighted in dark grey).

overlay were cast on top of the fabric. The mixture design designated Mixture 2 in Table 1 was used for casting the overlay beam for all specimens and has a w/cm of 0.42 with a target flexural strength of 650 psi and a target slump of 1–3 in. The difference in strength between the overlay beam and existing concrete beam is to simulate the effect of age on long term strength gain that would occur in the existing in-service pavement. All specimens were cured according to ASTM C192 (*11*).

The specimens representing the existing pavement and interlayer for the asphalt UBOL consisted of asphalt-PCC composite beams sawed from in-service pavements. This ensures that the mixture proportioning, density, particle orientation, and aging of the asphalt interlayers are typical of those found in the field. The asphalt-PCC composite beams were obtained from the Minnesota Department of Transportation (MnDOT), Michigan Department of Transportation (MDOT), and Pennsylvania Department of Transportation (PennDOT).

A summary of in-service pavements from which the UBOL specimens were harvested is provided in Table 2. Once the beams were brought to the laboratory, each specimen had to be cut to the desired dimensions. Sand patch testing (ASTM E965) was performed on the asphalt surface of each beam to characterize the surface roughness and the final beam dimensions were measured (*12*). The overlay was cast on top of the asphalt in the same manner as the fabric specimens using Mixture 2 from Table 1. Additional information about the test setup can be found in (*13*).

Developing a Relationship between Totsky k-value and Beam Response with ISLAB

The Totsky approach for modeling the interlayer behavior in overlays of existing PCC pavements can be employed using ISLAB (*14*). Figure 3 provides a representation of the model used to determine the Totsky *k*-value for the different interlayers. It should be noted that the simulated load is applied as a 0.25 in wide line-load along the beam depth to replicate the actual loading condition (indicated in black in Figure 3). Thus, the load contact area is 1.5 in^2. As the finite element model is static, a single load of 1 kip is applied to determine the response of the beam model.

In ISLAB, the notch at mid-span in the existing concrete is modeled by inserting a joint at mid-span. In the upper layer (the overlay), this joint fully transfers the load (the load-transfer efficiency is 100%). However, no load is transferred across the joint (load transfer efficiency is near 0%) in the lower layer (the existing concrete).

With the beam model, a factorial of cases is modeled to observe the response when different interlayer properties are used. In each case, only the Totsky interlayer *k*-value (k_{totsky}) is varied. The remainder of the properties used in the model include the following:

- Layer 1: $h_{OL} = 6\,\text{in}$, $E_{OL} = 4{,}255{,}000\,\text{psi}$ (average for all of the concrete made to cast the overlay beam portion of the specimens), Poisson ratio $\nu = 0.15$, and unit weight $\gamma = 0.087\,\text{lb/in}^3$
- Interlayer: k_{totsky} varied from 100 to 50,000 psi/in
- Layer 2: $h_{EX} = 6\,\text{in}$, $E_{EX} = 4{,}790{,}000\,\text{psi}$ (average for all of the concrete made to cast the existing beam portion of the specimens), Poisson ratio $\nu = 0.15$, and unit weight $\gamma = 0.087\,\text{lb/in}^3$
- Mesh details: Mesh elements are square (0.125 inches to a side) for the entire model, as illustrated in Figure 3.
- A static load of 1-kip is applied to determine a linear beam response associated with interlayer properties.

Figure 4 illustrates the final relationship determined for the modeled beam response and Totsky interlayer stiffness. Also included in the figure is an exponential relationship determined by transforming the variables and finding a linear least-squares fit. As shown in the figure, the R^2 value for the fitness of the exponential relationship is 0.99, thus the model adequately describes the relationship between the model response and the Totsky *k*-value for this range of values. With the relationship developed in ISLAB, interlayer Totsky *k*-values can be established for each beam specimen tested and therefore each type of interlayer system included in the laboratory study.

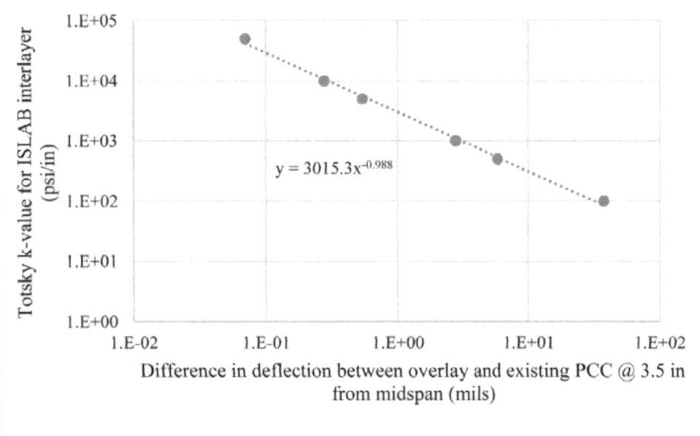

Figure 4. Developed relationship between the difference in layer deflection and Totsky *k*-value for the interlayer from ISLAB.

Interlayer Beam Response

Table 3 presents the beams for each interlayer and the corresponding Totsky *k*-value. Results for these beams are provided in Table 3. Please refer to Table 2 for the specimen designation. Given the response of the different interlayer beams under a 1-kip load in the lab, the modeled relationship provided in Figure 3 was used to infer an associated Totsky interlayer stiffness. The average and standard deviation for the different interlayers tested in the laboratory are presented in Table 4.

Hypothesis testing was performed to evaluate the effects of the different interlayers and determine if there is a statistical difference between the interlayers. Tukey's range test was used to compare all possible pairs of means (*15*). The null hypothesis is that the means of the two interlayers compared are equal while the alternative hypothesis is that the mean of one of the two interlayers differs from the other. Table 5 presents all pair-wise comparisons between each interlayer. The difference in means is the result of the subtraction of the averages of the two interlayers compared. The 95% confidence intervals on the difference between interlayers are also presented. The two interlayers are statistically different at 95% if the range of the confidence interval does not contain zero. As can be seen from Table 5, the means of the fabric interlayers are statistically different from each of the asphalt interlayers with the exception of the open graded asphalt from Minnesota. The difference with the open graded asphalt from Minnesota could be because the gradation was more open graded and contained more rounded aggregate particles, resulting in a lower strength and stiffness as compared to the other interlayers. No statistical difference was detected between any of the asphalt interlayers or between the fabric interlayers.

It should be noted that there does not appear to be a relationship between either the interlayer asphalt thickness and the inferred Totsky *k*-value or between asphalt stiffness and the Totsky *k*-value. Based on the model and the lab data, other factors, including the interlayer bond and perhaps the loading/support conditions, must be considered if the inferred Totsky *k*-value is to be considered beyond an average across all asphalt lab beams.

MnROAD Falling Weight Deflectometer Analysis

To supplement the use of the laboratory beam testing in establishing the Totsky interlayer *k*-value, an analysis

Table 3. Established Totsky *k*-values for Laboratory Specimens

Overlay PCC UBOL specimen*	*E* (psi)	*f'c* (psi)	Fabric type	Difference in deflection @ 1 kip (mils)	Totsky k-value (psi/in)
F15	4280000	5059	F15	8.27	411
	4280000	5059	F15	10.41	325
	4430000	4632	F15	12.33	274
F10	3880000	4512	F10	10.58	320
	4170000	5069	F10	7.76	439
	3880000	4512	F10	9.48	358
			Asphalt thickness		
MNDAU	3880000	4590	2.9	0.93	3824
	4480000	5106	2.8	2.32	1504
	4430000	4632	2.8	0.76	4698
MNDAM	4300000	4696	0.9	1.37	2581
	4480000	5106	1	1.25	2828
	4490000	4732	0.8	0.66	5431
MNONU	4480000	5106	1.7	1.52	2324
	4650000	5131	1.7	0.93	3824
	4430000	4632	1.8	2.3	1518
MIDAU	4230000	5106	1.1	0.65	5521
	4790000	5131	1	0.99	3584
	4430000	4632	1.3	1.17	3033
MIOAU	4710000	5013	1.8	1.28	2760
	4620000	5073	1.9	0.68	5263
	4490000	4632	1.8	1.32	2675
PADNU	4630000	4966	1.5	1.98	1766
	4340000	4824	1.4	1.3	2717
	4340000	4824	1.5	0.63	5690

Note: UBOL = unbonded concrete overlay of existing concrete pavements.

Table 4. Average and Standard Deviation of Totsky *k*-value for Each Interlayer Type

UBOL specimen*	Average Totsky *k*-value	Standard deviation
F15	337	63
F10	372	55
MNDAU	3342	1262
MNDAM	3613	1175
MNONU	2555	901
MIDAU	4046	966
MIOAU	3566	1095
PADNU	3391	1533

Note: UBOL = unbonded concrete overlay of existing concrete pavements.

was carried out using FWD data from MnROAD UBOLs. MnROAD Cells 105, 205, 304, 405, 505, and 605 are UBOLs constructed with either an open graded permeable asphalt stabilized stress relief course (PASSRC) or a non-woven geotextile fabric. It should be noted that the MNONU laboratory specimens were also harvested from an in-service pavement in Minnesota with a PASSRC interlayer. The designs of these cells are summarized in Table 6 below. The existing concrete pavement in cell 5 was constructed in 1993 and consisted of 7.1 in of PCC placed over 3 in of class 4 aggregate base over 27 in class 3 aggregate subbase over a clay subgrade (*16*, *17*). Cell 5 had 20 ft long by 13 ft (passing lane) or 14 ft (driving lane) wide panels and bituminous shoulders. FWD data was available for each cell except cell 105.

Thermocouple data was available for cells 205, 305, and 605. As the designs for cells 105, 405, and 505 are identical to 205, 305, and 605 respectively, the same temperature data was used for the similar cells. The temperature profile through the PCC overlay, as well as an approximate temperature of the interlayer at the time of FWD testing, was then established for each cell and test time. FWD testing at the joints in the wheelpath was used to establish the Load Transfer Efficiency (LTE) to be used in the ISLAB finite element model. The slab stiffness was obtained either directly from elastic modulus tests performed on specimens or through a correlation with strength. The layers beneath the existing PCC were modeled as a Winkler foundation with a *k*-value of 250 psi/in, which was backcalculated using FWD data collected before the placement of the overlay.

The Totsky formulation is ISLAB was then used to model the structure for a the case when FWD testing is

Table 5. Pair-Wise Interlayer Comparisons

Comparison[a]	Difference of mean Totsky k-value between interlayers	95% confidence interval of difference
F10 - F15	35	(−2762, 2833)
MNDAU - F15	**3006**	**(208, 5803)**
MNDAM - F15	**3277**	**(479, 6074)**
MNONU - F15	2218	(−579, 5016)
MIDAU- F15	**3709**	**(912, 6507)**
MIOAU - F15	**3229**	**(432, 6027)**
PADNU - F15	**3054**	**(257, 5852)**
MNDAU - F10	**2970**	**(173, 5768)**
MNDAM - F10	**3241**	**(444, 6039)**
MNONU - F10	2183	(−615, 4980)
MIDAU- F10	**3674**	**(876, 6471)**
MIOAU - F10	**3194**	**(396, 5991)**
PADNU - F10	**3019**	**(221, 5816)**
MNDAM - MNDAU	271	(−2526, 3069)
MNONU - MNDAU	−787	(−3585, 2010)
MIDAU - MNDAU	704	(−2094, 3501)
MIOAU - MNDAU	224	(−2574, 3021)
PADNU - MNDAU	49	(−2749, 2846)
MNONU - MNDAM	−1058	(−3856, 1739)
MIDAU - MNDAM	433	(−2365, 3230)
MIOAU - MNDAM	−47	(−2845, 2750)
PADNU - MNDAM	−223	(−3020, 2575)
MIDAU - MNONU	1491	(−1306, 4289)
MIOAU - MNONU	1011	(−1786, 3809)
PADNU - MNONU	836	(−1962, 3633)
MIOAU - MIDAU	−480	(−3278, 2318)
PADNU - MIDAU	−655	(−3453, 2142)
PADNU - MIOAU	−175	(−2973, 2622)

Note: Bold font indicates statistically significant comparisons.
[a]UBOL = unbonded concrete overlay of existing concrete pavements.

performed at midslab to establish the interlayer Totsky k-value that produces the closest measured deflection response. Mesh convergence was achieved by examining the deflection and overlay slab stress beneath the center slab load and the mesh size was 4 in. Three sensors were used to define the deflection, including directly under the load plate, and the sensors at $\pm$ 12 in from the applied FWD load. Slabs that exhibited cracking and had a corresponding center slab drop after the cracking had initiated were excluded from this analysis to attempt to isolate the effect of the interlayer on the resulting response. A batch of runs were then generated for Totsky interlayer k-value in increments of 100 psi/in. The FWD deflections were then matched to the Totsky k-value using linear interpolation to obtain the interlayer stiffness. The results of the Totsky k-value determination are presented in Figure 5. The range of interlayer k-values is 1180–8770 psi/in with an average value of 3900 psi/in for the cells with the PASSRC interlayer. The range of interlayer k-values is 135–900 psi/in with an average value of 425 psi/in for the cell with the nonwoven geotextile fabric interlayer. The Totsky k-values established from the laboratory testing are also plotted for comparison.

As can be seen in Figure 5, there is no apparent trend between the interlayer k-value and asphalt temperature, which is consistent with the laboratory data, in that there was no apparent trend between different asphalts with varying stiffness. Statistical testing was carried out to see if a statistically significant difference could be identified between the k-values obtained from the laboratory specimens and those found using the FWD testing at MnROAD. Student t-tests were carried out using the null hypothesis that the mean laboratory k-values are equal to the mean k-values obtained from the FWD testing. These results are summarized in Table 7 below. Table 7 shows that the difference between the Totsky k-values established using FWD results and the laboratory specimens is statistically significant for both asphalt and fabric interlayers. Finally, there is not a significant difference between the Totsky k-values established from the laboratory specimens and the FWD results for the fabric interlayer or the open graded Minnesota asphalt.

Table 6. UBOL MnROAD Cells

Cell	Construction date	Slab size[a] (length × width) (ft × ft)	Dowels (in)	Overlay concrete thickness (in)	Interlayer thickness (in)	Interlayer type	Existing concrete thickness (in)
105	10/8/08	15 × 14	None	4	1	Permeable asphalt (PASSRC)	7.5
205	10/8/08	15 × 14	None	4	1	Permeable asphalt (PASSRC)	7.5
305	10/8/08	15 × 14	None	5	1	Permeable asphalt (PASSRC)	7.5
405	10/8/08	15 × 14	None	5	1	Permeable Asphalt (PASSRC)	7.5 (cracked)
505	8/24/11	6 × 7	None	5	-	Fabric	7.5 (cracked)
605	8/24/11	6 × 7	None	5	-	Fabric	7.5

Note: PASSRC = permeable asphalt stabilized stress relief course.
[a]Sizes shown for driving lane. For sections 15 × 14, passing lane is 15 × 13. For sections 6 × 7, passing lane is 6 × 6.5. This matches the width of Cell 5 driving and passing lanes.

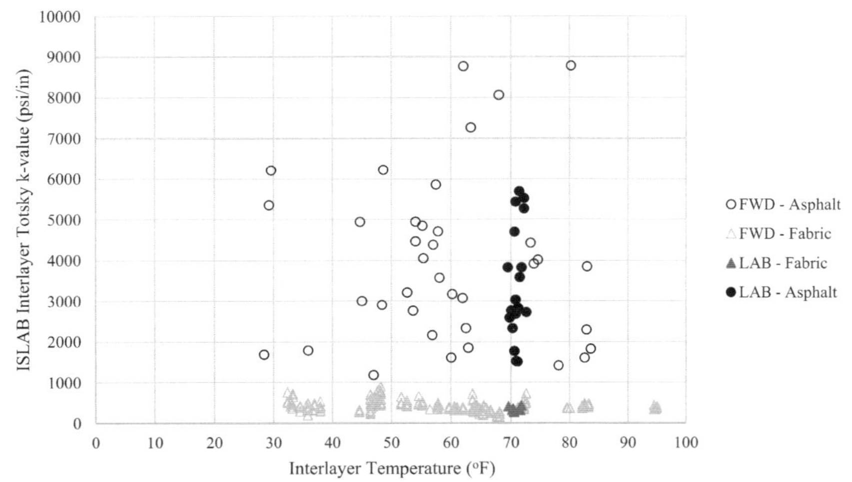

Figure 5. Interlayer Totsky k-value established for MnROAD cells 205, 305, 405 (asphalt) and 505, 605 (fabric) and the lab specimens.

Table 7. t-Tests Comparing FWD Totsky *k*-value Results

Comparison between means of established Totsky *k*-values	*p*-value of *t*-test for difference in means
Fabric LAB vs. MnROAD Fabric FWD	0.126
MNONU LAB vs. MnROAD Asphalt (PASSRC) FWD	0.137
MnROAD Fabric FWD vs. MnROAD Asphalt (PASSRC) FWD	<0.001

Note: MnROAD = Minnesota Road Research Facility; FWD = falling weight deflectometer; PASSRC = permeable asphalt stabilized stress relief course.

Sensitivity of Totsky *k*-value to Predicted Response

A sensitivity analysis was conducted using ISLAB to show the effect of the Totsky *k*-value on the response of the slab. The effect of the Totsky *k*-value and the overlay thickness on the resulting peak deflection from a midslab FWD load was examined. The overlay consists of 15 ft × 12 ft panels with a PCC elastic modulus of 4 million psi. The mesh size was 4 in. The joint LTE was set to 90%, and the existing pavement thickness was 8 in with a PCC elastic modulus of 5 million psi. The base and subgrade were modeled as a Winkler foundation with a *k*-value of 200 psi/in.

As can be seen from Figure 6, four different overlay thicknesses were considered, and the Totsky *k*-value was varied from 100 to 10,000 psi/in. As expected, the deflections for the established fabric *k*-value are greater than that for the asphalt *k*-value. In addition, the effect of the *k*-value diminished as the slab thickness increased. The range of asphalt Totsky *k*-values that were obtained from the laboratory and FWD analysis is roughly 1500–9000 psi/in. It can be seen for the analysis presented in Figure 6 that minimal variation occurs over the estimated deflections for this range, especially in the thicker overlays. This indicates that the sensitivity of the Totsky *k*-value to the calibrated asphalt stiffness is relatively low for the range of values obtained from the lab and the FWD analysis conducted.

Conclusions and Recommendations

Laboratory beam testing results and FWD tests from MnROAD cells were modeled in ISLAB to establish the interlayer Totsky *k*-value. From the laboratory testing, the only significant comparisons were that all asphalt interlayers, except for the open graded asphalt (PASSRC) interlayer used in Minnesota, were significantly different from the two fabric interlayers. The difference with the PASSRC (MNONU) could be because the gradation was more open graded and contained more rounded aggregate particles resulting in a lower strength and stiffness than the other interlayers. in addition, no apparent relationship exists between asphalt stiffness or thickness and the Totsky *k*-value for the asphalt interlayers included in this study. The Totsky *k*-values

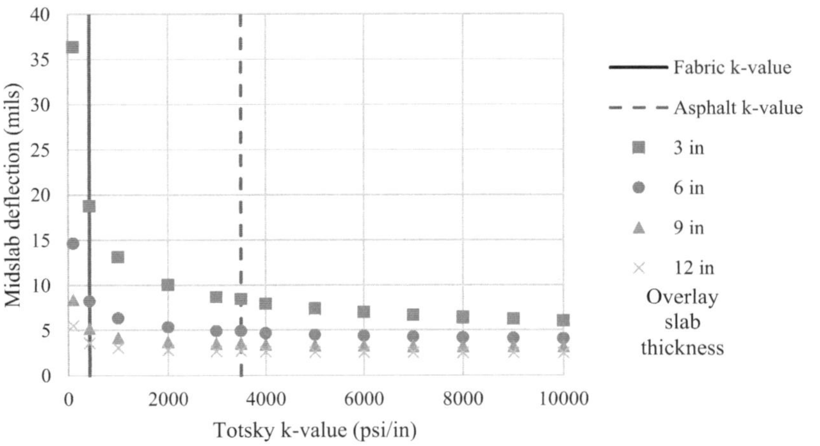

Figure 6. ISLAB sensitivity of Totsky *k*-value to predicted response.

obtained using the FWD data are not statistically different from those established using the laboratory specimens for the same interlayer type. The *k*-values established using FWD data were statistically different from one another for both the fabric and asphalt.

As there is no apparent trend between stiffness and asphalt type or stiffness and asphalt temperature, one *k*-value is recommended as an average for all asphalt interlayer types and temperatures. Averaging the results of both the laboratory and FWD investigation for asphalt interlayers produces an average value of approximately 3500 psi/in. This value is recommended for use in the development of a design procedure for UBOLs with an asphalt interlayer. No discernable difference was detected between fabrics having different weights; however, the fabric stiffness was shown to be statistically different from the asphalt stiffness. Therefore, one value is recommended as an average for all nonwoven geotextile fabrics. The average value for the laboratory and FWD results is 425 psi/in for the fabric interlayer.

Additional Considerations

It is important to remember that there are other characteristics of the interlayer system that also effect the performance of the UBOLs that were not considered in this investigation. This includes the drainage characteristics of the interlayer system. The fabrics and open graded HMAs provide increased drainage over dense graded HMA. Drainage through the interlayer prevents water pressure buildup. At the same time, sufficient resistance to stripping has to be provided so that the binder is not stripped away as water flows through the open graded HMA interlayer. An advantage to the fabric is that good drainage can be achieved without stripping concerns. The disadvantage is that the better the drainage characteristics the less stiff the interlayer tends to be, which might have an adverse effect on the fatigue life of the overlay. Friction and shear transfer, which develops at the interfaces is also a critical consideration. The interlayer system must be able to provide a slip plane to allow the overlay to move freely with respect to the existing pavement. However, field observations have indicated that some interlayer systems do not provide sufficient restraint to allow for joint activation. This leads to excessively long slabs with high curling stresses. The joints that do activate are excessively wide, which can result in the development of distress at the joints. Therefore, the interlayer system must provide a balance between allowing sufficient slip for free movement of the overlay and providing sufficient restraint for joint deployment. This study focused on solely defining the stiffness of the interlayer to be used in a design procedure but, as described above, many other factors contribute to the performance of the interlayer and must be considered.

Acknowledgments

The authors would like to thank the Georgia, Iowa, Kansas, Michigan, Minnesota, Missouri, North Carolina, and Oklahoma Departments of Transportation. These states participated in the FHWA Pooled Fund Study TPF 5-269 under which this work was performed. The authors would also like to thank the following people for their assistance in their research effort: Andrew Bennett, Benjamin Krom, and John Staton of MDOT; Thomas Burnham, Robert Strommen and Jack Herndon at the MnROAD Research Facility; Rob Golish from MnDOT; the McCrossan Company; Elwood Balog and James Foringer of PennDOT; and Golden Triangle Construction for their assistance. The supply of the nonwoven geotextile fabrics and technical expertise provided by Eric Little, formerly with Propex, was also greatly appreciated.

Author Contributions

The authors confirm contribution to the paper as follows: study conception and design: Vandenbossche and Khazanovich; data collection: Sachs; analysis and interpretation of results: Sachs and Tompkins; draft manuscript preparation: Sachs, Vandenbossche, Tompkins, and Khazanovich. All authors reviewed the results and approved the final version of the manuscript.

References

1. American Association of State Highway and Transportation Officials (AASHTO). *AASHTO Guide for Design of Pavement Structures*. Washington, D.C.: AASHTO, 1993.
2. Applied Research Associates (ARA), Inc., ERES Division. *NCHRP 1-37A Final Report: Guide for Mechanistic-Empirical Design of New and Rehabilitated Pavement Structures*. Transportation Research Board of the National Academies, Washington, D.C., 2004.
3. Departments of the Army, and the Air Force. *Rigid Pavements for Airfields Other than Army*. Technical Manual 5-824-3/Air Force Manual 88-6, Chapter 3, Washington, D.C., 1970.
4. Tayabji, S. D., and P. A. Okamoto. Thickness Design of Concrete Resurfacing. *Proc., 3rd International Conference on Concrete Pavement Design and Rehabilitation*, Purdue University, West Lafayette, Ind., 1985, pp. 367–379.
5. Rollings, R. S. *Design of Overlays for Rigid Airport Pavements*. Final Report, Report No. DOT/FAA/PM-87/19. Department of the Army, Geotechnical Laboratory, 1988.
6. Minnesota Department of Transportation (MnDOT). *Unbonded Concrete Overlay Design Procedures*. Minnesota Department of Transportation, Office of Materials and Road Research, 2010.
7. Totsky, O. N. *Behavior of Multi-Layered Plates and Beams on Winkler Foundation*. Stroitel'naya Mekhanika i Raschet Sooruzhenii, Moscow, 1981.
8. Khazanovich, L. *Structural Analysis of Multi-Layered Concrete Pavement Systems*. PhD thesis. University of Illinois, Urbana, 1994.
9. Khazanovich, L., and A. M. Ioannides. Structural Analysis of Unbonded Concrete Overlays under Wheel and Environmental Loads. *Transportation Research Record: Journal of the Transportation Research Board*, 1994. 1449: 174–181.
10. ASTM International. *Standard Test Method for Flexural Strength of Concrete (Using Simple Beam with Third-Point Loading)*. ASTM Standard C78, West Conshocken, Pa., 2015.
11. ASTM International. *Standard Practice for Making and Curing Concrete Test Specimens in the Laboratory*. ASTM Standard C192, West Conshocken, Pa., 2013.
12. ASTM International. *Standard Test Method for Measuring Pavement Macrotexture Depth Using a Volumetric Technique*. ASTM Standard E965, West Conshocken, Pa., 2015.
13. Sachs, S. G., J. M. Vandenbossche, K. Alland, J. DeSantis, and L. Khazanovich. Laboratory Investigation into the Effects of Different Interlayer Systems on Reflective Cracking in Unbonded Overlays of Existing Concrete Pavements (UBOLs). *Transportation Research Record: Journal of the Transportation Research Board*, 2016. 2591: 33–41.
14. Khazanovich, L., H. T. Yu, S. Rao, K. Galasova, E. Shats, and R. Jones. ISLAB 2000- Finite Element Analysis Program for Rigid and Composite Pavements: User's Guide. ERES Consultant, Urbana Champaign, Ill., 2000.
15. Montgomery, D. *Design and Analysis of Experiments*, 8th ed. John Wiley & Sons, Inc., New York, 2012.
16. Watson, M., and T. Burnham. *Construction Report for MnROAD Thin Unbonded Concrete Overlay Test Cell 5 (Sub-Cells 105 - 405) Final Report*. Minnesota Department of Transportation, Office of Materials and Road Research, June 2010.
17. *Standard Specifications for Construction*. Minnesota Department of Transportation, St. Paul, Minn., 2016.

The Standing Committee on Design and Rehabilitation of Concrete Pavements (AFD50) peer-reviewed this paper (18-04111).

Article

TRR
JOURNAL OF THE TRANSPORTATION RESEARCH BOARD

Transportation Research Record
2018, Vol. 2672(40) 264–274

DOI: 10.1177/0361198118796383
journals.sagepub.com/home/trr
SAGE

A Tale of Two Pavements: Forensic Investigation of an Unbonded Concrete Overlay and a Concrete Pavement Reconstruction on I-40 near Flagstaff, Arizona

Thomas J. Van Dam[1], Nicole Dufalla[1], and Linda Pierce[1]

Abstract

In 1990, the Arizona Department of Transportation (ADOT) constructed a Specific Pavement Studies (SPS)-6 experiment on eastbound Interstate 40 (I-40) east of Flagstaff, AZ, as part of the Long-Term Pavement Performance Program. The SPS-6 experiment was designed to evaluate treatment and rehabilitation options for an existing concrete pavement. Nineteen sections were constructed, one of which was a 10 in. thick unbonded concrete overlay (UBCOL). Additionally, ADOT planned to construct an 11 in. unbonded portland cement concrete (PCC) overlay on the adjacent I-40 westbound to investigate benefits of additional concrete thickness. The eastbound unbonded PCC overlay has significantly outperformed all other pavement sections and remains in good condition after 27 years of service. However, the westbound PCC section is severely distressed and in need of immediate rehabilitation. An investigation was conducted to evaluate why two similar unbonded PCC overlays performed with vastly different results. It was found that the westbound "unbonded PCC overlay" was not an overlay after all, but instead a 12 to 13 in.thick jointed plain concrete pavement (JPCP). The high level of distress in the westbound JPCP is likely related to the presence of a weak subgrade that infiltrated the aggregate base. Additionally, during coring, water filled the core holes suggesting that the subgrade in the westbound direction was saturated. Further, petrographic analysis indicated deicing chemicals contributed to corrosion of the dowel bars and deterioration of the concrete joints. This investigation demonstrates some advantages for using unbonded overlays as rehabilitation alternatives for existing concrete pavements.

The Arizona Department of Transportation (ADOT) Specific Pavement Studies 6 (SPS-6) project was constructed in 1990 on a 2.7 mile segment of a 10 mile ADOT rehabilitation project on Interstate 40 (I-40) eastbound, east of Flagstaff, AZ, extending from milepost 202.16 to milepost 204.84. The original pavement structure consisted of 8 to 9 in. of undoweled jointed plain concrete pavement (JPCP) on 7 in. of cement-treated base on existing subgrade material, as seen in Figure 1. The existing JPCP was severely distressed with spalled transverse and longitudinal joints and cracks (reported to be severe durability cracking), and shattered (broken into three or more pieces) slabs.

The objective of the Long-Term Pavement Performance Program (LTPP) SPS-6 projects was to evaluate the overall performance of specific maintenance and rehabilitation treatments on JPCP (*1*). Consistent with other SPS-6 sites, the average length of each experimental section was established to be approximately 500 ft. In addition to the eight core SPS-6 sections, the ADOT SPS-6 site includes eleven ADOT supplemental sections. The treatments evaluated are as follows (*2*):

- LTPP Core Sections:
  - Crack and seat existing JPCP using a nominal crack spacing of 3 × 3 ft followed by an asphalt concrete (AC) overlay.
  - Minimum JPCP restoration included partial-depth spall repair and full-depth patching.
  - Maximum JPCP restoration included removing and replacing AC patches, full-depth

[1]NCE (Nichols Consulting Engineers), Reno, NV

Corresponding Author:
Address correspondence to Thomas J. Van Dam: tvandam@ncenet.com

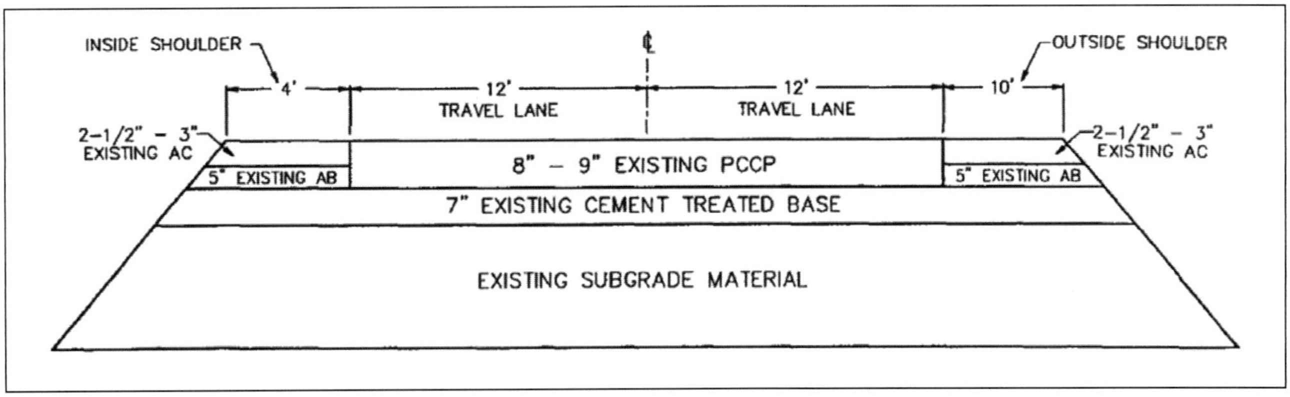

Figure 1. Existing JPCP structure prior to SPS-6 construction.

remove and replace existing distressed slabs, partial-depth spall repair, and milling and replacing existing AC shoulders.

- ADOT Supplemental Sections:
  - Rubblize JPCP into nominal 1 to 2 in. pieces, compact with a vibratory roller, followed by a gap-graded asphalt rubber AC (ARAC) overlay.
  - Unbonded portland cement concrete (PCC) overlay with 2 in. AC bond breaker, created from milling the existing 8 in. overlay to a thickness of 2 in. and using it as a bond breaker.
  - Crack and seat existing JPCP using a nominal crack spacing of 4 × 6 ft followed by an ARAC overlay.
  - Crack and seat existing JPCP using a nominal crack spacing of 4 × 6 ft, placement of non-woven geotextile, followed by an AC overlay.
  - ARAC overlay of existing JPCP (no pre-treatments).

In addition, ADOT investigated the effect of a marginal increase in UBCOL thickness on performance. Later in 1990, following the completion of the SPS-6 construction, ADOT constructed what was reported to be an 11 in. thick UBCOL on westbound I-40 directly opposite the SPS-6 10 in. thick unbonded PCC overlay on eastbound I-40. However, it was discovered in the course of this investigation that the UBCOL was not constructed; instead, the on-site investigation revealed the constructed section consists of 12 to 13 in. of doweled JPCP placed over an unknown depth of aggregate base. Despite extensive efforts to determine the reason for the discrepancy between the designed and constructed pavements, including contacting both current and former ADOT personnel, the exact reason for this change in pavement type remains unknown. But this happenstance did provide an excellent opportunity to contrast the performance of these two pavement types.

Forensic Investigation

In May 2017, a forensic investigation was initiated to review the long-term performance of the ADOT SPS-6 project to obtain potential long-term benefits of the studied treatments. The sections included in this forensic study were 040663, 040608, 040607, 040606, 040659, 040604, and 040603. Additionally, the forensic investigation included an evaluation of what was thought to be the slightly thicker UBCOL placed contemporaneously on westbound I-40 directly across the median from section 040663.

The forensic evaluation included a review of all available construction documentation, visual condition assessment, coring and dynamic cone penetrometer (DCP) testing, falling weight deflectometer (FWD) testing, and petrographic analysis of extracted concrete cores. All site investigation activities were conducted June 6–8, 2017.

Construction Documentation

Construction of the UBCOL on eastbound I-40 included milling off approximately 6 in. of an existing 8 in. thick AC overlay, leaving nominally 2 in. of AC as a bond-breaker. A 10 in. thick unbonded JPCP overlay was then placed on the milled surface. Since construction of the SPS-6 sections in 1990, ADOT has applied numerous treatments to the AC-surfaced sections, but there is no record or evidence of any treatments being applied to the UBCOL during its 27 years of service.

Information on the westbound ADOT JPCP section was more difficult to find. No construction records were available—it was confirmed that they had been destroyed—and almost every ADOT employee involved has since retired. However, the former resident engineer recalled that the change in pavement type occurred due to the installation of the weigh-in-motion (WIM) traffic device located at the western edge of the concrete pavement section, and the construction of JPCP pavement rather than an unbonded overlay greatly simplified the construction geometrics. Standard practice at that time dictated placing the JPCP on an unspecified thickness of aggregate base. Results from coring conducted for this investigation revealed that the fine-grained subgrade has fully infiltrated with the aggregate base. The westbound I-40 JPCP required an AC maintenance overlay and full-depth AC patch placed in 2013 in the right lane (truck lane) to improve ride quality due to severe slab cracking, displacement, and spalling. Additionally, subsealing using polyurethane was conducted in an attempt to stabilize settled/rocking slabs in multiple unknown locations.

Visual Condition Assessment

The ADOT SPS-6 sections were placed out-of-study in 2002 as most sections other than the UBCOL required overlays to maintain serviceability. At that time, the 10 in. thick unbonded PCC overlay exhibited the best performance among all sections in the SPS-6 experiment, with the only observed distresses being longitudinal joint seal damage, polished aggregate, and some longitudinal joint spalling (*1*). No visual distress surveys are reported for the westbound I-40 JPCP in 2002.

In May 2017, a condition assessment was completed in accordance with the LTPP distress identification procedures (*3*). The condition assessment summary for the eastbound SPS-6 UBCOL section and the westbound JPCP section are shown in Table 1. The condition survey for the westbound JPCP was conducted as an asphalt-surfaced roadway due to the presence of the AC maintenance overlay and patch which fully covered the surface. However, since the interest of this investigation is the performance of the underlying concrete pavement and the overlay is relatively thin, the cracking noted in Table 1 only includes transverse and longitudinal cracking visible as reflection cracks, and not cracking over joints.

The eastbound UBCOL is currently performing well, as seen in Figure 2. Current distresses include joint seal damage, polished aggregate Figure 2*c*), minor longitudinal joint spalling, and limited asphalt patching associated with the installation of a WIM station (Figure 2*e*). Figure 3 shows the unsealed joints and polished aggregate typical of the section in detail.

Conversely, the westbound JPCP test section is in poor condition, with typical distresses seen in Figure 4, exhibiting significant longitudinal and transverse slab cracking and a large, distressed full-depth asphalt patch. Further, the underlying cracks and joints are showing evidence of deterioration although these were not rated in the distress survey, which was conducted for an asphalt surface. Severe observations from this pavement are shown in Figure 5. Figure 5*a* shows extreme reflection cracking in the AC overlay at frequent intervals. A full-depth AC repair is shown in Figure 5*b* with distresses including rutting, shoving, and fatigue cracking. Finally, visible joint deterioration in the concrete pavement below the overlay is shown in Figure 5*c*.

Table 1. Condition Assessment Summary for Eastbound Unbounded Concrete Overlay and Westbound JPCP

Distress type	UBCOL	JPCP[a]
Transverse joint seal damage (count)	29	—
Transverse cracking (ft)	0	1,977
Longitudinal cracking (ft)	0	2,555
Spalling transverse joints (ft)	17	—
Asphalt patching (ft^2)	56	6,000
Longitudinal joint seal damage (ft)	869	—
Polished aggregate (ft^2)	526	—

[a]Due to AC maintenance patch, distress only includes transverse and longitudinal cracking and AC patching. Reflection cracking over joints is not counted.

Coring and DCP Testing

Three cores were extracted from the eastbound UBCOL and four cores from the westbound JPCP to verify thicknesses, assess the materials, conduct DCP testing, and to obtain samples for concrete petrographic analysis. The coring information is summarized in Table 2, including core location, tests to be conducted, general conditions, and DCP testing results. It is observed that the DCP calculated California Bearing Ratio (CBR) values vary with depth, and alternating soft and hard soil layers were observed during testing.

FWD Testing and Analysis

FWD testing was conducted on all sections evaluated as part of this forensic investigation to determine layer moduli and load transfer efficiency (LTE). LTE was measured at transverse joints.

Back Calculation of PCC Modulus and k-Value

The AREA method was used to estimate PCC elastic modulus and static k-value as described by Hall (*4*) and estimates of these values for the eastbound UBCOL and westbound JPCP are provided in Table 3. Although the backcalculated average k-value for the eastbound UBCOL is lower than the average k-value for the

Figure 2. Typical conditions observed in eastbound SPS-6 UBCOL: (*a*) begin section (looking east); (*b*) transverse joint (typical); (*c*) polished aggregate; (*d*) wheelpath wear (tine marks missing in wheelpaths); (e) AC patch of weigh-in-motion scale; and (*f*) end section (looking west).

westbound JPCP (196 lbf/in.^{2}/in. versus 345 lbf/in ^{2}/in , respectively), subgrade support for the UBCOL is relatively consistent, while the k-value for the JPCP is quite variable, ranging from a low of 85 lbf/in.^{2}/in. to a high of 598 lbf/in.^{2}/in. Similarly, the average PCC modulus for the two sections falls within typical ranges for PCC yet the PCC moduli for the JPCP range from approximately 124,000 lb/in.^{2} to 5,900,000 lb/in.^{2}. The relatively more consistent values for the eastbound UBCOL support the uniformly good observed pavement condition, whereas

Figure 3. Close up of joint seal damage and polished aggregates observed in eastbound SPS-6 UBCOL.

the variation in results for the westbound JPCP indicate a pavement structure with a high variability across the length of the pavement.

Load Transfer Efficiency

The FWD data was used to determine the LTE across the joints of the two sections and these values are summarized in Table 4. The load transfer values for the UBCOL were relatively uniform and satisfactory, with an average of 85% on both the approach and leave sides and a range from 75% to 93%. Conversely, the load transfer efficiencies measured on the westbound JPCP section averaged fair to good (60% on the approach side and 80% on the leave side), but exhibited significant variability on the approach side, ranging between 15% and 99%.

Petrographic Analysis

Five cores were sent to DRP Consulting, Inc., for petrographic analysis. As detailed in Table 5, the submitted cores included two cores from the eastbound UBCOL section (cores 0663A and 0663C) and three cores from the westbound JPCP section (cores 0699A, 0699B, and 0699C1). The petrographic analysis was conducted in accordance with ASTM C856, *Standard Practice for Petrographic Examination of Hardened Concrete*, which

Table 2. Core Description and DCP Results

Core no.	Core location	Tests	Core description	CBR range from DCP with depth[a]
UBCOL 0663A	Center of panel	DCP, petrographic analysis	9.6 in. PCC, good condition 3.1 in. AC, good condition, debonded from lower PCC 8.8 in. PCC, cracked in pieces	20–57 (<24 in.) >57 (24–36 in.)
UBCOL 0663B	Outside wheelpath, mid-panel	DCP	10.4 in. PCC, good condition 3.0 in. AC, good condition, debonded from lower PCC 8.5 in. PCC, good condition	17–85 (<24 in.) >65 (24–36 in.)
UBCOL 0663C	Outside wheelpath, transverse joint	Petrographic analysis	10.2 in. PCC, dowel bar corrosion and cracking 2.4 in. AC, stripping, debonded from PCC layers 8.5 in. PCC, cracking in lower lift	N/A
JPCP 0699A	Lane/shoulder longitudinal joint	DCP, petrographic analysis	13.1 in. PCC, cracked full-depth No detectable base Weak subgrade	20–85 (<24 in.) 3–85 (24–36 in.)
JPCP 0699B	Outside wheelpath, transverse joint	DCP, petrographic analysis	1.4 in. AC maintenance cold patch 12.8 in. PCC, good condition No detectable base Weak subgrade	28–85 (<24 in.) 20–48 (24–36 in.)
JPCP 0699C1	Outside wheelpath, mid-panel	Petrographic analysis	1.2 in. AC maintenance cold patch, some delamination 11.6 in. PCC, cracked full-depth No detectable base	N/A
JPCP 0699C2	Outside wheelpath, transverse joint	None	11.8 in. PCC, broken into several pieces No detectable base	N/A

[a]Depth <24 in. represents subgrade CBR from bottom of core to a depth of 24 inches.

Figure 4. Typical conditions observed in westbound JPCP section: (*a*) begin section (looking east); (*b*) underlying JPCP failure; (*c*) reflection cracking; (*d*) reflection cracking; (e) spalled reflection cracking; and (*f*) end section (looking west).

included visual and optical microscopical examination of the core surfaces and of polished slabs made from the submitted cores, and phenolphthalein staining of longitudinal sections of each core.

Table 5 indicates whether basic or comprehensive ASTM C856 testing was conducted. Basic petrography was performed all on cores and included examination of polished concrete slabs using a stereo optical microscope. Several specimens were then selected for comprehensive examination which entailed examining thin sections using a petrographic optical microscope. In addition, ASTM C1152 was used to determine acid-soluble chloride contents to better understand the degree to which chemical deicing salts had permeated into the concrete.

Figure 5. Severe conditions observed in the westbound JPCP lanes: (*a*) slab cracking observed in westbound JPCP; (*b*) rutting, shoving, and fatigue cracking of full-depth AC repair in westbound JPCP section; and (*c*) deterioration of transverse joints in the westbound JPCP.

Table 3. Summary of k-Value and PCC Modulus

	UBCOL		JPCP	
Statistic	k-value (lbf/in.^{2}/in.)	PCC modulus (lb/in.^{2})	k-value (lbf/in.^{2}/in.)	PCC modulus (lb/in.^{2})
Average	195	4,000,000	345	2,250,000
Std deviation	22	660,000	155	1,495,000
Median	198	3,925,718	345	2,341,626
Minimum	145	2,863,735	85	124,429
Maximum	236	5,468,356	598	5,902,325
25th percentile	177	3,380,000	225	1,030,000
75th percentile	210	4,340,000	481	3,080,000

Based on the petrographic analysis, two distinct concrete mixtures are observed in the extracted cores. The concrete observed in the overlay (top) of the eastbound UBCOL and in the new westbound JPCP were similar, having hardened cement paste containing hydrated portland cement and fly ash, a natural gravel coarse aggregate (siliceous in composition), and a natural sand fine aggregate from the same source. The concrete is air-entrained with an estimated air content range between 6% and 8%.

The concrete from the original pavement (the bottom concrete layer from the UBCOL section) contained no supplementary cementitious materials and had a significantly higher estimated water/cementitious materials ratio than the newer concrete. The coarse aggregate is a crushed fine-grained siliceous volcanic rock. The charac-

Table 4. Summary of LTE Results

Statistic	UBCOL	JPCP
Average	85/85	60/80
Std deviation	4/3	30/4
Median	86/86	61/80
Minimum	75/79	15/73
Maximum	93/90	99/86
25th percentile	83/84	30/77
75th percentile	87/87	91/84

Note: Values shown are for the approach/leave side of the joint in percentages.

Table 5. Summary of Samples and Petrographic Testing Schedule

Core No.	Layer	Testing
0663A	Overlay (top)	• ASTM C856 comprehensive petrography • ASTM C1152 acid soluble chloride
0663A	Original (bottom)	• ASTM C856 comprehensive petrography • ASTM C1152 acid soluble chloride
0633C	Overlay (top)	• ASTM C856 basic petrography
0633C	Original (bottom)	• ASTM C856 basic petrography
0699A	New JPCP	• ASTM C856 basic petrography
0699B	New JPCP	• ASTM C856 comprehensive petrography • ASTM C1152 acid soluble chloride
0699C	New JPCP	• ASTM C856 comprehensive petrography • ASTM C1152 acid soluble chloride

teristics of this aggregate suggest that it is not susceptible to freeze–thaw damage even though the reported distress affecting this pavement was durability cracking (D-cracking), which is the result of coarse aggregate freeze–thaw damage. The fine aggregate is a natural sand. The concrete is air-entrained with a total air content range estimated between 8% and 10%.

In both concrete types, the entrained air-void systems appear to be acceptable to protect the bulk concrete against freeze thaw damage and no clear evidence of cracking or microcracking near the exposed surface due to freeze–thaw damage was observed. However, the aggregates in both concrete mixtures contain an abundance of aggregate particles potentially susceptible to alkali–silica reaction (ASR). The newer concrete in the UBCOL and JPCP exhibited little evidence of ASR with

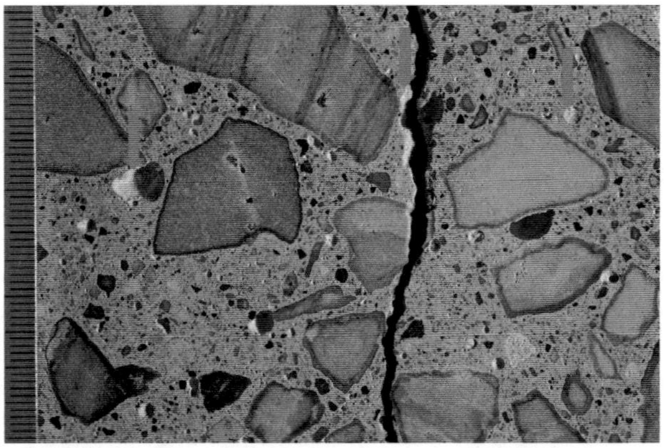

Figure 6. Gel deposits (arrows) in original pavement from Core No. 0663A. Note dark "reaction rims" common in concrete affected by ASR surrounding coarse aggregate particles in original (bottom) concrete from test section 040663.

rare deposits of gel observed in voids. No cracking, microcracking, or any other form of deterioration or damage was observed in association with ASR in these cores, suggesting that the use of fly ash may have had a mitigating effect. In the original older concrete found in the eastbound UBCOL section, gel deposits were found on large fracture surfaces and large reaction rims were observed around the coarse aggregate particles, as seen in Figure 6. However, there was no clear evidence of microcracking directly related to ASR as the deposits of gel may have followed pre-existing cracks or caused the cracking. Further, these cores do not show an abundance of microcracking related to ASR which is commonly observed when such reactions drive internal expansion.

Figure 7 presents the ASTM C1152 acid-soluble chloride testing results, revealing very high concentrations of chloride at the surface (top) of cores obtained away from a joint compared with the middle of the same core. These results are indicative of heavy exposure to chloride-based deicing chemicals. Further, in the one core subjected to chloride testing obtained from a joint (JPCP Core 0699C), the chloride content at the surface is almost identical to that at the middle of the core. This strongly suggests that chloride solutions are penetrating into the joint and being absorbed into the concrete along the joint sidewall. The corrosion of the dowel bar obtained from the UBCOL (Core No. 0663C) showed evidence of significant corrosion, as seen in Figure 8. Further, crystalline deposits that contain chloride were observed in the paste in this core and crystalline exudations were observed in the UBCOL Core No. 0663A after exposure to elevated temperature (105°F) and humidity (100% relative humidity) indicating that the potential expansion from ASR exists and can be seen in Figure 9.

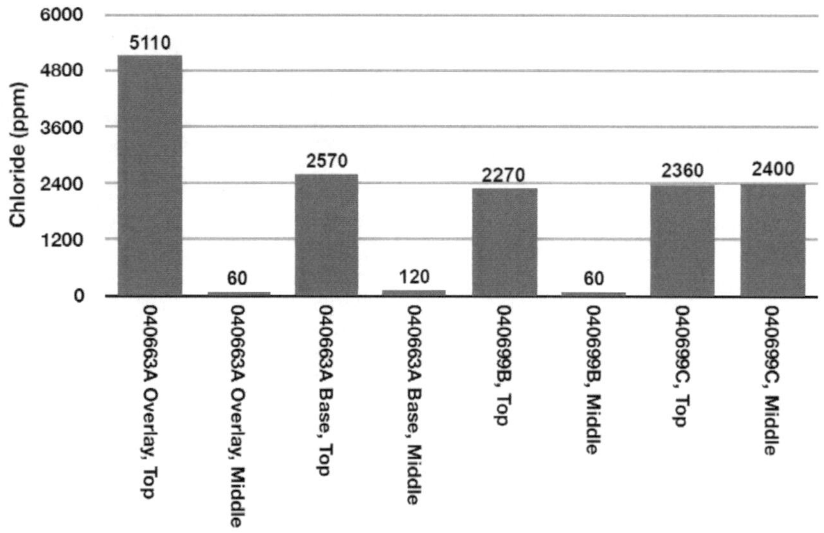

Figure 7. Bar chart summarizing ASTM C1152 acid-soluble chloride data.

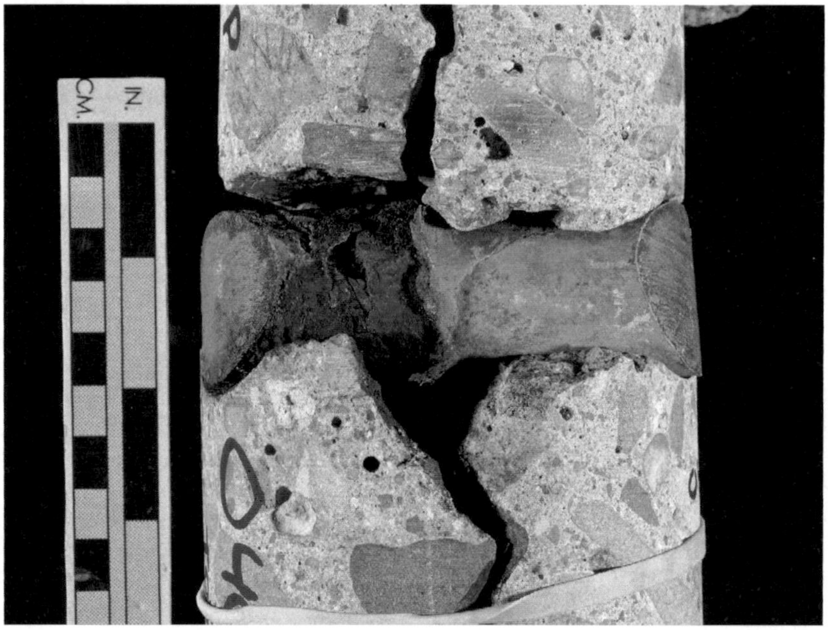

Figure 8. Corroded steel dowel in overlay of Core No. 0669C. Notice the green epoxy coating is no longer effective.

Conclusion

The following conclusions can be drawn from this forensic study of the concrete pavement sections associated with the ADOT SPS-6 site:

- Although a 10 in. thick UBCOL was constructed in the eastbound direction of I-40 as planned, the 11 in. thick UBCOL was not constructed as designed in the adjacent westbound direction;

Figure 9. Reflected light photomicrograph of the polished surface of Core No. 0699C (top). Shows chloride containing crystalline deposits observed as exudations after exposure to elevated temperature and humidity conditions.

instead, core and DCP results indicated that this section was constructed as a 12 to 13 in. JPCP placed over an unspecified thickness of untreated aggregate base, or possibly even on subgrade.

- The eastbound UBCOL has performed well after 27 years of service with minimal distress even though little to no maintenance has been conducted. In contrast, the westbound JPCP has performed poorly, suffering significant slab cracking and shattered slabs. An AC maintenance overlay was applied over the entire truck lane in 2003 and a large AC patch was needed to replace one or two slabs. The weak and saturated subgrade layer, as indicated by the flow of water into the core holes even after being vacuumed dry, contributed to the settlement and cracking (in most cases shattered) of the westbound JPCP section. Currently, the JPCP is in need of replacement.
- The average k-value and average PCC modulus estimates for the UBCOL and JPCP sections, in general, are within expected values. However, estimated values for the JPCP are quite variable, indicating areas of distressed to sound pavement structure. The average load transfer values at joints and cracks for both sections are good, but they are better and far more uniform for the UBCOL section. For the JPCP, the approach side joint load transfer efficiencies are highly variable, ranging from 15% to 99%.
- Petrographic analysis and acid-soluble chloride testing revealed that high levels of deicing have occurred. Further, the chlorides likely played a factor in the corrosion of the steel dowel bars and joint deterioration. It was also revealed that the use of an acceptable supplementary cementing material (SCM) in the new concrete can help mitigate the damaging effects of ASR, some evidence of which existed in the original concrete, which did not contain any SCM.

In 2016, the average annual daily traffic was reported to be 11,400 in the eastbound direction and 10,800 in the westbound direction and both directions had approximately 26% truck traffic (*5*). Under this high volume of truck traffic, the UBCOL has performed exceptionally well after 27 years of service in a harsh freeze–thaw environment. It has remained in service with little maintenance even though all the other ADOT SPS-6 test sections were removed from the study in 2002 and subsequently overlaid. The UBCOL constructed in the eastbound lane also significantly outperformed the contemporaneously constructed 12 to 13 in. thick JPCP in the adjacent westbound lane which was subjected to similar, although slightly lower, truck traffic. This study clearly illustrates the potential benefits offered by an UBCOL for the treatment of a deteriorated concrete pavement over other treatment options.

Acknowledgments

The project team acknowledge the Arizona Department of Transportation and the Federal Highway Administration, specifically Jack Springer, for supporting this project. In addition, the authors thank Chad Auker and Tod Skinner for their assistance with this investigation. Larry Scofield's initial review and discussion of need for this forensic study and review of the final report on this important project is also greatly appreciated. The comprehensive information stored in the LTPP database assisted the researchers in conducting this investigation.

Author Contributions

The authors confirm contribution to the paper as follows: Study conception and design: The broader study was conceived and designed under the auspices of the Federal Highway Administration's Long-Term Pavement Performance (LTPP) program. The specifics of this study to compare the unbonded concrete overlay to the newly constructed concrete section were devised by T. Van Dam and L. Pierce. Data collection: Field data was collected by a team of accredited LTPP evaluators. Concrete petrography was conducted by Dr David Rothstein of DRP, a Twining Company. T. Van Dam and N. Duffala participated in the field and collected supplemental data specific to this study. Analysis and interpretation of results: T. Van Dam, N. Duffala, and L. Pierce. Draft manuscript preparation: T. Van Dam and N. Duffala. All authors reviewed the results and approved the final version of the manuscript.

References

1. Puccinelli, J., S. Karamihas, K. Hall, and K. Senn. *Performance Evaluation of Arizona's LTPP SPS-6 Project: Strategic Study of Rehabilitation Techniques.* Report No. FHWA-10-396. Arizona Department of Transportation, Phoenix, AZ, 2010.
2. Austin Research Engineers. *SPS-6: Rehabilitation of Jointed Portland Cement Concrete Pavements.* Report No. FHWA-AZ92-379-II. Arizona Department of Transportation, Phoenix, AZ, 1992.
3. Miller, J. S., and W. Y. Bellinger. *Distress Identification Manual for the Long-Term Pavement Performance Program,* 4th ed. FHWA-RD-03-031. Federal Highway Administration, McLean, VA, 2003.
4. Hall, K. T. *SHRP Contract P-020: Backcalculation Solutions for Concrete Pavements.* Transportation Research Board of the National Academies, Washington, D.C., 1992.
5. Arizona Department of Transportation. *HPMS Location Report for Year 2016: Traffic Counts.* 2016. https://www.azdot.gov/docs/default-source/planning/2016-aadt-publication-interstate.pdf?sfvrsn = 2 Accessed May 10, 2017.

The Standing Committee on Design and Rehabilitation of Concrete Pavements (AFD50) peer-reviewed this paper (18-03256).

Article

TRR
JOURNAL OF THE TRANSPORTATION RESEARCH BOARD

Transportation Research Record
2018, Vol. 2672(40) 275–281

DOI: 10.1177/0361198118787632
journals.sagepub.com/home/trr
SAGE

A Long-Term Performance Evaluation of an Experimental Concrete Overlay

James Greene[1], Ohhoon Kwon[1], Abdenour Nazef[1], and Bouzid Choubane[1]

Abstract

Flexible pavements are often rehabilitated by milling distressed asphalt and placing new asphalt at a thickness that accounts for expected traffic growth and pavement life. However, there are many reported benefits to concrete overlays as a method to rehabilitate and preserve distressed asphalt pavements. In 1988, the Florida Department of Transportation designed and constructed an unbonded concrete overlay on US-1 between Daytona Beach and Titusville. The 1.9-mile concrete overlay was part of a larger 8-mile milling and resurfacing of a deteriorated asphalt pavement. The concrete overlay test sections were divided into three groups based on design thicknesses of 6, 7, and 8 inches. Each of these groups included subsections with three joint spacing levels and two dowel bar configurations consisting of standard 12-inch spacing and wheel path only. The overlay sections are still in service with no major rehabilitation effort. The primary distress is pavement roughness found in sections with wheel path dowels only. The section with the most cracked slabs had joint spacings of 20 ft, which is no longer recommended. This paper documents the experimental program and presents the analysis and findings.

Flexible pavements are often rehabilitated by milling distressed asphalt and placing new asphalt at a thickness that accounts for expected traffic growth and pavement life. However, there are many reported benefits to concrete overlays as a rehabilitation and preservation technique. In addition to being cost-effective, properly designed concrete overlays can be constructed quickly, can be easy to maintain, and can extend the life of the original pavement for 30 years or more (*1*). Concrete overlays are designed as bonded or unbonded. The type of design is primarily dictated by the condition of the existing asphalt pavement (*1*). In the United States, more than 200 concrete overlay projects have been constructed each decade since the 1980s. Of these projects, nearly 80% were unbonded concrete overlays and approximately 40% of these unbonded concrete overlays were constructed over an existing asphalt pavement rather than a concrete or composite pavement (*2*).

In 1988, the Florida Department of Transportation (FDOT) designed and constructed an unbonded concrete overlay on US-1 between Daytona Beach and Titusville. The 1.9-mile concrete overlay was part of a larger 8-mile milling and resurfacing of a deteriorated asphalt pavement. The concrete overlay test sections were divided into three groups based on design thicknesses of 6, 7, and 8 inches. Each of these groups included six 500 ft subsections with three joint spacing levels and two dowel bar configurations.

Concrete Overlay Types

Bonded concrete overlays may be used to add structural capacity and eliminate surface distresses. The existing asphalt pavement must be at least 3 inches thick and in good to fair condition with limited fatigue cracking and no raveling or stripping. Bonded concrete overlays range from 3 to 6 inches in thickness. The bond between the concrete and asphalt layers is critical. The asphalt base stiffness reduces the concrete stress and also provides resistance to differential movement of the asphalt and concrete (*3*). A MnROAD study found that distresses formed soon after concrete panels became unbonded (*4*). Research sponsored by the Colorado Department of Transportation recommended that bonded concrete overlays be considered as partially bonded composite pavement systems and the design procedure should account for the additional stress resulting from a partial rather than full bond (*5*). Concrete panel sizes ranging from 3 to 8 ft are recommended to reduce differential movement between the asphalt and concrete and curling/warping stresses (*1*, *5*).

[1]Florida Department of Transportation, State Materials Office, Gainesville, FL

Corresponding Author:
Address correspondence to Bouzid Choubane:
bouzid.choubane@dot.state.fl.us

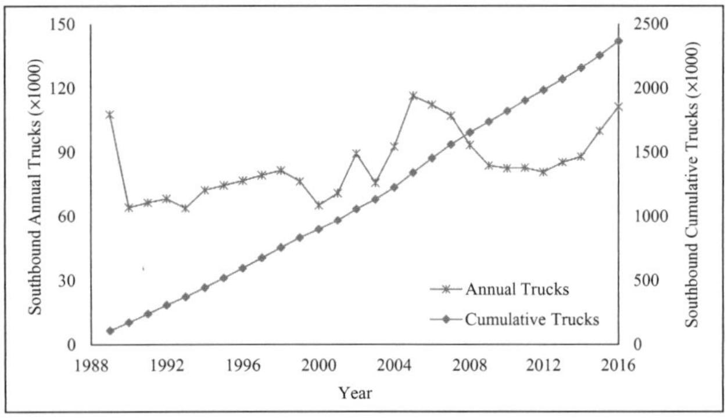

Figure 1. Truck traffic applied to the concrete overlay sections.

Figure 2. Application of lime slurry on the asphalt surface.

As indicated earlier, unbonded concrete overlays are more common and less complex since bonding of the asphalt and concrete is not essential to good performance. Unbonded concrete overlays are typically greater than 4 inches and are used to restore structural capacity to an asphalt roadway with moderate to significant deterioration. Since the bond is not critical the overlay may be designed as a new concrete pavement using several design methods. The asphalt layer primarily serves as a stable, non-erodible base (*3*). Asphalt milling and patching may be required to ensure the asphalt base provides adequate support.

Research Objectives

The primary objective of this investigation was to evaluate the effectiveness of an unbonded concrete overlay as a rehabilitation method for distressed asphalt pavements. To accomplish this study, the effect of different design factors such as overlay thickness, joint spacing, and dowel configuration was assessed after nearly 30 years of service. In addition to the primary objective, documentation of the lessons learned from this experimental project may provide better understanding of the relationship between design and long-term performance of concrete overlays.

Project Description

The concrete overlay experimental project is located in the southbound direction of a four-lane divided highway on US-1 between Daytona Beach and Titusville. The concrete overlay was constructed in 1988. Figure 1 shows the truck traffic trend since construction was completed. In 2015, the average annual daily traffic was over 6,000 vehicles with 4.4% trucks. More than 2.25 million trucks have used this roadway over the last 28 years.

The original asphalt pavement had significant cracking and some rutting. The asphalt pavement thickness was variable but consisted of approximately 5 inches of asphalt, on average, supported by an 8.5-inch limerock base and granular subgrade. Approximately 4 inches of asphalt was milled and a 1-inch leveling course was applied. All of the existing asphalt was removed from the control section. The asphalt surface was sprayed with a lime slurry, as shown in Figure 2, to cool the temperature of the asphalt surface. The joint load transfer system included an 18-inch dowel with standard spacing of 12 inches and an alternative configuration that consisted of three dowels spaced at 12 inches in each wheel path. Dowel diameters were 0.75 inches for the 6-inch slabs and 1 inch in diameter for the 7- and 8-inch slabs. Dowel baskets were fixed to the asphalt surface in standard and special wheel path configurations as shown in Figure 3. The concrete was placed using fixed forms and was manually distributed and vibrated. A wet burlap drag and transverse tining were used to texture the concrete surface. A white pigmented curing compound was used and the joints were cut and sealed after approximately 4 hours. The joints were sealed with a low modulus silicon.

A standard jointed plain concrete pavement (JPCP) with 14-ft joint spacing was constructed at the south end of the test sections to serve as a control. The original JPCP was designed with a 10-year design life based on the AASHTO 1986 Design Guide to accelerate damage and obtain relatively early results. Design factors included a modulus of subgrade reaction (K) of 240 pci, concrete flexural strength of 500 psi, and 1.2 million equivalent single-axle loads, which resulted in a slab thickness of 7 inches. Edge drains were also provided in all of the sections. Figure 4 presents the overlay project layout.

Pavement Performance Evaluation

The performance evaluation focused on the effects of different design factors within the concrete overlay sections compared with the control section. Performance parameters that were measured included bond assessment, pavement deflection, cracking, smoothness, and faulting. The initial performance indicators measured in 1988 were compared with measurements from recent performance evaluations conducted in 2015.

Figure 3. Standard (left) and special (right) dowel bar configurations.

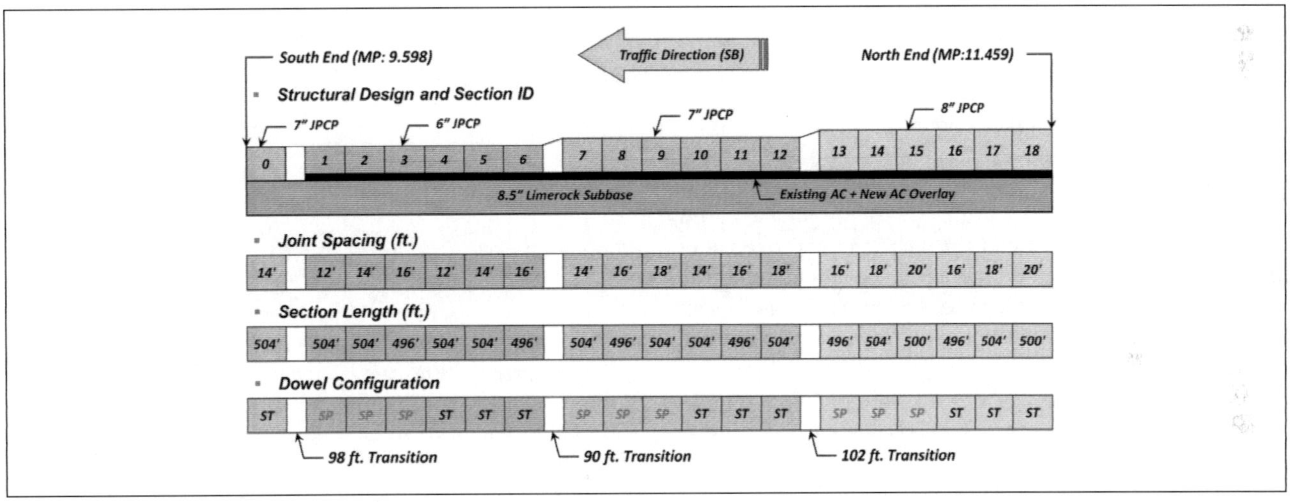

Figure 4. Concrete overlay and control sections.
Note: ST = standard dowel configuration; SP = alternative dowel configuration.

Concrete/Asphalt Bond

As mentioned previously, the asphalt surface was sprayed with a lime slurry to cool the asphalt during concrete placement. Researchers at the time initially thought this whitewash treatment may inhibit the bond between the asphalt and concrete. However, the bond shear strength assessed using test method Iowa 406-C after construction found that of six cores tested, all had bond shear strengths over 90 psi and had an average of over 400 psi. Out of the 19 cores taken during 2015, ten were found to have no bond. A large enough sample of cores was not taken to develop statistically sound correlations between design parameters and bonding condition. However, there was no apparent pattern of debonding due to slab thickness or dowel configuration. More often, cores were found debonded on slabs with longer joint spacing. For example, all three cores taken on slabs with 20-ft joint spacings were found to be debonded. Cores taken along the slab corners were debonded approximately 85% of the time while those along the slab center were debonded approximately 30% of the time. Of the cores that retained a bond, the average bond shear strength was 160 psi.

Pavement Deflection

Deflection measurements are often used to infer the structural capacity of a pavement system. Falling Weight Deflectometer (FWD) measurements were obtained during the day and night to account for potential curling effects. Slab surface temperatures measured during the 1988 and 2015 FWD surveys ranged from 92°F to 103°F during the day and from 78°F to 85°F at night. The FWD center deflection was used in the following

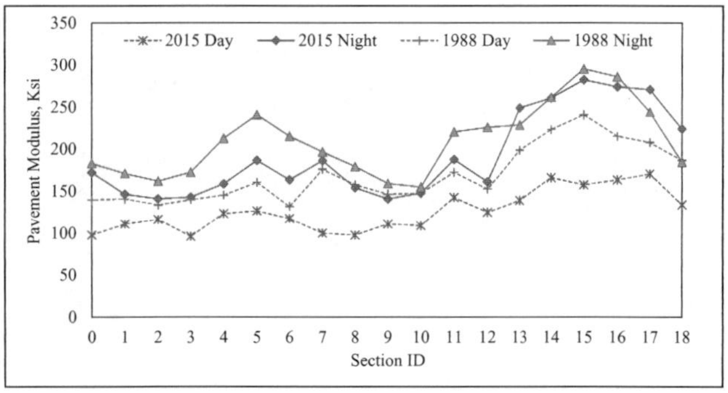

Figure 5. Pavement modulus.

equation to determine the composite modulus of the pavement system (*6*).

$$E_0 = \frac{1.5a\sigma_0}{D_0}$$

Where: E_0 = pavement composite modulus of the entire pavement system beneath the loading plate

a = radius of loading plate

σ_0 = peak pressure of the FWD loading plate

D_0 = peak center FWD deflection

Figure 5 shows the pavement composite modulus measured during the day and night in 1988 and 2015. On average, the daytime pavement modulus was approximately 20% less than the nighttime modulus in 1988 and approximately 40% less during 2015. The difference in day and nighttime pavement modulus shows the impact of curling as well as an indication of the bond deterioration between the concrete and asphalt base.

Joint deflections provide an indication of the effectiveness of a joint load transfer system. Deflections were measured at slab corners during both surveys while deflections measured along the outside wheel path joint were measured only in the 2015 survey. All deflection measurements were performed at night when slab corners and joints may be curled upward. It is difficult to find published deflection criteria, but there is some support for limiting corner deflections to 25 mm or less and differential deflections to 5 mm or less (*7*). Corner deflections indicated that a combination of slab thickness and dowel configuration helped control deflections. Corner deflections remained similar in 2015 compared with 1988 for the 8-inch slabs with standard dowels. However, as slab thickness decreased from 8 to 6 inches in the standard dowel sections, corner deflections increased by approximately 30% in 1988 and by more than 70% in 2015. Corner deflections remained approximately 20 mm for the special doweled sections regardless of thickness. Table 1 shows the average corner deflections for each survey year.

Joint load transfer efficiency (LTE) and joint differential deflections are shown in Table 2. Deflections were measured along the outside wheel path joints during the night. Sections

Table 1. Average Corner Deflection Measurements

Slab thickness (inches)	Corner deflections in special dowel sections (mm)		Corner deflection in standard dowel sections (mm)	
	1988	2015	1988	2015
6	18	25	17	23
7	20	20	14	17
8	20	22	12	11

Table 2. Load Transfer Properties

Thickness (inches)	% LTE	Differential deflection (mm)
Special dowel configuration		
6	48	6.7
7	46	7.5
8	43	6.1
Standard dowel configuration		
6	72	3.4
7	78	2.2
8	86	0.8
7 (control)	64	3.0

with a standard dowel configuration had average LTE values of approximately 80% and average differential deflections of 2 mm while the special dowel sections had average LTE values of 46% and differential deflections of 7 mm. Differential deflections also slightly decreased as slab thickness increased in the standard dowel sections. Despite having a standard dowel configuration, the control section had an average LTE of 64% and average differential deflection of 3 mm. The primary difference in this section and the overlay sections is the base type.

Cracking

A crack survey conducted in 1988 indicated that the only observed cracks were due to drying shrinkage. A crack survey performed in 2015 found that no cracks were observed in sections 5, 8, 9, 13, and 14, and most other sections had very little cracking. Of the 612 slabs that make up the experimental sections, cracks were observed in only 33 (5%) slabs. No section had more than 10% slabs with either transverse or longitudinal cracks. Section 18 was found to have the highest percentage of total cracked slabs (20%). This section is 8 inches thick with 20-ft joint spacing. Corner cracks were not observed in any of the 8-inch sections. No clear trends were found between dowel configuration and

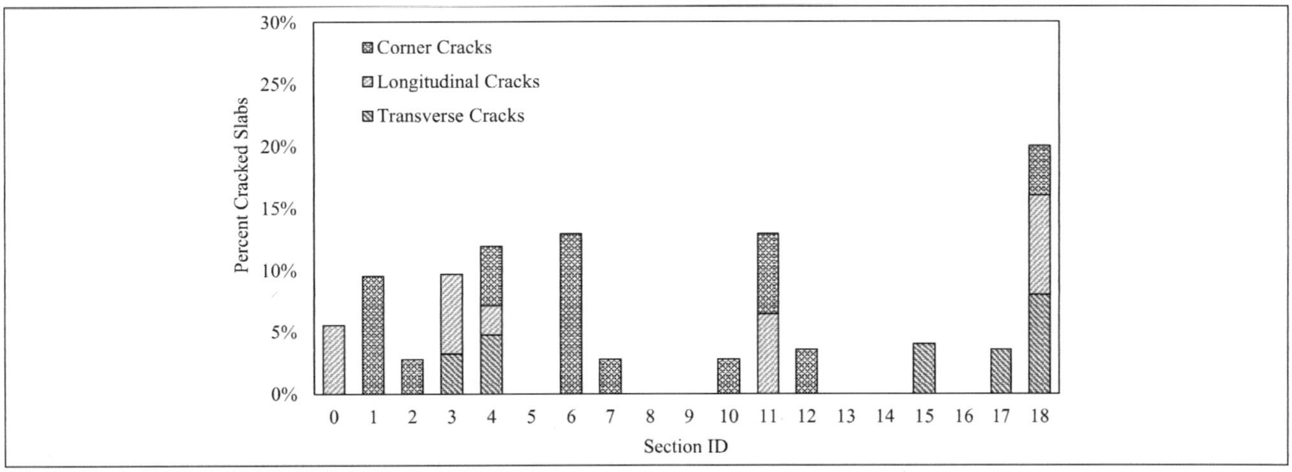

Figure 6. Percent cracked slabs.

Figure 7. Overview of the test sections.

cracking performance. Figure 6 summarizes the result of the 2015 crack survey. Figure 7 shows the beginning of the experimental project (Section 18) and Figure 8 shows typical distresses observed within the test sections.

Smoothness

Figure 9 shows the pavement smoothness in terms of international roughness index (IRI) measured during 1988 and 2015. Both sets of measurements were made during the daytime. The slab edges and corners are expected to be curled downward during the daytime and may represent a smoother condition compared with nighttime conditions where corners and edges are curled upwards. The initial smoothness measured in 1988 was approximately 70 inches/mile. Fault depths were measured with the high-speed profiler in 2015 using AASHTO R36 Method B. In general, sections with the special dowel configuration were rougher and had greater fault depths than those with the standard dowel configuration. After 28 years of service, the IRI of the special and standard dowel sections increased nearly 50 inches/mile and 10 inches/mile (60% and 15%) on average, respectively. There does not appear to be a relationship between smoothness and joint spacing or slab thickness. Interestingly, the control section experienced an increase in IRI of 55 inches/mile and greater faulting than other sections with a standard dowel configuration. The primary difference between the control (Section 0) and the comparable overlay sections with the same slab thickness and joint spacing is the base type. The limerock base in the control section is erodible compared with the asphalt base. Joint load transfer properties in Table 2 also confirm that LTE and differential deflections improved for the overlay section with an asphalt base compared with the control.

Findings and Conclusion

This project successfully demonstrated the effectiveness of a concrete overlay as a rehabilitation method for distressed asphalt pavements. After 28 years of service and more than 2.25 million trucks, the experimental concrete overlay sections on US-1 have outlived the design life by almost 20 years without the need for major repair or rehabilitation. While all of the sections performed well over the original design life, thicker sections, as one would expect, were found to have fewer cracked slabs at the end of the study. Higher corner deflections and decreased joint load transfer performance were measured on slabs with dowels placed only in the wheel path. Dowel configuration did not have an impact on cracking performance but did affect the overall pavement smoothness. The asphalt base in the concrete overlay sections served as a non-erodible base, which reduced faulting and slowed the increase in pavement roughness. Below are specific observations made during this study:

- Despite being designed as an unbonded overlay, a bond between the concrete overlay and asphalt was

Figure 8. Distresses observed in the test sections.

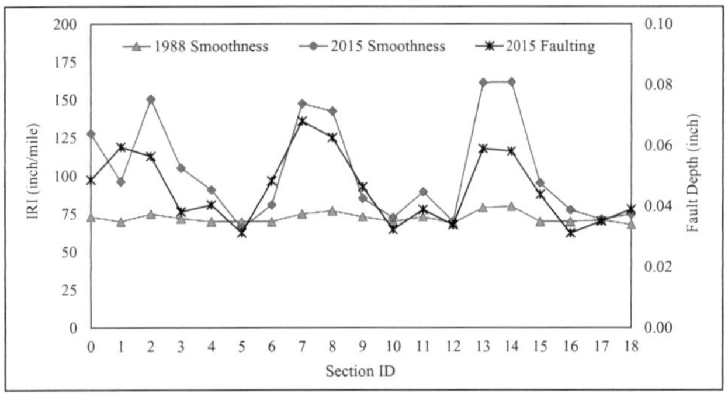

Figure 9. Pavement smoothness.

confirmed after construction. After 28 years of service, the bond was still evident in nine of 19 cores. Debonding occurred at the corners more often than the slab center. Joint spacing likely played a role in bond retention.

- Corner deflections may be controlled through a combination of slab thickness and dowel configuration. Corner deflections for the 8-inch overlay with standard dowels were reduced by nearly 50% compared with the 8-inch overlay with dowels only in the wheel paths.
- Sections with a standard dowel configuration had average LTE values of approximately 80% and differential deflections of 2 mm, while the sections with dowels in the wheel paths alone had average LTE values of 46% and differential deflections of 7 mm.
- In general, most sections exhibited good cracking performance. The section with 20-ft joint spacing had the most overall cracked slabs (20%). Of the 25 slabs within this section, two were found with transverse and longitudinal cracks and one was found with a corner crack. No corner cracks were observed in any 8-inch overlay sections.
- Dowel configuration and base type had the biggest impact on pavement smoothness. IRI increased by more than 60% for sections with dowels in the wheel path only and more than 15% for sections with standard dowels. IRI increased approximately 75% in the control section, which included standard dowels and a limerock base.

Unbonded concrete overlays with dowels in the wheel path only are an economical solution on low volume local roadways where pavement roughness thresholds may not be as strict and traffic shifts due to emergency evacuations or other events are minimal. All three slab thicknesses performed well, but corner cracks were eliminated when slab thickness increased to 8 inches. The section with 20-ft joint spacing was found to have the greatest number of cracks. Current FDOT guidelines limit joint spacing to 15 ft and validate this finding. Reduced joint spacing should be considered for slabs less than 7 inches thick, which may help in minimizing curling/warping stresses.

Author Contributions

The authors confirm contributing to the paper as follows: study conception and design: James Greene, Abdenour Nazef, and Bouzid Choubane; data collection: n/a; analysis and interpretation of results: James Greene, Ohhoon Kwon, Abdenour Nazef, and Bouzid Choubane; draft manuscript preparation: James Greene, Ohhoon Kwon, Abdenour Nazef, and Bouzid Choubane. All authors reviewed the results and approved the final version of the manuscript.

References

1. Harrington, D., and G. Fick. *Guide to Concrete Overlays: Sustainable Solutions for Resurfacing and Rehabilitating Existing Pavements*, 3rd ed. National Concrete Pavement Technology Center, Iowa State University, Ames, 2014.
2. Fick, G., and D. Harrington. *Technical Brief: Performance History of Concrete Overlays in the United States*. National Concrete Pavement Technology Center, Iowa State University, Ames, 2014.
3. Mateos, A., J. C. Paniagua, and F. Paniagua. *Development of Improved Guidelines and Designs for Thin Whitetopping: Literature Review*. Report No. UCPRC-TM-2015-01. California Department of Transportation, Sacramento, 2015.
4. Burnham, T. R. *Forensic Investigation Report for MnROAD Ultrathin Whitetopping Test Cells 93, 94, and 95*. Report MR/RC-2005-45. Minnesota Department of Transportation, Maplewood, 2005.
5. Sheehan, M. J., S. M. Tarr, and S. Tayabji. *Instrumentation and Field Testing of Thin Whitetopping Pavement in Colorado and Revision of the Existing Colorado Thin Whitetopping Procedure*. Report No. CDOT-DTD-R-2004-12. Colorado Department of Transportation, Denver, 2004.
6. Stubstad, R. N., Y. J. Jiang, and E. O. Lukanen. *Guidelines for Review and Evaluation of Back Calculation Results*. Report No. FHWA-RD-05-152. Federal Highway Administration, McLean, Va., 2006.
7. Snyder, M. *Guide to Dowel Load Transfer Systems for Jointed Concrete Roadway Pavements*. Iowa State University, Ames, 2011.

The Standing Committee on Design and Rehabilitation of Concrete Pavements (AFD50) peer-reviewed this paper (18-02722).

Article

TRR
JOURNAL OF THE TRANSPORTATION RESEARCH BOARD

Transportation Research Record
2018, Vol. 2672(40) 282–290

DOI: 10.1177/0361198118781146
journals.sagepub.com/home/trr
SAGE

Performance of Concrete Overlays on Illinois Interstates, 1967 through 2016

Laura B. Heckel[1] and Charles J. Wienrank[2]

Abstract

The Illinois Department of Transportation (IDOT) has been constructing concrete overlays of existing concrete pavements on the Interstate system for nearly 50 years. The construction of two bonded concrete overlays and eight unbonded concrete overlays have been documented to date. IDOT has also constructed one thin unbonded concrete overlay. An evaluation of performance history of the eleven pavement sections was undertaken. The following conclusions can be drawn from the evaluation of the concrete overlays in Illinois: 1) bonded concrete overlay performance on the Interstate system varied widely, and should only be used in the future judiciously, with engineering judgment; 2) unbonded concrete overlays performed consistently well, with all observed pavements exceeding or expected to exceed their design lives and design traffic factors; and 3) the thin unbonded overlay was too new to conduct a meaningful analysis of performance.

The Illinois Department of Transportation (IDOT) has been constructing concrete overlays on the Interstate system for nearly 50 years. Concrete overlays typically take one of two forms: bonded and unbonded. A bonded concrete overlay involves placing new concrete directly on the existing concrete, creating one thicker monolithic pavement slab capable of handling traffic for a normal design life, typically 20 years from the placement of the overlay. Bonded concrete overlays are thin, in the range of 3 to 4 in. (75 to 100 mm). The distresses of the existing concrete pavement will reflect through a bonded concrete overlay. Because of this, pavements in good condition but with insufficient structural capacity for the anticipated traffic levels are the best choices for bonded concrete overlays. A typical application would be in areas with an existing bare concrete pavement where truck traffic increased dramatically, such as in areas of new commercial or manufacturing development.

An unbonded concrete overlay involves placing a fully designed new pavement over the top of an existing pavement. The design assigns little or no structural value to the existing pavement. An interlayer is used between the existing concrete pavement and the new concrete pavement to cause the existing and new pavements to act independently, thereby reducing stresses between the two layers. An unbonded concrete overlay is a good choice for deteriorated pavements at the point of needing reconstruction. An unbonded concrete overlay eliminates the need for pavement removal and the construction of a new subgrade/subbase, significantly reducing the cost to construct a new pavement.

In Illinois, the construction of two bonded concrete overlays and eight unbonded concrete overlays on the Interstate system have been documented to date. IDOT has also constructed one thinner unbonded concrete overlay on the Interstate system, as an alternative to rehabilitation with a hot mix asphalt (HMA) overlay rather than as an alternative to reconstruction. The earliest concrete overlay on record was placed on I-70 in Bond County in 1967 during the construction of the Interstate system, using the original alignment of US-40. Construction of the most recent concrete overlay occurred in 2016. All concrete overlays on the Illinois Interstate system have been constructed over existing concrete pavements.

An evaluation of the construction and performance history of the eleven pavement sections is provided here. Performance measures include biennial condition rating survey (CRS) values, international roughness index (IRI) values, and accumulated traffic loadings. Structural performance was evaluated through the use of falling weight deflectometer (FWD) data. Maintenance and rehabilitation information for each pavement section was also obtained.

Project Locations

The concrete overlays are identified by route and county. Table 1 lists all concrete overlay projects, their county, original pavement design and construction date, and new pavement design and construction date. Four of the pavements have been rehabilitated or reconstructed since the concrete overlays were placed. They are denoted in the table by "No" and the year they were taken out of service in the "In service?" column of the table.

[1]Applied Research Associates, Inc., Champaign, IL
[2]Illinois Department of Transportation, Springfield, IL

Corresponding Author:
Address correspondence to Laura B. Heckel: lheckel@ara.com

Table 1. Concrete Overlay Locations, Original Construction, and New Construction

Route	County	Original pavement type	Original pavement year	New overlay type	New overlay year	In service?
Bonded concrete overlays						
I-80	Rock Island	8" CRCP	1965	4" concrete	1994–1995	No – 2010
I-88	Whiteside	8" CRCP	1976	3" concrete	1996	Yes
Unbonded concrete overlays						
I-70 WB	Bond	8" JRCP	1939	6", 7", 8" CRCP	1967	No – 1987
I-55	South Sangamon	7" JRCP	1933	8" CRCP	1970	No – 2001
I-55 NB	North Sangamon	10" JRCP	mid 1950s	9" CRCP	1976	No – 1997
I-74 WB	Knox	7" CRCP	1969	9" CRCP	1995	Yes
I-88	Whiteside	8" CRCP	1975	9" CRCP	2001	Yes
I-70	Clark	8" CRCP	1969	12" CRCP	2002	Yes
I-57	Jefferson	8" CRCP	1969	10" CRCP	2014	Yes
I-57	Union/Johnson	10" JRCP	1960–1961	9.75" JPCP	2016	Yes
Thin unbonded concrete overlay						
I-72	Sangamon	8" CRCP	1976	6' × 6' × 6" concrete	2015	Yes

Note: CRCP = continuously reinforced concrete pavement; JRCP = jointed reinforced concrete pavement; JPCP = jointed plain concrete pavement; 1 in. (") = 25.4 mm; 1 ft (') = 0.305 m.

Literature Review

Many reports have been written throughout the years regarding the performance of concrete overlays. The 1999 National Highway Cooperative Research Program (NCHRP) Report 415, *Evaluation of Unbonded Portland Cement Concrete Overlays*, compiled and analyzed data collected on unbonded concrete overlays of concrete pavements throughout the country (*1*). The report recommended best practices particularly relevant to overlays constructed with continuously reinforced concrete pavements (CRCP), including constructing an interlayer of at least 1 in. (25.4 mm) and constructing the new pavement at least 7 in. thick (178 mm).

In 2002, the Federal Highway Administration (FHWA) published *Portland Cement Concrete Overlays: State of the Technology Synthesis*, which compiles best design and construction practices as well as performance history on all types of concrete overlays (*2*). Specific to bonded concrete overlays, the FHWA paper recommends that a pavement section should only be selected for bonded concrete overlay if the existing pavement does not have a significant amount of distress. Bonded concrete overlays are most appropriate for a minimally distressed pavement with an unanticipated increase in traffic volume. More recently, in 2014, the National Concrete Pavement Technology Center published *Performance History of Concrete Overlays in the United States* (*3*). This report indicated that 45 states had constructed concrete overlays by the time of the research. All three reports indicated that properly designed and constructed concrete overlays of concrete pavements generally perform well.

IDOT has written several reports throughout the years on the construction and performance of concrete overlays. These reports were reviewed to identify concrete overlays in Illinois, to obtain construction information, as well as to gain any insights into the early performance of concrete overlays. Published in 1996, Physical Research Report No. 123, *Thin Bonded Concrete Overlay and Bonding Agents*, covered the construction of the I-80 bonded concrete overlay (*4*). Physical Research Report No. 134, *Performance of Thin Bonded Concrete Overlays in Illinois*, from 2000, reported on the construction of the I-88 section, and also contained preliminary performance data of both bonded concrete overlays, spanning the time period from October 1994 to September 1998 (*5*). Physical Research Report No. 143, published in 2002, *Bonded Concrete Overlay Performance in Illinois*, added to Report No. 134 and provided performance details through December 2001 (*6*).

Physical Research Report No. 80, *Continuously Reinforced Concrete Overlays on Existing Portland Cement Concrete Pavement*, provided details of the construction and the first 10 years' performance of the I-70 Bond County project, and was published in 1978 (*7*). Physical Research Report No. 101, *Performance Evaluation of Concrete Overlays*, covered the construction of both I-55 concrete overlays and the performance of the first three unbonded concrete overlays through 1988 (*8*). Physical Research Report No. 122, *Planning, Design and Construction of an Unbonded Concrete Overlay*, reported on the construction of the I-74 concrete overlay and was published in 1996 (*9*). Physical Research Report No. 140, published in 2002, *Performance of an Unbonded Concrete Overlay on I-74*, described the performance of the concrete overlay through 2001 (*10*).

Performance Measures

Several measures were used to quantify the performance of the concrete overlays. They are condition metrics, accumulated traffic, structural performance measures, and maintenance and rehabilitation activities.

Condition Metrics

IDOT performs CRS reviews on every Interstate pavement section every 2 years. Teams review video tapes of the pavement sections and identify up to five predominant distresses and corresponding severities, according to the CRS pavement distress codes contained in IDOT's *Illinois Highway Information Systems Roadway Information and Procedure Manual* (*11*). These distresses along with the IRI values and rutting values (for asphalt-surfaced pavements) are entered into a formula that weights each distress and surface characteristic, and subtracts those values from 9.0, the CRS value for a new pavement surface. The 2007 Illinois Center for Transportation report, *Update of Condition Rating Survey (CRS) Calculation/Prediction Models – Final Report*, detailed this process and the most recent equations used by IDOT CRS teams (*12*). All CRS and IRI values as well as predominant distresses were obtained from IDOT databases for the analysis.

IDOT uses performance models developed from past pavement CRS deterioration rates to predict future CRS values. Performance models found in the 2007 Illinois Center for Transportation report were used for this analysis (*12*). IDOT uses the same model for all new CRCPs, bonded concrete overlays of CRCPs, and unbonded overlays constructed with CRCP. The CRS results for all of the concrete overlays were compared with the CRCP performance model as one method of performance evaluation. The 2007 model and a terminal service condition of CRS = 4.5 were used to predict end of life for the overlays that have not been rehabilitated or reconstructed. A CRS value at or below 4.5 corresponds to "poor" condition, as defined in IDOT's *Condition Rating Survey Summary Report FY16, January 2017* (*13*).

Accumulated Traffic

An important factor in how long a pavement section will last is the amount of traffic it carries over its life span. Each pavement is designed to carry a certain quantity of traffic loadings, with a given reliability factor, over the course of its design life. The design traffic factor quantifies these expected loads. IDOT calculates design traffic factors by determining the current traffic, projecting it forward to the midpoint of the design period, and multiplying the traffic values by equivalency factors (*14*). For all but one of the concrete overlays, the design traffic factor was located in the project files, with the I-88 Whiteside County unbonded overlay being the exception. Its traffic factor was estimated from actual traffic values at the time of construction. Actual time to rehabilitation or predicted time to a CRS of 4.5 was compared with design life.

To determine the actual accumulated traffic loadings of a pavement, the historical traffic data were obtained from IDOT databases for each year of a pavement's life then converted to millions of equivalent single axle loads (MESALs), the same traffic factor measurement used for the design traffic factor. These values were summed to determine the cumulative MESALs. To compare the actual traffic loadings to the design traffic factor, *percent consumption* was calculated by dividing cumulative MESALs by the design traffic factor, then multiplying by 100.

Structural Performance Monitoring

FWD data were obtained from IDOT's Bureau of Research to analyze the structural performance of the concrete overlays. Data analyzed include deflection under the load (D0) measured in the center of the lane, deflection basin area, subgrade resilient modulus (E_{RI}), and load transfer efficiency (LTE) across transverse cracks/joints (LC or LJ LTE) and pavement/shoulder joints (TE LTE).

Lower deflections indicate thicker slabs and high levels of support, with a greater ability to carry traffic into the future without structural issues. Deflections under the load below 6 mil (0.15 mm) were considered excellent for concrete pavements. The deflection basin area represents the ability of the pavement to spread an applied load. Higher deflection basin areas indicate a greater load-spreading ability of the pavement. Areas of approximately 30 in. (762 mm) are typical for concrete pavements. Subgrade resilient modulus (E_{RI}) values were calculated using an equation developed by the University of Illinois (*15*). A higher E_{RI} represents a more stable subgrade. An E_{RI} greater than 10 ksi (69 MPa) was considered good support. The LTE was calculated by dividing the deflection on the unloaded side of the joint or crack by the deflection directly beneath the loaded plate. An LTE of 100% indicated ideal performance of the joint or crack in transferring load. A LTE less than 50% indicated the joint or crack was not performing its function of transmitting load.

Maintenance and Rehabilitation

Contract maintenance and rehabilitation information contained in IDOT's Illinois Highway Inventory System (IHIS), as well as information obtained by viewing videos of the pavement sections, were obtained to analyze the maintenance and rehabilitation performed on the pavements.

Bonded Concrete Overlays

Design

The bonded concrete overlay thicknesses were determined by calculating the recommended thicknesses from the AASHTO and NCHRP design procedures for bonded concrete overlays and adjusting the recommendations for site conditions (*4*). The I-80 overlay contained test sections with varying amounts of microsilica in the concrete mix as well as using or not using grout to enhance the bond between the existing and new concrete. The effects of each of these design

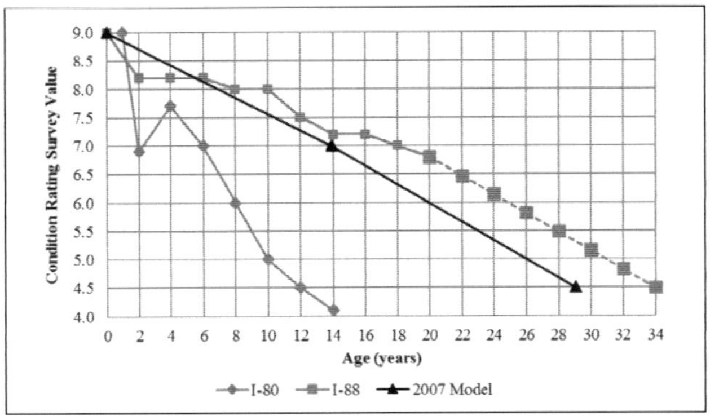

Figure 1. Bonded concrete overlay condition rating survey values vs. age.

parameters are not explored here, as they have been sufficiently analyzed in previous reports (*4–6*). The I-88 project used neither microsilica nor grout.

Condition Metrics

Figure 1 shows the actual and predicted CRS values for both bonded concrete overlays. Solid lines denote actual CRS values. Dashed lines represent predicted CRS values. The black line represents the 2007 CRS performance model for CRCPs. The CRS on I-80 dropped rapidly after construction, and reached a CRS of 4.1 in only 14 years, the last survey prior to reconstruction. The improvement in CRS between years 2 and 4 was due to a patching contract. This pavement significantly underperformed the 2007 model. The literature review revealed that the I-80 overlay was constructed on a pavement section with a lower condition level than research at the time advised, which was most likely a factor in the rapid deterioration of the pavement. Poor bonding between the old and new concrete was also noted, possibly caused by microsilica used in the concrete mix and reflective cracking of underlying distresses (*4–6*).

Twenty years after construction, I-88 was at a CRS of 6.8. The section had outperformed the 2007 model as of 2016. Furthermore, using the 2007 model, this section of I-88 would not reach the "poor" CRS rating of 4.5 until 34 years of service.

Two years following the 1995 construction of the bonded concrete overlay on I-80, the CRS recorded frequent medium-severity transverse cracking, in addition to infrequent low-severity longitudinal cracking and infrequent punchouts. Even with several patching contracts, discussed in the Maintenance and Rehabilitation section, the 2008 CRS revealed frequent high-severity transverse cracking, high-severity punchouts, and permanent patching of 12% to 18% of the pavement section.

The I-88 bonded concrete overlay followed a completely different pattern of distress development than the I-80 project. No distresses were recorded as part of the CRS evaluation until the pavement was 8 years old, when infrequent punchouts and a few permanent patches were recorded. The 2016 CRS evaluation, 20 years after construction, recorded infrequent medium-severity transverse cracks, low-severity centerline deterioration and longitudinal cracking, 1 to 3 punchouts/mi (0.6 to 1.8 punchouts/km), and less than 6% of the section had permanent patches.

Roughness data were obtained for both bonded concrete overlays. The IRI on I-80 rose very quickly from around 100 in./mi (1.58 m/km) in both directions initially to 175 in./mi (2.76 m/km) westbound and 300 in./mi (4.73 m/km) eastbound 14 years after construction. However, the roughness on I-88 has been hovering between about 65 and 80 in./mi (1.03 and 1.26 m/km) throughout its life.

Accumulated Traffic

Table 2 summarizes the traffic data for the bonded concrete overlays. Both pavements exceeded their design traffic factors. Interstate 80 carried 180% of its design traffic from the time the bonded overlay was constructed in 1994–1995 until the pavement section was reconstructed in 2010. During that time, ADT grew by 57% and annual MESALs grew by 84%. The significant increase in traffic may have contributed to the rapid deterioration of the I-80 section. I-88 had carried 135% of the design traffic factor and met its 20-year design life by 2016. It is still in service as of July 2017. From construction through 2016, ADT grew 7% and annual MESALs grew 13%.

Structural Performance

The results of the FWD analyses are contained in Table 3. Both bonded concrete overlays had very low deflections, high deflection basin areas and E_{RI} values, as well as very good LTE across transverse cracks throughout testing. The good results on I-80 indicated that the performance issues were not related to the pavement structure itself (pavement stiffness and subgrade support), but were most likely caused by the deteriorated condition of the original pavement at the time of overlay and the high traffic volumes. The I-88 bonded concrete overlay did not indicate any structural issues as of the most recent testing in October 2016.

Maintenance and Rehabilitation

Significant early cracking required a substantial amount of patching soon after the I-80 section was completed. Information from Physical Research Report No. 143 and IHIS indicated patching contracts were awarded in 1998, 2000, 2001, 2002, and 2004 (*6*). No additional patching contracts were recorded between 2005 and 2010. Due to the extensive patching and low CRS values, IDOT reconstructed this section in 2010, 15 years after construction. IDOT awarded patching contracts for the I-88 bonded concrete overlay in 2003, 2008, 2011, and 2016, with total patching by 2016 of about 3%. No rehabilitation of this section has been performed.

Table 2. Bonded Concrete Overlay Traffic Data

Route	Traffic years	Cumulative MESALs	Percent trucks	Design life (years)	Design traffic factor (MESALs)	Percent consumed
I-80 Rock Island Co.	1995–2009	35.3	32.5	20	19.5	180.8
I-88 Whiteside Co.	1996–2016	17.7[a]	32.8	20	13.1	134.9

Note: MESALs = millions of equivalent single axle loads.
[a]Cumulative MESALs as of 2016, as the pavement section is still in service.

Table 3. Bonded Concrete Overlay Structural Performance Data

Date	Center deflection (mil)	Center area (in.)	Center E_{RI} (ksi)	LC LTE (%)
I-80				
10/1996	2.5	NC	NC	87.6
04/1999	2.7	31.4	13.3	87.1
03/2001	2.8	31.4	13.3	92.3
09/2006	2.7	31.2	13.8	93.0
I-88				
10/1996	2.2	30.0	16.6	91.5
08/1998	2.3	NC	NC	89.7
10/2016	2.5	31.2	14.8	92.5

Note: E_{RI} = subgrade resilient modulus; LC LTE = leave side of crack load transfer efficiency; NC = not collected; 1 in. = 25.4 mm; 1 mil = 0.0254 mm; 1 ksi = 6.89 MPa.

Because of the disparity in performance between the I-80 and I-88 bonded concrete overlays, IDOT has not constructed another bonded concrete overlay on the Interstate system. IDOT's Physical Research Report No. 143 recommended "extreme caution" if considering a bonded concrete overlay (*6*). If a new bonded concrete overlay was desired, care should be taken to select a pavement still in excellent to good condition but under-designed due to an unanticipated increase in truck traffic.

Unbonded Concrete Overlays

Design

IDOT designed the pavement thicknesses for each project by calculating the design traffic factor, determining the thickness required if the pavement had been new construction, and subtracting 1 in. (25 mm) of thickness as credit for the underlying pavement. Illinois typically uses the existing HMA overlay as an interlayer. The minimum interlayer on the unbonded concrete overlays placed to date was 3 in. (75 mm), whereas the maximum interlayer thickness was 5 in. (125 mm), after the removal of 2.75 in. (70 mm) of existing HMA overlay.

Seven of the eight unbonded overlays were designed for a 20-year life. The eighth project, I-70 in Clark County, was designed for a 30-year life. To obtain a 30-year life, the percentage of longitudinal steel was increased from 0.7% to 0.8%. Standard reinforcing uses #4 (#13) bars at 4-ft (1.2 m) centers transversely to support the longitudinal steel. This contract added an extra #4 (#13) bar at 2 ft (0.6 m) to provide additional stabilization to the longitudinal steel. The tie bars between lanes and the mainline pavement and the shoulders were #8 (#25) bars (as opposed to #6 (#19) bars), 30-in. (762 mm) long, on 24-in. (610 mm) centers (as opposed to 30-in. (762 mm) centers for centerline joints). All steel was epoxy-coated and the freeze-thaw requirements were narrowed to ensure a more durable aggregate. IDOT continues to monitor this project for the 30-year life aspects in addition to the nature of the project as an unbonded concrete overlay.

Condition Metrics

The 2007 performance models were used to predict future CRS values for unbonded concrete overlays as well. Figure 2 depicts the actual CRS data in solid lines and predicted data in dashed lines. The black line represents the 2007 prediction model for all CRCP pavements. The I-70 Bond County section and both I-55 sections are depicted until the final CRS prior to rehabilitation with HMA. Occasional improvements in CRS are due to changes in the predominant distresses selected by the CRS teams.

The I-70 overlay in Bond County and the I-55 section in North Sangamon County underperformed the 2007 model slightly. These projects were overlaid with HMA after 20 and 21 years of service, respectively. The I-55 section in South Sangamon County did not perform as well as the 2007 model initially, but performed quite well after the 10th year and was eventually overlaid with HMA after 31 years of service. The unbonded concrete overlays constructed between 1995 and 2002 are all performing significantly better than the 2007 model. It is too early to reliably predict the performance of the two newest unbonded concrete overlays, both on I-57.

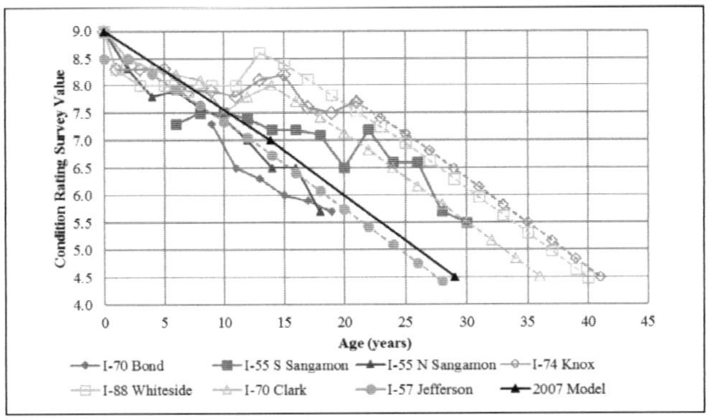

Figure 2. Unbonded concrete overlays condition rating survey values vs. age.

The I-57 Jefferson County project started at a CRS of 8.5 due to some construction issues not related to the overlay design.

No records of distresses exist for I-70 in Bond County. No distress records exist for either I-55 overlay until 1994. By the time the I-55 overlay in South Sangamon County was resurfaced in 2002, it was showing high-severity centerline deterioration, frequent low-severity longitudinal cracking, low-level D-cracking, infrequent punchouts, and 6% to 12% of the section had been patched. The I-55 unbonded overlay in North Sangamon County had medium-level D-cracking, medium-level centerline deterioration, frequent narrow longitudinal cracks, frequent popouts, and less than 6% patching of the pavement during its 1994 survey.

The CRS of the pavement section on I-74 in Knox County recorded no distresses until 2002, when centerline deterioration was logged. The 2016 CRS survey, with the pavement age at 21 years, recorded infrequent low-severity longitudinal cracking, 1–3 punchouts/mi (0.6–1.8 punchouts/km), infrequent popouts, and less than 6% of area patched.

Records of the I-88 unbonded concrete overlay indicated very little distress of the pavement throughout its 14-year life. The CRS in 2016 identified infrequent low-severity longitudinal cracking.

The CRS team documented no distresses on the 2002 overlay on I-70 in Clark County until 2008. The most recent survey noted infrequent low-severity transverse cracking and 4 or more low-severity punchouts/mi (2.5 punchouts/km).

The CRS have recorded no distresses to date on the I-57 unbonded concrete overlay in Jefferson County, although patches were noted on a video survey. These patches were the result of construction defects. The I-57 section in Union and Johnson Counties, the first JPCP overlay, will be surveyed in 2018.

All available roughness data was obtained for the unbonded concrete overlays. The overlay on I-55 in South Sangamon County was stable throughout its life at nearly 140 in./mi (2.20 m/km), while the roughness on the newer overlays was consistently lower, between 60 and 80 in./mi (0.95 and 1.26 m/km) since construction.

Accumulated Traffic

Table 4 contains a summary of the traffic data for the unbonded concrete overlays. The first three unbonded concrete overlays have been rehabilitated, thus their cumulative MESALs and percent consumed reflect the total traffic carried before rehabilitation. These three unbonded concrete overlays exceeded their design lives, and accumulated traffic from 174% to 222% of design traffic. The oldest unbonded concrete overlay still in service, I-74 Knox County, reached 100% design traffic consumption in 2016, and is not yet on IDOT's *FY 2018-2023 Proposed Highway Improvement Program* for rehabilitation (*16*).

Table 5 shows the age and the percent of design traffic consumed for each pavement section when it is predicted to reach a CRS "poor" rating of 4.5. The predicted percent traffic consumed was based on extrapolated traffic values using historical growth rates on each route. All four unbonded concrete overlays constructed between 1995 and 2014 were predicted to exceed their design lives (20 or 30 years) and were predicted to achieve between 102% and 235% of design traffic. I-57 in Union/Johnson Counties was not included since no starting CRS exists from which to predict future performance.

Structural Performance

Two of the unbonded concrete overlays have been tested with an FWD over their lives. The results of the testing are included in Table 6. Both sections had average deflections under 2 mil (0.05 mm), deflection basin area values around 32 in. (813 mm), E_{RI} values between 17 and 20 ksi (117 and 138 MPa), and LC LTE above 90% throughout all testing. All of these values were within the limits of good performance defined in the Performance Measures section. TE LTE was 54% on I-74; however, no distresses such as lane-shoulder separation were observed. Mainline to shoulder load transfer on I-70 was nearly 92%.

Maintenance and Rehabilitation

The three oldest unbonded concrete overlays were rehabilitated after 20 to 31 years of service. No maintenance records prior to rehabilitation exist for these pavement sections.

The I-70 section in Clark County included a 5-year warranty. Before the warranty expired, punchouts totaling 0.1% of the pavement area required patching. The punchouts were found to be design-related, possibly caused by full-thickness shoulders tied to the CRCP with #8 (#25) tie bars. No additional maintenance has been performed on this section.

The centerline and shoulder joints on the I-74 concrete overlay were sealed by contract maintenance in 2014. No contract maintenance or rehabilitation had been performed on I-88, or either I-57 unbonded concrete overlay, as of July 2017.

Overall, the performance of unbonded concrete overlays throughout their almost 50-year history has been very good.

Table 4. Unbonded Concrete Overlay Traffic Data

Route	Traffic years	Cumulative MESALs	Percent trucks	Design life (years)	Design traffic factor	Percent consumed
I-70 Bond Co.	1967–1987	23.4	37.6	20	13.4	174.7
I-55 S. Sangamon Co.	1970–2001	38.1	21.5	20	17.1	222.0
I-55 N. Sangamon Co.	1976–1997	27.5	25.6	20	15.8	174.1
I-74 Knox Co.	1995–2016	25.3[a]	29.1	20	24.0	104.7
I-88 Whiteside Co.	2001–2016	13.7[a]	35.1	20	16.2[b]	84.7
I-70 Clark Co.	2002–2016	45.3[a]	49.8	30	116.6	38.8
I-57 Jefferson Co.	2014–2016	9.6[a]	30.7	20	80.0	12.0
I-57 Union/Johnson Co.	N/A[c]	N/A[c]	19.1	20	32.8	N/A[c]

Note: MESALs = millions of equivalent single axle loads.
[a]Cumulative MESALs as of 2016, as the pavement section is still in service.
[b]Estimated from 1995 traffic data with a 0.5% growth rate.
[c]Constructed in 2016, so no traffic had accumulated in 2016.

Table 5. Predicted Life to CRS 4.5

Pavement section	Predicted age at CRS 4.5	Year CRS 4.5 reached	Predicted % consumed at CRS 4.5
I-74 Knox County	41	2036	208
I-88 Whiteside Unbonded	40	2041	235
I-70 Clark County	36	2038	102
I-57 Jefferson County	28	2042	161

Note: CRS = condition rating survey.

Table 6. Unbonded Concrete Overlay Structural Performance Data

Date	Center deflection (mil)	Center area (in.)	Center E_{RI} (ksi)	LC LTE (%)	TE LTE (%)
I-74 Knox County					
7/1995	2.0	31.0	17.4	NC	NC
8/1996	2.2	31.6	15.1	NC	NC
7/1997	2.4	31.4	14.3	92.6	NC
4/1999	2.1	31.0	17.8	92.6	NC
7/2001	2.3	30.8	16.3	91.6	NC
9/2016	1.8	31.7	17.1	91.9	53.9
I-70 Clark County					
9/2016	1.1	32.1	20.2	90.4	91.7

Note: E_{RI} = subgrade resilient modulus; LC LTE = leave side of crack deflection load transfer efficiency; TE LTE = true edge (pavement/shoulder joint) load transfer efficiency; NC = not collected; 1 mil = 0.0254 mm; 1 in. = 25.4 mm; 1 ksi = 6.89 MPa.

All sections have exceeded or are expected to exceed their design lives as well as their design traffic factors. Roughness has not been an issue, nor has the structural performance of the sections. Maintenance has been minimal and project rehabilitation has not been earlier than expected.

Thin Unbonded Concrete Overlay

Design

The thin unbonded concrete overlay was designed as a 6-in. (150-mm) concrete overlay sawed to create 6-ft × 6-ft (1.8-m × 1.8-m) panels. The design was chosen as an alternative approach to rehabilitation with HMA, rather than as an alternative to reconstruction. The overlay was constructed on the existing 8-in. (200-mm) CRCP. The existing HMA overlay was completely removed due to stripping. A 1.25-in. (32-mm) HMA interlayer was constructed in the eastbound direction and a geotextile was placed as the interlayer for the westbound direction. Synthetic macro fibers were used in the concrete mix. Tie bars at the construction joints and longitudinal centerline joints comprised the only reinforcing steel in the project.

Condition Metrics

The unbonded concrete overlay on I-72 has had one CRS rating since construction, in 2016. The survey returned a CRS value of 8.7 in both directions with no distresses noted.

Table 7. I-72 Sangamon County Structural Performance Data

Date	Direction	Center Deflection (mil)	Center area (in.)	Center E_{RI} (ksi)	LJ LTE (%)
10/2016	EB	2.4	32.3	13.8	89.9
10/2016	WB	3.9	32.5	9.2	78.2

Note: E_{RI} = subgrade resilient modulus; LJ LTE = leave side of joint load transfer efficiency; 1 mil = 0.0254 mm; 1 ksi = 6.89 MPa; 1 in. = 25.4 mm.

The eastbound IRI was 48 in./mi (0.75 m/km) and the westbound IRI was 62 in./mi (0.98 m/km), both good values.

Accumulated Traffic

Cumulative traffic data through 2016 for the thin unbonded concrete overlay on I-72 in Sangamon County totaled 1.5 MESALs. The thickness of this section was not designed for an anticipated amount of traffic, so calculations for design traffic factor and percent design traffic consumed were not appropriate.

Structural Performance

The deflection testing results for the thin unbonded concrete overlay are contained in Table 7. The deflection data were separated by direction of travel due to the difference in interlayer materials used in the two directions. Deflection testing across the longitudinal mid-lane joint was not performed.

The average deflection was higher in the westbound than in the eastbound direction, but both were similar to the other concrete overlays. The area values were similar, indicating a similar ability to disperse the stresses from loadings through the pavement. The E_{RI} was lower in the westbound than in the eastbound direction. Both E_{RI} values were acceptable. The LTE across the transverse joints was higher in the eastbound than in the westbound direction. The differences in structural performance data between the directions will be monitored over time to determine if they persist.

Maintenance and Rehabilitation

A field survey performed in the fall of 2016, one year after construction, corroborated IDOT database records indicating no maintenance activities were required or performed.

Summary and Conclusion

This study was undertaken to examine performance of concrete overlays on the Interstate system in Illinois. Two bonded concrete overlays, eight unbonded concrete overlays, and one thin unbonded concrete overlay were identified from previous reports and through available construction documents.

The study evaluated performance by examining the CRS and corresponding distresses, the IRI, accumulated traffic, structural performance through the use of a FWD, and maintenance and rehabilitation history of each pavement section. Based on the totality of the analyses, the following conclusions can be drawn:

- Bonded concrete overlay performance varied widely, and should only be used in the future with extreme caution. Consistent with the FHWA recommendation to select existing pavements with little distress prior to placement of a bonded concrete overlay, the pavement section in Illinois with more distress prior to overlay did not perform well, whereas the less-distressed pavement performed quite well.
- Unbonded concrete overlays performed consistently well, with all observed pavements exceeding or expected to exceed their design lives and design traffic factors. All of Illinois' unbonded overlays placed to date were designed at or above the NCHRP report's recommendations for interlayer thickness and new CRCP pavement thickness. As a result, good performance has been attained.
- The thin unbonded overlay was too new to conduct a meaningful analysis of performance.

Because unbonded concrete overlays have performed well, they should be strongly considered whenever an existing pavement reaches the end of its service life and is in need of reconstruction. All concrete overlay projects will continue to be monitored into the future.

Acknowledgments

This research study was funded by the Illinois Department of Transportation. The authors thank IDOT's Bureau of Research for performing testing of the pavements and the Bureau of Programming for providing historical data. The authors also thank Robert Rescot, John Senger, and LaDonna Rowden for manuscript review.

Author Contributions

The authors confirm contribution to the paper as follows: study conception and design: L.H, C.W.; data collection: L.H.; analysis and interpretation of results: L.H., C.W.; draft manuscript preparation: L.H. All authors reviewed the results and approved the final version of the manuscript.

References

1. ERES Consultants, Inc. *NCHRP Report 415: Evaluation of Unbonded Portland Cement Concrete Overlays*. Transportation Research Board of the National Academies, Washington, D.C., 1999.
2. Smith, K. D., H. T. Yu, and D. G. Peshkin. *Portland Cement Concrete Overlays: State of the Technology Synthesis*. Publication FHWA-IF-02-045. FHWA, U.S. Department of Transportation, 2002.
3. Fick, G., and D. Harrington. *Technical Brief: Performance History of Concrete Overlays in the United States*. National Concrete Pavement Technology Center, Iowa State University, 2014.

4. Rowden, L. R. *Thin Bonded Concrete Overlay and Bonding Agents*. Physical Research Report No. 123. Illinois Department of Transportation, 1996.
5. Volle, T. H. *Performance of Thin Bonded Concrete Overlays in Illinois*. Physical Research Report No. 134. Illinois Department of Transportation, 2000.
6. Winkleman, T. J. *Bonded Concrete Overlay Performance in Illinois*. Physical Research Report No. 143. Illinois Department of Transportation, 2002.
7. Dhamrait, J. S., and D. R. Schwartz. *Continuously Reinforced Concrete Overlays on Existing Portland Cement Concrete Pavement*. Physical Research Report No. 80. Illinois Department of Transportation, 1978.
8. Lippert, D. L., and J. B. DuBose. *Performance Evaluation of Concrete Overlays*. Physical Research Report No. 101. Illinois Department of Transportation, 1988.
9. McNeal, A. F. *Planning, Design and Construction of an Unbonded Concrete Overlay*. Physical Research Report No. 122. Illinois Department of Transportation, 1996.
10. Heckel, L. B. *Performance of an Unbonded Concrete Overlay on I-74*. Physical Research Report No. 140. Illinois Department of Transportation, 2002.
11Illinois Department of Transportation. *Illinois Highway Information Systems Roadway Information and Procedure Manual, Appendix C*. Illinois Department of Transportation, 2014.
12. Heckel, L., and Y. Ouyang. *Update of Condition Rating Survey (CRS) Calculation/Prediction Models*. Final Report. ICT R27-13, Illinois Center for Transportation, 2007.
13. Illinois Department of Transportation. *Condition Rating Survey Summary Report FY16*. 2017. http://www.idot.illinois.gov/Assets/uploads/files/Transportation-System/Reports/OP&P/Travel-Stats/FY%202016%20Condition%20Rating%20Survey%20Summary%20Report.pdf. Accessed July 27, 2017.
14. Illinois Department of Transportation. *Bureau of Design and Environment Manual*. Illinois Department of Transportation, 2016.
15. Thompson, M. R. ILLI-PAVE Based NDT Analysis Procedures. In *Nondestructive Testing of Pavements and Backcalculation of Moduli* (A. J. Bush, III, and G. Y. Baladi, ed.), ASTM STP 1026, ASTM, Philadelphia, 1989, pp. 487–501.
16.Illinois Department of Transportation. *2018-2023 Proposed Highway Improvement Program*. http://www.idot.illinois.gov/MYP2018-202317. Accessed June 30, 2017.

The Standing Committee on Design and Rehabilitation of Concrete Pavements (AFD50) peer-reviewed this paper (18-00960).

Article

TRR
JOURNAL OF THE TRANSPORTATION RESEARCH BOARD

Transportation Research Record
2018, Vol. 2672(40) 291–303

DOI: 10.1177/0361198118801347
journals.sagepub.com/home/trr
SAGE

Cold Central Plant Recycled Asphalt Pavements in High Traffic Applications

David H. Timm[1], **Brian K. Diefenderfer**[2], **and Benjamin F. Bowers**[2]

Abstract

Cold central plant recycling (CCPR) is gaining wider use in the U.S. for rehabilitating existing asphalt pavements or for new construction. Although it is used widely in lower traffic volume situations, CCPR use in high volume pavements remains an open question when considering its structural capacity and expected performance. A project completed in 2011 on I-81 in Virginia indicated CCPR may be suitable for high-volume traffic applications and was further evaluated with the construction of three CCPR test sections at the National Center for Asphalt Technology Test Track in 2012. These sections are now approaching 20 million equivalent single axle load applications and this paper documents their structural and surface performance thus far. The structural characterization indicates healthy pavements with no significant increases in measured pavement response or decreases in backcalculated moduli over time. Performance has been excellent with no cracking observed on any section, rut depths less than 0.3 inches and ride quality that has remained almost unchanged. Perpetual pavement analyses were also conducted and found that the section with a cement-stabilized base layer supporting the CCPR layer met the criteria and is likely perpetual. The other two sections, without the cement-stabilized base, did not meet the criteria and may develop bottom-up cracking. Data from the I-81 and Test Track sections enabled the Virginia Department of Transport (VDOT) to proceed with a design-build project on I-64 that will feature CCPR with a cement-stabilized base and full-depth reclamation (FDR). It is estimated that nearly 170,000 tons of reclaimed asphalt pavement will be used with over $10 million in savings.

Pavement recycling is a series of processes that includes cold in-place recycling (CIR), cold central-plant recycling (CCPR), and full-depth reclamation (FDR). Pavement recycling is effective at rehabilitating existing pavements or constructing new pavements while reducing the construction costs, environmental impacts, and construction time (*1–3*). Although the use of pavement recycling is becoming more common in several US states, its use is not consistent or widespread throughout the country. Several reasons for this exist and include the absence of defined materials characterization inputs for engineered structural design, a lack of rapidly applied quality control and/or assurance procedures, and a shortage of long-term performance data in the literature that documents the structural performance of existing sections. Studies related to the first two of these reasons are either underway or recently completed as part of work sponsored by the National Cooperative Highway Research Program (*4–5*); this paper addresses the third reason by describing research using CCPR and FDR conducted by the Virginia Department of Transportation (VDOT) and the National Center for Asphalt Technology (NCAT).

CCPR is a process in which newly milled or existing stockpiled reclaimed asphalt pavement (RAP) is combined with a recycling agent (and chemical additives, if used) to create a paving material that can be placed with conventional asphalt paving equipment. Great cost and energy savings are realized because the RAP particles are not reheated. These processes occur at a mobile plant which can be located at or near the paving project or RAP source stockpile. Also, because CCPR is not performed in-place the underlying foundational materials can be stabilized (using a process such as FDR), if needed, before placement of the CCPR material. Additionally, the CCPR process can be used to place multiple lifts for thicker applications. It is through this process that CCPR can be used to produce a recycled base course for reconstruction, new construction, lane widening, shoulder strengthening, and other projects.

[1]Auburn University, Auburn, AL
[2]Virginia Transportation Research Council, Charlottesville, VA

Corresponding Author:
Address correspondence to David H. Timm: timmdav@auburn.edu

Typical layer thicknesses for CCPR range from 2 to 6 inches (*6*) but multiple lifts may be used to increase the total thickness of the recycled layer.

VDOT has completed construction and is conducting research projects subjecting recycled pavements to high truck volumes to study their long-term structural performance, including CCPR. In 2011, VDOT completed construction on a portion of Interstate-81 in western Virginia that carries more than 6,000 trucks per day. CCPR and FDR were used in combination to reconstruct the right lane on this 3.7 mile-long project. Along with using CIR in the left lane, this project was the first time these three processes were used together on the interstate system in the US (*7*). In 2012, VDOT sponsored three test sections at the NCAT Pavement Test Track to further study CCPR pavements under high truck traffic conditions and included embedded instrumentation in the sections (*8*). The sections performed well through the first research cycle subjecting them to 10 million equivalent single axle loads (ESALs) of trafficking and were left in place for a second research cycle that would apply an additional 10 million ESALs.

As CCPR pavements have rarely been subjected to such high traffic levels, it is critical to provide updates on their condition and overall performance as agencies look to using CCPR more frequently in high traffic conditions. Additionally, the pavements showed signs that they may be perpetual (i.e., no deep structural distress development) and therefore an analysis was warranted to determine if they are predicted to be perpetual based on modern design criteria. Finally, the results of the research on the I-81 project and the test sections at the NCAT Test Track have given VDOT the confidence to use pavement recycling on other high profile projects (*9*). Providing discussion of this particular example may help other agencies begin using CCPR more routinely in high-volume applications.

Objectives

Given the needs described above, the objectives of this paper include the following:

1. Document the performance and structural condition of the VDOT sections at the NCAT Test Track through two test cycles approaching 20 million ESALs.
2. Evaluate the VDOT Test Track sections under perpetual pavement criteria.
3. Discuss real-world implementation, including cost comparisons, of CCPR pavements in high truck traffic volume situations.

Scope of Work

To accomplish the objectives, data collected from the three VDOT sections at the NCAT Test Track were compiled over two test cycles. The data included routine surface performance measurements (i.e., rutting, cracking, and roughness) in addition to structural evaluation through frequent falling weight deflectometer (FWD) testing and backcalculation. Frequent measurements using embedded instrumentation were also used for structural characterization. The strain measurements were used as part of a perpetual pavement analysis in addition to simulations conducted using the design software, PerRoad. Real world implementation was demonstrated through the I-64 reconstruction and widening project near Williamsburg, VA. Cost analyses, using data from VDOT, were also conducted that compared the recycled versus conventional approaches.

Full-Scale Testing and Evaluation

NCAT Test Track Facility

As described previously, the NCAT Test Track is a 1.7 mile closed-loop full-scale flexible pavement test facility located in Opelika, AL (*8*). Operating on two-year test cycles, the 46 test sections are loaded with approximately ten million ESALs per cycle. During this time, pavement response, deflection and performance measurements were made on a routine basis. Precise climate records and traffic data, applied by a fleet of triple-trailer trucks, were also kept during the test cycle. On average, the test site receives approximately 53 inches of rain per year with mid-depth pavement temperatures ranging from 40 °F to 120 °F. The CCPR materials were subjected to typical testing for recycled materials during mixture design that included an assessment of the retained indirect tensile strength. The materials used in this study passed the requirements of 70% retained indirect tensile strength (ITS) and a dry ITS of 45 psi. The sections in this experiment were constructed during the summer of 2012 and trafficking began in October 2012. Ten million ESALs were applied between October 2012 and October 2014. A second test cycle applying another ten million ESALs began in October 2015.

Test Track Cross Sections

The pavement cross-sections in this study are shown in Figure 1. Each section featured a stone-matrix asphalt (SMA) surface and Superpave dense-graded asphalt concrete (AC) layers above the CCPR layer. Sections N3 and N4 were constructed on top of a crushed granite aggregate base layer and S12 was built on a cement-stabilized base layer simulating FDR. The FDR layer is

Table 1. As-Built Layer Properties (*8*)

Section	N3-6"AC	N4-4"AC	S12-4"AC SB
Layer description	Lift 1-19 mm NMAS SMA with 12.5% RAP and PG 76-22 binder		
Binder content, %	6.1	6.0	6.1
Air voids, %	4.3	4.7	4.2
Layer description	Lift 2-19 mm NMAS Superpave with 30% RAP and PG 67-22 binder		
Binder content, %	4.6	4.6	4.7
Air voids, %	7.1	7.4	6.7
Layer description	Lift 3-19 mm NMAS Superpave with 30% RAP and PG 67-22 binder		
Binder content, %	4.4	NA	NA
Air voids, %	6.4	NA	NA
Layer description	CCPR-100% RAP with 2% foamed 67-22 and 1% type II cement		
Layer description	Crushed granite aggregate base		6" Crushed granite aggregate base and 2" subgrade stabilized in-place with 4% Type II cement
Layer description	Subgrade – AASHTO A-4 Soil		

Note: AC = Superpave dense-graded asphalt concrete; NMAS = nominal maximum aggregate size; CCPR = cold central plant recycling; RAP = reclaimed asphalt pavement: SMA = stone-matrix asphalt; AASHTO = American Association of State Highway and Transportation Officials; SB = stabilized base.

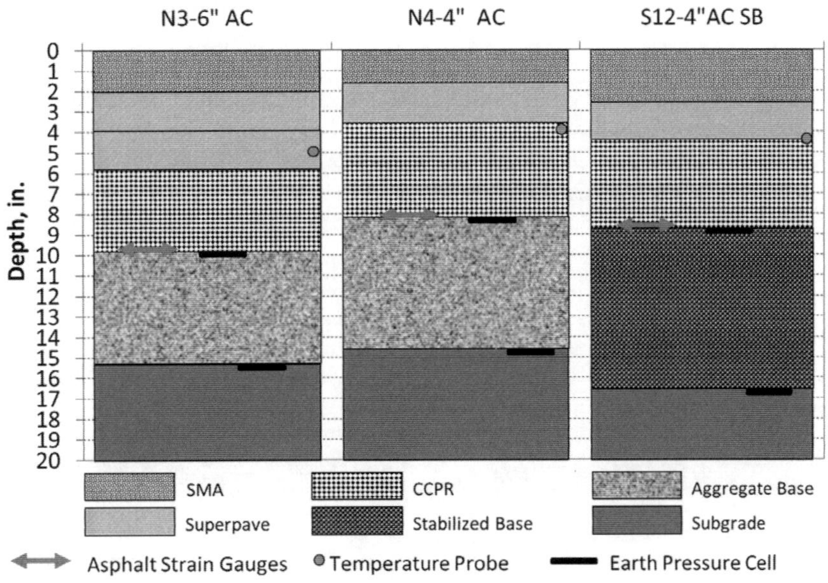

Figure 1. Average test track as-built thicknesses and depth of instrumentation (*8*).

described as simulated because, although traditional FDR construction techniques and equipment were used, no bound materials were included in the FDR process. All three sections were constructed on the same subgrade native to the Test Track and classified as an A-4 soil (*10*). Sections N3 and N4 were designed to evaluate the difference in performance between 4 inches and 6 inches of AC over 5 inches of CCPR. Sections N4 and S12 were designed to determine the performance differences between the aggregate base (6 inches) and the stabilized base (SB) layers (8 inches). Table 1 contains the as-built layer properties and further details of the mix design and construction were previously documented (*8*), but it is important to point out that the CCPR layers were placed as a single lift.

Figure 1 also shows the types and depths of instrumentation used in the sections. Asphalt strain gauges were installed to measure bending at the bottom of the

CCPR. Earth pressure cells (EPCs) were installed at the bottom of the CCPR and the top of the subgrade to measure vertical stress distributions through the pavement. Finally, temperature probes were installed at approximately mid-depth of the AC/CCPR layer to capture environmental effects on the pavement.

The three recycled sections on the Test Track were designed to mimic the cross section of a project placed in Virginia on Interstate 81 in 2011 (*7*, *11*). It is possible to determine the expected service lives of the three Test Track sections by following procedures in the American Association of State Highway and Transportation Officials (AASHTO) 1993 Pavement Design Guide (*12*) and using VDOT's typical design parameters for asphalt pavements (*13*, *14*). Structural numbers for Sections N3, N4, and S12 were calculated as 5.1, 4.2, and 5.5, respectively. Using these structural numbers and inputs for designing an interstate-style pavement, the ESALs that could be carried assuming a soil support of 9400 psi (about 1/3 of the FWD measured values, [*10*]) yields ESAL values ranging from about 3 to 16 million.

Test Track Traffic

Trafficking of the sections began on October 22, 2012 and was applied with five triple-trailer trucks each having a steer axle (11,000 lb), a drive tandem axle (40,000 lb), and five single axles (20,750 lb/axle). This truck configuration corresponded to approximately 10 ESALs per truck pass. During the first two year study (2012–2014), approximately 10 million ESALs were applied. The second test cycle (2015–2017) was to apply an additional 10 million ESALs. There was approximately a one-year break, during Test Track forensic and reconstruction activities, during which no traffic was applied to the sections. At the time of writing this paper (mid-October, 2017), the total traffic for the two cycles was approximately 19 million ESALs.

Test Track Performance Characterization

During each two-year cycle, the sections were manually inspected weekly for signs of cracking in addition to manual rut depth and ride-quality measurements made with an automated road profiling vehicle. These measurements and observations were conducted at different intervals, so the data sets are current through slightly different dates and ESAL totals. For 19 million ESALs, no cracking was observed in any of the three CCPR test sections. This excellent cracking performance was further evaluated according to perpetual pavement criteria as described later in this paper.

Figure 2 shows the rutting performance through to the August 7, 2017 (18.3 million ESALs). Sections N3 and N4 (without the SB) were nearly indistinguishable with maximum rutting of approximately 0.3 inches. S12 (SB) had maximum rut depths of approximately 0.25 inches. As the measured rutting in all three sections

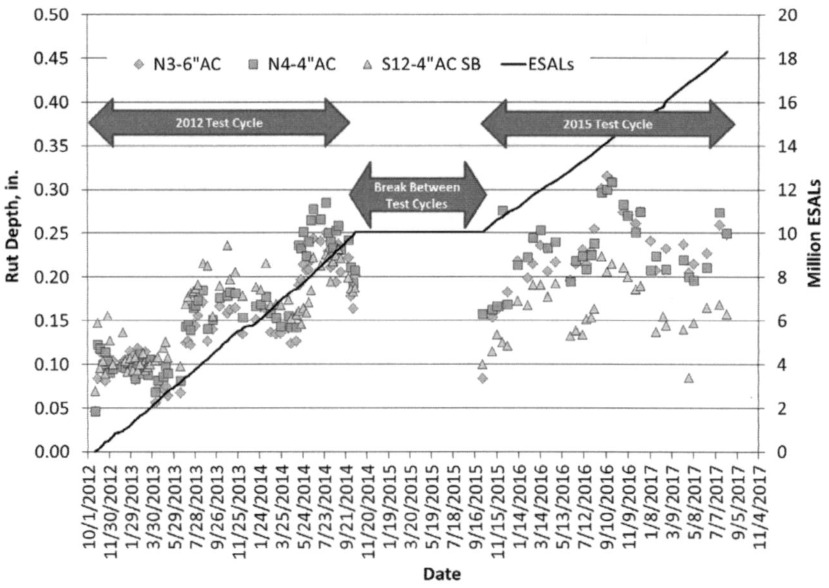

Figure 2. Test track sections—rutting performance.

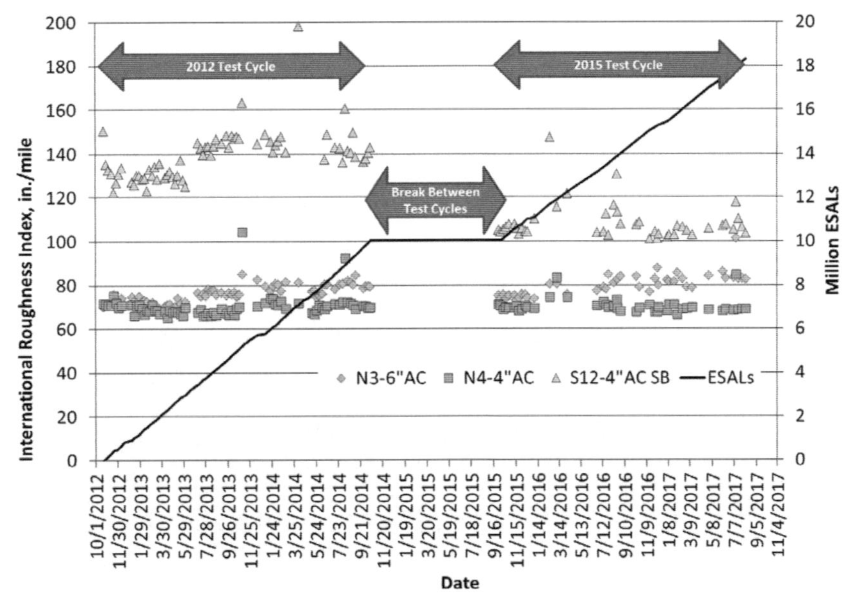

Figure 3. Test track sections—ride quality.

was well below the failure threshold of 0.5 inches, they each exhibited excellent rutting performance with the SB section showing a slight advantage.

Figure 3 plots pavement smoothness, expressed as the International Roughness Index (IRI), versus time through 18.3 million ESALs (August 7, 2017). As noted previously, Section S12 (SB) had an initial roughness of nearly double that of the other sections (*8*). This section had a localized low spot approximately 40 to 60 feet into the section that was identified immediately after construction which produced relatively high IRI values. In addition, the transition from the previous section (S11) into S12 was notably rough at the start of the 2012 research cycle. Between the 2012 and 2015 test cycles, the transition zone was milled and inlaid which appreciably improved the ride quality data. Overall, it is important to note that the IRI of S12 has not changed significantly over time.

Test Track Structural Characterization

As described previously (*8*), FWD testing was conducted on each section multiple times per month during each research cycle. Each round of data consisted of deflections collected at twelve locations corresponding to between, inside and outside wheelpaths at four longitudinal locations in each section. Although four drop-heights were used, with three replicates at each height, only data from the 9-kip load-level are presented here. Following previously-documented procedures that utilized EVERCALC 5.0 for backcalculation (*8*), the data shown in Figure 4 shows the backcalculated AC moduli normalized to 68 °F for both test cycles, through to July 10, 2017 (17.9 million ESALs). It is important to note that the AC and CCPR layers were combined into a single layer for the purposes of backcalculation, as was previously documented (*8*).

Visual inspection of Figure 4 does not indicate appreciable changes in the modulus over time. If there were significant aging or damage occurring over the two test cycles, one would expect to see increases or decreases in temperature-normalized moduli, respectively. Trendlines were fit to each series in Figure 4, but were not included as they all resulted in R^2 values below 0.03, meaning there was no appreciable change attributable to time or traffic application. In other words, the sections appear to be structurally healthy and have not experienced significant aging detectable through backcalculation. Additionally, the apparent increase in modulus for S12 may be partly owing to the presence of the SB layer which may artificially increase the AC modulus during backcalculation. The slight increase over time in the S12 AC modulus may also be attributed to curing of the SB.

Test Track Pavement Response Measurements

Asphalt strain gauges and earth pressure cells installed at the time of construction were routinely monitored throughout the two research cycles to track changes in pavement response over time. Previous documentation

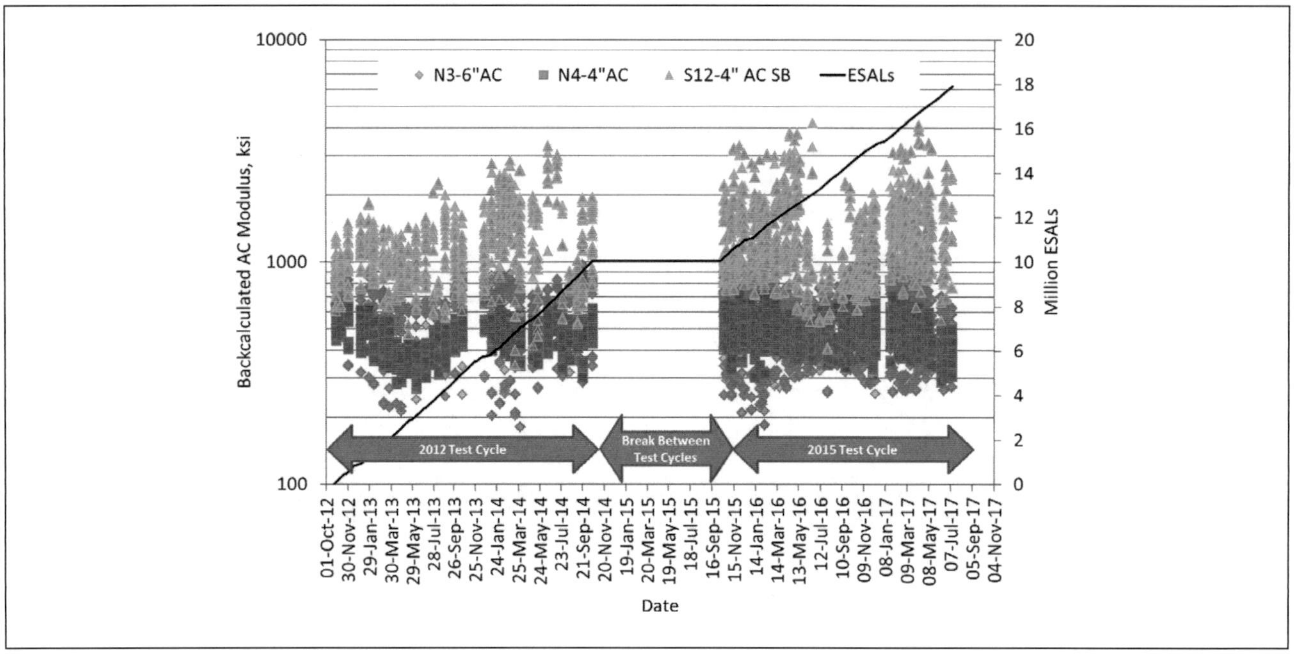

Figure 4. Back calculated AC/CCPR modulus normalized to 68 °F vs. date.

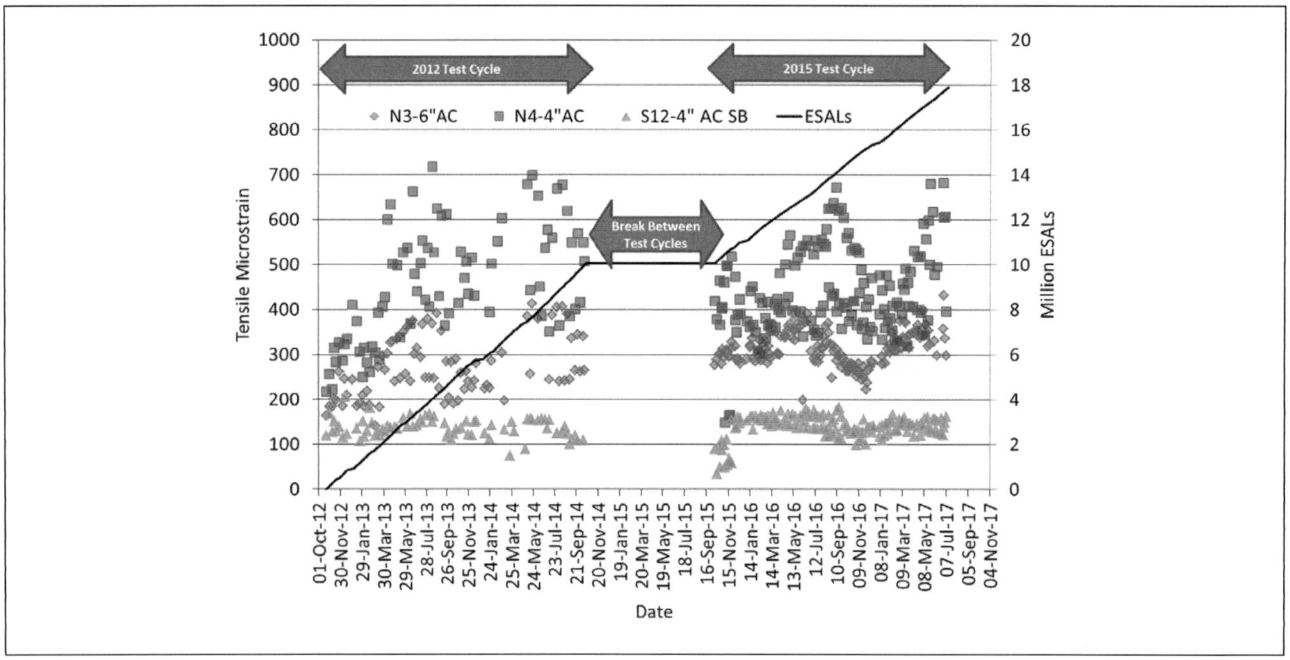

Figure 5. Tensile microstrain normalized to 68 °F vs. date.

described the data collection and reduction processes from these gauges in great detail (*8*). Of particular interest are the strain measurements made at the bottom of the CCPR layer as they are indicative of fatigue cracking performance. Figure 5 plots strain measurements, normalized to 68 °F, made over the two research cycles following the aforementioned procedures (*8*). The data are current through to July 7, 2017 (17.9 million ESALs).

Figure 5 supports the conclusions from Figure 4 that the sections all appear to be structurally healthy with no

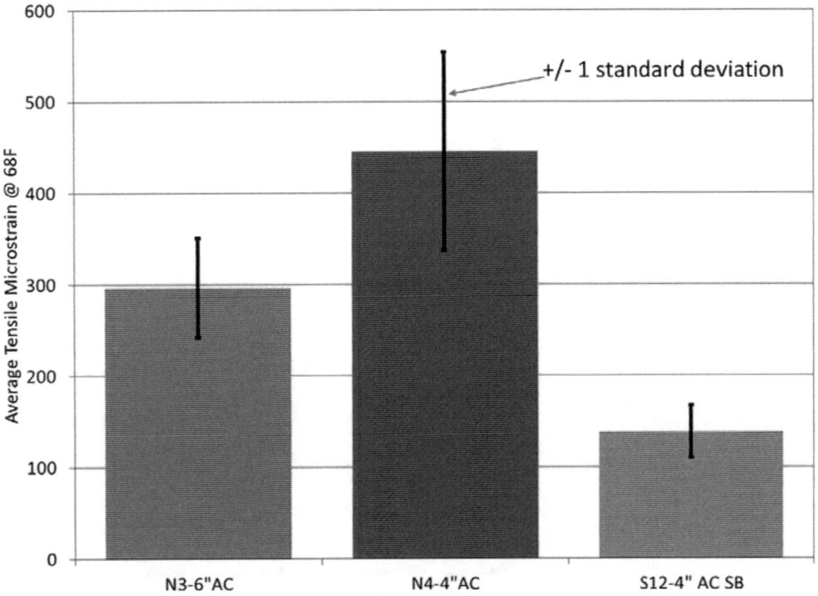

Figure 6. Average tensile microstrain and standard deviation normalized to 68 °F.

appreciable changes in strain response over the two test cycles. Interestingly, there does appear to be an initial increase in strain response for section N4 at the very beginning of the experiment in 2012, but it quickly leveled off and no upward or downward trends have been observed since then. Trendlines were also fit to the series in Figure 5 but were not included as the R^2 were again very low (<0.2). This indicates, as observed in the backcalculated moduli data above, that there has been no appreciable change attributable to time or traffic application.

Differences in strain magnitude are evident between the sections, as expected, owing to differences in AC thickness and the presence of the SB layer. The average strain values and standard deviation, computed from Figure 5, are plotted in Figure 6. Tukey–Kramer analysis ($\alpha = 0.05$) showed that the difference in all mean values was statistically significant so the observed 33% decrease in strain from N4 to N3 and the 69% decrease from N4 to S12 are both practically and statistically significant.

Perpetual Pavement Analysis

AS the CCPR sections at the Test Track exhibited excellent performance through two research cycles, analyses were conducted to evaluate whether these sections may be considered as perpetual pavements with respect to bottom-up fatigue cracking. Studying the fatigue performance of a pavement section subjected to high truck traffic volumes is a common practice for traditionally-constructed asphalt pavements. However, it is not yet known if the CCPR materials used in this study are expected to ultimately fail by fatigue or some other mechanism.

The Test Track sections were evaluated with respect to recently-developed criteria that utilize control strain distributions in the evaluation process. The first, a measured strain distribution, was developed from Test Track data using measured strain data from several sections that either did or did not experience bottom-up fatigue cracking (*15*, *16*). This control distribution serves as an upper limit to the strain response above which cracking may be expected and below which perpetual performance is expected. The second, a simulated strain distribution, was also developed from NCAT Test Track data, but relied on simulated strain levels in the design software, PerRoad, and was further validated with field-documented perpetual pavements (*17–19*). It is important to emphasize that these criteria were not developed from CCPR or cement-stabilized base pavement sections so it is currently unclear whether they are truly applicable. However, they do serve as a checkpoint against conventional flexible pavements and further trafficking and monitoring of the sections will help to further validate whether they may be applied, or new criteria are needed.

To evaluate the sections against the measured strain distribution criteria, a cumulative strain distribution was generated from measured data for each section and plotted with the control distribution as shown in

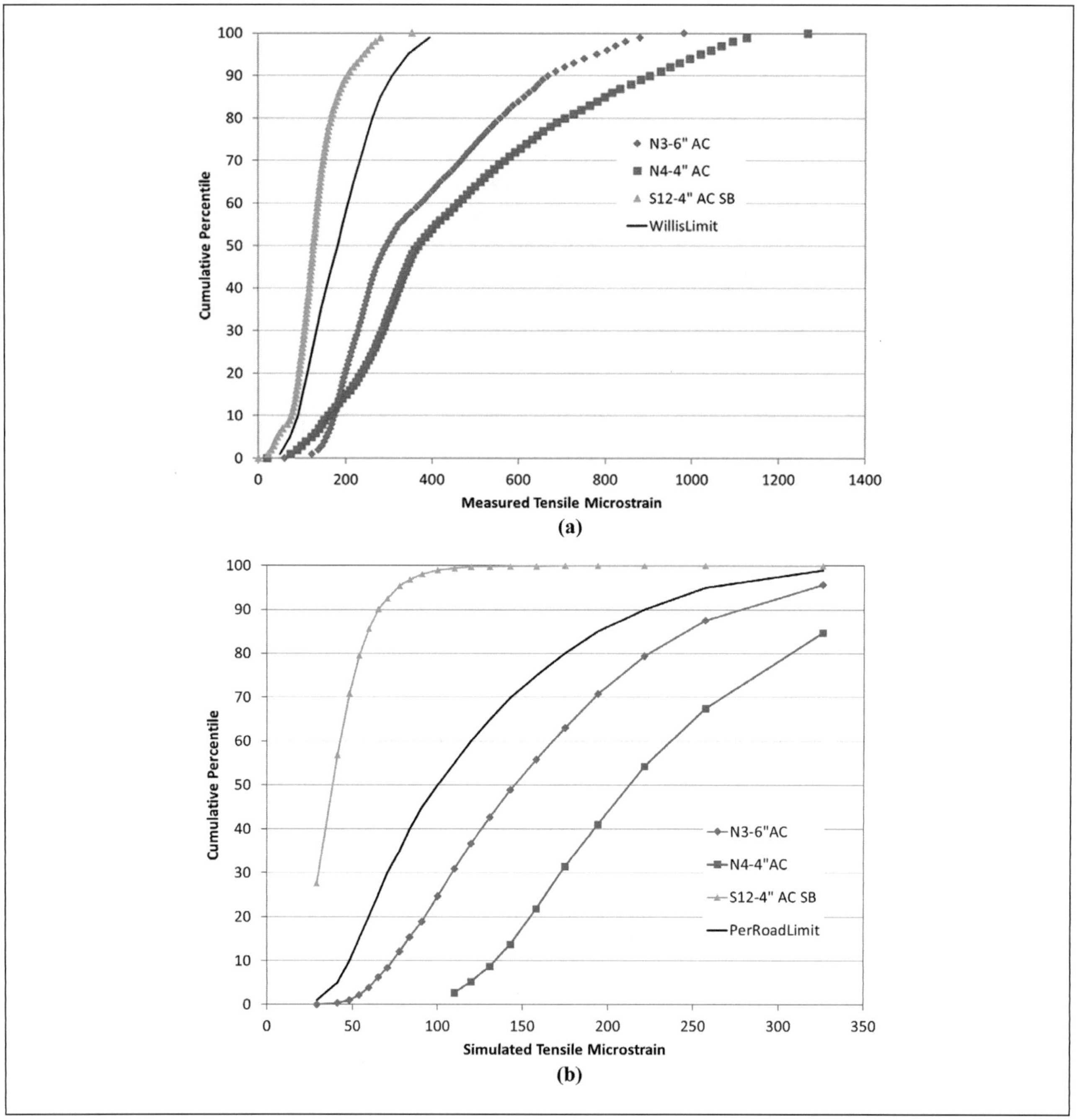

Figure 7. Perpetual pavement analysis cumulative strain distributions (*a*) measured strains; and (*b*) PerRoad simulated strain.

Figure 7*a*. Simulated strain distributions were generated through the PerRoad software by entering relevant input parameters (i.e., layer moduli, layer thicknesses, and traffic conditions) and utilizing the Monte Carlo features in PerRoad to produce cumulative distributions as illustrated in Figure 7*b*. Both analyses resulted in the same conclusion that section S12 – with the cement-stabilized base layer – is expected to be perpetual as its strain distribution is less than the control distribution. The other two sections both exceed the control distribution, and according to the criteria, are expected to experience bottom-up cracking at some point. Furthermore, both analyses show the benefit of the additional 2 inches of AC resulting in lower strain levels for N3 relative to N4.

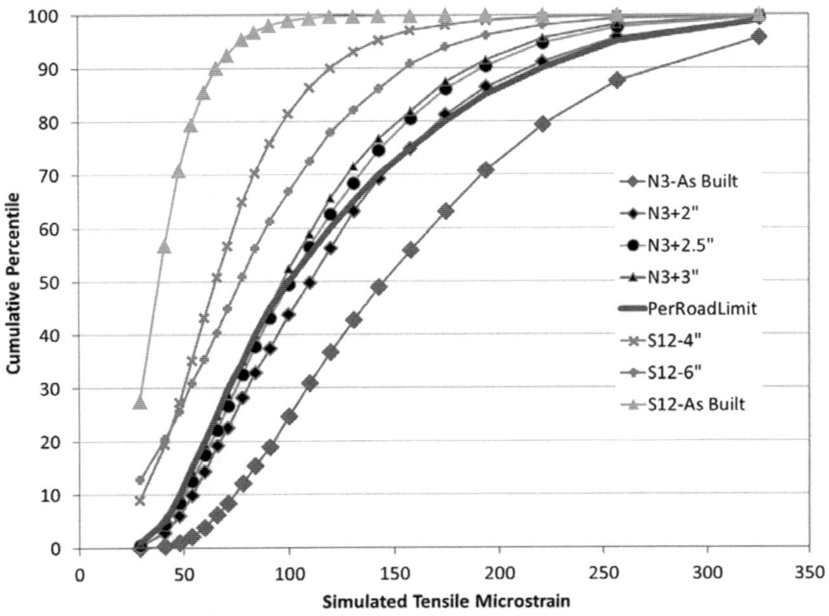

Figure 8. Additional perpetual strain analyses.

At this point, it is unknown whether the sections exceeding the strain limits will truly develop bottom-up cracking as the criteria were not developed from CCPR sections. Likewise, cracking could develop in S12, perhaps from cracking of the cement-stabilized layer reflecting through the CCPR and AC layers. However, application of the current perpetual criteria indicate that S12 is likely perpetual while the others are not.

Further perpetual analyses were conducted with PerRoad to evaluate sections N3 and S12 to quantify the changes in thickness needed to bring them closer to the perpetual design limit (i.e., optimized perpetual design). In section N3, this required additional thickness while S12 required thickness reductions. Figure 8 summarizes the results of the analyses in which the as-built cross-sections were first analyzed followed by incrementally increasing (N3) or decreasing (S12) the AC thickness. For example, the N3 + 2" series represents the simulation in which an additional 2 inches were added to the as-built AC/CCPR layer thickness. The figure shows that an additional 2.5 to 3 inches of AC/CCPR, bringing the total AC/CCPR depth to 11.75 to 12.25 inches, will move N3 into the non-cracking side of the perpetual limit.

Figure 8 also shows that a dramatic decrease in the AC/CCPR thickness (up to 6 inches) still leaves the strain distribution well to the left of the perpetual limit. An obvious concern with this cross section, which focuses only on controlling strain at the bottom of the AC/CCPR layer, is that covering the SB layer with only three or less inches of AC/CCPR would potentially lead to cracking of the SB layer. Additional mechanistic simulations were conducted with the layered elastic program, WESLEA for Windows, to examine the horizontal strain levels through the depth of varying S12 cross-sections. These simulations, discussed in depth below, used the average modulus values from the PerRoad analysis and a single wheel load of 9,000 lb with 100 psi of contact pressure.

Figure 9 summarizes the results for the three simulated S12 cross-sections. In each case, the neutral axis (point of zero strain) of the cross-section lies within the AC/CCPR layer with the bottom of the CCPR and SB layers both in tension. The as-built cross-section has nearly equivalent strain levels at the bottom of both layers, but the SB layer experiences significantly higher strain levels as the AC/CCPR thickness is reduced. Changing from as-built to 6-inch results in a 40% increase in the tensile strain at the bottom of the CCPR layer, but the SB tensile strain increases by almost a factor of three. The strain (or stress) tolerance of the SB layer is not currently known, but this analysis highlights the potential risk of placing too little AC/CCPR over the SB layer in which it would be forced to carry significantly greater tensile loadings. Measurement of the fatigue tolerance of the SB should be studied in the future.

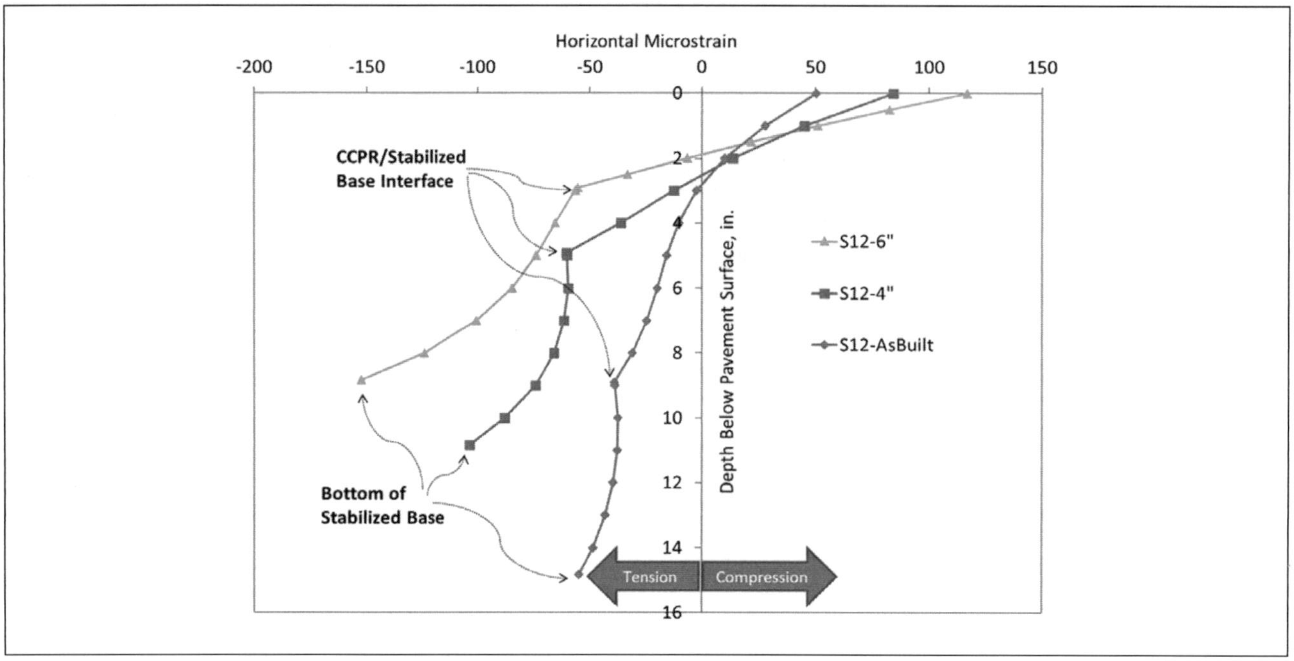

Figure 9. S12 Horizontal strain versus depth simulations.

Application of Results

In 2016, VDOT awarded a design-build contract to reconstruct and widen a portion of Interstate 64 near Williamsburg, VA using FDR and CCPR as major components of the new pavement sections. The first phase of the project will consist of building a new travel lane and full-width shoulder to the inside of the existing roadway in both directions. These new lanes will be an asphalt pavement having a 6-inch-thick CCPR base placed on top of 12 inches of cement treated base produced from crushed concrete. The second phase of the project will consist of replacing the existing two lanes and outside shoulder of the jointed concrete pavement with an asphalt pavement having a 6-inch CCPR base on top of a 12-inch-thick FDR foundation. The project length is 7.08 miles and this section of I-64 carries approximately 37,500 vehicles per day (in each direction) with about 8.5% trucks. Considering typical RAP contents for the surface asphalt layers and the CCPR layer, it is estimated that nearly 200,000 tons of RAP will be used on the project.

VDOT gained the confidence to move forward with this innovative pavement solution from the excellent performance results observed at the NCAT Test Track sections in conjunction with the performance of Virginia's Interstate 81 pavement recycling project (*7*, *11*). Given the current ESAL counts for the NCAT sections (approaching 20 million ESALs), it would take approximately 22 years on the I-64 project to reach the same ESAL values.

Cost Analysis of Recycled Versus Conventional Designs

Figure 10 shows the pavement cross-sections for the original and recycled designs. The two designs are structurally equivalent and use the VDOT typical layer coefficients (*13*, *14*). The calculated structural numbers are 7.08 and 7.06 for the original and recycled designs, respectively. A significant cost savings using the recycling design was identified by comparing local material prices (shown in Table 2) for the original design to the as-bid prices for the recycled design. The costs for the original and recycled sections were calculated as \$88 per square yard and \$42 per square yard, respectively. For this project, the cost savings using the recycled design is greater than \$10 million.

Conclusions and Recommendations

The following conclusions and recommendations are drawn from this study:

- The three recycled pavement test sections at the NCAT Test Track are examples of new or reconstructed pavement structures built with high percentages of recycled materials. The results of this

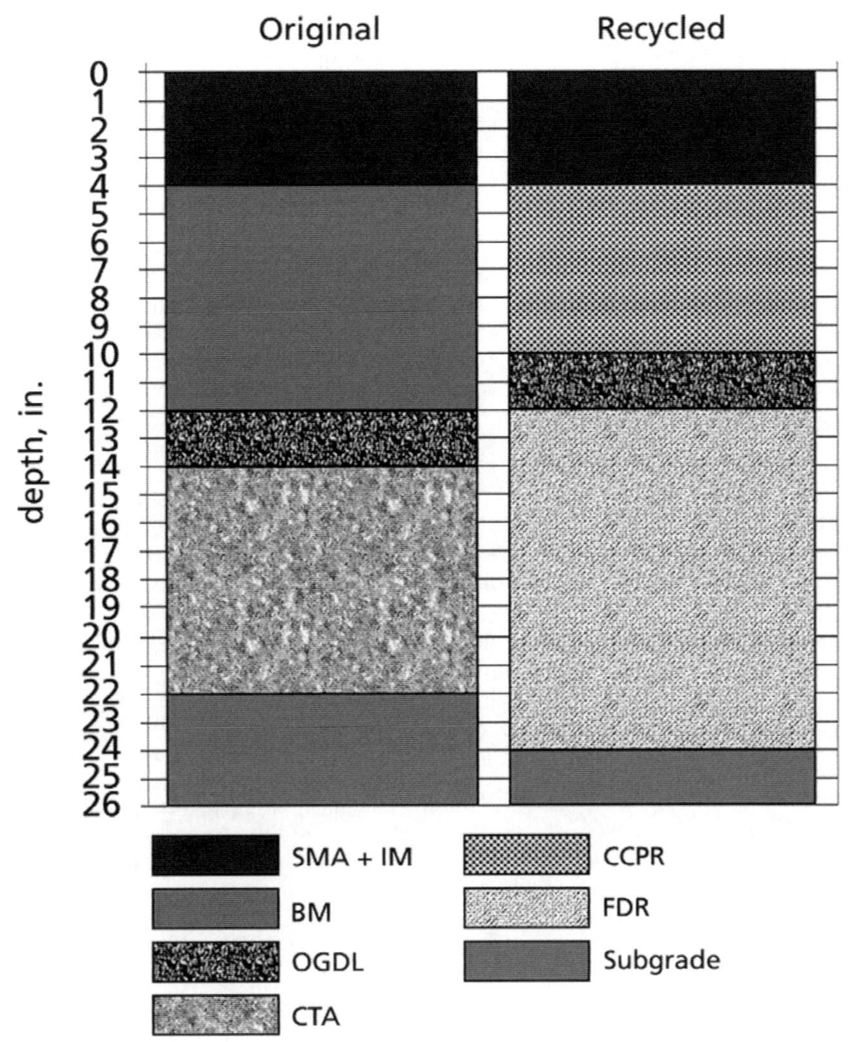

Figure 10. Original and recycled designs for I-64 project.

Table 2. Local Material Prices Used in Cost Analysis

Material	Unit	Unit Cost ($)
Asphalt surface (SMA)	Tons	106
Asphalt intermediate (SMA)	Tons	93
Asphalt base	Tons	82
Cement treated aggregate	Tons	44
CCPR	Tons	45
FDR	Square yard	12

Note: SMA = stone-matrix asphalt; CCPR = cold central plant recycling; FDR = full-depth reclamation.

study show that they have outperformed their designed service lives based on the current traffic level of 19 million ESALs and probably much longer based on performance and structural characterizations.

- CCPR layers may be treated like conventional asphalt concrete materials in pavement design and modeling as supported by the measured and simulated strain data presented in the perpetual pavement analysis in addition to the backcalculated layer properties using layered elastic analysis. This is consistent with other studies of similar materials that found that CCPR materials could be modeled in the same way as asphalt concrete materials (*20, 21*).

- Additional AC thickness was more effective at reducing AC strain than the cement stabilized layer. Although the cement-stabilized base produced a greater overall strain reduction (69% reduction) than additional AC thickness (33% reduction), when the thickness of the respective layers is considered the additional AC reduced the strain by 14.9% per inch of AC, while the SB was 8.8% per inch of base.
- Based on perpetual strain analysis, section S12 – with the cement-stabilized base layer – is expected to be perpetual as its strain distribution is less than the control distribution. This assumes that the previously-developed criteria may be applied to CCPR pavements with a cement-stabilized layer. Section S12 should be left in place for the next cycle of trafficking to validate this assumption and expectation.
- Based on perpetual strain analysis, sections N3 and N4 exceed the control strain distribution, and are expected to experience bottom-up cracking at some point. Again, it is currently unknown whether the criteria apply to CCPR pavements, and this warrants further trafficking and monitoring in the next test cycle.
- By using the information learned from the NCAT Test Track, a recycled design for a new or reconstructed pavement can be constructed with significant cost savings compared to the original design.

Acknowledgments

The authors acknowledge the assistance of Andy Babish, Rob Crandol, Affan Habib, David Kaulfers, and Mike Wells, VDOT Materials Division; Chaz Weaver, VDOT Staunton District; Troy Deeds, Donnie Dodds, Benjamin Earl, and Kevin McGhee, VTRC; Trenton Clark and Richard Schreck, Virginia Asphalt Association; Ken Lanford, Lanford Brothers Co.; Larry Roberts and David Stowell, Slurry Pavers; Marlin Hewett, B & S Contractors; Brent Moore, Branscome, Inc.; Charles Schwartz, University of Maryland; Alex Apeagyei; University of East London; Rennie Shenmugam, Loudon International; Mike Marshall, Wirtgen America; Buzz Powell, Jennifer Still, and Randy West, NCAT; and Wangyu Ma, Auburn University. The authors also acknowledge the efforts of Michael Brown, Richard Ferron and Al Soltis. The authors acknowledge Linda Evans, VTRC, for assistance with the editorial process.

Author Contributions

The authors confirm contribution to the paper as follows: study conception and design: Timm, Diefenderfer, and Bowers; data collection: Timm, Diefenderfer, and Bowers; analysis and interpretation of results: Timm, Diefenderfer, and Bowers; draft manuscript preparation: Timm, Diefenderfer, and Bowers. All authors reviewed the results and approved the final version of the manuscript.

References

1. Bemanian, S., P. Polish, and G. Maurer. Cold In-Place Recycling and Full-Depth Reclamation Projects by Nevada Department of Transportation: State of the Practice. *Transportation Research Record: Journal of the Transportation Research Board*, 2006. 1949: 54–71.
2. Thenoux, G., A. Gonzalez, and R. Dowling. Energy Consumption Comparison for Different Asphalt Pavements Rehabilitation Techniques Used in Chile. *Resources, Conservation and Recycling*, Vol. 49, 2007, pp. 325–339.
3. Stroup-Gardiner, M. *NCHRP Synthesis 421: Recycling and Reclamation of Asphalt Pavements Using In-Place Methods*. Transportation Research Board of the National Academies, Washington, D.C., 2011.
4. *NCHRP Project 09-51: Material Properties of Cold In-Place Recycled and Full-Depth Reclamation Asphalt Concrete for Pavement Design*. Transportation Research Board of the National Academies, Washington, D.C., 2017.
5. *NCHRP Project 09-62: Rapid Tests and Specifications for Construction of Asphalt-Treated Cold Recycled Pavements*. Transportation Research Board of the National Academies, Washington, D.C., 2017.
6. Asphalt Recycling and Reclaiming Association. *Basic Asphalt Recycling Manual*. Publication No. FHWA-HIF-14-001. Asphalt Recycling and Reclaiming Association, Annapolis, MD, 2014.
7. Diefenderfer, B. K., A. K. Apeagyei, A. A. Gallo, L. E. Dougald, and C. B. Weaver. In-Place Pavement Recycling on I-81 in Virginia. *Transportation Research Record: Journal of the Transportation Research Board*, 2012. 2306: 21–27.
8. Timm, D. H., M. Diaz-Sanchez, and B. Diefenderfer. Field Performance and Structural Characterization of Full-Scale CCPR Pavements. Presented at the 94th Annual Meeting of the Transportation Research Board, Washington, D.C., 2015.
9. Virginia Department of Transportation. *Interstate 64 Widening – Segment II*. http://www.i64widening.org/learn_more/segment_2.asp. Accessed July 24, 2017.
10. Taylor, A. J., and D. H. Timm. *Mechanistic Characterization of Resilient Moduli for Unbound Pavement Layer Materials*. Report No. 09-06. National Center for Asphalt Technology, Auburn University, Auburn, AR, 2009.
11. Diefenderfer, B. K., B. F. Bowers, and A. K. Apeagyei. Initial Performance of Virginia's Interstate 81 In-Place Pavement Recycling Project. *Transportation Research Record: Journal of the Transportation Research Board*, 2015. 2524: 152–159.
12. American Association of State Highway and Transportation Officials. *Guide for Design of Pavement Structures*. American Association of State Highway and Transportation Officials, Washington, D.C., 1993.

13. Virginia Department of Transportation. *Notice of Revision to Materials Division's Manual of Instructions Chapter VI.* Memorandum 386-15. Virginia Department of Transportation, Richmond, VA, 2015.
14. Virginia Department of Transportation, Materials Division. *Guidelines for 1993 AASHTO Pavement Design.* Virginia Department of Transportation, Richmond, VA, 2000.
15. Willis, J. R., and D. H. Timm. Development of Stochastic Perpetual Pavement Design Criteria. *Journal of the Association of Asphalt Paving Technologists*, Vol. 79, 2010, pp. 561–596.
16. Willis, J. R., and D. H. Timm. *Field-Based Strain Thresholds for Flexible Perpetual Pavement Design.* Report No. 09-09. National Center for Asphalt Technology, Auburn University, Auburn, AR, 2009.
17. Castro, A. J., N. Tran, M. M. Robbins, D. H. Timm, and C. Wagner. Further Evaluation of Limiting Strain Criteria for Perpetual Asphalt Pavement Design. *Transportation Research Record: Journal of the Transportation Research Board*, 2017. 2640: 41–48.
18. Robbins, M. M., N. Tran, D. H. Timm, and J. R. Willis. Adaptation and Validation of Stochastic Limiting Strain Distribution and Fatigue Ratio Concepts for Perpetual Pavement Design. *Journal of the Association of Asphalt Paving Technologists*, Vol. 84, 2015, pp. 181–217.
19. Tran, N., M. M. Robbins, D. H. Timm, J. R. Willis, and C. Rodezno. *Refined Limiting Strain Criteria and Approximate Ranges of Maximum Thicknesses for Designing Long-Life Asphalt Pavements.* Report No. 15-05. National Center for Asphalt Technology, Auburn University, Auburn, AR, 2015.
20. Diefenderfer, B. K., B. F. Bowers, C. W. Schwartz, A. Farzaneh, and Z. Zhang. Dynamic Modulus of Recycled Pavement Mixtures. *Transportation Research Record: Journal of the Transportation Research Board*, 2016. 2575: 19–26.
21. Schwartz, C. W., B. K. Diefenderfer, and B. F. Bowers. *NCHRP Research Report 863: Material Properties of Cold In-Place Recycled and Full-Depth Reclamation Asphalt Concrete.* Transportation Research Board of the National Academies, Washington, D.C., 2017.

The Section on Pavements (AFD00) peer-reviewed this paper (18-01535).

Article

TRR
JOURNAL OF THE TRANSPORTATION RESEARCH BOARD

Transportation Research Record
2018, Vol. 2672(40) 304–314

DOI: 10.1177/0361198118788426
journals.sagepub.com/home/trr
SAGE

Calibrating the Mechanistic–Empirical Pavement Design Guide Rutting Models using Accelerated Pavement Testing

Yu Tian[1], Jusang Lee[2], Tommy Nantung[2], and John E. Haddock[3]

Abstract

This paper presents a new method for the local calibration of the rutting models in the *Mechanistic–Empirical Pavement Design Guide* (MEPDG) for Indiana's pavements using accelerated pavement testing (APT). The study focused on the verification, calibration, and validation of the rutting models in the MEPDG using both field and APT sections. In this study, a new calibration methodology was developed that uses layer rutting distribution. Also, a procedure was developed to provide the most realistic simulations of APT conditions (climate, traffic, and aging) using virtual weather station data generation, a special traffic configuration, and falling weight deflectometer evaluation. The primary benefits of APT are that it uses test sections with high distress levels and low measurement errors and it provides findings about the contributions of each layer to the total rut depth (i.e., pavement surface rutting). The accuracy of the MEPDG's rutting prediction models was improved significantly following implementation of the new calibration process. The sum of squared errors (SSE) was reduced by 73% and the standard error of estimates was reduced by 38%. No significant differences were found between the predicted and measured asphalt layer, subgrade, and total rut depths at the 95% confidence level. When compared with the conventional calibration method, which uses the total rut depth without knowing the rut depth of each layer, the proposed method provides more reliable predictions for both asphalt mixture and subgrade rutting, most especially for the latter. The bias and SSE of the predicted subgrade rutting were improved by 46% and 62%, respectively.

In the early 2000s, a new pavement design approach based on mechanistic methods, the *Mechanistic–Empirical Pavement Design Guide* (MEPDG), was introduced as part of NCHRP Project 1-37A to serve as a replacement for the 1993 AASHTO pavement design method. However, to use the MEPDG correctly, it must first be calibrated for local conditions. Since the introduction of the MEPDG, many state and local agencies as well as consultants have worked on MEPDG local calibration.

Several challenges have been encountered when calibrating the MEPDG for Indiana practice, most especially when calibrating the rutting models. First, the pavement sections used for local calibration should exhibit a level of distress that is at least 50% of the design failure criterion. Otherwise, the accuracy and bias of the predictions might not be well defined at the pavement failure trigger values. Second, the Superpave volumetric mix design procedure has been used by most U.S. state highway agencies since 2005 (*1*), and the Indiana Department of Transportation (INDOT) has been using Superpave mixtures for all Indiana pavements ever since. Because the final product of this study will be implemented by INDOT for future pavement design and analysis to maintain a high level of prediction for Superpave pavements, it was decided that only Superpave-designed pavements would be selected for this study. However, most such pavements have been in service for less than 10 years and show only limited rutting distress in Indiana.

The third challenge, as indicated by the MEPDG developers (*2*), is a limitation of the rutting models themselves. These models were developed using total rut depths measured at the pavement surface, as these data were all that were available from the Long-Term Pavement Performance (LTPP) program database. Trench data, or any other information that could be

[1]Key Laboratory of Road and Traffic Engineering of the Ministry of Education, Tongji University, Shanghai, P.R. China
[2]Office of Research and Development, Indiana Department of Transportation, West Lafayette, IN
[3]Purdue University, Lyles School of Civil Engineering, West Lafayette, IN

Corresponding Author:
Address correspondence to Jusang Lee: jlee@indot.in.gov

used to derive layer-wise rutting, were not available to the MEPDG developers (*2*). Without knowing the individual layer rutting information, it is impossible to calibrate the models precisely, because there is no way to determine from the data how much of the total rutting occurred in the unbound layers and how much occurred in the asphalt layers. Given this limitation, the MEPDG rutting models were calibrated globally based on the assumption that the contribution of each of the layers to the total rut depth measured at the surface would be similar to that predicted by the MEPDG. However, this assumption is questionable and is likely one of the major sources of error in MEPDG rutting prediction results. The results of several local calibration efforts have indicated that the globally calibrated models over-predicted the results (*2–7*).

Objectives

The primary objectives of this research are to (1) evaluate the prediction performance of the MEPDG rutting models, (2) develop a new method based on accelerated pavement testing (APT) techniques for calibrating local MEPDG rutting models, and (3) provide well calibrated MEPDG models that are ready for INDOT implementation. Note: The version of the MEPDG used in this study is Pavement ME 2.0.

APT Rutting Performance

Benefits

APT has provided potential solutions to overcome the aforementioned difficulties. The benefits of using APT are:

- Although field sections used for calibration typically have 10 or fewer years of in-service life and often exhibit "fair" to "good" performance, APT sections can be made to exhibit distress levels that are close to the rutting failure design criterion by accelerating traffic load applications.
- The use of APT sections can reduce the required number of field sections needed for calibration, as the APT sections lead to lower standard error of estimates (SEE) values due to well-controlled test conditions and measurement procedures.
- Using a layer-wise rutting measurement technique, that is, the mid-depth profile monitoring system (MPMS), which was developed as part of this study, the distributed rutting data were employed in the calibration process to fine-tune the two rutting models, that is, one for the unbound layers and one for the asphalt layers, individually.

Accelerated Pavement Testing

This study used a full-scale APT facility designed and built by Purdue University and INDOT researchers and personnel. The test sections were constructed using a typical INDOT full-depth asphalt pavement structure with open graded (OG) drainage layer. The four APT test sections consisted of two different pavement structures and two different surface course materials, one a conventional Superpave-designed dense-graded (dense) mixture, the second a stone matrix asphalt (SMA) mixture. Details about the APT test sections can be found in a previous publication (*8*).

APT Parameters

APT loading was applied using a one-half standard axle with dual tire assembly. This study employed a tire size of 11R22.5, loading of 4.08 tonnes (9,000 lb.), and tire pressure of 0.69 MPa (100 psi). The target load application was 50,000 passes, which produced about 10.16 mm (0.4 in.) of total rut depth at the pavement surface. This rut depth was targeted based on the rutting failure criterion used in INDOT practice. The tire wandering feature was not activated in order to further accelerate the rutting progression. Embedded pavement sensors indicated that the pavement temperature during testing was 47°C (117°F) at a depth of 3.8 cm (1.5 in.). This target temperature was based on the design high pavement temperature derived from the mean average maximum air temperature for the hottest seven-day period within a 20-year period obtained from a weather station near the field sites used in the calibration process.

Mid-Depth Profile Monitoring System

The rutting that occurred within each layer was measured using the MPMS. It consists of a series of monitoring holes that are drilled into the pavement surface and reach down to the various pavement layer interfaces. An automated laser profiler installed beside the loading wheels is used to monitor the change in elevation at the bottom of each monitoring hole to determine the layer deformations. Details about the MPMS can be found in a previous publication (*8*). It was found, for typical INDOT full-depth flexible pavements, regardless of pavement thickness and surface layer material, the subgrade deformation remained at a very low level in this study. Only about 10% of the rutting occurred in the subgrade.

Calibration of Rutting Models

Calibration Database

This study used full-depth flexible pavements constructed with Superpave-designed asphalt mixtures in an attempt

Table 1. Pavement Structure and Material Properties

ID	Route name	Pavement structure and material properties						
			Surface	Intermediate	Upper base	OG	Lower base	SG
1	APT Test Lane 1	Thickness (in.)	1.5	2.5	3	2.5	3	12.5
		NMAS (mm)	9.5	19	19	19	19	N/A
		PG	70-22	70-22	64-22	76-22	64-22	N/A
2	APT Test Lane 1	Thickness (in.)	1.5	2.5	3	2.5	3	12.5
		NMAS (mm)	9.5 (SMA)	19	19	19	19	N/A
		PG	70-22	70-22	64-22	76-22	64-22	N/A
3	APT Test Lane 3	Thickness (in.)	1.5	2.5	6	2.5	3	15.5
		NMAS (mm)	9.5 (SMA)	19	19	19	19	N/A
		PG	70-22	70-22	64-22	76-22	64-22	N/A
4	APT Test Lane 4	Thickness (in.)	1.5	2.5	6	2.5	3	15.5
		NMAS (mm)	9.5	19	19	19	19	N/A
		PG	70-22	70-22	64-22	76-22	64-22	N/A
5	Previous APT Lane 1	Thickness (in.)	2	3	12			16
		NMAS (mm)	9.5	19	19			N/A
		PG	70-22	70-22	64-22			N/A
6	Previous APT Lane 2	Thickness (in.)	2	3	9			16
		NMAS (mm)	9.5	19	19			N/A
		PG	70-22	70-22	64-22			N/A
7	SR-43	Thickness (in.)	1.5	2.5	4	2.8	3	12
		NMAS (mm)	9.5	19	19	19	19	N/A
		PG	76-22	76-22	64-22	76-22	64-22	N/A
8	SR-28	Thickness (in.)	1.5	2.5	4	3.4	4	12
		NMAS (mm)	9.5	19	19	19	19	N/A
		PG	70-22	70-22	64-22	76-22	64-22	N/A
9	US-41	Thickness (in.)	1.5	2.5	4	2.3	3	12
		NMAS (mm)	9.5	19	19	19	19	N/A
		PG	76-22	76-22	64-22	76-22	64-22	N/A
10	SR-43	Thickness (in.)	1.5	2.5	11.5			12
		NMAS (mm)	9.5	19	19			N/A
		PG	70-22	64-22	64-22			N/A
11	US-40	Thickness (in.)	1.5	2.5	4	3.2	4.5	12
		NMAS (mm)	9.5	19	19	19	19	N/A
		PG	76-22	76-22	64-22	76-22	64-22	N/A
12	I-465	Thickness (in.)	1.5	2.5	5.5	3.6	5.5	12
		NMAS (mm)	9.5 (SMA)	19	19	19	19	N/A
		PG	76-22	76-22	64-22	76-22	64-22	N/A
13	US-421	Thickness (in.)	1.5	2.5	4	3.5	4	12
		NMAS (mm)	9.5	19	19	19	19	N/A
		PG	70-22	70-22	64-22	76-22	64-22	N/A
14	US-31	Thickness (in.)	2	3	12			16
		NMAS (mm)	9.5	19	19			N/A
		PG	70-22	70-22	64-22			N/A

to calibrate the MEPDG rutting models. Only a few full-depth in-service pavement sections were available for calibration purposes. After a thorough search of the INDOT project database, eight field projects were identified for use in the calibration scheme. The selected field sections were three state roads (SRs), four U.S. highways, and one interstate (I) highway. The sections were constructed between 2007 and 2009 and the pavement thicknesses varied from 31.8 cm to 43.2 cm (12.5 in. to 17 in., respectively). The average annual daily traffic varied from 332 to 14,463 vehicles per day. Additionally, four APT sections, as previously described, and two APT sections from a previous APT study were selected and combined with the field sections to create an MEPDG local calibration database. Table 1 lists these pavement structures and material properties, that is, pavement thickness, nominal maximum aggregate size (NMAS), and performance grade (PG) for each of the pavement layers (surface, intermediate, upper base, OG, lower base, and subgrade) for each of the test sections.

MEPDG Level 3 input data were used in this study. Level 1 input requires laboratory characterization of the materials, but the laboratory tests typically are not required for construction. Thus, to be consistent with the future implementation of this calibration product, Level 1 input was excluded from the scope of this study. For Level 3 input, the MEPDG estimates the dynamic modulus values of hot mix asphalt mixture using the gradation and binder grade, and the resilient modulus values of unbound material using soil classifications.

All the required input parameters were prepared based on contractors' surveys and data extraction from the INDOT pavement management database, including weather station data, site-specific traffic counts and parameters, construction information such as job mix formulas, pavement structures, construction logs, and quality assurance/quality control data.

APT Simulations

In order to take advantage of APT techniques, several considerations must be taken into account. First, the MEPDG simulates realistic environmental conditions based on weather station data and then predicts daily and seasonal temperature and moisture variations in pavements, whereas APT typically utilizes controlled climate conditions that challenge even the most realistic simulations obtained via the MEPDG. Second, the MEPDG simulates more asphalt aging than actually occurs during APT, as reported by Von Quintus et al. (*5*). In order to provide the most realistic simulations of APT conditions in the MEPDG simulations, a procedure was developed in this study to consider the conditions that are unique to APT, that is, climate, traffic, and aging.

A virtual weather station was generated to represent the APT climate conditions. By adjusting the corresponding climatic parameters, that is, shortwave absorptivity, percentage of sunshine, precipitation, and air temperature, the pavement temperature and moisture content could be kept constant during the entire MEPDG simulation period.

Traffic was simulated for APT using the special axle configuration feature offered by the MEPDG to customize and define the wheel assembly for APT. The dual tires were simulated using two tires, each with a tire load of 2.04 tonnes (4,500 lb.), the dual tire spacing was 34.29 mm (13.5 in.), and the tire inflation pressure was 0.69 MPa (100 psi). Lateral wander standard deviation was set to zero so that the load could be applied the same way as for the APT sections.

With regard to the aging condition, once asphalt mixtures are placed and exposed to the environment, they begin to age as the asphalt binder oxidizes. As a result, the viscous properties of the asphalt binder change over time and, thus, so do the mixture properties (*9*). Asphalt mixtures tend to become stiffer over time, which plays a vital role in the material's behavior and the pavement's long-term performance. The MEPDG incorporates the

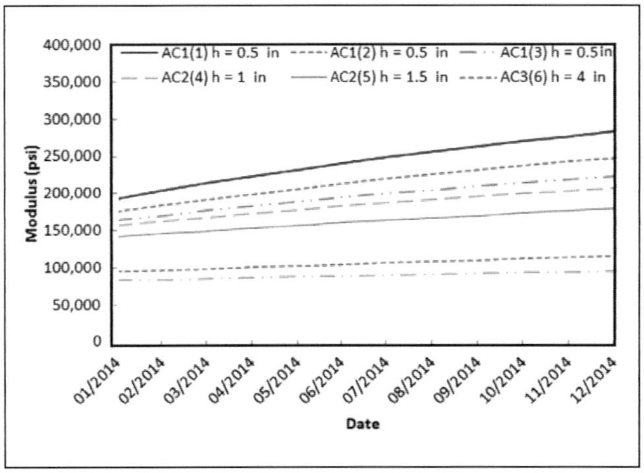

Figure 1. MEPDG GAS-simulated aging effect for APT Lane 1.

effects of aging using the Global Aging System (GAS) developed by Mirza and Witczak (*10*). The GAS predicts the binder viscosity at any time and at any depth in the pavement system. The predicted viscosity is then incorporated into the determination of the dynamic modulus values.

Although the GAS simulates a realistic aging condition for in-service pavement analysis, such a model challenges the most faithful simulations of APT. APT usually is performed within a few months of pavement placement whereas MEPDG simulations usually are conducted over a much longer analysis period. As a result, more asphalt aging is simulated by the MEPDG than actually occurs during APT. The minimum analysis period in the MEPDG is one year, which means that a minimum of one year of aging must occur, even for APT sections that are constructed and tested within only a few months.

Figure 1 shows a preliminary MEPDG GAS analysis run using APT Lane 1 parameters; the aging-induced evolution of the modulus value of each sublayer is plotted. AC1(1), AC1(2), and so forth represent the sublayers used in the MEPDG analysis. During the one-year analysis period, about a 50% increase in the asphalt layer modulus value occurred.

Falling weight deflectometer (FWD) testing was conducted in this study to account for the different aging levels in the actual APT and MEPDG simulations. The FWD back-calculated modulus value was used to quantify the actual amount of increase in the APT asphalt layer in situ modulus value. This amount of increase was then compared with the GAS-simulated aging effect, as shown in Figure 1. Then, the analysis period (referred to as the 'aging analysis period'), in accordance with the amount of increase in the MEPDG simulation, could be determined. As an example, the in situ modulus value of

APT Lane 1 was found to increase by 27% after the APT loading was completed using FWD back-calculation. Specifically, Figure 1 indicates that an average modulus value increase of 27% occurred within the first six months; that is, the aging analysis period is six months.

The aging analysis period was then used in the MEPDG simulations. Actual APT load applications were applied within the aging analysis period instead of the MEPDG analysis period, and the simulation results at the end of the aging analysis period were used. Using the same APT Lane 1 example, the total APT load application is 50,000 passes, which must be applied during the six-month period (i.e., the same as the aging analysis period determined earlier). Thus, 8,333 passes per month were used in the MEPDG one-year simulation. Finally, the simulation results for the first six-month period were used, although one-year simulation was performed.

Calibration Procedure

Local calibration is a process that adjusts the calibration coefficients of the rutting models in order to eliminate or minimize bias and SEE. Two rutting models are embedded in the MEPDG to predict the rutting that will occur in the unbound material layers and asphalt layers, respectively. The model for the unbound material layer is shown in Equation 1 and that for the asphalt layer is shown in Equation 2.

$$\delta_a(N) = \beta_{s1} k_1 \varepsilon_v h \left(\frac{\varepsilon_0}{\varepsilon_r}\right) e^{-\left(\frac{\rho}{N}\right)^{\beta}} \quad (1)$$

where

δ_a = permanent deformation for a given layer (inch)

N = number of load repetitions

β_{s1} = calibration factor

k_1 = 2.03 for granular material, 1.35 for fine-grained material

ε_v = vertical resilient strain in the layer as obtained from a mechanistic model

ε_r = resilient strain imposed in a laboratory test to obtain material properties

h = layer thickness (inch)

$\varepsilon_0, \rho, \beta$ = material properties, which are correlated with the layer water content and layer resilient modulus

$$\frac{\varepsilon_p}{\varepsilon_r} = K_Z \beta_{r1} 10^{k_{r1}} (T)^{k_{r2}\beta_{r2}} (N)^{k_{r3}\beta_{r3}} \quad (2)$$

where

ε_r = incremental resilient strain at the mid-depth of a thickness increment

ε_p = plastic strain calculated at the mid-depth of a thickness increment

T = temperature at the mid-depth of a thickness increment (°F)

N = number of axle load applications of a specific axle type and load interval within a specific time period

$\beta_{r1}, \beta_{r2}, \beta_{r3}$ = local calibration coefficients, all equal to 1 for the globally calibrated model

k_{r1}, k_{r2}, k_{r3} = plastic deformation factors

K_Z = depth function

The calibration coefficients β_{r2} *and* β_{r3} in the models are used to adjust the nonlinear effect of temperature and the number of traffic load repetitions for the MEPDG predictions, respectively. These coefficients were fine-tuned by executing the MEPDG using all the predetermined potential values of β_{r2} *and* β_{r3} to determine the values that result in the lowest sum of squared errors (SSE). Coefficients β_{r1} and β_{s1}, on the other hand, are direct multipliers. In other words, they are linearly proportional to the predictions and thus could be optimized externally without MEPDG execution. This optimization was performed using the Microsoft Excel Solver optimization routine based on the generalized reduced gradient method.

Figure 2 illustrates that the field and APT sections were combined to form the calibration database. The uniqueness of each section was taken into account by performing the most realistic MEPDG simulations using the actual parameters, that is, traffic, material properties, and climate conditions. For the APT sections, the MEPDG simulations were conducted according to the procedure developed in this study.

Ten percent of the measured total rut depth was distributed into the subgrade layers for the field sections. According to a previous finding (*8*), for typical Indiana full-depth flexible pavements (i.e., pavement thickness greater than 12.5 in.), about 10% of the total rut depth is in the subgrade; this percentage remains relatively constant for different pavement structures and materials.

Then, β_{s1} was optimized externally to minimize the SSE between the predicted and measured subgrade rut depths. For the asphalt layer, β_{r1}, β_{r2}, and β_{r3} must be optimized simultaneously. A large factorial of every possible combination of β_{r2} and β_{r3} was determined based on previous research findings and experience gained from the MEPDG execution (*11*, *12*). Twenty combinations were used in this study, ranging from 1.7 to 2.0 and 0.1 to 0.5 for β_{r2} and β_{r3}, respectively. The interval for each combination is 0.1. It should be noted that the calibration process is very time-consuming, and the MEPDG was executed for the 14 sections for each β_{r2} and β_{r3} combination. The total number of MEPDG executions was 280. The total running time was 70 hours, with 15 minutes per run. According to previous studies (*11*, *12*), an interval of 0.1 is sufficient to provide satisfactory results; besides, a further decrease of the interval by 50%

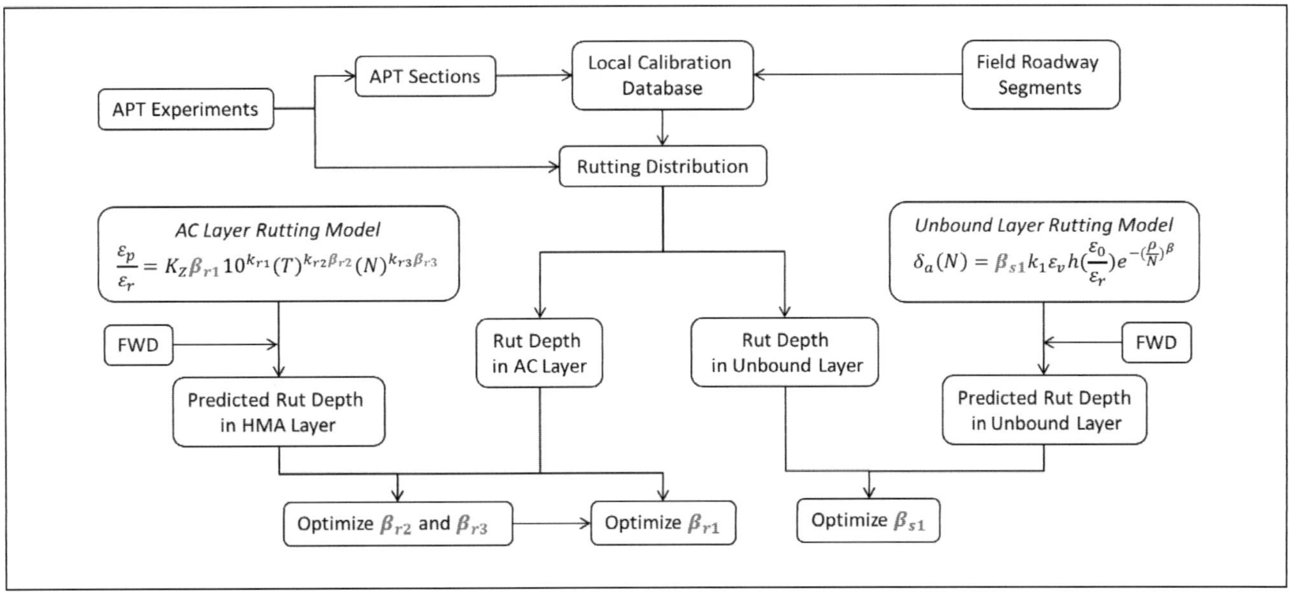

Figure 2. Calibration flow chart.

Table 2. Local Calibration Factors

β_{s1}	β_{r1}	β_{r2}	β_{r3}
0.115	0.075	1.900	0.400

would result in a 400% increase in the MEPDG runtime.

Finally, the predicted rut depths of the asphalt layer and subgrade were exported and prepared for external optimization. β_{r1} was then optimized for each combination of β_{r2} and β_{r3}. Table 2 lists these calibration factors.

Model Validation

Model validation is necessary because it assesses the capability of the calibrated models to produce robust and accurate predictions of pavement performance. A successful validation can be achieved if the bias and precision statistics obtained from the validation process are similar to those obtained from the calibration process, or, alternatively, a statistical test should be performed with the null hypothesis that no significant difference exists between the predicted and observed values.

Validation typically requires an additional set of roadway sections that is independent of the set used for the calibration. However, in this study, the number of available sections was extremely limited and an additional independent set of sections was not available. As suggested by AASHTO (*13*), a jack-knife resampling approach can be used for a small database. As shown in Figure 3, an n - 1 jack-knife validation process employs a sample with n roadway sections divided first into two groups. One group has n -1 sections that are used for the 1st model calibration, and then the 1st calibrated model is applied to the remaining section from which the standard error (e_1) is calculated. Next, another section is withheld from n roadway sections and the new n - 1 sections are used for the 2nd model calibration. The 2nd calibrated model is applied to the withheld segment for the standard error (e_2) calculation. This process is repeated until all n sections have been withheld and n values of standard error are obtained. Those standard errors can be used to calculate the jack-knife statistics.

In this study, the database contained 14 sections; therefore, calibration was performed 14 times in order to obtain the validation statistics. For each calibration, the jack-knifed standard errors were computed from the sections that were not used in the calibration process. Thus, the jack-knife statistics are considered to be independent evaluations of the model predictions.

Verification, Calibration, and Validation Results and Discussion

Verification of Globally Calibrated Models

Before proceeding to the calibration process, the accuracy of the globally calibrated models was evaluated for local practice. When the MEPDG was first developed, the global models were calibrated using the LTPP program database. The global calibration factors almost

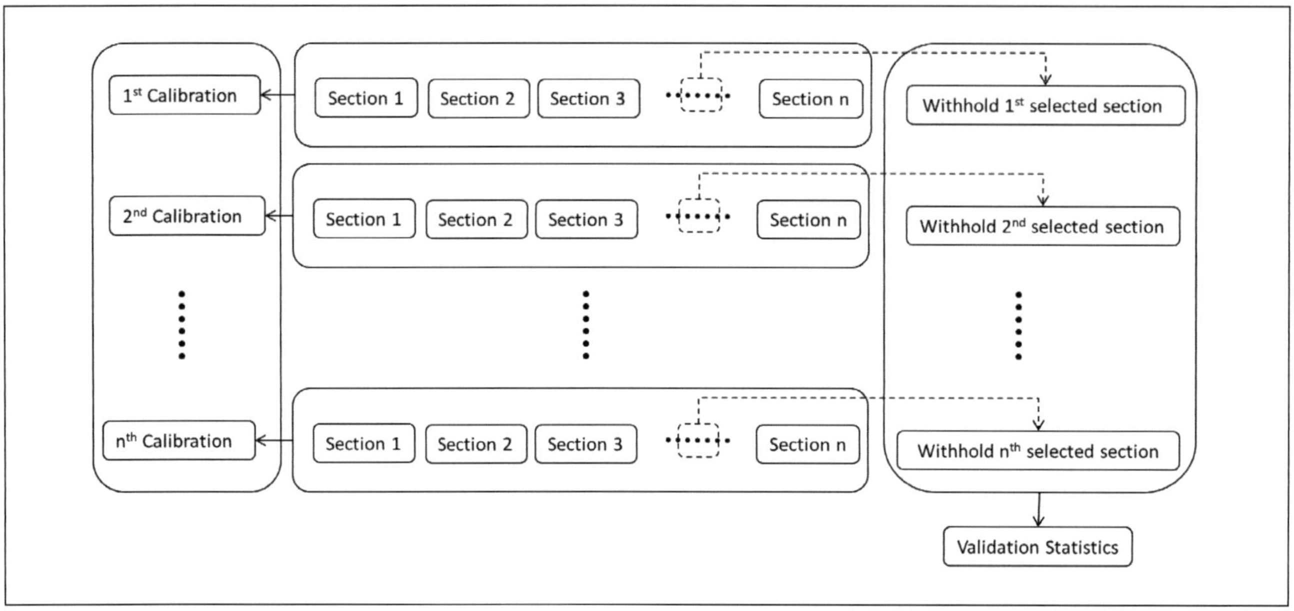

Figure 3. Validation flow chart.

certainly do not reflect local conditions, however. Furthermore, when the MEPDG was first calibrated using the LTPP program database, newly developed mix design methods such as the Superpave mix design method, and unconventional mixtures such as SMA mixtures, were not incorporated into the database. The discrepancies between global and local conditions, old and new mixture design methods, and conventional and unconventional mixtures can introduce bias and error into distress predictions.

The MEPDG, with its inherent global models, was executed in this study to analyze preselected field and APT sections. Figure 4 presents comparisons between the predicted and actual rut depths in the globally calibrated models for the total rut depth, asphalt layer rut depth, and subgrade rut depth, respectively. Extensive dispersion and poor correlations are evident between the predicted and actual values for the total, asphalt layer, and subgrade rut depths. The total rut depth was overpredicted by 1.6 mm (0.06 in.). This trend was due mostly to the overestimation of the subgrade rutting (i.e., bias is 3.2 mm or 0.13 in.). For the asphalt layer rutting, the predicted values are lower than the actual values with an average of 1.6 mm (0.06 in.). The prediction accuracy for both the field and APT sections are poor, especially for the two APT sections that experienced the most rutting. The poor performance of the globally calibrated MEPDG models for Indiana pavements further proves the need for this study.

Statistical analyses at the 95% confidence level were performed to evaluate the predictions. As shown in Table 4, the null hypotheses for both the asphalt layer and subgrade predictions were rejected, indicating significant differences between the predictions and actual values. Although the null hypothesis for the total rut depth was accepted, the *p*-value is very close to the critical value of 0.05.

Calibration

Figure 5 presents comparisons of the predicted rut depths and predicted residual errors for the globally calibrated and locally calibrated models. The notable bias that is evident in the globally calibrated models for the rut depths in the asphalt layer and subgrade was corrected in the local model. Precision also improved after the calibration; the residual errors are less scattered, especially for the subgrade, and they are evenly distributed around the zero line, indicating unbiased or low bias predictions.

As shown in Table 4, before the calibration, the statistical test results showed that significant differences existed between the predictions and measured values for both the asphalt layer and the subgrade rut depths. After calibration, all of the hypotheses were accepted, with a clear tendency toward significance at the 95% confidence level. The coefficients of correlation improved overall to above 0.6 except for the subgrade rut depth predictions. The precision of the predictions also increased, as indicated by the reduction of the SEE by 38% and reduction of the SSE by 73%. The bias of the predictions also was reduced, as indicated by the reduction of the bias by 87%.

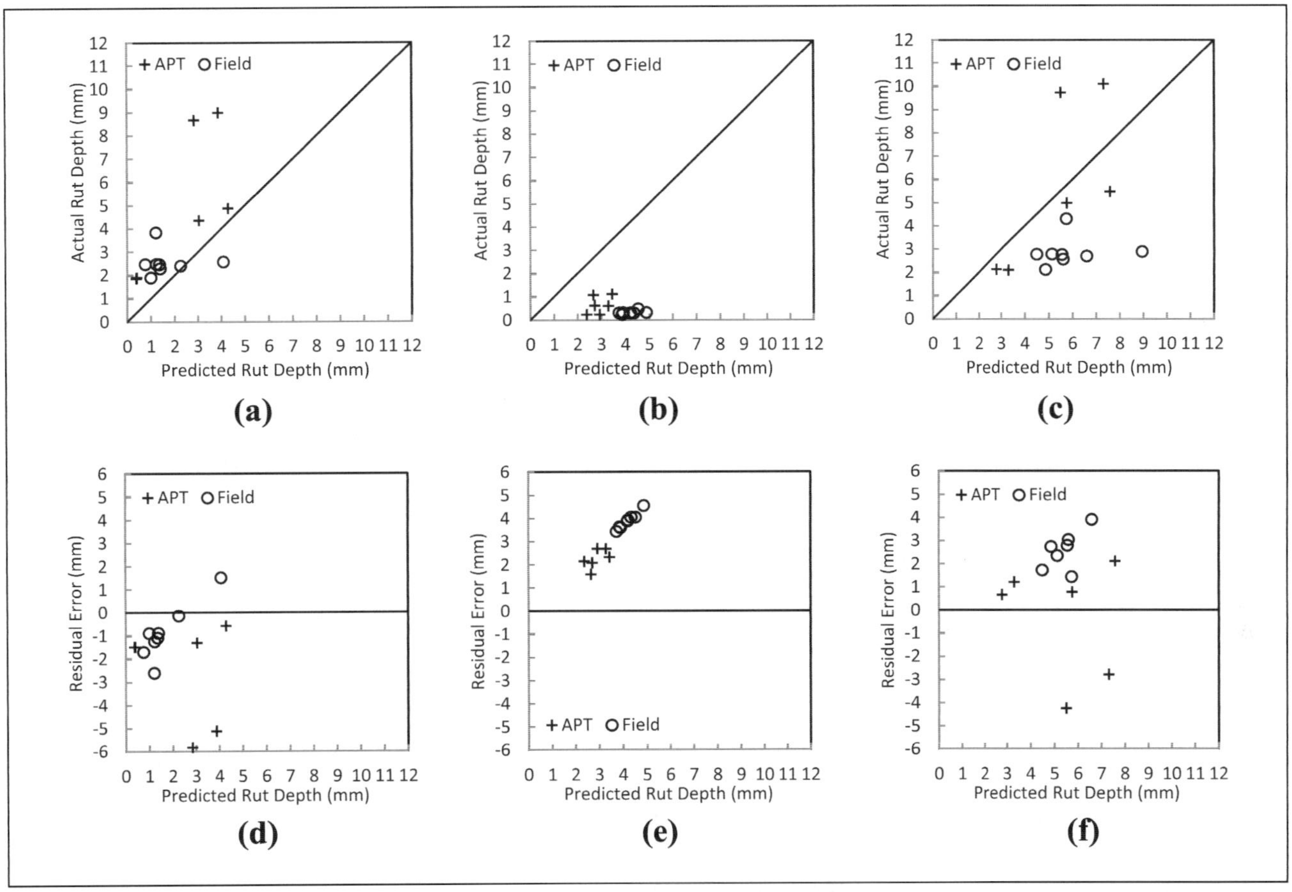

Figure 4. Predicted rut depths (*a*, *b*, *c*) and prediction errors (*d*, *e*, *f*) of globally calibrated models for asphalt layer (*a*, *d*), subgrade (*b*, e), and total rut depths (*c*, *f*).

Evaluation of Conventionally Calibrated Local Models

The rutting models were calibrated conventionally based on the assumption that the contribution of each of the layers to the total rut depth measured at the surface would be similar to that predicted by the MEPDG. Due to the fact that only the total rut depths measured at the pavement surface were available for calibration, the two models were calibrated simultaneously. In other words, compared with the proposed calibration method used in this study, where the two rutting models (i.e. Equations 1 and 2) are calibrated separately using the pavement layer-wise deformation, the conventional calibration method optimizes the rutting models and the four calibration factors (i.e., $\beta_{r1}, \beta_{r2}, \beta_{r3}, and\ \beta_{s1}$) all together by simply assuming that the summation of Equations 1 and 2 equals the total rut depth. Such an approach could produce multiple optimal values with similar SSE due to the under-determination of the system. Moreover, this approach focuses only on minimizing the standard error of the total rut depth, which ignores the prediction accuracy for individual layers. In this research, for comparative purposes, the MEPDG rutting models were calibrated using the conventional approach and the same test sections. Table 3 presents a comparison of the calibration statistics for both methods, that is, the conventional calibration method and the new proposed calibration method. Although the statistics for the conventional method that were applied for the total rut depth predictions are similar to those of the proposed method, the proposed method provided more reliable predictions for both the asphalt and subgrade layers. For the subgrade predictions in particular, the bias and SSE were improved by 46% and 62%, respectively. The conventional calibration method appears to provide results that ignore the effects and contributions of the subgrade and asphalt layers with regard to total rut depth and thus could be problematic for pavement design and analysis.

Validation

A similar trend between the calibration and validation results for the new proposed calibration model in terms of bias and precision was found. Table 4 compares the

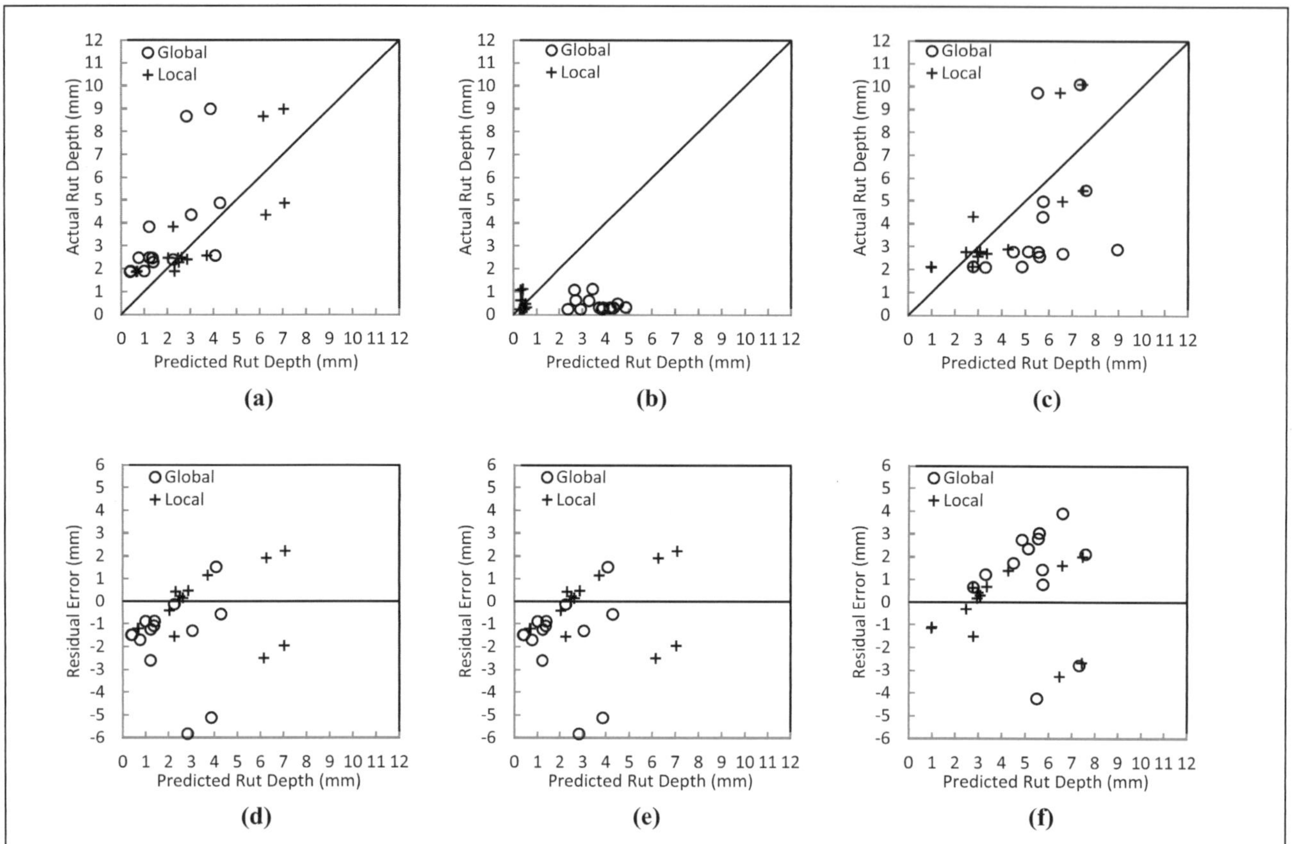

Figure 5. Predicted rut depths (*a, b, c*) and prediction errors (*d, e, f*) of globally calibrated and locally calibrated models for asphalt layer (*a, d*), subgrade (*b, e*), and total rut depths (*c, f*).

Table 3. Comparison of Calibration Statistics for Conventional and Proposed Methods

Layer	Model	Bias (mm)	SSE (mm²)	S_e
Asphalt layer	Conventional	0.4	27.7	1.41
	Proposed	-0.2	26.0	1.40
Subgrade	Conventional	-0.3	4.2	0.46
	Proposed	-0.2	1.5	0.30
Total	Conventional	0.0	29.9	1.52
	Proposed	-0.2	32.4	1.60

statistics obtained from the globally calibrated models, locally calibrated models, and model validation. The validation statistics are similar overall to those of the locally calibrated models. Although the SSE values obtained from validation are relatively higher than those obtained from the locally calibrated models, they are still lower than those obtained from the globally calibrated models. The coefficients of correlation remained at the same level for validation. The same hypotheses also were accepted for the asphalt layer, subgrade, and total rut depths at the 95% confidence level, with p-values similar to those obtained from the calibrated models.

Findings and Conclusions

In this research, a methodology was developed to calibrate the MEPDG prediction models using a database that contains both APT sections and field roadway sections and accounts for the rutting in individual pavement layers. The following findings and conclusions were drawn from this study:

- For the globally calibrated MEPDG models, the predicted rut depths of the asphalt layer and subgrade were found to be significantly different from the actual measurements. The total rut depths were over-predicted due mostly to the overestimation of the subgrade rutting.
- The accuracy of the MEPDG prediction models was improved after the implementation of the new calibration process. The SSE was reduced by 73%.

Table 4. Statistics for Globally Calibrated and Locally Calibrated Models and Model Validation

Layer	Model	Mean predicted (mm)	Mean actual (mm)	Bias (mm)	SSE (mm^2)	S_e (mm)	R^2	H_0:
Asphalt layer	Global	2.0	3.6	-1.6	83.0	1.9	0.38	Rejected, $p = 0.035$
	Local	3.5		-0.2	26.0	1.4	0.66	Accepted, $p = 0.850$
	Validation	3.5		-0.1	58.2	2.1	0.39	Accepted, $p = 0.911$
Subgrade	Global	3.6	0.5	3.2	153.2	0.9	0.13	Rejected, $p = 0.000$
	Local	0.4		0.0	1.5	0.3	0.13	Accepted, $p = 0.656$
	Validation	0.4		0.0	1.7	0.4	0.24	Accepted, $p = 0.656$
Total	Global	5.7	4.1	1.6	119.7	2.6	0.14	Accepted, $p = 0.078$
	Local	3.9		-0.2	32.4	1.6	0.66	Accepted, $p = 0.828$
	Validation	3.5		-0.1	67.0	2.3	0.38	Accepted, $p = 0.887$

The SEE was reduced by 38%, indicating that the precision of the predictions was improved. No significant difference was found between the predicted and actual rut depths.

- A comparison of the proposed calibration method with the conventional calibration method shows that the new method provides more reliable predictions for both the asphalt and subgrade layers. For the subgrade predictions in particular, the bias and SSE were improved by 46% and 62%, respectively.
- The model validation, which employed jack-knife resampling, confirmed that the calibrated models are able to provide accurate and statistically sound pavement performance predictions.
- The proposed calibration method as well as the developed procedure for using APT sections as a supplement for MEPDG calibration can be used also by other states for their own MEPDG implementation efforts.
- The product of this study already has been implemented by INDOT for full-depth flexible pavement design and analysis.

Acknowledgments

This research work was supported by the Indiana Department of Transportation and Joint Transportation Research Program.

Author Contributions

The authors confirm contributing to the paper as follows: study conception and design: Yu Tian, Jusang Lee, Tommy Nantung, John E. Haddock; data collection: Yu Tian; analysis and interpretation of results: Yu Tian, Jusang Lee; draft manuscript preparation: Yu Tian, Jusang Lee, John E. Haddock. All authors reviewed the results and approved the final version of the manuscript.

References

1. Zhou, J., J. M. Signore, and J. T. Harvey. *Superpave Implementation Phase I: Determining Optimum Binder Content*. UCPRC-TM-2012-03. University of California Pavement Research Center, 2012.
2. Applied Research Associates, Inc. *NCHRP Project 1-37A: Guide for Mechanistic-Empirical Design of New and Rehabilitated Pavement Structures*. Transportation Research Board of the National Academies, Washington, D.C., 2004.
3. Hong, F., and D.-H. Chen. Calibrating Mechanistic-Empirical Design Guide Permanent Deformation Models Based on Accelerated Pavement Testing. *Journal of Testing and Evaluation*, 2008.
4. Kim, Y. R., F. M. Jadoun, T. Hou, and N. Muthadi. *Local Calibration of the MEPDG for Flexible Pavement Design*. HWY-2007-07. NC State University, 2011.
5. Von Quintus, H., J. Mallela, R. Bonaquist, C. W. Schwartz, and R. L. Carvalho. *NCHRP Report 719: Calibration of Rutting Models for Structural and Mix Design*. Transportation Research Board of the National Academies, Washington, D.C., 2012.
6. Souliman, M. I., M. S. Mamlouk, M. M. El-Basyouy, and C. E. Zapata. (2010). *Calibration of the AASHTO MEPDG for Flexible Pavement for Arizona Conditions*. Proc., 89th Annual Meeting of the Transportation Research Board, Washington, D.C., 2010.
7. Wu, Z., and X. Yang. *Evaluation of Current Louisiana Flexible Pavement Structures Using PMS Data and New M-E Pavement Design Guide*. Louisiana Department of Transportation and Development, 2012.
8. Tian, Y., J. Lee, T. NanTung, and J. Haddock. Development of a Mid-depth Profile Monitoring System for Accelerated Pavement Testing. *Construction & Building Materials*, Vol. 140, 2017.
9. Cong, P., N. Liu, Y. Tian, and Y. Zhang. Effects of Long-Term Aging on the Properties of Asphalt Binder Containing Diatoms. *Construction & Building Materials*, Vol. 123, 2016.
10. Mirza, M. W., and M. W. Witczak. Development of a Global Aging System for Short and Long Term Aging of Asphalt Cements. *Journal of the Association of Asphalt Paving Technologists*. 1995.
11. El-Basyouny, M. M. *Calibration and Validation of Asphalt Concrete Pavements Distress Models for 2002 Design Guide*. PhD dissertation, Arizona State University, 2004.
12. Jadoun, F. *Calibration of the Flexible Pavement Distress Prediction Models in the Mechanistic Empirical Pavement*

Design Guide (MEPDG) for North Carolina. PhD dissertation. North Carolina State University, 2011.

13. American Association of State Highway and Transportation Officials. *Guide for the Local Calibration of the Mechanistic-Empirical Pavement Design Guide*. American Association of State Highway and Transportation Officials, 2010.

The Standing Committee on Full-Scale Accelerated Pavement Testing (AFD40) peer-reviewed this paper (18-02761).

The contents of this paper reflect the views of the authors, who are responsible for the facts and the accuracy of the data presented here. The contents do not necessarily reflect the official views or policies of the sponsors. These contents do not constitute a standard, specification, or regulation.

Article

TRR
JOURNAL OF THE TRANSPORTATION RESEARCH BOARD

Transportation Research Record
2018, Vol. 2672(40) 315–325

DOI: 10.1177/0361198118758058
journals.sagepub.com/home/trr
SAGE

Development of a Structural Capacity Prediction Model Based on Traffic Speed Deflectometer Measurements

Zia Uddin Ahmed Zihan[1], Mostafa A. Elseifi[1], Kevin Gaspard[1], and Zhongjie Zhang[2]

Abstract

The importance of incorporating pavement structural conditions in the selection of maintenance and rehabilitation strategies along with functional indices has been recognized by state agencies. To measure in-service pavement structural capacity, surface deflection under a defined load has been typically used. The traffic speed deflectometer (TSD) has emerged as a continuous deflection-measuring device as it operates at traffic speed and reduces lane closure and user delays. The present study developed a nonlinear regression model to predict pavement structural number (SN) based on surface deflections measured by the TSD along with the total pavement layer thickness and traffic volume. The proposed model was successfully developed and validated with SN calculated based on TSD and falling weight deflectometer deflection data obtained from two testing programs in Louisiana and Idaho. The model was further validated with respect to its prediction of in-service structural capacity loss and deficiency. Based on statistical measures and the model's ability in identifying structurally deficient sections, results showed satisfactory accuracy of the model and supports its use for network-level decision-making processes in the pavement management system.

In-service pavement conditions are typically described by several functional factors such as surface distresses, roughness, and rutting, which do not necessarily assess structural damage of a pavement. As roads are being subjected to loads higher than the design traffic loads and to extreme weather events, the increasing rate of deterioration necessitates the use of a structural capacity indicator in a pavement management system (PMS) for effective rehabilitation and maintenance decision-making. Structural capacity is a valuable input in the design of asphalt concrete (AC) overlays and for identifying structurally deficient pavements. It is also important in selecting treatment methods and making pavement rehabilitation decisions. Pavement deflection under a given static or moving load is a fast and reliable method to evaluate pavement structural capacity. Deflection is also an important measurement that is used in pavement deterioration models (*1–4*).

Pavement deflection is typically measured by applying a defined load using the falling weight deflectometer (FWD); this method allows the measurement of deflections with a high accuracy but it requires lane closures causing traffic delays and safety concerns. This has limited the use of FWD to project level applications and has led to the introduction of traffic-speed deflection devices including the traffic speed deflectometer (TSD) and the rolling wheel deflectometer (RWD) (*5*). The TSD is the focus of this study as a potential device in assessing the in-service structural capacity of a road along its entire length and without lane closures. The use of the TSD is promising for pavement evaluation at the network level because it can measure deflection at traffic speeds, which enables large spatial coverage and can generate a continuous deflection profile rather than measuring pavement deflection at discrete points (*6*). Another advantage of TSD is, unlike RWD, it allows complete measurement of the deflection basin (*7*).

This study developed a nonlinear regression model for the prediction of pavement structural number (SN) and structural deficiency at 0.01-mile interval. The model was successfully validated based on data points obtained from TSD and FWD measurements in Louisiana and Idaho. The validation and performance evaluation of the model was conducted by comparing its prediction with the SN calculated from FWD deflection measurements. Extracted cores and functional indices were also used for evaluation of the model's ability in identifying structurally deficient pavements.

[1]Department of Civil and Environmental Engineering, Louisiana State University, Baton Rouge, LA
[2]Pavement Geotechnical Research Administrator, Louisiana Transportation Research Center, Louisiana State University, Baton Rouge, LA

Corresponding Author:
Address correspondence to Mostafa A. Elseifi: elseifi@lsu.edu

Background

The need for considering structural conditions along with functional conditions has been recognized in the past decade by state agencies, which supported the incorporation of structural conditions in PMS network-level decision-making processes. The TSD, a Doppler-laser based continuous deflection measurement device, has emerged as a promising approach to measure vertical deflection velocity continuously along a road section. The TSD consists of an articulated truck that uses a rear axle up to 22,000 lbs to load the pavement structure. The operational speed of the device is up to 60 mph; the TSD concept is based on the measurement of the deflection velocity rather than the absolute deflection of the road surface (*6, 7*). In Australia and New Zealand, a research study found a strong correlation between TSD and FWD deflection measurements (*8*).

The development of FWD in the past four decades allowed practitioners to determine detailed structural attributes of in-service pavements (*1*). Research studies have developed methodologies to evaluate the structural conditions of in-service pavement and its SN based on surface deflections measured using FWD and RWD as described in Table 1. A pooled fund TSD study has developed a methodology for estimating the effective structural number (SN_{eff}) from TSD measurements using Rohde's (1994) method (see Table 1) and introduced thresholds for predicting pavement structural conditions based on derived parameters from TSD measured deflections (*9, 10*). As the use of FWD is limited at the network level and with the acceptance of continuous deflection measurement devices in many countries, a model is needed to predict SN from continuous devices such as TSD. The present study aimed at developing a SN prediction model based on TSD deflections, which would allow the identification of structurally deficient pavements. Such information would benefit state agencies in network-level decision-making processes and in avoiding inaccurate selection of maintenance and rehabilitation (M&R) activities.

Objectives

The main objective of the present study was to develop a structural capacity indicator model to predict in-service SN based on TSD deflections at 0.01-mile intervals of a road section. A secondary objective of the study was to evaluate the proposed model's ability in identifying structurally deficient locations along a road segment. Furthermore, the importance of incorporating structural capacity along with the functional indices in PMS decision-making was evaluated using functional and structural indices.

Data Description and Methodology

In 2016, a TSD device operated by the Australian Road Research Board (ARRB) was used to measure vertical deflection velocity, horizontal speed of the vehicle, air temperature, and pavement surface temperature in six parishes of District 05 in Louisiana. Measurements were conducted for 13 control sections at 0.01-mile intervals. FWD measurements were also collected for the same control sections at 0.1-mile intervals for the evaluation and comparison with TSD measurements. TSD and FWD measurements were also conducted in Idaho for one road segment that was 13.4 miles in length. TSD measurements were reported at 0.006-mile intervals and FWD measurements were reported at 100 feet (0.02-mile) intervals.

Collected raw measurements (vertical deflection velocity and actual horizontal speed) of the TSD device were used to calculate the deflection basin at each milepost according to the methodology known as "area under the curve (AUTC)" proposed by Muller and Roberts (*15*). According to this method, the vertical deflection velocity is divided by the actual speed of the vehicle to get the deflection slope; slopes are then plotted against TSD sensor locations. Afterwards, the plotted curve is numerically integrated assuming the deflection slope is zero at locations 0 and 137.8 in. from the load. The deflection value was then calculated at the selected locations with adequate curve fitting using the piecewise cubic Hermite function as suggested by the AUTC method. Surface deflections were calculated at nine locations (i.e., 0, 8, 12, 18, 24, 36, 48, 60, 72 in. from the center of the load). Temperature corrections for both TSD and FWD surface deflections were conducted to a reference temperature of 20°C according to the methodology developed by Kim and Park (*16*) and asphalt mid-depth temperature was calculated using Bell's equation (*17*).

A nonlinear model was developed based on TSD deflections. Afterwards, the model's ability in identifying structurally deficient locations based on percentage loss in in-service SN was evaluated by comparing the prediction with extracted cores and surface distress data. The core reports, traffic data, and functional indices data were collected from the Louisiana Department of Transportation and Development (LaDOTD) PMS.

Model Development

The proposed model was developed based on TSD surface deflection measurements calculated from the deflection slope by the AUTC method. Initially, measured deflections at nine offset distances referred to as D_0, D_8, D_{12}, D_{18}, D_{24}, D_{48}, D_{60}, and D_{72} were used as independent variables along with the corresponding pavement total thickness (T_{th}), and the average daily traffic (ADT). The SN calculated from the AASHTO 1993 method based on FWD and TSD deflections was used as the dependent variable in the development of the model. To ensure

Table 1. Developed SN Models Based on FWD and RWD Measurements

Method	SN models	Description
SN_{eff} prediction methodology based on TSD using Rhode's equation (*9*)	$SIP = D_0 - D_{1.5Hp}$ $SN_{eff} = K_1\, SIP^{k_2}\, H_p^{k_3}$	SIP = structural index of pavement D_0 = peak deflection $D_{1.5Hp}$ = deflection at distance 1.5 times pavement depth H_p = pavement depth For asphalt pavements, k_1 = 0.4728, k_2 = −0.4810, and k_3 = 0.7581
RWD based SN model (*11*)	$SN = -14.72 + 27.55*\left(\frac{AC_{th}}{D_0}\right)^{0.04695} - 2.426*\ln(SD) + 0.29*\ln(ADTTPLN)$	AC_{th} = asphalt layer(s) thickness D_0 = Avg. RWD deflection (mils) SD = standard deviation of RWD deflection ADTPLN = average annual daily traffic per lane (vehicle per day) $SN_{RWD0.1}$ = SN based on RWD measurements defined each 0.16 km (0.1 mile)
FWD center deflection based SN model (*12*)	$SN = 6.3763 - 0.3364 d_0 + 0.0062* d_0^{\,2} - 0.0805*D + 0.01*D^2 - 0.0008(d_0 * D)) - 0.4115 \log(EAL) + 0.1438* (\log(EAL))^2 + 0.0836*ETCR - 0.0091** EFCR + 0.0004*EFCR^2 - 0.4061*Rut$	SN = pavement structural number d_0 = center deflection (mils) D = pavement depth (in.) EAL = equivalent standard daily traffic EFCR/ETCR = equivalent fatigue/ transverse cracking Rut = Rut depth (in.)
SN model based on the derived parameters of FWD deflections (*13*)	$SN = e^{5.12} A_{UPP}^{-0.78} BLI^{0.31}$	A_{upp} = area under pavement profile BLI = base layer index (surface curvature index, SCI)
Backcalculated Moduli AASHTO NDT Method (*14*)	$SN = \sum_{i=1}^{n} h_i a_g \left(\frac{E_i}{E_g}\right)^{1/3}$	a_g = layer coefficient of standard materials E_i = layer resilient modulus E_g = layer resilient modulus of standard materials h_i = layer thickness (in.)
AASHTO Method II (*14*)	$D_0 = \frac{1.5P}{\pi lr}\left(\frac{(0.0045HP)^3}{SN^3}\left[1 - \frac{1}{(1+(HP/lr)^{1/2}}\right] + \frac{1}{E_s\left(1+\frac{40000SN^3}{lr^2 E_{SG}^{2/3}}\right)^{1/2}}\right)$	D_0 = peak FWD deflection (in.) HP= layer thickness (in.) lr = load radius (in.) E_{SG} = the subgrade modulus
Jameson's formula (*13*)	$SN = 1.69 + \frac{842.8}{(D_0 - D_{1500})} + \frac{42.94}{D_{900}}$	D_0 = the peak deflection (μm) D_{900} = deflection at 900 mm from loading (μm) D_{1500} = deflection at 1,500 mm from loading (μm)
Asgari's formula (*13*)	$SNC = a0(D_0)^{a1}$	$a0$, $a1$ = Asgari co-efficients D_0 = Center deflection
The Wimsatt formula (*13*)	$SN_{eff} = 0.0045(D)E_p^{0.333}$	D = total layer thickness E_p= existing pavement modulus of the layers above subgrade

accuracy, several statistical analyses were conducted; that is, pairwise correlation, significance testing using regression analysis, and multicollinearity testing among all the independent variables.

Table 2. Pearson Correlation Coefficients

	SN_{FWD}	D_0	D_8	D_{12}	D_{18}	D_{24}	D_{36}	D_{48}	D_{60}	D_{72}	T_{th}	ADT
SN_{FWD}	1.00	−0.55	−0.36	−0.20	−0.002	0.16	0.30	0.34	0.35	0.39	0.82	0.44
D_0	−0.55	1.00	0.95*	0.87*	0.73*	0.57	0.38	0.30	0.22	0.11	−0.28	−0.27
D_8	−0.36	0.95*	1.00	0.98*	0.89*	0.78*	0.61*	0.53	0.45	0.33	−0.14	−0.20
D_{12}	−0.20	0.87*	0.98*	1.00	0.96*	0.88*	0.75*	0.67*	0.59	0.47	−0.03	−0.14
D_{18}	−0.002	0.73*	0.89*	0.96*	1.00	0.97*	0.89*	0.83*	0.75*	0.65*	0.11	−0.08
D_{24}	0.16	0.57	0.78*	0.88*	0.97*	1.00	0.96*	0.92*	0.86*	0.77*	0.21	−0.05
D_{36}	0.30	0.38	0.61*	0.75*	0.89*	0.96*	1.00	0.99*	0.95*	0.88*	0.28	−0.03
D_{48}	0.34	0.30	0.53	0.67*	0.83*	0.92*	0.99*	1.00	0.98*	0.93*	0.29	−0.03
D_{60}	0.35	0.22	0.45	0.59	0.75*	0.86*	0.95*	0.98*	1.00	0.95*	0.29	−0.03
D_{72}	0.39	0.11	0.33	0.47	0.65*	0.77*	0.88*	0.93*	0.95*	1.00	0.30	−0.02
T_{th}	0.82	−0.28	−0.14	−0.03	0.11	0.21	0.28	0.29	0.29	−0.30	1.00	0.37
ADT	0.44	−0.27	−0.20	−0.14	0.08	−0.05	−0.03	−0.03	−0.03	−0.02	0.37	1.00

Note: Highly collinear relationships are marked with an asterisk.

Pairwise Correlation

Pairwise correlation was conducted among the independent variables to avoid using collinear or multicollinear independent variables in the model, which may increase the variance of the estimated regression coefficients. All the possible independent variables that may have an influence on the prediction of the SN were evaluated using pairwise correlation analysis. The correlation coefficient is an indication of the level of collinearity among the independent variables. The coefficient is called the Pearson's correlation coefficient, which ranges from −1 to +1. A large absolute value indicates high collinearity between the variables. The positive or negative sign represents the positive or negative relationship between the variables (*18, 19*). The results of pairwise correlation analysis is presented in Table 2. A Pearson's coefficient that is greater than 0.6 was considered highly collinear in the analysis. As shown in this table, most of the deflection measurements were correlated, which was expected as surface deflections tend to increase or to decrease concurrently with the exception of far distance deflections (i.e., D_{60} and D_{72}), which may indicate weakness in the underlying layers and the subgrade. It is worth noting that the final model did not include D_{60} or D_{72} because the use of these variables would limit the application of the model as some TSD surveys do not measure far distance deflections from the load. Moreover, the prediction accuracy was satisfactory with the use of D_{48}, which was deemed more reasonable than the use of D_{60} or D_{72}, even though they showed a better correlation with SN_{FWD}. Based on the results of the analysis, collinear variables were not used. Different combinations of noncollinear variables were considered in the regression analysis.

Regression Analysis

To assess the significance of the independent variables (D_0, D_8, D_{12}, D_{18}, D_{24}, D_{48}, D_{60}, D_{72}, T_{th}, and ADT) on the prediction of the dependent variable (SN), regression analysis was conducted. Independent variables with no significance on the dependent variable were removed from the model to avoid overfitting of the dependent variable. When overfitting occurs, the regression model becomes tailored to fit the random noise in the data set rather than reflecting the actual trends in the measurements. Regression analysis was conducted on several combinations of independent variables using SAS 9.4 software. At a confidence level of 95%, a *P*-value less than 0.05 would represent a significant effect. The combination of independent variables that had significant effect on the dependent variable is presented in Table 3.

Table 3. Results of Regression Analysis and Multicollinearity Test

Variable	Pr > |*t*|	Interpretation	Variance inflation (VIF)
Intercept	<.0001	Significant	0
D_0	<.0001	Significant	1.35
D_{48}	<.0001	Significant	1.35
T_{th}	<.0001	Significant	1.22
ADT	<.0001	Significant	1.49

Variance Inflation Factor

To further filter out the multicollinear independent variables, a second statistical factor known as the variance inflation factor (VIF) was used. Even after pairwise correlation analysis between two variables, there is a possibility of multicollinearity resulting from the combination of one variable with more than one variable. To address this issue, the most used statistical factor is the VIF. Because of multicollinearity, an inflation can occur in the standard error, which is measured by VIF. A maximum VIF value of 5 to 10 is recommended in the literature (*20*). In this study, the VIF values for the selected independent variables were within the acceptable range (Table 3), which indicates that no multicollinear independent variables have been used in the model.

Nonlinear Regression Model

A nonlinear regression analysis was conducted using SAS 9.4 to predict the SN of a pavement, which is referred as SN_{TSD} based on the statistically significant TSD deflections (D_0 and D_{48}), ADT, and total pavement thickness (T_{th}) including bound and unbound layers. The nonlinear relationship was obtained based on an iterative process to achieve the best prediction accuracy and was continued until the highest R^2 and lowest RMSE were obtained. The iterative process included use of different function assigned to the independent variables. About 70% of the data points from Louisiana and 30% of the data points from Idaho were used in the development phase to fit the model; the remaining data points from Louisiana and Idaho were used to validate the fitted model. The model demonstrated an acceptable accuracy with a coefficient of determination (R^2) of 0.93 in the development phase and with a root mean square error (RMSE) of 0.85 as shown in Figure 1*a*.

An independent TSD and FWD data set was used to validate the model. The use of Idaho data points in the model validation demonstrated the model's compatibility with different climatic regions and construction practices. The model performed satisfactorily in the validation phase with an R^2 of 0.88 and with a RMSE of 1.06 as shown in Figure 1*b*. A good agreement was also found when the average SN_{TSD} for each road section was compared with the average SN_{FWD}; see Figure 1*c*. The proposed model is presented in Equation 1:

$$SN_{TSD} = 18.67*e^{(-0.013*D_0)} + 8.65*(D_{48})^{0.11} + 0.18*(T_{th}) + 0.31*\ln(AADT) - 24.28 \quad (1)$$

where

SN_{TSD} = SN based on TSD measurements;

D_0 = deflection of pavement under loaded tire or center deflection (mils);

D_{48} = deflection at 48 in. distance from center deflection (mils);

T_{th} = total layer thickness of pavement (in.); and

ADT= average daily traffic (vehicles per day).

Longitudinal Profile Comparison

A continuous SN profile was developed from the proposed model at short intervals of 0.01 mile. Longitudinal profiles for SN_{TSD} obtained from the model were compared for both the Louisiana and Idaho control sections. The longitudinal profiles for one Louisiana section and one Idaho section are shown in Figure 2, *a* and *b*, respectively. Higher variability was noted in the Idaho section, possibly because of higher roughness and cracking at the surface as suggested by the lower SN predicted for this section. Past studies concluded that both FWD and RWD test methods resulted in a greater average deflection and scattering in sites in poor conditions (*21*).

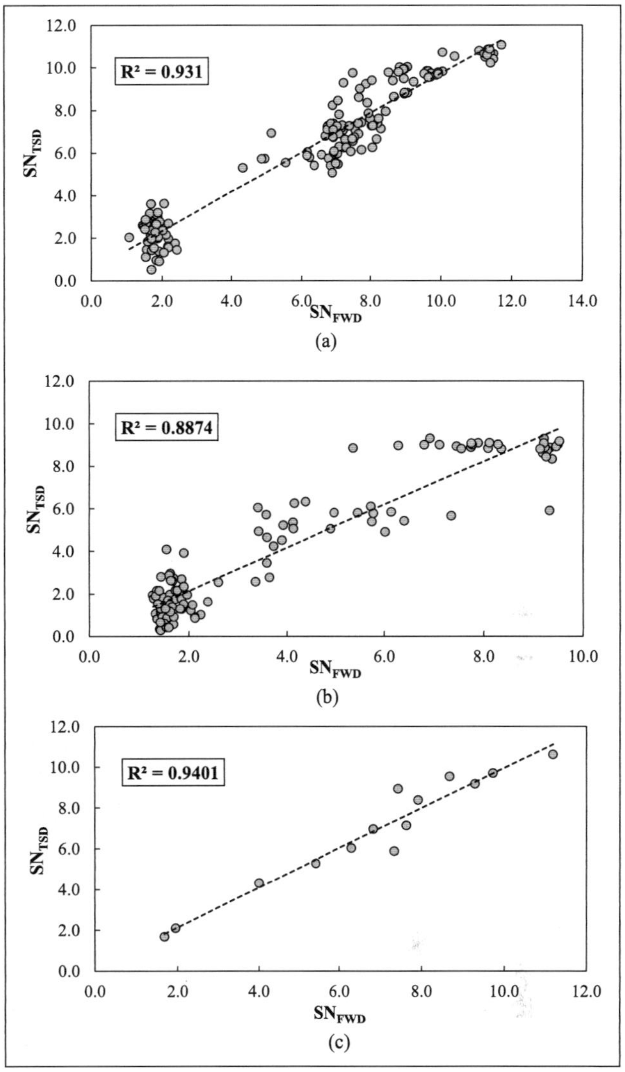

Figure 1. (*a*) Model development, (*b*) model validation, and (*c*) average SN comparison for each section.

Sensitivity Analysis

A sensitivity analysis was conducted for the proposed model's output as a function of the model's inputs. Sensitivity of the dependent variable (SN_{TSD}) was tested for different input parameters, which were varied within their maximum and minimum values. The average of each of the input parameters was used as the baseline in the sensitivity analysis. From the results of the sensitivity analysis, it was found that the predicted SN_{TSD} was most sensitive to D_0 among all other parameters and the least sensitive to ADT. The change in the predicted SN_{TSD} with the varying input parameters is shown in Figure 3.

Analysis of Variance

An analysis of variance (ANOVA) was conducted with the measurements from the Louisiana control sections to evaluate

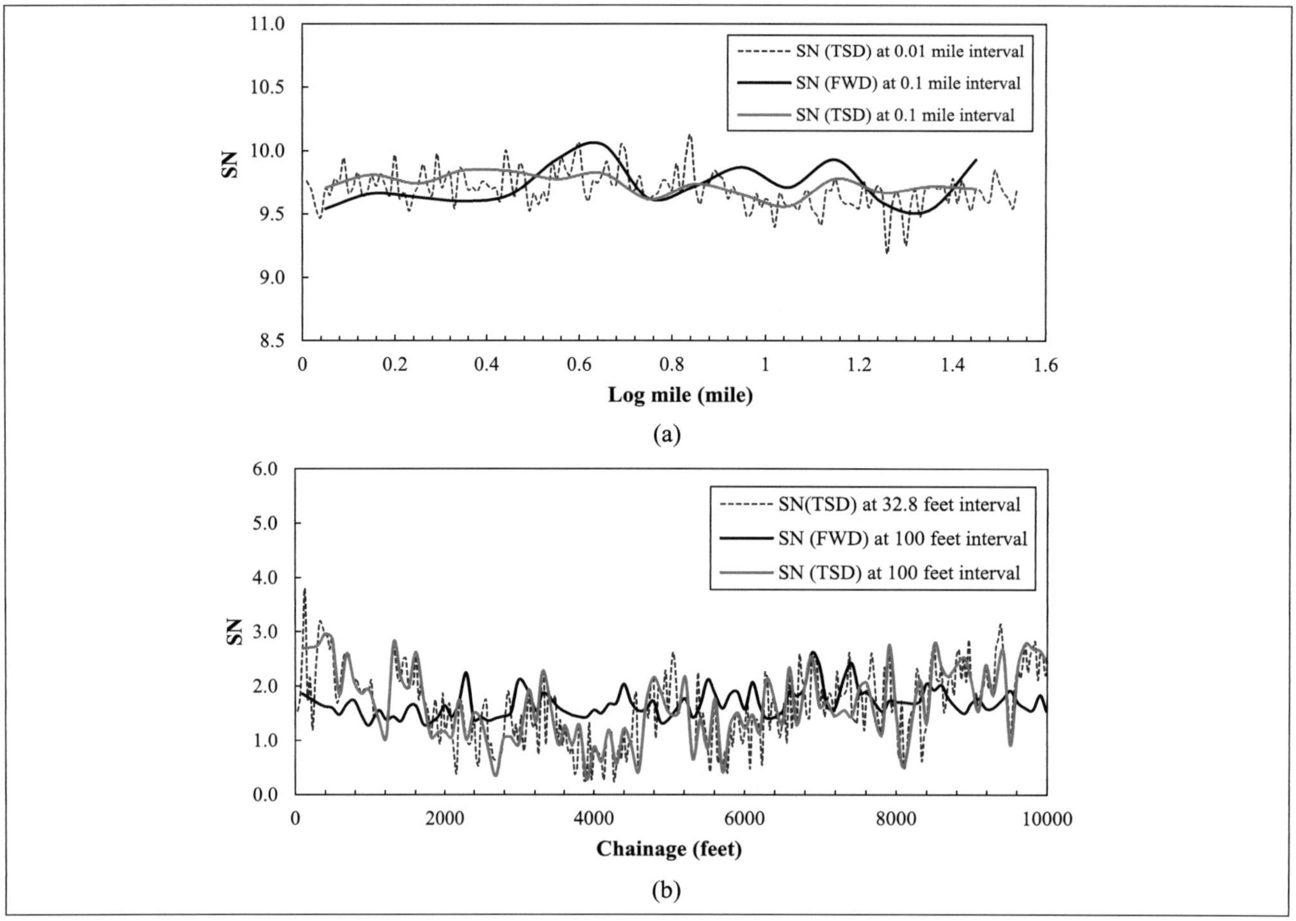

Figure 2. (*a*) Longitudinal comparison of SNTSD and SNFWD for Louisiana and (*b*) longitudinal comparison of SN_{TSD} and SN_{FWD} for Idaho.

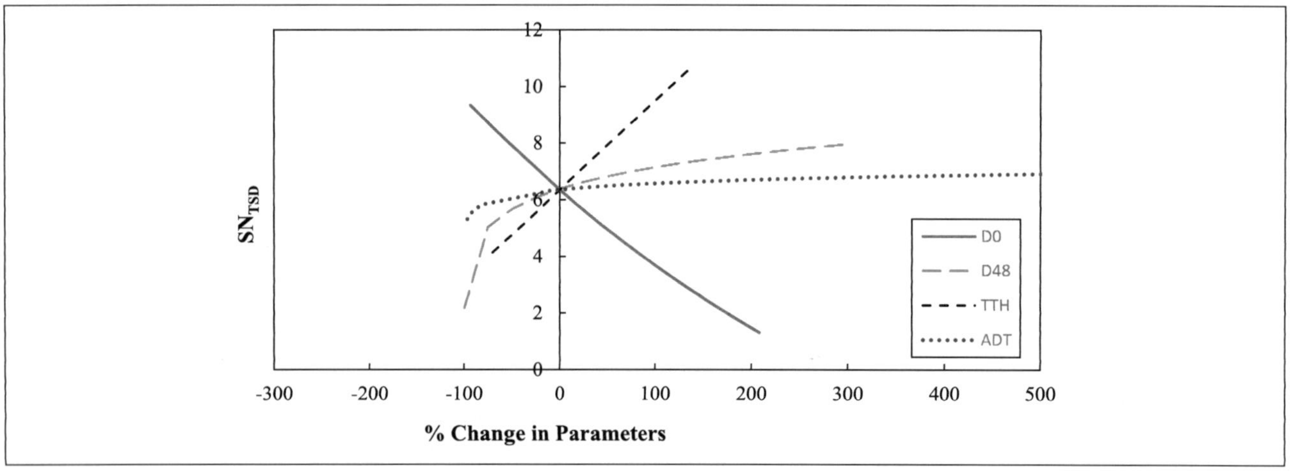

Figure 3. Sensitivity analysis for the SN_{TSD} model.

the degree of influence of functional indices on in-service structural capacity at a confidence level of 95%. Functional indices considered in the comparison with SN_{TSD} were alligator cracking (ALCR), random cracking (RNDM), patching

Table 4. Significance of Functional Indices on SN

Road Category	Functional indices (Pr > |t|)					
	ALCR	RNDM	PTCH	RUT_IND	RUFF	PCI
Thick (treated)	*0.7457**	0.0002	*0.4741**	*0.5430**	*0.7276**	*0.2646**
Thick (untreated)	0.0448	*0.5146**	*0.3604**	*0.3329**	0.0179	*0.4810**
Medium (treated)	<0.0001	<.0001	<.0001	0.0212	*0.7933**	<.0001
Medium (untreated)	*0.4792**	*0.6867**	NA	*0.7065**	*0.7834**	*0.8272**
Thin (treated)	*0.2205**	*0.6867**	NA	NA	*0.1645**	*0.2347**

Note: Nonsignificant relationships are marked with an asterisk. NA = not available.

(PTCH), rutting index (RUT_IND), roughness (RUFF), and pavement condition index (PCI). Road sections were divided into five categories based on asphalt layer(s) thickness and type of base layer (treated and untreated) as shown in Table 4 (*11*). Thick sections were those that had an AC layer greater than 6 in.; medium sections were those with AC layers between 3 and 6 in.; and thin sections were those with AC layers less than 3 in. The correlation between each of the aforementioned functional indices and structural indices (SN_{TSD}) was evaluated using the *P*-value obtained from the statistical analysis. As a 95% confidence limit was used, a *P*-value less than 0.05 would represent significant correlation between the condition indices and vice versa. As shown by the results in Table 4, a nonsignificant statistical relation was found between functional indices and SN in four of the five road categories. Therefore, one may assume that considering a structurally based index in a PMS would allow for the identification of road segments that are in need of structural repair and that are not currently identified by the functional indices.

Performance Evaluation of the Developed Model

The model's precision and adequacy were evaluated by comparing the percentage loss in SN_{TSDeff} and SN_{FWDeff}. The proposed model's ability in identifying structurally deficient sections was also evaluated as compared to structural deterioration identified from extracted cores and from FWD. As cores were only available for the Louisiana data set, this analysis was exclusively conducted for the Louisiana road sections.

Loss in In-Service SN Calculation

To determine the percentage loss in in-service SN, the AASHTO design SN during construction was calculated using Equation 2:

$$SN = a_1 * D_1 + a_2 * m_2 * D_2 + a_3 * m_3 * D_3 \quad (2)$$

where

a_1 = asphalt layer coefficient, a_2 = base layer coefficient, a_3 = subbase layer coefficient;

D_1 = asphalt layer thickness (in.), D_2 = base layer thickness (in.), and D_3 = subbase layer thickness (in.); and

m_2 = base layer drainage coefficient and m_3 = subbase layer drainage coefficient.

The values of the layer coefficients were selected in accordance with LaDOTD design standards: $a_1 = 0.42$; $a_2 = 0.28$ for treated (cement stabilized) base and 0.07 for untreated base; $a_3 = 0.11$ for cement treated subbase and 0.04 for untreated subbase. The values of m_1 and m_2 was considered 1.0 in all cases.

After calculating the design SN of the sections during construction, SN_{TSD} was corrected according to the findings of a study conducted by Wu and Gaspard to account for design and construction practices in Louisiana as follows (*22, 23*); see Equation 3:

$$SN_{eff} = 2.58 \ln(SN_{FWD}) - 0.77. \quad (3)$$

As the SN_{TSD} model was developed and validated based on SN calculated from FWD deflections, the predicted SN_{TSD} of the road sections were also adjusted through the same model using Equation 4:

$$SN_{TSDeff} = 2.58 \ln(SN_{TSD}) - 0.77. \quad (4)$$

The loss in SN_{TSDeff} was then calculated at every 0.1-mile interval using Equation 5:

$$\text{Loss in SN}(\%) = \frac{\text{Design SN} - SN_{TSDeff}}{\text{Design SN}} * 100. \quad (5)$$

The percentage loss in SN_{TSDeff} was compared with SN_{FWDeff} at each extracted core location. It is noted that the only control sections considered in this comparison were the ones that had the cores extracted at almost the same location to ensure precise evaluation of the SN model. The average SN over 1.5 miles was also calculated and compared. As shown in Figure 4, the estimated percentage loss from the model was in good agreement with the percentage loss predicted from SN_{FWD}.

Analysis of the Extracted Cores

In this section, the model's efficiency in identifying structural deficient sections was evaluated. Past studies by the

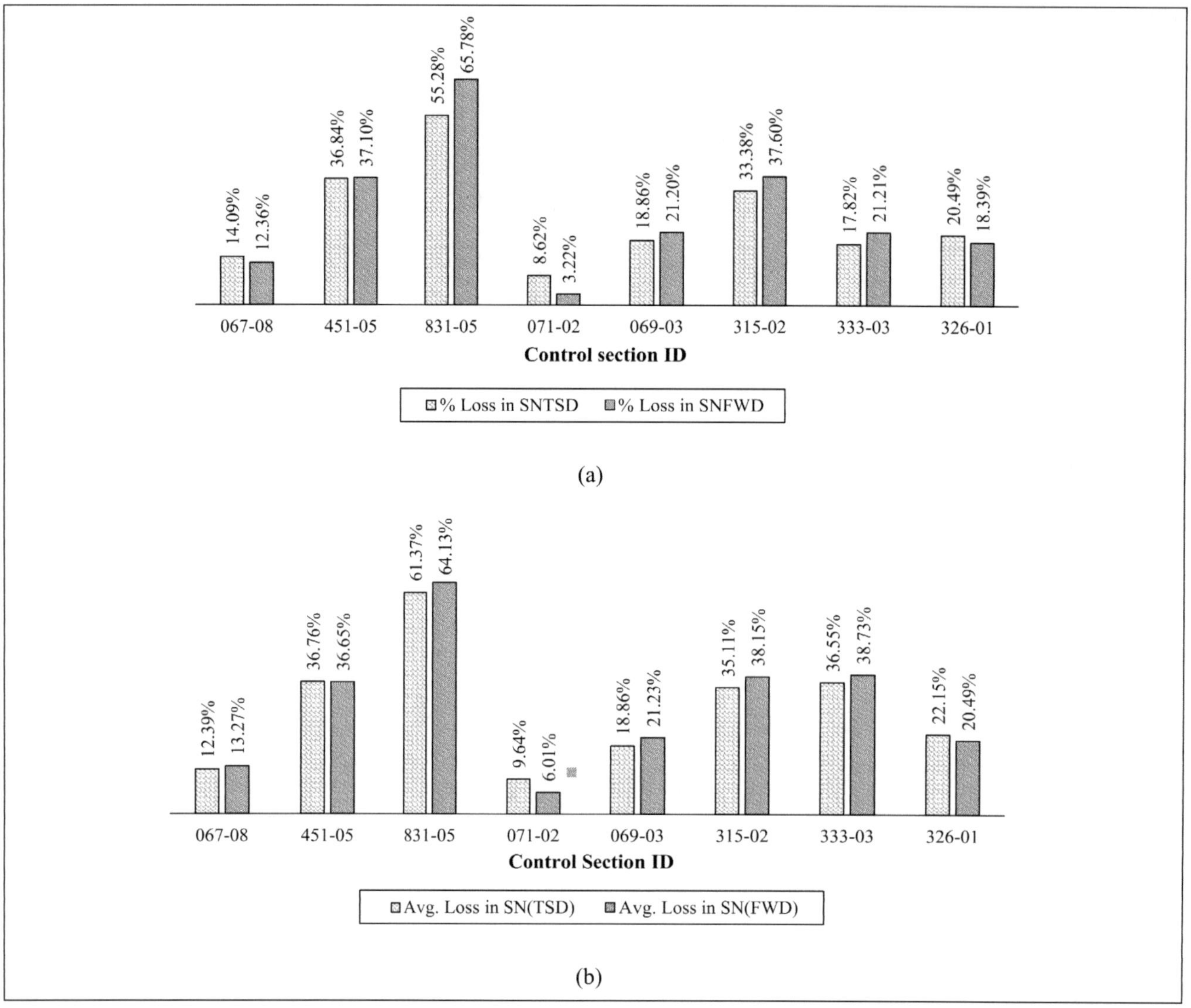

Figure 4. (*a*) Comparison of loss in in-service SN at core location and (*b*) comparison of loss in average in-service SN.

authors used 50% loss of AASHTO SN as the threshold to identify structurally deficient locations (*11*). In the present study, the model's competency in identifying structurally deficient locations was also based on a 50% loss in structural capacity. The extracted cores were compared with the estimated loss in SN (%) along with the functional indices at the same locations. A detailed evaluation of four typical road section is presented; see Figure 5.

Control Section 831-05. The control section is located on route LA 821 with a length of 8.18 miles located at Lincoln parish in Louisiana District 05. The total layer thickness of this control section obtained from core reports was found to be 13 in. consisting of three AC layers of 5 in. and a granular base layer of 8 in. After assessment of the extracted core, deterioration (stripping) was detected in the third asphalt layer, which was 2 in. thick. The percentage loss in SN_{TSDeff} at the core location and the average SN (%) loss over 1.5 miles was found to be 55.3% and 61.4%, respectively. The average PCI, a combined functional index, for this control section was 92.5 over 1.5 mile and the PCI at the core location was 90.3 indicating excellent functional conditions. Given the average SN loss (%) is greater than 50%, this control section was identified as structurally deficient even with a sound PCI rating.

Control Section 315-02. The control section is located on route LA 143 with a length of 9.26 miles located at Ouachita parish in Louisiana District 05. The total layer thickness of the control section obtained from core reports was 23 in. consisting of two AC layers of 9.5 in. on top of a cement stabilized sand clay gravel base layer of 13.5 in.

Figure 5. Analysis of the extracted cores: (*a*) control section 831-05, (*b*) control section 315-02, (*c*) control section 333-03, and (*d*) control section 451-08.

After assessment of the extracted core, debonding was detected between the bottom AC layer and the underlying base layer. It is to be noted that poor drainage conditions were also detected in this road section. The average percentage loss in SN_{TSDeff} over 1.5 miles and at the core location was 33.4% and 35.1%, respectively. The average PCI for this control section was found to be 92.4 over 1.5 miles and 96.7 at the core location indicating excellent functional conditions. Although the predicted structural capacity loss was less than 50%, the model predicted a loss in SN possibly related to the detected debonding between the AC and the base layers.

Table 5. Model's Performance Evaluation Based on Extracted Cores

Control section	Average PCI	Type of deterioration in cores	Note	Average loss in SN_{TSDeff} (%)	Remarks on model's efficiency
831-05	92.4	Stripping	N/A	61.4	Identified structurally deficient section
315-02	92.3	Separation	Cement stabilized base layer	33.4	Predicted loss in structural capacity because of debonding
333-03	67.9	Stripping	New overlay applied since core extraction	36.6	Reasonable estimation as new overlay was applied
451-08	95.78	Stripping	Core extracted from 13.5 miles	31.66	Identified stripping deficiency

Note: N/A = not applicable.

Control Section 333-03. The control section is located on Route LA 582 with a length of 6.83 miles located at E Carroll parish in Louisiana District 05. The total layer thickness of this control section obtained from core reports was 18 in. consisting of five AC layer of 9.5 in. and a granular base layer of 8.5 in. After assessment of the extracted core, deterioration (stripping) was found in the bottom AC layer that was 1.5 in. thick. The average percentage loss in SN_{TSD} over 1.5 miles and at the core location was calculated as 36.5% and 17.8%, respectively. On further assessment of the control section, it was found that a new overlay had been applied since the core extraction and functional survey explaining the adequate structural capacity of the control section. Therefore, the model's estimated loss in SN can reasonably be justified. The average PCI in 2015 for this control section was found to be 67.9 over 1.5 miles and the PCI at the core location was 73.6, which was before the new overlay.

Control Section 451-08. The control section is located at Madison parish with a length of 32.97 miles in Louisiana District 05. The total layer thickness obtained from core reports was 23.3 in. consisting of two AC layers of 5.2 in., a concrete layer of 8 in., an AC layer of 4 in., on top of a granular base layer of 6.0 in. The average PCI for this control section was 95.8; therefore, this section can be considered functionally sound. An average loss in in-service SN_{TSD} over 1.5 miles was calculated at 31.7%. Although not structurally deficient, stripping was detected along the length of the road section at a distance of 13.5 miles. Therefore, the model's estimated loss in SN is considered reasonable.

Summary of the Model's Structural Efficiency Prediction

The proposed model can reasonably estimate the average loss in in-service SN (%) as compared with the extracted cores as discussed in the previous sections. Even though functionally sound, several control sections were identified as structurally deficient in relation to Avg. loss in SN_{TSDeff} as summarized in Table 5. This implies the necessity of considering a structural index along with functional indices in a PMS, which was also demonstrated using the ANOVA analysis presented in Table 5.

Summary and Conclusions

The main objective of the present study was to develop a structural capacity indicator model to predict in-service SN based on TSD deflections at 0.01-mile intervals of a road section. A secondary objective of the study was to evaluate the proposed model's ability in identifying structurally deficient sections along a road segment. Based on the results of the analysis, the following conclusions may be drawn:

- A statistical nonlinear regression model was developed for predicting in-service pavement structural capacity. The model showed an acceptable prediction accuracy with a satisfying R^2 and RMSE value in the development and validation phases. The model was successfully developed and validated with SN calculated based on TSD and FWD deflection data obtained from two contrasting data sets from Louisiana and Idaho.
- The importance of considering structural indices along with functional indices was demonstrated based on ANOVA analysis and extracted cores.
- The estimated percentage loss in structural capacity from the model was in good agreement with the percentage loss calculated from FWD.
- Core samples showed that sections that were predicted to be structurally deficient suffered from asphalt stripping and debonding problems. Yet, some of these sections were in very good conditions according to their functional conditions.

In summary, results demonstrate that the developed model is a valuable tool that could be used in a PMS at the network level to predict pavement structural conditions based on TSD testing with an acceptable level of accuracy and to complement currently adopted functional indices.

Acknowledgments

The financial support of the Louisiana Transportation Research Center (LTRC) is greatly appreciated. The authors also acknowledge the help of Nadarajah Sivaneswaran from the FHWA Turner-Fairbank Highway Research Center in providing the Idaho data.

References

1. Zofka, A., M. Graczyk, and J. Rafa. *Qualitative Evaluation of Stochastic Factors Affecting the Traffic Speed Deflectometer Results*. Presented at 94th Annual Meeting of the Transportation Research Board, Washington, D.C., 2014.
2. Elseifi, M., and O. Elbagalati. *Assessment of Continuous Deflection Measurement Device in Louisiana–Rolling Wheel Deflectometer*. Final report 581, Project 14-2P, Interim Report, 2013–2016, Louisiana Department of Transportation and Development, Baton Rouge, LA, 2017.
3. Zofka, A., J. Sudyka, M. Maliszewski, and P. Harasim. Alternative Approach for Interpreting Traffic Speed Deflectometer Results. *Transportation Research Record: Journal of the Transportation Research Board*, 2014. 2457: 12–18.
4. Katicha, S. W., J. Bryce, G. Flintsch, and B. Ferne. Estimating "True" Variability of Traffic Speed Deflectometer Deflection Slope Measurements. *Journal of Transportation Engineering*, Vol. 141, No. 1, 2014, p. 04014071.
5. Flintsch, G. W., B. Ferne, B. Diefenderfer, S. Katicha, J. Bryce, and S. Nell. Evaluation of Traffic Speed Deflectometers. *Transportation Research Record: Journal of the Transportation Research Board*, 2012. 2304: 37–46.
6. Chai, G., S. Manoharan, A. Golding, G. Kelly, and S. Chowdhury. Evaluation of the Traffic Speed Deflectometer Data Using Simplified Deflection Model. *Proc., 6th Transport Research Arena*, April 18–21, Elsevier B.V., doi: 10.1016/j.trpro.2016.05.444, 2016.
7. Elbagalati, O., M. Mousa, M. A. Elseifi, K. Gaspard, and Z. Zhang. *Development of a Methodology to Backcalculate Pavement Layer Moduli Using the Traffic Speed Deflectometer*. Presented at 96th Annual Meeting of the Transportation Research Board, Washington, D.C., 2017.
8. Roberts, J., U. Ai, T. Toole, and T. Martin. *Traffic Speed Deflectometer: Data Review and Lessons Learnt*. Austroads Project No. AT1730 and Publication No. AP-T279-14, Austroads Publication Online, October 2014.
9. Katicha, S., G. Flintsch, S. Shrestha, and S. Thyagarajan. *Demonstration of Network Level Structural Evaluation with Traffic Speed Deflectometer*. Final report. Prepared for the Federal Highway Administration, USDOT, 2017.
10. Rohde, G. T. Determining Pavement Structural Number from FWD Testing. *Transportation Research Record: Journal of the Transportation Research Board*, 1994. 1448: 61–68.
11. Elbagalati, O., M. A. Elseifi, K. Gaspard, and Z. Zhang. Prediction of In-service Pavement Structural Capacity Based on Traffic-Speed Deflection Measurements. *Journal of Transportation Engineering*, Vol. 142, No. 11, 2016, p. 04016058.
12. Gedafa, D. S., M. Hossain, R. Miller, and T. Van. Network-Level Flexible Pavement Structural Evaluation. http://krex.ksu.edu, 2014.
13. Schnoor, H., and E. Horak. Possible Method of Determining Structural Number for Flexible Pavements with the Falling Weight Deflectometer. *Abstracts of the 31st Southern African Transport Conference (SATC 2012)*, 2012. Proceedings ISBN Number: 978-1-920017-53-8.
14. American Association of State Highway and Transportation Officials (AASHTO). *AASHTO Guide for Design of Pavement Structures*. Published by the American 7 Association of State Highway and Transportation Officials, Washington, D.C., 1993.
15. Muller, W. B., and J. Roberts. Revised Approach to Assessing Traffic Speed Deflectometer (TSD) Data and Field Validation of Deflection Bowl Predictions. *International Journal of Pavement Engineering*, Vol. 14, No. 4, 2013, pp. 388–402.
16. Kim, Y., H. Richard, and G. Park. *Use of Falling Weight Deflectometer Multi-Load Data for Pavement Strength Estimation*. Final report FHWA/NC/2002-006. Department of Civil Engineering, North Carolina State University, Raleigh, 2002.
17. Lukanen, E. O., R. Stubstad, and R. Briggs. *Temperature Predictions and Adjustment Factors for Asphalt Pavement*. Report No. FHWA-RD-98-085. Federal Highway Administration, Washington, D.C., 2000.
18. Miller, I., and J. E. Freund. *Mathematical Statistics: With Applications*. India: Pearson Education, 2004.
19. Freund, R. J., W. J. Wilson, and P. Sa. *Regression Analysis*. Academic Press, London, 2006.
20. Hair, J. F., Jr., R. E. Anderson, R. L. Tatham, and W. C. Black. *Multivariate Data Analysis with Readings*. Prentice Hall, New Jersey, 1995.
21. Elseifi, M. A., A. Abdel-Khalek, K. Gaspard, Z. Zhang, and S. Ismail. Evaluation of Continuous Deflection Testing Using the Rolling Wheel Deflectometer in Louisiana. *Journal of Transportation Engineering*, Vol. 138, No. 4, 2012, pp. 414–422.
22. Wu, Z., and K. Gaspard. *Mechanistic Flexible Pavement Overlay Design Program*. Report FHWA/LA. 08/454, Baton Rouge, LA, 2009.
23. Wu, Z., Z. Zhang, and C. Abadie. Determining Structural Strength of Existing Asphalt Layer Using Condition Survey Data. *International Journal of Pavement Engineering*, Vol. 14, No. 7, 2013, pp. 603–611.

The Standing Committee on Pavement Structural Modeling and Evaluation peer-reviewed this paper (18-00998).

Article

TRR
JOURNAL OF THE TRANSPORTATION RESEARCH BOARD

Transportation Research Record
2018, Vol. 2672(40) 326–335

DOI: 10.1177/0361198118758025
journals.sagepub.com/home/trr
SAGE

Development of Domain Analysis to Predict Multi-Axial Flexible Airfield Pavement Responses Due to Gear and Environmental Loadings

Angeli Gamez[1], Jaime A. Hernandez[1], and Imad L. Al-Qadi[1]

Abstract

Flexible pavement design procedures use maximum mechanistic strains to predict service life via empirical transfer functions. The conventional method of using predefined point locations for potential damage may not accurately represent realistic pavement scenarios. For instance, flexible airfield pavement analysis mainly considers the critical strain at the bottom of the asphalt concrete (AC), which may not characterize near-surface cracking potential. In lieu of point strains, domain analysis, a new method, accounts for the multi-axial behavior of pavements, as inherently excited by three-dimensional (3-D) and nonuniform aircraft tire–pavement contact stresses. Initially applied on highway pavements considering truck tire loading, this approach is an initial breakthrough for implementing domain analysis on flexible airfield pavements; in this study, A-380 and F-16 landing gear tire loads were considered. As anticipated, speed and temperature had significant influence on cumulative domain stress and strain ratios. The decrease in speed and increase in temperature not only increased the cumulative ratios up to 1.81, but nonlinearity of the problem became more prevalent at worst loading conditions (8 kph and 45°C). Minimal difference in ratios for F-16 cases suggests that the National Airport Pavement Test Facility pavement structure became less sensitive to conditions under low loads. Point response analysis revealed that critical strains were not significantly influenced by the tire-inflation pressure, for example, tensile strain at the bottom of the AC only increased up to 13.6% (considering 8 kph speed), whereas domain analysis quantified the increase with respect to 3-D stress or strain states.

Conventional flexible airfield pavement design procedure utilizes the maximum vertical strain at the top of the subgrade and maximum horizontal strain at the bottom of asphalt concrete (AC) layers to predict pavement structural life. Similar to highway pavement design, transfer functions relate mechanistic strains to empirical equations to determine the total number of load cycles that a pavement structure can withstand prior to failure (*1*). The Federal Aviation Administration (FAA) has developed the publicly available software FAARFIELD to design pavements considering airplane characteristics, including loading gear type and geometry, tire-inflation pressure, and traffic volume. Although FAARFIELD implements a multilayered elastic and three-dimensional (3-D) finite element (FE)-based design procedure, several inputs are simplified. For instance, AC is characterized as an elastic layer and for the aircraft tire loading, the contact pressure is assumed equivalent to the tire-inflation pressure (uniformly distributed over a circular or rectangular contact area).

FE modeling is a robust tool used by the scientific community to simulate complex problems (e.g., airfield pavement analysis). In contrast to analytical models, 3-D FE pavement modeling capabilities include multi-axial tire–pavement contact stresses, viscoelastic and nonlinear material characterization, material damping, layer interface properties, dynamic analysis, infinite boundary, and continuously moving load (*2–5*). All the aforementioned considerations provide a step forward in simulating a realistic loaded pavement scenario.

Furthermore, flexible pavement test sections have been studied to determine the effect of numerous variables on pavement behavior, including aircraft tire load, temperature, traveling speed, and tire-inflation pressure. Garg and Hayhoe reported their findings from the National Airport Pavement Test Facility (NAPTF), that the increase in temperature and decrease in speed excite the viscous deformation more significantly than elastic deformation (*6*). The increase in the

[1]University of Illinois at Urbana-Champaign, Illinois Center for Transportation, Rantoul, IL

Corresponding Author:
Address correspondence to Angeli Gamez: agamez2@illinois.edu

temperature from 11.1°C to 22.2°C resulted in an increase in strains ranging from 100% to 120%. In addition, Al-Qadi and co-authors measured in-situ responses of an airport runway with actual airplane traffic (*7*), wherein the data were later used to validate a 3-D FE airfield pavement model (*4*). The magnitude of actual aircraft loads was a clear governing factor to trigger high levels of measured stresses and strains. Roginski indicated that high tire pressure had minimal adverse effect on pavement responses within the upper layer (*8*). A later study by Wang and co-authors determined that for an A-380 tire, an increase in tire-inflation pressure from 1.45 to 1.69 MPa caused 9% to 11% increase in maximum shear and compressive strains within the AC layer (9); however, these responses are not considered within the conventional design method.

A recent study dealing with truck tires revealed that point strains cannot adequately represent the multidimensional state of stresses and strains within a loaded pavement structure (*10*). Although an increase in tire-inflation pressure is expected to directly influence pavement behavior as the tire–pavement contact stress distribution is altered, the maximum strains did not change significantly. To address this issue, a new method, domain analysis, was developed to account for 3-D and nonuniform pavement responses that directly manifest from applied tire–pavement contact stresses (*5,11–17*). This paper aims to implement the domain analysis on an FE airfield pavement model considering the influence of varying AC surface temperatures, traveling speeds, and tire-inflation pressures for A-380 and F-16 tire loads. Utilizing the same FE airfield pavement output database, critical strain responses at point locations were extracted and compared with the 3-D cumulative stress and strain ratios. Details of the FE model and the domain analysis procedure are described in later sections. The paper concludes by highlighting the advantages of the domain analysis outputs over point responses.

FE Model

A NAPTF flexible airfield pavement structure was simulated using Abaqus. Previously constructed in the High Tire Pressure test area at NAPTF (*18*), the pavement structure consists of 125 mm of AC (P401), 200 mm of crushed stone base (P209), 150 mm of Econocrete (P306), and 810 mm of uncrushed stone (P154) (*19*). These material notations are used in later sections. Regarding material properties, AC was assumed as linear viscoelastic, while the supporting layers below the crushed stone base were characterized as linear elastic. On the other hand, a nonlinear, isotropic, and stress-dependent model was implemented for the crushed stone base layer due to high stress levels excited by aircraft tire loads.

In addition, a mesh sensitivity analysis was completed to determine optimum discretization of the selected pavement structure with the criteria of balancing an acceptable level of accuracy with feasible computational effort. The FE pavement model considered AC temperature profiles, layer interface properties, infinite boundary elements for far-field region behavior (*20*), continuously moving load (*2*), and nonuniform contact stresses. Details of the FE model, including material properties and mesh sensitivity analysis, are discussed elsewhere (*21*).

Analysis Matrix

Three-dimensional and nonuniform contact stresses from a single landing gear of the A-380 and F-16 aircrafts were considered in this study. The tire loads for A-380 and F-16 were assumed constant at 262 and 72 kN, respectively. The numerical matrix included three traveling speeds (8, 140, and 270 kph) and three AC surface temperatures (–12, 21, and 45°C). Tire-inflation pressures also varied; however, appropriate values were assumed for each aircraft tire. The A-380 landing gear tire-inflation pressures were assumed to be 1.45 and 1.69 MPa; for the F-16 landing gear, tire-inflation pressures were assumed to be 2.14 and 2.41 MPa. A total of 36 numerical simulations were conducted using computing resources from the Extreme Science and Engineering Discovery Environment (*22*). It is important to recall that the analysis is constrained to consider the AC layer as a viscoelastic material, although under the worst loading conditions (slow speed, high tire-inflation pressure, and high load) AC materials tend to exhibit more viscoplastic behavior. Given that the FE model considers a 3-D moving load, stress-dependency of the base material, and dynamic analysis, incorporating viscoplasticity as a part of the structural analysis would be computationally infeasible at the current time.

Tire–Pavement Contact

Representing the tire–pavement contact stresses for the A-380 and F-16 as 3-D and nonuniform, the distribution of vertical contact stresses along each rib σ_z was calculated using (*23–25*):

$$\sigma_z(x) = \frac{\alpha L}{lb}\left(1+\frac{1}{2n}\right)\left[1-\left(2\frac{x}{l}-1\right)^{2n}\right] \quad (1)$$

where L is the applied load; l is the contact length; b is the rib width; x is the distance along contact length; α is the percentage of L carried by a rib; and n is the shape parameter of $\sigma_z(x)$. The maximum vertical contact stresses were assumed to occur at the center of the contact length (this might be slightly inaccurate for high speed) by assigning the shape parameter as follows:

$$n = \frac{1}{2\left(\frac{lb\sigma_{z,\max}}{\alpha L}-1\right)} \quad (2)$$

where $\sigma_{z,\max}$ is the maximum vertical contact stress along the corresponding rib.

Results reported within the literature defined the contact area details. The contact patch of the A-380 tire comprised five ribs, wherein the assumed width of the exterior and middle ribs were 90 and 70 mm, respectively, and the width of the other ribs was 45 mm. A contact length of 560 mm was implemented for all ribs (*26*). Contact area details for the F-16 were based on results reported by Lindsley and co-authors (*27*). Five 240 mm long ribs composed the contact patch; the widths for the exterior and middle ribs were 30 mm and for the other ribs were 20 mm. To determine the distribution of σ_z along the contact length, the ratio between the maximum σ_z and tire-inflation pressure was assumed. The ratios for A-380 cases were 1.8 for the two exterior ribs, and 1.1 for the three interior ribs; the F-16 cases considered 1.7 and 1.3 for the exterior and interior ribs, respectively.

Variations of the longitudinal and transverse contact stresses were defined based on experimental measurements from truck tires (*28*). Transverse contact stresses σ_y had the same shape as σ_z; however, the magnitude was assigned as $0.40\sigma_z$ and applied only at the rib edges. Longitudinal contact stresses consisted of two skewed parabolas with a common point at x=0.85l, and peak values of $\sigma_{x,\max}=0.20\sigma_{z,\max}$ and $\sigma_{x,\min}=-0.15\sigma_{z,\max}$ at x=0.20l and x=0.85l, respectively (*23*). The contact stresses along the vertical, longitudinal, and transverse directions for a representative rib of F-16 are presented in Figure 1. One could clearly observe the effect of tire-inflation pressure on the peak σ_z and shape of σ_z versus x curve wherein higher tire-inflation pressure resulted in larger and better-defined maximum vertical contact stresses. Details of the A-380 tire–pavement contact stresses are presented in a different publication (*23*). It is worth noting that the generated contact stress inputs reflect a free-rolling tire condition and are anticipated to change when other scenarios are considered. Special cases considering turning, braking, and accelerating conditions, which generate significantly different contact stresses mainly in the longitudinal direction (*29*), were not included as they are outside the scope of this study.

FE Point Responses

FAA flexible pavement design relies upon two point responses: tensile strain at the bottom of the AC $\varepsilon_{11,ac}$ and vertical strain on top of the subgrade $\varepsilon_{22,sg}$ (*1*). The aforementioned strain responses relate to fatigue cracking within the AC and rutting within the subgrade, respectively; however, multiple critical pavement responses are associated with other pavement distresses. In particular, the transverse strain at the bottom of the AC $\varepsilon_{33,ac}$ influences bottom-up fatigue cracking; transverse surface strain $\varepsilon_{33,sf}$ affects near-surface cracking; and shear strain in each layer is related to rutting ($\varepsilon_{23,ac}$, $\varepsilon_{23,bs}$, $\varepsilon_{23,P306}$, $\varepsilon_{23,P154}$, and $\varepsilon_{23,\text{subg}}$ for the vertical shear strain in the AC, base, P306, P154, and subgrade, respectively).

The aforementioned strain responses were extracted from airfield pavement simulations. Figure 2 presents the impact

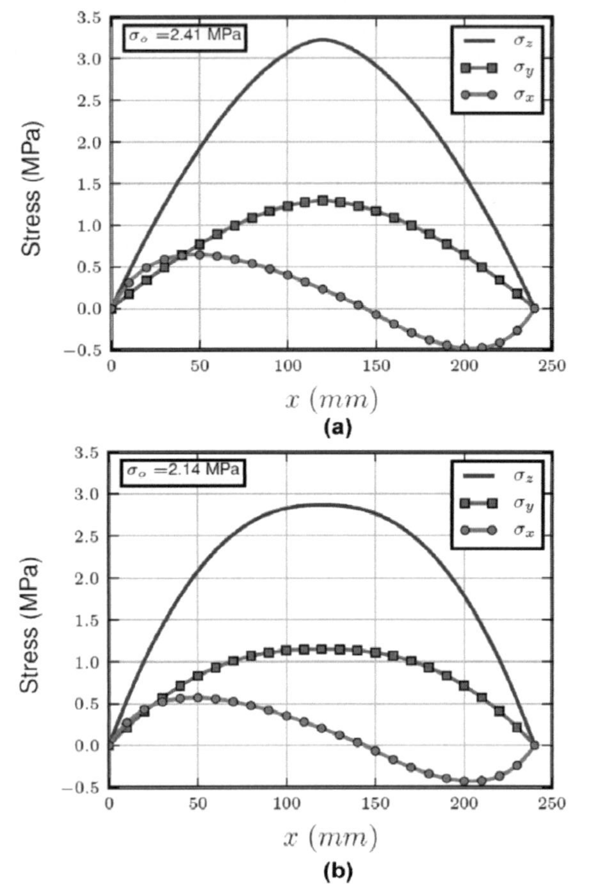

Figure 1. Three-dimensional contact stresses along the center rib of F-16 considering tire-inflation pressures of: (*a*) 2.14 MPa and; (b) 2.41 MPa.

of tire-inflation pressure and speed on critical pavement responses, excited by F-16 tire loading (the same analysis scheme was applied for A-380 loading cases). The point responses along the horizontal axis are grouped and each bin consists of six bars; each bar represents a case considering a combination of speed and tire-inflation pressure. All the strain responses for each load case are normalized to a reference case, which considers the speed of 8 kph and tire-inflation pressure of 2.14 MPa (as indicated on the vertical axis description). For example, one could notice that the first bar, for each bin (left-most) within Figure 2, equates to unity as this is the reference case.

Expected trends, resulting from changes in temperature, speed, and tire-inflation pressure, were observed for both aircraft tire loads. First, as temperature increased from −12°C to 45°C, AC stiffness decreased, thus causing significantly higher strains. Second, in general, strains are reduced as speed increases because of the stiffening effect of speed on AC. And third, increasing tire-inflation pressure resulted in

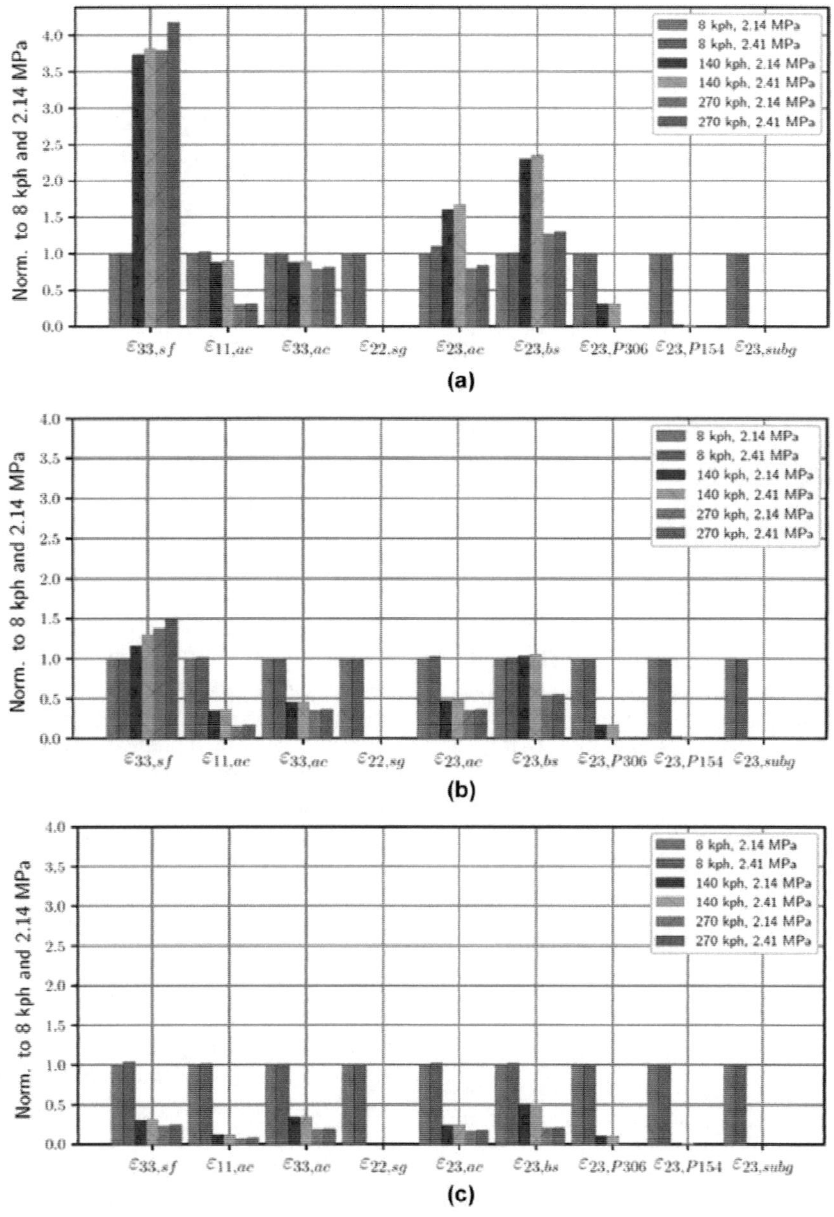

Figure 2. Critical pavement responses for F-16 considering AC surface temperatures of: (*a*) −12°C; (*b*) 21°C and; (*c*) 45°C.

higher pavement responses, although the difference between cases seemed irrelevant for multiple cases.

Specifically, changes in the tire-inflation pressure did not have significant influence on the tensile strain at the bottom of the AC, which is one of the two main design inputs considered by FAARFIELD. The increments in tire-inflation pressures were 13.4% and 12.6% for A-380 and F-16, respectively. At the lowest speed, $\varepsilon_{11,ac}$, for A-380 increased by 13.6% following the increase of tire-inflation pressure. The corresponding change for F-16 was much lower, wherein the greatest change was 3.1%. This smaller effect on F-16 is attributed to 27.5% lower tire loading than the A-380 cases on the same airfield pavement structure.

Strain responses are extracted from a single point location aligned with a specific direction. Furthermore, the traditional design procedure predetermines the location of potential damage initiation (only at two locations: bottom of AC and top of subgrade), which may not be true for all cases. Consequently, calculating damage using transfer functions unrealistically assumes negligible interaction between various damage mechanisms. These drawbacks, along with limited capability of point responses to capture the influence of tire-inflation pressures and near-surface responses, are addressed by the proposed method: domain analysis. Details of the new analysis scheme are presented in the following sections.

Domain Analysis Procedure

A new procedure, domain analysis, quantifies the multi-axial stress or strain states using a single scalar parameter. The proposed analysis scheme includes four main steps: (1) calculate multi-axial stress and strain states of a pavement structure; (2) discretize the pavement to identify critical zones; (3) compare stress and strain states with respect to the failure criteria; and (4) combine the failure potential of critical regions to obtain a scalar called cumulative ratio. A brief description of each step is presented; however, details of the domain analysis can be found elsewhere (*10*).

Multi-Axial Stress and Strain States

Stresses and strains are calculated using the simulated loaded airfield pavement structure. A subdomain is extracted from the FE pavement model to focus the analysis on critical regions, which envelops a volume with dimensions 1240 × 962 mm^2 in the plan view and a depth of 2035 mm. Principal stresses and strains were utilized to compute the hydrostatic stress, p_σ, and shear stress indicator, q_σ, which can be calculated by

$$p_\sigma = \frac{1}{3}(\sigma_1 + \sigma_2 + \sigma_3) \tag{3}$$

$$q_\sigma = \sqrt{\frac{1}{2}[(\sigma_1 - \sigma_2)^2 + (\sigma_2 - \sigma_3)^2 + (\sigma_1 - \sigma_3)^2]} \tag{4}$$

where σ_1, σ_2, and σ_3 are the maximum, intermediate, and minimum principal stresses, respectively. In the following, the same form of equations can represent the hydrostatic strain, p_ε, and shear strain indicator, q_ε:

$$p_\varepsilon = \frac{1}{3}(\varepsilon_1 + \varepsilon_2 + \varepsilon_3) \tag{5}$$

$$q_\varepsilon = \sqrt{\frac{2}{9}\left[(\varepsilon_1 - \varepsilon_2)^2 + (\varepsilon_2 - \varepsilon_3)^2 + (\varepsilon_1 - \varepsilon_3)^2\right]} \tag{6}$$

where ε_1, ε_2, and ε_3 are the maximum, intermediate, and minimum principal strains, respectively. The hydrostatic and shear stress or strain indicator at each element centroid within the subdomain were plotted in a $p-q$ diagram.

Domain Discretization

Relative to point strains, domain analysis aims to represent the spread of $p-q$ coordinates within the 3-D subdomain. Another key part of the proposed method is to identify critical locations of potential pavement damage. Nine zones were identified, wherein the horizontal boundary is a function of the pavement layer thickness and the vertical boundary is defined by the tire width of a specific load case with an addition of 50 mm to both right and left sides to capture localization at tire edges. This addition corresponded to previous observations wherein high shear values localized near the tire edges (*30, 31*). Moreover, the horizontal boundary was confined within 100 mm from the surface, where near-surface cracking is generally observed (*13*). For simplicity, this horizontal boundary at the near-surface was mirrored to create zones at the bottom of the layer. Figure 3 presents the nine zones generated along the vertical cross-section.

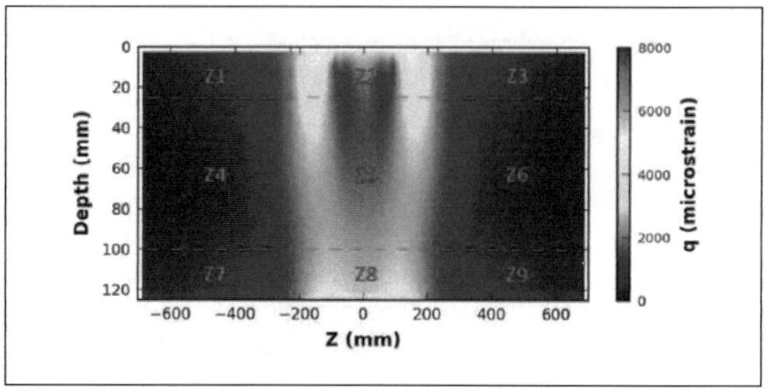

Figure 3. Zoning process.

In particular, six critical zones were considered to govern the analysis: (1) Z2 within the AC relating to near-surface cracking and rutting; (2) Z8 within the AC relating to bottom-up fatigue cracking; and (3) Z2 within the base, (4) Z2 within the Econocrete, (5) Z2 within the uncrushed stone base, and (6) Z2 within the subgrade related to rutting.

Inclusion of Failure Criteria

Using failure criteria, damage potential is quantified with respect to the current $p-q$ state. The Modified Drucker–Prager Cap Model was selected as it accounts for tension cutoff, compression, and shear. It should be noted that any failure surface could be adapted within the domain analysis. Furthermore, the Cartesian coordinates of the $p-q$ diagram were transformed into polar coordinates. The polar coordinate system was regionalized into 30 sectors, stemming from six radii and seven equally spaced angle boundaries. The rationale in the transformation is that polar sectors closest to the failure surface are penalized more than the ones farther away. The weights are defined by the stress or strain ratio (i.e., current stress or strain state of the sector midpoint divided by the allowable stress or strain state). A weight of 1.0 indicates that the midpoint sector coincides with the yield surface and failure is reached.

Cumulative Ratio

In the final step, a single scalar is calculated by considering geometry and adding all stress or strain states within the subdomain. A homogenizing factor was applied to account for the geometric difference between load cases along the vertical cross-sectional area. Using individual element area, a_{jl}, of element j within zone l and the total area, A_l, the homogenizing factor, a_{jl} / A_l is calculated.

Furthermore, another geometric difference could arise along the traveling direction. The element length, l_{elem}, is divided by the total length of the subdomain, L_{sub}, to obtain the volumetric factor, $l_{\text{elem}} / L_{\text{sub}}$. Using the vector magnitude of each $p-q$ coordinate from the origin and weight factors, the $p-q$ point cloud may be combined into a cumulative value, coined as the cumulative stress $C\sigma$ or cumulative strain $C\varepsilon$, which is calculated as follows:

$$C\sigma,\varepsilon_{\text{case}} = \frac{\sum_{l=1}^{z}\sum_{j=1}^{e}\sum_{i=1}^{s}\left|(pq)_{\sigma,\varepsilon}\right|_{jl} \times a_{jl} \times l_{\text{elem}} \times w_i}{A_l \times L_{\text{sub}}} \quad (7)$$

where $C\sigma,\varepsilon_{\text{case}}$ is the cumulative stress or strain of the specific load case; $\left|(pq)_{\sigma,\varepsilon}\right|_{jl}$ is the vector magnitude of the element j for a total of e elements within the zone l for a total of z zones; and w_i is the weight of the specific sector i for a total of s sectors.

The cumulative ratio is calculated as follows:

$$CRS = \frac{C\sigma_{\text{case}}}{C\sigma_{\text{ref}}} \quad (8)$$

$$CRE = \frac{C\varepsilon_{\text{case}}}{C\varepsilon_{\text{ref}}} \quad (7)$$

where CRS is the unitless cumulative ratio in the stress domain; CRE is the unitless cumulative ratio in the strain domain; $C\sigma_{\text{ref}}$ is the cumulative stress of the reference load case; and $C\varepsilon_{\text{ref}}$ is the cumulative strain of a reference load case. A sensitivity analysis of the zoning limits was completed to evaluate their influence on the ratios. Although the cumulative stress or strain values were directly influenced, the cumulative ratios did not vary as significantly as the cumulative values. The trend of the ratios was also maintained despite the variation in zoning limits, which may be due to the fact that the two cases being compared (that form the ratio) utilize the exact same partitioning scheme. Future work may include determining a meaningful use of the cumulative values, along with the ratios.

Results and Discussion

Domain Analysis of A-380 Tire Load Cases

In contrast to point responses, the domain analysis accounts for the inherent 3-D response of a loaded pavement structure. For instance, Figure 4 illustrates the complex response of the stress-dependent base considering a single A-380 landing gear tire load and AC surface temperature of 45°C. Note that

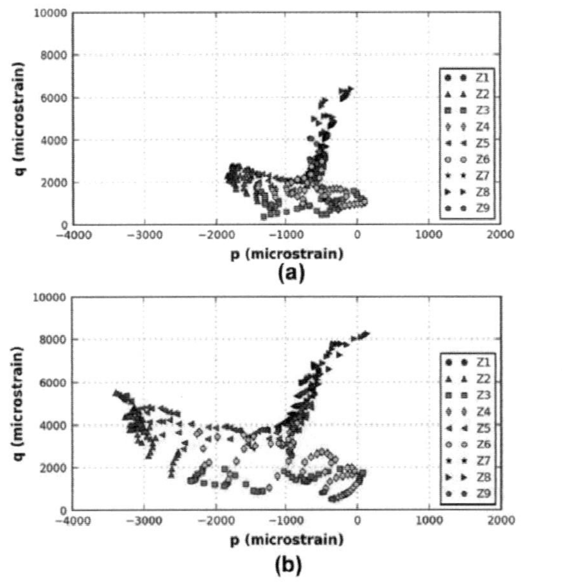

Figure 4. Case of A-380, $p-q$ diagrams of loading cases with 8 kph, 45°C, and two tire-inflation pressure values: (*a*) 1.45 MPa and; (*b*) 1.69 MPa.

the $p-q$ diagrams were extracted at a specific vertical cross-section along the subdomain contact length, where the maximum values were located. It is evident from the $p-q$ cloud that an increase in the tire-inflation pressure by 17% led to significantly higher levels of compression and shear values.

The outputs of the domain analysis include CRS and CRE values for each critical zone, and the weighted average of all the ratios from the selected six critical zones. The weighted average was calculated as a single metric to compare the numerical simulations. It is noteworthy that the loading cases were categorized with respect to the surface temperature and were normalized to the simulation considering 8 kph speed and 1.45 MPa tire-inflation pressure. Hence, under the same AC temperature, CRS and CRE define the potential pavement damage relative to the normalizing load case.

A sample plot of the CRS and CRE values within the six critical zones is presented in Figure 5, considering an AC surface temperature of 21°C for varying tire speeds and inflation pressures. As the loading condition worsened, that is, decrease in speed and increase in tire-inflation pressure, the CRS at the near-surface and bottom zones of AC steadily increased from 0.20 to 1.36 (Figure 5*a*). Furthermore, the supporting layers below the granular base experienced significantly low stress or strain levels for cases considering tire speeds greater than 8 kph. Cumulative ratios within $Z2$ of P306, P154, and subgrade layers tended to zero, which indicates that the principal stresses and strains were also zero.

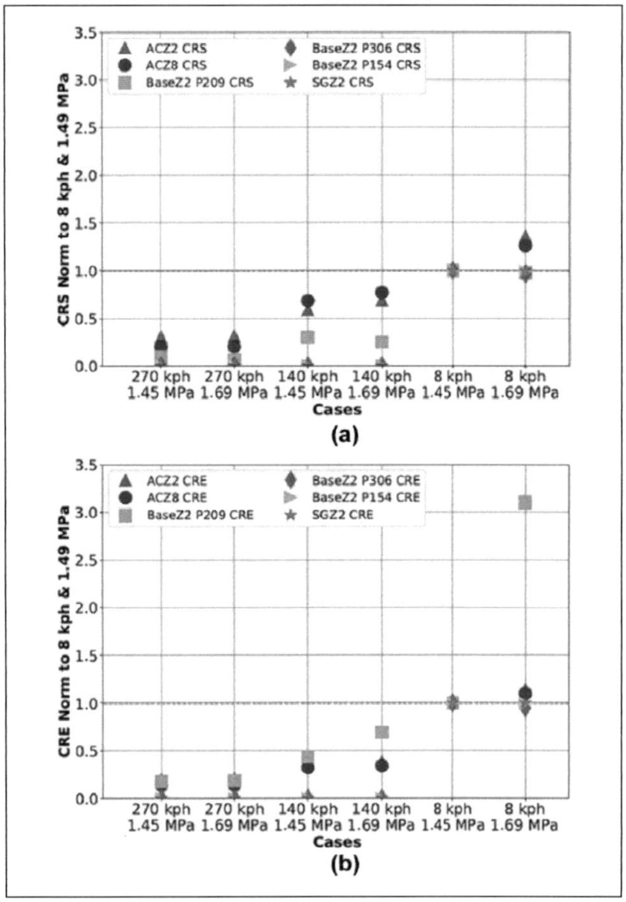

Figure 5. (*a*) CRS and (*b*) CRE within critical zones of A-380 cases considering 21°C AC surface temperature.

Most of the *CRE* values within the AC were significantly lower than those of the loading case under 8 kph and 1.45 MPa (Figure 5*b*). Under high speeds, AC layers tend to behave elastically. Conversely, as the loading speed reduced to 8 kph, the viscous components governed, leading to high strains. On the other hand, the base layer was relatively more sensitive to the changes in loading conditions as the *CRE* ranged from 0.17 to 3.1. Comparing the *CRE* trends with respect to AC surface temperatures, the cases with –12°C revealed a steady increase, whereas the trend became highly nonlinear as the surface temperature further increased to 45°C.

An increase in the weighted average of the *CRS* was evident as the surface temperature increased from –12°C to 45°C, while the weighted average of the *CRE* decreased relative to the case under 8 kph and 1.45 MPa (Figure 6). This coincides with the rationale that at high temperatures, the stress state of the pavement structure becomes less sensitive to other parameters as temperature governs. The nonlinearity effect was also apparent as the disparity between *CRS* and *CRE* increased while the AC layer was subjected to higher temperatures. For instance, the *CRE* values for the cases considering 45°C varied between 0.05 and 1.7, wherein the considerable increase was observed under the speed of 8 kph. A clear change in the ratios as the tire-inflation pressure increased for each combination of speed and tire-inflation pressure is also presented in Figure 6. The importance of these findings supports the fact that the domain analysis appropriately captures the impact of the analysis variables on 3-D pavement responses.

Domain Analysis of F-16 Tire Load Cases

The same analysis scheme was implemented for F-16 tire loading cases (Figure 7). As expected, the ratios increased with the decrease in speed and increase in tire-inflation pressure. In contrast to the A-380 cases, reducing the tire load to 72 kN for the F-16 cases led to minimal differences in the ratios when the tire-inflation pressure changed by 12.6%. This elucidates that the considered airfield pavement structure becomes less sensitive to the applied tire–pavement contact stress distribution under low load (given the assumptions described earlier).

However, as nonlinearity begins to govern with increasing temperature, a slight increase in the ratios was observed as the tire-inflation pressures changed. At low speed of 8 kph and AC surface temperature of 45°C, a *CRS* increment of 5.82% resulted within the near-surface region of the AC layer. Corresponding to the same loading case, the *CRE* within the crushed stone base increased by 6.08% as the tire-inflation pressure was altered from 2.14 to 2.41 MPa. Furthermore, the max increase in the weighted average *CRS* and *CRE* values were 4.56% and 2.20%, respectively, for the worst loading scenario.

Summary and Conclusions

Initially applied on highway pavements, the domain analysis aims to capture the direct influence of 3-D and nonuniform tire–pavement contact stresses on pavement responses. As damage initiation may not always be trivial to locate, the proposed method allows the localized region (high stress or strain levels) to be highlighted and inherently penalized via the use of weights, which are defined by the selected failure criteria. Moreover, the combined influence of compression/tension and shear are accounted for in this weighting scheme.

Even though this method has been applied on low-volume and interstate highway flexible pavements considering truck tire loading, the presented study is an initial breakthrough in implementing the proposed analysis on airfield flexible pavements loaded with A-380 and F-16 landing gear. As anticipated, speed and temperature had significant influence on *CRS* and *CRE* ratios for both A-380 and F-16 tire loading cases. Not only did the decrease in speed and increase in temperature increase the cumulative ratios, but the nonlinearity of the problem also became more prevalent at the worst conditions (8 kph and 45°C). The minimal difference

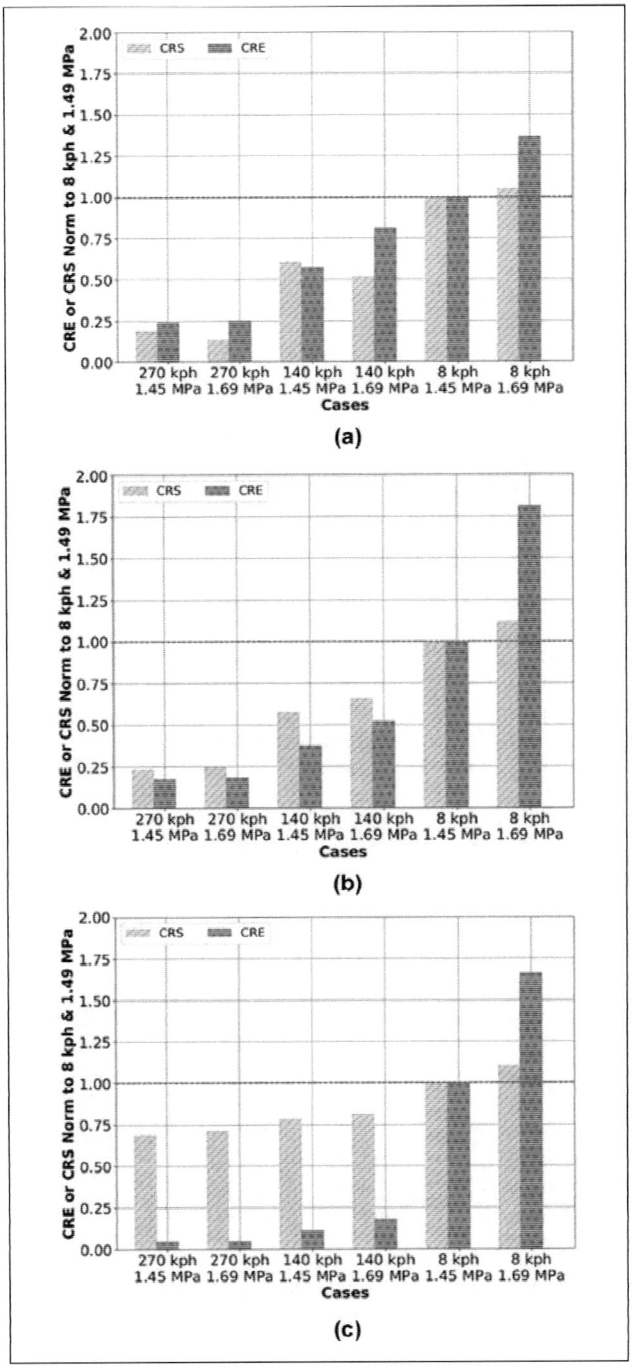

Figure 6. Weighted average of CRS and CRE for an airfield pavement loaded with A-380 tire considering AC surface temperatures of: (*a*) −12°C; (*b*) 21°C and; (*c*) 45°C.

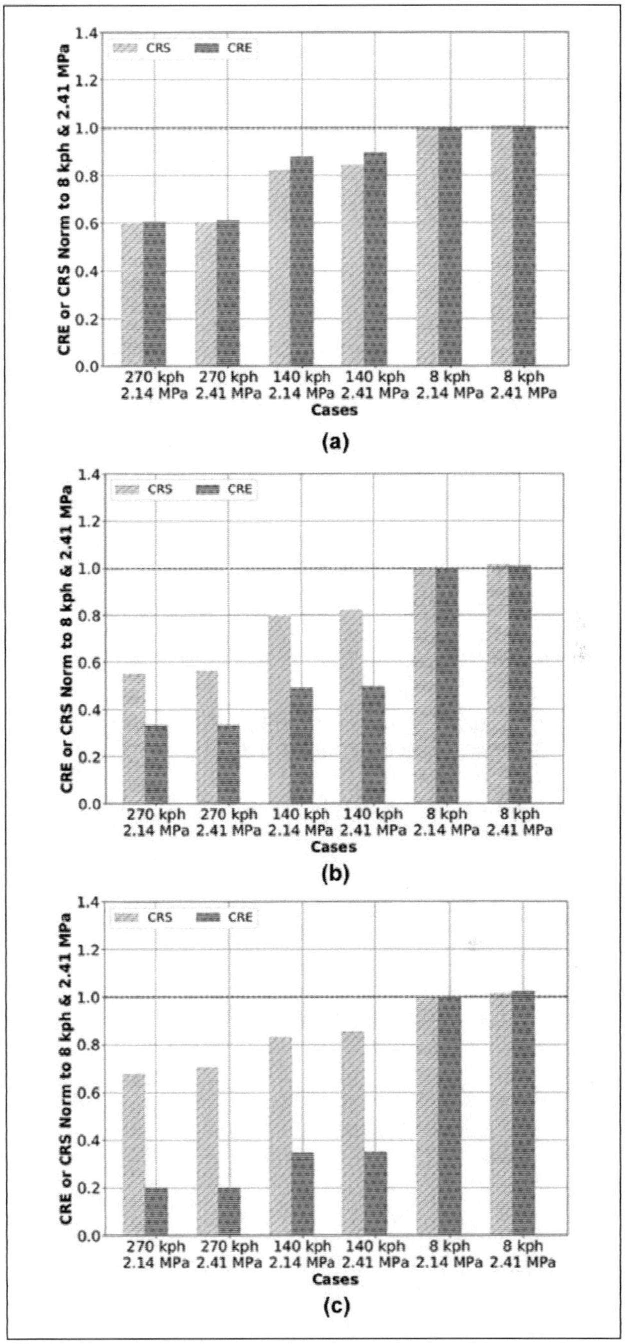

Figure 7. Weighted average of CRS and CRE for an airfield pavement loaded with F-16 tire considering AC surface temperatures of: (*a*) −12°C; (*b*) 21°C and; (*c*) 45°C.

in ratios for the F-16 loading cases suggests that the selected airfield pavement structure becomes less sensitive to conditions under low loads.

In contrast to point responses, the critical tensile strain at the bottom of the AC under the slowest speed was minimally influenced by the increase in tire-inflation pressure, wherein the strain increments were 13.6% and 3.1% for A-380 and F-16 cases, respectively. The cumulative ratios, on the other hand, increased up to 81% considering the same loading cases. This is a clear example of the limited capability of point responses to illustrate the influence of tire-inflation pressures, while the domain analysis determined the increase in 3-D stress or strain states via a single scalar. Furthermore, minimal differences in point strains directly influence pavement damage prediction, provided that transfer functions use critical strains as inputs.

Potential adaptation of the presented concepts into flexible airfield pavement design procedures will lead to better representation of actual pavement responses and evolution of pavement damage. Moreover, domain analysis could be easily applied to other scenarios considering varying pavement structures, tire types and configurations, environmental conditions, and tire rolling conditions. Another future parameter to consider is wander, wherein additional FE runs may not be required. Instead, the same output database could be utilized, and wander could be taken into account by shifting the subdomain. Different failure criteria could also be integrated into the domain analysis engine. Lastly, the next step forward is to relate domain analysis results to field measurements with the goal of improving the link between mechanistic responses and damage prediction.

Acknowledgments

This project was conducted in cooperation with the Illinois Center for Transportation. The authors are indebted to Navneet Garg and David Brill for providing the pavement material properties required for the numerical simulations. This study used the Extreme Science and Engineering Discovery Environment (XSEDE), which is supported by National Science Foundation grant number ACI-1548562. We also would like to thank Mahidhar Tatineni for his unyielding assistance in using XSEDE resources.

References

1. Advisory Circular 150/5320-6F. *Airport Pavement Design and Evaluation*. US Department of Transportation, Federal Aviation Administration, 2016.
2. Yoo, P., I. L. Al-Qadi, M. Elseifi, and I. Janajreh. Flexible Pavement Responses to Different Loading Amplitudes Considering Layer Interface Condition and Lateral Shear Forces. *International Journal of Pavement Engineering*, Vol. 7, No. 1, 2006, pp. 73–86.
3. Al-Qadi, I. L., H. Wang, P. J. Yoo, and S. Dessouky. Dynamic Analysis and In Situ Validation of Perpetual Pavement Response to Vehicular Loading. *Transportation Research Record: Journal of the Transportation Research Board*, 2008. 2087: 29–39.
4. Wang, H., I. L. Al-Qadi, S. Portas, and M. Coni. Three-Dimensional Finite Element Modeling of Instrumented Airport Runway Pavement Responses. *Transportation Research Record: Journal of the Transportation Research Board*, 2013. 2367: 76–83.
5. Hernandez, J. A., A. Gamez, and I. L. Al-Qadi. Effect of Wide-Base Tires on Nationwide Flexible Pavement Systems: Numerical Modeling. *Transportation Research Record: Journal of the Transportation Research Board*, 2016. 2590: 104–112.
6. Garg, N., and G. F. Hayhoe. Asphalt Concrete Strain Responses at High Loads and Low Speeds at the National Airport Pavement Test Facility (NAPTF). In *Advancing Airfield Pavements* (W. G. Buttlar and J. E. Naughton, ed.), American Society of Civil Engineers, Reston, V.A., 2001, pp. 1–14.
7. Al-Qadi, I. L., S. Portas, M. Coni, and S. Lahouar. Runway Instrumentation and Response Measurements. *Transportation Research Record: Journal of the Transportation Research Board*, 2010. 2153: 162–169.
8. Roginski, M. J. Effects of Aircraft Tire Pressures on Flexible Pavements. In *Proc., International Conference on Advanced Characterization of Pavement and Soil Engineering*, Athens, Greece, 2007, pp. 1473–1481.
9. Wang, H., M. Li, and N. Garg. Simulation of NAPFT High Tire Pressure Tests with Advanced Finite Element Modeling. In *Proc., 2014 FAA Worldwide Airport Technology Transfer Conference*, Galloway, New Jersey, 2014, pp. 1–13.
10. Gamez, A., J. A. Hernandez, H. Ozer, and I. L. Al-Qadi. Development of Domain Analysis for Determining Potential Pavement Damage. *ASCE Journal of Transportation Engineering, Part B: Pavements*, 2017, accepted for publication.
11. Yoo, P. J., and I. L. Al-Qadi. Effect of Transient Dynamic Loading on Flexible Pavements. *Transportation Research Record: Journal of the Transportation Research Board*, 2007. 1990: 129–140.
12. Al-Qadi, I. L., and P. J. Yoo.Effect of Surface Tangential Contact Stresses on Flexible Pavement Response (With Discussion). *Journal of the Association of Asphalt Paving Technologists*, Vol. 76, 2007, pp. 663–695.
13. Yoo, P. J., and I. L. Al-Qadi. The Truth and Myth of Fatigue Cracking Potential in Hot-Mix Asphalt: Numerical Analysis and Validation. *Journal of Association of Asphalt Paving Technologists*, Vol. 77, 2008, pp. 549–590.
14. Wang, H., and I. L. Al-Qadi. Combined Effect of Moving Wheel Loading and Three- Dimensional Contact Stresses on Perpetual Pavement Responses. *Transportation Research Record: Journal of the Transportation Research Board*, 2009. 2095: 53–61.
15. Xue, W., and E. Weaver. Influence of Tyre Configuration on Pavement Response and Predicted Distress. *International Journal of Pavement Engineering*, Vol. 16, No. 6, 2015, pp. 538–548.
16. Gungor, O. E., J. A. Hernandez, A. Gamez, and I. L. Al-Qadi, Quantitative Assessment of the Effect of Wide-Base Tires on Pavement Response by Finite Element Analysis. *Transportation Research Record: Journal of the Transportation Research Board*, 2016. 2590: 37–43.
17. Gungor, O. E., I. L. Al-Qadi, A. Gamez, and J. A. Hernandez. In-Situ Validation of Three- Dimensional Pavement Finite Element Models. In *The Roles of Accelerated Pavement Testing in Pavement Sustainability* (J. P. Aguiar-Moya, A. Vargas-Nordcbeck, F. Leiva-Villacorta, and L. G. Loria-Salazar, ed.), Springer, New York, 2016, pp. 145–159.
18. Song, I., and N. Garg. High Tire Pressure and Temperature Effects on Hot Mix Asphalt Concrete Permanent Deformation Using Customized Asphalt Pavement Analyzer. In *Proc., 2010 FAA Worldwide Technology Transfer Conference, Atlantic City, New Jersey*, 2010, pp. 1–14.
19. Advisory Circular 150/5370-10G. Standards for Specifying Construction of Airports. US Department of Transportation, Federal Aviation Administration, 2014.

20. Hjelmstad, K. D., Z. Qiuhai, and J. Kim. Elastic Pavement Analysis Using Infinite Elements. *Transportation Research Record: Journal of the Transportation Research Board*, 1997. 1568: 72–76.

21. Hernandez, J. A., A. Gamez, and I. L. Al-Qadi, Domain Analysis for Airfield Pavement: Moving Forward from Point Responses. In *Tenth International Conference on the Bearing Capacity of Roads, Railways and Airfields*, Athens, Greece, 2017, pp. 1693–1700.

22. Towns, J., T. Cockerill, M. Dahan, I. Foster, K. Gaither, A. Grimshaw, V. Hazlewood, S. Lathrop, D. Lifka, G. D. Peterson, et al. XSEDE: Accelerating Scientific Discovery. *Computing in Science & Engineering*, Vol. 16, No. 5, 2014, pp. 62–74.

23. Hernandez, J. A., and I. L. Al-Qadi, Airfield Pavement Response Caused by Heavy Aircraft Takeoff: Advanced Modeling for Consideration of Wheel Interaction. *Transportation Research Record: Journal of the Transportation Research Board*, 2015. 2471: 40–47.

24. Hernandez, J., A. Gamez, I. L. Al-Qadi, and M. De Beer. Analytical Approach for Predicting Three-Dimensional Tire–Pavement Contact Load. *Transportation Research Record: Journal of the Transportation Research Board*, 2014. 2456: 75–84.

25. Guo, K., and D. Lu. UniTire: Unified Tire Model for Vehicle Dynamic Simulation. *Vehicle System Dynamics*, Vol. 45, No. S1, 2007, pp. 79–99.

26. Wang, H., and I. L. Al-Qadi. Impact of Non-Uniform Aircraft Tire Pressure on Airfield Pavement Responses. In *Transportation and Development Institute Congress 2011: Integrated Transportation and Development for a Better Tomorrow*, Chicago, Illinois, 2011, pp. 844–853.

27. Lindsley, N. J., J. P. Medzorian, M. K. Lay, and W. W. Macy. *Effects of Flywheel Curvature on Aircraft Tire Footprint Behavior*. Technical Paper 952020. SAE, Warrendale, Pennsylvania, 1995.

28. Hernandez, J. A., I. L. Al-Qadi, and M. De Beer. Impact of Tire Loading and Tire Pressure on Measured 3D Contact Stresses. In *Airfield and Highway Pavement 2013: Sustainable and Efficient Pavements* (I. L. Al-Qadi and S. Murrell, ed.), American Society of Civil Engineers, Reston, V.A., 2013, pp. 551–560.

29. Hernandez, J. A., and I. L. Al-Qadi. Hyperelastic Modeling of Wide-Base Tire and Prediction of Its Contact Stresses. *Journal of Engineering Mechanics*, Vol. 142, No. 2, 2015, p. 04015084.

30. Hernandez, J. A., A. Gamez, and I. L. Al-Qadi, Effect of Wide-Base Tires on Nationwide Flexible Pavement Systems: Numerical Modeling. *Transportation Research Record: Journal of the Transportation Research Board*, 2016. 2590: 104–112.

31. Wang, H., and I. Al-Qadi. Near-Surface Pavement Failure under Multiaxial Stress State in Thick Asphalt Pavement. *Transportation Research Record: Journal of the Transportation Research Board*, 2010. 2154: 91–99.

The Standing Committee on Pavement Structural Modeling and Evaluation peer-reviewed this paper (18-01291).

The contents of this report reflect the view of the authors, who are responsible for the facts and the accuracy of the data presented herein. The contents do not necessarily reflect the official views or policies of the Illinois Center for Transportation. This paper does not constitute a standard, specification, or regulation.

Article

TRR
JOURNAL OF THE TRANSPORTATION RESEARCH BOARD

Transportation Research Record
2018, Vol. 2672(40) 336–347

DOI: 10.1177/0361198118758010
journals.sagepub.com/home/trr
SAGE

Improved Backcalculation Procedure for Continuously Reinforced Concrete Pavement

Yating Zhang[1] **and Jeffery Roesler**[2]

Abstract

Falling weight deflectometer (FWD) testing is effective in evaluating the structural response of in-situ concrete pavements through the backcalculated pavement layer parameters. Specifically, the FWD data can be used to backcalculate the foundation layer and concrete stiffness or the soil layer stiffness, effective slab thickness, and slab–base interface condition. Since continuously reinforced concrete pavement (CRCP) has closely spaced transverse cracks, the traditional backcalculation assumption of an infinite slab can lead to significant errors in the backcalculated results. In this paper, solutions for backcalculated modulus of subgrade reaction (k-value), elastic modulus of concrete (E), and effective thickness (h_{eff}) for different crack spacing have been derived from 2-D finite element analysis. AASHTO sensor configuration (0, 12, 24, 36 in.) was recommended for CRCP with crack spacing ≥6 ft, and an alternative solution for crack spacing of 4 and 5 ft was proposed with $AREA_{24}$. Crack load transfer efficiency (LTE) across transverse cracks had limited impact on backcalculated results if the LTE was >80%. As expected, the backcalulation values were sensitive to the load plate's longitudinal position relative to the transverse crack especially for crack spacings smaller than 8 ft. The proposed backcalculation method was applied to a field CRCP test section with different crack spacing, reinforcement ratio, and base types.

Falling weight deflectometer (FWD) testing is an effective method to evaluate in-situ pavement responses and estimate the pavement's load capacity and future performance. FWD tests have been conducted on many continuously reinforced concrete pavements (CRCPs) with almost all the backcalculation following jointed plain concrete pavement (JPCP) procedures. Tayabji et al. conducted FWD investigations on 23 CRCP projects across the United States and utilized ILLI-BACK to backcalculate modulus of subgrade reaction (k-value) and slab rigidity ($D = \ell^4 k$) (*1,2*). Li and White demonstrated that FWD sensor configuration had a significant impact on backcalculated results by comparing two closed-form algorithms, ILLI-BACK and NUS-BACK (*3*). Khazanovich et al. developed an improved AASHTO methodology called Best Fit Method and validated with field deflection testing data on two JPCP sections and one CRCP section from the Long-Term Pavement Performance program in the United States and Canada, indicating the improved model worked with a hypothetical joint spacing of 15 ft for CRCP (*4*). Kohler and Roesler demonstrated the feasibility of an FWD device to characterize the crack width of in-situ CRCP given at least five distinct load-level drops at the transverse cracks (*5*). Roesler and Huntley evaluated the structural performance over time of CRCP containing crushed concrete aggregates from the backcalculated elastic modulus of the concrete (E) and k-value using the traditional FWD equations for concrete pavements as well as the transverse crack load transfer efficiency (LTE) (*6*). Choi et al. completed an extensive field investigation of 27 CRCP test sections in Texas with various slab thicknesses, pavement ages, and environmental conditions (*7*). As expected, mid-slab deflections decreased with increased crack spacing, confirming that in some circumstances correction factors must be considered for more accurate CRCP backcalculation.

The traditional FWD backcalculation procedure for concrete pavements is based on an infinite slab assumption, as described in the AASHTO Guide (*8*). For jointed plain concrete with standard slab sizes, this assumption is generally satisfied. Since long-term transverse crack spacings in CRCP are typically 4–6 ft given the current standards for steel reinforcement and base friction, the assumption of infinite slab seems to be violated for existing CRCP backcalculations. Ioannides et al. stated the minimum non-dimensional slab

[1]Zhejiang University, Hangzhou, China
[2]University of Illinois at Urbana-Champaign, Urbana, IL

Corresponding Author:
Address correspondence to Yating Zhang: crystal_zyt@zju.edu.cn

size (L/ℓ) for interior loading was 3.5 and 8 in order to meet the Westergaard infinite slab assumptions for stresses and deflections, respectively (*9*). Later, Crovetti developed a series of finite-sized slab deflection correction factors based on a range of L/ℓ values (*10*). Unfortunately, the study primarily focused on JPCP and did not cover a sufficient range to include all expected L/ℓ for all types of concrete pavements. King and Roesler proposed a backcalculation procedure for the bonded concrete overlays of asphalt pavement system based on smaller panel sizes, with a smaller slab length to radius of relative stiffness ratio (L/ℓ) ranging from 1.1 to 5.3 given a slab aspect ratio of one (*11*). Ironically, no study has been found in the literature justifying backcalculation of CRCP layer properties or effective thickness with infinite slab equations. Furthermore, existing finite-sized slab solutions (*10,11*) will not directly work for CRCP because of their larger slab aspect ratios, for example, 0.25–1.0.

Objectives

The objective of this study is to theoretically determine when the infinite slab assumption is valid for CRCP in terms backcalculated *k*-value, *E*-value, and effective thickness (h_{eff}), and, when this assumption is not applicable, to develop an improved CRCP backcalculation procedure based on finite-size slab panels with larger slab aspect ratios and check validity with field data. In addition, the sensitivity of the LTE across the transverse cracks and sensor configuration on the backcalculated results needs to be determined to set reasonable limits on the backcalculation equations. Finally, the sensitivity of the FWD plate location has to be investigated to quantify the backcalculation error when plate is longitudinal offset from the mid-panel location.

Traditional Rigid Pavement Backcalculation

The *AREA* algorithm (*12*) has been widely used to quantify the deflection basin generated from FWD testing:

$$AREA = \frac{1}{2d_0}\left[d_0 r_1 + \left(\sum_{i=1}^{n-1} d_i\left(r_{i+1} - r_i\right)\right) + d_n\left(r_n - r_{n-1}\right)\right] \quad (1)$$

where d_i is measured deflections, n is number of FWD sensors minus one, and r_i are distances between center of the loaded plate and the ith sensor. *AREA* is the normalized deflection basin relative to the deflection measured at the center of the loading plate.

Ioannides et al. found that there is a unique relationship between *AREA* and the radius of relative stiffness (ℓ) for a given radius of loaded area (a) based on the assumption of an infinite concrete slab (*13*). For rigid pavements, the most common *AREA* equations are the AASHTO (0, 12, 24, 36 in.) and SHRP (*14*) (0, 8, 12, 18, 24, 36 and 60 in.) sensor configurations, named *A4* and *A7*, respectively. Those regression equations to relate ℓ and *A7* (*15*), and ℓ and *A4* (*16*), as shown in the following equations:

$$\ell_{A7} = \left[\frac{\ln\left(\frac{60 - A7}{289.708}\right)}{(-0.698)}\right]^{2.566} \quad (2)$$

$$\ell_{A4} = \left[\frac{\ln\left(\frac{36 - A4}{1812.279}\right)}{(-2.5559)}\right]^{4.387} \quad (3)$$

With the theoretical relationship between ℓ and *AREA* known from Equations 2 to 3, Westergaard's (*17*) equation for calculating the non-dimensional deflection at the slab interior ($\Delta_{int}*$) under a circular plate with fixed a/ℓ ratio can be determined using

$$\Delta_{int}* = \frac{1}{8}\left[\left[1 + \left(\frac{1}{2\pi}\right)\left(\ln\left(\frac{1.7810725a}{\ell}\right) - 1.25\right)\left(\frac{a}{\ell}\right)^2\right]\right] \quad (4)$$

given an infinite slab assumption, maximum plate load (P), and measured center plate deflection (d_0).

With the calculated non-dimensional interior deflection ($\Delta_{int}*$), the modulus of subgrade reaction (k) can be determined by rearranging

$$\Delta_{int}* = \frac{k d_0 \ell^2}{P} \quad (5)$$

Given the foundation layer *k*-value, the concrete's stiffness (E) or slab thickness (h) can be calculated from the radius of relative stiffness

$$\ell = \left[\frac{Eh^3}{12\left(1-\upsilon^2\right)k}\right]^{1/4} \quad (6)$$

This closed-form backcalculation process summarized above (*18*) has been extensively applied to evaluate the structural response and capacity of concrete pavement for roads and airfields including jointed plain, jointed reinforced, and CRCPs.

Feasibility of Traditional Backcalculation Procedure for CRCP

The unique relationship between *AREA* and radius of relative stiffness ℓ in Equations 2 and 3 and the Westergaard interior solution in Equations 4 and 5 is based on the assumption of infinite slab size where the edges and joints do not affect the deflection basin in an interior-loaded slab. Given transverse crack spacing develops fully in CRCP after the first 2–3 years (*19*) at spacings <8 ft, the feasibility of applying the

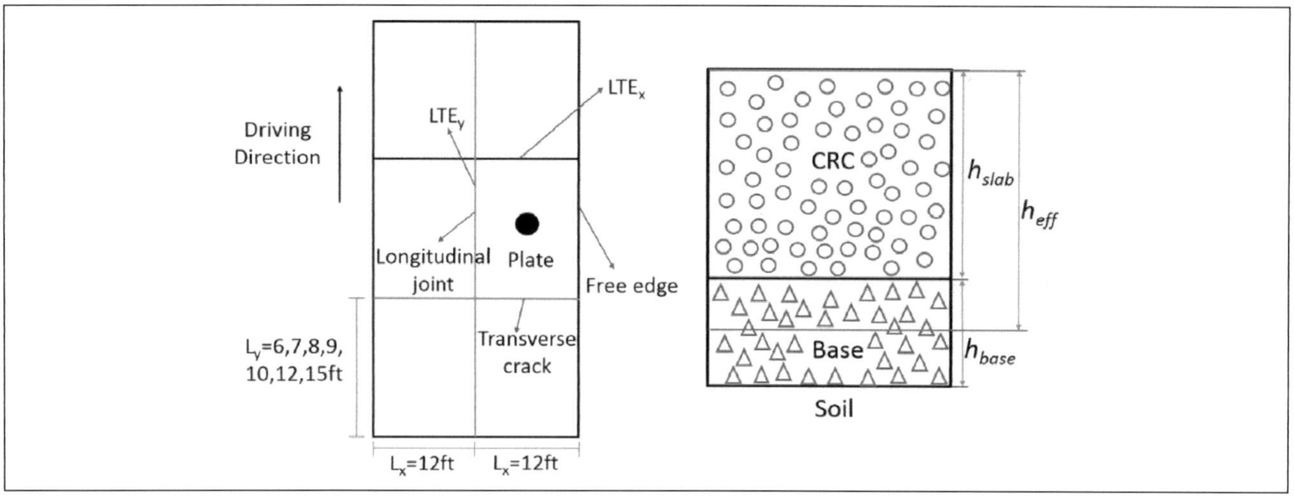

Figure 1. Layout of two-dimensional finite model and CRCP cross-section.

traditional backcalculation formulas has to be evaluated with numerical modeling.

Finite-Sized Slab Model for CRCP

As shown in Figure 1, a six-slab system was used to model a two-lane CRCP in ILLI-SLAB (*20,21*), a 2-D finite element analysis software. The FWD plate was positioned at the center of one of the central slabs, equidistant from the transverse cracks and 6 ft from the free edge and longitudinal joint. To determine the error introduced in CRCP backcalculation with the traditional backcalculation procedure, the following inputs were assumed for the concrete and subgrade properties and FWD test parameters.

1. Modulus of elasticity (E): 5,000,000 psi;
2. Poisson's ratio (υ): 0.15;
3. Modulus of subgrade reaction (k-value): 100 psi/in.;
4. Load (P): 9,000 lbs;
5. Radius of plate (a): 5.9 in.

The width of each slab is fixed at 12 ft with the slab length ranging from 6 ft to 15 ft (slab aspect ratios from 0.5 to 1.2) to simulate average crack spacing at different pavement ages. Slab panels <6 ft (72 inches) will not be able to accurately calculate *AREA* (*A4*) because the last sensor is at a 36 inch offset from plate (Equation 3). The thickness of concrete slab varied from 2 in. to 10 in. with a 0.5-in. increment for every combination of slab size and crack LTE. For the theoretical analyses, no base layer was assumed to simplify comparisons. Transverse crack LTE were chosen to be 100, 90, 80, 50 and 0% in both of transverse (LTE_x) and longitudinal joints (LTE_y) with $LTE_x = LTE_y$ for each case. The range of ℓ-values produced for the chosen slab thickness, concrete properties, and soil stiffness is 13.59–45.44 inches with the corresponding non-dimensional slab (L/ℓ) ranged from 1.58 to 13.25. Theoretical deflections at the pre-set sensor configurations were extracted from ILLI-SLAB in order to perform the backcalculation comparisons. Equation 1 (*A4*) and Equation 3 were used for backcalculation when crack spacing was <10 ft. Equation 1 (*A7*) and Equation 2 were used when crack spacing was ≥10 ft.

Comparison of Backcalculated k and E-Values and Effective Thickness (h_{eff})

The backcalculated values of modulus of subgrade reaction (k-value) and the concrete's modulus of elasticity (E) can be used to examine the accuracy of the traditional algorithm given that the inputs to the numerical model are known. The effective thickness (h_{eff}) concept was suggested by Ioannides et al. to describe the bond/interface condition between concrete slab and base layer (*22*). By rearranging Equation 6 after backcalculating k-value and assuming E-value and Poisson's ratio (υ), h_{eff} can be obtained as shown in

$$h_{eff} = \sqrt[3]{\frac{12\ell^4\left(1-\upsilon^2\right)k}{E}} \quad (7)$$

In this analysis, the base layer is automatically combined with the concrete slab into a single thickness value with the h_{eff} representing the combined slab–base bond condition, slab–base thickness, and base stiffness. Assuming the pavement system has a base layer, the effective thickness is always greater than slab thickness. If the backcalculated h_{eff} is higher than the in-situ thickness of concrete slab, it implies that there is a partial to full bond between concrete slab and base layer or an unbonded, very stiff/thick base layer. If h_{eff} is the same or very close to the thickness of concrete slab, the interface between the concrete slab and base is behaving unbonded and/or there is a low stiffness base (e.g., aggregate base) present. If no base layer is present, then the h_{eff} represents the backcalculated thickness of the CRCP slab.

Figure 2, *a–c* present the *k* and *E*-values and h_{eff} determined from the traditional backcalculation procedure for various crack spacing (*L*), CRCP thickness, and LTE of 90% given the deflection basins from ILLI-SLAB. Figure 2*a* indicates that the backcalculated *k*-value varies significantly with changes in crack spacing and slab thickness. For a given thickness, the error between the input (100 psi/in.) and backcalculated value increases when crack spacing decreases and slab thickness increases, that is, for small L/ℓ values. Likewise the error between the input and backcalculated *E*-value (Figure 2*b*) follows similar trend as *k*-values. In both comparisons, errors are higher when L/ℓ decreases, which occurs when crack spacing decreases or slab thickness increases. For typical CRCP with 10-in. thickness and 6-ft crack spacing, errors between the actual and backcalculated *k* and *E*-values can be >50%.

Figure 2*c* demonstrates that the transverse crack spacing affects the accuracy of the backcalculated h_{eff}. The error between input thickness and h_{eff} increases with decreasing panel size and increasing panel thickness. The largest h_{eff} error reaches 15% at 6 ft crack spacing and 10 in. thickness and starts becoming significant for crack spacing <8 ft. For crack spacing >8 ft, the error in h_{eff} remains <7.6% for all analyzed thicknesses. The effective thickness error for 10 in. slab and 6 ft crack spacing is 15%, which is less than the error in backcalculated *k* and *E*-values. However, an error in thickness by 15% is worse than an error in *k* or *E*-value of 50% based on Westergaard's (*23*) interior slab stress (σ),

$$\sigma = \frac{0.316P}{h^2}\left[4\log\left(\frac{\ell}{\sqrt{1.6a^2+h^2}-0.675h}\right)+1.069\right] \quad (8)$$

given the change in stress for h_{eff}, *k*-value, and *E*-value being 22.4, 6.5, and 3.8%, respectively, indicating higher accuracy in h_{eff} is more important than accuracy in *k* and *E*-values.

Given an unacceptable error in h_{eff} for typical CRCP crack spacing and thicknesses with the traditional backcalculation procedure for rigid pavements as well as large errors for *k* and *E*-values, alternative equations are needed to consider different L/ℓ values present in-situ CRCP. Therefore, a revised CRCP backcalculation method for finite-sized panels with different slab aspect ratios is needed to accommodate in-situ CRCP responses especially if assessment of the composite behavior between the slab and base layer is desired.

CRCP Backcalculation Procedure Based on Finite-Sized Slabs

In order to limit the error introduced by the analysis assumption, an improved backcalculation method for CRCP was developed using the theoretical deflection basin data from the previous section. The relationship between *AREA* and the radius of relative stiffness and a/ℓ ratio and the non-dimensional deflection (Δ_{int}*) was defined for each crack spacing. Furthermore, the LTE across the cracks and FWD sensor configuration were also considered.

Relationship Between AREA and Radius of Relative Stiffness (ℓ)

The defined *AREA* equations with crack spacing ranging from 4 ft to 15 ft, based on Equation 1, are

$$AREA_{24} = 6\left(1+2\frac{d_{12}}{d_0}+\frac{d_{24}}{d_0}\right)(\text{For CS}<6\text{ft}) \quad (9)$$

$$AREA_{36} = 6\left(1+2\frac{d_{12}}{d_0}+2\frac{d_{24}}{d_0}+\frac{d_{36}}{d_0}\right) \quad (10)$$
$$(\text{For 6f} \le \text{CS} < 8\text{f}t)$$

$$AREA_{48} = 6\left(1+2\frac{d_{12}}{d_0}+2\frac{d_{24}}{d_0}+2\frac{d_{36}}{d_0}+\frac{d_{48}}{d_0}\right) \quad (11)$$
$$(\text{For 8ft} \le \text{CS} < 10\text{ftt})$$

$$AREA_{60} = 6\left(1+2\frac{d_{12}}{d_0}+2\frac{d_{24}}{d_0}+2\frac{d_{36}}{d_0}+2\frac{d_{48}}{d_0}+\frac{d_{60}}{d_0}\right) \quad (12)$$
$$(\text{For CS} \ge 10\text{ft})$$

for each slab size with the applicable number of sensors. Theoretical deflections were extracted for calculating *AREA* from sensors at 0 in. (d_0), 12 in. (d_{12}), 24 in. (d_{24}), 36 in. (d_{36}), 48 in. (d_{48}), 60 in. (d_{60}) offsets from the center of the loading plate.

Figure 3*a* presents the relationship between $AREA_{36}$ and ℓ for 6 ft × 12 ft slab with multiple-crack LTE as an example. This unique relationship for a given combination of crack spacing and crack LTE can now be statistically defined for all CRCP cases of interest.

Relationship Between a/ℓ and Non-Dimensional Deflection (Δ_{int})*

After determining ℓ with the appropriate *AREA* term (e.g., Figure 3*a*), Δ_{int}* can be calculated using Equation 5. Figure 3*b* shows an example of relationship between Δ_{int}* and a/ℓ for 6 ft × 12 ft slab with five-crack LTEs. The data can be statistically fitted to relate Δ_{int}* and a/ℓ given any combination of crack spacing and crack LTE.

Effect of Transverse Crack LTE on Backcalculation

Based on field investigations (*1,7*), LTE for CRCP is typically in the range of 80–100%, mostly around 90%, even

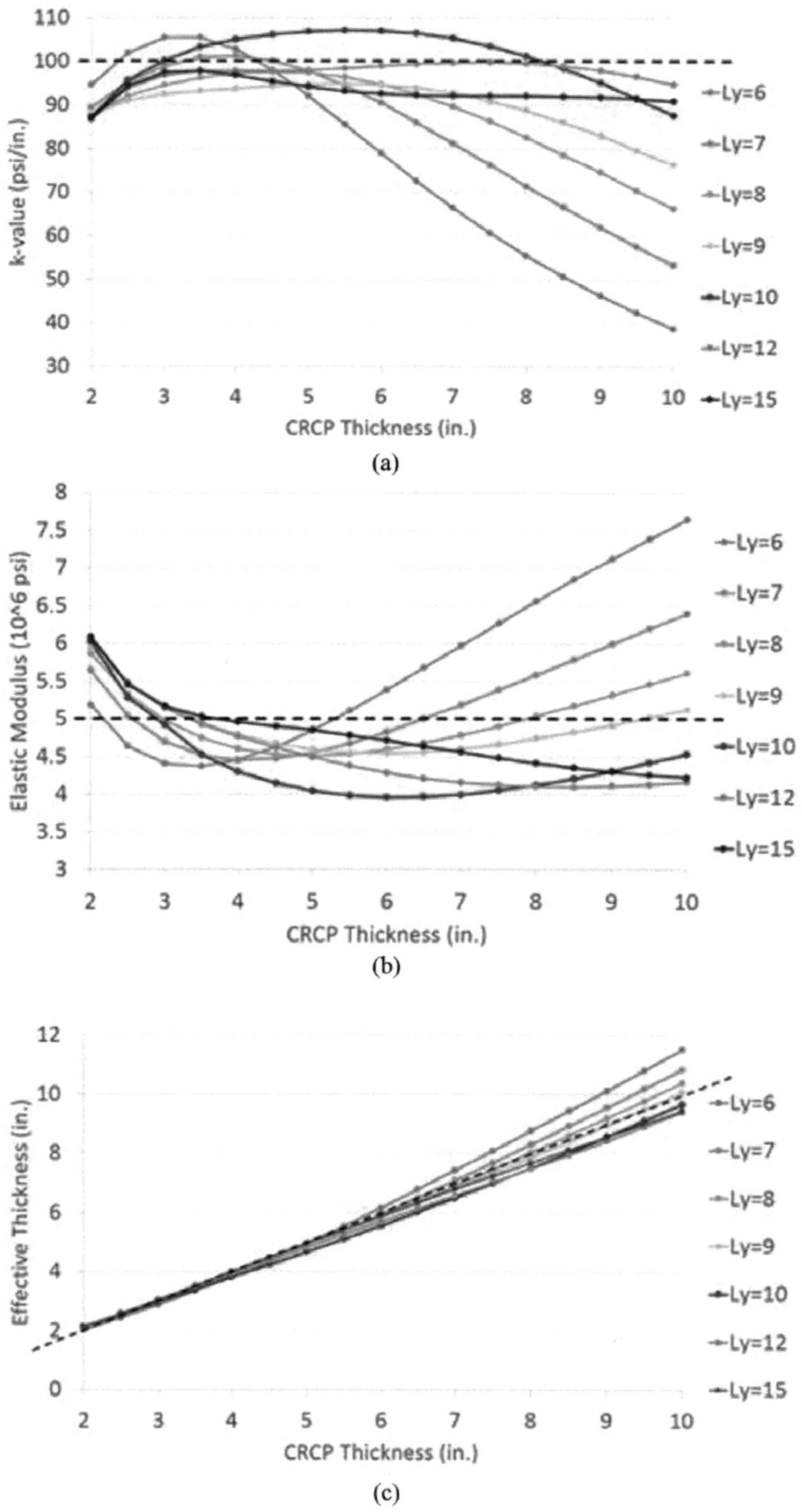

Figure 2. Comparison of traditional CRCP backcalculation process with infinite slab assumption for (*a*) *k*-value, (*b*) *E* and (*c*) h_{eff} given ILLI-SLAB deflections for different slab thickness and crack spacing (slab size) and fixed LTE = 90%.

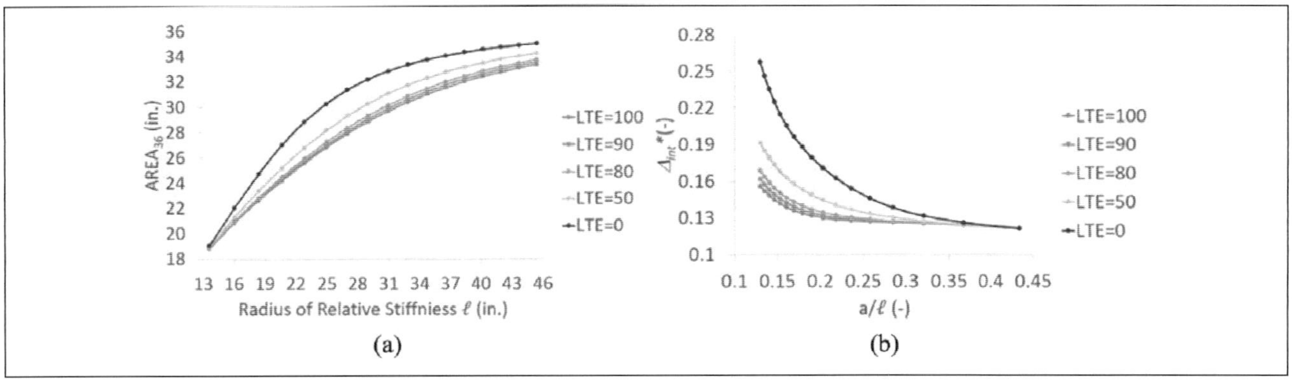

Figure 3. (*a*) $AREA_{36}$ versus radius of relative stiffness and (*b*) non-dimensional deflection (Δ_{int}*) versus a/ℓ for 6 ft × 12 ft slab.

Table 1. Percent Difference Errors in Backcalculated Results (h_{eff}, *k*-value, *E*) Compared with Input Values for 10 inch Thickness and Multiple Slab Sizes and LTE

	Load transfer efficiency (LTE)								
	100%			90%			80%		
Slab size (ft)	h_{eff}	*k*-value	*E*	h_{eff}	*k*-value	*E*	h_{eff}	*k*-value	*E*
15 × 12	0.15	1.9	0.44	0.10	0.24	0.29	0.020	1.5	0.059
12 × 12	0.42	1.6	1.3	0.61	1.4	1.8	1.0	4.9	3.2
10 × 12	0.047	2.2	0.14	0.27	2.1	0.80	0.90	7.1	2.7
9 × 12	0.54	2.4	1.6	0.20	2.6	0.61	0.42	8.3	1.3
8 × 12	1.1	3.1	3.2	0.70	2.6	2.1	0.059	9.1	0.18
7 × 12	1.1	5.0	3.1	0.53	1.3	1.6	0.26	8.3	0.78
6 × 12	1.6	7.4	4.7	0.67	0.14	2.0	0.35	7.3	1.1
5 × 12	2.0	8.6	6.0	0.68	0.40	2.0	0.50	7.6	1.5
4 × 12	2.9	10	8.5	0.71	0.18	2.1	0.99	9.0	3.0

after 20 years of service life. According to Figure 3, *a* and *b*, the relationship between *AREA* and ℓ, Δ_{int}* and a/ℓ does not change significantly when LTE is >80%. Table 1 lists errors in backcalculated h_{eff}, *k* and *E*-values for a range of CRCP panel size given a slab thickness of 10 in. and multiple-crack LTE values with predicted equations assuming 90% of LTE. As expected, the smaller the crack spacing, the greater the error. The influence of LTE over 80% on backcalculated results is limited, with the largest error being around 3, 10, and 8.5% for h_{eff}, *k* and *E*, respectively. Therefore, in order to simplify the procedure, the proposed equations for backcalculating ℓ and Δ_{int}* can be applied to in-situ crack LTE over 80% with all statistically derived relationships assuming transverse crack LTE = 90%.

Sensitivity of Sensor Configurations

As shown in Equations 9 to 12, the number of sensors applicable for various crack spacing is different. When analyzing an entire CRCP project with various crack spacing, requiring multiple sensor configurations increases the complexity of the procedure. Therefore, given an acceptable range of error, it is better to use the sensor configuration from AASHTO design guide (0, 12, 24, 36 in.) for CRCP. To assess the impact of sensor configuration, the percent difference in h_{eff}, *k*, and *E*-value was compared between $AREA_{36}$ and $AREA_{60}$ relative to the actual value (see Table 2) for crack spacing >6 ft. The percent difference between $AREA_{36}$ and $AREA_{24}$ for crack spacing <6 ft was also compared, listed in Table 3. According to backcalculated results, the sensor configuration from AASHTO Guide can be used to backcalculate CRCP at crack spacing ≥6 ft, given the greatest difference in h_{eff}, *k*-value, and *E* being <0.51, 3.3, and 1.5%, respectively, for ILLISLAB-generated deflections. For crack spacing <6 ft, $AREA_{24}$ should be used since the error in h_{eff} with $AREA_{36}$ cannot be neglected.

Finite-Sized Slab Backcalculation Method for CRCP

Based on the sensitivity results for transverse crack LTE and sensor configuration, statistical regression equations for predicting ℓ and Δ_{int}* were derived from theoretical (CRCP) FWD deflection basins to backcalculate more accurate

Table 2. Percent Difference Errors in Backcalculated Results (h_{eff}, k-Value, E) Compared with Input Values from $AREA_{36}$ and $AREA_{60}$ for Several Slab Sizes and 10 inch Slab Thickness

	Sensor configurations					
	$AREA_{36}$			$AREA_{60}$		
Slab size (ft)	h_{eff}	k-value	E	h_{eff}	k-value	E
15 × 12	0.12	0.31	0.37	0.10	0.24	0.29
12 × 12	0.47	1.0	1.4	0.61	1.4	1.8
10 × 12	0.31	2.2	0.94	0.27	2.1	0.80
9 × 12	0.076	3.3	0.23	0.20	2.6	0.61
8 × 12	0.51	3.1	1.5	0.70	2.6	2.1

Table 3. Percent Difference Errors in Backcalculated Results (h_{eff}, k-Value, E) Compared with Input Values from $AREA_{36}$ and $AREA_{24}$ for 10 inch Thickness and Multiple Slab Sizes

	Sensor configurations					
	$AREA_{36}$			$AREA_{24}$		
Slab size (ft)	h_{eff}	k-value	E	h_{eff}	k-value	E
5 × 12	4.0	18	12	0.68	0.40	2.0
4 × 12	2.8	2.7	8.2	0.71	0.18	2.1

pavement layer properties, that is, k-value, E-value and h_{eff}. For crack spacing ranging from 4 ft to 15 ft, the recommended fitted equations are summarized in Table 4 (ℓ versus $AREA_{36}$, ℓ versus $AREA_{24}$) and Table 5 (Δ_{int}* versus a/ℓ) with the statistical indicator (ADJ R^2) evaluating the accuracy of these fitted equations. Once ℓ and Δ_{int}* are calculated from Equations 8 to 10, then k-value and E-value or h_{eff} can be calculated from Equations 5 through 7.

Sensitivity of Plate Position on Backcalculated Properties

Transverse crack spacing on CRCP sections have been shown to follow a Weibull distribution (*5,24*) and, as a result, assuming an average transverse crack spacing and uniform drop pattern, for example, 100 ft, for FWD testing would result in the plate not being dropped in the middle of each CRCP slab panel. Therefore, the acceptable level of FWD plate offset from the middle of the panel needs to be determined with respect to the backcalculated CRCP properties given the set of finite-sized slab equations for CRCP. In order to perform this sensitivity, various plate offsets have been selected for multiple CRCP panel sizes given the formulas in Tables 4 and 5. Figure 4 shows the offsets configuration for a 15 ft × 12 ft slab given deflection basin measured only behind plate. Based on the improved CRCP backcalculation procedure for using h_{eff}, the influence of various FWD plate location offsets were evaluated relative to an exact center drop.

Table 4. Regression Coefficient for ℓ versus $AREA_{24}$ or $AREA_{36}$ for Multiple CRCP Slab Sizes

	Regression coefficient			
Slab size	A	B	C	ADJ R^2
[1]15 ft × 12 ft	2819	−2.976	4.812	0.9997
[1]12 ft × 12 ft	637.6	−1.805	3.699	0.9998
[1]10 ft × 12 ft	74.26	−0.3753	1.852	0.9994
[1]9 ft × 12 ft	44.39	−0.1513	1.365	0.9996
[1]8 ft × 12 ft	33.50	−0.06352	1.065	0.9997
[1]7 ft × 12 ft	30.68	−0.03551	0.9117	1
[1]6 ft × 12 ft	34.43	−0.03594	0.8874	0.9997
[2]5 ft × 12 ft	26.53	−0.1182	1.118	0.9996
[2]4 ft × 12 ft	69.44	−0.3424	1.432	0.9995

With general equation form in Equations 13 and 14:

$$^{1}\ell_{36} = \left(\frac{\ln\left(\frac{36 - AREA_{36}}{A} \right)}{B} \right)^{C}, AREA_{36} = 6\left(1 + 2\frac{d_{12}}{d_0} + 2\frac{d_{244}}{d_0} + \frac{d_{36}}{d_0} \right) \quad (13)$$

$$^{2}\ell_{24} = \left(\frac{\ln\left(\frac{24 - AREA_{24}}{A} \right)}{B} \right)^{C}, AREA_{24} = 6\left(1 + 2\frac{d_{12}}{d_0} + \frac{d_{24}}{d_0} \right) \quad (14)$$

Table 5. Regression Coefficient for Δ_{int}* Versus a/ℓ for Multiple CRCP Slab Sizes

	Regression coefficient			
Slab size	A	B	C	ADJ R^2
15 ft × 12 ft	0.06840	6.744	0.1130	0.9999
12 ft × 12 ft	0.05906	2.954	0.09949	0.9939
10 ft × 12 ft	0.09264	1.040	0.05721	0.9864
9 ft × 12 ft	0.07014	1.461	0.08018	0.9740
8 ft × 12 ft	0.05260	5.651	0.1153	0.9349
7 ft × 12 ft	0.2253	16.95	0.1235	0.9506
6 ft × 12 ft	0.6071	21.56	0.1240	0.9908
5 ft × 12 ft	0.7299	19.60	0.1226	0.9995
4 ft × 12 ft	0.7290	16.08	0.1200	0.9998

General equation form in Equation 15:

$$\Delta_{int}^* = A \cdot \exp\left(-B \cdot \frac{a}{\ell} \right) + C \quad (15)$$

Figure 5, *a* and *b* provide two examples of backcalculated h_{eff} for crack spacing of 15 ft and 8 ft with multiple plate offsets from the center of slab in two directions, given positive and negative offsets representing plate closer to the approach and leave side of crack, respectively. This allows for analysis of plate drops occurring anywhere longitudinally between two transverse cracks. As expected, the difference between the backcalculated CRCP thickness and input thickness varied more significantly for greater plate offsets, shorter slab sizes,

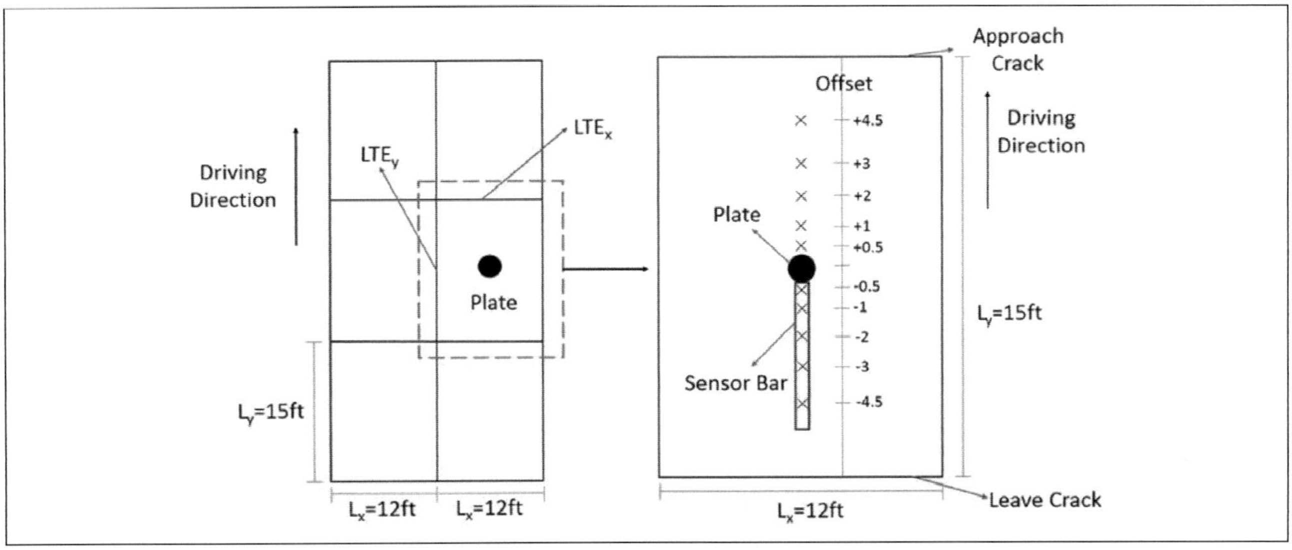

Figure 4. Plate offset configuration for 15 ft by 12 ft slab given deflection basin measured only behind plate.

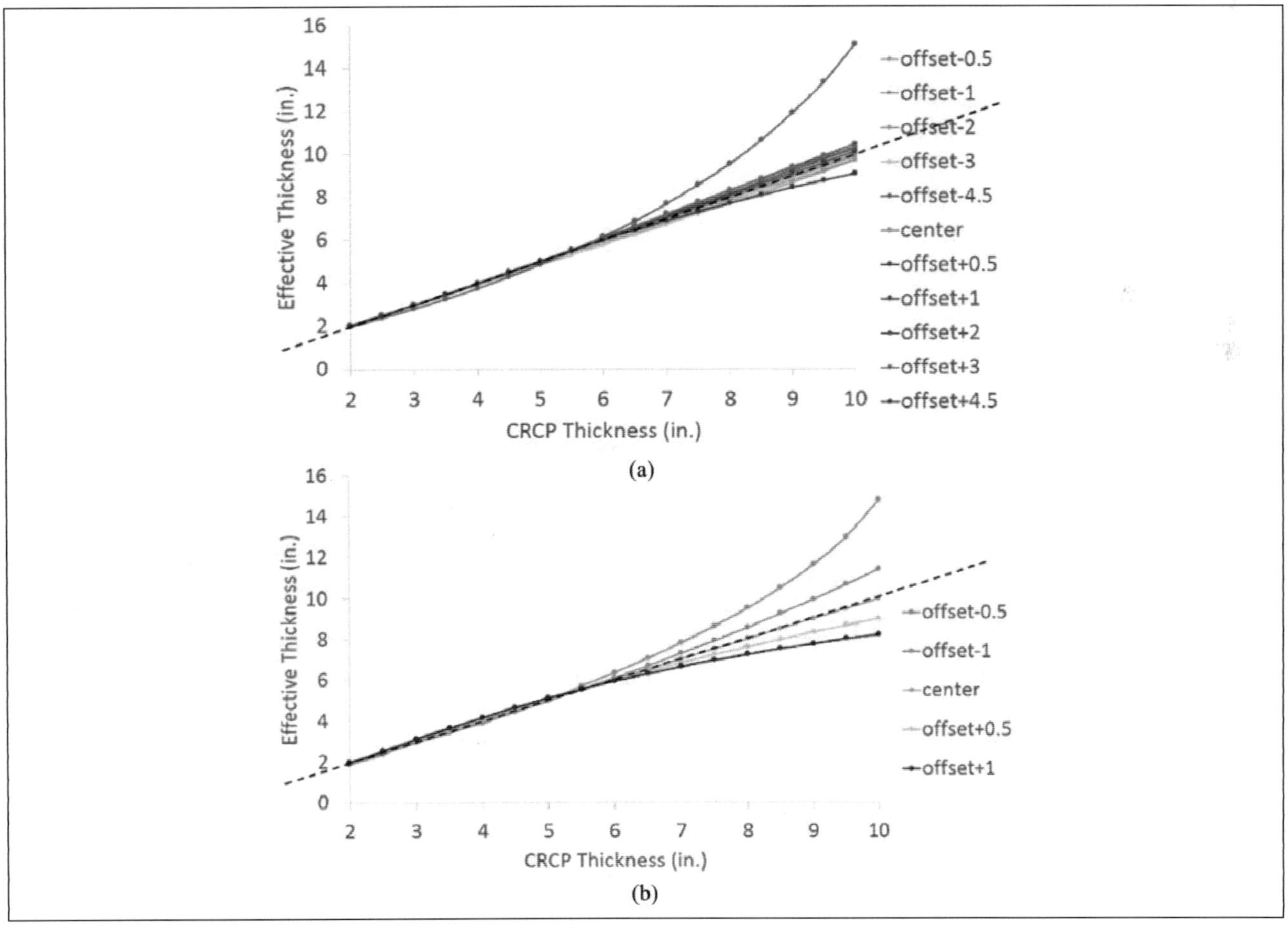

Figure 5. Backcalculated h_{eff} at several plate offset locations for (*a*) 15 ft × 12 ft and (*b*) 8 ft × 12 ft slab where positive offset means plate is closer to the approach side of crack and negative offset means plate is closer to the leave side of crack.

Table 6. Percent Error in h_{eff} Between Center Plate Drop and Plate Offsets (ft)

Crack spacing	Designed thickness	Plate offset (negative offset: plate closer to leave crack; positive offset: plate closer to the approach)										
		−4.5	−3	−2	−1	−0.5	0	0.5	1	2	3	4.5
6 ft	9 in.					36	0	20				
	9.5 in.					47	0	22				
	10 in.					69	0	23				
8 ft	9 in.				29	10	0	7.4	14			
	9.5 in.				37	12	0	8.6	16			
	10 in.				49	15	0	9.7	17			
10 ft	9 in.			33	7.2	2.6	0	1.8	3.7	9.8		
	9.5 in.			43	9.5	3.6	0	2.7	5.2	12		
	10 in.			56	12	4.7	0	3.5	6.8	15		
12 ft	9 in.		34	5.7	0.21	0.50	0	0.80	1.5	0.61	6.1	
	9.5 in.		43	8.5	0.79	0.040	0	0.41	0.73	0.95	8.4	
	10 in.		56	12	1.9	0.50	0	0.013	0.20	2.7	11	
15 ft	9 in.	32	1.0	3.6	2.5	1.3	0	1.4	2.6	4.5	4.0	6.2
	9.5 in.	41	0.36	3.2	2.4	1.3	0	1.4	2.7	4.6	3.8	7.6
	10 in.	51	1.9	2.7	2.3	1.2	0	1.4	2.8	4.5	3.2	9.0

Table 7. Summary of Illinois Tollway CRCP Test Sections

Section	IE	ID	IC	IB	IA
Slab thickness (in.)	10.5	10.5	10.5	10.5	10.5
Base type	2 in.WMA + 4 in. CTB	2 in.WMA + 4 in. CTB	2 in.WMA + 4 in. CTB	Microsurfacing + 6 in. CTB	Microsurfacing + 6 in. CTB
Average d_0 (10^{-3} in.)	1.83	1.87	1.72	1.65	1.62
Average k-value (psi/in.)	466	449	408	433	517
Average E (10^6 psi)	10.6	10.8	13.3	13.6	12.0
Average h_{eff} (in.)	13.5	13.5	14.5	14.6	14.0
LTE (%)	89.6	90.3	90.7	89.8	89.4

Note: WMA = warm-mixed asphalt; CTB = cement-treated base.

and larger slab thickness (decreased L/ℓ). When the CRCP thickness was <6 in., plate offset had little impact on h_{eff}; however, when thickness was >9 in., plate offsets could have a significant impact on h_{eff} depending on the magnitude of offset.

Table 6 summarizes the error between the backcalculated h_{eff} for center-loaded CRCP panels and panels with the plate offset from center for crack spacing of 8, 10, 12 and 15 ft. For a smaller crack spacing, for example, ≤8 ft, it is important to make sure the plate drops at the center of the slab panel during FWD testing in order to guarantee the accuracy of the backcalculated CRCP properties. For a larger crack spacing, 12 or 15 ft, dropping at small offsets (several feet) does not have a strong impact on the backcalculated results.

Field Application of Improved CRCP Backcalculation Procedure

This improved backcalculation method for CRCP was applied to a CRCP highway near Chicago, constructed in August 2016. The CRCP test roadway was divided into five sections with different base types and steel contents as seen in Table 7. The design thickness of concrete slab and base are 10.5 and 6 in., respectively. The FWD tests were conducted 10 months after the roadway construction and an observer assisted during testing to minimize any plate offset form the center of the panel. Average measured d_0 and transverse crack LTE as well as the average backcalculated k-value, E, and h_{eff} given improved method are presented in Table 7. The concrete stiffness is assumed to be 5×10^6 psi for the backcalculation of h_{eff}. Figure 6 presents the backcalculated k-value, E, and h_{eff} along with crack spacing (ft) for each location in the CRCP section given average drop distance being approximately 40 ft.

Figure 6*a* shows that backcalculated k-value with traditional method is lower than the results with the improved method, which is consistent with the theoretical results described in Figure 2*a*. For E-value, the backcalculated results with the traditional method are lower than those with the improved method given crack spacing higher than 9 ft shown in Figure 6*b*, similar to the trend in Figure 2*b*. Overall,

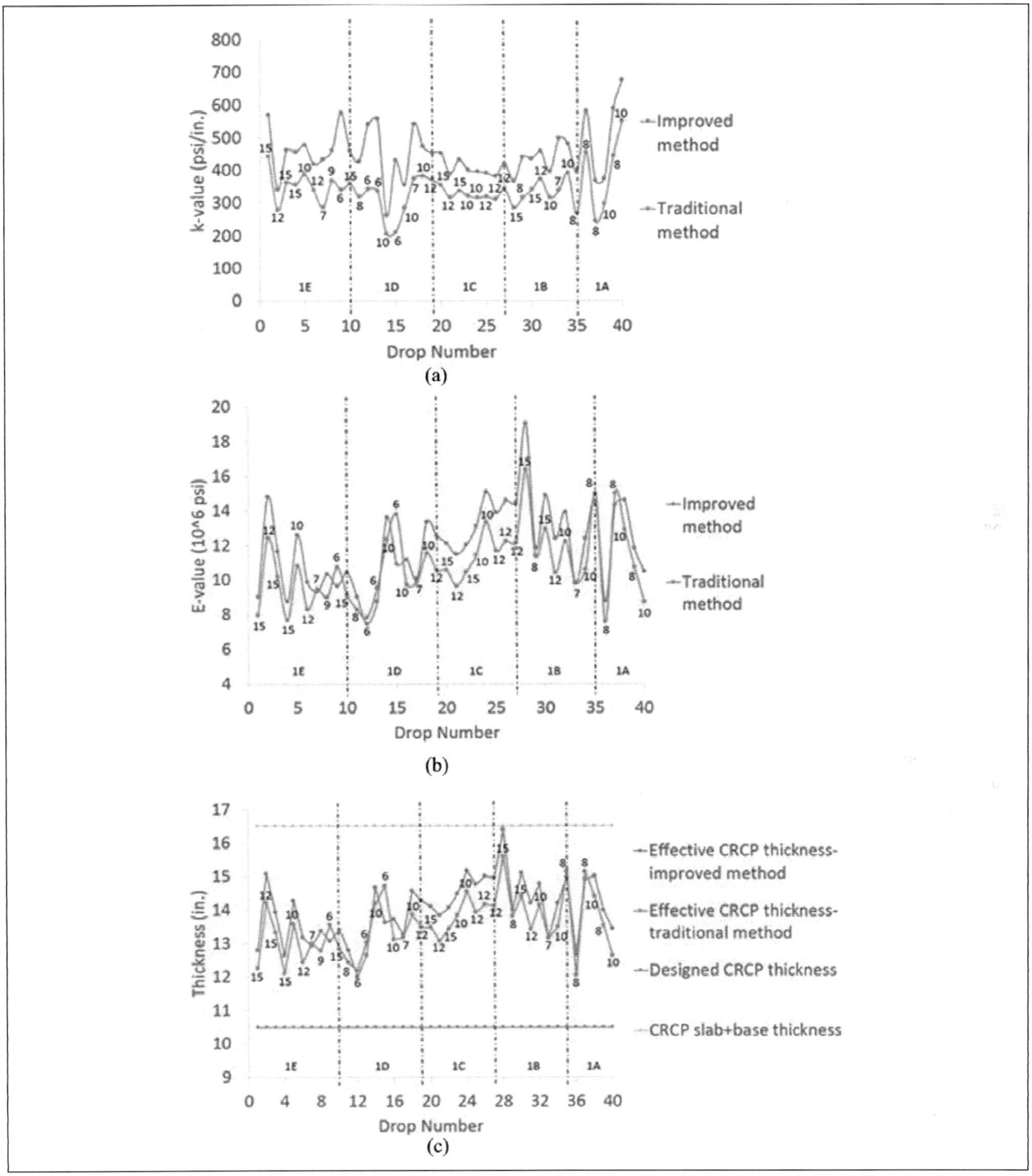

Figure 6. CRCP (*a*) *k*-value, (*b*) *E* and (*c*) effective slab thickness along the test section with crack spacing (ft) of each drop being marked.

the average h_{eff} is around 3–4 in. higher than designed thickness of the concrete slab according to average values in Table 7. Given the base layers are not the same stiffness as the CRCP, the effective thickness is not expected to reach 16.5 in. even with full slab–base interface bond. Both base types evaluated demonstrated partial bond with the CRCP. There is not any huge difference in h_{eff} from two methods shown in Figure 6*c* given an average difference being 0.57 in. This is because the range of crack spacing in Section 1 for FWD test was between 6 ft and 15 ft, mostly around 10–12 ft with a mean value of 10.5 ft. The difference in h_{eff} with traditional and improved backcalculation only becomes significant when crack spacing is <8 ft according to the feasibility study of traditional method in Figure 2*c*.

There are peaks and valleys in the backcalculated values shown in Figure 6*c* when viewing all drop locations, which can indicate variation in the slab thickness, base thickness, and/or slab–base interface condition. Given a large fluctuation from one location to another, this can be caused by longitudinal plate offset from the center. The plate offset during this FWD test was controlled to be <±0.5 ft, which can cause error in h_{eff} ranging from 0.013% to 69% given different crack spacing for a 10 in. slab (see Table 6). Therefore, the control of longitudinal plate offset plays an important role in achieving accurate backcalculation results. Despite these variations, backcalculated *k*-value, *E*, and h_{eff} reported in Table 7 and Figure 6 are within reasonable ranges, suggesting the improved backcalculation procedure for CRCP provides enhanced capabilities compared with the traditional method.

Conclusions

The traditional backcalculation procedure for rigid pavements was shown to have limitations when applied to CRCP because of the assumption of an infinite slab condition. This assumption was especially problematic when backcalculating effective thickness, which can be used to evaluate the slab–base interface condition of the CRCP structure. An improved backcalculation method for CRCP was proposed for finite-sized slabs with various slab aspect ratios and non-dimensional slab sizes (L/ℓ). The unique relation of ℓ and *AREA* as well as Δ_{int}* and a/ℓ were developed for multiple CRCP crack spacing. The theoretical analysis demonstrated that LTE had negligible influence on backcalculation results when it was >80%, giving the largest error of 3, 10 and 8.5% for h_{eff}, *k*-value and *E*, respectively. Similarly, the sensor configuration had limited influence on backcalculated h_{eff}, *k*, and *E*-values given $AREA_{48}$ versus $AREA_{36}$ and $AREA_{60}$ versus $AREA_{36}$ for panel sizes >6 ft. Therefore, an average LTE of 90% was recommended for CRCP backcalculation with the standard AASHTO sensor configuration of 0, 12, 24, 36 in. for CRCP crack spacing ≥6 ft. For crack spacing of 4 and 5 ft, a backcalculation procedure with three sensors (0, 12, 24 in.) was proposed. A sensitivity study for plate location was conducted to investigate the impact of longitudinal plate offset from the center of the panel on h_{eff}. The theoretical analysis demonstrated that it is important to drop the plate at the center of the CRCP panel given smaller crack spacing (≤8 ft) to achieve accurate backcalculation results compared with larger crack spacing (12, 15 ft), for which longitudinal plate offsets by a few feet did not have significant effect on the backcalculated properties. The new backcalculation method was applied to a test section on the Illinois Tollway. Overall, there is a good frictional bond condition in the CRCP test section with the h_{eff} 3–4 in. over the design CRCP thickness.

Acknowledgments

This publication is partially based on the results of Illinois State Toll Highway Authority Project RR-14-9168, *Innovative Structural and Material Design for Continuously Reinforced Concrete Pavement* and from financial support of the *China Scholarship Council*. The authors would like to acknowledge the support of Steve Gillen from the Illinois State Toll Highway Authority and Dan Gancarz of Applied Research Associates.

References

1. Tayabji, S. D., P. J. Stephanos, J. S. Gagnon, and D. G. Zollinger. *Performance of Continuously Reinforced Concrete Pavements Volume II: Field Investigations of CRC Pavements*. FHWA-RD-94-179. PCS/Law Engineering, Beltsville, Maryland, 1998.
2. Tayabji, S. D., D. G. Zollinger, J. R. Vederey, and J. S. Gagnon. *Performance of Continuously Reinforced Concrete Pavements Volume III: Analysis and Evaluation of Field Test Data*. FHWA-RD-94-180. PCS/Law Engineering, Beltsville, Maryland, 1998.
3. Li, S., and T. D. White. Falling-Weight Deflectometer Sensor Location in the Backcalculation of Concrete Pavement Moduli. *Journal of Testing and Evaluation*, Vol. 28, No. 3, 2000, pp. 166–175.
4. Khazanovich, L., S. D. Tayabji, and M. I. Darter. *Backcalculation of Layer Parameters for LTPP Test Sections, Volume I: Slab on Elastic Solid and Slab on Dense-Liquid Foundation Analysis of Rigid Pavements*. FHWA-RD-00-086. ERES Consultants, Columbia, Maryland, 2001.
5. Kohler, E., and J. Roesler. Crack Spacing and Crack Width Investigation from Experimental CRCP Sections. *International Journal of Pavement Engineering*, Vol. 7, No. 4, 2006, pp. 331–340.
6. Roesler, J., J. Huntley, and A. Amirkhanian. Performance of Continuously Reinforced Concrete Pavement Containing Recycled Concrete Aggregates. *Transportation Research Record: Journal of the Transportation Research Board*, 2011. 2253: 32–39.
7. Choi, P., D. Kim, B. Lee, and M. C. Won. Evaluation of Structural Responses of Continuously Reinforced Concrete Pavement (CRCP) using Falling Weight Deflectometer. *Canadian Journal of Civil Engineering*, Vol. 43, No. 1, 2016, pp. 28–39.
8. American Association of State Highway and Transportation Officials. *AASHTO Guide for Design of Pavement Structures*. AASHTO, Washington, D.C., 1993.
9. Ioannides, A. M., M. R. Thompson, and E. J. Barenberg. Westergaard Solutions Reconsidered. *Transportation Research*

Record: Journal of the Transportation Research Board, 1985: 1043: 13–23.

10. Crovetti, J. A. *Design and Evaluation of Jointed Concrete Pavement Systems Incorporating Free Draining Base Layers*. PhD dissertation. University of Illinois, Urbana-Champaign, 1994.
11. King, D., and J. R. Roesler. Backcalculation Procedure for Bonded Concrete Overlays of Asphalt Pavement. *Transportation Research Record: Journal of the Transportation Research Board*, 2014. 2457: 72–79.
12. Hoffman, M. S., and M. R. Thompson. *Mechanistic Interpretation of Nondestructive Pavement Testing Deflections*. UILU-ENG-81-2010. University of Illinois, Urbana-Champaign, 1981.
13Ioannides, A. M. Dimensional Analysis in NDT Rigid Pavement Evaluation. *Journal of Transportation Engineering*, Vol. 116, No. 1, 1990, pp. 23–36.
14Strategic Highway Research Program. *Strategic Highway Research Program: Layer Moduli Backcalculation Procedure-Software Selection*. Strategic Highway Research Program, Washington D. C., 1991.
15. Hall, K. T., M. I. Darter, T. E. Hoerner, and L. Khazanovich. *LTPP Data Analysis Phase I: Validation of Guidelines for k-Value Selection and Concrete Pavement Performance Prediction*. FHWA-RD-96-198. ERES Consultants, Champaign, IL, 1997.
16. Hall, K. T. *Backcalculation Solution for Concrete Pavements. Technical Memo Prepared for SHRP Contract P-020, Long-Term Pavement Performance Data Analysis*. National Research Council, Washington, D.C., 1992.
17. Westergaard, H. M. Stresses in Concrete Runways of Airports. *Highway Research Board Proceedings*, Vol. 19, 1939, pp. 197–202.
18. Ioannides, A. M., E. J. Barenberg, and J. A. Lary. Interpretation of Falling Weight Deflectometer Results Using Principles of Dimensional Analysis. *Proc., 4th International Conference on Concrete Pavement Design and Rehabilitation*, Purdue University, 1989, pp. 231–247.
19.McGhee, K. H. Experience with Continuously Reinforced Concrete Pavements in Virginia. *Transportation Research Record: Journal of the Transportation Research Board*, 1974. 485: 14–24.
20. Tabatabaie, A. M., and E. J. Barenberg. Finite Element Analysis of Jointed or Cracked Concrete Pavements. *Transportation Research Record: Journal of the Transportation Research Board*, 1978. 671: 11–18.
21. Khazanovich, L. *Structural Analysis of Multi-Layered Concrete Pavement Systems*. PhD dissertation. University of Illinois, Urbana-Champaign, 1994.
22. Ioannides, A. M., L. Khazanovich, and J. L. Becque. Structural Evaluation of Base Layers in Concrete Pavement Systems. *Transportation Research Record: Journal of the Transportation Research Board*, 1992. 1370: 20–28.
23Westergaard, H. M. Stresses in Concrete Pavements Computed by Theoretical Analysis. *Public Roads*, Vol. 7, 1926, pp. 25–35.
24.Selezneva, O., M. Darter, D. Zollinger, and S. Shoukry. Characterization of Transverse Cracking Spatial Variability: Use of Long-Term Pavement Performance Data for Continuously Reinforced Concrete Pavements Design. *Transportation Research Record: Journal of the Transportation Research Board*, 2003. 1849: 147–155.

The Standing Committee on Pavement Structural Modeling and Evaluation peer-reviewed this paper (18-02277).

The contents of this paper reflect the view of the authors, who are responsible for the facts and accuracy of the data presented here.

Article

TRR
JOURNAL OF THE TRANSPORTATION RESEARCH BOARD

Transportation Research Record
2018, Vol. 2672(40) 348–359

DOI: 10.1177/0361198118758675
journals.sagepub.com/home/trr
SAGE

Application of Traffic Speed Deflectometer for Network-Level Pavement Management

Shivesh Shrestha[1], Samer W. Katicha[1], Gerardo W. Flintsch[1], and Senthilmurugan Thyagarajan[2]

Abstract

In this paper, the traffic speed deflectometer (TSD), a device used for network level structural evaluation, is assessed. TSD testing was performed in nine states on a total of 5,928 miles (some repeated) during three time periods: November 2013, May to July 2014, and June to September 2015. This paper presents (1) the results of repeatability and comparison of the TSD with the falling weight deflectometer (FWD), (2) the results of the comparison of TSD measurements with typical pavement management system (PMS) data, and (3) an approach that can be implemented by State Highway Agencies (SHAs) to incorporate indices derived from TSD data into their PMS decision-making process. The results show that repeated TSD measurements follow similar trends and the TSD measurements and FWD measurements on the same pavement sections follow similar trends as well. Comparing TSD measurements with PMS surface condition data confirmed that the TSD provided valuable information about the structural condition of the tested pavement sections that cannot be derived from the already available pavement surface condition as part of an agency's PMS. An example of how TSD information can be used to refine the triggered maintenance treatment category as part of a network-level PMS analysis is presented for a roughly 75-mile section of I-81 south in Virginia.

Most of the State Highway Agencies (SHAs) rely on a surface condition index to summarize the overall condition of their pavements and assess rehabilitation needs. The surface condition index does not fully take into account the structural condition of the pavement, which is an important driving cause of pavement deterioration. This can cause less than optimal treatment selection at the network level as it has been shown that the correlation between the surface condition and structural measurements of pavement response is weak (*1, 2*) and the rate of deterioration of pavement sections is affected by the structural condition (*3*). Therefore, incorporating the pavement structural condition along with the pavement surface condition in the pavement management decision-making process could lead to better-informed decisions, more cost-effective pavement rehabilitation, and preservation strategies (*4–7*).

Some SHAs have started using deflection measurements from the falling weight deflectometer (FWD) to enhance their network level pavement management processes (*8, 9*). The FWD is a stationary deflection-measuring device that requires traffic control and can only take stationary measurements at discrete points limiting the density of points where data are collected. This makes the device inefficient for network level data collection. The Second Strategic Highway Research Program 2 (SHRP 2) R06(F) project investigated most of the deflection measuring devices and identified the TSD as a device that can effectively measure the structural condition of the pavements at the network level (*10*). The Federal Highway Administration (FHWA) initiated a pooled fund study to carry out a field demonstration of the traffic speed deflectometer (TSD) in the nine participating SHAs: California Department of Transportation (Caltrans), Georgia Department of Transportation (DOT), Idaho DOT, Illinois DOT, Nevada DOT, New York DOT, Pennsylvania DOT, South Carolina DOT, and Virginia DOT. This paper provides an example of how the results of TSD testing could be implemented within a pavement management system (PMS).

Objective

The objective of this paper is to evaluate the TSD and show how information derived from TSD measurements can be

[1]Center for Sustainable Transportation Infrastructure, Virginia Tech Transportation Institute, Blacksburg, VA
[2]ESC Inc., Turner Fairbank Highway Research Center, McLean, VA

Corresponding Author:
Address correspondence to Shivesh Shrestha: shivesh@vt.edu.

implemented in a PMS. This is achieved in three tasks. The first task evaluates the repeatability of the TSD and shows how TSD measurements compare with FWD measurements. The second task shows that the information derived from the TSD is different from the typical information recorded from a PMS. The third task shows an example of how TSD information can be implemented in a PMS.

Approach

The paper has three main sections. The first section presents an overview of the TSD testing performed in the nine SHAs that participated in the pooled fund study and the processing of TSD measurements. The second section presents the results of the evaluation of the technology and the last one provides an example on how the information collected with the TSD can be used to enhance pavement management.

TSD Data Collection and Processing

This section presents details on the participating states, the total number tested miles in each state, the month and year of testing, and the data processing approach.

The TSD data are first processed to calculate deflection indices from the deflection slopes measured by the TSD. The deflection indices are then corrected to a reference temperature. The calculated indices presented in this paper are the surface curvature index (SCI300) and the effective structural number (SN_{eff}). The SCI300 was temperature corrected using the method developed by Rada et al. *(11)* whereas for SN_{eff} the approach described in Lukanen et al. *(12)* which corrects for center deflection is used. SCI300 thresholds to classify the pavement condition into good, fair, and poor were then determined based on a large database of simulated pavement sections by Rada et al. *(11)*.

TSD Evaluation

The TSD was evaluated with respect to two aspects:

1. *TSD repeatability and comparison with FWD measurements*: Short-term repeatability is presented for measurements performed on the same pavement in consecutive days and long-term repeatability is presented for measurements performed on the same section in consecutive years. Comparison with FWD is performed on more than 70 miles of I-81 in Virginia.
2. *Comparison of TSD measurements with data from a PMS*: Fatigue cracking on interstate sections in Virginia is compared with TSD SCI300. Also, the SN_{eff} from the Pennsylvania Department of Transportation (PennDOT) PMS is compared with SN_{eff} calculated from TSD measurements. This comparison is informative as the SN_{eff} from the PMS is based on layer thicknesses, layer coefficients, and layer age whereas the SN_{eff} from the TSD represents actual measured field structural condition.

Implementation

The last part of the effort presented a possible approach that can be used to implement TSD information into the PMS. The approach presented is a simplified version based on the Virginia Department of Transportation (VDOT) approach that incorporated FWD measurements into the PMS decision process for interstate roads. The main feature of the approach is that the treatment selection is first obtained based on the functional condition (surface condition) without considering the structural condition. The information from the structural condition is then used to adjust the selected treatment. For example, the adjustment could be to use a lighter treatment, keep the same treatment, or use a heavier treatment depending on whether the structural condition is good, fair, or poor. The main feature of the approach is that it modified the tried and trusted decision process of a SHA currently used with the data currently available in its PMS.

TSD Data Collection and Processing

Nine SHAs participated in the pooled fund study, with testing performed over three years: in November 2013, between May and July in 2014, and between June and September in 2015. Two data collection cycles were performed in all SHAs except Idaho and Virginia, which joined the pooled fund during the 2015 data collection cycle. In total, 5,928 lane miles (with some repeated) of testing were performed. The details for each state are shown in Table 1.

Calculation of the Deflection Indices

The TSD measures the deflection slope of the deflection basin rather than pavement deflection. The deflection can be obtained from the deflection slope by integration as follows:

$$d(x) = \int_{x}^{\infty} s(y)\,dy \tag{1}$$

where $s(y)$ is the slope at distance y measured from the applied load and $d(x)$ is the deflection at distance x measured from the applied load.

Greenwood Engineering, the manufacturer of TSD and who provided data collection services for this study, uses a parameterized model for the shape of the deflection slope developed by Pedersen (*13*) to obtain deflections from the deflection slope by optimizing the model parameters to fit the deflection slope data. The calculated deflections provided by Greenwood Engineering were normalized with the measured dynamic load during testing to a lead of 9,000 lb. Structural indices were then calculated from the normalized

Table 1. Participating States and Total Tested Lane Miles

Participating SHA	Testing month and year	Tested lane miles
Caltrans	June 2014 and August 2015	980
Georgia DOT	May 2014 and July 2015	646
Idaho Transportation Department	September 2015	1,040
Illinois DOT	June 2014 and September 2015	400
Nevada DOT	June 2014 and August 2015	352
New York DOT	November 2013 and July 2014	595
Pennsylvania DOT	July 2014 and June 2015	567
South Carolina DOT	May 2014 and July 2015	726
Virginia DOT	June 2015	622
Total		5,928

deflections. The SCI300 and SN_{eff} are used in this paper. The SCI300 is calculated as follows:

$$SCI300 = D0 - D300 \tag{2}$$

where $D0$ is the deflection at the point of load application (mid-point between the dual tires) and $D300$ is the deflection 11.81in. (300 mm) away from the point of load application.

The SN_{eff} is calculated using the method of Rohde (*14*) as follows:

1. Determine the structure index SIP of the pavement,

$$SIP = D0 - D_{1.5Hp} \tag{3}$$

where $D_{1.5Hp}$ is the deflection at a lateral distance of 1.5 times the pavement depth, H_p, the total thickness of all layers above the subgrade.

2. Determine the existing pavement SN_{eff} as

$$SN_{eff} = k_1 SIP^{k_2} H_p^{k_3} \tag{4}$$

where, for asphalt pavements, $k_1 = 0.4728$, $k_2 = -0.4810$, and $k_3 = 0.7581$. $D0$ used in the calculation was corrected to a reference temperature of 68°F using the procedure described in Lukanen et al. (*12*).

Temperature Correction of SCI300

The temperature-correction procedure used was developed by Rada et al. (*11*). The procedure corrects for the strain at the bottom of the asphalt. The strain is estimated using

$$\varepsilon = a(SCI300)^b \tag{5}$$

where the parameters a and b are given in Table 2.

Once the strain is estimated, it is temperature corrected using a procedure that relates the dynamic modulus to the strain. The steps for this procedure are as follows (*11*):

Table 2. Relationship between SCI300 and Maximum Tensile Strain at the Bottom of the Asphalt Layer

Asphalt concrete layer thickness (in.)	Parameter	
	a	*b*
3–4	52.438	0.9620
4–5	50.814	1.0200
5–6	53.725	1.0240
6–7	59.704	0.9870
7–8	62.539	0.9520
8–9	64.595	0.9120
9–10	65.645	0.8820
10–11	65.656	0.8373
11–12	64.639	0.8103
12–13	63.058	0.7895
13–14	60.592	0.7479
14–15	58.494	0.7594
15–16	55.386	0.7285
3–6 (thin)	57.818	0.9270
6–9 (medium)	63.202	0.9350
9–16 (thick)	62.538	0.8412

1. Compute the asphalt layer dynamic modulus at the test temperature, E_f, based on the calculated strain (from SCI300 using Equation 5) using the following equation:

$$E_f = c \times \varepsilon^d \tag{6}$$

where c and d are model parameters that depend on the asphalt layer thickness. When the thickness is not known, default values are provided.

2. Compute a temperature-correction factor, T_c, for the dynamic modulus as follows:

$$T_c = 19.791\left(e^{-0.043T_r} - e^{-0.043T_f}\right) \tag{7}$$

where T_r is the reference temperature (typically 70°F) and T_f is the asphalt temperature during the test.

Table 3. Thresholds for SCI300 (TSD)

Road category	Asphalt concrete layer thickness (in.)	Annual traffic (million ESAL)	Threshold for fatigue cracking at wheelpath (%)	Threshold for poor		Threshold for fair	
				N_f (million ESAL)	SCI300 (mil)	N_f (million ESAL)	SCI300 (mil)
Interstate	> 9	1.4	10	2.8	3.7	7.0	2.7
Primary	6–9	0.2	10	0.4	6.2	1.0	4.9
Secondary	3–6	0.07	45	0.14	9.7	0.35	7.3

3. Compute the dynamic modulus, E_r, at the selected reference temperature as follows:

$$E_r = \frac{E_f}{1-T_c}. \quad (8)$$

4. Compute the strain, ε_r, at the selected reference temperature by rearranging Equation 6 as follows:

$$\varepsilon_r = \left(\frac{E_f}{c}\right)^{\frac{1}{d}}. \quad (9)$$

5. Calculate the temperature-corrected TSD index using the inverse of Equation 5.

The asphalt temperature T_f is taken as the mid-depth temperature and calculated from the measured surface temperature using the BELLS3 equation (*12*):

$$T_f = 0.95 + 0.892 \times IR + \left[\log(d) - 2.65\right] \times \left[\begin{array}{l}-0.448 \times \mathrm{IR} + 0.621 \times (1\mathrm{day}) + \\ 1.883\sin(\mathrm{hr}18 - 15)\end{array}\right] + 0.042\mathrm{IR}\ \sin(\mathrm{hr}18 - 13.5) \quad (10)$$

where
T_f = pavement temperature at mid-depth d, °C,
IR = pavement surface temperature, °C,
log = base 10 logarithm,
d = mid-depth of the asphalt concrete layer, inches (in.),
1 day = average air temperature the day before testing, °C,
sin = sine function on an 18-hr clock system, with 2π radians equal to one 18-hr cycle,
hr18 = time of day, in a 24-hr clock system, but calculated using an 18-hr asphalt concrete temperature rise-and-fall time cycle.

The previous day's average air temperature was obtained at the closest weather station from the National Center for Environmental Information weather site, https://gis.ncdc.noaa.gov, and used in the BELLS3 equation to calculate the mid-depth temperature. The computed mid-depth temperature was used with the temperature-correction procedure.

Determination of Default Thresholds for Pavement Structural Condition Classification

A methodology based on the number of remaining equivalent single axle loads (ESALs) was used to arrive at a preliminary estimate for the threshold between good/fair and fair/poor segments. The remaining ESAL thresholds used in the report are only for illustrative purposes and it is expected that they will be revised based on the experience gained from TSD implementation efforts by individual SHAs.

Thresholds were obtained for three road categories—interstate, primary, and secondary roads— based on asphalt layer thickness as shown in Table 3. The thresholds given in Table 3 are based on a large database of simulated pavement sections by Rada et al. (*11*). The responses from these simulated pavements were used to determine the number of load repetitions to failure, N_f, computed using the Asphalt Institute equation (*15*):

$$N_f = C \times 0.00432 \left(\frac{1}{\varepsilon_t}\right)^{3.291} \left(\frac{1}{E}\right)^{0.854}. \quad (11)$$

where C is the calibration coefficient, ε_t is the tensile strain at the bottom of the asphalt layer (load corresponding to 9,000 lbf loaded dual-tire configuration with 13.5-in tire spacing and 116 psi tire pressure), and E is the stiffness of asphalt mixture (psi).

The value of C was set to 13.3 for interstate and primary roads and 18.4 for secondary roads corresponding to the failure criteria of 10% and 45% of wheelpath cracking, respectively (*16*). To convert remaining ESALs to remaining life, the following default levels of annual ESAL were considered for the three road categories:

- Interstate: 1.4 million ESAL—equivalent to about 6,500 annual daily truck traffic (ADTT) (or 2,000 singles, 4,000 doubles, and 500 trains or triples)
- Primary: 0.2 million ESAL—equivalent to about 950 ADTT (or 700 singles, 220 doubles, and 30 trains or triples)
- Secondary: 0.07 million ESAL—equivalent to about 375 ADTT (or 300 singles, 75 doubles)

Thresholds were set so that a pavement with 2 years or less of life remaining is considered "poor," between 2 and 5 years "fair," and 5 or more years "good." The two strain levels that separate good from fair and fair from poor can be obtained from Equation 11. These strains were then used to determine the appropriate SCI300 using Equation 5, which are given in Table 3.

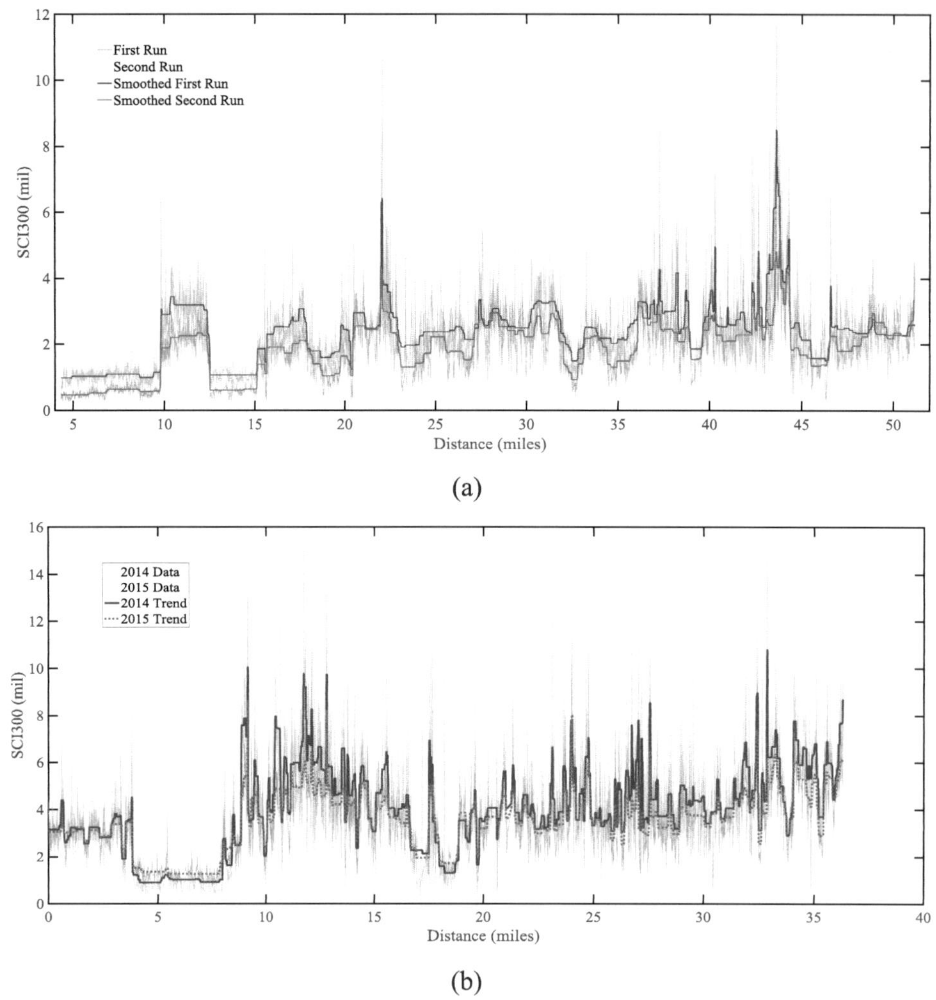

Figure 1. Illustration of short- and long-term repeatability: (*a*) short-term repeatability on US 29 south in Virginia performed on two consecutive days and (*b*) long-term repeatability on Route 144 in Pennsylvania.

TSD Evaluation

Repeatability

This section presents examples of short-term and long-term repeatability. Figure 1*a* shows an example of short-term SCI300 repeatability of tests (corrected to a reference temperature of 70°F) performed in Virginia. The repeated runs followed similar trends. A smoothed SCI300 is also shown in the figure to better highlight the trends. Figure 1*b* provides an example of long-term repeatability of sections tested in Route 144 Pennsylvania. The repeated SCI300 (corrected to a reference temperature of 70°F) followed similar trends in Pennsylvania. These results are in agreement with previous studies that have analyzed the repeatability of the measurements from the TSD (*17, 18*).

Reproducibility

Figure 2*a* shows the comparison between the *D*0 measurements from the TSD and FWD on a 74-mile section of I-81 South in Virginia. Both the TSD and FWD followed similar trends along the length of the tested section. TSD measurements were not corrected for temperature in this comparison.

Between milepost 214 and 218, the TSD and the FWD significantly disagreed with each other; the *D*0 measured from the TSD (in black) shows that the pavement section is significantly stronger than that represented by the *D*0 from the FWD (in orange). This section between mileposts 214 and 218 was rehabilitated (using a variety of recycling methods) in 2011, and the FWD was tested in 2007 and the TSD was tested in 2015 on the section. Thus, the TSD was able to

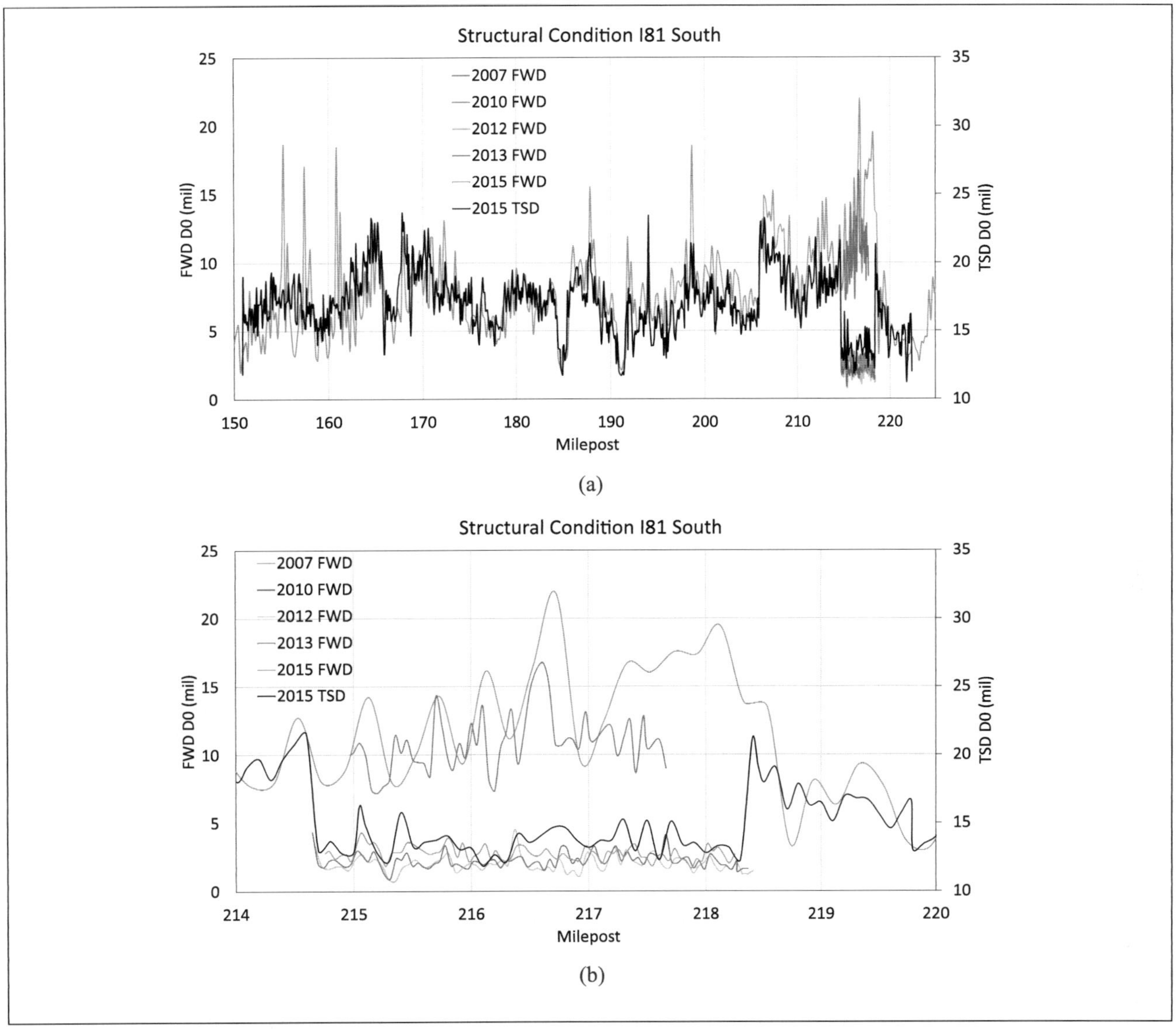

Figure 2. Comparison of TSD and FWD *D0* on I-81 South: (*a*) comparison of the entire section and (*b*) details of comparison.

identify the pavement section that was structurally stronger compared to others around it.

Figure 2*b* shows details of the comparison between milepost 214 and 218. FWD data between 2007 and 2010 show that the structural condition of the section did not change significantly in the 3- to 4-year period. (If any change is observed, the data suggest that the pavement slightly strengthened during that time period, although no treatment was applied, with the average temperature-corrected 2010 *D0* being 10.9 mils and the average temperature-corrected 2007 *D0* being 12.2 mils.) The rehabilitation work improved the structural condition of the pavement section as reflected in the decreased *D0* of FWD measurements performed in 2012, 2013, and 2015. The TSD-calculated *D0* from the 2015 test follows the trend of the FWD results of 2012, 2013, and 2015, suggesting that the improved structural condition of the pavement section was also captured by the TSD measurements.

The TSD and FWD are two different devices that use different mechanisms to compute the pavement deflection. The TSD measures the pavement deflection slope under maximum load, whereas the FWD measures the maximum deflection measurement as shown in Figure 3. Thus, the TSD and FWD do not exactly measure at the same instance even though the sensors are placed at the same distance from the applied load. The TSD and FWD also use different loading mechanisms; the TSD uses dual tire loading configurations to impart a dynamic load to the pavement whereas the FWD uses impulse load on a circular plate to apply the load on the pavement. Because of these reasons, the TSD and FWD do not always have the same

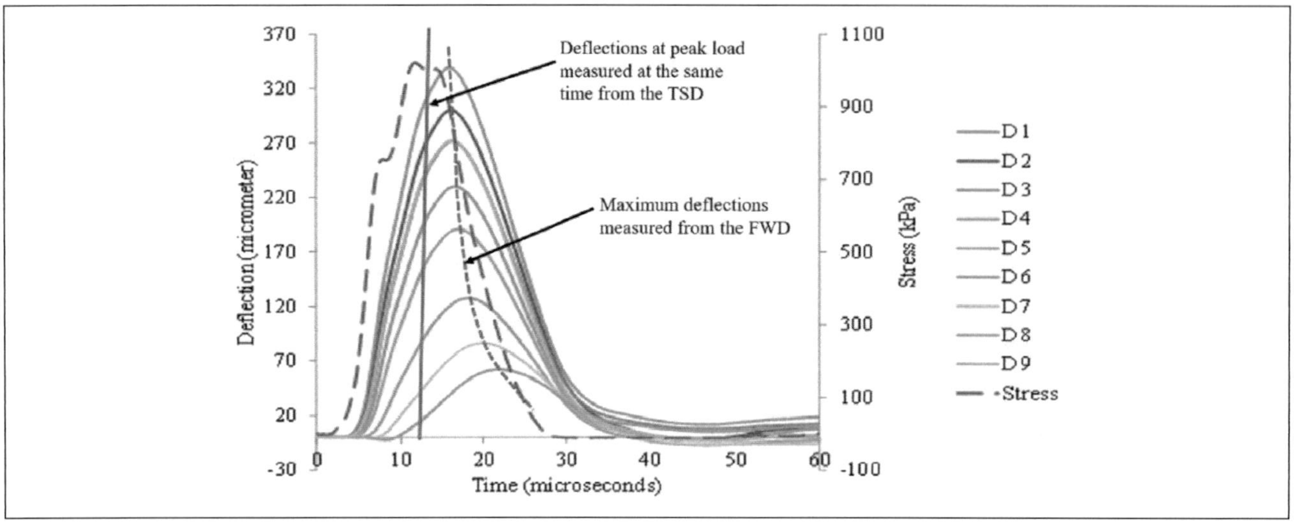

Figure 3. Deflection measurement from the FWD and TSD (19).

deflection measurements. However, there are cases in which the TSD and the FWD measurements are almost identical. When comparing the TSD and FWD measurements it is important to take these points into consideration and focus on the trends rather than the actual values themselves.

Comparison of TSD Measurements with PMS Data

Comparison with Detailed Fatigue Cracking Data. The fatigue cracking data were obtained from the Virginia PMS. Virginia records the fatigue cracking at three severity levels from 1 to 3, 1 being the least severe and 3 being the most severe. Figure 4*a* shows the SCI300 data measured by the TSD along with the fatigue cracking data obtained from the PMS for I-81 South. Pavements sections between milepost 205 and 215 have higher SCI300 values meaning the pavement is structurally weaker compared to its neighboring sections whereas the fatigue cracking on this pavement section seems to be very low. The pavements between mileposts 200 to 203 have a very severe fatigue cracking (severity level 3) but the pavement is structurally stronger compared to the section between mileposts 205 and 215.

Figure 4*b* shows the severity level 1 fatigue cracking in the three interstate roads I-81, I-64, and I-95. I-95 had significantly higher fatigue cracking compared to I-81 and I-64. More than 90% of I-64 and I-81 sections had practically no fatigue cracking compared with 50% of I-95 sections. Between 2011 and 2012, most of I-95 had received a 3- to 4.5-in. mill and overlay, but fatigue cracking reappeared rather quickly. On the other hand, some sections on I-81 and I-64 received a mill and overlay treatment and are still performing well with regard to fatigue cracking. Furthermore, I-81 sections that have not received a mill and overlay between 2007 and 2014 are still generally performing well with regard to fatigue cracking. Figure 4*c* shows the distribution of temperature-corrected TSD SCI300 for the three roads. The I-95 distribution shows higher SCI300 values and as such is a weaker pavement. This sheds light on why more cracking is observed on I-95. Although a large portion of it had been recently resurfaced, these portions are still relatively weak, which is leading to fatigue cracking in a short time period. A subpopulation on I-95 has low SCI300 values, but this consists of a strong composite pavement. The Spearman rank correlation between SCI300 and severity level 1 fatigue cracking was 0.20 for I-81 and 0.16 for I-95, and 0.40 for both combined. These results suggest that, in general, weaker pavement sections will exhibit more cracking (I-95 weaker than I-81); however, this is a very broad trend and observed cracking cannot be a substitute for pavement structural condition measurements (for both I-81 and I-95, cracking does not correlate well with structural condition). Also, surface cracking is a lagging indicator, which limits cost-effective alternatives that would otherwise have been possible if the structural deterioration is identified well before the fatigue cracks appear on the surface.

Comparison with Layer Coefficient Estimated Structural Number. PennDOT calculates their SN_{eff} based on the construction history of the pavement. The estimated deterioration is based on age assuming a fixed loss in structure with time as shown in Table 4. The structural coefficient used in calculating the SN_{eff} for flexible pavements are categorized based on the type of pavement layer and pavement age: 32A representing pavements ≤9 years and 32B representing pavements ≥10 years.

The comparison between TSD-calculated SN_{eff} and SN_{eff} estimated based on the American Society of State Highway and Transportation Officials (AASHTO) 1993

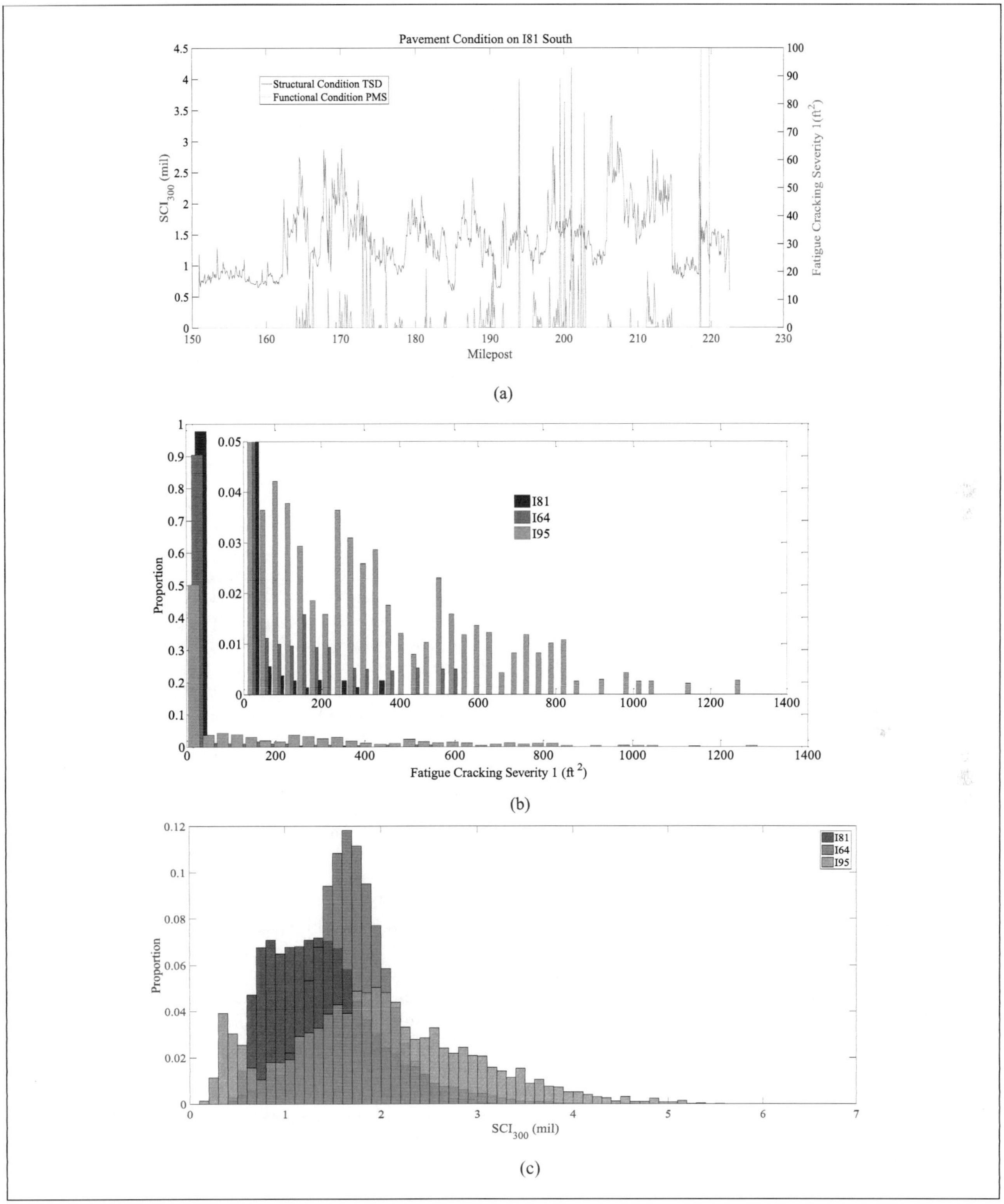

Figure 4. Comparison of TSD measurements with PMS data: (*a*) comparison of TSD SCI300 and severity level 1 fatigue cracking on I-81 South; (*b*) observed frequency of severity level 1 fatigue cracking on tested sections in Virginia; and (*c*) observed frequency of TSD SCI300 on tested interstate sections.

Table 4. Structural Layer Coefficients Used by PennDOT

Pavement layer	Structural coefficient 32A	32B
Surface course; new construction, reconstruction, or overlay		
Superpave 0.37 in., 0.49 in., 0.74 in., 0.98 in. (Wearing and Binder Courses)	0.44	0.3
FB-1, FB-2 (Wearing and Binder Courses)	0.2	0.2
FJ-1, FJ-1C, FJ-4 (Wearing Courses)	0.5	0.3
Base course; new construction, or reconstruction		
Superpave 0.98 in. Base Course	0.4	0.3
Superpave 1.47 in. Base Course	0.4	0.3
Crushed Aggregate (CABC)	0.14	0.14
Crushed Aggregate, Type DG (CABCDG)	0.18	0.18
Aggregate–Bituminous (ABBC)	0.3	0.3

(*20*) layer coefficient approach was performed for data from Pennsylvania and Idaho. The Pennsylvania PMS records a calculated SN_{eff} for their pavement sections based on layer thickness, layer structural coefficient, and layer structural coefficient reduction as a function of layer age. The calculation is similar to the AASHTO 1993 design method except that the layer coefficients are reduced as the layer ages. Figure 5*a* shows a significant discrepancy between the TSD-calculated SN_{eff} and the Pennsylvania PMS-calculated SN_{eff} between mile 12 and mile 25. This shows that the PMS-calculated SN_{eff} does not reflect the actual pavement structural condition.

The SN_{eff} approach used in the PMS makes assumptions about how the structural condition of the pavement will deteriorate; however, the actual field deterioration can be different than assumed. On the other hand, the TSD takes measurements that are directly related to the actual structural condition of the pavement.

Figure 5*b* reinforces the argument that the structural condition cannot be obtained from currently available PMS data. The figure shows an example of a road in Idaho where detailed layer thickness and composition were collected using ground penetrating radar (GPR) in conjunction with the TSD testing (both at 10-m intervals). The layer thickness and composition were used to calculate the AASHTO 1993 SN_{eff} using a layer coefficient of 0.44 for the asphalt layer and 0.14 for the base layer, which represents the as-constructed SN_{eff}. The SN_{eff} was also calculated from the TSD data using the Rohde (*14*) method (Equation 4). Again, the trend between TSD SN_{eff} and AASHTO 1993 is not the same indicating non-uniform structural deterioration from the initial construction. This shows the importance of getting actual field measurements to characterize the structural condition and not relying on design parameters that are not representative of the true current structural condition.

Implementation into the PMS Decision Process

VDOT uses a set of pavement management decision matrices with distresses as inputs and treatment activities as outputs. The matrices are grouped based on the following roadway classifications: interstates, primary routes, secondary routes, and unpaved roads, in addition to the following pavement types: bituminous-surfaced (BIT), bituminous-surfaced composite pavements (with jointed concrete pavement below the surface, BOJ), bituminous-surfaced composite pavements (with continuously reinforced concrete pavement below the surface, BOC), continuously reinforced concrete (CRC), and jointed concrete pavements (JCP). Additionally, updated cost estimates per mile for each treatment are available for each road category. The decision process is a two-phase approach (Figure 6*a*). In 2008, this two-phase approach was modified to include structural condition and truck traffic volumes, and the enhanced decision tree was integrated into the process. One of the main features of the approach is that the addition of the pavement structural information did not alter the core of the decision process already in place but provided an additional step that can be used when pavement structural condition is available. If structural information becomes unavailable, the decision process can revert to the core process already in place. VDOT currently uses the following five treatment categories (from do nothing to heavier treatments): do nothing (DN), preventive maintenance (PM), corrective maintenance (CM), rehabilitation maintenance (RM), and reconstruction (RC). At the preliminary treatment selection stage, one of these five categories is selected based on the surface condition index and the decision matrices. In the enhanced decision process, based on the structural condition (and traffic level and construction history), the selected preliminary treatment can be either retained, or modified to a heavier or lighter treatment.

This section presents an example of how the structural condition will affect the triggered maintenance category at the network-level analysis for the tested I-81 section in Virginia. The approach is based on the Criticial Condition Index (CCI) recommended treatment used by VDOT. It should be noted that in the actual VDOT decision process, the different distresses—for example, cracking, rutting, and so forth—are used to trigger the maintenance category and the CCI is used as an additional filter. Here the process is simplified for a clearer exposition. This example uses data from I-81 and therefore the triggered categories for the interstate road system are used. The different treatment categories are codified with a numerical value as follows: DN = 1, PM = 2, CM = 3, RM = 4, and RC = 5. After the triggered maintenance category is obtained from the CCI at the first stage, the structural condition information is used to potentially bump up the triggered maintenance category. The structural condition information used is the one obtained from the SCI300 with thresholds determined from

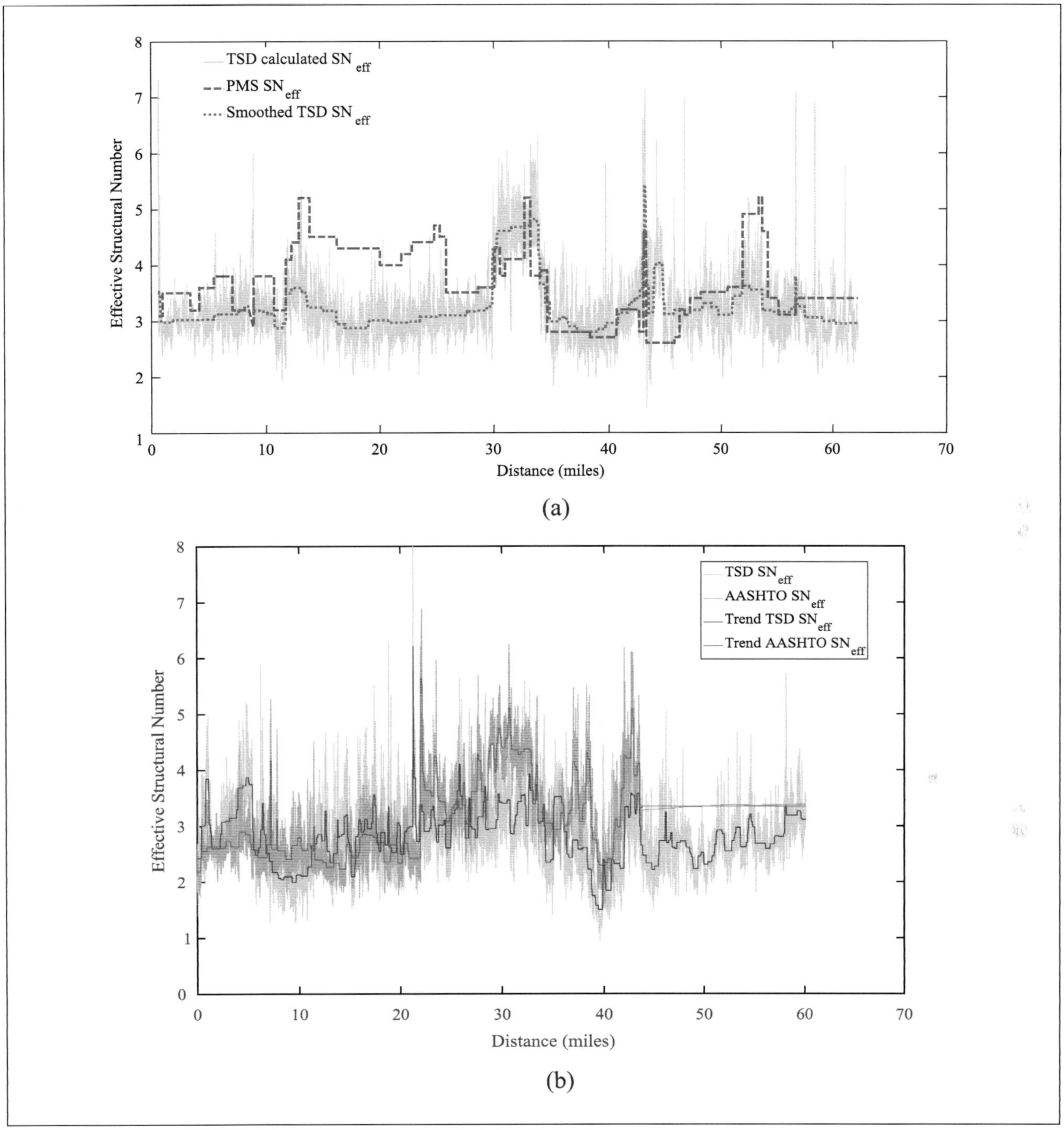

Figure 5. SN_{eff} comparisons: (*a*) TSD SN_{eff} and Pennsylvania PMS SN_{eff} and (*b*) TSD-calculated SN_{eff} and AASHTO layer thickness and coefficient SN_{eff}.

Table 3. The approach to bump up the triggered maintenance category is as follows: if all the SCI300 measurements in the pavement section have good structural condition, keep the maintenance category triggered by the CCI with a maximum category of 3; if there are SCI300 measurements within the pavement section that are classified as fair, bump up the maintenance category triggered by the CCI by one not to exceed a maintenance category of 5; if there are SCI300 measurements within the pavement section that are classified as poor, bump up the maintenance category triggered by the CCI by two not to exceed a maintenance category of 5 and a minimum category of 4.

In total, there are 747 0.1-mile sections tested in I-81. The triggered maintenance categories obtained before and

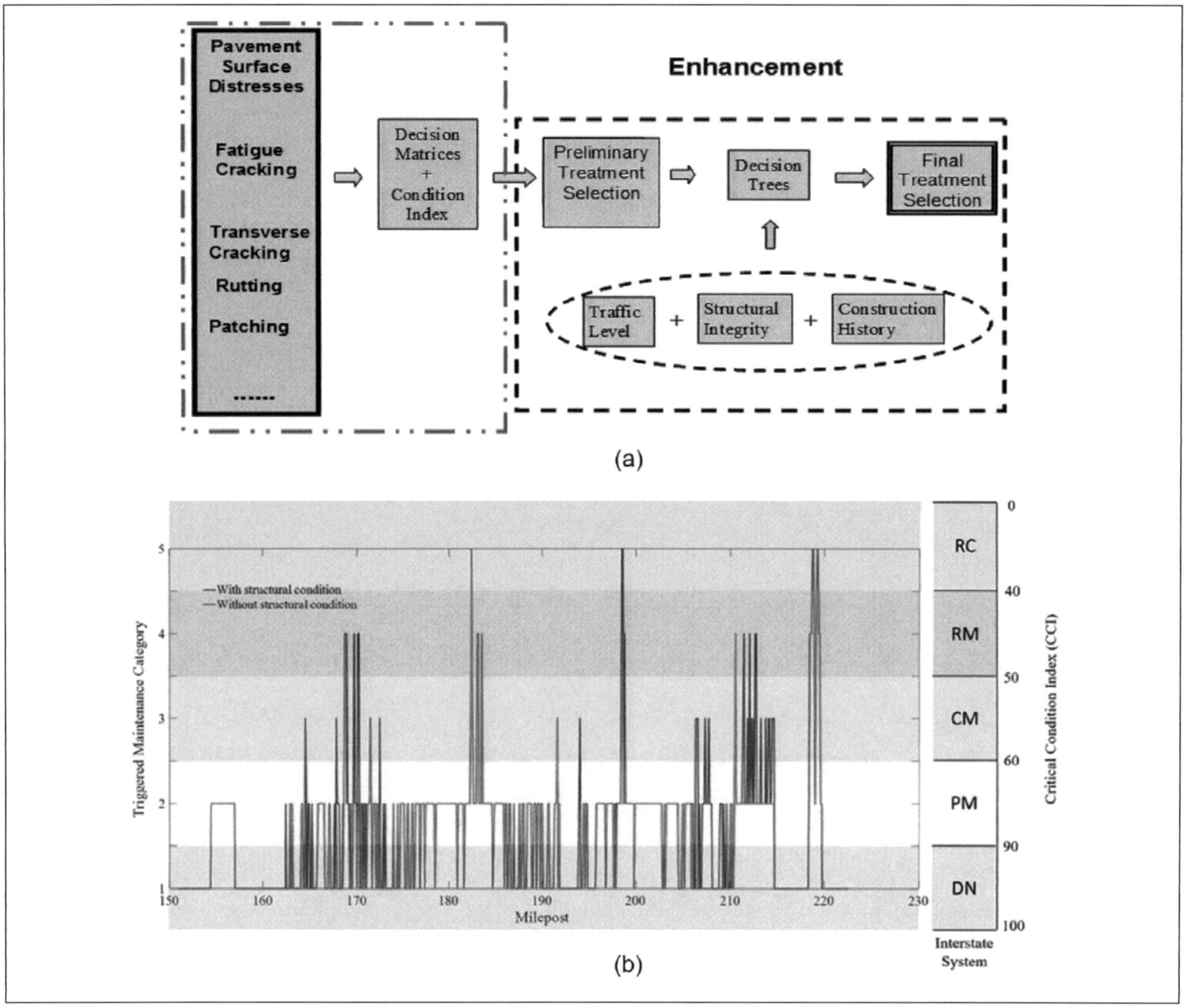

Figure 6. Application of the TSD data into PMS decision process: (*a*) DOT two-phase decision process (*21*) and (*b*) triggered maintenance category before and after structural condition consideration.

after taking into account the structural condition are shown in Figure 6*b*. Of the 747 sections, 66 (8.8%) had the triggered maintenance category bumped up, 50 by one and 16 by two. The total number of sections in each maintenance category before and after incorporating the structural condition are DN: 384 before, 345 after; PM: 338 before, 347 after; CM: 6 before, 25 after; RM: 14 before, 23 after; RC: 5 before, 7 after. The number of sections in the DN categories decreased, whereas the number of sections in all other categories increased. Again, this example is simplified, and other factors such as traffic level and pavement age can be considered (as is done by VDOT). It is expected that different SHAs will tailor their approach for their specific needs.

Conclusion

This paper summarizes the results and findings from the TSD pooled fund study. Based on the analysis, the following conclusions are made:

1. The short- and long-term repeatability of the TSD is generally good. Repeated measurements performed on consecutive days or in two different years followed similar trends. However, work is needed to more accurately account for the effect of temperature on TSD measurements.
2. A comparison of the TSD with the FWD on a section of I-81 in Virginia showed that the two devices followed

similar trends along the tested road. Rehabilitation work performed on parts of the tested road suggests that the structural improvement is reflected in FWD testing. TSD measurements followed the trend of FWD measurements, which suggests that the structural improvement is also picked up by TSD measurements.

3. Comparison between TSD indices and PMS surface condition data (e.g., cracking) suggests little relationship. Detailed investigation between TSD SCI300 and PMS-recorded fatigue cracking shows that structurally weaker pavement sections in general exhibit more fatigue cracking. However, this is a broader trend and observed fatigue is not a robust indicator of the pavement structural condition.
4. A framework to incorporate the TSD-measured structural condition within a SHA's PMS is presented. The framework is based on the current two-stage approach followed by VDOT. In the first stage, the triggered maintenance category in the PMS is based on surface condition. In the second stage, the triggered maintenance category is combined with structural condition information to determine the final recommended maintenance category.

Acknowledgments

This study was possible through the FHWA pooled fund project "Demonstration of Network Level Pavement Structural Evaluation with Traffic Speed Deflectometer" DTFH61-11-D-00009-T-13008. We thank Dr. Nadarajah Sivaneswaran of the FHWA for his guidance and support throughout the project. We also acknowledge Greenwood Engineering for collecting and providing the TSD data and Brian Diefenderfer for his guidance with the Virginia PMS data. Finally, we also thank all the states that participated in this study.

References

1. Flora, W. F. *Development of a Structural Index for Pavement Management: An Exploratory Analysis*. Purdue University, West Lafayette, 2009.
2. Bryce, J., G. Flintsch, S. Katicha, and B. Diefenderfer. Developing a Network-Level Structural Capacity Index for Asphalt Pavements. *Journal of Transportation Engineering*, Vol. 139, No. 2, 2012, pp. 123–129.
3. Katicha, S. W., S. Ercisli, G. W. Flintsch, J. M. Bryce, and B. K. Diefenderfer. *Development of Enhanced Pavement Deterioration Curves*. PhD Dissertation, Virginia Tech, 2016.
4. Ferne, B., P. Langdale, M. Wright, R. Fairclough, and R. Sinhal. Developing and Implementing Traffic-Speed Network Level Structural Condition Pavement Surveys. *Proc., 9th International Conference on the Bearing Capacity of Roads, Railways and Airfields*, Trondheim, Norway, 2013.
5. Lister, N., C. Kennedy, and B. Ferne. The TRRL Method for Planning and Design of Structural Maintenance. *Proc., 5th International Conference on the Structural Design of Asphalt Pavements, University of Michigan*, Ann Arbor, Michigan, 1982, pp. 709–725.
6. Zaghloul, S., Z. He, N. Vitillo, and J. Kerr. Project Scoping Using Falling Weight Deflectometer Testing: New Jersey Experience. *Transportation Research Record: Journal of the Transportation Research Board*, 1998. 1643: 34–43.
7. Steele, D. A., C. A. Beckemeyer, and T. P. Van. Optimizing Highway Funds by Integrating RWD Data into Pavement Management Decision Making. *Proc., 9th International Conference on Managing Pavement Assets*, VTechWorks, Washington, D.C., 2015.
8. Flintsch, G., and K. McGhee. *Managing the Quality of Pavement Data Collection*. Transportation Research Board of the National Academies, Washington, D.C., 2009.
9. Katicha, S., G. Flintsch, and B. Ferne. Optimal Averaging and Localized Weak Spot Identification of Traffic Speed Deflectometer Measurements. *Transportation Research Record: Journal of the Transportation Research Board*, 2013. 2367: 43–52.
10. Flintsch, G., S. Katicha, J. Bryce, B. Ferne, S. Nell, and B. Diefenderfer. *SHRP 2 Report S2-R06F-RW-1: Assessment of Continuous Pavement Deflection Measuring Technologies*. Transportation Research Board of the National Academies, Washington, D.C., 2013.
11. Rada, G., S. Nazarian, B. Visintine, R. V. Siddharthan, and S. Thyagarajan. *Pavement Structural Evaluation at the Network Level*. FHWA-DTFH61- 12-C-00031. FHWA, U.S. Department of Transportation, 2015.
12. Lukanen, E. O., R. Stubstad, and R. Briggs. *Temperature Predictions and Adjustment Factors for Asphalt Pavement*. FHWA-RD-98-085. FHWA, McLean, Va., 2000.
13. Pedersen, L. *Viscoelastic Modeling of Road Deflections for Use with the Traffic Speed Deflectometer*. PhD dissertation. Technical University of Denmark, Lyngby, 2013.
14. Rohde, G. T. *Determining Pavement Structural Number from FWD Testing. Transportation Research Record: Journal of the Transportation Research Board,* 1994. 1448: 61–68.
15. Asphalt, I. *Research and Development of the Asphalt Institute's Thickness Design Manual (MS-1) Ninth Edition*. Asphalt Institute, College Park, Maryland, 1982.
16. Finn, F., C. Saraf, R. Kulkarni, K. Nair, W. Smith, and A. Abdullah. The Use of Distress Prediction Subsystems for the Design of Pavement Structures. *Volume I of Proc., 4th International Conference on Structural Design of Asphalt Pavements*, Ann Arbor, Michigan, August 22–26, 1977, pp. 3–38.
17. Shrestha, S., S. Katicha, and G. Flintsch. *Field Demonstration of the Traffic Speed Deflectometer*. Presented at the World Conference on Pavement and Asset Management, Milan, Italy, 2017.
18. Katicha, S. W., and G. W. Flintsch. Field Demonstration of the Traffic Speed Deflectometer in New York. Presented at the *Transportation Research Board*, 2015.
19. Chatti, K., M. E. Kutay, N. Lajnef, I. Zaabar, S. Varma, and H. S. Lee. *Enhanced Analysis of Falling Weight Deflectometer Data for Use with Mechanistic-Empirical Flexible Pavement Design and Analysis and Recommendations for Improvements to Falling Weight Deflectometer*. DTFH61-11-C-00026. FHWA, U.S. Department of Transportation, 2013.
20. American Association of State Highway and Transportation Officials (AASHTO). Guide for design of pavement structures. Washington, D.C., 1993.
21. Chowdhury, T. *Supporting Document for the Development and Enhancement of the Pavement Maintenance Decision Matrices Used in the Needs-Based Analysis*. Virginia Department of Transportation, Richmond, 2008.

The Standing Committee on Pavement Structural Modeling and Evaluation peer-reviewed this paper (18-05514).

Article

TRR
JOURNAL OF THE TRANSPORTATION RESEARCH BOARD

Transportation Research Record
2018, Vol. 2672(40) 360–370

DOI: 10.1177/0361198118758637
journals.sagepub.com/home/trr
SAGE

Development of Artificial Neural Networks for Predicting the Response of Bonded Concrete Overlays of Asphalt for Use in a Faulting Prediction Model

John W. DeSantis[1], Julie M. Vandenbossche[1], Kevin Alland[1], and John Harvey[2]

Abstract

Transverse joint faulting is a common distress in bonded concrete overlays of asphalt pavements (BCOAs), also known as whitetopping. However, to date, there is no predictive faulting model available for these structures. To account for conditions unique to BCOA, a computational model was developed using a three-dimensional finite element program, ABAQUS, to predict the response of these structures. The model was validated with falling weight deflectometer (FWD) data from existing field sections at the Minnesota Road Research Facility (MnROAD) as well as at the University of California Pavement Research Center (UCPRC). A large database of analyses was then developed using a fractional factorial design. The database is used to develop predictive models, based on artificial neural networks (ANNs), to rapidly estimate the structural response at the joint in BCOA to environmental and traffic loads. The structural response will be related to damage using the differential energy concept. Future work includes the implementation of the developed ANNs in this study into a faulting prediction model for designing BCOA.

Bonded concrete overlays of asphalt pavements (BCOA), also known as whitetopping, consist of a Portland cement concrete (PCC) overlay placed on an existing asphalt or composite pavement. The existing asphalt is often milled before the overlay to remove any surface distress and to increase the surface area, which may promote greater bond between the PCC and asphalt. This technique is intended as a cost-effective rehabilitation solution for marginally distressed in-service asphalt or composite pavements. These structures typically have smaller panel sizes than traditional jointed plain concrete pavements (JPCP) to reduce curling and warping stresses and bending stresses produced by applied loads (*1–3*).

Faulting has been observed to be a common distress in BCOAs (*4,5*). However, faulting is not considered in any of the current design procedures for BCOA. Faulting in BCOAs, like other PCC-surfaced pavements, is caused by the pumping mechanism. Pumping in a BCOA can develop at either the bottom of the overlay slab within the asphalt layer or below the asphalt in the granular layer, depending on the pavement structure (*4*). The development of pumping, and therefore faulting, is related to the structural response of the pavement to applied loads. Therefore, it is necessary to estimate the response of the pavement to environmental and traffic loadings to predict faulting.

A computational model is used to estimate the response of BCOAs to these loadings. The accuracy of this model is confirmed using falling weight deflectometer (FWD) data collected from in-service BCOA structures. The computational model is then used in the development of artificial neural networks (ANNs). The ANNs can very accurately reproduce the same deflections as the computational model, given a certain set of required inputs. These ANNs can then be used to develop predictive equations to rapidly estimate the response of the BCOA for use in a faulting model, similar to those within Pavement Mechanistic-Empirical (ME) for JPCP faulting and stress prediction in the BCOA-ME (*6–9*).

Effect of Depth of Joint Deployment

A better understanding of joint behavior for BCOAs is necessary to accurately model the response of these structures. Joints can deploy through just the PCC layer, both the overlay and the underlying asphalt layer, or can have both depths

[1]University of Pittsburgh, Pittsburgh, PA
[2]Department of Civil and Environmental Engineering, University of California Pavement Research Center, Davis, CA

Corresponding Author:
Address correspondence to Julie M. Vandenbossche: jmv7@pitt.edu

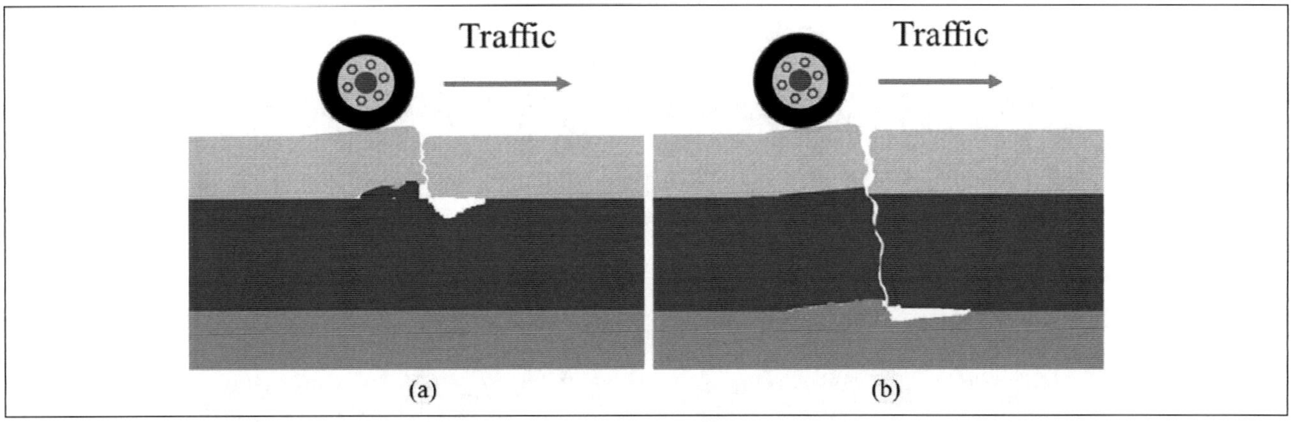

Figure 1. Faulting following pumping of the different layers (*4*). (a) Joint depth is PCC only. (b) Joint depth is through PCC and asphalt.

of deployment along the same overlay. The different depths of joint deployment can be estimated based on the flexural stiffness ratio (FSR) between the overlay and the asphalt layer and the panel size (*4,10*). The depth of joint deployment dictates the layer that will be eroded and pumped. If the transverse joint in the overlay does not propagate through the asphalt layer, then pumping of the asphalt material occurs at the top of the asphalt layer, as shown in Figure 1*a*. The interfacial bond can deteriorate over time, and water can seep in along the interface and become trapped. This contributes to stripping and the deterioration of the structural integrity of the asphalt, which leads to pumping and faulting (*10*). If the transverse joint propagates through the asphalt layer, then the PCC and asphalt act as a monolithic structure and faulting develops because of pumping of the granular material below the asphalt or the subgrade material, as shown in Figure 1*b*. The joint depth also plays an important role in the magnitude of curling, as well as the load transfer efficiency (LTE) at the joint. A fully bonded BCOA with a continuous asphalt layer will decrease the curling stresses because of moment transfer, whereas when a full-depth joint is present, the moment transfer is reduced and can lead to higher curling stresses. A continuous asphalt layer will provide more continuous support across the joint and will provide a higher LTE than when a full-depth joint is present.

A computational model for BCOAs must be able to account for both depths of joint deployment. A continuous asphalt layer, owing to joint deployment only through the overlay, is difficult to model in pavement-specific finite element modeling (FEM) programs, such as ISLAB (*11*). Therefore, the computational model was developed in a multipurpose 3D FEM software, ABAQUS, which is able to model both depths of joint deployment (*12*).

Development of Computational Models

The computational model consists of at least one full lane-width; an eight-slab system was used for partial lane-width panels (6 × 6 ft and 8 × 8 ft), and a four-slab system was used for full lane-width panels (12 × 12 ft). The eight-slab system consists of two adjacent panels in the transverse direction and four panels in the longitudinal direction. The four-slab system consists of four full lane-width panels in the longitudinal direction. The effects from the adjacent lane were negligible for the full lane-width panels on account of the critical response location being the interior corner. Therefore, a single lane was used. Both slab systems include a shoulder that is 6 ft wide and is either tied PCC or asphalt. Figure 2 presents the 3D FEM model.

Interface Between Layers

The interface between the PCC (elastic solid) and the asphalt (elastic solid) layer was fully bonded. Composite action is essential for the performance of BCOA. This was achieved by using a "tie" constraint, which treats adjacent nodes as rigidly connected to one another at the interface between the PCC and asphalt layers (*12*). A dense liquid foundation beneath the asphalt was simulated using a "linear" interface interaction with a rigid, fully constrained body (*12*). This allows the asphalt to become unsupported, unlike the "Foundation" support interaction in ABAQUS (*12*). The stiffness of the interaction is defined by a modulus of subgrade reaction, k-value (psi/in).

Joint Behavior

Another important feature of the computational model is joint behavior. When joint deployment through the asphalt is expected, the asphalt layer is discontinuous at the transverse joint and no interaction is assumed within the asphalt layer across adjacent slabs. The asphalt layer is modeled as continuous beneath all longitudinal joints within the lane width, and all transverse joints when the joint does not deploy through the asphalt. When joint deployment is only through the PCC for the eight-slab system, adjacent joints are modeled only through the PCC layer. On the other hand, when the joint of interest is

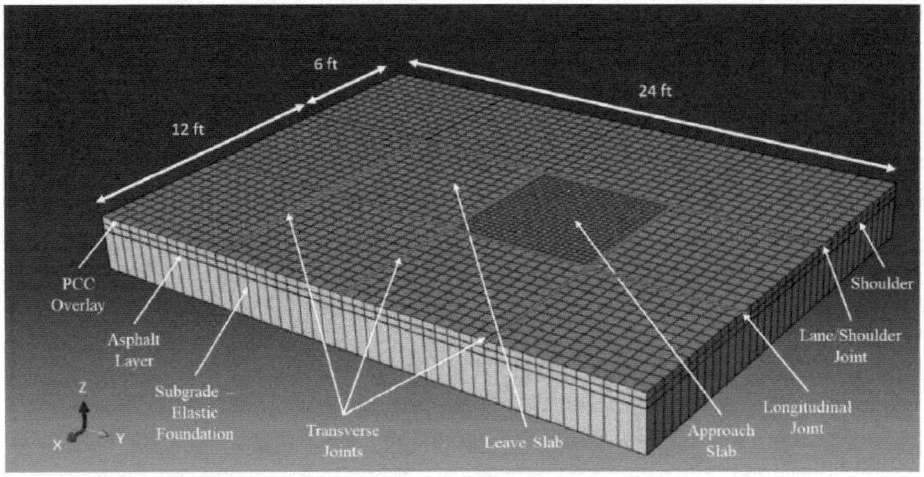

(a)

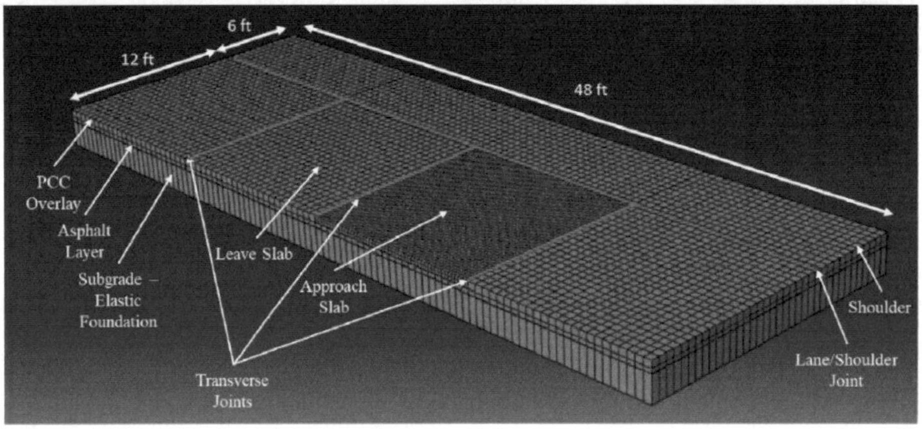

(b)

Figure 2. Model configuration. (a) eight-slab system: 6 × 6 ft panels. (b) four-slab system: 12 × 12 ft panels.

modeled as a full-depth joint, the adjacent joints are assumed to only propagate through the PCC layer. For the four-slab system, all FEM analyses contained full-depth joints for every joint. The longer panels have larger joint width movement than the smaller panel sizes. Also, these panel sizes typically have thicker overlays with a high FSR (*10*). These factors contribute to the likelihood of adjacent full-depth joints. These assumptions on joint deployment were based on an extensive study that was performed to better understand joint performance in BCOAs (*10*). The asphalt remains continuous for both joint depths for all longitudinal joints within the driving lane. However, the asphalt layer is not continuous across the lane/shoulder joint.

Aggregate interlock across the PCC joints is addressed by using shear springs. The spring stiffness per unit area is calculated for a given LTE based on Equation 1, developed by Crovetti (*13*):

$$AGG^* = \left(\frac{\frac{1}{LTE} - 0.01}{0.012} \right)^{-1/0.849} \quad (1)$$

where: AGG^* is the nondimensional joint stiffness of the transverse joint; and LTE is the corresponding LTE to the assigned joint stiffness. The stiffness of the individual springs is calculated based on the tributary area. The longitudinal joints in the driving lane are modeled with the same LTE as the defined LTE for the transverse joints. The lane/shoulder joint is defined as 90% LTE when it is a tied PCC shoulder, and 0% when it is an asphalt shoulder. For both shoulder conditions, the asphalt layer is not continuous across the lane/shoulder joint and is assumed to provide no support. The tied PCC shoulder was assumed to be 90% LTE to account for the discontinuity in the asphalt. The full-depth asphalt shoulder is also assumed to provide no support. A

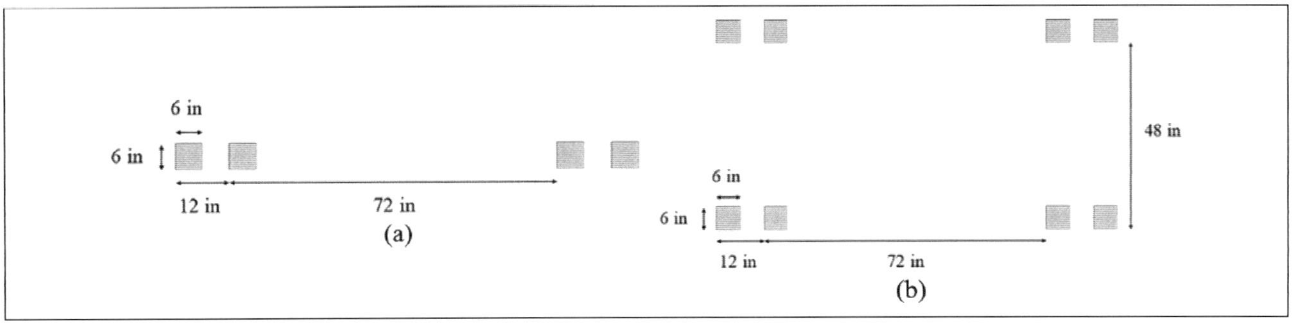

Figure 3. Axle configurations. (a) Single axle configuration. (b) Tandem axle configuration.

Table 1. Mesh Convergence Study Parameters

Joint depth	Panel size, L × W (ft)	PCC thickness (in)	Asphalt thickness (in)	Modulus of subgrade reaction (psi/in)
PCC only	6 × 6	3	7	100
PCC only	6 × 6	3	7	315
PCC only	6 × 6	5	7	100
PCC only	6 × 6	5	7	315
PCC only	12 × 12	3	7	100
PCC only	12 × 12	3	7	315
Full-depth	12 × 12	6	7	100
Full-depth	12 × 12	6	7	315

"hard" contact surface interaction is also applied to all joint faces to allow compression at the joint faces if they come into contact with each other.

Wheel and Thermal Loads

Loading is performed in two steps: thermal loading and traffic loading. In the first step, an effective linear temperature gradient (ELTG) is applied, which accounts for curling caused by temperature gradients, drying shrinkage, and built-in curl. The ELTG is applied by defining nodal temperatures at each of the five layers of nodes in the PCC slab. The temperature at the bottom of the PCC is set to 0°F. The ELTG varies linearly from the bottom to the top of the PCC. All nodes in the asphalt layer are set to 0°F. Traffic is applied as a uniform pressure evenly distributed over the tire footprint. Single and tandem axle loads were considered. The dual tire footprint can be seen in Figure 3.

Finite Element Mesh

Accurate prediction of structural response depends on using an appropriate FEM mesh in the computational models. Quadratic brick elements have been shown to predict pavement behavior well (*14*). Twenty node quadratic brick elements (C3D20) were used in ABAQUS. These elements are similar to elements used in pavement-specific FEM, such as EverFE (*15*). This node configuration has been shown to be able to provide a high level of accuracy, along with an acceptable computational time demand (*16*).

The mesh size was selected using a mesh convergence study. Both depths of joint propagation were examined with different mesh fineness. Two structures were examined for both the 6-ft panel size and 12-ft panel size models. Table 1 presents the different structures examined in the convergence study. Mesh convergence was achieved using 3 × 3 inch by half of the layer thickness for the loaded slab, and 6 × 6 inch by the layer thickness for the remainder of the model. Because second-order elements are used, one element along the thickness for each layer is sufficient for modeling flexural models (*16*).

Model Validation

It is essential that the computational model accurately predicts the pavement response. To validate the FEM model, four test sections from two testing facilities were used. Two BCOA sections from the Minnesota Road Research Facility (MnROAD), Cells 60 and 97, and two sections from the University of California Pavement Research Center (UCPRC) Heavy Vehicle Simulator (HVS) test sections, Sections B and F, were used. The parameters used for the validation sections can be seen in Table 2. The parameters for the MnROAD sections were obtained from construction

Table 2. Pavement Parameters for Validation Sections

Parameter	Cell 60	Cell 97	Section B	Section F[a]
PCC thickness (in)	5	6	4.5	4.5
Asphalt thickness (in)	7	7	4.5	1 (RHMA) 3.5 (HMA)
PCC modulus of elasticity (10^6 psi)	4.6	4.7	7.54	7.54
Modulus of subgrade reaction, *k*-value (psi/in)	315	315	720	720
Asphalt modulus of elasticity (10^6 psi)	0.35	0.35	0.475	0.95 (RHMA) 0.55 (HMA)
Asphalt testing temperature (°F)	90	90	95	85
Asphalt testing load frequency (Hz)	25	25	20	20
Panel size, L × W (ft)	5 × 6	10 × 12	6 × 6	6 × 6

[a]Section F consisted of a 1-inch newly placed rubberized hot-mixed asphalt (RHMA) layer on top of an older 3.5-inch hot-mixed asphalt (HMA) layer.

reports and research summary publications (*8,17–21*). The asphalt stiffness at the time of the FWD testing was established using a mastercurve derived using laboratory data and temperatures measured from thermocouples embedded at mid-depth in the asphalt layer. FWD testing at mid-slab was used to backcalculate the k-value of the subgrade. The parameters corresponded well with previous research conducted on the BCOA sections at MnROAD (*17–21*). The asphalt stiffness for the UCPRC sections was estimated using FWD and laboratory test data to develop mastercurves at a frequency of 20 Hz.

Figure 4 presents the deflections measured with the FWD and predicted from the computational model as a 9-kip load is applied in the outer wheelpath adjacent to the transverse joint. Section B at UCPRC was tested using a 13.5-kip load. For Cell 60, the asphalt layer was expected to be continuous through the joint early in the service life. For Cell 97, both depths of joint deployment are expected. Early in the service life the asphalt layer is likely still continuous and later in the service life the asphalt layer was not. For Sections B and F, the asphalt is likely continuous during the time of testing based on measured deflections and visual inspection.

From the plots shown in Figure 4, it is evident that the computational models can accurately predict the response of the BCOA structures to applied loads.

Predictive Model

Performing a computational analysis of all loading configurations and environmental conditions for all analysis periods for developing and utilizing a faulting prediction model is computationally prohibitive. Therefore, ANNs are used to produce responses rapidly, similar to the methodology employed in Pavement ME and linear regression stress prediction models in the BCOA-ME (*6,8,9,22*).

Critical Response Parameters

The critical responses from the model are to be used to calculate the differential energy of subgrade deformation, shown in Equation 2. The critical response to be obtained from the computational model is the deflection basin on the approach and leave slab after the environmental loading step and the traffic loading step. The deflection basin is defined as 2 × 6 ft wide on both sides of the transverse joint, adjacent to the lane/shoulder joint. The volume of the deflection basin is calculated as the sum of the nodal deflections in the basin multiplied by the tributary area corresponding to the nodes. The same number of nodal deflections are used for the approach and leave slab basins. The deflection basin can be seen in Figure 5. When the joint depth is through the PCC only, the response is recorded at the bottom of the PCC layer. When the joint is full-depth, the response is recorded at the bottom of the asphalt layer. In Pavement ME, the critical response to determine differential energy for concrete pavement design are the deflections at the corners on both the loaded (approach slab) and unloaded (leave slab) sides of the joint (*6*). In this study, corner deflections, full lane-width deflection basins, triangular deflection basins and 2 × 6 ft deflection basins are all considered. The deflection basin was used instead of deflections at the corner, because for BCOA the maximum deflection is not always at the corner of the slab. This is especially true when the wheelpath occurs at mid-slab. Selection of the 2 ×6 ft basin allowed for consideration of the basin in the area most heavily influenced by the load. The selection of this basin area also provided improved accuracy between predicted and measured performance during the calibration process:

$$\mathrm{DE} = \frac{1}{2}\mathrm{k}\left(B_{\delta L}^2 - B_{\delta UL}^2\right) \tag{2}$$

where: DE is the differential energy of subgrade deformation; k is the modulus of subgrade interaction; $B_{\delta L}$ is the deflection basin on the loaded slab; and $B_{\delta UL}$ is the deflection basin on the unloaded slab.

Partial Factorial Analysis

An accurate predictive model requires the computational model to be analyzed over an inference space with 10

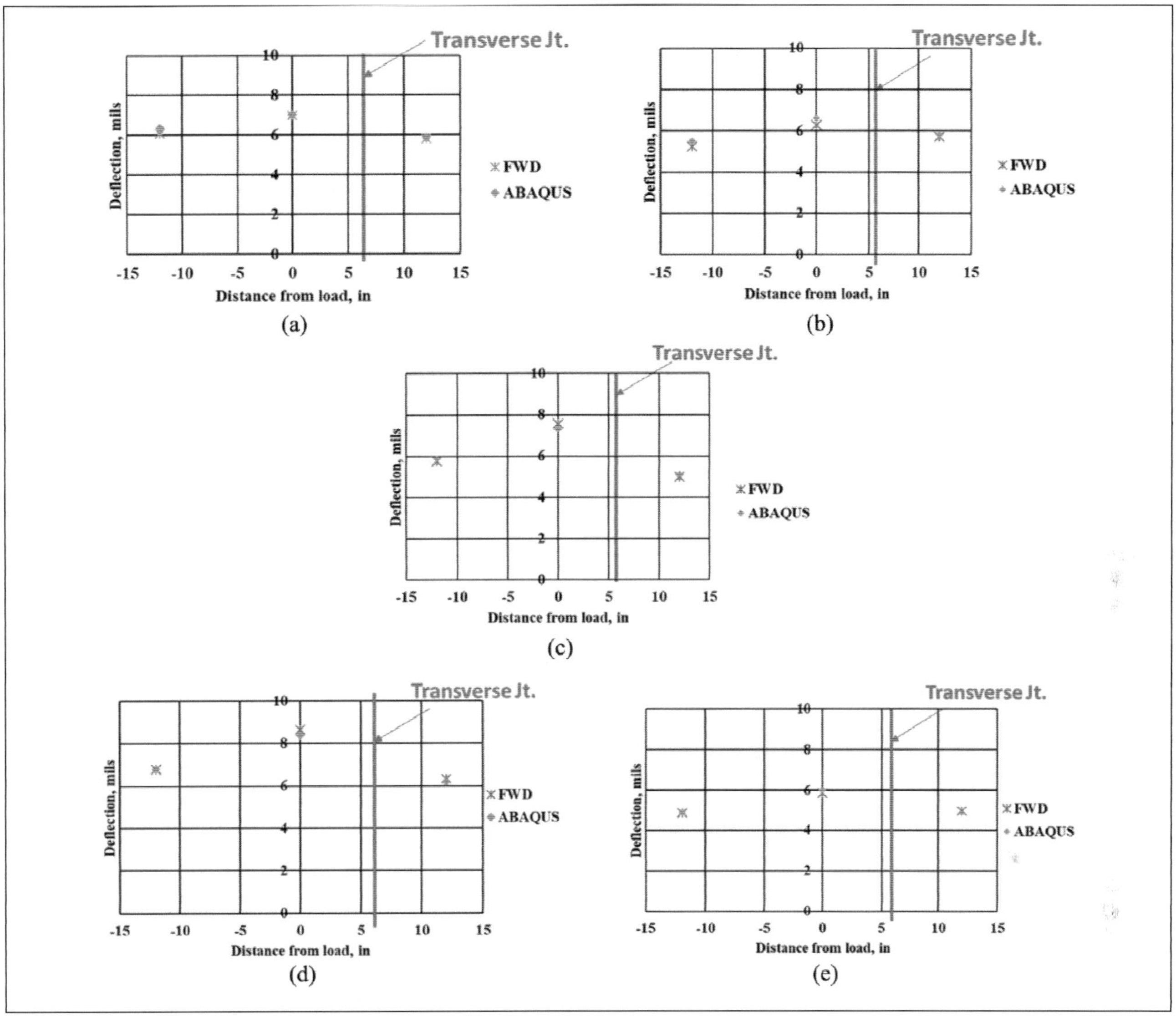

Figure 4. Model validation. (a) Cell 60: continuous asphalt (9-kip load). (b) Cell 97: continuous asphalt (9-kip load). (c) Cell 97: noncontinuous asphalt (9-kip load). (d) Section B: continuous asphalt (13.5-kip load). (e) Section F: continuous asphalt (9-kip load).

parameters. The computational model needs to evaluate a large range of parameters for populating the database used to develop the ANNs. All parameters considered, along with their corresponding values, can be seen in Table 3. A full factorial analysis would result in a total of approximately 105,000 FEM analyses. To account for the different joint depths, there are 18 possible combinations examining joint depth, PCC thickness, and asphalt thickness. However, only 12 of these combinations are feasible based on the FSR established by DeSantis et al. (*4*). Three structures were analyzed considering the joint depth through the PCC, three structures were analyzed considering a full-depth joint, and three structures were analyzed considering both joint depths separately, resulting in 12 combinations. This reduces the total number of analyses to 23,328 per joint spacing for the partial lane-width panels and 11,664 for the large panels because only full-depth joints were considered for the large panels. To further reduce the number of analyses required, a fractional factorial is used (*23*). Additional analyses were not performed to account for the three levels of asphalt stiffness. Instead, each level of asphalt stiffness (low, medium, or high) was assigned to one-third of the analyses in a full factorial of the other parameters. This reduces the required number of analyses by two-thirds. The total amount of FEM analyses conducted in ABAQUS was 19,440 (7,776 analyses for each of the partial lane-width panels, and 3,888 analyses for the full lane-width panels) using the fractional factorial design.

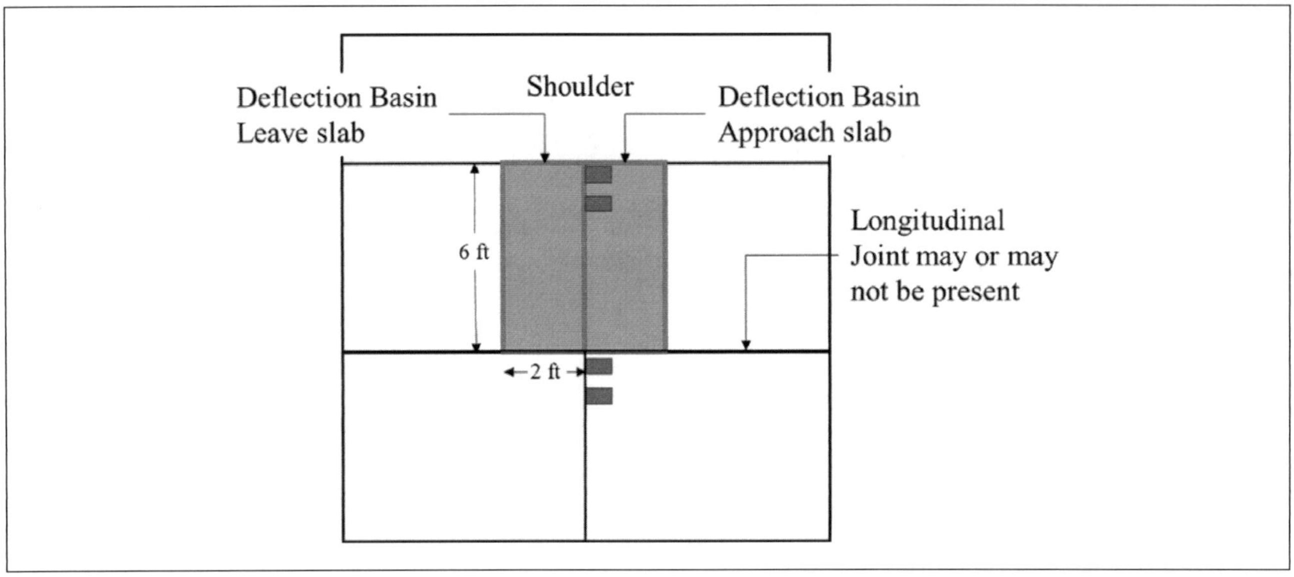

Figure 5. Deflection basin definition.

Table 3. Overall Design Matrix

Parameter	Parameter values		
Joint propagation depth	PCC only	PCC and asphalt (0% LTE through asphalt)	
PCC thickness (in)	3.5	5.5	8
PCC modulus of elasticity (psi)	4.0E+06		
PCC Poisson's ratio	0.18		
PCC CTE (in/in/°F)	5.5E−06		
PCC joint spacing (ft)	6	8	12
PCC temp. difference (°F)	−12	0	24
Asphalt thickness (in)	3.5	5.5	7.5
Asphalt modulus of elasticity (psi)	1.0E+05	8.0E+05	3.0E+06
Asphalt Poisson's ratio	0.35		
Modulus of subgrade reaction, k-value (psi/in)	100	250	400
Shoulder width (ft)	6		
Lane shoulder LTE (%)	0 (Asphalt)	90 (tied PCC)	
Transverse joint LTE (%)	50	70	95
Wheel wander (in)	0	6	18
Single axle (kip)	18	30	
Tandem axle (kip)	36	60	

Development of ANNs

The ANNs are developed to predict the deflection basins for both the loaded and unloaded side of the joint. To train and test the ANNs using the factorial analyses, the neural network toolbox in MATLAB® was used (*24*). A total of 15 ANNs were trained, 10 ANNs for the medium size panels, and 5 networks for the large panels. These networks are separated into loaded or unloaded slab, axle type, joint depth, and temperature. Because of symmetry of the temperature loading condition, only one ANN is developed for both the loaded and unloaded sides of the joint (*25*). Each of the ANNs with each of their predictors are shown in Equations 3–5.

$$NN_{\Sigma L,A,JD}\left(\begin{matrix}\text{JTSpace}, l_{eff}, q_i^*, \text{k},\\ \text{LTE}_{Joint}, LTE_{shoulder}, s, \Phi\end{matrix}\right)\Big| \qquad (3)$$

Table 4. Predictability of ANNs

ANNs	Slab size	Deflection basin location	Axle (A)	Joint depth (JD)	R^2
$NN_{\Sigma L,A,JD}$	Medium	Loaded	Single	PCC only	0.999
$NN_{\Sigma UL,A,JD}$	Medium	Unloaded	Single	PCC only	0.999
$NN_{\Sigma L,A,JD}$	Medium	Loaded	Tandem	PCC only	0.999
$NN_{\Sigma UL,A,JD}$	Medium	Unloaded	Tandem	PCC only	0.999
$NN_{\Sigma L,A,JD}$	Medium	Loaded	Single	Full-depth	0.999
$NN_{\Sigma L,A,JD}$	Medium	Unloaded	Single	Full-depth	0.999
$NN_{\Sigma L,A,JD}$	Medium	Loaded	Tandem	Full-depth	0.999
$NN_{\Sigma L,A,JD}$	Medium	Unloaded	Tandem	Full-depth	0.999
$NN_{\Sigma T,JD}$	Medium	Loaded – (temp. only)	–	PCC only	0.999
$NN_{\Sigma T,JD}$	Medium	Loaded – (temp. only)	–	Full-depth	0.999
$NN_{\Sigma L,A,JD}$	Large	Loaded	Single	Full-depth	0.999
$NN_{\Sigma L,A,JD}$	Large	Unloaded	Single	Full-depth	0.999
$NN_{\Sigma L,A,JD}$	Large	Loaded	Tandem	Full-depth	0.999
$NN_{\Sigma L,A,JD}$	Large	Unloaded	Tandem	Full-depth	0.999
$NN_{\Sigma T,JD}$	Large	Loaded – (temp. only)	–	Full-depth	1.00

$$NN_{\Sigma U,A,JD}\left(\begin{matrix}\text{JTSpace}, l_{eff}, q_i^*, \text{k}, \\ \text{LTE}_{Joint}, LTE_{shoulder}, s, \Phi\end{matrix}\right) \tag{4}$$

$$NN_{\Sigma T,JD}\left(\begin{matrix}\text{JTSpace}, l_{eff}, q_i^*, \text{k}, \\ \text{LTE}_{Joint}, \text{LTE}_{shoulder}, s, \Phi\end{matrix}\right) \tag{5}$$

where:

$NN_{\Sigma L,A,JD}$ is the neural network for the sum of the 2 × 6-ft deflection basin for the loaded slab for axle type A (1 for single, 2 for tandem) for joint depth JD (0 for PCC only, 1 for full-depth);

$NN_{\Sigma U,A,JD}$ is the neural network for the deflection basin for the unloaded slab for axle type A (1 for single, 2 for tandem) for joint depth JD (0 for PCC only, 1 for full-depth);

$NN_{\Sigma T,JD}$ is the neural network for the deflection basin for the condition in which only temperature is present for joint depth JD (0 for PCC only, 1 for full-depth);

JTSpace is the joint spacing of the overlay (inches);

l_{eff} is the effective radius of relative stiffness of the overlay (inches) using bonded plate theory, and can be seen in Equation 6.

$$l_{eff} = \sqrt[4]{\frac{\text{Plate Stiffness}}{\left(1-\mu_{PCC}^{\;2}\right)k}} \tag{6}$$

where:

Plate Stiffness is determined using bonded plate theory (psi × in^3) (*26*);

μ_{PCC} is Poisson's ratio of the PCC;

k is the modulus of subgrade reaction (psi/in); and

q_i^* is the adjusted load/pavement weight ratio, and can be seen in Equation 7.

$$q_i^* = \frac{P_i}{A * \gamma_{eff} * h_{eff}} \tag{7}$$

where:

P_i is the axle load (lb.);

A is the parameter for axle type (1 for single, 2 for tandem axles);

γ_{eff} is the effective unit weight (pci);

h_{eff} is the effective height (inches); and

LTE_{Joint} is the LTE of the joint (%) (this parameter was selected over AGG* so the inputs were evenly distributed in the inference space);

$\text{LTE}_{shoulder}$ is the lane/shoulder LTE (%);

s is wheel wander offset from the lane/shoulder joint (inches)

Φ is Korenev's nondimensional temperature gradient, which is shown in Equation 8.

$$\Phi = \frac{2\alpha_{pcc}\left(1+\mu_{pcc}\right)l_{eff}^2}{h_{eff}^2}\frac{k}{\gamma_{eff}} * \Delta T \tag{8}$$

Where:

α_{pcc} is the coefficient of thermal expansion for the PCC overlay (inches/inches/°F); and

ΔT is the temperature difference in the overlay (°F).

The training of ANNs can have relatively high variability because of the possibility of local minima in the objective function (*27*). To reduce this variability, 10 ANNs were trained for each predictive model with different semirandom starting conditions. The model prediction was obtained by taking a robust average of the 10 ANNs and discarding the two highest and lowest estimates. Each ANN was trained using two hidden layers of 20 neurons each, which

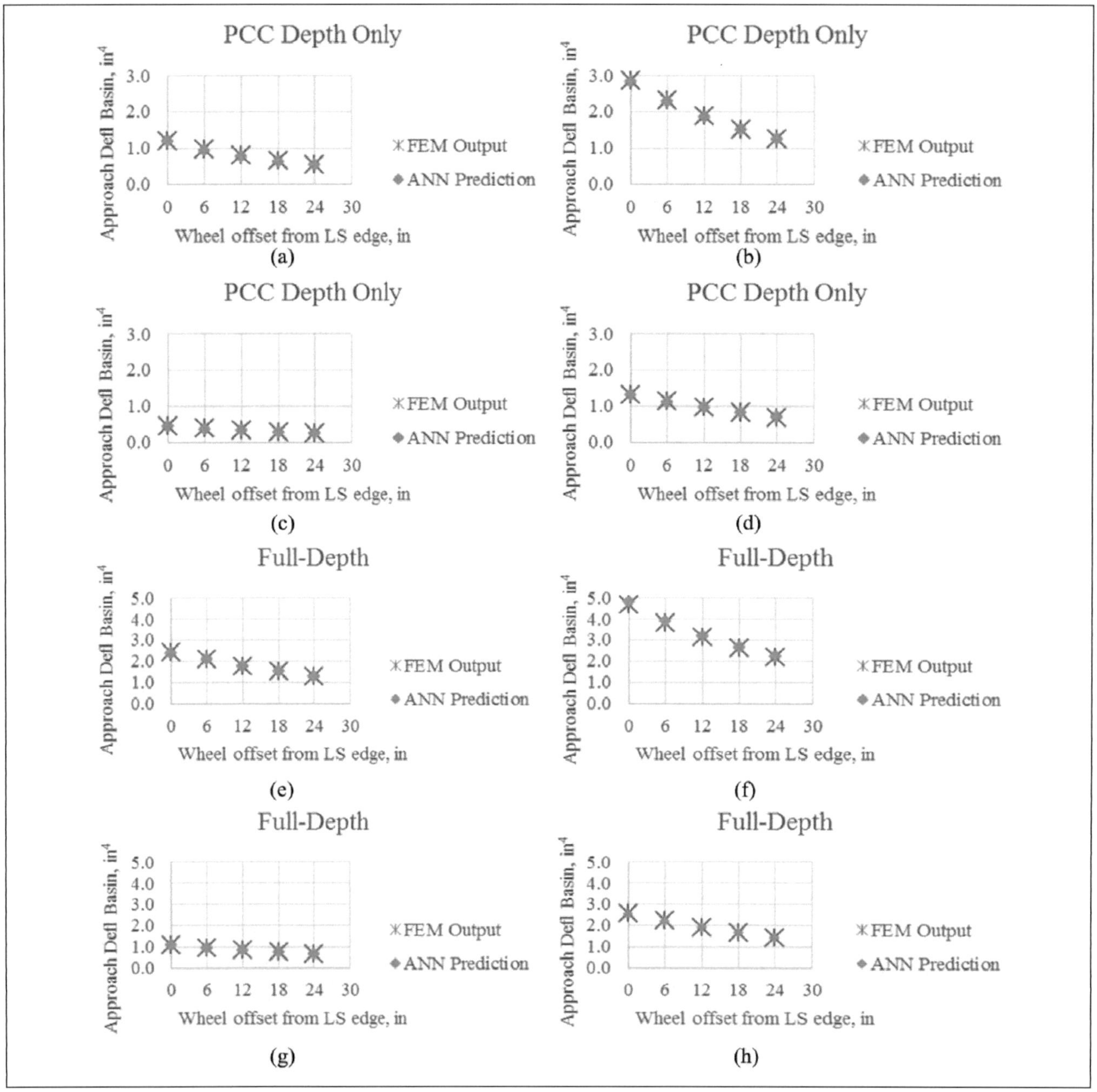

Figure 6. ANN validation. (*a*) Single axle: PCC = 3.5 inches, asphalt = 3.5 inches. (*b*) Tandem axle: PCC = 3.5 inches, asphalt = 3.5 inches. (*c*) Single axle: PCC = 8 inches, asphalt = 7.5 inches. (*d*) Tandem axle: PCC = 8 inches, asphalt = 7.5 inches. (*e*) Single axle: PCC = 3.5 inches, asphalt = 3.5 inches. (*f*) Tandem axle: PCC = 3.5 inches, asphalt = 3.5 inches. (*g*) Single axle: PCC = 8 inches, asphalt = 7.5 inches. (*h*) Tandem axle: PCC = 8 inches, asphalt = 7.5 inches.
Note: ANN = artificial neural network; FEM = finite element model; LS = lane shoulder.

was found to provide enough flexibility for accurate prediction. Overfitting was prevented by using weight decay selected using the Bayesian regularization training algorithm. This method was selected over early stopping algorithms, such as Levenberg–Marquardt optimization, since computational time for training was not a concern. To evaluate model performance, 85% of the data were used in the training set and the remaining 15% in the testing set. The performance of the test sets for each predictive model can be seen in Table 4.

A small validation was conducted on the effects of wheel offset for two structures using the medium-sized panel

ANNs. The joint spacing, PCC stiffness, asphalt stiffness, and k-value for both structures was 6 ft, 4E+06 psi, 8E+05 psi, and 100 psi/inch, respectively. Both depths of joint propagation were also considered. The comparisons can be seen in Figure 6.

Summary

Currently available BCOA design procedures only have the capability of predicting the performance life using fatigue cracking as the failure criteria. Therefore, there is a great need to be able to incorporate faulting as a failure criteria into the design process for BCOAs. To help address this, models were developed for predicting the response of BCOA to environmental and traffic loading conditions without the need to perform an FEM analysis for each loading condition and each different pavement structure. These BCOA response prediction models can now be incorporated into a framework for predicting the development of faulting in BCOA.

To accomplish this, first a computational model was developed using 3D FEM software, ABAQUS, to accurately predict the behavior of BCOA. The model was validated with sections at MnROAD and the UCPRC testing facility. The computational model included two models. One accounts for a joint that only propagates through the PCC and the other accounts for when the joint propagates through both the PCC and asphalt layers. The critical response for each model is the deflection basin on the approach and leave slabs. When the joint is only through the PCC, the response is recorded at the bottom of the PCC layer. When the joint is full-depth, the response is recorded at the bottom of the asphalt layer. The deflection basins are to be used to more accurately represent the difference in energy density on both sides of the joint as compared to just consideration of the corner deflections.

Finally, a fractional factorial was performed for a range of parameters, resulting in 19,440 FEM analyses that were used to train ANNs. The use of ANNs allow very accurate and quick predictions. These ANNs were developed to estimate the mechanistic response of BCOAs using a defined set of inputs. These estimates can be used in conjunction with performance data to produce a mechanistic empirical predictive faulting model for BCOA.

Acknowledgments

This paper describes research activities that were requested and sponsored by the California Department of Transportation (Caltrans). This sponsorship is gratefully acknowledged.

References

1. Mack, J. W., L. W. Cole, and J. P. Mohsen. Analytical Considerations for Thin Concrete Overlays on Asphalt. *Transportation Research Record: Journal of the Transportation Research Board*, 1993. 1388: 167–173.
2. Mack, J. W., L. D. Hawbaker, and L. W. Cole. Ultrathin Whitetopping: State-of-the-Practice for Thin Concrete Overlays of Asphalt. *Transportation Research Record: Journal of the Transportation Research Board*, 1998. 1610: 39–43.
3. Wu, C. L., S. M. Tarr, T. M. Refai, M. A. Nagi, and M. J. Sheehan. *Development of Ultrathin Whitetopping Design Procedure*. Research and Development Series No. 2124. Portland Cement Association, Skokie, Illinois, 1998.
4. DeSantis, J. W., J. M. Vandenbossche, K. Alland, S. Sachs, T. Burnham, and A. Montenegro. Joint Performance in Bonded Concrete Overlays of Asphalt. *Proc., 11th International Conference for Concrete Pavements (ICCP)*, San Antonio, TX, August 2016.
5. National Pavement and Concrete Technology Center. Faulting Study on BCOA Pavements in Iowa. In *Concrete Overlay Performance on Iowa's Roadways, 2017*. National Pavement and Concrete Technology Center, Ames, IA.
6. ARA, Inc., ERES Consultants Division. *Guide for Mechanistic-Empirical Design of New and Rehabilitated Pavement Structures*. Final report, NCHRP Project 1-37A. Transportation Research Board of the National Academies, Washington, D.C., 2004. http://www.trb.org/mepdg/guide.htm.
7. Khazanovich, L., M. Darter, and H. T. Yu. Mechanistic-Empirical Model to Predict Transverse Joint Faulting. *Transportation Research Record: Journal of the Transportation Research Board*, 2004. 1896: 34–45.
8. Vandenbossche, J. M., N. Dufalla, and Z. Li. Bonded Concrete Overlay of Asphalt Mechanistic-Empirical Design Procedure. *International Journal of Pavement Engineering*, Vol. 18, No. 11, 2017, pp. 1–12. https://doi.org/10.1080/10298436.2016.1141410.
9. Li, Z., and J. M. Vandenbossche. Redefining the Failure Mode for Thin and Ultra-Thin Whitetopping with a 1.8-×1.8-m (6-×6-ft) Joint Spacing. *Transportation Research Record: Journal of the Transportation Research Board*, 2013. 2368: 133–146.
10. DeSantis, J. W., S. Sachs, and J. M. Vandenbossche. Faulting Development in Concrete Overlays and Pavements. *International Journal of Pavement Engineering* (submitted Nov 2017, In Review).
11. Khazanovich, L., H. T. Yu, S. Rao, K. Galasova, E. Shats, and R. Jones. *ISLAB2000: Finite Element Analysis Program for Rigid and Composite Pavements. User's Guide*. ERES Consultants, Champaign, IL, 2000.
12. ABAQUS Inc. *ABAQUS Software Version 6.13: Analysis User's Manual*. ABAQUS Inc., Providence, RI, 2013.
13. Crovetti, J. A. *Design and Evaluation of Jointed Concrete Pavement Systems Incorporating Open-Graded Permeable Bases*. Ph.D. dissertation. University of Illinois at Urbana, Champaign, 1994.
14. Guo, E., L. Ricalde, and I. Kawa. *FAA Finite Element Design Procedure for Rigid Pavements*. Report DOT/FAA/AR-07/33,. Federal Aviation Administration, Washington, D.C., 2007.
15. Davids, W., G. Turkiyyah, and J. Mahoney. EverFE: Rigid Pavement Three-Dimensional Finite Element Analysis Tool. *Transportation Research Record: Journal of the Transportation Research Board*, 1998. 1629: 41–49.

16. Kuo, C. M. *Three-Dimensional Finite Element Analysis of Concrete Pavement*. Ph.D. dissertation. University of Illinois at Urbana, Champaign, 1994.
17. Burnham, T. R. *Construction Report for MnROAD Thin Whitetopping Test Cells 60-63*. Report MN/RC-2006-18. Office of Minnesota Road Research, St. Paul, MN, 2006.
18. Vandenbossche, J. M., and A. J. Fagerness. Performance, Analysis, and Repair of Ultrathin and Thin Whitetopping at Minnesota Road Research Facility. *Transportation Research Record: Journal of the Transportation Research Board*, 2002. 1809: 191–198.
19. Barman, M., F. Mu, and J. M. Vandenbossche. *Development of a Rational of Mechanistic-Empirical Based Design Guide for Thin and Ultra-thin Whitetopping: Task 1 Report*. University of Pittsburgh, Pittsburgh, PA, 2011.
20. Mu, F., and J. M. Vandenbossche. *Development of a Rational of Mechanistic-Empirical Based Design Guide for Thin and Ultra-Thin Whitetopping. Task 2 Report: Review and Selection of Structural Response Models*. University of Pittsburgh, Pittsburgh, PA, 2011.
21. Li, Z., J. M. Vandenbossche, and N. Dufalla. Structural Model for Longitudinal Cracking in Bonded Whitetopping with a 1.83-m × 1.83-m (6-ft × 6-ft) Joint Spacing. *American Society of Civil Engineering Journal of Transportation Engineering, Part B: Pavements*, Vol. 143, No. 4, 2017, p. 04017015. https://doi.org/10.1061/JPEODX.0000015.
22. Li, Z., N. Dufalla, F. Mu, and J. M. Vandenbossche. Bonded Concrete Overlay of Asphalt Pavements Mechanistic-Empirical Design Guide (BCOA-ME). *Theory Manual, FHWA TFP Study*, Vol. 5, 2013, p. 165.
23. Montgomery, D. C. *Design and Analysis of Experiments*, 8th ed. John Wiley & Sons, Inc., New York, 2013.
24. The MathWorks. *MATLAB and Statistics Toolbox Release*. The MathWorks, Inc., Natick, MA, 2013.
25. Sachs, S. *Development of a Joint Faulting Model for Unbonded Concrete Overlays of Existing Concrete Pavements Through a Laboratory and Numeric Analysis*. Ph.D. dissertation. University of Pittsburgh, 2017.
26. Khazanovich, L. *Structural Analysis of Multi-Layered Concrete Pavement Systems*. Ph.D. dissertation. University of Illinois, Urbana, 1994.
27. Ripley, B. D. *Pattern Recognition and Neural Networks*. Cambridge University Press, Cambridge, 1996.

The Standing Committee on Pavement Structural Modeling and Evaluation peer-reviewed this paper (18-06069).

The contents of this paper reflect the views of the authors and do not necessarily reflect the official views or policies of the State of California or the Federal Highway Administration. This paper does not represent any standard or specification.

Article

TRR
JOURNAL OF THE TRANSPORTATION RESEARCH BOARD

Transportation Research Record
2018, Vol. 2672(40) 371–381

DOI: 10.1177/0361198118758636
journals.sagepub.com/home/trr
SAGE

Identification of Potential Location and Extent of Localized Interface Debonding in the Wheelpath of Asphalt Pavements

David Hernando[1], Jeremy A. M. Waisome[1], Jian Zou[1], and Reynaldo Roque[1]

Abstract

Debonding between asphalt layers is usually modeled as a global or "smeared" phenomenon across the entire lane. However, field evidence indicates debonding may be limited to a portion of the interface. The objective of this study was to determine the potential location and extent of localized interface debonding in asphalt pavements so that more realistic interface conditions and pavement responses can be employed in future performance predictions. A parametric study was conducted to locate stress states potentially conducive to interface debonding. Factors considered included asphalt concrete (AC) layer thickness, AC-to-base stiffness ratio, interface compliance, tire size, and traffic wander. The parametric study showed existence of a zone of high shear stress coupled with low confinement for a broad range of depths (1–3 in. below the surface) and extending to 2 in. from the tire edge. Given the drop in confinement immediately outside the tire edge and that shear stress magnitude in this zone was similar to shear strength values reported in the literature, it was concluded that the repetition of these critical stress state conditions can cause localized debonding of an interface located about 2 in. below the pavement surface. Existence of a potential zone of localized interface debonding around the edge of a tire can promote a debonded strip below the wheelpath, which is consistent with field observations. The width of the debonded strip can extend to 42 in. Future research efforts should examine the stress redistribution associated with the presence of a debonded strip below the wheelpath.

The structural response of asphalt pavements is greatly affected by the bonding conditions of the multiple layers that make up the pavement structure. Achieving good bond at the interface between asphalt layers is desirable because it reduces bending stress induced by traffic loads by making the pavement act as a monolithic structure. Many studies have investigated the effect of the presence of an interface on pavement response and the potential affect on pavement distress, namely cracking, by means of smeared bonding conditions (*1–6*). This approach assumes the same properties and same response (either fully bonded, fully debonded or with some degree of compliance) across the entire interface.

One of the most widely used models to simulate smeared bonding conditions between asphalt layers is that derived by Uzan (*7*). In Uzan's model, a linear relationship between shear stress and the relative displacement between the two surfaces in contact at the interface is defined by a parameter called interface reaction modulus. Debonding between asphalt layers is introduced by means of a low or even a null (complete slip) interface reaction modulus applied to the entire interface. However, field studies indicate that debonding is not a global or "smeared" phenomenon that occurs across the entire lane, but rather a local phenomenon limited to a portion of the interface. Field cores extracted by Muench and Moomaw (*8*) in Washington State showed interface debonding only in cracked areas of the wheelpath. Cores removed outside the wheelpath exhibited no debonding or distress. In addition, they reported the average depth to the debonded interface was 1.8 in. Likewise, investigation on section trenches performed by the University of Florida during a cracking study also found presence of localized debonding limited to the wheelpath (unfortunately, those trenches are no longer available and no graphic proof could be included). Furthermore, interface debonding concurrent with longitudinal cracking was observed on wheelpath cores extracted as part of a Superpave monitoring project (*9*), as shown in Figure 1.

Field evidence indicates that debonding may be limited to a portion of the interface, which contradicts the way debonding has been traditionally modeled when analyzing the structural

[1]Department of Civil and Coastal Engineering, University of Florida, Gainesville, FL

Corresponding Author:
Address correspondence to David Hernando: dhernando@ufl.edu

Figure 1. Wheelpath core exhibiting interface debonding and longitudinal cracking (I-10, Duval County) [reference measurement in inches (1 in. = 2.54 cm)].

response of asphalt pavements (Figure 2). Therefore, there is a need to further investigate the potential location and extent of interface debonding so that more realistic interface conditions can be introduced to compute stresses and strains for pavement performance prediction.

Objective and Scope

The objective of this study was to determine the potential location and extent of localized interface debonding in asphalt pavements.

Some clarification on the differences between potential distress mechanisms resulting from deficient interface bonding conditions, mainly slippage and cracking, is warranted (Figure 3). First, slippage is associated with single events where high contact stresses from braking reach bond strength (i.e., braking-induced shear), whereas cracking is thought to be related to progressive interface bond breakdown under repeated (multiple events) shear stress induced by bending of the pavement structure (i.e., bending-induced shear). Second, slippage is restricted to the very top of the pavement; it results from failure of the uppermost interface located directly beneath the surface layer. Conversely, interface bond breakdown may occur at an interface deeper in the pavement structure. Finally, slippage is a local distress that may appear in braking zones, while interface bond breakdown could occur anywhere and according to Muench and Moomaw (*8*) "could potentially be much more prevalent on highways and thus represent a greater concern." Therefore, the scope of this study was bending-induced shear rather than braking-induced shear.

Research Approach

A parametric study was conducted to locate stress states potentially conducive to interface debonding. Uzan et al. (*1*) showed that interfaces follow a Mohr-Coulomb failure law; that is, shear resistance increases with the level of confinement. As illustrated in Figure 4, for a given shear stress, stress states of lower normal pressure will fall closer to the failure envelope and will be potentially more susceptible to interface failure. Therefore, the analysis focused on locating stress states of high shear in combination with low normal pressure.

The multilayer linear elastic program BISAR was employed to conduct axisymmetric analyses of multiple pavement structures and obtain responses at selected locations. Factors considered included asphalt concrete (AC) layer thickness, AC-to-base stiffness ratio (SR), interface conditions (level of compliance associated with the presence of an interface), tire size and traffic wander (lateral displacement of vehicles within the lane). Figure 5 summarizes the various pavement sections selected for analysis. Note that pavement sections were labeled according to their corresponding SR, that is, the ratio of AC stiffness to base stiffness. Also, an additional section with a stiffness gradient through the depth of the AC layer was included to study the effect of field oxidative aging.

All materials were considered linear elastic and were defined by their elastic modulus and Poisson's ratio. A broad range of AC elastic moduli was covered from 200,000 psi, which represents an asphalt mixture on a hot summer day, to 800,000 psi, which is more illustrative of a late Fall/early Spring evening. Two stiffness values were considered for a granular base, 25,000 and 40,000 psi, while subgrade stiffness was fixed at 15,000 psi. Poisson's ratio values of 0.35, 0.40 and 0.45 were assumed for AC, base, and subgrade layers, respectively.

Furthermore, two AC thicknesses were evaluated: 4 in., which represents a thin asphalt layer; and 8 in., which can be found on interstate pavements and major highways with a relatively long service life and rehabilitation history. An 18-kip single axle load with a uniform tire pressure of 100 psi was selected, which resulted in a circular static tire imprint with a radius of 5.35 in.

Stress states obtained from BISAR analyses were compared to stress states of failure reported in the literature to identify areas where the onset of debonding is more likely.

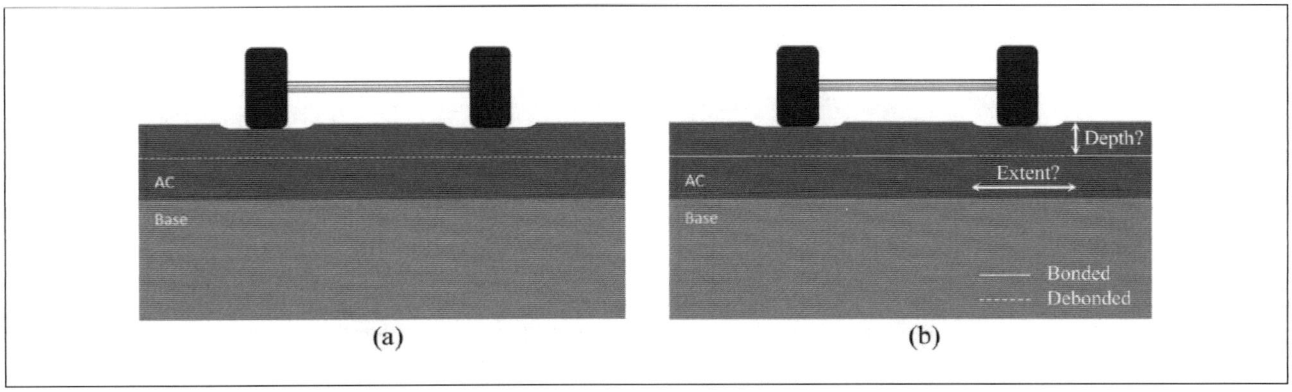

Figure 2. Difference between (*a*) smeared interface debonding traditionally used in modeling and (*b*) localized interface debonding observed in the field.

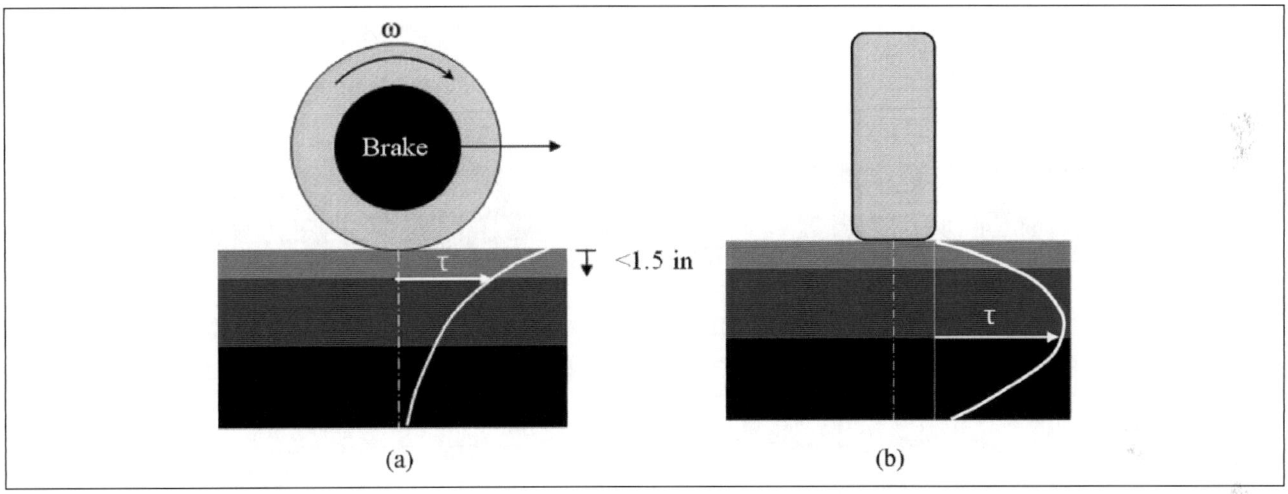

Figure 3. Schematic illustration of (*a*) braking-induced shear on a longitudinal section and (*b*) bending-induced shear on a transverse section (1 in. = 2.54 cm).

Results and Discussion

Effect of Pavement Structure

This section examines the effect of AC layer thickness and AC-to-base SR on the potential location and extent of debonding between asphalt layers. All layers were considered fully bonded at this stage (i.e., no poor bonding conditions or tracking of tack coat).

Stress distribution results for the pavement structures with a 4-in. thick AC layer are shown in Figure 6. Figure 6*a* illustrates the horizontal shear stress through the depth of the AC layer under the edge of the tire (i.e., at a radial distance of 5.35 in.). High shear stress was observed for a broad range of depths around the center of the layer (1–3 in.), with a maximum at a depth slightly <2 in. This finding is consistent with the depth at which debonding was observed in the field by Muench and Moomaw (*8*) and Roque et al. (*9*). Maximum shear stress occurred for SR = 32, closely followed by the section with a stiffness gradient. This was expected as shear stress is a function of bending, which increases with SR. Figure 6*b* shows the horizontal shear stress distribution along a horizontal plane located 2 in. below the surface. The maximum shear stress was under the edge of the tire for all cases. Figure 6*b* also illustrates the existence of an area of high vertical normal stress below the tire (confined zone), but confinement drastically decreased below 30 psi outside the edge of the tire. More importantly, a zone of relatively high shear stress (between 70 psi and 25 psi) coupled with low vertical normal stress (<30 psi) was identified immediately next to the edge of the tire. Note that for a normal stress of 30 psi, Uzan et al. (*1*) reported interface shear strength values between 90 psi and 20 psi. Likewise, Mohammad et al. (*6*) measured shear strength values below 40 psi for a maximum confinement pressure of 20 psi. Thus, stress states obtained in this study seemed to agree with failure conditions reported in the literature, which substantiates that localized interface debonding is likely to occur.

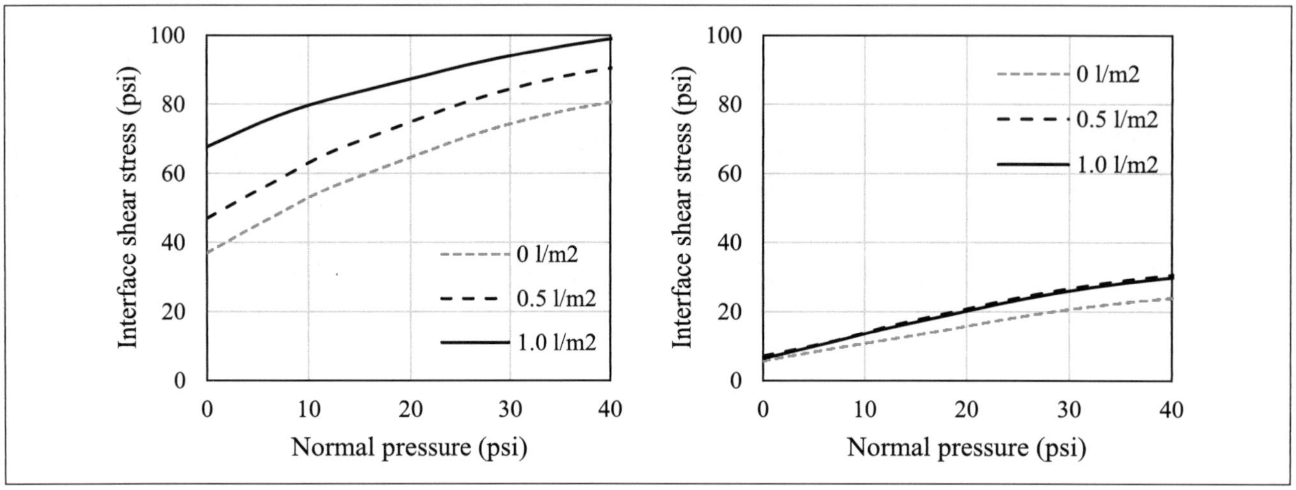

Figure 4. Interface shear strength as a function of normal pressure for multiple tack coat application rates and testing temperatures: (*a*) 25°C and (*b*) 55°C (*1*) (1 psi = 6.89 kPa).

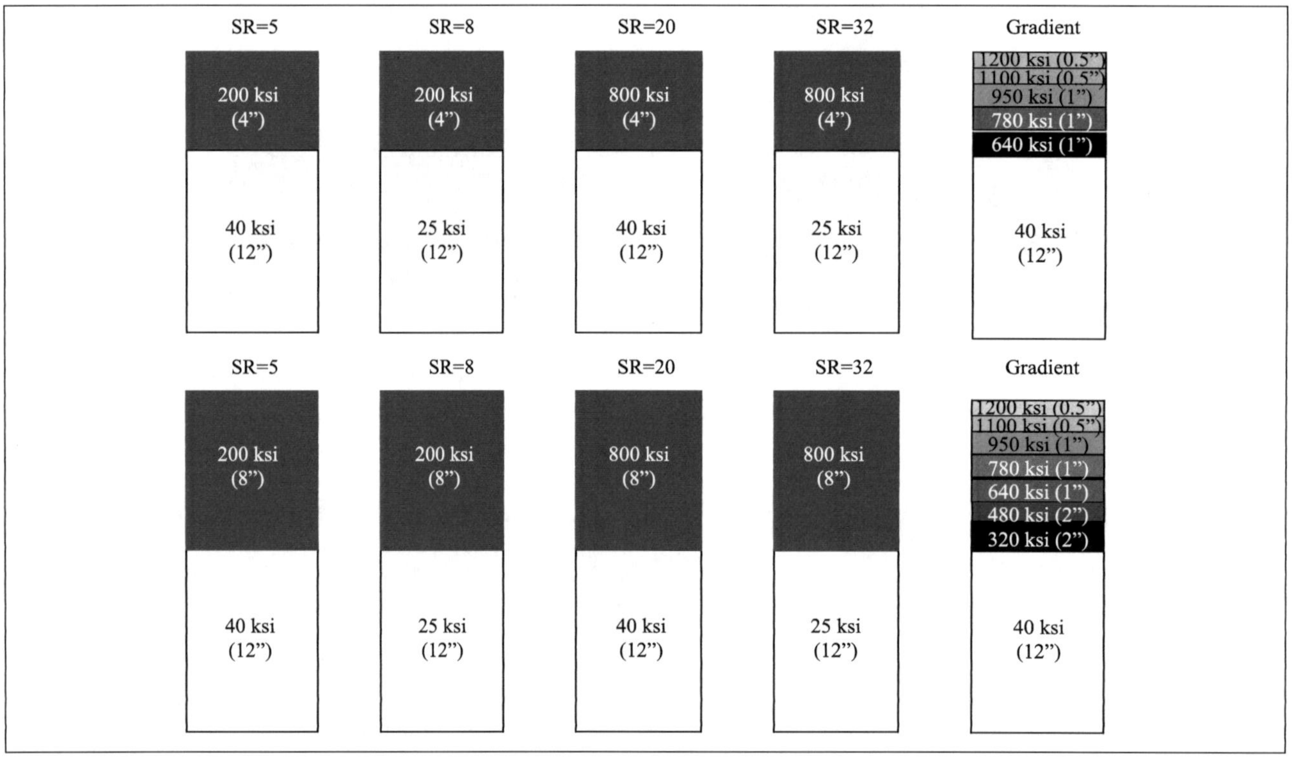

Figure 5. Pavement sections selected for evaluation of structural characteristics (layer thickness shown in brackets, 1 ksi = 6.89 MPa, 1 in. = 2.54 cm).

Figure 7 depicts the stress distribution for the pavement structures with an 8-in. thick AC layer. Horizontal shear stress magnitude was lower than in the case of a 4-in. AC layer given the reduction in bending resulting from a thicker AC layer. Furthermore, Figure 7*a* shows the existence of an area of high shear stress between 1 in. and 3 in. from the surface, with a maximum at a depth of about 2 in., which agreed with the results obtained for a 4-in. AC layer. In other words, horizontal shear stress tended to peak below the edge of the tire at a depth of 2 in. regardless of AC thickness for the wide range of pavement structures analyzed. Figure 7*b* plots the horizontal shear stress distribution and average

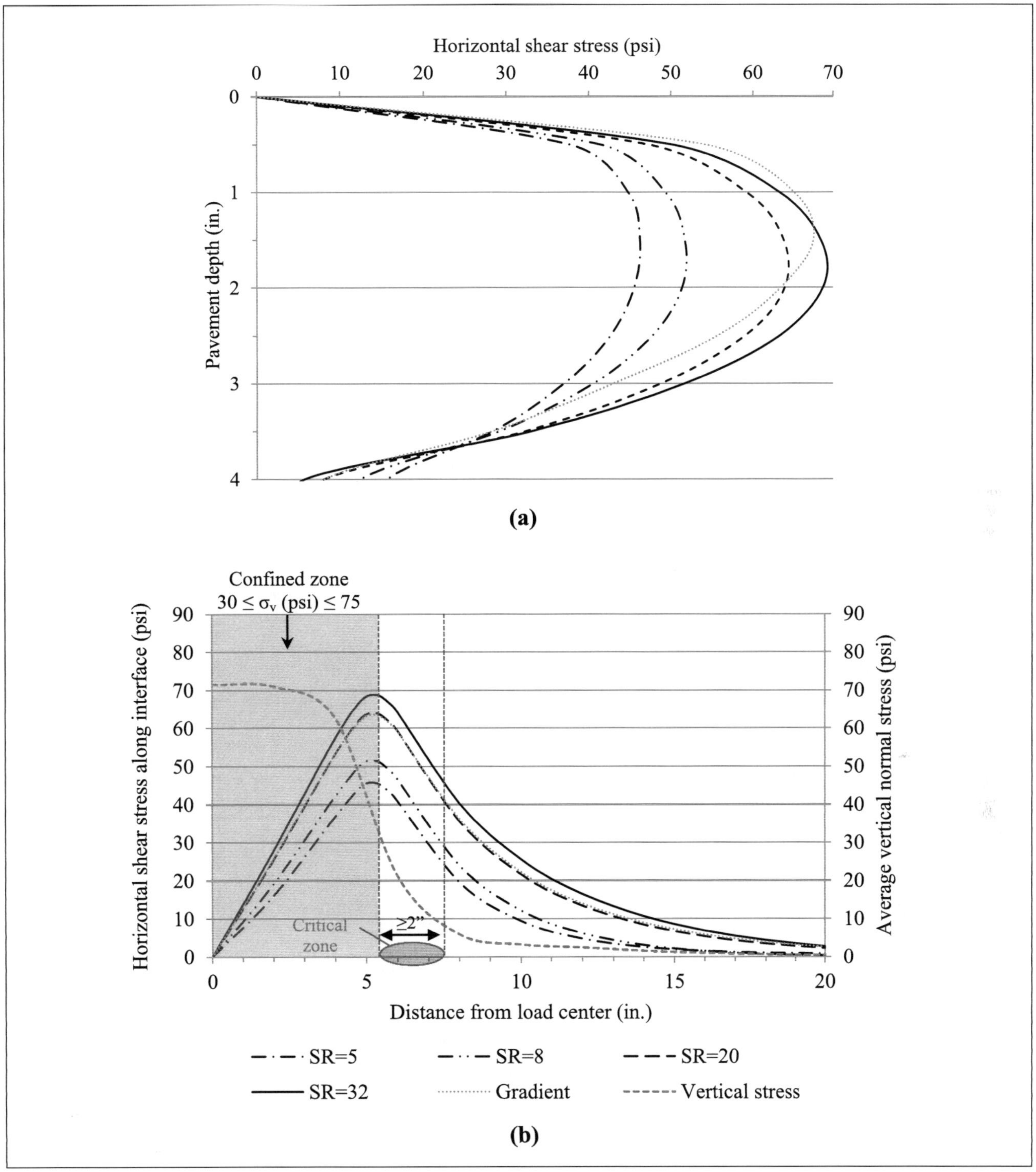

Figure 6. Stress distribution of pavement structures with a 4-in. AC layer: (*a*) horizontal shear stress through the depth of the asphalt layer under the tire edge and (*b*) horizontal shear stress and average vertical stress at a depth of 2 in. (1 psi = 6.89 kPa, 1 in. = 2.54 cm).

vertical normal stress along this critical depth of 2 in. Like the 4-in. AC layer case, an area of high vertical normal stress was found below the tire (confined zone). Also, confinement dropped outside the edge of the tire where high shear stress still existed (shear stress outside the confined zone ranged from 45 psi to 25 psi).

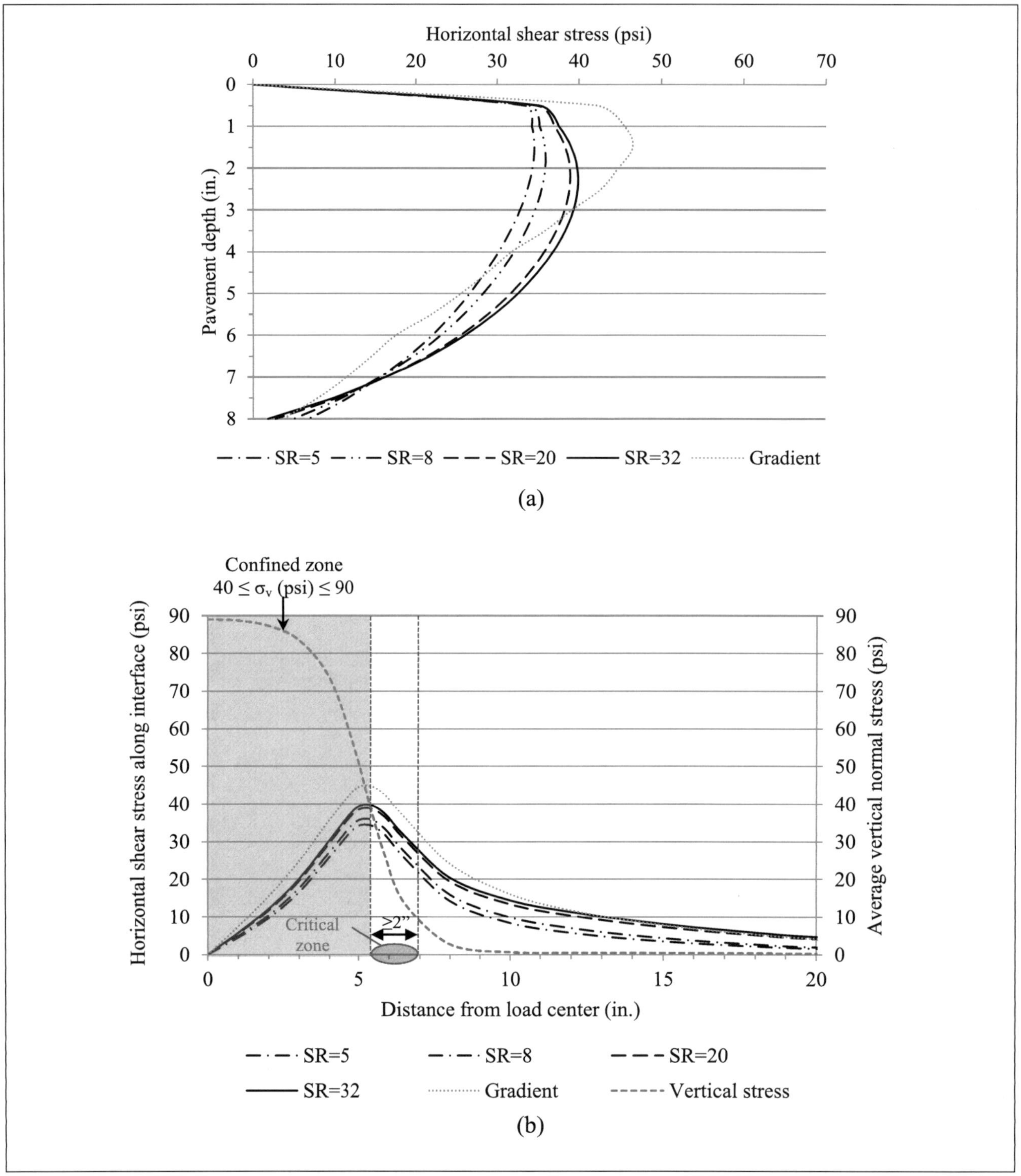

Figure 7. Stress distribution of pavement structures with an 8-in. AC layer: (*a*) horizontal shear stress through the depth of the asphalt layer under the tire edge and (*b*) horizontal shear stress and average vertical stress at a depth of 2 in. (1 psi = 6.89 kPa, 1 in. = 2.54 cm).

Results in Figures 6 and 7 identified the existence of a zone of stress states of relatively high shear (from 25 psi to 70 psi) coupled with low normal pressure at a depth of 2 in. and extending to at least 2 in. from the edge of the tire. Given the drop in confinement immediately outside the tire edge and that shear stress magnitude in this zone was similar to

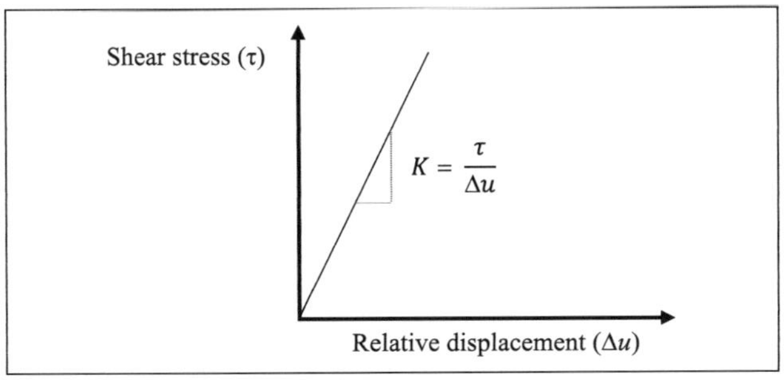

Figure 8. Interface model for evaluation of smeared bonding conditions (7).

shear strength values reported in the literature, it is believed that the repetition of these critical stress state conditions can cause localized debonding of an interface located about 2 in. below the pavement surface.

Effect of Interface Conditions

Most studies in the literature employed the model derived by Uzan (*7*) to simulate interface bonding conditions between asphalt layers. In Uzan's model (Figure 8), a linear relationship between shear stress (τ) and the relative displacement between the two surfaces in contact at the interface (Δu) is defined by a parameter called interface reaction modulus (K). The interface reaction modulus can account for different levels of interface compliance associated with the presence of a bonding agent between two asphalt layers. For example, harder bonding agents applied at a low rate will tend to be less compliant (higher K value) than softer bonding agents at a high application rate (lower K value).

Field experiences discussed by Uzan et al. (*1*) suggested the state of adhesion in real conditions is an intermediate one between full adhesion and zero adhesion, depending on material properties and construction quality. In other words, there is some inherent compliance associated with the interface. As a reference, Uzan et al. (*1*) recommended a K value of 10 N/mm^3 to model an interface in the upper part of an asphalt pavement subjected to a tire pressure of 75–100 psi at normal highway speed and 25°C.

An axisymmetric analysis with the multilayer linear elastic program BISAR was performed in this section to evaluate the effect of interface conditions following Uzan's model. Figure 9 plots the results obtained for the pavement structure with a 4-in. AC layer and a SR of 32 (this structure was chosen for yielding the greatest shear stress found in Figures 6 and 7). An interface with varying reaction modulus was introduced at the critical depth of 2 in. Trends shown in Figure 9 agreed with observations reported by Uzan (*7*): K values above 100 N/mm^3 yielded similar results to the fully bonded case, and K values below 1 N/mm^3 tended to represent the fully debonded scenario. The horizontal shear stress distribution along the interface followed the same pattern regardless of K value: shear stress started at zero under the center of the load because of the symmetry conditions, peaked below the edge of the tire, and gradually reduced away from the tire. In addition, interface conditions were found to have a minimal effect on vertical stress distribution: in all cases, high vertical stress existed below the tire but dropped outside the edge of the tire. Figure 9 shows that, even when smeared bonding interface conditions were considered by way of the interface reaction modulus model, a 2-in. zone of high shear stress coupled with low confinement where localized debonding can potentially develop was still present.

It should be pointed out that in the case of excessively compliant interfaces, for example interfaces with K values below 1 N/mm^3, the maximum horizontal shear stress in the unconfined portion of the interface would barely reach 10 psi. Under these conditions, development of localized interface debonding seems rather unlikely. Figure 10 depicts that an excessively compliant interface or, in an extreme scenario, a fully debonded interface would result in high tension above the interface, which could theoretically lead to bottom-up cracking under the center of the tire. Nevertheless, it is important to keep in mind that the area below the tire is subjected to high vertical stress (confined zone), which allows for the development of horizontal frictional resistance along the interface. The development of horizontal frictional resistance reduces the tensile stress above the interface and, consequently, reduces the likelihood of bottom-up cracking below the center of the tire. The high tensile stress traditionally predicted under the center of a tire in a debonded pavement is the result of an excessively simplified model in which the frictional resistance resulting from the high vertical stress below the tire is not captured.

Consideration of Tire Load Travel

The two previous sections identified the presence of a critical zone extending to 2 in. from the edge of a static tire where interface debonding can potentially occur. However, tire loads in the field do not remain static but rather travel with traffic flow. Figure 11 depicts a plan view of a truck axle on a roadway lane (circular tire imprints were considered for illustration purposes). As a truck travels the roadway, the critical zone of high shear coupled with low confinement found around the edge of the tire (represented as a red ring in Figure 11) moves longitudinally along both wheelpaths. Note that the critical ring-shaped zone around the tire describes a strip when moved longitudinally. Consequently, the presence of a critical stress state zone around a tire can promote a debonded strip below both wheelpaths. This mechanism, which to the authors' knowledge has not been reported in the literature, is consistent with the presence of localized debonding in the wheelpath found in field studies by Muench and Moomaw (*8*) and Roque et al. (*9*).

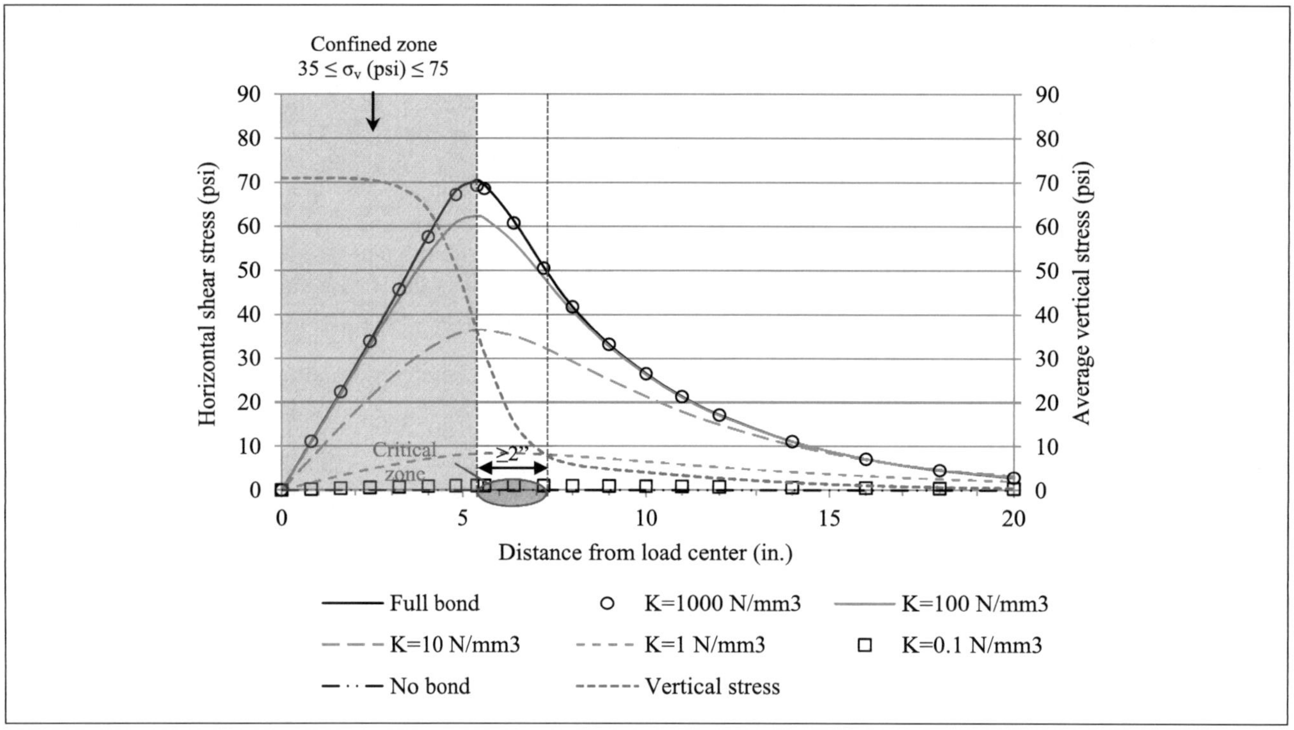

Figure 9. Horizontal shear stress and average vertical stress at a depth of 2 in. for different interface reaction modulus (*K*) values (4-in. AC layer pavement structure, stiffness ratio = 32) (1 psi = 6.89 kPa, 1 in. = 2.54 cm).

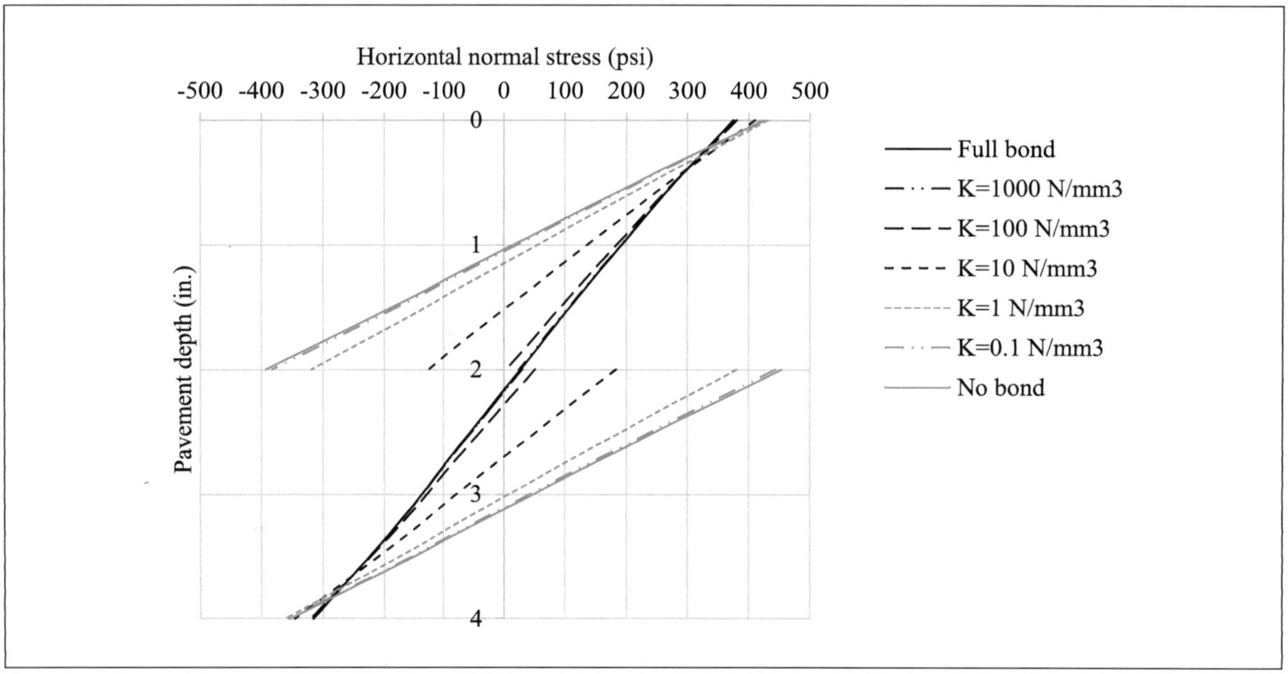

Figure 10. Horizontal normal stress under the center of a tire through the depth of the AC layer for different interface reaction modulus (*K*) values (4-in. AC layer pavement structure, stiffness ratio = 32) (negative values denote tension, 1 psi = 6.89 kPa, 1 in. = 2.54 cm).

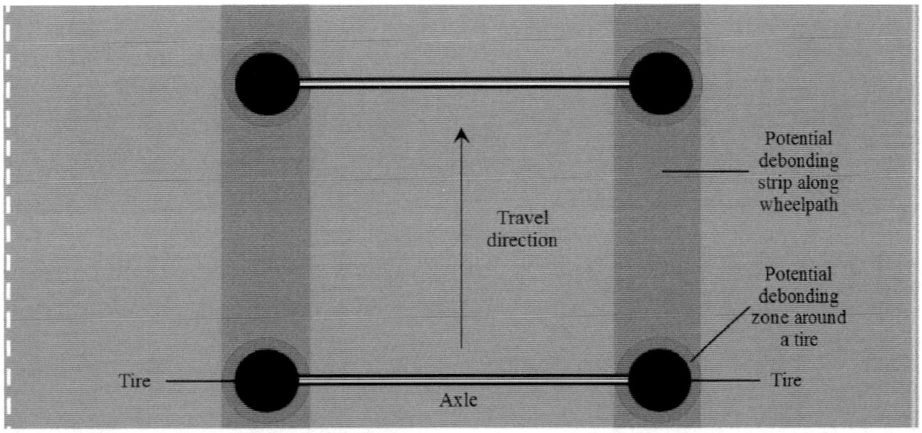

Figure 11. Schematic representation of development of a debonded strip along the lane wheelpath.

Table 1. Approximate Contact Width of Different Radial Tires (*12–14*) (1 in. = 2.54 cm)

Category	Tire denomination	Contact width (in.)
Conventional	275/80R22.5	8
First-generation wide-base (super-single)	385/65R22.5	11.5
	425/65R22.5	12.5
New-generation wide-base	445/50R22.5	17.8
	455/55R22.5	18

Effect of Tire Size and Traffic Wander

Within the last two decades, trucking companies have gradually introduced wide-base radial tires because of economic and environmental benefits such as improved fuel efficiency, reduced emissions, and increased load capacity. Compared to a conventional radial tire with a typical contact width of 8 in., wide-base tires range from 12 in. (first generation) up to 18 in. (new generation), as illustrated in Table 1. Thus, tire type has a significant effect on the extent of the potentially debonded strip along the wheelpath. Furthermore, debonding extent is also affected by the lateral movement of the tire within the lane, referred to as traffic wander. Field studies showed that the lateral position of wheel loads within a lane follows a normal distribution with a standard deviation of about 10 in. (*10, 11*). Since 68% of a normal distribution lies within one standard deviation of the mean, it is reasonable to believe that the width of the potentially localized debonding strip along the wheelpath can extend up to 10 in. to each side. Thus, for normal traffic wander conditions, defined by a standard deviation of 10 in., the width of the debonded strip could extend up to 42 in. when new-generation wide-base tires are dominant (Figure 12).

Conclusions

The parametric study conducted proved existence of a zone of high shear stress coupled with low confinement for a broad range of depths (1–3 in. below the surface, with the more critical condition being about 2-in. deep) and extending to 2 in. from the tire edge. Given the drop in confinement immediately outside the tire edge and that shear stress magnitude in this zone was similar to shear strength values reported in the literature, the repetition of these critical stress state conditions can cause localized debonding of an interface located about 2 in. below the pavement surface. This location was independent of AC thickness although shear stress magnitude was greater for sections with a thinner AC layer. Structures with a high AC-to-base SR or a stiffness gradient in AC layer resulting from field oxidative aging were more susceptible to debonding (greater shear stress at similar confinement levels). Existence of a potential zone of localized interface debonding around the edge of a tire can promote a debonded strip below the wheelpath as truck traffic travels the roadway, which is consistent with field observations. The width of the debonded strip can range from 12 in. for conventional tires with no wander to 42 in. in the case of new-generation wide-base tires and a traffic wander standard deviation of 10 in.

Future Research

It is recommended that future research efforts examine the stress redistribution associated with the presence of a

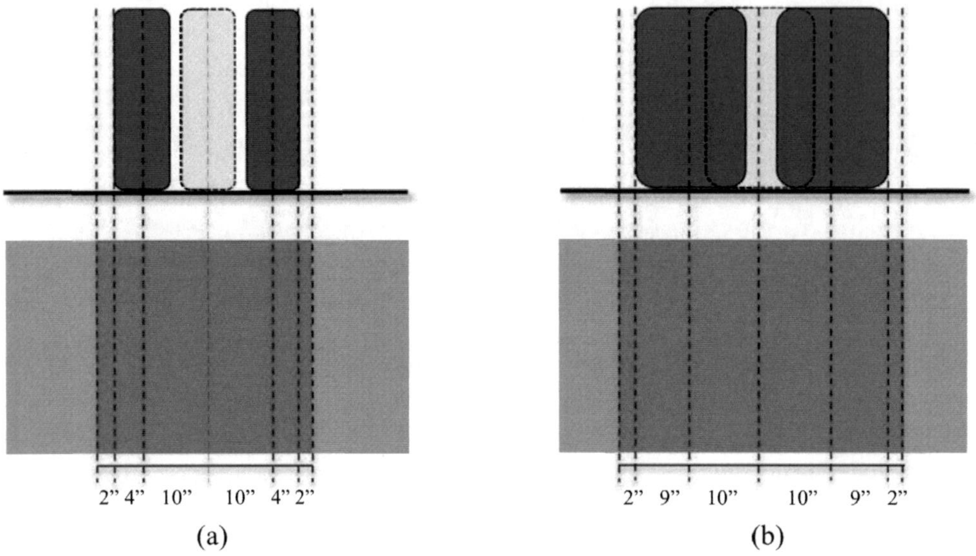

Figure 12. Potential extent of localized debonded strip along the wheelpath considering traffic wander and tire width: (*a*) conventional radial tire (8-in. wide) and (*b*) new-generation wide-base tire (18-in. wide) (1 in. = 2.54 cm).

debonded strip below the wheelpath. An emphasis should be placed on the study of progressive interface debonding along the transverse direction versus initiation of a crack through the asphalt layers that meet at the interface.

Acknowledgments

The authors thank the Florida Department of Transportation for supporting this study.

References

1. Uzan, J., M. Livneh, and Y. Eshed. Investigation of Adhesion Properties Between Asphaltic Concrete Layers. *Journal of the Annual Meeting of the Association of Asphalt Paving Technologists*, Vol. 47, 1978, pp. 495–521.
2. Romanoschi, S. A., and J. B. Metcalf. Effects of Interface Conditions and Horizontal Wheel Loads on the Life of Flexible Pavement Structures. *Transportation Research Record: Journal of the Transportation Research Board*, 2001. 1778: 123–131.
3. Kruntcheva, M. R., A. C. Collop, and N. H. Thom. Properties of Asphalt Concrete Layer Interfaces. *ASCE Journal of Materials in Civil Engineering*, Vol. 18, No. 3, 2006, pp. 467–471.
4. Maina, J. W., M. De Beer, and K. Matsui. Effects of Layer Interface Slip on the Response and Performance of Elastic Multi-layered Flexible Airport Pavement Systems. *Proc., 5th International Conference on Maintenance and Rehabilitation of Pavements and Technological Control*, Utah, 2007, pp. 145–150.
5. Willis, J. R., and D. H. Timm. Forensic Investigation of Debonding in Rich Bottom Pavement. *Transportation Research Record: Journal of the Transportation Research Board*, 2007. 2040: 107–114.
6. Mohammad, L. N., M. A. Elseifi, A. Bae, N. Patel, J. Button, and J. A. Scherocman. *NCHRP Report 712: Optimization of Tack Coat for HMA Placement*. Transportation Research Board of the National Academies, Washington, D.C., 2012.
7. Uzan, J. Influence of the Interface Condition on Stress Distribution in a Layered System. *Transportation Research Record: Journal of the Transportation Research Board*, 1976. 616: 71–73.
8. Muench, S. T., and T. Moomaw. *De-Bonding of Hot Mix Asphalt Pavements in Washington State: An Initial Investigation*. Report no. WA-RD 712.1. Washington State Department of Transportation, Olympia, Wash., 2008.
9. Roque, R., S. Chun, J. Zou, G. Lopp, and C. Villiers. *Continuation on Superpave Projects Monitoring*. FDOT Final Report BDK-75-977-06. University of Florida, Gainesville, Fla., 2011.
10. Erlingsson, S., S. Said, and T. McGarvey. Influence of Heavy Traffic Lateral Wander on Pavement Deterioration. *Proc., 4th European Pavement and Asset Management Conference*, Malmö, Sweden, 2012.
11. Timm, D., and A. Priest. *Wheel Wander at the NCAT Test Track*. Report 05-02. National Center for Asphalt Technology, Auburn, Ala., 2005.
12. Myers, L. A., R. Roque, B. Ruth, and C. Drakos. Measurement of Contact Stresses for Different Truck Tire Types to Evaluate Their Influence on Near-Surface Cracking and Rutting. *Transportation Research Record: Journal of the Transportation Research Board*, 1999. 1655: 175–184.
13. Greene, J., U. Toros, S. Kim, T. Byron, and B. Choubane. *Impact of Wide-Base Single Tires on Pavement Damage*. Research Report 09-528. Florida Department of Transportation, State Materials Office, Gainesville, Fla., 2009.

14. Wang, G., and R. Roque. Impact of Wide-Based Tires on the Near-Surface Pavement Stress States Based on Three-Dimensional Tire-Pavement Interaction Model. *Road Materials and Pavement Design*, Vol. 12, No. 3, 2011, pp. 639–662.

The Standing Committee on Pavement Structural Modeling and Evaluation peer-reviewed this paper (18-06110).

Article

TRR
JOURNAL OF THE TRANSPORTATION RESEARCH BOARD

Transportation Research Record
2018, Vol. 2672(40) 382–394

DOI: 10.1177/0361198118796728
journals.sagepub.com/home/trr
SAGE

Finite Element Studies of Skid Resistance under Hot Weather Condition

Tianchi Tang[1], Kumar Anupam[1], Cor Kasbergen[1], Reginald Kogbara[2], Athanasios Scarpas[1,3], and Eyad Masad[2,4]

Abstract

The skid resistance of a pavement surface is an important characteristic that influences traffic safety. Previous studies have shown that skid resistance varies with temperature. However, relatively limited work has been carried out to study the effect of temperature on skid resistance in hot climates. Recent developments in computing and computational methods have encouraged researchers to analyze the mechanics of the tire-pavement interaction phenomenon. The aim of this paper is to develop a thermo-mechanical tire pavement interaction model that would allow more robust and realistic modeling of skid resistance using the Finite Element (FE) method. The results of this model were validated using field tests that were performed in the State of Qatar. Consequently, the validated FE model was used to quantify the effect of factors such as speed, inflation pressure, wheel load, and ambient temperature on the skid resistance/braking distance. The developed model and analysis methods are expected to be valuable for road engineers to evaluate the skid resistance and braking distance for pavement management and performance prediction purposes.

The skid resistance of an asphalt pavement is an important characteristic that influences driving safety conditions. Research studies show that there is a relationship between low skid resistance of a pavement surface and traffic accidents (*1*). Moreover, the accident risk may increase dramatically when the friction numbers decrease below certain thresholds (*2*). Therefore, it is important for road researchers to understand the fundamental mechanisms of tire-pavement interaction.

Studies have shown that the skid resistance is significantly influenced by various pavement surface characteristics (*3*). It has also been demonstrated through experimental and mathematical modeling that temperature and tire related factors such as load, speed, and inflation pressure also influence skid resistance (*4–6*). It was observed that skid resistance decreases with an increase in temperature (*7*, *8*). This relationship can be rationalized by the fact that temperature-dependent viscoelastic properties of tire rubber have a significant influence on its hysteretic friction. The lack of proper consideration of the effect of temperature on skid resistance measurements would lead to erroneous maintenance management decisions and pavement performance predictions.

To properly assess the surface condition of a pavement, highway agencies and authorities around the world conduct regular skid resistance measurements. One of the popular ways to measure skid resistance is by the use of continuous friction measuring equipment (CFME). This is because CFMEs are designed to test the friction around 10%–20% slip condition, which is similar to that used in vehicles equipped with anti-lock braking systems (ABS) (*9*). GripTester is one such device that measures the longitudinal friction along the wheel track and operates at a 15% fixed slip ratio. The device has the capability for skid testing at a wide range of speeds (from 5 to 130 km/h) (*10*). Although the GripTester could provide reliable data, uncertainties related to temperature and vehicle speed variations exist. To handle these uncertainties, statistical based approaches are often adopted by road agencies. The challenge with these approaches, however, is that they tend to ignore the fundamental mechanisms involved in skid resistance and are only valid for the section(s) where they were calibrated. Moreover, several field tests

[1]Section of Road Engineering, Faculty of Civil Engineering & Geosciences, Delft University of Technology, Delft, The Netherlands
[2]Mechanical Engineering Program, Texas A&M University at Qatar, Education City, Doha, Qatar
[3]Khalifa University of Science and Technology, Abu Dhabi, United Arab Emirates
[4]Zachry Department of Civil Engineering, Texas A&M University, College Station, TX

Corresponding Author:
Address correspondence to Tianchi Tang: t.tang@tudelft.nl

are needed to obtain such statistical-based transfer functions, which require enormous time and resources, apart from temperature-related uncertainties.

Recent developments in computational methods and computing power have allowed researchers to develop and analyze complex tire-pavement interaction models using the Finite Element (FE) method. Such models, on validation, can help pavement engineers and authorities evaluate several scenarios for the selection of pavement materials. In general, FE simulations of the tire-pavement interaction include three main aspects, namely, modeling of pavement roughness and structure, modeling of tire structure and modeling of tire-pavement contacts (*11*). Tanner (*12*) presented a numerical model for the solution of the frictional contact problems for the tire-pavement interaction. Davis (*13*) proposed a numerical model primarily to study tire mechanical properties. Johnson et al. (*14*) used a FE model to simulate the tire loading. In recent years, Cho et al. (*15*) presented a tire-pavement interaction model using FE tools. The majority of these models simulated the pavement sub model as a rigid flat plate, which does not reflect a real field situation.

Wollny et al. (*16*) proposed an inelastic pavement layered structure in their coupled tire-pavement interaction FE model. However, the pavement surface texture was not considered in this model. Ong and Fwa (*17–20*) presented a FE model where a pavement with grooved surfaces was considered to study skid resistance. However, the major limitation of their model is the use of non-rotating tire sub model. Based on the above review of models, it can be seen that very few studies have evaluated the effect of temperature primarily because of the complexity of a thermo-mechanical tire-pavement interaction mechanism.

Recently, researchers (*21–23*) used an innovative staggered interaction FE model to determine the temperature developments for a rolling/skidding tire by considering actual pavement surface morphologies (obtained on the basis of computed tomography imaging scans). However, these models were developed for normal temperature conditions and real size test tires like PIARC, ASTM, and Goodyear and so forth. As mentioned earlier, in this research program, GripTester measurements were carried out under field conditions in Qatar. To the best of the authors' knowledge, a thermo-mechanical model to study the tire-pavement interaction mechanism in hot weather is still not available to researchers. Therefore, the aim of this paper is to develop such a model for skid resistance evaluation using a GripTester, and to validate the FE model on the basis of field tests.

Scope of Study

The aim of this study is to develop a thermo-mechanical tire-pavement interaction model and to validate the model using actual field tests. The model would then be used to:

- propose a mechanics-based methodology to transfer the skid resistance measurement from a GripTester tire to a passenger car tire;
- analyze safety aspects (i.e., braking distance) related to skid resistance;
- evaluate the influence of temperature on skid resistance and pavement safety aspects; and
- evaluate the influence of the pavement roughness characteristics on skid resistance and pavement safety aspects.

Description of the FE Model

Numerical Modeling of Pavement Structure

The asphalt pavement section was designed and constructed in line with Qatar Construction Specifications (QCS), which essentially adopt the Marshall mix design method. Pen 60–70 base bitumen and gabbro were used in the asphalt mixtures of the surface course of the pavement section. The asphalt specimens were extracted from the aforementioned section and later used to capture surface morphologies and internal structures with a laser profilometer and an X-ray tomographer. The output images were then imported into Simpleware software (*24*). The image segmentation and filtering technique available in Simpleware were used to distinguish the three phases of the mixture namely air voids, aggregates, and asphalt binder. Once the post-segmentation process was completed, the microstructure FE meshes for the pavement surface were generated with aggregates modeled as an elastic material and asphalt binder as a viscoelastic material, as shown in Figure 1.

Numerical Modeling of Tire Structure

Recent advances in computational methods and computing power bring the modeling of the tire-pavement interaction to a new level of accuracy accounting for the tire cross-section geometry and other components (*25–27*). The standard GripTester tire (ASTM E 1844) was simulated using the commercial software Abaqus (*28*) and an in-house software at TU Delft. The geometrical details of the GripTester tire can be found ASTM E 1844 procedure.

The modeling of the GripTester tire starts with the development of a 2D FE mesh of its cross section as obtained from ASTM E 1844. By revolving the 2D FE mesh, the 3D FE mesh of the tire body and tread are generated. As the contact section of the tire with the pavement is the tread surface, it is not necessary to mesh the tire body and tread with the same mesh density. At the

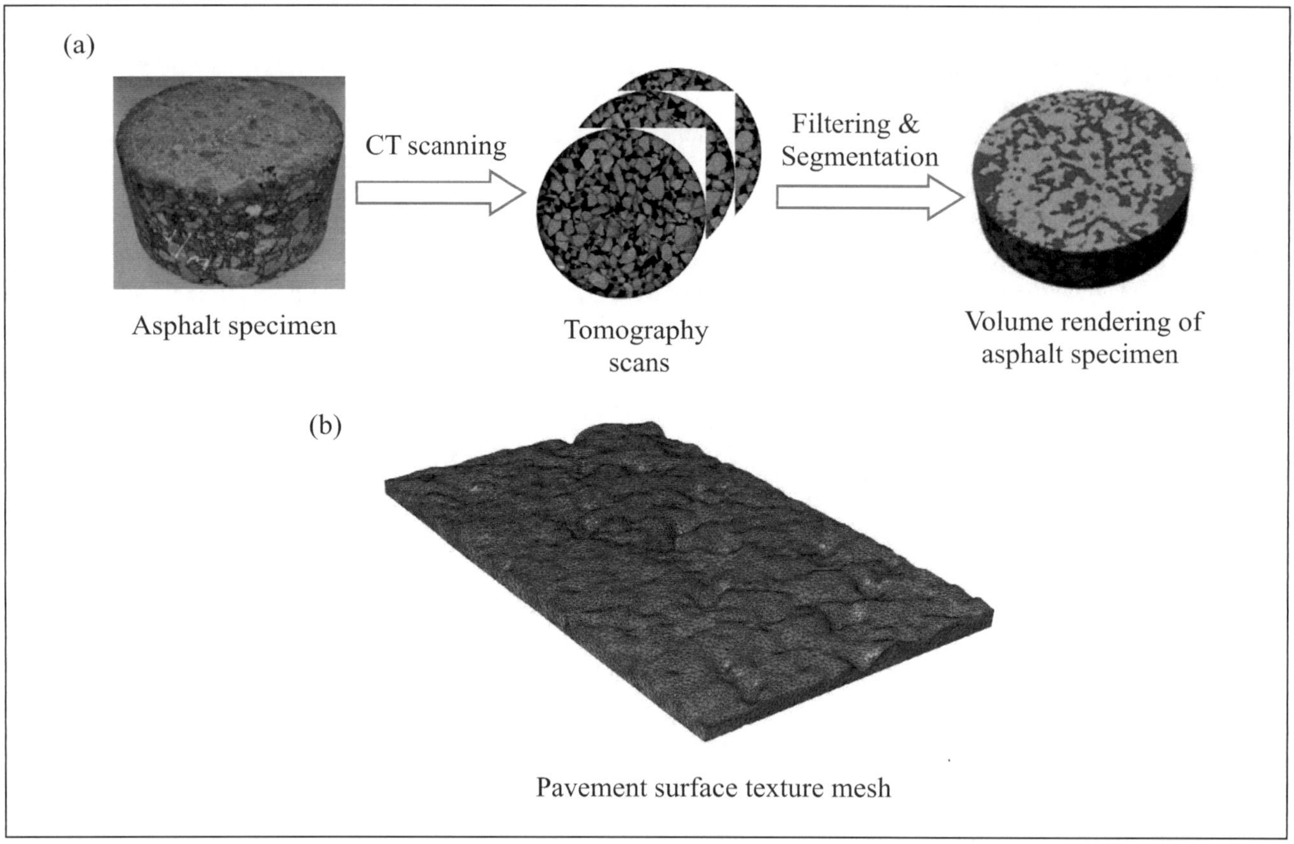

Figure 1. Generation of asphalt surface FE mesh: (*a*) procedure to obtain asphalt surface morphology and (*b*) resulting FE asphalt surface meshes from Simpleware.

last step, the whole tire model is completed by assembling the tire body and tread parts together using surface-surface constraint. The developed FE tire mesh is illustrated in Figure 2.

It is generally believed that the hysteresis energy loss as the tire rotates under the weight of the vehicle is the main cause of energy loss associated with skid resistance. To obtain the dissipation energy, it is necessary to model the tire rubber as a viscoelastic material. The rheological material characteristics of the tire rubber are obtained by means of Dynamic Shear Rheometer (DSR) tests. The details of the test procedure can be found in a previous paper (*29*). Furthermore, static load-deflection tests were performed to obtain the stiffness characteristics of the GripTester tire, as shown in Figure 3*a*. The comparison of the aspect ratio of the tire footprints between static load-deflection test and FE footprint prediction at a load of 150, 200, 300 and 400 N, respectively, is shown in Figure 3*b*.

Description of the Tire-Pavement Interaction Model

To properly model the interfacial interaction between a tire and the underlying road surface in the context of a large deformation framework, the use of a robust 3D contact algorithm that handles contact problems is necessary. Therefore, the so-called surface-to-surface contact algorithm, originally developed by Puso and Laursen (*30*, *31*), is implemented at the tire-pavement interaction interface. The modeling and the development of the tire-pavement interaction model can be broadly divided into two steps:

1. In the first step, static analysis of the developed 3D FE model of the GripTester tire rolling over smooth pavement is carried out. To do the rolling simulation, a steady-state analysis is invoked with the application of Arbitrary Lagrangian Eulerian (ALE) formulation (*32*).
2. In the second step, the history of stresses, strains, and displacements retained from the first step is imported. In the analysis, the FE test tire is forced to roll or slip at the desired speed on a FE mesh of an asphalt pavement surface, as shown in Figure 4. Due to the roughness of the textured pavement surface, the contact pressure that develops at the tire–pavement interface changes with

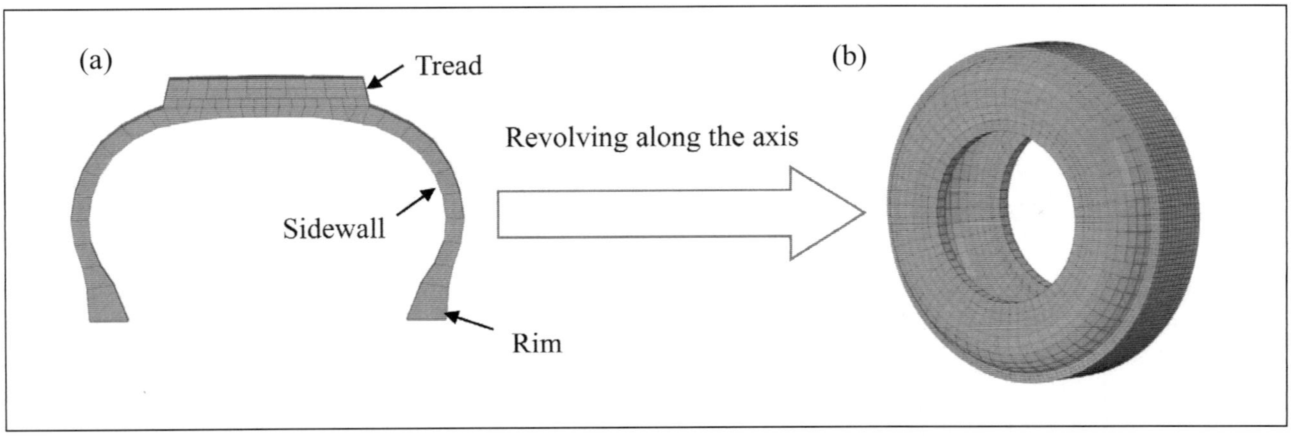

Figure 2. GripTester tire structure: (*a*) cross-section view of Goodyear tire and (*b*) 3-D view of Goodyear tire mesh.

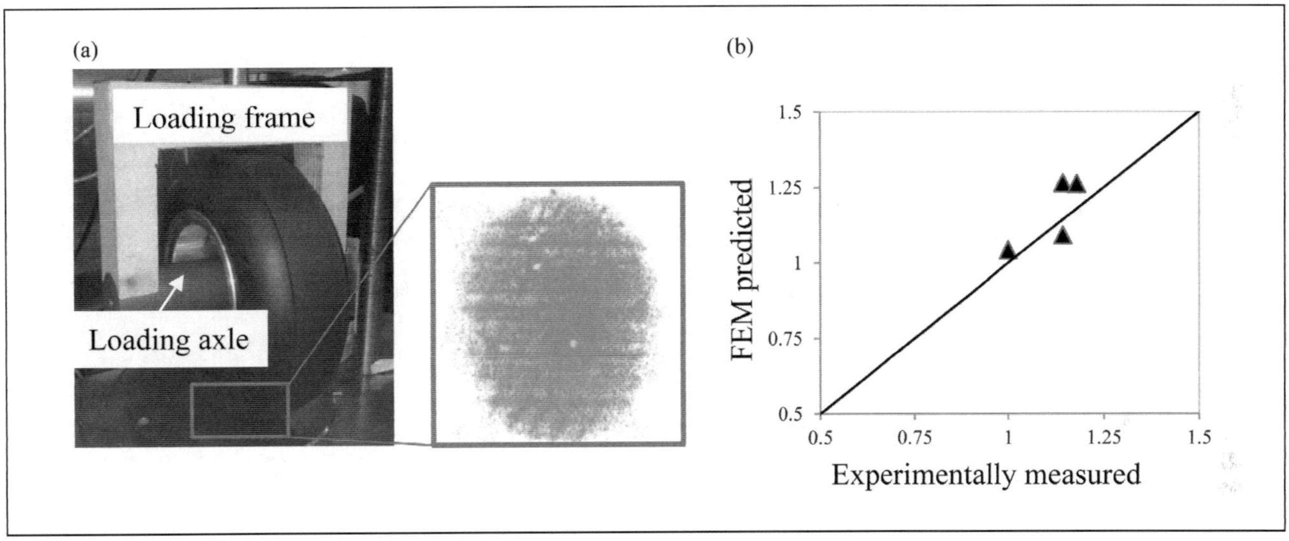

Figure 3. Validation of the FE GripTester: (*a*) static-load test to obtain the footprint and (*b*) comparison of the aspect ratio of the foot prints between FE predicted results and test results.

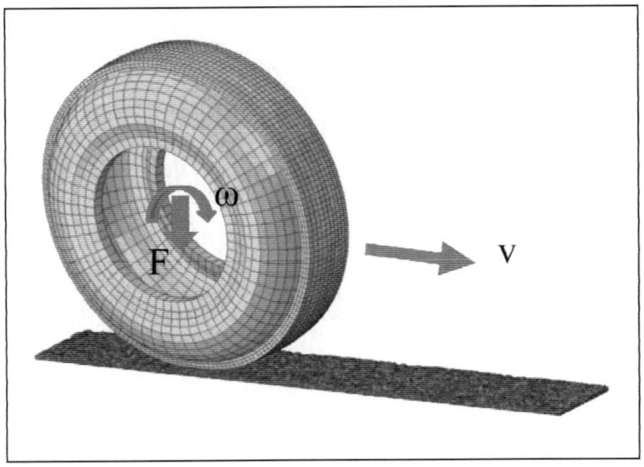

Figure 4. Coupled tire-pavement model.

the rolling or slipping of the tire. The output (stresses and strains) of the tire in this analysis is then exported for the computation of skid resistance. This will be discussed in the next section.

Computation of Skid Resistance

A detailed description of the computation procedure can be found in Srirangam et al. (*22*). However, for ease of understanding, a brief description is presented in this section. In this approach, the skid resistance is calculated based on the theory of hysteresis energy dissipation (Equations 1 to 11). The stresses and strains obtained from Step 2 in the previous section are used to compute the energy dissipation in a separate module. The dissipation losses at each Gauss point of all the cross-section

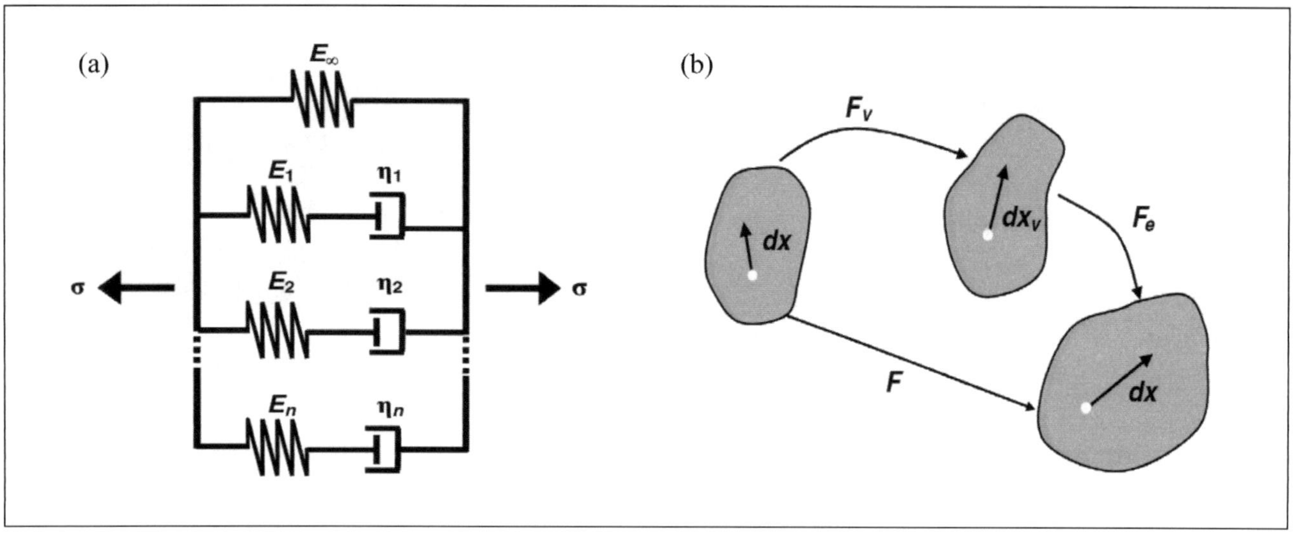

Figure 5. (*a*) One-dimensional schematic diagram of viscoelastic material model and (*b*) multiplicative decomposition of deformation gradient.

elements are computed from the circumferential strain cycles of the deformation analysis. The strain cycles and the linearized viscoelastic model are then used to calculate the energy loss in each circumferential ring of elements. The energy loss is summed for all rings to give a total energy loss. Six total stress and six viscous strain components are extracted at each Gauss point of every element in each circumferential group to calculate the energy loss.

The multiplicative decomposition concept was used to derive energy dissipation rates shown in Figure 5. The total deformation of a Maxwell component can be split into its elastic and viscous part as follows:

$$F = F_e F_v \tag{1}$$

Here F, F_e, and F_v are the total, elastic, and viscous deformation gradient, respectively.

The Lagrange–Green strain is defined as a function of the right Cauchy–Green deformation tensor $C = F^T F$:

$$\begin{aligned} E &= \frac{1}{2}(C - I) \\ &= \frac{1}{2}(F_V^T F_V^T F_e F_V - I) \\ &= \frac{1}{2}F_V^T(F_V^T F_e - F_V^{-T} F_V^{-T})F_V \\ &= \frac{1}{2}F_V^T(C_e - b_V^{-1})F_V \\ &= F_V^T\left(\frac{1}{2}(C_e - I) + \frac{1}{2}\left(I - b_V^{-1}\right)\right)F_V \\ &= F_V^T(\overline{E}_e + \overline{E}_v)F_V \end{aligned} \tag{2}$$

Here $\overline{E}_e$ and $\overline{E}_v$ are the elastic and viscous Lagrange–Green strain in the intermediate configuration.

The viscous Lagrange–Green strain in the reference configuration is derived via a pull-back operation over the viscous deformation gradient:

$$\begin{aligned} E_V &= F_V^T \overline{E}_v F_V \\ &= F_V^T \frac{1}{2}(I - b_V^{-1})F_V \\ &= \frac{1}{2}(F_V^T F_V - F_V^T b_V^{-1} F_V) \\ &= \frac{1}{2}(C_V - I) \end{aligned} \tag{3}$$

Consequently, the viscous Lagrange–Green strain rate in the reference configuration is defined as:

$$\begin{aligned} \dot{E}_V &= \frac{1}{2}\dot{C}_V \\ &= \frac{1}{2}(\dot{F}_V^T F_V - F_V^T \dot{F}_V) \\ &= \frac{1}{2}(F_V^T F_V^{-T} \dot{F}_V^T F_V - F_V^T \dot{F}_V F_V^{-1} F_V) \\ &= \frac{1}{2}F_V^T(L_v^T + L_v)F_V \\ &= F_V^T d_v F_V \end{aligned} \tag{4}$$

where L_v is the spatial velocity gradient tensor and d_v is the viscous rate of deformation tensor. The tensor d_v is defined as a function of L_v:

$$d_v = \frac{1}{2}(L_v^T + L_v) = L_v = \mathbb{C}_v^{-1} : \sum\nolimits_e \tag{5}$$

with

$$L_v = \mathbb{C}_v^{-1} : \sum\nolimits_e = L_v^T \quad (6)$$

where $\mathbb{C}_v$ is a double symmetric fourth-order tensor representing the viscosity, the expression of $(\bullet)^{-1}$ denotes the inverse of the tensor, $\sum_e = C_e S_e$ is the Mandel stress tensor and $S_e$is the second Piola–Kirchhoff tensor in the intermediate configuration. The rate of viscous work, which has to be always positive, can now be expressed as:

$$\begin{aligned}\dot{W}_V &= S : \dot{E}_V \\ &= (F_V^{-1} S_e F_V^{-T}) : F_V^T d_v F_V \\ &= F_V^{-1} S_e : F_V^T d_v F_V F_V^{-1} \\ &= S_e : F_V^{-1} F_V d_v \\ &= S_e : d_v \\ &= S_e : \mathbb{C}_v^{-1} : \sum_e \\ &= S_e : \mathbb{C}_v^{-1} : C_e S_e \\ &= C_e S_e : \mathbb{C}_v^{-1} : S_e \geq 0\end{aligned} \quad (7)$$

The viscous work itself can be calculated via the following incremental formula:

$${}^{t+\Delta t}W_V = {}^{t}W_V + (S_e : L_v) \cdot \Delta t, \quad {}^{0}W_V = 0 \quad (8)$$

The energy loss ΔW per unit volume per cycle can be written as:

$$\Delta W = \int_0^T S_e : L_v dt \quad (9)$$

The contraction of the stress tensor and strain tensor is integrated over one tire revolution to obtain the dissipated energy. Dissipated energy:

$$\Delta W = \sum_{m=1}^{M} \sum_{k=1}^{K} \int_0^T S_e^{km} : L_v^{km} dt \quad (10)$$

where $k = 1, 2, ..., K$ is the number of the sections in the 3D tire model and $m = 1, 2, ..., M$ is the number of elements in each section of the 3D tire model.

Then the skid resistance of a given tire is computed by:

$$F_{skid} = \frac{\Delta W}{2\pi R_e} \quad (11)$$

where R_e is the effective rolling radius of the tire.

Field Tests and Model Validation

Field tests were carried out to validate the developed FE model by collecting the necessary information such as temperature in the tire cross section, vehicle speed, ambient temperature and pavement temperature. A new temperature measurement system was developed to calibrate the FE model. This system contains small thermal sensors which were embedded into the test tire at various prescribed locations. A cabinet containing a data-logger was used to collect the temperature information from the sensors and transmit it wirelessly to a WiFi-enabled laptop placed in the skid resistance-measuring device. The detailed description of the instrumentation is beyond the scope of this study.

Field Tests

The field tests were carried out on a pavement that consists of three different sections in Qatar. The difference between the sections consists of the asphalt binder type and mixture design in the wearing course. The test procedure and data collection were carried out in accordance with the European Technical Specification CEN/TS 15901-7. Figure 6 shows a GripTester with the temperature measurement system installed on the test tire. The temperature measuring cabinet was installed on the wheel rim. Then, the tire was mounted and firmly anchored to the stub axle. The horizontal drag force and vertical load were measured by two strain gauges to determine the friction number. The operating and temperature conditions for the field test are summarized in Table 1.

The skid measurements were recorded at 5m intervals. Figure 7 shows that the friction number was mostly in the range of 0.2 to 0.3 with an average value of 0.24. Some high values were recorded as indicated in the data spikes at the points of transition from one section to another or because of other irregularities of the pavement. It is noted that the recorded values are below acceptable skid resistance limits (*33*).

Model Validation

The validation of the FE GripTester tire-pavement interaction model was achieved by evaluating the skid resistance at three different pavement sections (A, B, and C respectively) from the same road mentioned previously. Figure 8 shows a comparison of the predicted model results against the experimental data. The difference between the simulation results and experimental data for sections A, B, and C are 0.05 (17%), 0.05 (16%), and 0.07 (25%), respectively.

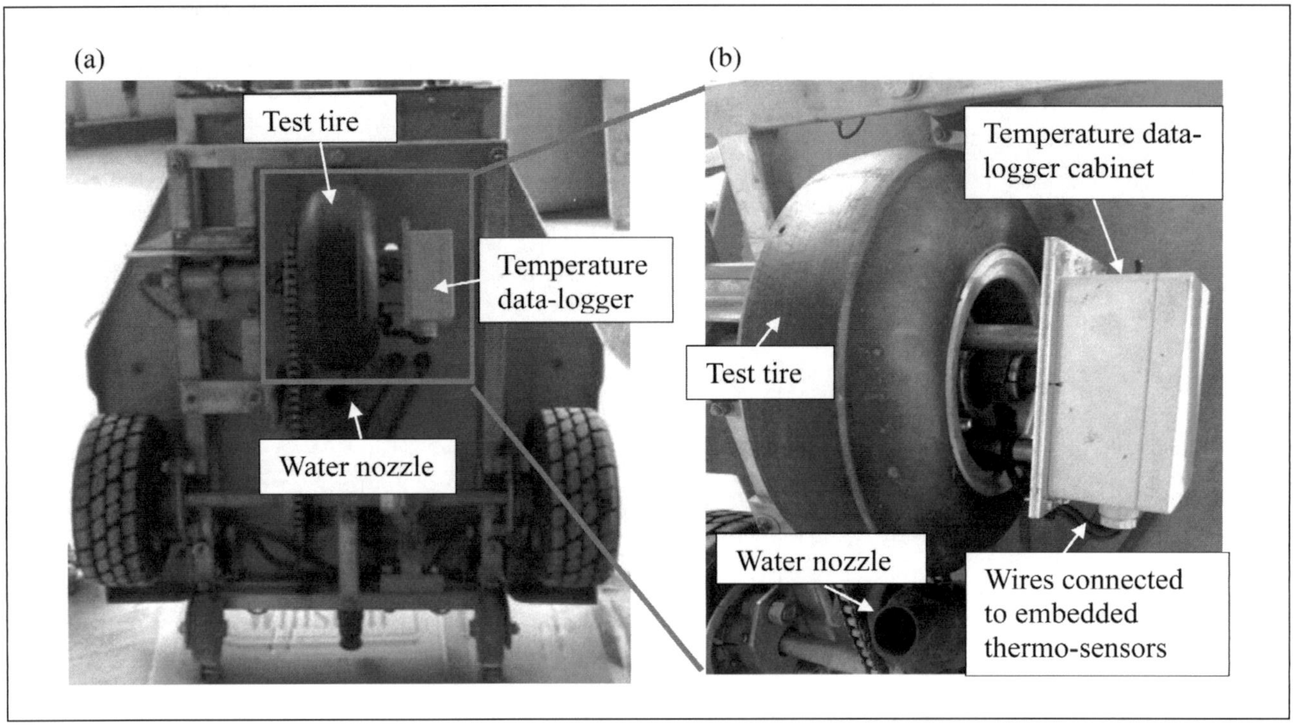

Figure 6. GripTester: (*a*) components of a GripTester and (*b*) zoom-in view of instrumented GripTester tire.

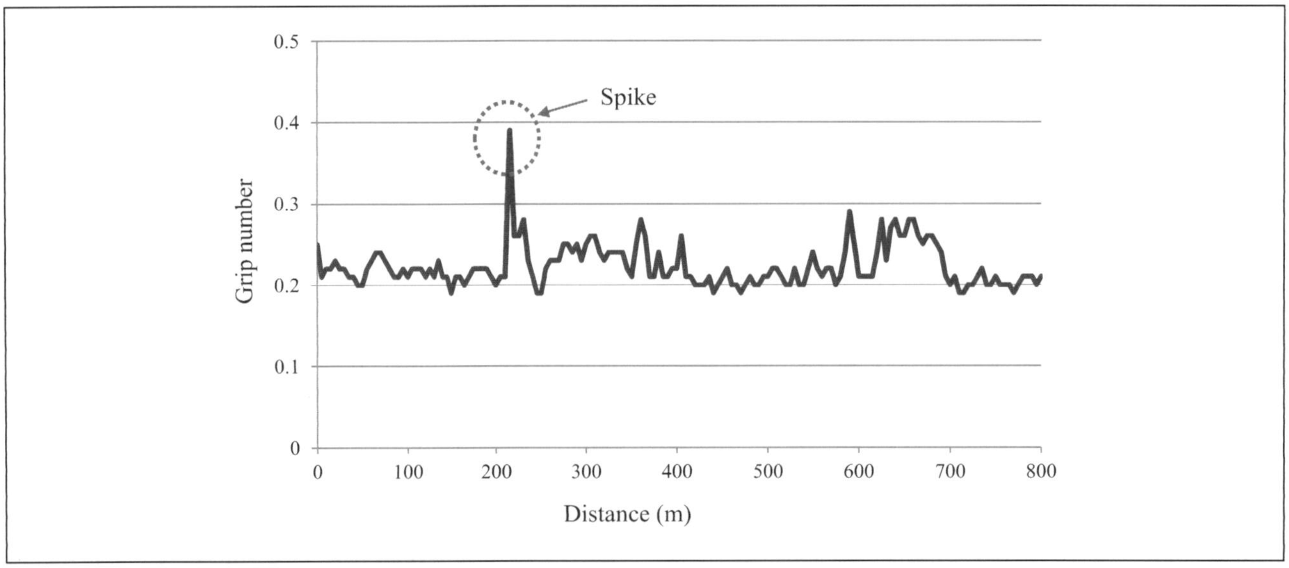

Figure 7. An example of GripTester measurements.

Table 1. Measurement Conditions

Tire inflation pressure	140 kPa
Axle load	210 N
Test speed	50 km/h
Ambient temperature	34°C
Pavement temperature	44°C

The model was also able to predict the development of temperature correctly. On average, the temperature recorded inside the tire material was 40°C whereas the model predicted a temperature of 43°C. This shows a good agreement between the measured and predicted temperatures, considering the unavoidable variations of field conditions and possible measurement errors.

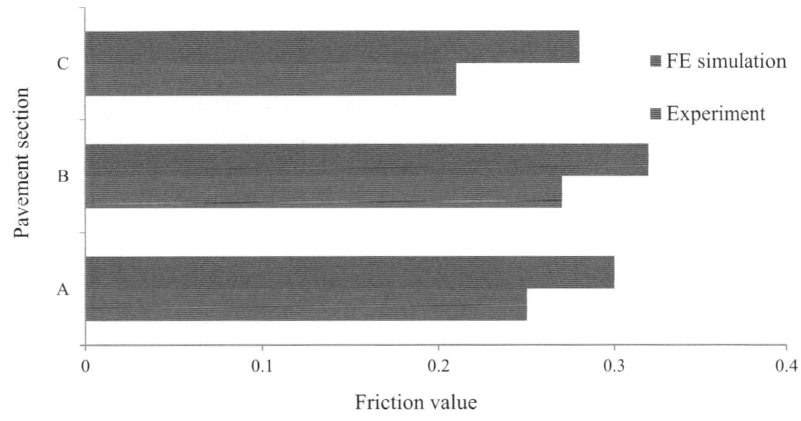

Figure 8. A comparison of GripTester measurements and FE results.

Braking Distance Calculations from PIARC Tire

To predict the friction values of a real scale tire (PIARC tire) based on test results obtained from GripTester, the following steps were taken:

1. The developed model was used to simulate the interaction between the GripTester and the pavement surface accounting for the surrounding temperature conditions. Friction and temperature development profiles were recorded as outputs from the simulation.
2. The surrounding temperature conditions were kept the same as in Step 1, and the Grip tire was replaced by the PIARC tire.
3. The PIARC tire was run at a prescribed maximum speed of V_i (120 km/h) and the speed was gradually reduced in smaller steps to a prescribed lower limit (20 km/h) by repeating the above-mentioned steps. A relationship between the coefficient of friction and the speed was established which was utilized as an input for the following step.
4. To carry out braking simulation, the PIARC tire was rolled over a plain pavement where the tire-pavement interaction interface was provided with the friction law according to the earlier obtained friction-speed relationship. This simulation was carried out under various conditions to study the tire-pavement interaction mechanisms.

Results and Discussion

In a field measurement, slight variations in the operating conditions (test speed, inflation pressure, water flow volume etc.) are to be expected. Furthermore, if the measurements are conducted at different times of the day, the ambient temperature during the test does not remain constant. These variations may result in an unreliable interpretation of skid resistance data. The FE model was used to study the effect of the variation in speed and ambient temperature on friction number.

Influence of Test Speed and Ambient Temperature on GripTester Results

In order evaluate the effect of speed on the friction number, test speeds of the GripTester were varied between 40 and 100 km/h with an interval of 20 km/h as shown in Figure 9*a*. Results show that the friction number decreases from 0.3 to 0.23 (21%) with an increase of test speed from 40 to 100 km/h. This can be attributed to the fact that the increase in speed causes less energy dissipation.

The effect of ambient temperatures from 5 to 50°C on skid resistance are shown in Figure 9*b*. The results show that an increase in ambient temperature will cause a decrease in friction number. This is because higher ambient temperature increases the overall tire temperature, which lowers the material's mechanical properties. This leads to the decrease of energy dissipation and therefore, lower friction number. Such analysis is useful in obtaining correction factors of friction measurements accounting for temperature variations.

Influence of Vehicle Operating Parameters on Pavement Safety

In the earlier section, a methodology was proposed to predict the friction of a PIARC test tire, based on field test results obtained from a GripTester. It is expected

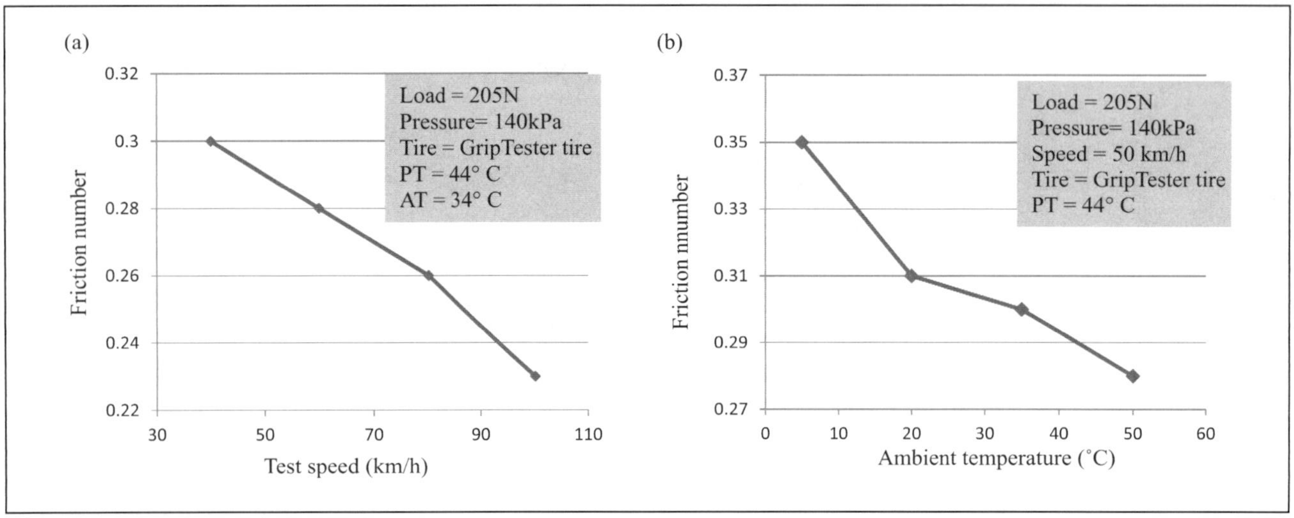

Figure 9. Results of influential factors on the friction number: (*a*) test speed and (*b*) ambient temperature.

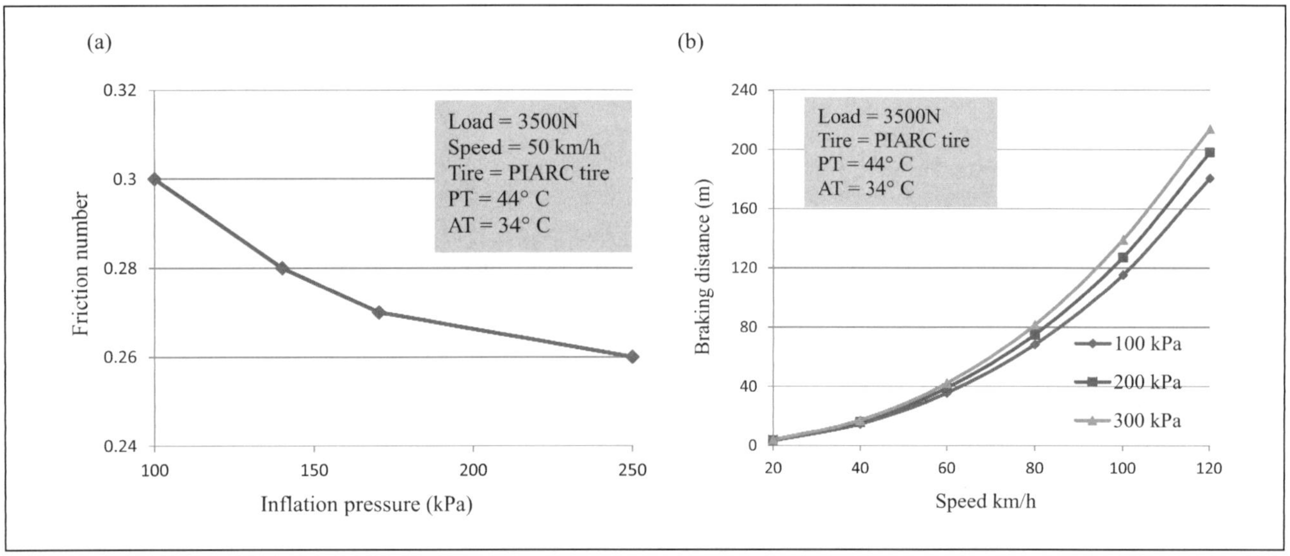

Figure 10. Results of effect of inflation pressure on (*a*) the skid resistance and (*b*) the braking distance.

that a PIARC tire would better represent a passenger car tire in terms of geometrical attributes and operating conditions. In this subsection, further analysis will be carried out based on the friction obtained by a PIARC tire.

As discussed in earlier sections, the skid resistance is influenced by many factors and thus the available braking distance is also affected by these variations. Since the interpretation of the results is a lot easier in terms of the braking distances, the effects of vehicle speed, inflation pressure, wheel load and ambient temperature on the skid resistance and braking distance will be discussed. It was observed that the developed model is able to compute the braking distance directly; a further post process analysis is not needed.

Effect of Inflation Pressure on the Skid Resistance/Braking Distance. Various inflation pressures: 100, 140, 170, and 250 kPa were considered to evaluate the effect of tire pressure on skid resistance. The results are shown in Figure 10*a*. Under the prescribed conditions, results show that the skid resistance decreased from 0.3 to 0.26 (13%) with an increase of inflation pressure from 100 to 250 kPa. This effect can be attributed to a decrease of contact surface with an increase in pressure.

Figure 10*b* shows that the braking distances increase with an increase in inflation pressure; however, their rates of increment depend on the slipping condition of the vehicle. The braking distance is calculated at a speed range from 20 to 120 km/h for three different inflation

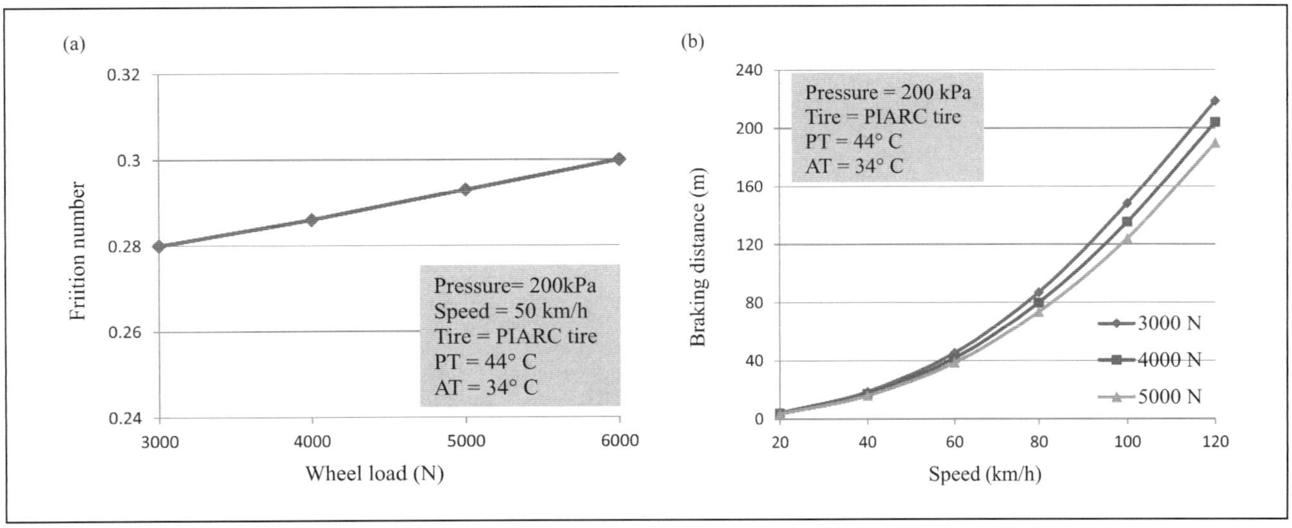

Figure 11. Results of effect of wheel load on (*a*) the skid resistance and (*b*) the braking distance.

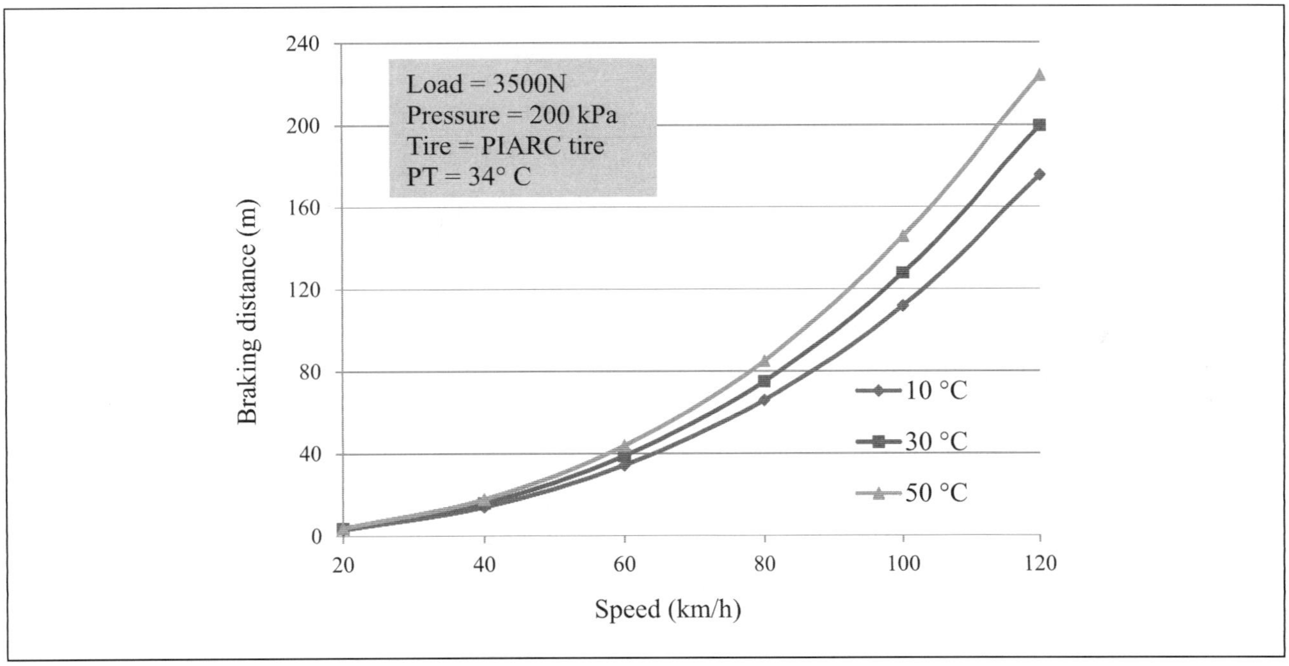

Figure 12. Results of effect of temperature on the braking distance.

pressures: 100, 200, and 300 kPa. The percentage decrease in braking distance was observed to be 18% at a speed of 120 km/h, when the tire inflation pressure changed from 100 to 300 kPa. This highlights the importance of maintaining proper inflation pressure in the tire.

Effect of Vehicle Loading on the Skid Resistance/Braking Distance. The wheel load was varied from 3000 to 6000 N at an interval of 1000 N. The results are shown in Figure 11*a*. It can be observed that as the tire load increases, the skid resistance increased from 0.28 to 0.3 (7%) indicating that the effect of load is not significant. This result can be attributed to the higher deformation of the tire body caused by increasing tire load, which increases the energy losses and ultimately causes an increase in the friction force.

Figure 11*b* shows the effect of wheel loads on the braking distance. Wheel loads ranging from 3000 to 5000 N were considered. The plots show that the braking

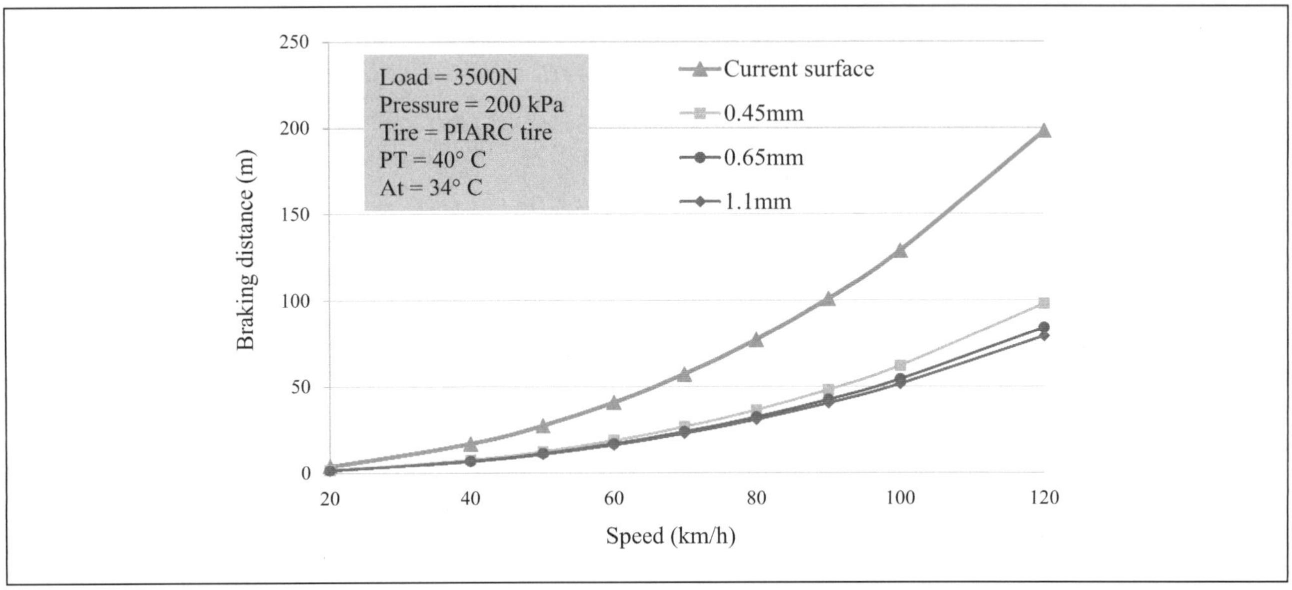

Figure 13. Results of effect of MPD on the braking distance.

distance decreases with the increase of the wheel load. At the speed of 120 km/h, the predicted braking distance decreases from 218 to 189 m (15%) with an increase of wheel load from 3000 to 5000 N.

Effect of Ambient Temperature on the Braking Distance. Figure 12 shows the effect of ambient temperatures on the braking distance. The results show that at constant load, inflation pressure, and speed, the values of braking distance increase with an increase in the ambient temperature. An increase of the braking distance from 224 to 175 m (28%) was observed for an increasing ambient temperature from 10 to 50°C.

Effect of MPD on the Braking Distance. Figure 13 demonstrates the effect of MPD on the braking distance. Three pavement surfaces with different MPD values of 0.45, 0.65, and 1.1 mm were obtained by magnifying the scales of the pavement mesh. The results were compared with the current pavement surface. It can be clearly seen that braking distances increase with decreasing MPD. The results show that for a given speed of 120 km/h, braking distance reduces from 198 to 79 m for the MPDs considered. Specifically, the difference between the braking distances for MPD values of 0.45 and 0.65 mm is 6%, while that for MPD values of 0.65 and 1.1 mm is 17%. This may be attributed to the fact that a contact between the rubber tire and many macro-asperity contact regions of the pavement surface will result in hysteresis energy loss leading to higher skid resistance. It implies that proper maintenance for better pavement roughness could significantly reduce the braking distance. Such findings could be important for the road authorities to enable them make maintenance/rehabilitation planning.

On the basis of Figures 9–12, it can be concluded that the skid resistance/braking distance at the chosen section is lower than the acceptable limits in Europe and the USA. The pronounced role of the temperature can be clearly seen in these plots. The findings of this study indicate that failure to maintain tire pressures and adhere to speed limits increase the accident risk even more as the temperature increases. Figure 13 implies that maintainence of the pavement roughness is of great importance in terms of braking distance.

Conclusions

This paper presents the development and application of a FE thermo-mechanical tire pavement interaction model. This model simulates the condition of a GripTester tire running over an asphalt pavement surface obtained from a test section. The model was validated using field measurements carried out in Qatar. The model capabilities were illustrated by analyzing the effects of variations of speed, tire inflation pressure, static wheel load and temperature on the skid resistance/braking distance. It was demonstrated that the model can be used to develop correction factors for the effects of temperature on friction measurements. In addition, the paper presented a finite element approach to evaluate skid resistance obtained using different tire sizes and configurations. The results showed that different tires would result in different friction values for the same surface under the same vehicle operating conditions.

Based on both field data and FE analysis, it was concluded that the rise in temperature could have a significant influence on reducing skid resistance and increasing braking distance. This reduction in skid resistance is a potential safety risk for vehicles traveling on pavement surfaces with elevated temperatures. The results also demonstrated that braking distance could be improved with better surface roughness. This information could be vital to road agencies to develop maintenance/rehabilitation strategies.

Acknowledgments

This publication was made possible by an NPRP award (NPRP No. 7-482-2-184: Thermo-Mechanical Tire-Pavement Interaction: Computational Modeling and Field Measurements) from the Qatar National Research Fund (QNRF-a member of The Qatar Foundation).

Author Contributions

The authors confirm contribution to the paper as follows: study conception and design: Kumar Anupam, Tianchi Tang, Athanasios Scarpas and Eyad Masad; data collection: Reginald Kogbara; analysis and interpretation of results: Tianchi Tang, Kumar Anupam and Cor Kasbergen; draft manuscript preparation: Tianchi Tang. All authors reviewed the results and approved the final version of the manuscript.

References

1. Long, K., H. Wu, Z. Zhang, and M. Murphy. *Quantitative Relationship between Crash Risks and Pavement*. Texas Department of Transportation, Austin, TX, 2014.
2. Hall, J. W., K. L. Smith, L. Titus-Glover, J. C. Wambold, T. J. Yager, and Z. Rado. Guide for Pavement Friction. *Final Report for NCHRP Project*, Vol. 1, 2009, p. 43.
3. Clark, J. D., and D. J. Schuring. Load, Speed and Inflation Pressure Effects on Rolling Loss Distribution in Automobile Tires. *Tire Science and Technology*, Vol. 16, No. 2, 1988, pp. 78–95.
4. Schuring, D. J., and S. Futamura. Rolling Loss of Pneumatic Highway Tires in the Eighties. *Rubber Chemistry and Technology*, Vol. 62, 1990, pp. 315–367.
5. Mahmoud, E., and E. Masad. Experimental Methods for the Evaluation of Aggregate Resistance to Polishing, Abrasion and Breakage. *Journal of Materials in Civil Engineering*, Vol. 19, 2007, pp. 977–985.
6. Elkin, B. L., K. J. Kercher, and S. Gullen. Seasonal Variation of Friction of Bituminous Surfaces in Indiana. *Transportation Research Record: Journal of the Transportation Research Board*, 1979. 777: 50–58.
7. Al-Mansour, A. Effects of Pavement Skid Resistance on Traffic Accidents. *Journal of Engineering Research*, Vol. 3, 2006, pp. 75–78.
8. Anupam, K., S. K. Srirangam, A. Scarpas, and C. Kasbergen. Influence of Temperature on Tire–Pavement Friction: Analyses. *Transportation Research Record: Journal of the Transportation Research Board*, 2013. 2369: 114–124.
9. Najafi, S., G. W. Flintsch, and K. K. McGhee. Assessment of Operational Characteristics of Continuous Friction Measuring Equipment (CFME). *International Journal of Pavement Engineering*, Vol. 14, 2013, pp. 706–714.
10. *Findlay Irvine Ltd*, 2017. http://www.findlayirvine.com. Accessed July 26, 2017.
11. Kogbara, R. B., E. A. Masad, E. Kassem, A. Scarpas, and K. Anupam. A State-of-the-art Review of Parameters Influencing Measurement and Modeling of Skid Resistance of Asphalt Pavements. *Construction and Building Materials*. Vol. 114, 2016, pp. 602–617.
12. Tanner, J. A. *Computational Methods for Frictional Contact with Applications to the Space Shuttle Orbiter Nose-Gear Tire*. NASA Technical Paper No. 3574. NASA, Langley Research Center, Hampton, VA, 1996.
13. Davis, P. A. *Quasi-Static and Dynamic Response Characteristics of F-4 Bias-Ply and Radial-Belted Main Gear Tires*. NASA Technical Paper 3586. NASA, Langley Research Center, Hampton, VA, 1997.
14. Johnson, A. R., J. A. Tanner, and A. J. Mason. *Quasi-Static Viscoelastic Finite Element Model of an Aircraft Tire*. NASA/TM-1999-209141. NASA, Langley Research Center, Hampton, VA, 1999.
15. Cho, J. R., H. W. Lee, J. S. Sohn, G. J. Kim, and J. S. Woo. Numerical Investigation of Hydroplaning Characteristics of Three-Dimensional Patterned Tire. *European Journal of Mechanics and Solids*, Vol. 25, 2006, pp. 914–926.
16. Wollny, I., R. Behnke, K. Villaret, and M. Kaliske. Numerical Modeling of Tyre–Pavement Interaction Phenomena: Coupled Structural Investigations. *Road Materials and Pavement Design*. Vol. 17, 2016, pp. 563–578.
17. Ong, G. P., and T. F. Fwa. Effectiveness of Transverse and Longitudinal Pavement Grooving in Wet-Skidding Control. *Transportation Research Record: Journal of the Transportation Research Board*, 2007. 2005: 172–182.
18. Ong, G. P., and T. F. Fwa. Transverse Pavement Grooving against Hydroplaning, I–Simulation Model. *Journal of Transportation Engineering*, Vol. 132, No. 6, 2006, pp. 441–448.
19. Ong, G. P., and T. F. Fwa. Transverse Pavement Grooving against Hydroplaning, II–Design. *Journal of Transportation Engineering*, Vol. 132, No. 6, 2006, pp. 449–457.
20. Ong, G. P., and T. F. Fwa. Modeling Skid Resistance of Commercial Trucks on Highways. *Journal of Transportation Engineering*, Vol. 136, No. 6, 2010, pp. 510–517.
21. Srirangam, S. K., K. Anupam, C. Kasbergen, and A. Scarpas. Study of Influence of Operating Parameters on Braking Friction and Rolling Resistance. *Transportation Research Record: Journal of the Transportation Research Board*, 2015. 2525: 79–90.
22. Srirangam, S. K., K. Anupam, A. Scarpas, and A. Köster. Influence of Temperature on Tire-Pavement Friction-1: Laboratory Tests and Finite Element Modeling. Presented at 92nd Annual Meeting of the Transportation Research Board, Washington, D.C., 2013.

23. Anupam, K., S. K. Srirangam, A. Scarpas, and C. Kasbergen. Influence of Temperature on Tire–Pavement Friction: Analyses. *Transportation Research Record: Journal of the Transportation Research Board*, 2013. 2369: 114–124.
24. *Simpleware*. ScanIP 32-bit (Version 4.3) + CAD Module 32-bit (Version 1.3). Synopsys, Inc., Mountain View, CA, 2011.
25. Gall, R., T. Tabaddor, D. Robbins, P. Majors, W. Sheperd, and S. Johnson. Some Notes on the Finite Element Analysis of Tires. *Tire Science and Technology*, Vol. 23, 1995, pp. 175–188.
26. Meschke, G., H. J. Payer, and H. A. Mang. 3D Simulations of Automobile Tires: Material Modeling, Mesh Generation and Solution Strategies. *Tire Science and Technology*, Vol. 25, 1997, pp. 154–176.
27. Cho, J. R., K. W. Kim, W. S. Yoo, and S. I. Hong. Mesh Generation considering Detailed Tread Blocks for Reliable 3D Tire Analysis. *Advances in Engineering Software*, Vol. 35, 2004, pp. 105–113.
28. *ABAQUS (Version 6.13) User's Manual*. Dassault Systemes Simulia Corp., Johnston, RI, 2013.
29. Zegard, A., F. Helmand, T. Tang, K. Anupam, and A. Scarpas. Rheological Properties of Tire Rubber using Dynamic Shear Rheometer for FEM Tire-Pavement Interaction Studies. Presented at 8th International Conference on Maintenance and Rehabilitation of Pavements, Singapore, 2016.
30. Puso, M. A., and T. A. Laursen. A Mortar Segment-to-Segment Contact Method for Large Deformation Solid Mechanics. *Computer Methods in Applied Mechanics and Engineering*, Vol. 193, 2004a, pp. 601–629.
31. Puso, M. A., and T. A. Laursen. A Mortar Segment-to-Segment Contact Method for Large Deformation Solid Mechanics. *Computer Methods in Applied Mechanics and Engineering*, Vol. 193, 2004b, pp. 4891–4913.
32. Braess, H., and P. Wriggers. Arbitrary Lagrangian-Eulerian Finite Element Analysis of Free Surface Flow. *Computer Methods in Applied Mechanics and Engineering*, Vol. 190(1–2), 2000, pp. 95–110.
33. American Association of State Highway Transportation Officials. *A Policy on Geometric Design of Highways and Streets (6th ed.)*. AASHTO, Washington, D.C., 2010.

The Standing Committee on Pavement Surface Properties and Vehicle Interaction (AFD90) peer-reviewed this paper (18-04085).

The statements made here are solely the responsibility of the authors.

TRR
JOURNAL OF THE TRANSPORTATION RESEARCH BOARD

Article

Transportation Research Record
2018, Vol. 2672(40) 395–407

DOI: 10.1177/0361198118777384
journals.sagepub.com/home/trr
SAGE

Field Investigation of Relationship between Pavement Surface Texture and Friction

Sareh Kouchaki[1], Hossein Roshani[2], Jorge A. Prozzi[3], Natalia Zuniga Garcia[1], and Joaquin Bernardo Hernandez[4]

Abstract

Proper tire–pavement interaction is essential for the safety of motorists. Pavement surface texture is a major contributing factor to tire–pavement friction. This study performed a series of statistical analyses of field-measured friction and texture data to find the texture–friction correlation. Three test sections with different pavement types were selected within the state of Texas. Data were collected at three locations in the right wheel path and three locations in the center of the lane for each test section. To measure the texture data, the researchers used the circular track meter (CTM) and a prototype measurement device developed in-house and consisting of a line laser scanner (LLS). Friction measurements were obtained with the dynamic friction tester (DFT) and Grip-Tester. The mean profile depth (MPD) was calculated by using the measured texture data. The relationship between the MPD values and the friction numbers obtained from the Grip-Tester and DFT was investigated at speeds of 50 and 70 km/h (31.1 and 43.5 mph). The repeatability and reliability of both the developed LLS prototype and the Grip-Tester were also evaluated, as well as the effect of test speed on friction measurement. The results indicated a strong positive correlation between the texture and friction data. In addition, the developed LLS prototype was able to scan the pavement surface texture more reliably and precisely than the CTM in terms of vertical and horizontal resolution. The Grip-Tester showed promising results compared with the DFT with regards to the friction measurement.

The direct force developed in the tire–pavement interface is known as *skid resistance*, a property defined by the properties of the tire, the vehicle speed, and the pavement condition and texture. Pavement texture is determinant of the resistance of the pavement surface to a vehicle sliding and skidding (*1*, *2*). The extent of skid resistance on any given pavement is dependent on the design of the surface texture—specifically its micro- and macro-texture—as the texture can affect the skid resistance, splash-and-spray, rolling resistance, and tire wear (*3*). Pavement design can adjust surface pavement properties to provide the safety needed (*4*).

Pavement surface texture is influenced by many factors, such as aggregate type and size, mixture gradation, and texture orientation, among others. Pavement texture is the result of the deviations of the surface layer from an actual planar surface (*5*). The World Road Association has categorized pavement texture by a range of wavelength (λ) and amplitudes (A). The standard specification organizations, such as the American Society for Testing Materials (*5*), International Organization for Standardization (*6*), and German Institute for Standardization (DIN on ISO 13473-1), accepted and incorporated these definitions. The ISO 13473-1 refined the terms incorporating typical amplitudes (*7*) as follows:

- Micro-texture (λ < 0.5 mm, A = 1–500 μm) (where λ is wavelength and A is amplitude)
- Macro-texture (0.5 mm < λ < 50 mm, A = 0.1–20 mm)
- Mega-texture (50 mm < λ < 500 mm, A = 0.1–50 mm)

[1]Department of Civil, Architectural, and Environmental, The University of Texas at Austin, Austin, TX
[2]City of San Antonio, Streets Engineering & Infrastructure Management, Transportation and Capital Improvements (TCI), TX
[3]International Center for Partnered Pavement Preservation, Department of Civil, Architectural, and Environmental, The University of Texas at Austin, Austin, TX
[4]Center for Transportation Research, The University of Texas at San Antonio, Austin, TX

Corresponding Author:
Address correspondence to Sareh Kouchaki: sareh.kouchaki@utexas.edu

Micro-texture refers to the small-scale texture of the aggregate surface, which controls the contact between the tire rubber and the pavement surface. Micro-texture is a function of aggregate particle mineralogy, petrology, and source (natural or manufactured), and is affected by the environmental effects and the action of traffic (*8*). *Macro-texture* refers to the large-scale texture of the pavement surface resulting from the aggregate particle size and arrangement. In asphalt pavements, the mixture properties (aggregate shape, size, and gradation), which define the type of mixture, control the macro-texture (*9*). Mega-texture has wavelengths in the same order of size as the tire/pavement interface. Examples of mega-texture include ruts, potholes, and major joints and cracks. It affects vibration in the tire walls but not the vehicle suspension, and it is therefore strongly associated with noise and rolling resistance (*9*).

Pavement friction is the result of a complex interplay between two principal frictional force components: adhesion and hysteresis. Although there are other components of pavement friction, such as tire rubber shear, they are insignificant when compared with the adhesion and hysteresis force components (*3*, *9*). Thus, friction can be viewed as the sum of adhesion and hysteresis. *Adhesion* is the friction that results from the small-scale bonding/interlocking of the vehicle tire rubber and the pavement surface. It is a function of the interface shear strength and contact area (*9*, *10*). The *hysteresis* component of the frictional forces results from the energy loss created within the tire as it responds the texture. Because the adhesion force is developed at the tire/pavement interface, it is most responsive to the micro-level asperities (micro-texture) of the aggregate particles. In contrast, the hysteresis force developed within the tire is most responsive to the macro-level asperities (macro-texture) formed in the pavement surface. Thus, in principle, adhesion governs the overall friction on smooth-textured and dry pavements, while hysteresis is the dominant component on wet and rough-textured pavements (*3*, *9*).

In 2015, Torbruegge and Wies in Germany studied the correlation between wet skid resistance and pavement texture. Their findings showed no correlation between the mean texture depth (MTD) as a surface texture characteristic and skid resistance as measured by British pendulum (*11*). Similar results were obtained by Gunaratne et al. in which no significant correlation was found between friction and MTD. However, Gunaratne et al. observed a strong correlation between friction and the texture frequency characteristics, such as the power spectral density calculated by fast Fourier transform (*12*).

The precise nature of the relationship between friction and texture remains unknown, although several attempts to reveal it have been made using methods such as Fourier, wavelet, and fractal analysis (*13–15*). These models' degree of complexity and the difficulty in finding the coefficients relating to the calculations result in complications for road engineers.

Goal and Objectives

Understanding the texture–friction relationship is valuable for transportation agencies and engineers for proper pavement maintenance and better design of pavement, especially when higher skid resistance is needed. A few studies have been conducted in this regard, but the effects of pavement texture on the friction produced at the pavement surface are still not fully understood. This study's primary goal is to find the correlation between field-measured texture and friction data. For this purpose, several objectives were defined:

- Develop an accurate texture measurement device, called a line laser scanner (LLS) prototype, and evaluate its performance by comparing its results to those of the circular track meter (CTM).
- Use the Grip-Tester to collect continuous friction data at traffic speed and compare it with the dynamic friction tester (DFT) at two different speeds.
- Calculate the mean profile depth (MPD) values of texture data and perform statistical analysis to establish correlations between texture and friction.

Experimental Methods and Theoretical Evaluations

Measurement of Skid Resistance

There are three main skid resistance measuring principles: longitudinal friction coefficient (LFC), transverse, and stationary or slow-moving (*16*, *17*). The LFC measurement principle aims to measure friction when a vehicle is traveling forward in a straight line. When the brakes are applied, both the angular velocity of the wheel and the overall velocity of the vehicle decrease. When the braking force on the wheels is too strong, the wheels are "locked" and consequently slide over a surface. LFC measurement devices have slip ratios that simulate the wheel slipping process when a wheel slides over a surface. More specifically, the slip ratio compares the vehicle's velocity with the angular velocity of the wheel. When the slip ratio is 0, the angular velocity of the wheel is the same as the velocity of the vehicle (i.e., no slip between wheel and surface). When the slip ratio is 1, there is no angular velocity in the wheel (i.e., wheel is fully locked and slides over the surface). LFC devices can have either a fixed or a variable slip ratio (*15*, *17*).

The stationary or slow-moving measurement principle is used in compact devices usually found in the laboratory or still testing. One such device uses a pendulum arm; others, such as the DFT, use a rotating head. Both devices use rubber sliders that use the surface friction to slow down the pendulum or the rotating head (*16–18*).

Table 1 summarizes LFC measurement devices commonly used around the world. LFC devices can measure between 5 and 140 km/h (5.1 and 87 mph) and can come in the form of compact trailers or large-capacity trucks. LFC devices use a watering system, requiring vehicles to carry a water tank for measurement purposes. Transverse friction measurement devices are usually larger, more expensive, and can take measurements for longer distances since greater water capacity increases maximum measurement distance. While inexpensive and easy to use, stationary or slow-moving devices must be operated manually (*18*).

Among the measurement devices presented in Table 1, the Grip-Tester was selected because it showed the most promise for measuring pavement friction because of its wider range of test speed, better repeatability and reproducibility (depending on the operating speed), greater efficiency in water usage, and commercial availability (*18*). The Grip-Tester is a friction-measuring trailer that uses a fixed slip wheel to simulate anti-lock braking on a wet road surface. This is achieved by the Grip-Tester having three wheels: two have patterned treads and are connected by a drive axle, while the third is a smooth tread tire. The drive axle, connecting the two drive wheels, has a 27-tooth sprocket; it is connected by a transmission chain to the measuring wheel that has a 32-tooth sprocket (*19*). The difference in sprocket teeth creates a 15% continuous slippage on the measurement tire. To create wet road conditions, a watering system supplies a constant water thickness specified by the user to the measuring wheel.

The axle connecting the measuring wheel outputs the dynamic friction by using strain gages to measure the horizontal drag force and vertical load force. The information is used to estimate the grip number (GN), or coefficient of friction, in real-time (*19*). As shown in Equation 1, the GN is the ratio between the fraction of tractive drag force (F_d) and the load force (Q).

$$GN = \frac{F_d}{Q} \tag{1}$$

The Grip-Tester (Figure 1*a*) needs a vehicle capable of towing the trailer as well as a water tank to supply water to the measuring wheel.

Along with the Grip-Tester, the DFT was used to measure the friction of the pavement at different speeds. This machine is specified under ASTM E1911. The DFT (shown in Figure 1, *b* and *c*) consists of a rotating disk and three rubber pads attached to the bottom of the disk. The disk is pushed by an electric motor to rotate until it reaches the target speed set by an operator. At the same time, water is applied to the pavement surface. The disk drops and the rubber pads come into contact with the wet area. Each rubber pad is loaded vertically at 11.8 N (2.65 lb). The friction force developed between pads and the pavement slows down the disk. The DFT measures the friction coefficient continuously until the disk stops completely (*13*, *20*). The coefficient of friction (μ) is the ratio between the friction force or horizontal torque force (F) and the applied vertical load on the rubber sliders of the DFT (*21*).

$$\mu = \frac{F}{Q} \tag{2}$$

Measurement of Pavement Surface Texture

Circular Track Meter. CTM is a common static method used to measure the macro-texture of the pavement surface. According to the ASTM E2157, the CTM (Figure 2, *a* and *b*) consists of a laser-displacement sensor that rotates over a circle with a diameter of 284.5 mm (11.2 in.) and measures the profile of pavement surface texture. The profile includes 1024 points scanned at an interval of 0.87 mm (0.034 in.). Using the instructions provided by ASTM E 1845, the measured profile then is sectioned into eight equal parts and the highest profile peak in each part is measured (*23*, *24*).

Developed LLS Prototype. LLS was utilized in this research study to scan the pavement texture. This device is two-dimensional non-contact laser sensor that projects blue light in a horizontal line. Small changes in the height caused by the texture irregularities can also be captured using this scanner system (*25*). A prototype (Figure 3*a*) was developed to enable LLS to capture three-dimensional data. For this purpose, the LLS was mounted on a linear motion controller (linear stage) to travel over the surface and scan it at a given interval. Figure 3*c* illustrates the picture of the triangulation system. As a result of the angle between the camera and the laser, the system could miss data from vertical edges.

Calculation of MPD

In 1992 the Permanent International Association of Road Congresses (PIARC) performed an experiment using 41 friction and texture measuring devices in Spain and Belgium. The results of that experiment showed the MPD parameter as the best pavement macro-texture indicator (*23*). The MPD is estimated from the surface height profile following ASTM E1845. The first step is to censor the inclination slope of the height profile such

Table 1. Comparison of the LFC Measuring Devices (16–18)

Device name	Theoretical water film thickness (WFT), speed, tire type, measurement interval (interval)	Assembly	Device picture
ADHERA	TWFT: 1 mm Speed: 40–120 km/h Tire type: smooth PIARC tire 165R15 Interval: 20 m Country/countries of use: France	Assembly: Trailer that can be hooked up to vehicle Commercially available: No	
BV-11	TWFT: 1 mm Speed: 20–160 km/h [C] Tire type: Trelleborg tire T49 Interval: 20 m Country/countries of use: England, Sweden, and Finland	Assembly: Trailer that can be hooked up to vehicle Commercially available: Yes	
Grip-tester	TWFT: 0.5 mm Speed: 5–100 km/h Tire type: smooth ASTM tire, 254 mm diameter Interval: 10 or 20 m Country/countries of use: United States, United Kingdom, and others	Assembly: Trailer that can be hooked up to vehicle Commercially available: Yes	
ROAR DK	WFT: 0.5 mm Speed: 60–80 km/h Tire type: ASTM 1551 Interval: 5m + Country/countries of use: Denmark	Assembly: Trailer that can be hooked up to a vehicle Commercially available: No	
ROAR NL	TWFT: 0.5 mm Speed: 50–70 km/h Tire type: ASTM 1551 Interval: 5–100 m Country/countries of use: Netherlands	Assembly: Three-axle tanker truck with two measuring systems mounted at the rear of the chassis (tank capacity is about 12,000 liters) Commercially available: No	
RWS NL	TWFT: 0.5 mm Speed: 50–70 km/h Tire type: PIARC smooth 165 R15 Interval: 5–100 m Country/countries of use: Netherlands	Assembly: Trailer that can be hooked up to a vehicle Commercially available: No	
Skiddometer BV-8	TWFT: 0.5 mm Speed: 40–80 km/h TT: AIPCR ribbed, 165 R15, with four longitudinal grooves Interval: 30–50 m Country/countries of use: Sweden	Assembly: Trailer that can be hooked up to a vehicle Commercially available: Yes	
SRM	TWFT: 0.5 mm Speed: 40–80 km/h TT: AIPCR ribbed, 165 R15, with four longitudinal grooves Interval: 20 m Country/countries of use: Germany	Assembly: The test wheels are mounted on the back of a tanker vehicle at the approximate location of a vehicle tire paths Commercially available: No	

(continued)

Table 1. (continued)

Device name	Theoretical water film thickness (WFT), speed, tire type, measurement interval (interval)	Assembly	Device picture
TRT	WFT: 0.5 mm Speed: 40–140 km/h Tire type: smooth ASTM Interval: Typically, 20 m Country/countries of use: Czech Republic	Assembly: The measuring equipment is under a specially equipped vehicle Commercially available: No	
SRT-3	WFT: 0.5 mm Speed: 30–90 km/h Tire type: Tire with thread (200 kPa) Interval: 100 m Country/countries of use: Poland	Assembly: Trailer that can be hooked up to a vehicle Note: SRT-3 is more sensitive to micro-texture changes than to macro-texture changes Commercially available: No	
IMAG	WFT: 1.0 mm Speed: up to 140 km/h Tire type: PIARC smooth profile tire Interval: N/A Country/countries of use: France, Germany, and Poland	Assembly: Trailer that can be hooked up to a vehicle Commercially available: Yes	
ICC	WFT: 1.0 mm Speed: 65–80 km/h Country/countries of use: United States, and others	Assembly: Trailer that can be hooked up to vehicle Commercially available: Yes	

that a regression line is calculated and then subtracted from the height profile and so the initial profile will be converted to a zero-mean profile. As a second step (Figure 4), the height profile is divided into two segments with the length of $\frac{L}{2}$ for which the maximum height is detected. By using Equation 3, MPD is calculated as the average of the two determined maximum heights.

$$MPD = \frac{Max\,Height\,of\,1st\,Segment + Max\,Height\,of\,2nd\,Segment}{2} \quad (3)$$

Field Data Collection

Test Sections and Pavement Type. Three test sections, each more than five years old, were selected. Each section had established measurement data for the DFT, CTM, and LLS. The sections chosen provide a variety of mix designs, ensuring variation of surface texture for the test. The test sections are located in Bastrop, Bryan, and Fort Worth:

- Bastrop:
  - Mix design: porous friction course (PFC)
  - Average annual daily traffic (AADT): 13,972
- Bryan:
  - Mix design: Dense-Graded Type C
  - AADT: 5,843
- Fort Worth:
  - Mix design: Dense-Graded Type D
  - AADT: 57,385

At each location, six samples were taken: three in the right wheel path and three in the center of the lane, with a 12.5 m (41 ft) distance between the locations (Figure 5). Within each location, one CTM and one DFT measurements were obtained. The CTM and DFT testing followed ASTM E2157 and ASTM E1911 procedures, which state the equipment should be placed in the same location.

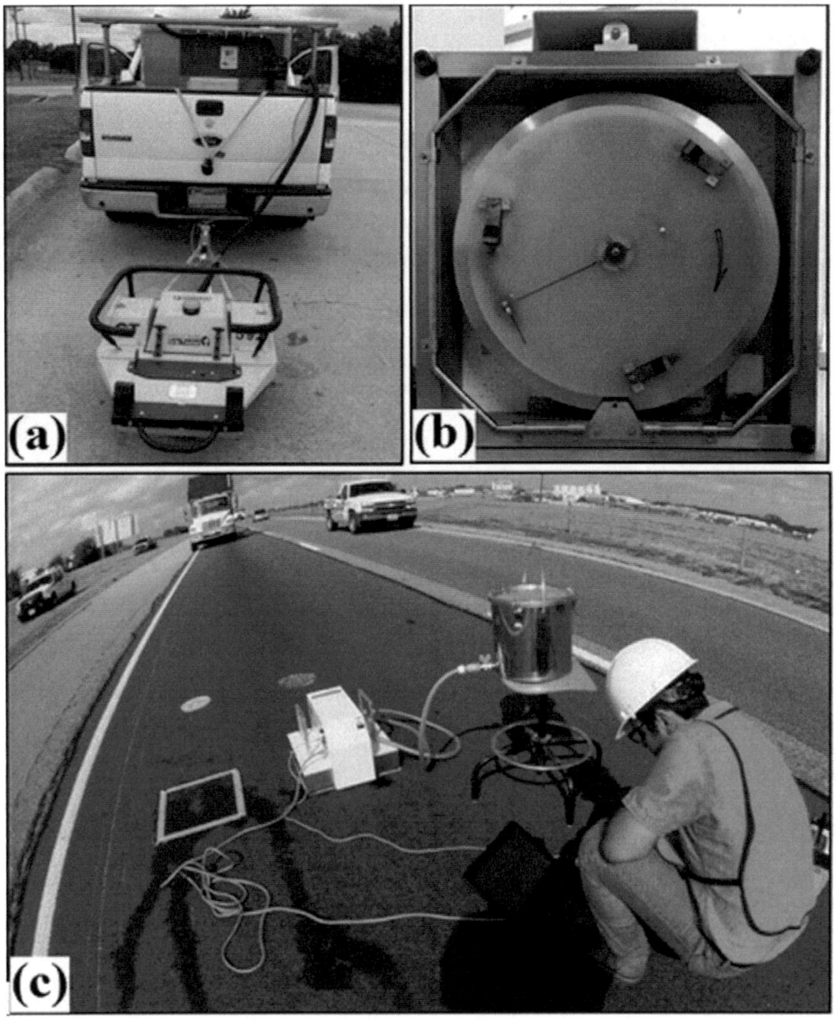

Figure 1. (*a*) the Grip-Tester attached to a vehicle, (*b*) bottom view of the DFT, and (*c*) the DFT connected to a water tank and digital command control (DCC) measuring the friction at the lane center path.

Figure 2. (*a*) the CTM powered by a battery measuring the texture at the right wheel path and (*b*) bottom view of the CTM.

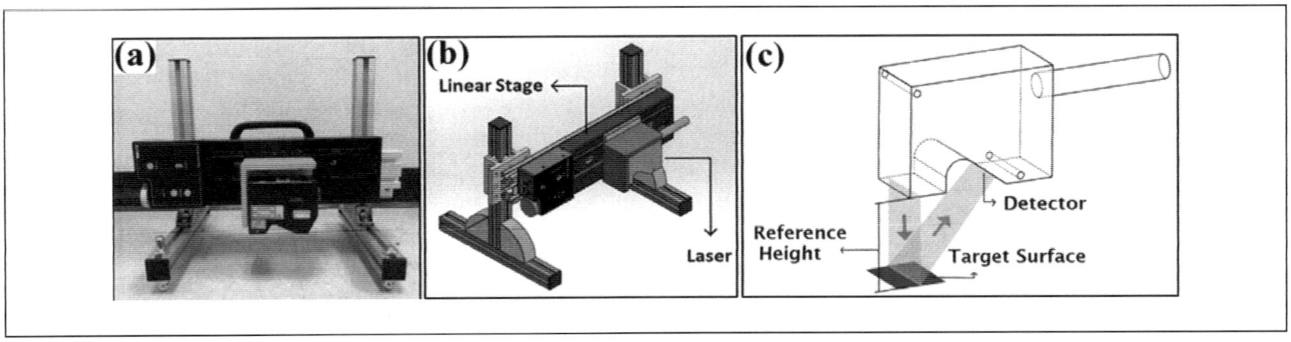

Figure 3. (*a*) a front view of the LLS prototype, (*b*) a 3-D schematic view of the LLS, and (*c*) the triangulation system of laser.

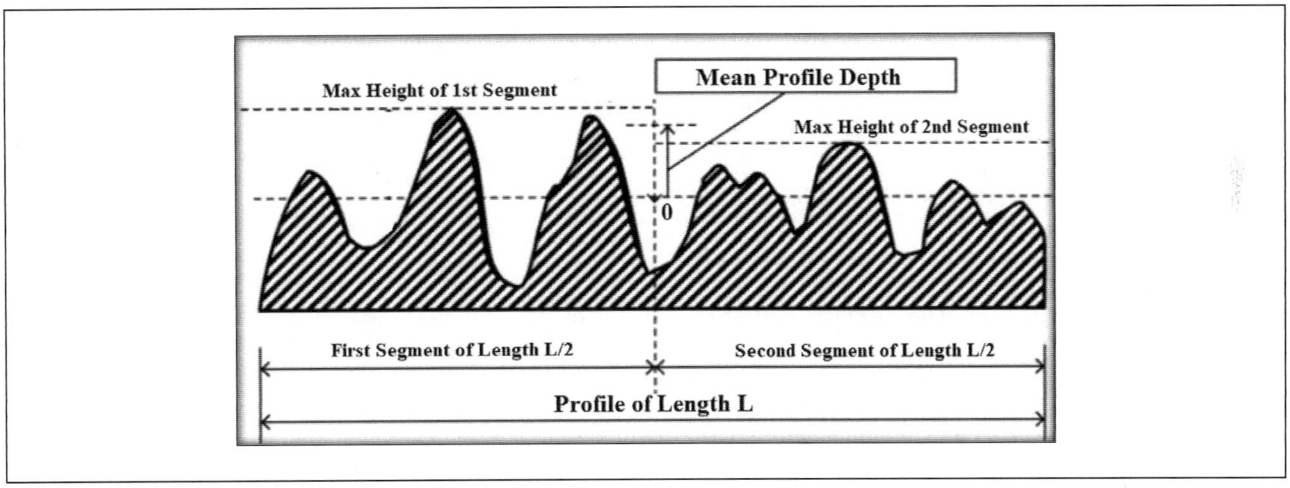

Figure 4. Graphical illustration of MPD calculation (*26*).

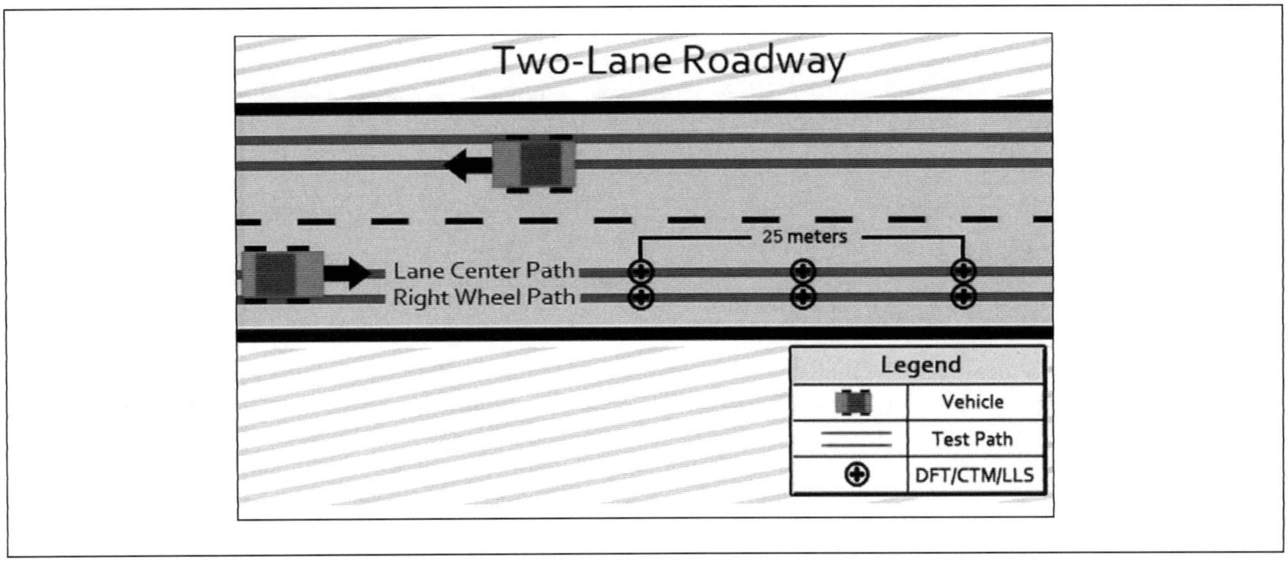

Figure 5. Illustration of test section and test location.

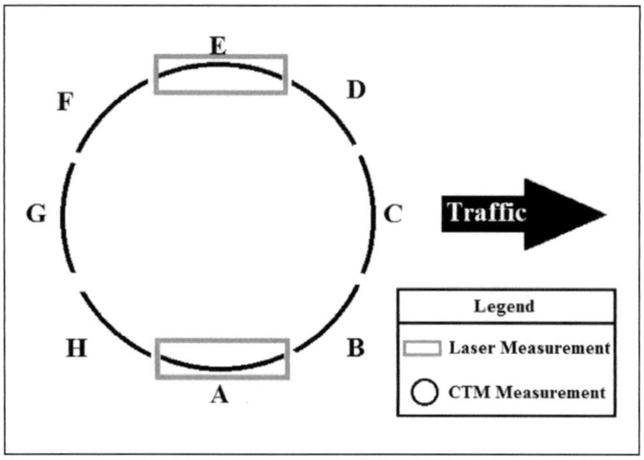

Figure 6. The scanned areas by the CTM and the LLS.

To maintain consistency in the surface measurements, the LLS was placed in the same locations described for the CTM. Note, however, that these devices operate differently owing to the mechanism of motion in each: the CTM measures in a circle and the LLS measures linearly. For the CTM, eight segments of 111.5 mm (4.39 in.) arcs are scanned by the laser, which results in a full circle. Among the segments, two of them are parallel to the direction of traffic. To ensure an equitable comparison with the CTM's measurements, the LLS was configured to scan a 120 mm (4.72 in.) length at a linear speed of 8 mm/s (0.31 in./s) with a sampling frequency of 1 kHz. The area scanned by the LLS was the same as the two segments of the CTM that were parallel to the direction of traffic.

For the Grip-Tester, the test consisted of two target speeds (50 km/h and 70 km/h) to evaluate the dependency of speed on friction. The speed needs to be maintained within 5% of the target speed. The 70 km/h (43.5 mph) target was selected to enable a direct comparison with the DFT equipment, which has a speed range of up to 80 km/h (49.7 mph).

Results and Discussions

Repeatability of Developed LLS Prototype

The LLS results were compared with those from the CTM for both segments A and E, shown in Figure 6, to investigate the repeatability and reliability of the developed LLS. The MPD, as a surface characteristic, was calculated for all test locations. Then, single values of MPD for the right wheel path and the center were obtained by averaging the three respective samples. This was done for both LLS and CTM as graphed in Figure 7, *a* and *b*—the -R and -C represent the right wheel path and lane center path, respectively. As this figure illustrates, the MPD values obtained by the LLS, denoted as MPD-by-LLS, are very close to the MPD values of CTM, denoted as MPD-by-CTM. There were no observable biases between the MPD values with which to draw a general conclusion.

Repeatability of the developed LLS prototype was evaluated by calculating the average and standard deviation of three MPD-by-LLS and MPD-by-CTM values obtained at each test section from their respective test paths. Then, the standard deviation of each group of data was compared. For example, the three MPD-by-LLS values for the pavement surface under the right wheel path at the Bryan test section were compared with each other with respect to their standard deviation. Because of the short distance between test locations, similar texture characteristics were expected at those locations. As Figure 7*b* indicates, for segment A, the standard deviations of MPD-by-LLS data were significantly lower than MPD-by-CTM in the entire test sections. Similarly, Figure 7*d* shows that the standard deviations of MPD-by-LLS data, obtained from segment E, were notably lower in three test sections (Bastrop-C, Bastrop-R, and Bryan-R). Comparable MPD values were observed at the rest of the sections. According to these results, one can suggest that the developed LLS prototype is capable of providing an accurate and precise measurement of the pavement surface texture.

Grip-Tester Results

Figure 8, *a* and *b* show the GN of all test locations plotted against the DFT values at two different speeds of 50 and 70 km/h (31.1 and 43.5 mph). As these figures illustrate, the GN values have a strong relationship with DFT values, with R^2 of 0.93 and 0.91 at the two different speeds. It can be concluded that this relationship is independent of the test speed.

These findings indicate that the Grip-Tester is a reliable alternative for the DFT and does not require traffic control; therefore, it is safer and more efficient. Besides, the Grip-Tester enables engineers to obtain continuous friction data over a long pavement section.

Based on the literature, no given speed has been specified by agencies to perform the test. Therefore, understanding the effect of speed on GN would help researchers and practitioners interpret the friction results in different cases when the test speed is a variable. The results plotted in Figure 8*c* show that the GN decreased when the speed increased. This finding is reasonable but contrasts with the results of a study conducted at the University of Auckland by Wilson et al. in 2013, in which they stated that at speeds less than 75 km/h, the GN may not be affected by test speeds (*27*).

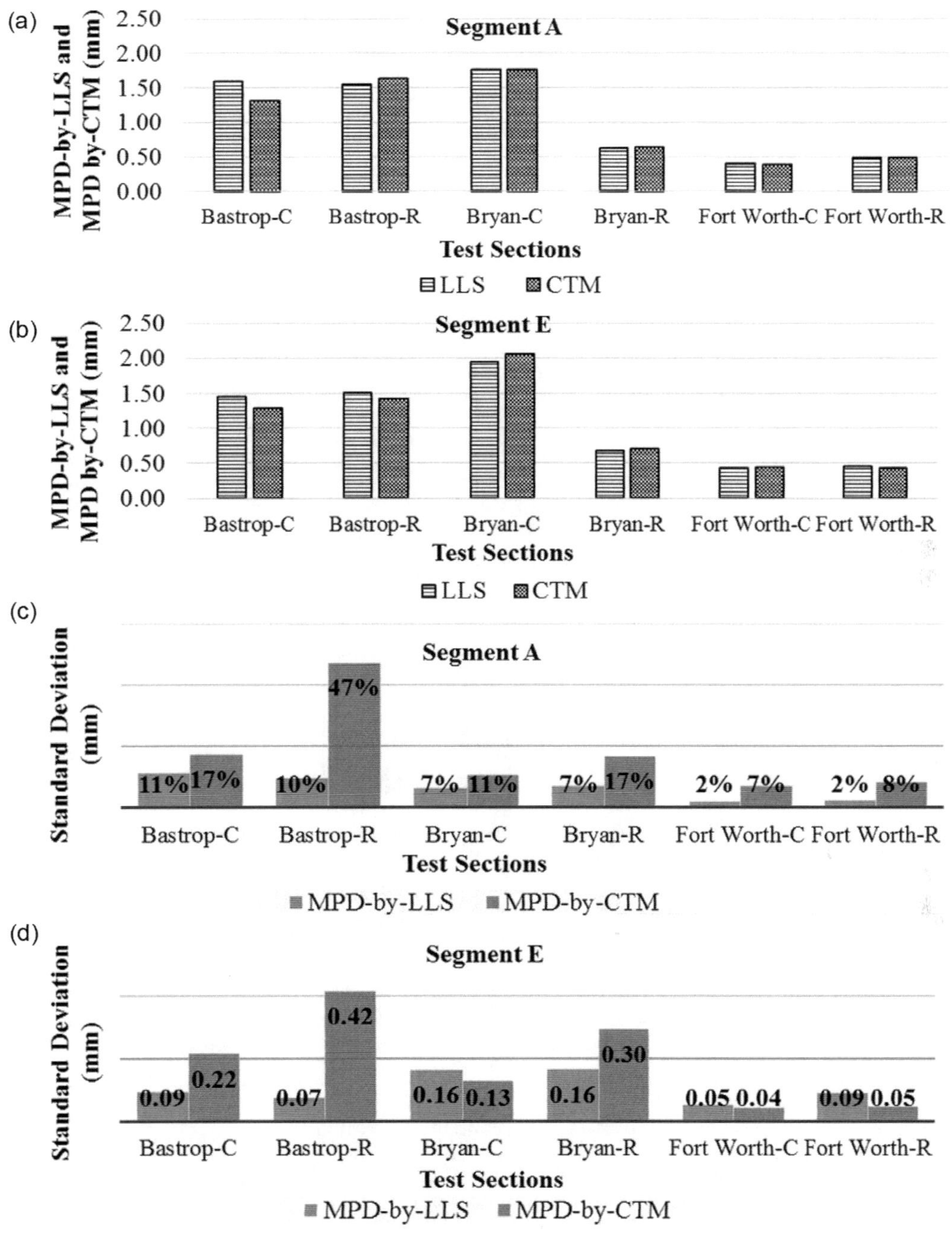

Figure 7. (*a*) and (*b*) show the comparison of mean MPD values obtained from developed LLS prototype and CTM in segments A and E, respectively; (*c*) and (*d*) show the standard deviation of MPD values measured by developed LLS prototype and CTM at six test sections in segments A and E, respectively.

According to Figure 7*a* and *c*, the MPD value of the section Bryan-R, which is about 0.70 mm (0.027 in.), is lower than Bastrop-R with MPD of 1.50 mm (0.06 in.). However, interestingly, in Figure 8*c*, the GN at Bryan-R is higher than that of Bastrop-R. It should be mentioned that, in these test locations, the water film thickness of 0.5 mm (applied by Grip-Tester) can only cover a portion of surface texture depth. The PFC pavement is an

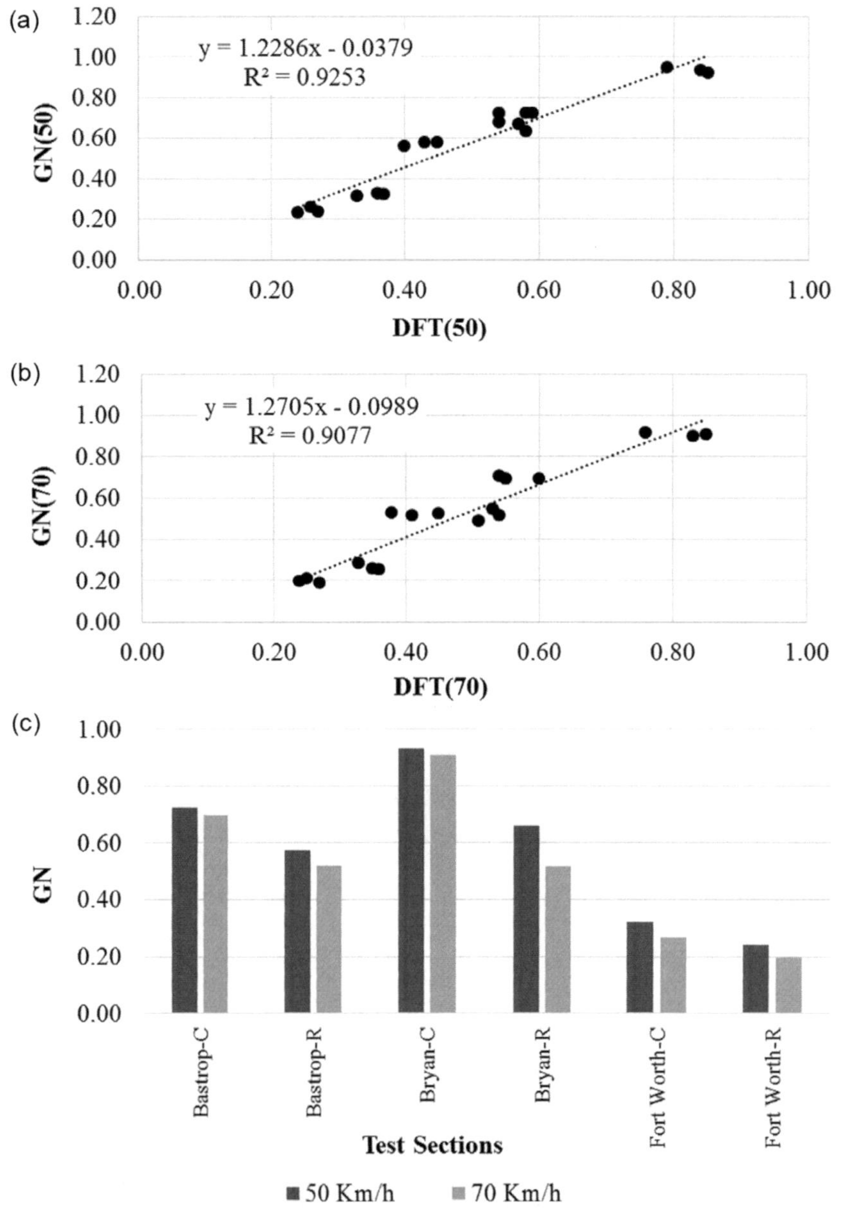

Figure 8. Correlation between GN and DFT values at two speeds of (*a*) 50 km/h, (*b*) 70 km/h, and (*c*) the GN at different test sections.

open-graded friction course that allows drainage of water through the asphalt layer. The surface pavement type of the Bastrop test section is PFC; therefore, the lack of small-size aggregates results in higher MPD values. On the other hand, the mix design of the Bryan test section is dense-graded type C, which contains all ranges of aggregate size and thus results in a lower MPD value but a greater tire–pavement contact area. In this case, it can be suggested that the adhesion factor is decisive in friction.

Although a PFC mix design increases the surface MPD (and thus causes more hysteresis), PFC's design also decreases the chance of the aggregates being fully exposed and coming into contact with the tire (thus decreasing the contact area and adhesion). A smooth surface is associated with more contact area and better adhesion, but it significantly decreases the MPD, which results in lower hysteresis. Therefore, the balance between adhesion and hysteresis in the pavement surface

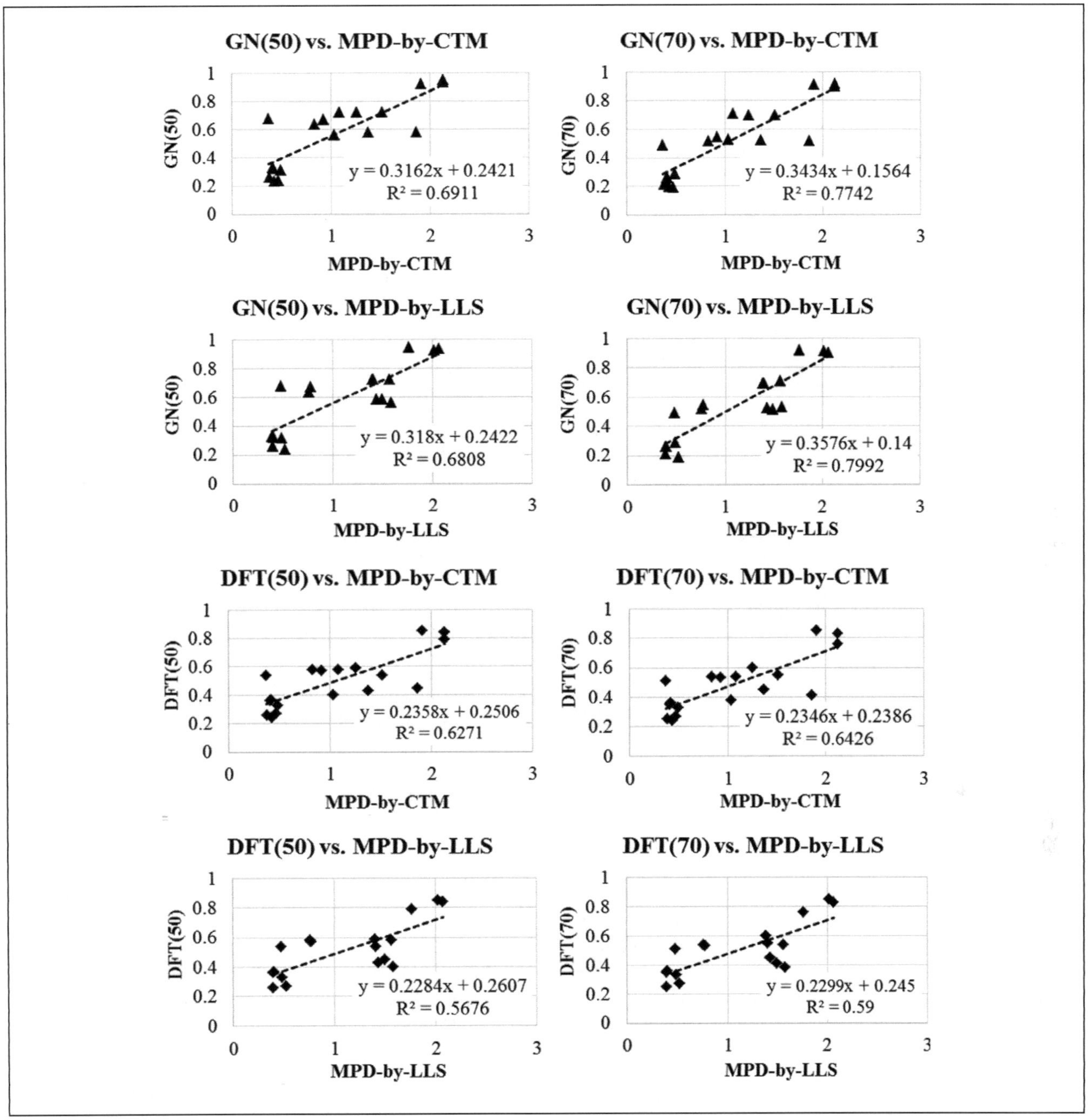

Figure 9. Plotted texture–friction data points including GN, DFT values, and MPD values.

should be optimized in order to maximize the surface friction and skid resistance. This is achievable by using an appropriate mix design providing both macro-texture and micro-texture (*15*).

The lowest friction was observed in the Fort Worth test sections where the MPD values were below 0.46 mm (0.018 in.). Considering the water film thickness of 0.5 mm (0.019 in.) applied by the Grip-Tester for wetting purpose, it can be concluded that the surface texture was fully covered by water. In this case, the water trapped in the V- or U-shaped areas of the pavement surface bears the load from the tire and prevents the tire penetrating the spaces between the aggregates—thus reducing the contribution of the aggregate surface to the friction force by decreasing the tire–aggregate contact area (*28*).

Relationship between the Texture and Friction

The GN and DFT values both represent the friction properties of the pavement surface; however, the MPD values obtained by the LLS and CTM reflect the texture characteristics of the surface. To explore the relationship between the friction and texture, those two groups were plotted against each other in four different forms: GN vs. CTM, GN vs. LLS, DFT vs. CTM, and DFT vs. LLS. For this purpose, the data obtained from the segment E, shown in Figure 6, were considered.

Using statistical analysis, the correlation coefficient (R) was used to evaluate the relationships. The R ranges from -1 to $+1$, where negative and positive signs represent the negative or positive linear correlation respectively.

The statistical analysis shows that the relationship between the GN values and MPDs is strongly linear with high correlation coefficients (Figure 9). For the Grip-Tester performed at 50 km/h, the $R = 0.83$ was found for GN-CTM, and $R = 0.82$ for GN-LLS. In addition, higher correlation coefficients were observed at the speed of 70 km/h. The R values of 0.88 and 0.89 show a strong linear relationship between GN as a friction parameter, and MPDs of CTM and LLS, respectively.

From the results provided in Figure 9, the Grip-Tester data have a better relationship with texture compared with the DFT data in terms of the calculated R^2 values. Thus, it can be suggested that friction and skid resistance could be estimated by means of texture measurement data with high confidence.

Conclusions

This study investigated the relationship between the texture and friction of pavement surfaces using field-measured data. During this study, a texture scanner prototype—the LLS—was developed to measure the texture data along with the CTM. The MPD was adopted as a pavement texture indicator. DFT and Grip-Tester were employed for friction measurement. The efficiency and repeatability of the developed LLS were explored through comparison with the CTM results.

The results of the repeatability analysis and standard deviation showed the reliability of the developed LLS prototype for texture measurement. In all the test sections, the MPD calculated from the LLS data was similar to that calculated from the CTM data. Therefore, it is recommended that the developed LLS prototype should be considered as an efficient alternative to the CTM device, especially when three-dimensional texture data are needed.

Regardless of the test speed and pavement type, the friction number obtained by the Grip-Tester showed a strong linear correlation with DFT measurements. On account of the reliable results of the Grip-Tester, its wider range of friction measurement, and the lack of need to control traffic, this device is recommended for use by state Departments of Transportation for friction measurement. The quality of the friction data obtained by a Grip-Tester seem to be somewhat sensitive to the test speed, because the GN decreased slightly when the test speed increased from 50 km/h to 70 km/h (31.1 mph to 43.5 mph). This finding implies that keeping the test speed fixed at different test sections is important to achieving comparable results.

This study's statistical analysis findings pointed to a strong positive linear correlation between texture and friction in pavements. The highest linear correlation coefficients (R) were observed between the GNs obtained at 70 km/h (43.5 mph) and texture data obtained using either the LLS or the CTM. The authors of this study recommend running the Grip-Tester at higher speeds in cases when the texture–friction relationship analysis is the point of interest. As a further recommendation, accurate friction prediction models could be developed based on the texture measurement data to eliminate the need for direct friction measurement.

Author Contributions

The authors confirm contribution to the paper as follows: Study conception and design: Sareh Kouchaki, Hossein Roshani, Joaquin Bernardo Hernandez, Jorge A. Prozzi; Data collection: Sareh Kouchaki, Hossein Roshani, Joaquin Bernardo Hernandez; Natalia Zuniga Garcia; Analysis and interpretation of results: Sareh Kouchaki, Hossein Roshani, Natalia Zuniga Garcia; Draft manuscript preparation: Sareh Kouchaki, Hossein Roshani, Joaquin Bernardo Hernandez, Jorge A. Prozzi; All authors reviewed the results and approved the final version of the manuscript.

References

1. Asi, I. M. Evaluating Skid Resistance of Different Asphalt Concrete Mixes. *Journal of Building and Environment*, Vol. 42, 2007, pp. 325–329.
2. Wang, H., and Z. Wang. Evaluation of Pavement Surface Friction Subject to Various Pavement Preservation Treatments. *Journal of Construction and Building Materials*, Vol. 48, 2013, pp. 194–202.
3. Henry, J. J. *NCHRP Synthesis of Highway Practice 291: Evaluation of Pavement Friction Characteristics*. TRB, National Research Council, Washington, D.C., 2000.
4. Hoerner, T. E., and K. D. Smith. *High Performance Concrete Pavement: Pavement Texturing and Tire-Pavement Noise*. Report No. FHWA-DTFH61-01-P-00290. FHWA, U.S. Department of Transportation, Washington, D.C., 2002.
5. American Society for Testing and Materials. *Standard Terminology Relating to Vehicle-Pavement Systems*. ASTM

E867. American Society for Testing and Materials, Philadelphia, Pennsylvania, 2012.

6. International Organization for Standardization. *Characterization of Pavement Texture by Use of Surface Profiles.* ISO 13473-1. International Organization for Standardization, 1997.
7. Sandberg, U. *Influence of Road Surface Texture on Traffic Characteristics Related to Environment, Economy, and Safety: A State-of-Art-Study Regarding Measures and Measuring Methods.* VTI notat53A-1997. Swedish National Road Association, Linkoping, Sweden, 1998.
8. Kogbara, B. R., E. A. Masad, E. Kassem, A. T. Scarpas, and K. Anupam. A State-of-the-Art Review of Parameters Influencing Measurement and Modeling of Skid Resistance of Asphalt Pavements. *Journal of Construction and Building Materials*, Vol. 114, 2016, pp. 602–617.
9. Hall, J. W., K. L. Smith, L. Titus-Glover, L. D. Evans, J. C. Wambold, and T. J. Yager. *NCHRP Project 01-43: Guide for Pavement Friction.* Transportation Research Board, Washington, D.C., 2009.
10. Gunaratne, M., N. Bandara, J. Medzorian, M. Chawla, and P. Ulrich. Correlation of Tire Wear and Friction to Texture of Concrete Pavements. *Journal of Materials in Civil Engineering*, Vol. 12, 2000, pp. 46–54.
11. Stefan Torbruegge, S., and B. Wies. Characterization of Pavement Texture by Means of Height Difference Correlation and Relation to Wet Skid Resistance. *Journal of Traffic and Transportation Engineering*, Vol. 2, 2015, pp. 59–67.
12. Gunaratne, M., N. Bandara, J. Medzorian, M. Chawla, and P. Ulrich. Correlation of Tire Wear and Friction to Texture of Concrete Pavements. *Journal of Materials in Civil Engineering*, Vol. 12, No. 1, 2000.
13. Huang, H., and J. Pan. Speech Pitch Determination Based on Hilbert–Huang transform. *Journal of Signal Processing*, Vol. 86, 2006, pp. 792–803.
14. Rado, Z., and M. Kane. An Initial Attempt to Develop an Empirical Relation between Texture and Pavement Friction Using the HHT Approach. *Wear*, Vol. 309, 2014, pp. 233–246.
15. Kane, M., I. Artamendi, and T. Scarpas. Long-Term Skid Resistance of Asphalt Surfacings: Correlation between Wehner– Schulze Friction Values and the Mineralogical Composition of the Aggregates. *Wear*, Vol. 303, 2013, pp. 235–243.
16. Descornet, G., B. Schmidt, M. Boulet, M. Gothie, M. T. Do, J. Fafie, M. Alonso, P. Roe, R. Forest, and H. Viner. *HERMES Project: Harmonization of European Routine and Research Measuring Equipment for Skid Resistance.* FEHRL Report, 2006/01. Forum of European National Highway Research Laboratories.
17. Do, M. T., and P. Roe. *Report on State-of-the-Art of Test Methods.* TYROSAFE Project Deliverable D04, 2008, Forum of European National Highway Research Laboratories.
18. Andriejauskas, T., V. Vorobjovas, and V. Mielonas. Evaluation of Skid Resistance Characteristics and Measurement Methods. *Proc., 9th International Conference of Environmental Engineering*, Vilnius, Lithuania, 2014.
19. Thomas, L. *Grip-Tester MK2 D-type Maintenance Manual.* Issue 4, Findlay, Irvine Limited, Scotland, 2008.
20. Masad, E., A. Rezaei, A. Chowdhury, and P. Harris. *Predicting Asphalt Mixture Skid Resistance Based on Aggregate Characteristics.* FHWA/TX-09/0 5627-1. FHWA, Texas A&M University, 2009.
21. Austroads. *Guide to the measurement of road surface skid resistance.* Austroads Incorporated, Sydney, Australia, 2005.
22. Wambold, J. C., C. E. Antle, J. J. Henry, and Z. Rado. *International PIARC Experiment to Compare and Harmonize Texture and Skid Resistance Measurements.* PIARC Technical Committee on Surface Characteristics, Madrid, 1995.
23. Abbas, A., M. E. Kutay, H. Azari, and R. Rasmussen. Three-Dimensional Surface Texture Characterization of Portland Cement Concrete Pavements. *Journal of Computer-Aided Civil and Infrastructure Engineering*, Vol. 22, 2007, pp. 197–209.
24. Mataei, B., H. Zakeri, M. Zahedi, and F. M. Nejad. Pavement Friction and Skid Resistance Measurement Methods: A literature Review. *Journal of Civil Engineering*, Vol. 6, 2016, pp. 537–565.
25. KEYENCE, *Measurement Sensors.* http://www.keyence.com/products/measure/index.jsp. Accessed January 1, 2017.
26. American Society for Testing and Materials. *Standard Practice for Calculating Pavement Macrotexture Mean Profile Depth.* ASTM E1845. American Society for Testing and Materials, Philadelphia, Pennsylvania, 2015.
27. Wilson, D. J., B. Jacobsen, and B. Chan. *The Effect of Road Roughness (and test speed) on GripTester Measurements.* NZ Transport Agency Research Report 523. NZ Transport Agency, Wellington, New Zealand, 2013.
28. Meegoda, J. N., and G. Shengyan. Evaluation of Pavement Skid Resistance Using High Speed Texture Measurement. *Journal of Traffic and Transportation Engineering*, Vol. 2, No. 6, 2015, pp. 382–390.

The Standing Committee on Pavement Surface Properties and Vehicle Interaction (AFD90) peer-reviewed this paper (18-03767).

Article

TRR
JOURNAL OF THE TRANSPORTATION RESEARCH BOARD

Transportation Research Record
2018, Vol. 2672(40) 408–417

DOI: 10.1177/0361198118781392
journals.sagepub.com/home/trr
SAGE

Evaluation of Hydroplaning Risk on Permeable Friction Course using Tire–Water–Pavement Interaction Model

Yangmin Ding[1] and Hao Wang[1]

Abstract

Wet weather-related hazards such as hydroplaning can be reduced with the proper use of permeable friction course (PFC). At low rainfall intensities, PFC provides quick drainage of water and better skid resistance. However, at higher rainfall rates, the entire volume of runoff cannot be discharged within the porous layer, causing drainage to occur on pavement surface. Water flow on the road surface can result in hydroplaning of tires. The objective of the study is to evaluate hydroplaning risk of multi-lane roadways with PFC using a fluid–structure interaction model. A comprehensive three-dimensional grooved tire–water–pavement interaction model was developed to predict hydroplaning speeds on different pavement surfaces, rainfall intensities, and rutting depths for the passenger car tire with anti-lock braking system. The results demonstrate that PFC can effectively reduce hydroplaning risk for two-lane roadways under light rain rate to moderate rain rate as compared with impervious pavements. The hydroplaning risk becomes more apparent as the number of traffic lanes increases or with the presence of pavement rutting. However, hydroplaning risk on roadways with more than six traffic lanes under heavy rainfall intensity can still exist on PFC. The study results can be useful for both driver and transportation agencies to improve driving safety in wet weather.

Permeable friction courses (PFCs) are defined as asphalt mixtures designed with high total air void contents compared with traditional hot mix asphalt (*1*). PFCs are gaining in popularity because of their ability to provide better road safety during rain events, such as reduced risk for hydroplaning, less splash and spray, and improved visibility of pavement markings.

PFC consists primarily of open-graded aggregates bonded together by polymer-modified binder or resin to form a composite matrix with interconnected voids (*2*, *3*). At low rainfall intensities or at the onset of rainfall, surface runoff enters the pores in the PFC and flows toward the shoulder. However, the pore space can be filled with water at high rainfall rates or with continuous rainfall, which can cause water to flow both on pavement surface and within the PFC. In addition, PFCs have been shown to clog over time, which will weaken the subsurface drainage capacity, thus forcing more flow on pavement surface (*4*). The flow on pavement surface under these conditions is referred to as overland flow or sheet flow, which tends to increase water film thickness along the drainage path. From the safety perspective, water flow over pavement surface can reduce skid resistance and result in hydroplaning when vehicles abruptly enter standing water after driving on relatively dry pavement (*5*, *6*).

Experimental studies have been conducted to predict hydroplaning speeds on non-permeable pavements. These studies have revealed valuable knowledge on the mechanisms and factors affecting tire hydroplaning, and proposed empirical equations for determining hydroplaning speeds (*7–12*). However, experimental tests are limited to specific types of tires and water film depths because they are complex and time consuming. On the other hand, numerical simulation is more flexible and efficient in obtaining hydroplaning speeds at different conditions after model calibration and validation. The previous studies focused on hydroplaning risk either under locked wheel conditions or free-rolling conditions

[1]Department of Civil and Environmental Engineering, Rutgers, The State University of New Jersey, Piscataway, NJ

Corresponding Author:
Address correspondence to Hao Wang: hwang.cee@rutgers.edu

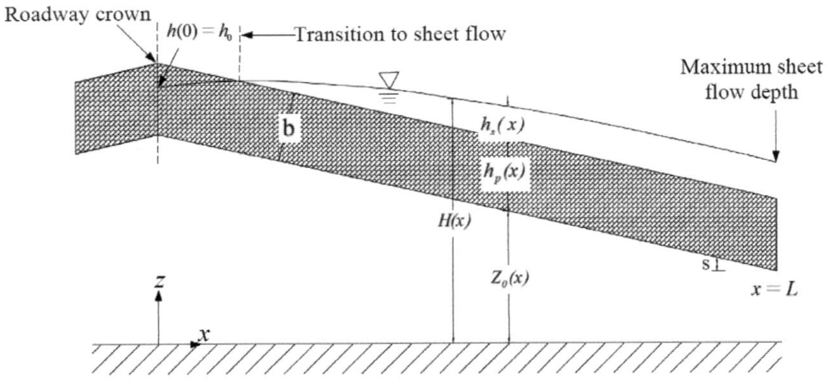

Figure 1. Cross-section along flow path of PFC with no-flow boundary (*4*).

on impervious pavements (*13–18*). The water film thickness in these models is often arbitrarily assumed, and is not directly related to rainfall intensity and pavement geometric characteristics. Limited studies have focused on the potential safety issues under wet weather on PFCs (*19*, *20*). These studies considered simplified tire geometry (bias passenger car tire with smooth treads), and the water film thickness ignored the subsurface flow in porous pavements.

The research presented in the current paper shares many attributes with previous studies to evaluate hydroplaning speed using an integrated fluid–structure interaction model. Its original contribution comes from several aspects that have yet to be addressed in the previous work. The hydroplaning simulation was conducted using more realistic tire profile, operation condition at braking, and water film depth. The water film depth was determined from a hydraulic model that combines surface and subsurface flow. The hydroplaning model predicts hydroplaning speed on multi-lane permeable pavements at different rain rates.

Objective

The objective of the study is to evaluate hydroplaning risk of multi-lane roadways with PFCs at different rainfall intensities. A comprehensive three-dimensional (3D) fluid–structure interaction numerical model was developed to predict hydroplaning speeds at different water film depths. The grooved tire profile with anti-lock braking system (ABS) reflects the real braking conditions of modern passenger car tires. The water film depths along the traffic lanes were computed from a hydrologic model that couples the surface and subsurface flow of PFCs. The hydroplaning risk was compared between permeable and impervious asphalt pavement surfaces. The effect of rutting on hydroplaning speeds was investigated.

Sheet Flow of Permeable Friction Course

Mathematical Model of Sheet Flow

Sheet flow occurs when the subsurface of the PFC becomes saturated and runoff flows onto the pavement surface. Figure 1 presents the sketch of a cross-section along the flow path on the roadway crown. The pavement has slope s, with the thickness of the PFC layer b in cm. The elevation of the bottom of the PFC layer with respect to a datum is $Z_0(x)$. The saturated thickness within PFC is $h_p(x)$, whereas the sheet flow depth on the surface is $h_s(x)$. The total head of water at any points is $H(x)$. In order to predict water film depths for hydroplaning simulation, the model proposed by Charbeneau and Barrett (*4*) was used, as shown in Equation 1. The equation can be used to determine the sheet flow depth along the drainage length at any specific rainfall intensity, pavement slope, and hydraulic conductivity.

$$h_s(x) = \left[\frac{K_{\mathrm{DW}}\nu(rx - Ts)}{8gs}\right]^{1/3} \tag{1}$$

Where, K_{DW} is the Darcy–Weisbach coefficient; ν is the kinematic viscosity in cm^2/s; r is rainfall intensity in cm/h; T is transmissivity of PFC defined as $T = Kb$ in cm^2/s; K is hydraulic conductivity of PFC in cm/s; b is PFC layer thickness in cm; g is the acceleration due to gravity in cm/s^2; x is measured from the roadway crown in cm (no-flow boundary); and s is slope.

Two critical steps are involved in using Equation 1 to estimate the sheet flow depth of PFC:

1. Determine the value of the Darcy–Weisbach coefficient (also called Darcy friction factor) K_{DW}, which is dependent on channel shape and frictional loss. For laminar flow in smooth channels, the value of K_{DW} is approximately 24 for

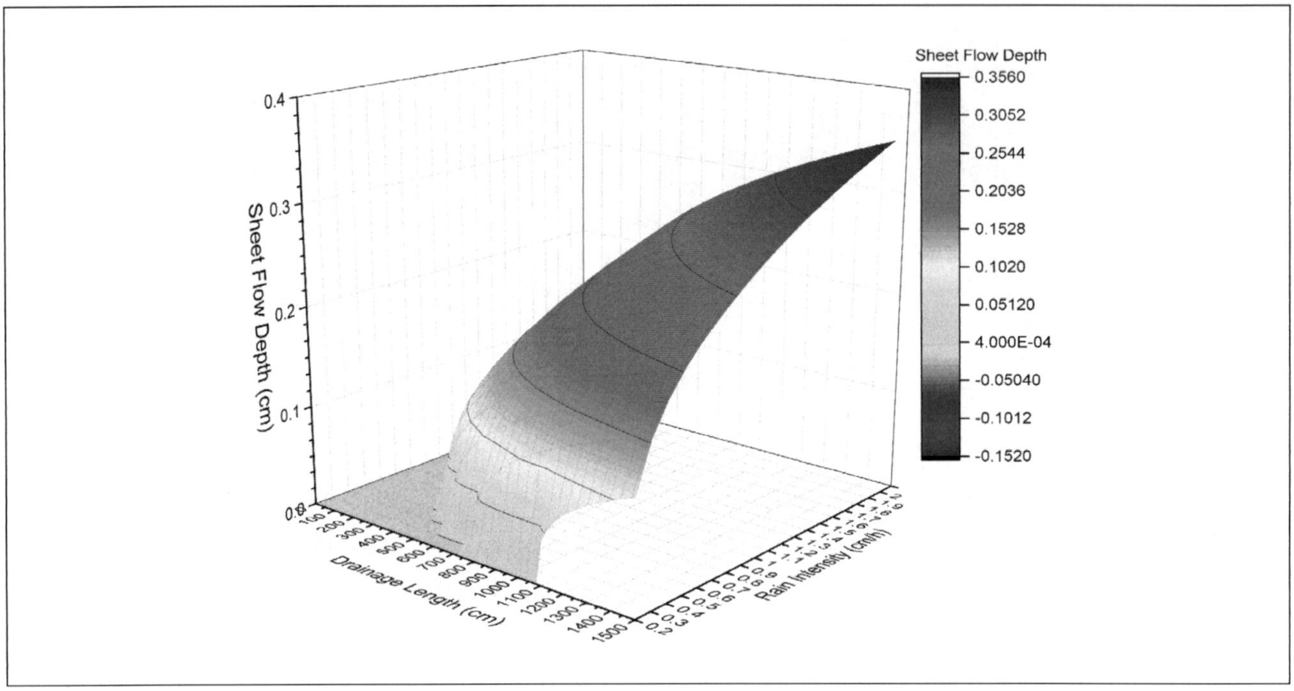

Figure 2. Sheet flow depths along drainage path.

rectangular channels (*21*). However, for the flow in PFC, as a result of the roughness of PFCs and rainfall effect, the value of K_{DW} is much larger (*4*). In this study, the recommended value (K_{DW} = 900) used in Charbeneau and Barrett's model was used.

2. Determine the boundary of the upstream edge of PFC. Depending on the conditions near the upstream edge of PFC, the boundary can be a no-flow boundary or drainage boundary. In the no-flow boundary, the origin of the x axis is the roadway crown, whereas in the drainage boundary, the coordinate system in the sheet flow model of PFC is located at the drainage divide. In this study, the no-flow boundary is used to calculate the maximum sheet flow depth along the drainage length.

Water Depth on Different Traffic Lanes

Geometric cross-sections of roadways describe the position and number of traffic lanes, lane widths, and cross-slopes, as well as the presence or absence of shoulders, curbs, and other roadway features. In general, cross-slopes of 1.5–4% are recommended for public roadways in the United States (*22*). Most public roads have at least two lanes, one for traffic in each direction. For rural and urban arterials, typical lane widths may range from 3.05 to 3.66 m (10 to 12 ft) and 3.66 m (12 ft) lanes should be used on roads with higher speed limits, higher traffic volumes, or significant numbers of trucks and other large vehicles. In this study, the width of lane is assumed as 3.66 m (12 ft), and the width of inner and outer wheel path are defined as 0.76 m (2.5 ft), with their center lines 0.76 m (2.5 ft) away from the lane edge (*23*).

The water film depths along the drainage length are calculated under the following conditions: slope$s$ = 0.02, thickness of PFCb = 3cm, hydraulic conductivity of PFC K = 1 cm/s, kinematic viscosity v = 0.01cm^2/s. The water film depths as a function of rainfall intensity and drainage length are presented in Figure 2. It is obvious that the sheet flow depth increases along the drainage path. For example, at the rainfall rate of 0.2 cm/h, there is no sheet flow on the first lane. However, there is about 1 mm sheet flow depth on the fourth lane. At the rainfall rate of 2 cm/h, the sheet flow depth on the first lane is around 1.9 mm, whereas the sheet flow depth on the fourth lane is about 3.5 mm.

In addition, the analysis shows that the PFC thickness mainly affects the location (traffic lane) where water film starts accumulating on pavement surface. When the porous layer is saturated and sheet flow already occurs on pavement surface, the water film depth only decreases slightly as the PFC thickness increases. Since this study focused on hydroplaning speeds caused by the existence of water film, the PFC thickness was kept unchanged in the analysis.

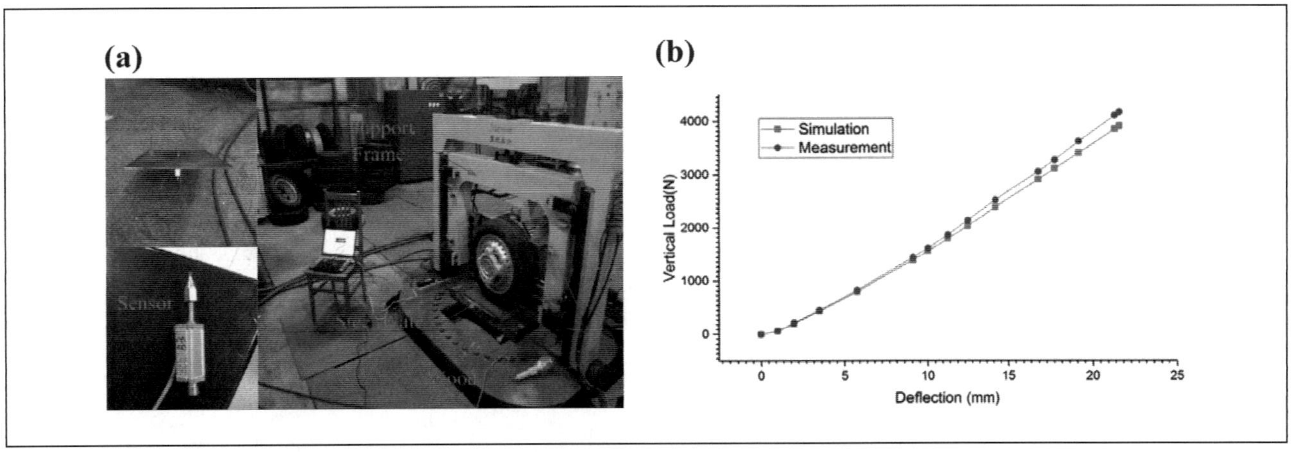

Figure 3. Tire deflection test: (*a*) test setup; (*b*) calculated and measured tire deflections.

Numerical Fluid–Structure Interaction Model

Hydroplaning Phenomenon

When a vehicle drives into thick water films at high speeds, the tire is subjected to pressure from the flowing fluid which causes deformation in the tire. In return, the deformed tire alters the fluid field. The interaction between rolling tire and flowing fluid repeats until the entire fluid–structure reaches steady state. This kind of interaction is called fluid–structure interaction. Depending on the reference frame, the tire mesh can move with the material or be fixed in space. The vertical load of the tire is transferred to the fluid through the interaction between tire and fluid. As the load transfer goes on, part of the water volume will flow into tire grooves; at the same time, because of the hydrodynamic pressure against the deformable tire structure, lift force arises and increases, which will result in the penetration of water under the tire contact patch, therefore reducing contact area at the tire–pavement interface. At a specified tire speed, the lift force reaches a certain level causing the tire to lose contact with the road surface, and full dynamic hydroplaning occurs.

Tire Finite Element Model

A passenger car tire model P205/45R16 with four 9.9 mm-wide grooves is considered for the hydroplaning simulation. The tire has an inflation pressure of 187 kPa and vertical load of 3,767 N. The choice of tire model parameters depends on balancing the type of behavior desired in the model with the information available. In the analysis of tire hydroplaning, the focus is on rainwater flow and global contact forces, not aiming at local tire deformation. The water flow and the resulting contact forces are more sensitive to the tread block geometry and the total kinetic energy of tire (*24*). In this case, the fiber-reinforced rubber part is modeled using solid rebar elements. The steel bead cords in the underlying rubber matrix are modeled as elastic elements. The tread and sidewalls constructed from pure rubber are modeled by the neo-Hookean model (*25*, *26*).

The tire model was validated using the predicted tire deflections against the measurements from static loading test for the same tire, as shown in Figure 3*a*. As presented in Figure 3*b*, the predicted tire deflections match well with the measured data. The simulated tire contact area was found within 5% error of the measured one. This indicates that the constructed tire model is suitable for finite element analysis.

The simulation of the finite element tire model includes five steps. In the first step, an internal inflation pressure is applied on the axisymmetric tire model. The second step is generation of 3D wedges with tire structure and tread. The third step is generation of a 3D full treaded tire. In the fourth step, the tire rim is mounted under inflation pressure. In the last step, the inflated tire reaches a stationary rolling condition under a vertical load at a constant traveling speed. The rolling tire is then imported to the non-stationary fluid–structure interaction model for hydroplaning analysis. Figure 4*a* presents the assembling process of grooved tire model.

Fluid Model

In this study, the Coupled Eulerian–Lagrangian (CEL) method is used to capture the hydroplaning phenomenon. With CEL, the flowing fluid can be modeled effectively using Eulerian analysis and the tire structure can be treated using traditional nonlinear Lagrangian analyses. In the CEL method, material is tracked as it flows

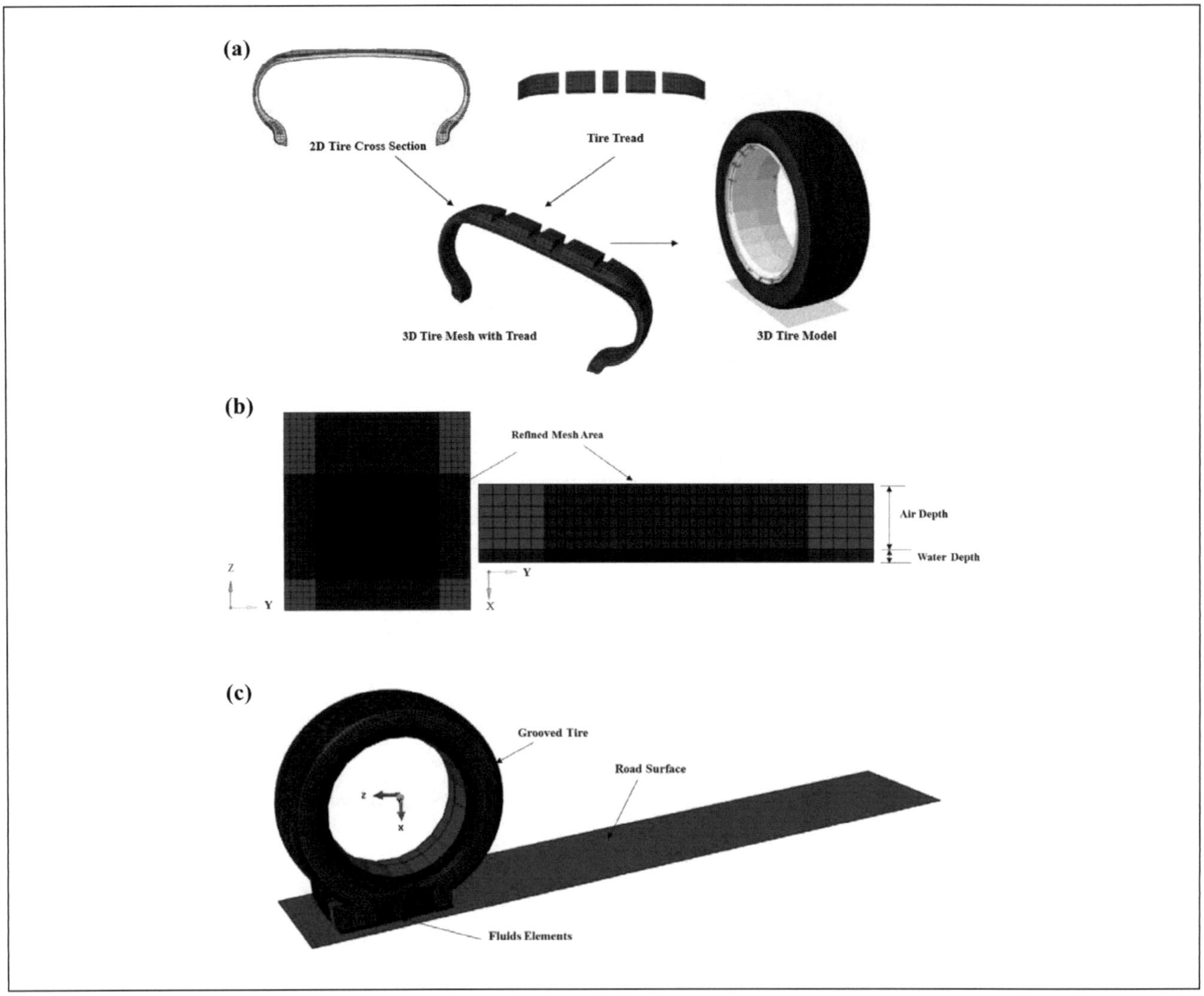

Figure 4. Illustration of (*a*) 3-D grooved tire model, (*b*) fluid model, and (*c*) fluid–structure interaction model for hydroplaning analysis.

through the mesh by computing its Eulerian volume fraction within each element (*24*). In the current study, the volume fraction $F(x;t) \in [0, 1]$ of water is used to track the flow boundary,

$$\frac{\partial F}{\partial t} + V \cdot \nabla F = 0 \quad (2)$$

By definition, if the element is fully filled with water, its volume fraction $F = 1$; if no water is present in an element, then the volume fraction $F = 0$. In the simulation of tire hydroplaning, both water and air voids are treated as Eulerian elements. If the sum of all material volume fractions in an element is less than one, that is, $0 < F < 1$, the air void will automatically fill the remainder of the element.

The Eulerian mesh of fluid (water and air) is presented in Figure 4*b*. The dimension of the fluid domain should be large enough to enclose the entire trajectory of interest, for example, the tire contact patch. In this study, the fluid domain has a width of 300 mm, length of 320 mm, and the depth of the air space above water film is fixed at 50 mm.

Hydroplaning Model

Modern cars are equipped with ABS to prevent the tire from locking up and skidding on the roadway. The concept of ABS is such that during braking, each tire continues to turn, thus reaching maximum braking force. Full braking or traction occurs when all the contact points at

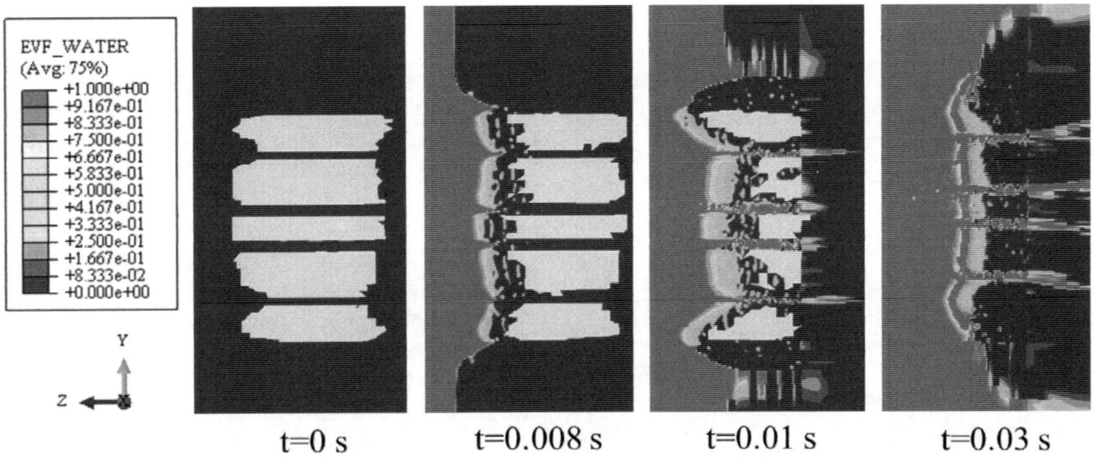

Figure 5. Evolution of water penetration into tire contact patch with time.

the tire contact patch are slipping. When hydroplaning occurs, a tire equipped with ABS will continue to rotate with an angular velocity. In this study, the ABS is realized by applying an exponential decay friction model with a slip ratio of 18% at a given angular velocity (*27*, *28*).

In the analysis, water is treated as an incompressible, isotropic, and Newtonian fluid with a constant viscosity. The only body force considered in the simulation is gravity, and the temperature change of fluid and its influence on the tire is negligible. The hydroplaning analysis is conducted in a moving frame of reference, in which the water and road surface are moving toward the tire structure at a given speed, and the tire rolls at a fixed location with angular velocity. Figure 4*c* shows the fluid–structure interaction model for hydroplaning simulation.

Figure 5 shows the contour plots of tire contact patches at hydroplaning speed of 72 km/h on water film depth of 10 mm. The EVF_WATER shown in the plot denotes the volume fraction for water, where the areas occupied by water (EVF = 1) appear as a uniform color form the top of contour spectrum, and the areas unoccupied by water presents as a different color from the bottom of contour spectrum. The boundary of water appears in a range of colors as EVF transitions from one to zero. The white area in Figure 5 indicates the contact patch of tire ribs at the tire–pavement interface. At t = 0, the tire has full contact with road surface. At t = 0.008 s, when rainwater interacts with the tire, water begins to occupy the tire contact patch. The contact patch gets smaller as water flow into tire grooves and becomes prominent. At t = 0.01 s, rainwater starts to drain from the left and right sides of tire in addition to drain through the grooves. As time goes on, at a certain time (t = 0.03 s), the hydrodynamic force reaches a certain level which is high enough to lift up the tire, and the tire loses contact with road surface.

Model Validation

There are many empirical equations that have been developed to predict hydroplaning speed. However, these existing empirical models are often limited to a certain type of tire or water film depth. For example, the NASA equation does not take into account of water film depth (*29*). The Gallaway equations are based on specific tire profiles (*30*, *31*). To validate the hydroplaning model proposed in this study, the predicted hydroplaning speeds were compared with the analytical results from previous work and experimental measurements reported in the literature.

To validate the numerical method of hydroplaning, the predicted hydroplaning speeds of 195/65R15 tires were compared with experimental data measured in the study of Okano and Koishi (*32*). Three different tread patterns of P195/65R15 tire were considered in the analysis, including smooth tread (no groove), narrow groove (groove width is 9 mm), and wide groove (groove width is 18 mm). The groove depth is 8 mm and groove spacing is 50 mm. The vertical load applied on the tire is 4 kN and the tire inflation pressure is 200 kPa. The water film depth is kept at 10 mm. In the experiment, the hydroplaning phenomenon was captured by the visual images of water flow around the contact area when tires run on water film. The comparisons of hydroplaning speeds obtained from the developed models and experimental measurements are summarized in Table 1. In addition, the comparisons of hydroplaning speeds obtained from

Table 1. Comparison of Hydroplaning Speeds with Experiment

Tire model	Hydroplaning speed (km/h) Experiment (*32*)	Current simulation	Difference (%)
Smooth tire (no grooves)	77.0	75.6	−1.8
9 mm groove width	82.6	80.0	−3.1
18 mm groove width	87.0	85.0	−2.3

Table 2. Comparison of Hydroplaning Speeds with Previous Model

Water film depth (mm)	Hydroplaning speed (km/h) Previous model (*33*)	Current simulation	Difference (%)
5	82.2	85.5	4
10	70.3	72.0	2.8
15	65.2	68.0	4.3
20	62.2	63.7	2.4

the developed model and previous work are summarized in Table 2. In the work by Oh et al., the combination of finite difference method and finite element method was used to solve hydroplaning speeds of P205/45R16 tire at different water film depths (5, 10, and 15 mm) (*33*). Good agreements were found between the predicted hydroplaning speed from the developed model in this study and the measurements and model results from previous work. These validations indicate that the proposed fluid–structure interaction model properly captures the hydroplaning phenomenon.

Results and Discussion

Hydroplaning Risk on Different Traffic Lanes

This subsection evaluates the hydroplaning risk of each traffic lane under different rainfall intensities. The maximum water film depths along the drainage length are calculated under the following conditions: slope $s = 0.02$, thickness of PFC $b = 3$cm, hydraulic conductivity of PFC $K = 1$ cm/s, and kinematic viscosity $\nu = 0.01$ cm^2/s. The hydroplaning simulations were performed for the 205/45R16 tire with a constant inflation pressure of 187 kPa and a vertical load of 3,767 N. From the maximum water depth presented in Table 3, the hydroplaning speeds under different water depths can be computed by the developed fluid–structure interaction model.

Figure 6 compares the hydroplaning speeds at different traffic lanes on permeable pavements. It can be seen that as water film depth increases with drainage length along the traffic lane under a given rainfall rate, the hydroplaning risk increases with the increase of lane numbers. For example, under light rain rate, the runoff enters the pores of PFC, and there is no sheet flow on the cross-sections with two lanes and four lanes. However, there is still hydroplaning potential on the third lane and fourth lane on the cross-sections with six lanes and eight lanes. In addition, it is obvious that the hydroplaning risk increases with the increase of rainfall intensity. At light to moderate rainfall intensities, the hydroplaning risk for a two-lane cross-section is relatively low; however, at heavy rainfall intensity, the hydroplaning risk increases dramatically, especially for roadway cross-sections with greater number of lanes.

The results show that PFC provides better road safety for two-lane and four-lane cross-sections at the same rain rates. The simulation results provide significant safe driving guidance for both drivers and transportation agencies. For example, during wet weather, it is advised that drivers travel on the innermost lane and avoid frequent changing of traffic lanes, especially driving from an inner lane to an outer lane which will increase the risk of hydroplaning. For transportation agencies, the roadway locations prone to hydroplaning can be treated with reduced speed limits and slippery pavement warning signs as low-cost interim measures to improve road safety. For example, at light rain rate, roadways with eight lanes should be operated at speeds no higher than

Table 3. Maximum Sheet Flow Depths (mm) at Different Lanes

Rain rate (cm/h)	Type	2 lanes	4 lanes 1st	4 lanes 2nd	6 lanes 1st	6 lanes 2nd	6 lanes 3rd	8 lanes 1st	8 lanes 2nd	8 lanes 3rd	8 lanes 4th
<0.25	Impervious	1.1	1.1	1.4	1.1	1.4	1.6	1.1	1.4	1.6	1.8
(light rain)	Permeable	–	–	–	–	–	0.9	–	–	0.9	1.3
0.25–1	Impervious	1.7	1.7	2.2	1.7	2.2	2.6	1.7	2.2	2.6	2.9
(moderate rain)	Permeable	1.2	1.2	2.0	1.2	2.0	2.4	1.2	2.0	2.4	2.7
1–5	Impervious	3.0	3.0	3.8	3.0	3.8	4.4	3.0	3.8	4.4	4.9
(heavy rain)	Permeable	2.8	2.8	3.7	2.8	3.7	4.3	2.8	3.7	4.3	4.8

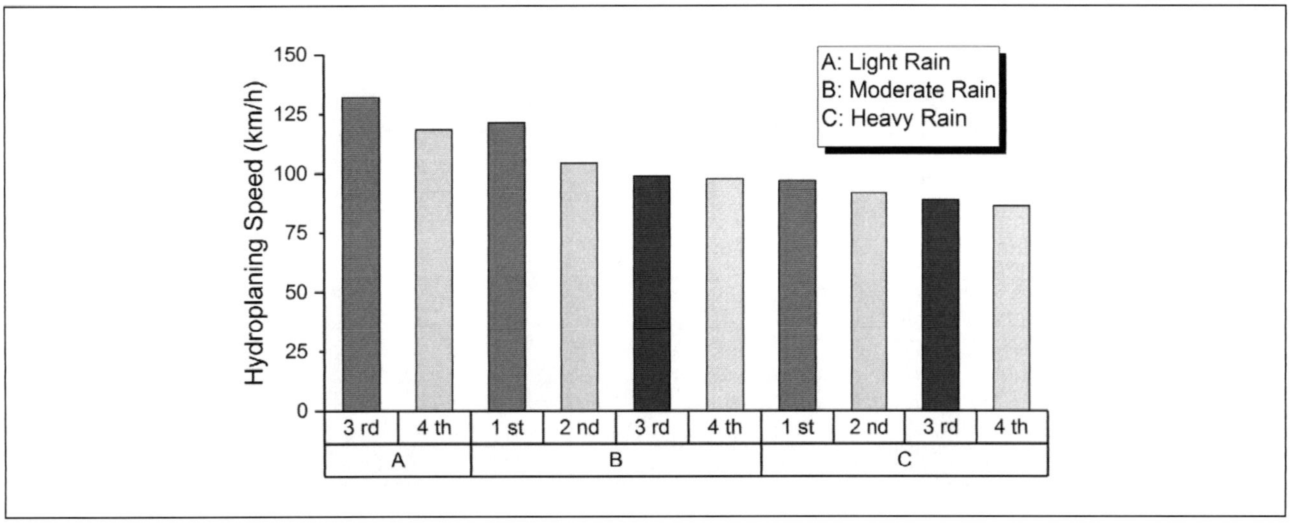

Figure 6. Hydroplaning speeds at each traffic lane on permeable pavements under different rain rates.

Table 4. Hydroplaning Speeds for Permeable and Impervious Pavements at Eight-Lane Roadway Cross-Section

	Permeable	Impervious
Rain rate (cm/h)	Hydroplaning speed range (km/h)	
<0.25 (light rain)	≤118	≤107
0.25–1 (moderate rain)	98~121	96~109
1–5 (heavy rain)	86~97	86~95

118 km/h; but the same roadway should be operated at speeds no high than 97 km/h and 86 km/h during moderate rain and heavy rain rates, respectively.

Comparison between Permeable and Impervious Pavements

This section compares hydroplaning risk between permeable and impervious asphalt pavements. For conventional impervious asphalt pavement, the hydraulic conductivity ranges from $8.5 \times 10^{-7}$to 10^{-4}cm/s (*6, 34*). In this paper, the hydraulic conductivity of 5.04×10^{-5}cm/s is assumed for impervious asphalt pavement. The maximum sheet flow depths at different lanes for permeable and impervious asphalt pavements were also compared in Table 3.

Table 4 shows the comparison of critical hydroplaning speeds at different rain rates for permeable and impervious pavements for an eight-lane roadway cross-section. The model results show that the PFC layer has reduced potential for hydroplaning during light rain and moderate rain conditions as compared with the impervious pavement surface, especially at the innermost traffic lane. For example, at light rain rate, there is no sheet flow on the first two lanes for an eight-lane cross-section; however, the sheet flow on the impervious pavement surface still has hydroplaning potential; at moderate rain rate, the hydroplaning speed for the innermost lane is 121 km/h, which is 11% higher than that on the impervious pavement surface. However, it is worth noting that the anti-hydroplaning potential of PFC is not significant in a heavy rain event, showing that in areas with frequent heavy rain rates, caution is still needed for drivers traveling on PFC.

Effects of Rutting on Hydroplaning Risk

Rutting on pavement surface can be particularly hazardous in wet weather conditions, because ruts can be filled with water and thus cause vehicle hydroplaning. Rutting depths can be classified into three severity levels: low severity (below 6 mm), moderate severity (7–12 mm), and high severity (greater than 13 mm) (*35*). In this section, a moderate-severity rutting (9 mm) and a high-severity rutting (14 mm) are adopted to analyze the effect of rutting on hydroplaning risk of permeable and impervious asphalt pavements. The first lane of roadway cross-section is considered, as it is experienced regardless of lane numbers.

The water film depth at the wheel path is calculated under the following conditions: $s = 0.02$, $b = 3$cm, $K = 1$ cm/s, hydraulic conductivity of impervious pavement $K = 5.04 \times 10^{-5}$cm/s, and $\nu = 0.01$cm^2/s. The maximum water film depths for permeable and impervious pavements at moderate rain rates are 10.2 mm and 10.7 mm at moderate-severity rutting, and 15.2 mm and 15.7 mm at high-severity rutting, respectively.

Figure 7 shows that pavement rutting significantly increases hydroplaning risk on both permeable and

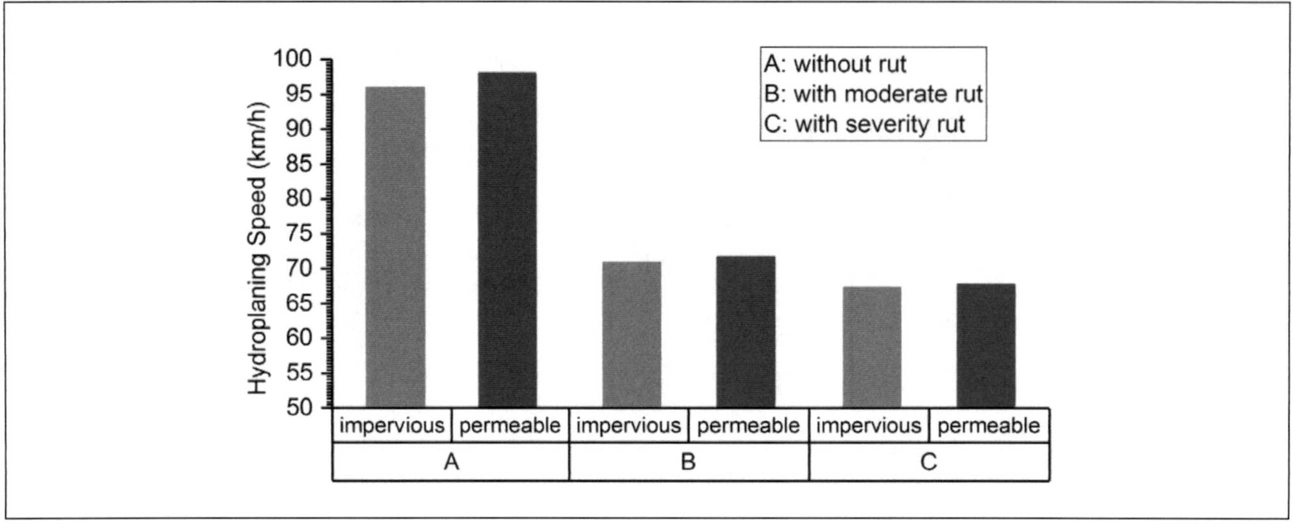

Figure 7. Effects of rutting on hydroplaning speeds at moderate rain rate.

impervious pavements. For example, the hydroplaning speeds on impervious pavements with moderate and severe rutting decrease to 70.9 km/h and 67.3 km/h, respectively, as compared with 96 km/h on pavements without rutting. This indicates that pavement with moderate to severe rutting should be given special attention, especially during wet weather conditions.

Conclusion

This paper presented an analytical framework to evaluate the hydroplaning risk of multi-lane roadways with PFC using a fluid–structure interaction model. For this purpose, the water film depths along the drainage length were predicted from a mathematical model which combines sheet flow and flow inside PFC. The hydroplaning speed was calculated using the developed grooved tire–water–pavement interaction model. The hydroplaning model can successfully identify the locations on multi-lane roadways that have high susceptibility to hydroplaning.

The results demonstrate that PFC can effectively reduce hydroplaning risk for two-lane roadways under light to moderate rain rates as compared with impervious pavements. The findings show that hydroplaning risk becomes more apparent as the number of traffic lanes increases or with the presence of pavement rutting. Therefore, the hydroplaning risk on roadways with more than six traffic lanes under heavy rainfall intensity can still exist on PFC.

The quantification of hydroplaning speed using the developed model can help transportation agencies reduce wet weather safety risk by setting measures depending on rain rates and cross-section design. The study findings can guide the project selection of PFCs at locations with high hydroplaning risk. It is apparent, according to this study, that roadways with two-lane to four-lane cross-sections under light to moderate rain rates can benefit most from the implementation of PFC. Future research should consider the requirement of PFC thickness under different geometry and weather conditions.

References

1. Alvarez, A. E., A. E. Martin, and C. Estakhri. Drainability of Permeable Friction Course Mixtures. *Journal of Materials in Civil Engineering*, Vol. 22, No. 6, 2009, pp. 556–564.
2. Wang, H., Y. Ding, G. Liao, and C. Ai. Modeling and Optimization of Acoustic Absorption for Porous Asphalt Concrete. *Journal of Engineering Mechanics*, Vol. 142, No. 4, 2016, p. 04016002.
3. Chen, J., X. Ma, H. Wang, P. Y. Xie, and W. Huang. Experimental Study on Anti-Icing and Deicing Performance of Polyurethane Concrete as Road Surface Layer. *Construction and Building Materials*, Vol. 161, 2018, pp. 598–605.
4. Suresha, S. N., G. Varghese, and U. R. Shankar. Laboratory and Theoretical Evaluation of Clogging Behaviour of Porous Friction Course Mixes. *International Journal of Pavement Engineering*, Vol. 11, No. 1, 2010, pp. 61–70.
5. Charbeneau, R. J., and M. E. Barrett. Drainage Hydraulics of Permeable Friction Courses. *Water Resources Research*, Vol. 44, No. 4, 2008, pp. 1–10.
6. Chen, J. Q., H. Wang, and H. Z. Zhu. Investigation of Permeability of OGFC Considering Effects of Anisotropy and Two-Dimensional Flow. *Construction and Building Materials*, Vol. 145, 2017, pp. 218–325.
7. Horne, W. B., U. T. Joyner, and T. W. Leland. *Studies of the Retardation Force Developed on an Aircraft Tire Rolling in Slush or Water*. NASA Technical Notes NASA-TN-D-552, L-1260. National Aeronautics and Space Administration, Washington, D.C., 1960.
8. Sinnamon, J. F., and J. T. Tielking. *Hydroplaning and Tread Pattern Hydrodynamics*. UM-HSRI-PF-74-10.

Highway Safety Research Institute, The University of Michigan, 1974.
9. Ivey, D. L. *Truck Tire Hydroplaning-Empirical Verification of Horne's Thesis*. Presentation to the Technical Seminar on Tire Service and Evaluation, ASTM Committee F-9, Akron, Ohio, 1984.
10. Horne, W. B., T. J. Yager, and D. L. Ivey. Recent Studies to Investigate Effects of Tire Footprint Ratio on Dynamic Hydroplaning Speed. In *The Tire Pavement Interface* (M. G. Pottinger and T. J. Yager, eds.), Special Technical Publication 929, ASTM International, West Conshohocken, Pa., 1986, pp. 26–46.
11. Anderson, D. A., R. S. Huebner, J. R. Reed, J. C. Warner, and J. J. Henry. *NCHRP Project 1-29: Improved Surface Drainage of Pavement: Final Report*. HRB, National Research Council, Washington, D.C., 1998.
12. Gunaratne, M., Q. Lu, J. S. Yang, J. Metz, W. Jayasooriya, M. Yassin, and S. Amarasiri. *Hydroplaning on Multi Lane Facilities*. BDK84 977-14. Florida Department of Transportation Research, 2012.
13. Ong, G. P., T. F. Fwa, and J. Guo. Modeling Hydroplaning and the Effects of Pavement Micro-Texture. *Transportation Research Record: Journal of the Transportation Research Board*, 2005. 1905: 166–176.
14. Ong, G. P., and T. F. Fwa. Transverse Pavement Grooving against Hydroplaning, I-Simulation Model. *Journal of Transportation Engineering*, Vol. 132, No. 6, 2006, pp. 441–448.
15. Fwa, T. F., and G. P. Ong. Wet-Pavement Hydroplaning Risk and Skid Resistance: Analysis. *Journal of Transportation Engineering*, Vol. 134, No. 5, 2007, pp. 184–190.
16. Kumar, S. S., K. Anupam, T. Scarpas, and C. Kashbergen. Study of Hydroplaning Risk on Rolling and Sliding Passenger Car. *Procedia-Social and Behavioral Sciences*, Vol. 53, No. 3, 2012, pp. 1019–1027.
17. Kosgolla, J. V. *Numerical Simulation of Sliding Friction and Wet Traction Force on a Smooth Tire Sliding on a Random Rough Pavement*. Graduate Theses and Dissertations, South Florida University, 2012.
18. Zhou, H., G. Wang, Y. Ding, J. Yang, and H. Zhai. Investigation of the Effect of Dimple Bionic Nonsmooth Surface on Tire Antihydroplaning. *Applied Bionics and Biomechanics*, Vol. 2015, Article ID 694068, 2015.
19. Zhang, L., G. P. Ong, and T. F. Fwa. Developing an Analysis Framework to Quantify and Compare Skid Resistance Performance on Porous and Nonporous Pavements. *Transportation Research Record: Journal of the Transportation Research Board*, 2013. 2369: 77–86.
20. Zhang, L., T. Fwa, G. P. Ong, and L. J. Chu. Analysing Effect of Roadway Width on Skid Resistance of Porous Pavement. *Road Materials and Pavement Design*, Vol. 17, No. 1, 2016, pp. 1–14.
21. Chow, V. T. *Open-Channel Hydraulics*. McGraw-Hill, New York, 1959.
22. American Association of State Highway and Transportation Officials. *A Policy on Geometric Design of Highways and Streets*. American Association of State Highway and Transportation Officials, Washington, D.C., 2001.
23. Miller, J. S., and Y. B. William. *Distress Identification Manual for the Long-Term Pavement Performance Program*. U.S. Department of Transportation, 2003, pp. 90–91.
24. Rugonyi, S., and K. J. Bathe. On Finite Element Analysis of Fluid Flows Fully Coupled with Structural Interactions. *Computer Modeling in Engineering and Sciences*, Vol. 2, No. 2, 2001, pp. 195–212.
25. Wang, H., I. L. Al-Qadi, and I. Stanciulescu. Simulation of Tire-Pavement Interaction for Predicting Contact Stresses at Static and Rolling Conditions, *International Journal of Pavement Engineering*, Vol. 13, No. 4, 2012, pp. 310–321.
26. Ding, Y. M., and H. Wang, FEM-BEM Analysis of Tire-Pavement Noise on Porous Asphalt Surfaces with Different Textures. *International Journal of Pavement Engineering*, 2017, pp. 1–8. (Published online).
27. Zhou, H., G. Wang, Y. Ding, J. Yang, C. Liang, and J. Fu. Effect of Friction Model and Tire Maneuvering on Tire-Pavement Contact Stress. *Advances in Materials Science and Engineering*, Volume 2015, Article ID 632647, 2015.
28. Wang, H., I. L. Al-Qadi, and I. Stanciulescu. Effect of Surface Friction on Tire-Pavement Contact Stresses during Vehicle Maneuvering. *Journal of Engineering Mechanics*, Vol. 140, No. 4, 2014, p. 04014001.
29. Horne, W. B., and R. C. Dreher. *Phenomena of Pneumatic Tire Hydroplaning*. NASA Technical Note D-2056. National Aeronautics and Space Administration, Washington, D.C., 1963.
30. Gallaway, B. M., R. E. Schiller, and J. G. Rose. *The Effects of Rainfall Intensity, Pavement Cross Slope, Surface Texture and Drainage Length on Pavement Water Depths*. Texas Transportation Institute, College Station, 1971.
31. Gallaway, B. M. *Pavement and Geometric Design Criteria for Minimizing Hydroplaning*. Publication FHWA-RD-79-31. FHWA, U.S. Department of Transportation, Washington, D.C., 1979.
32. Okano, T., and M. A. Koishi. A New Computational Procedure to Predict Transient Hydroplaning Performance of a Tire. *Tire Science and Technology*, Vol. 29, No. 1, 2001, pp. 2–22.
33. Oh, C. W., T. W. Kim, H. Y. Jeong, K. S. Park, and S. N. Kim. Hydroplaning Simulation for a Straight-grooved Tire by Using FDM, FEM and an Asymptotic Method. *Journal of Mechanical Science and Technology*, Vol. 22, No. 1, 2008, pp. 34–40.
34. Kanitpong, K., C. Benson, and H. Bahia. Hydraulic Conductivity (Permeability) of Laboratory-Compacted Asphalt Mixtures. *Transportation Research Record: Journal of the Transportation Research Board*, 2003. 1767: 25–32.
35. Hicks, R. G., S. Seeds, and D. G. Peshkin. *Selecting a Preventive Maintenance Treatment for Flexible Pavements*. FHWA-IF-00-027. Federal Highway Administration, Washington, D.C., 2000.

The Standing Committee on Pavement Surface Properties and Vehicle Interaction (AFD90) peer-reviewed this paper (18-03525).

Article

TRR
JOURNAL OF THE TRANSPORTATION RESEARCH BOARD

Transportation Research Record
2018, Vol. 2672(40) 418–429

DOI: 10.1177/0361198118781137
journals.sagepub.com/home/trr
SAGE

Effects of the International Roughness Index and Rut Depth on Crash Rates

Michael Mamlouk[1], Mounica Vinayakamurthy[1], B. Shane Underwood[2], and Kamil E. Kaloush[1]

Abstract

Pavement distresses directly affect ride quality, and indirectly contribute to driver distraction, vehicle operation, and accidents. In this study, analysis was performed on highways in the states of Arizona, North Carolina, and Maryland to investigate the relationship between accident rate and pavement ride quality (roughness) and rut depth. Two main types of data were collected: crash data from the accident records and International Roughness Index (IRI) and rut depth data from the pavement management system database in each state. Crash rates were calculated using the U.S. Department of Transportation method, which is the number of accidents per 100 million vehicle-miles of travel. Sigmoidal function regression analysis was performed to study the relationship between crash rate and both IRI and rut depth. In all cases, the crash rate did not show substantial increases until an IRI value of 210 inches/mile or a critical rut depth of 0.4 inches. When the IRI or rut depth increased above these values the crash rate increased. This is a key conclusion that provides empirically derived thresholds for IRI and rut depth to reducing the accident rate.

Several factors affect highway accident rates, such as human factors, vehicular causes, environment, roadway geometry, traffic volume, pavement condition, and their combinations. Studies show that the majority of accidents are caused by human factors such as distraction, alcohol, stress, physical deficiency, and age (*1*). Although pavement condition is not a major factor that affects accidents, the exact role of this factor is less studied and therefore less well understood. On one hand, a good quality pavement may reduce accident rates by reducing driver distraction and providing good quality vehicle operation conditions. Conversely, it may be argued that when the pavement condition is poor, drivers tend to be more cautious and reduce speed, which in turn might reduce the accident rate.

Pavement surface irregularities and rutting (Figure 1) directly affect ride quality, and indirectly contribute to driver distraction, vehicle operation, and accidents (*2*). A pavement that has many surface defects (potholes, corrugations, very severe fatigue cracking, etc.) can cause a vehicle to lose control when braking or turning, especially under adverse environmental conditions (*3*). When pavement surface irregularities increase, the contact area between vehicle tires and pavement decreases, resulting in lower brake friction (*4*). Also, these irregularities can contribute to greater vehicle instability as different friction forces may exist on the two sides of the vehicle. Moreover, vehicles bouncing up and down on extremely rough pavements may result in vehicles losing their loads, causing accidents (*5*). Similarly, rutting may result in a driver needing to exert extra effort to get out from the wheel path (if the rut depth is large), thus leading to

Figure 1. Surface irregularities and rutting may contribute to driver distraction, substandard vehicle operation, and accidents (right photo courtesy of Atkins http://www.atkinsglobal.com).

[1]Civil, Environmental and Sustainable Engineering Program, Arizona State University, Tempe, AZ
[2]Deprtment of Civil, Construction and Environmental Engineering, North Carolina State University, Raleigh, NC

Corresponding Author:
Address correspondence to Michael Mamlouk: mamlouk@asu.edu

uncertain and, in some cases, uncontrolled lateral vehicle movement. In addition, rutting is more hazardous in wet weather when water accumulates in the rut path and leads to hydroplaning and loss of vehicle control. The problem can be further exaggerated when human factors, such as distraction, alcohol, stress, physical deficiency, and age, are combined with pavement distresses.

The majority of the studies dealing with the effect of pavement condition on safety are related to skid resistance (*6, 7*). Only a limited number of studies have focused on exploring the relationship between accident frequency and roughness or rutting. These studies showed that increasing road roughness, in general, increases the rate of accidents. Very limited information is available to determine the pavement condition level the agency needs to maintain in order to actively reduce accident risk. Transportation agencies have been looking for the appropriate roughness and rut depth thresholds below which pavement-related accidents can be reduced.

Objectives

The main objective of this study is to determine the international roughness index (IRI) and rut depth thresholds that correlate to an increased accident rate. General trends that relate accident rates to roughness and rut depth are developed. Accident data and pavement conditions from three states in different geographic locations and climatic conditions are collected. Accident severity levels are separated and related to roughness or rut depth.

Literature Review

Pavement roughness can be defined as the deviations of a surface from a true planer surface with characteristic dimensions that affect vehicle dynamics, ride quality, dynamic loads, and drainage (*8*). The IRI is the most common parameter used to measure ride quality. Note that IRI does not measure roughness per se, since roughness is related to the micro/macro surface texture. IRI, therefore, measures the response to pavement roughness, or technically "ride quality." Depending on the body of literature and length scale of study, "roughness" can have multiple physical definitions. For example, in safety analysis, roughness is effectively interpreted as something in the order of millimeter to micrometer scales. The IRI is certainly affected by these surface irregularities, but much less so than it is affected by distortions related to driver comfort, between 2 and 20 m in wavelength (axle hop and body bounce). As such it is not a measure of roughness in the same way that roughness is interpreted in the safety literature.

King investigated the effect of road roughness on traffic speed and road safety in Southern Queensland, Australia (*9*). The study found a strong relationship between higher crash rates and increased pavement roughness. Crash rates involving light vehicles were more affected by increasing roughness than crashes involving heavy freight vehicles. Considering different crash severity levels, crashes resulting in hospitalizations and property damage had the strongest increase in crashes over a small increase in roughness. The study also found that speed is reduced when roughness increases. The study recommended that traffic authorities managing rural roads need to reduce roughness to an IRI value of 120 in./mile in order to provide a safer road environment.

Cairney and Bennet performed a study to determine the relationship between pavement surface characteristics and roadway crashes in Victoria, Australia (*10*). The authors measured pavement properties with a multi-laser profilometer and linked them to crash data with the help of the Global Positioning System (GPS). They found that there was good correlation between roughness and crash rate. However, no clear relationship could be found between rutting and crash rate.

In a study by Graves et al., the authors found that a disproportionate number of crashes were associated with certain pavement conditions, thus suggesting that they are correlated (*11*). The analysis was performed in Alabama and further suggested data mining could be a useful technique in the analysis process. Li et al. examined the relationship between crash severity and factors indicating the pavement condition in Texas (*12*). Results indicated that crashes of higher severity occurred on roads with poor pavement condition compared with roads with fair pavement condition. It was also noted that very good pavement conditions might induce speeding behaviors and therefore could have caused more severe crashes, especially on non-freeway arterials and during favorable driving conditions. Purposefully laying down slightly rougher pavements on high-speed roadways could potentially reduce high-severity crashes. A more recent study by Li indicated that pavement with poor surface conditions are responsible for higher crash rates (*13*).

Tehrani and Cowe Falls explored the relationship between the IRI value and number of collisions in the province of Alberta, Canada (*14*). The study indicated that the sections with high IRI values have more crashes in comparison to those with low IRI values. Also, the results indicated that there was a good correlation between rut depth and number of crashes in 1 km segments.

Cenek et al. performed a study to develop statistical models predicting the correlation between rut depths and fatal and injury crashes on New Zealand's state highway network (*15*). The results indicated that there was an increase in crash rate where the rut depth is 10 mm or higher, possibly as a result of water accumulation in the ruts. The authors concluded that the crash rate, for dry crashes in particular, has decreased slightly in sections where the rut depth is slightly higher than the normal range.

Chan et al. performed a study to understand the relationship between accident frequency and pavement condition using IRI, rut depth and present serviceability index (PSI) as parameters for pavement condition (*3*). The results show that IRI and PSI were significant in all types of models, whereas the rut depth model performed well in predicting the accidents that occurred during night time only.

Hu et al. developed mathematical relationships between IRI and driving comfort and safety (driving workload) (*16*, *17*). The authors developed threshold IRI values on road segments at different risk levels for driving comfort and safety. They also concluded that standard IRI values for pavement maintenance are beyond the comfort and safety threshold for both car and truck drivers.

In summary, it can be suggested that roughness and rutting can be contributing factors for traffic safety and crash occurrence. The literature suggests that pavement roughness has good correlation with crash rate and affects crash severity. The contribution of rut depth to traffic safety is not well defined. No guidelines are currently available in the literature to assist highway maintenance authorities to maintain their pavement at a certain level in order to minimize crash occurrences.

Data Collection and Processing

Data were collected from Interstate, U.S. and State roads in Arizona, North Carolina, and Maryland. The data were obtained for both flexible and rigid pavements without distinction between the years 2013 and 2015. Note that rigid pavement sections were automatically excluded from the accident-rutting correlations since no rutting data were typically available in the database for these sections.

Two main types of data were collected: crash data from the accident records and IRI and rut depth data from the pavement management system (PMS) databases. Data were brought together from the national Highway Performance Monitoring System (HPMS) public data release and open source data available from the three states.

The specific data items that were used in the study were crash frequency and severity, average annual daily traffic (AADT), IRI, and rut depth. As the duration of analysis is 1 year for the study, the data were sorted and separated for each year in each state. Crash data and PMS data were matched and merged together on the basis of location. For Arizona and North Carolina, data matching was performed by taking road name and milepost as common criteria. However, for Maryland, data were matched using latitude and longitude or geographic information system (GIS) coordinates as the matching criteria. The PMS data available in the shape file format were extracted using ArcGIS and were converted into csv files for use in further analysis. GIS coordinates along with the route ID which give information about road name and milepost were obtained from the shape files. After obtaining the filtered data, SQL queries were used to correlate the data and obtain the necessary results. SQL queries were written for all the crashes and for each severity level separately. After matching the data using SQL, Microsoft Excel was used to perform further analysis, and grouping the data on the basis of IRI and rut depth.

During the initial screening of the data, the crash occurrences that state other factors, such as weather condition, as the major cause of the crash were removed prior to the analysis. However, in most cases the contributing factor for crash occurrence was not reported. One mile segments are used for analysis and 1 year is taken as the period of analysis. The PMS data in Arizona and North Carolina were provided for each mile post. However, in Maryland, the PMS data were provided for every 0.1 miles. In order to maintain uniformity throughout the analysis, the Maryland PMS data were averaged to every mile.

Different states used different devices to collect IRI and rut depth data. In Arizona, the profilometer was used to measure both IRI and rut depth (*18*). In North Carolina, the profiler was used to measure IRI values and the profilometer to measure rut depth (*19*). In Maryland, the Automatic Road Analyzer (ARAN) was used to measure both IRI and rut depth data (*20*).

Accident data were reported at five levels of severity in Arizona and North Carolina and three levels in Maryland. In Arizona, the five levels are damage without injury, minor injury, non-incapacitating injury, incapacitating injury, and fatality. In North Carolina, the five levels are damage without injury, injury level C, injury level B, injury level A, and fatality. In Maryland, the three levels are property damage, physical injury, and fatality. In this study, the severity levels of the three states will be referred to as levels 1–5 for consistency.

Analysis of Results

Data Summary

Table 1 presents the crash frequency data used for the analysis divided into different severity levels. It can be noticed that the number of crashes used for analysis for Maryland is considerably lower than the other states. This is because crash data obtained from Maryland did not cover the whole state, as a portion of data was not publicly available. Therefore, the Maryland crash rates were used to study their trend with IRI and rut depth data, but not necessarily to show the actual crash rate values. Note also that severity levels 2–4 (physical injury) are combined in the state of Maryland. Also, no fatality crashes were reported in this Maryland sample of data.

Table 1. Summary Statistics of Crash Data

State (Year)	All severities	Severity level 1	Severity level 2	Severity level 3	Severity level 4	Severity level 5
Arizona (2013)	31,514	21,748	4,473	4,149	838	306
Arizona (2014)	32,570	22,809	4,454	4,296	767	243
North Carolina (2015)	97,612	67,601	20,625	6,702	835	601
Maryland* (2014)	807	607			204	-

Note: *Partial crash data were obtained.

Table 2. Average IRI and Rut Depth Values of Crash and Non-Crash Segments

State (Year)	Percent length of crash segments	Average IRI (in./mile)		Average rut depth (in.)	
		Non-crash segments	Crash segments	Non-crash segments	Crash segments
Arizona (2013)	36.6	86	72	–	–
Arizona (2014)	40.6	85	72	0.06	0.06
North Carolina (2015)	47.8	124	103	0.14	0.14
Maryland (2014)	–	–	133	–	0.15

Crash vs. Non-Crash Segments

The highways used in the analysis can be divided into crash and non-crash 1-mile segments. Crash segments are the road networks on which at least one accident has occurred in the study period. On the other hand, non-crash segments can be defined as the road networks on which no crashes have occurred during the study period.

Table 2 shows the percent of length of crash segments relative to the length of the whole pavement network in the different states/years. The table also shows the average IRI and rut depth values of crash and non-crash segments. Some data in the table are missing because the authors were not be able to obtain them within the time and budget available to the project.

The table shows that the length of crash segments represent 37–48% of the total length of the network in different states using 1-mile segments. The table also shows that IRI and rut depth values are different in different states. The difference could be because of the actual differences in pavement conditions or because of other reasons such as type of data measured, type of measuring equipment, data processing method, sampling method, unit length of pavement section, and number of runs of measuring devices (*21*).

Comparing crash and non-crash segments, the results show that the average ride quality and rutting values of crash and non-crash segments in each state/year are close to each other. In fact, the average IRI values of non-crash segments were slightly higher than those of crash segments. This suggests that ride quality and rutting are not the only factors affecting number of crashes but possibly in combination with other factors such as traffic volume, other pavement distresses, and so forth.

Crash Rate

For each state and each crash severity level, the IRI and rut depth values were broken down to categories of 50 inches/mile and 0.1 inches, respectively, and the corresponding number of crashes was obtained. The reason for separating the data into these bins was to obtain a reasonable number of IRI and rut depth categories. For each category, the crash rate was determined using the U.S. Department of Transportation method (*22*), which can be calculated using the formula mentioned below.

$$R = \frac{C \times 100,000,000}{V \times 365 \times L \times N} \quad (1)$$

where,

R = crash rate for the road segment per 100 million vehicle-miles of travel (VMT)

C = average number of crashes in the study period

V = average annual traffic volume entering the study area daily (AADT)

L = length of the road segment in miles

N = number of years of data

As crash rates were calculated using 1-mile road segments and 1-year periods, both L and N are equal to 1. In the rest of the analysis, only crash segments were used since non-crash segments have crash rates of zero. If both crash and non-crash segments are used, the large number of crash rates of zero will dominate the analysis and skew the results.

Ride Quality Analysis

The relationships between crash rates and IRI values were investigated. Several curve-fitting functions were tried such

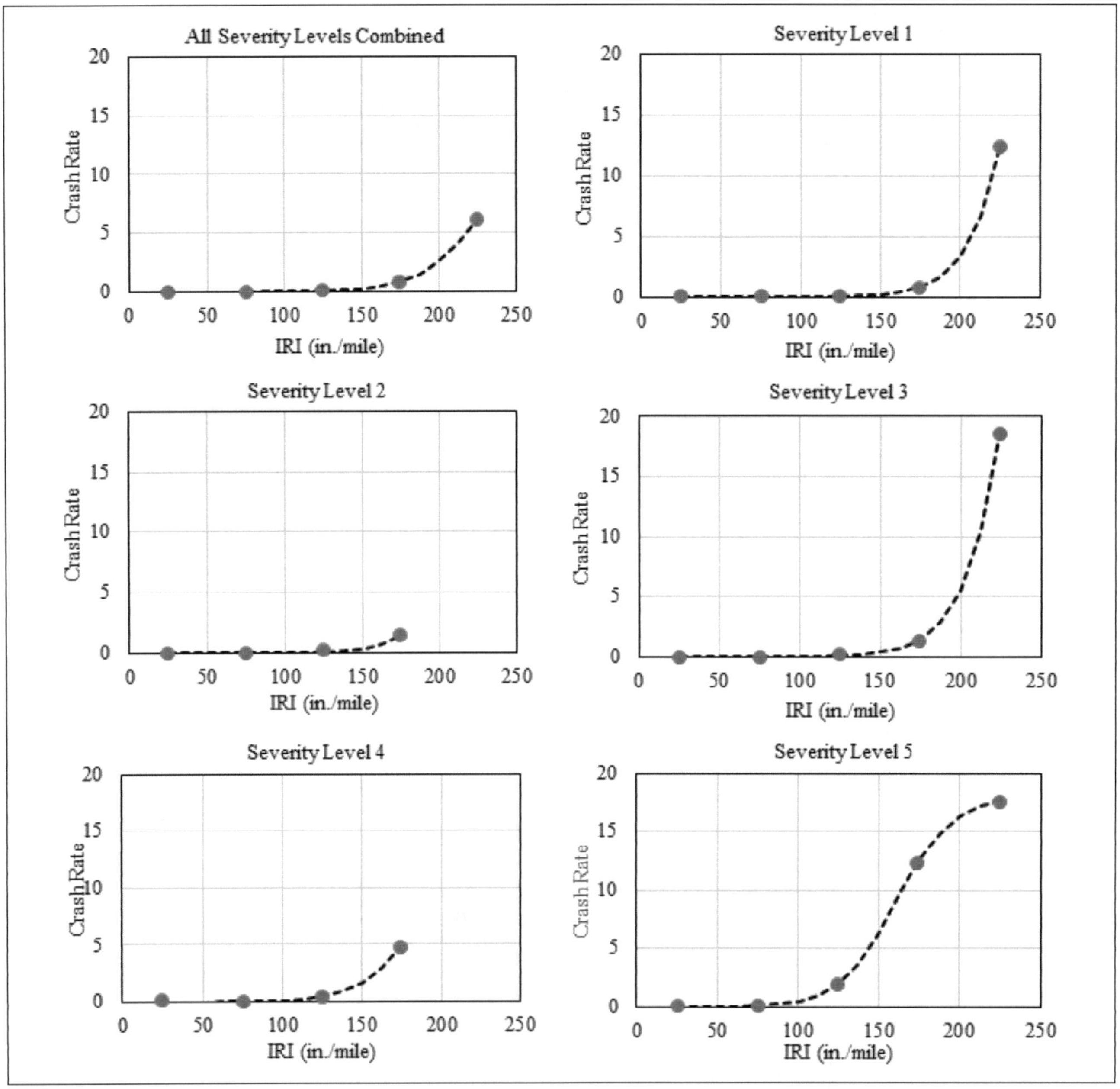

Figure 2. Relationship between crash rate and IRI values for different severity levels (Arizona 2013).

as exponential, power, and so forth. The sigmoidal function provided the best fit among other functions. Sigmoidal function models were developed between the average crash rates and the average IRI values for each crash severity level. During the analysis, data points that are obviously outside the typical range were removed from the analysis. Equation 2 shows the sigmoidal function used in each category.

$$\log R = \delta + \frac{\alpha}{1 + e^{\beta + \gamma(\log D)}} \tag{2}$$

where,

R = crash rate for the road segment per 100 VMT
D = average IRI (in./mile)
δ = minimum logarithmic value of R
$\delta+\alpha$ = maximum logarithmic value of R
β, γ = parameters describing the shape of the sigmoidal function

Figures 2–5 show the relationship between crash rate and IRI values for all severity levels combined and for individual severity levels for Arizona, North Carolina, and Maryland, respectively. The number of one-mile segments used to plot these graphs are 2267, 3230, 5821, and 120 for Arizona 2013, Arizona

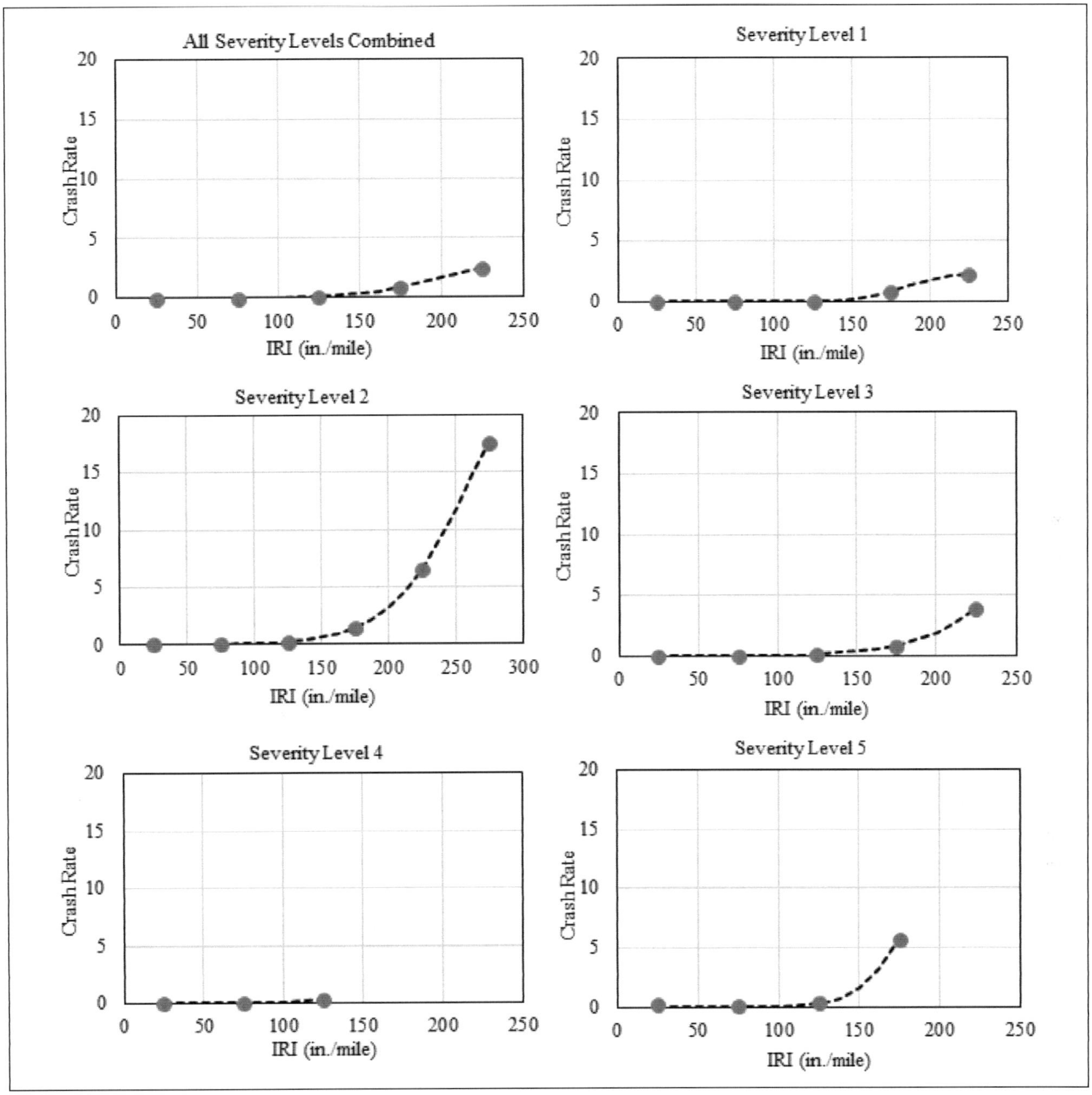

Figure 3. Relationship between crash rate and IRI values for different severity levels (Arizona 2014).

2014, North Carolina 2015, and Maryland 2014, respectively. The graphs show an average crash rate value in each category of IRI of 50. In all cases, the crash rate does not basically increase up to a certain IRI value, above which crash rate starts to increase. This trend was noticed for various crash severity levels. This phenomenon suggests that if the IRI value is kept below a certain value, crash rate can be reduced.

Rutting Analysis

Similar to IRI data, the rut depth values were broken down to categories of 0.1 inches for each state and each crash severity level. The relationships between crash rates and rut depth values were investigated. Sigmoidal function models (similar to Equation 2, except that D is the average rut

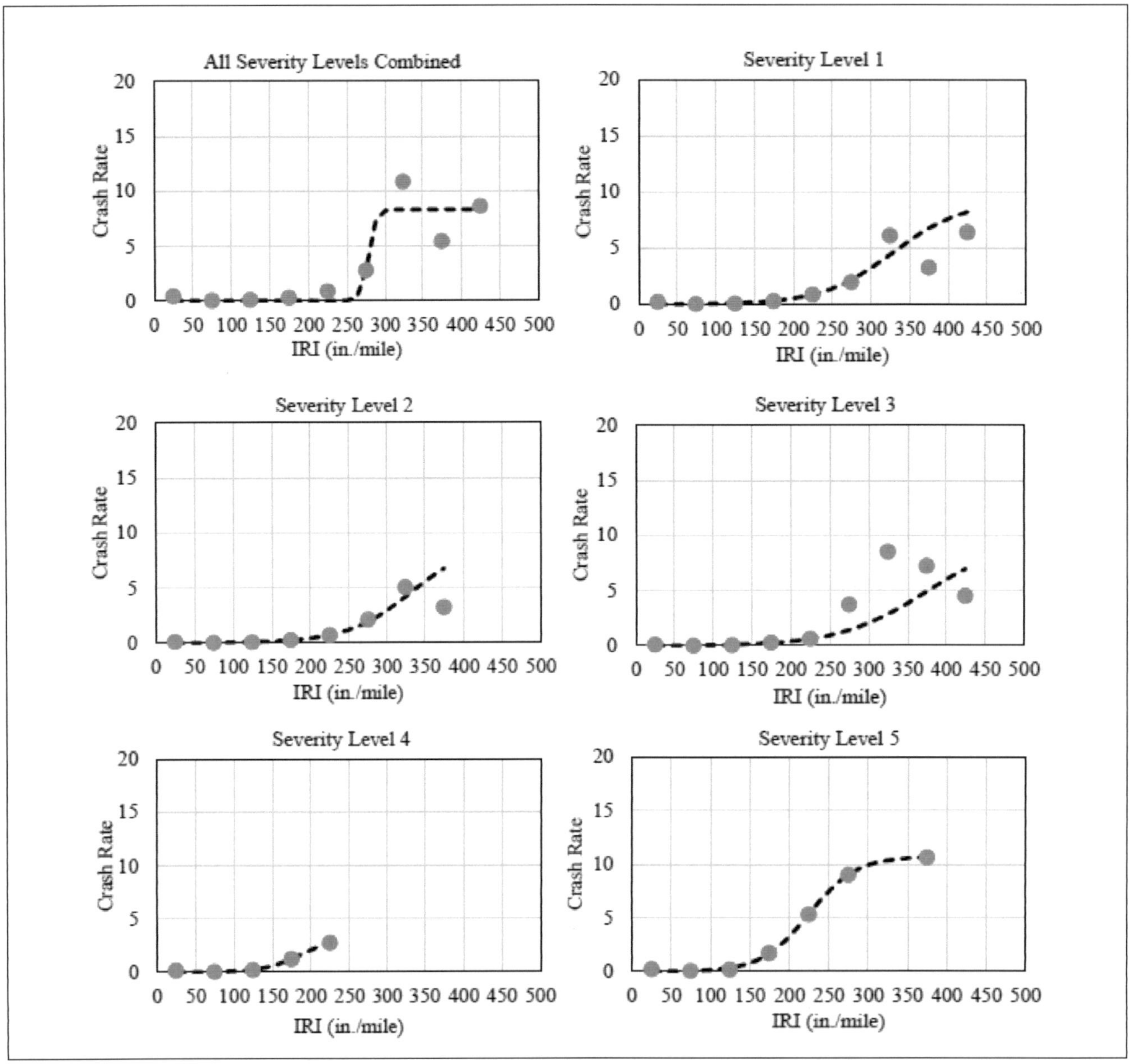

Figure 4. Relationship between crash rate and IRI values for different severity levels (North Carolina 2015).

depth in inches) were developed between the average crash rates and the average rut depth values for each crash severity level.

Figures 6–8 show the relationship between crash rate and rut depth values for all severity levels combined and for individual severity levels for Arizona, North Carolina, and Maryland, respectively. Similar to the IRI graphs above, these graphs show an average crash rate value in each category of rut depth of 0.1 inches. The number of 1-mile segments used to plot these graphs are 1841, 5922, and 111 for Arizona 2014, North Carolina 2015, and Maryland 2014, respectively. In all cases, the crash rate does not basically increase up to a certain rut depth value, above which crash rate starts to increase. This trend was noticed for various crash severity levels. This phenomenon suggests that if the rut depth is kept below a certain value, crash rate can be reduced.

Critical Pavement Condition

The aforementioned results show that the crash rate is not greatly affected by either IRI or rut depth at low values, but

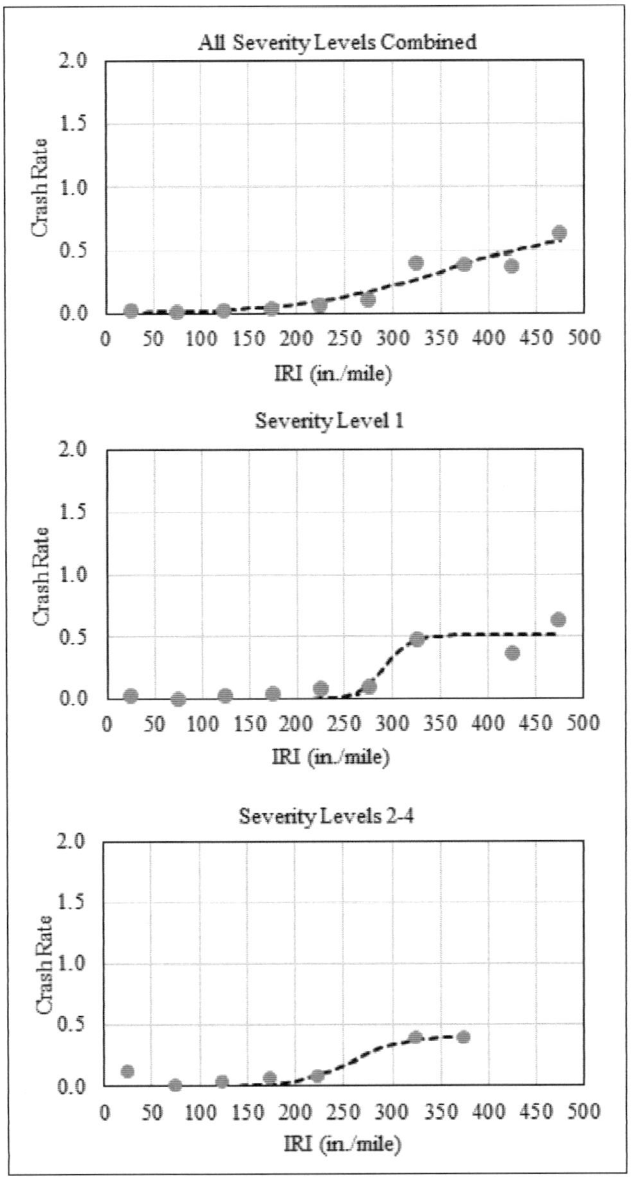

Figure 5. Relationship between crash rate and IRI values for different severity levels (Maryland 2014). Partial crash data were used.

at some critical value the crash rate increases. To objectively define this critical threshold the sigmoidal function was used to estimate the IRI or rut depth values when crash rate becomes non-zero. The second derivative (R'') of the sigmoidal function, Equation 3, was first used to identify the inflection point of the function.

$$R'' = \frac{(-\alpha \times \beta)^2\left[\left(e^{D\times\beta} - e^{\gamma}\right)\left(e^{(D\times\beta+\gamma)}\right)\right]}{\left(e^{D\times\beta} + e^{\gamma}\right)^3} \tag{3}$$

Where D is ether the average IRI value in inches/mile or rut depth in inches.

Next, the slope of the sigmoidal function at this point was taken and extrapolated until it intercepted the horizontal axis, as shown in Figure 9. The slope of the line at this point was determined by taking the first derivative (R') of the sigmoidal function, Equation 4. The intersection of the tangent line with the x-axis was selected as the critical IRI or rut depth.

$$R' = \frac{\alpha \times \beta \times e^{(D\times\beta+\gamma)}}{\left(e^{D\times\beta} + e^{\gamma}\right)^2} \tag{4}$$

In a few cases, the curves ended slightly below the inflection point because of the low accident data available. In such cases, the procedure for determining the critical IRI and rut depth values was kept the same for consistency. The amounts of error involved in these cases were within the rounding errors of IRI and rut depth used in the study.

The critical IRI and rut depth values were determined for the cases of all severity levels combined for each state as shown in Table 3. The table shows some variations in critical IRI and rut depth values for the different states. These differences could be because of differences in the type of measuring equipment, data processing methods, sampling method, or number of runs of measuring devices. The table also shows a difference in the critical IRI values of Arizona in 2013 and 2014, which could be within the measurement errors. Considering all values in the table, the critical IRI value can be taken as the average of all values for all cases. In such a case, the critical IRI value is 210 inches/mile, whereas the critical rut depth value is 0.4 inches. These average values, however, need to be used with caution because of the differences between states stated above. This is a key conclusion that provides empirically derived thresholds for safety concerns.

As crashes are typically caused by several combined factors, one of the limitations of this study is the fact that the effects of IRI and rut depth on crash rates were investigated separately and without considering other potential contributing factors. Future studies need to combine the effect of IRI and rut depth with other factors. Other contributing factors include human factors, vehicular malfunction, environmental factors, and roadway geometry.

Conclusion

The following conclusions can be drawn from the results of this study.

1. IRI and rut depth values of crash and non-crash segments in each state–year combination were close to each other. This suggests that ride quality and rutting are not the only factors affecting number of crashes, but possibly in combination with other factors such

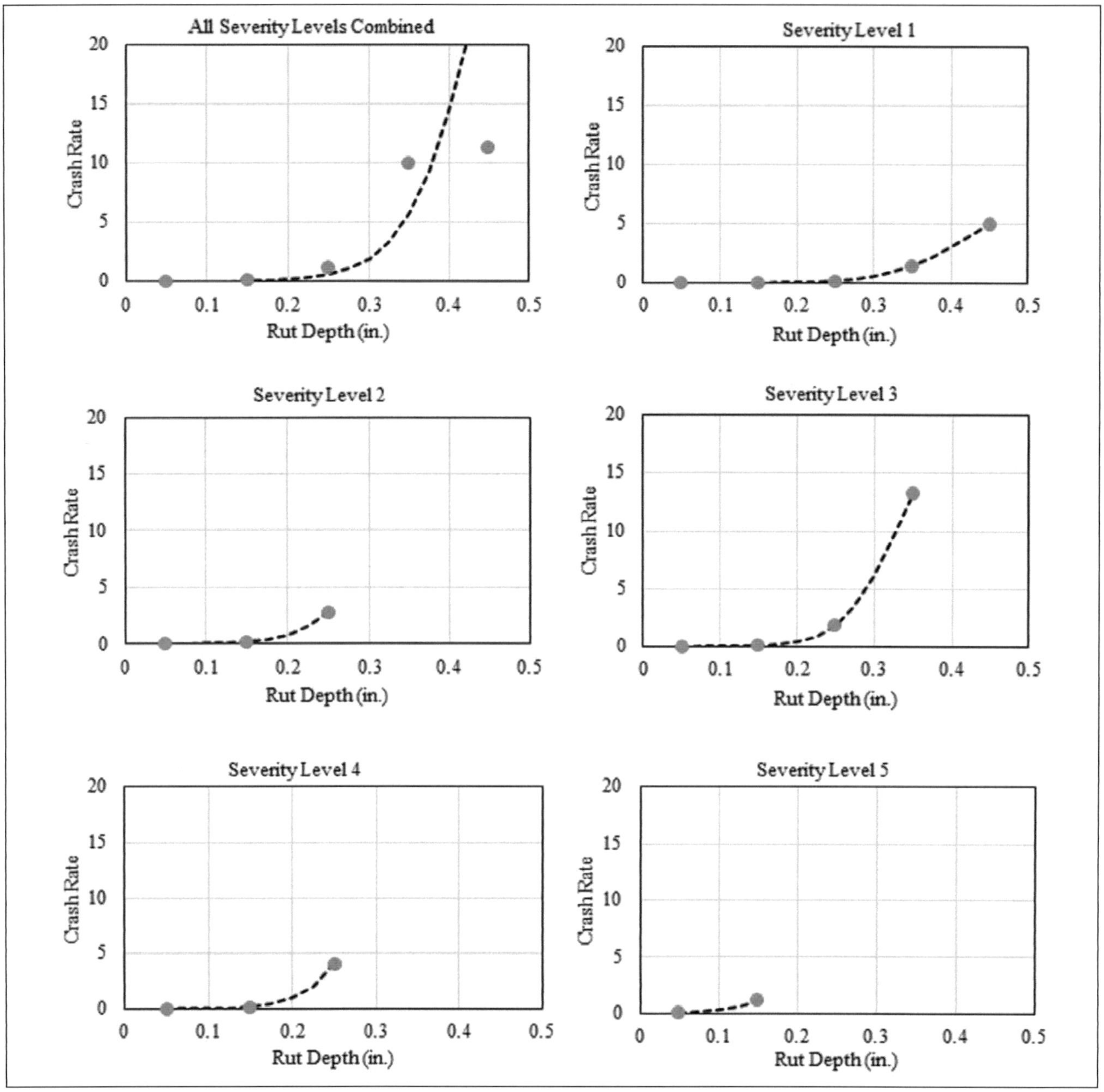

Figure 6. Relationship between crash rate and rut depth for different severity levels (Arizona 2014).

as traffic volume, other pavement distresses, and others.

2. There is a unique relationship between IRI and crash rate in all cases, indicating that crash rate does not basically increase up to a certain IRI value, above which crash rate starts to increase. This phenomenon occurred for individual crash severity levels as well as for all crash severity levels combined.
3. Similar to ride quality, crash rate does not essentially increase up to a certain rut depth value, above which crash rate starts to increase. This phenomenon occurred for individual crash severity levels and for all crash severity levels combined.
4. The critical IRI values above which crash rate starts to increase varied from one state to another because of the difference in the measurements in each state resulting from factors such as measurement equipment, data processing methods, sampling method, or number of runs of measuring devices. The average critical IRI value for all three

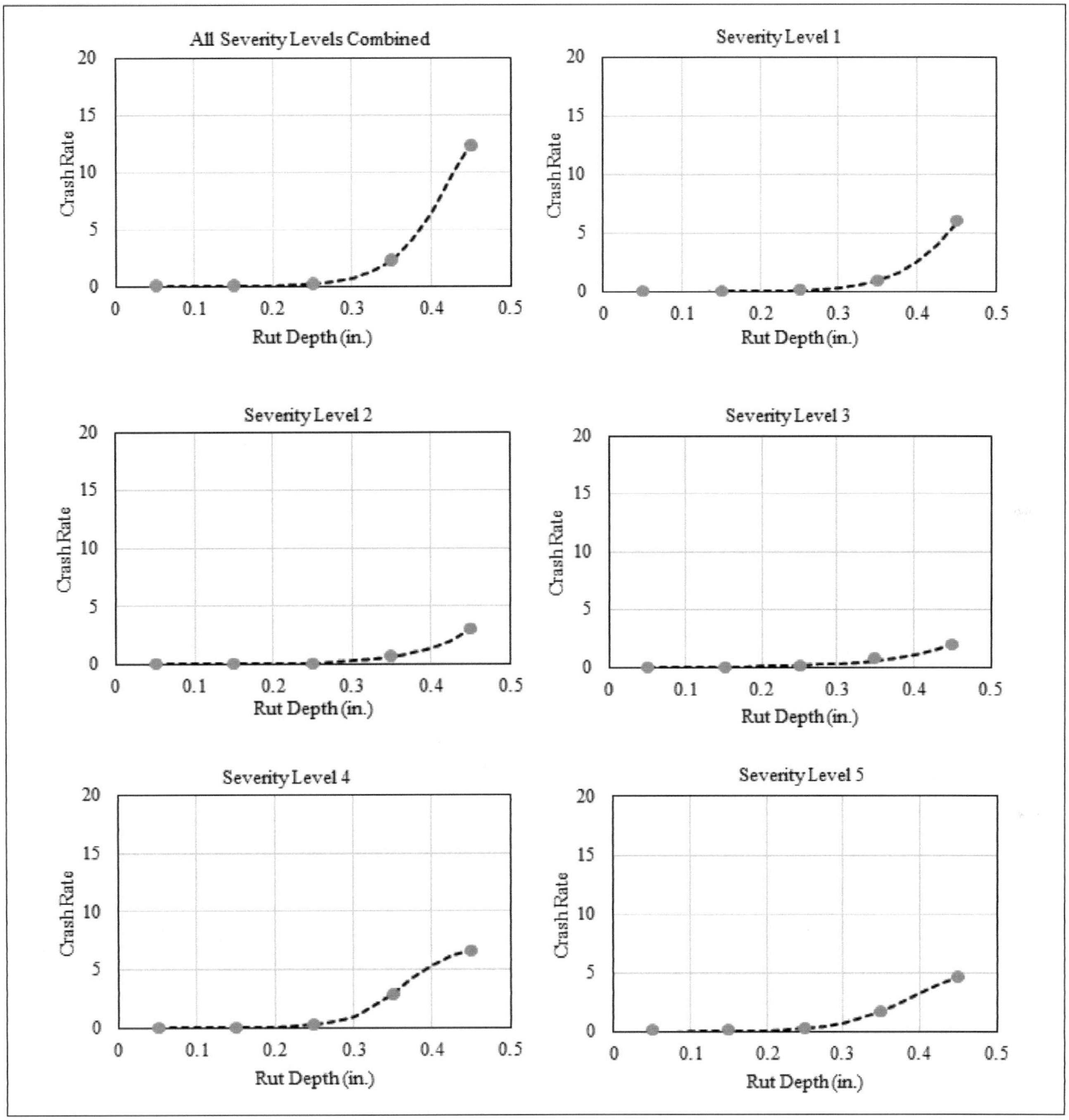

Figure 7. Relationship between crash rate and rut depth for different severity levels (North Carolina 2015).

states above which crash rate starts to increase is 210 inches/mile.

5. The average critical rut depth value above which crash rate starts in increase was almost the same for all three states, with an average value of 0.4 inches.

Note that the average threshold values of IRI and rut depth concluded in this study need to be used with caution because of the differences in measurements between states, as stated above. It should be also noted that crashes are rare and uncommon events. Therefore, more studies with large amounts of crash data are needed to confirm the results of

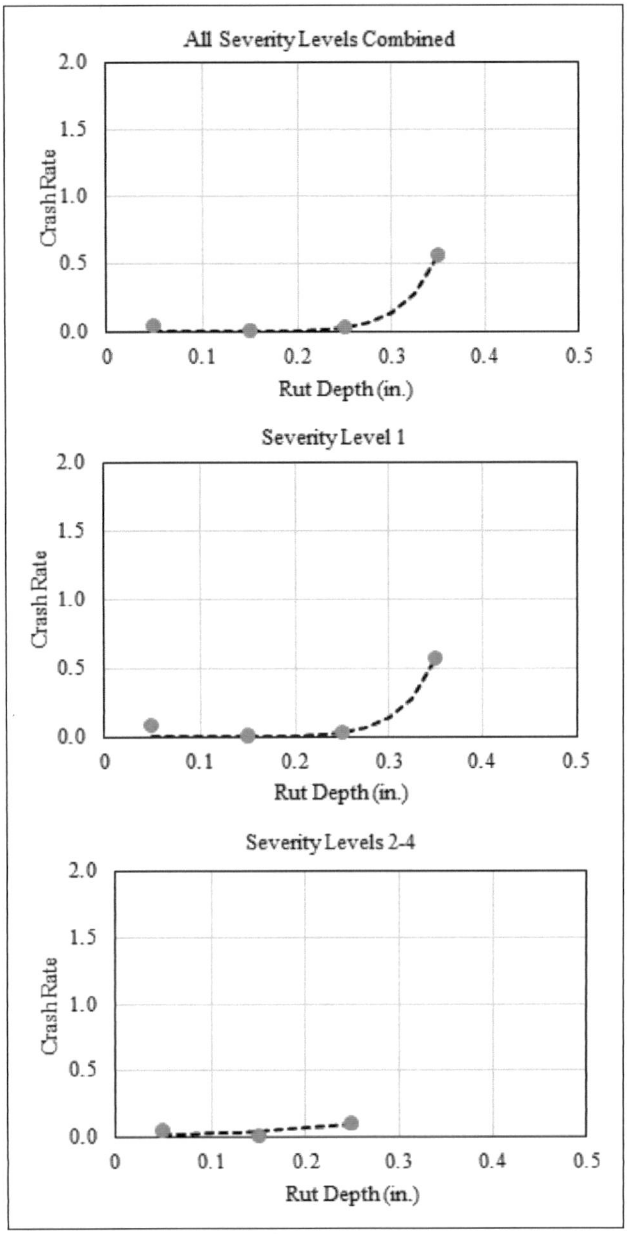

Figure 8. Relationship between crash rate and rut depth for different severity levels (Maryland 2014). Partial crash data were used.

this study. Future studies need to combine the effect of IRI and rut depth with other contributing factors such as human behavior, vehicular malfunction, environmental factors, and roadway geometry.

Acknowledgments

This project was funded by the National Transportation Center @ Maryland (NTC@Maryland), one of the five National Centers that were selected in this nationwide competition, by the Office of the Assistant Secretary for Research and Technology (OST-R), U.S. Department of Transportation (US DOT). The cost share contributed by Arizona State University is greatly appreciated. Appreciation is also given to transportation engineers in Arizona, Maryland, and North Carolina for answering queries and providing data used in this study.

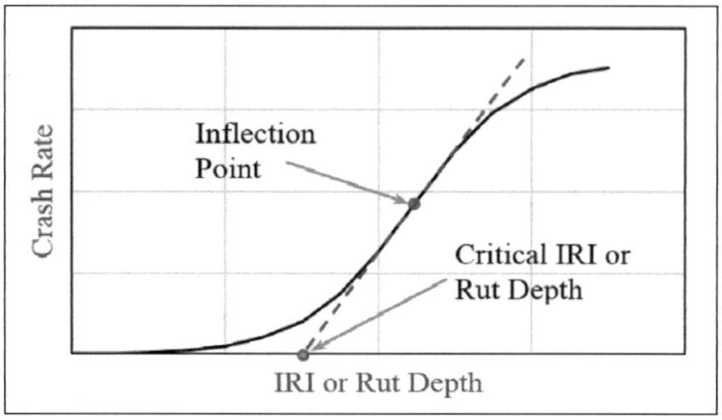

Figure 9. Determination of critical IRI or rut depth to minimize crash rate.

Table 3. Critical IRI and Rut Depth Values in Different Cases to Minimize Crash Rate

State	Year	Critical IRI (in./mile)	Critical rut depth (in.)
Arizona	2013	192	–
	2014	152	0.35
North Carolina	2015	268	0.35
Maryland	2014	208	0.4
Average (rounded)		210	0.4

References

1. Garber, N.J., and L.A. Hoel. *Traffic and Highway Engineering*, 5th ed. Cengage Learning, Stamford, CT, 2015.
2. Transportation Research Board. *Transportation Research Circular No. E-C134: Influence of Roadway Surface Discontinuities on Safety: State of the Art*. Transportation Research Board of the National Academies, Washington D.C., 2009.
3. Chan, C. Y., B. Huang, X. Yan, and S. Richards. Investigating Effects of Asphalt Pavement Conditions on Traffic Accidents in Tennessee Based on the Pavement Management System (PMS). *Journal of Advanced Transportation*, Vol. 44, No. 3, 2010, pp. 150–161.
4. Wambold, J. C., A. D. Brickman, W. H. Park, and J. Ingram. Effect of Road Roughness on Vehicle Braking. In *Highway Research Record 471*, HRB, National Research Council, Washington, D.C., 1973, pp. 76–82.
5. Burns, J. C. Roughness and Roadway Safety. *Transportation Research Record: Journal of the Transportation Research Board,* 1981. 836: 8–14.
6. Hall, J. W., K. L. Smith, L. Titus-Glover, J. C. Wambold, T. J. Yager, and Z. Rado. *NCHRP Project 1-43: Guide for Pavement Friction*. Transportation Research Board of the National Academies, Washington, D.C., 2009.

7. Noyce, D. A., H.U. Bahia, J.M. Yambo, and G. Kim. *Incorporating Road Safety into Pavement Management: Maximizing Asphalt Pavement Surface Friction for Road Safety Improvements*. Draft Literature Review and State Surveys, Midwest Regional University Transportation Center (UMTRI), Madison, Wis., 2005.
8. Gillespie, T. D., and M. Sayers (eds.). Development, Implementation, and Application of the Reference Quarter-Car Simulation. In *Measuring Road Roughness and Its Effects on User Cost and Comfort*. ASTM International, West Conshohocken, PA, 1985.
9. King, B. *The Effect of Road Roughness on Traffic Speed and Road Safety*. PhD dissertation. University of Southern Queensland, Australia, 2014.
10. Cairney, P., and P. Bennet. Relationship between Surface Characteristics on Selected Rural Roads in Victoria Rural Roads. *Proc., 23rd ARRB Conference*. ARRB Group, Vermont South, Victoria, Australia, 2008. http://worldcat.org/isbn/1876592540.
11. Graves, S. J., R. Daniel, and A. D. Michael. Mining and Analysis of Traffic Safety and Roadway Condition Data. UTCA Report 04310. The University of Alabama, Tuscaloosa, AL, July 2005. http://utca.eng.ua.edu/files/2011/08/04310fnl.pdf.
12. Li, Y., L. Chunxiao, and D. Liang. Impact of Pavement Conditions on Crash Severity. *Accident Analysis & Prevention*, Vol. 59, 2013, pp. 399–406.
13. Li, Y. Safety Impact of Pavement Conditions. *Transportation Research Record: Journal of the Transportation Research Board*, 2014. 2455: 77–88.
14. Tehrani, S., and L. Cowe Falls. A Comparison of IRI and Collision Rates in the Province of Alberta. *Canadian Society for Civil Engineering. Canadian Society for Civil Engineering*, Vol. 1024, 2012, p. 1.
15. Cenek, P. D., R. J. Henderson, M. Forbes, R. B. Davies, and A. Tait. *The Relationship between Crash Rates and Rutting*. Research Report No. 545. New Zealand Transport Agency, 2014.
16. Hu, S., and X. Qiao-Jun. Safety Evaluation Method of Freeway Pavement Condition. *Journal of Transportation Systems Engineering and Information Technology*, Vol. 12, No. 2, 2013, pp. 130–135.
17. Hu, J., X. Gao, R. Wang, and S. Sun. Research on Comfort and Safety Threshold of Pavement Roughness. *Transportation Research Record: Journal of the Transportation Research Board*, 2017. 2641: 149–155.
18. Arizona Department of Transportation. Vehicles of ADOT: Profilometer. Phoenix, AZ. https://www.azdot.gov/media/blog/posts/2015/01/23/vehicles-of-adot-profilometer. Accessed May 31, 2018.
19. North Carolina Department of Transportation. High Speed Inertial Profilometer. Raleigh, NC. https://connect.ncdot.gov/projects/construction/Structural%20Design%20AGCDOT%20Joint%20Bridge%20Design%20Commi/Pavement%20Design%20and%20Data%20Collection%20Section.pdf. Accessed May 31, 2018.
20. Maryland State Highway Administration Automatic Road Analyzer (ARAN). Baltimore, MD. http://www.roadware.com/news-and-events/News-Release-Maryland_ARAN9000_FINAL_Apr-11-16.pdf. Accessed on May 31, 2018.
21. Mamlouk, M., and C. Zapata. Necessary Assessment of Use of State Pavement Management System Data in Mechanistic-Empirical Pavement Design Guide Calibration Process. *Transportation Research Record: Journal of the Transportation Research Board*, 2010. 2153: 58–66.
22. Federal Highway Administration, U.S. Department of Transportation. *Roadway Safety Information Analysis: A Manual for Local Rural Road Owners*. Washington, D.C. https://safety.fhwa.dot.gov/local_rural/training/fhwasaxx1210/s3.cfm. Accessed on May 31, 2018.

The Standing Committee on Pavement Surface Properties and Vehicle Interaction (AFD90) peer-reviewed this paper (18-00421).

Article

TRR
JOURNAL OF THE TRANSPORTATION RESEARCH BOARD

Transportation Research Record
2018, Vol. 2672(40) 430–439

DOI: 10.1177/0361198118776114
journals.sagepub.com/home/trr
SAGE

Numerical and Experimental Predictions of Texture-Related Influences on Rolling Resistance

Dmytro A. Mansura[1], Nicholas H. Thom[2], and Hartmut J. Beckedahl[1]

Abstract

To overcome rolling resistance (RR) a typical vehicle on average consumes 4152 MJ/119 L of fuel annually as a result of both vehicle and pavement factors. A slight improvement in surface texture arrangement may therefore decrease fuel consumption bringing substantial long-term socio-economic benefits. This aligns with ever-tighter limits on CO_2 in the USA (163 g/km until 2025) fostering sustainable construction/exploitation of tires/pavements. This paper describes a multi-scale 3-D numerical methodology to calculate micro-distortional RR and contact indentations of surface aggregates into visco-elastic tread compound accounting for loading, velocity, temperature, and compound properties. It consists of a micro-scale tread block single aggregate model and a macro-scale car tire finite element model, rolling in steady-state mode over a rigid smooth surface. The surface texture is idealized in terms of hemispherical indenters. The micro-distortional RR estimates are based on contact force and energy lost per single stone. The computed contact/normal forces peak significantly due to visco-elastic effects at the beginning of the tire–surface contact phase, followed by a gradually relaxing stress region with a sudden release at the end of the interaction. The contact forces appear to be of a reasonable distribution and magnitude. It is found that micro-distortional RR is higher on a rougher and sparsely packed surface compared with a smoother and more tightly packed case. To determine the total tire-related RR, macro-distortional RR can then be added. The predictions were qualitatively confirmed and adjusted against real bituminous mixes by experimental testing, showing a reasonable agreement.

The harmful influence on the quality of life that stems from our ever-expanding vehicle fleet is well acknowledged by environmental agencies worldwide. This particularly concerns heavily trafficked countries/regions such as the USA and the EU, which have recently made coordinated steps in cutting carbon emissions from cars, vans, and trucks by specifying carbon footprint thresholds (e.g., 95 g/km per car by 2021 in the EU (*1*); 163 g/mile for both cars and trucks by 2025 in the USA (*2*)). Even with more fuel-efficient vehicles, low rolling resistance (RR) tires (*3*) and, since 2014, low RR wearing course (*4*) being produced to enhance sustainability, little or no progress in reducing carbon footprints has been noted to date (CO_2 has risen by 26% in the past 25 years) mainly because of the continuing increase in vehicle numbers on the roads (*5*).

Given the fact of air pollution, finite fossil fuel reserves, and that the price of fuel is still increasing, tire and pavement research institutes continually aim at designing and manufacturing tires and highways to be as environmentally friendly and cost effective as modern technologies allow. While the traditional means of minimizing fuel use are to modify aerodynamics, transmission, and tire characteristics, pavements are now seen as a further way of lowering energy consumption. It has been long established that pavements affect fuel consumption just as tires do, through RR, and decreasing RR by 10% may save 1% to 4% of fuel (*6*).

Generally, RR comprises a set of simultaneous compressive and tangential energy loss mechanisms and varies in magnitude from 25 N to 80 N per car tire. A typical vehicle uses on average 4152 MJ/119 L of fuel per year to overcome RR as a result of both tire and pavement factors, of which weak pavement structure and rough texture have been found to lead to extra fuel consumption

[1]Pavement Research Centre, Faculty of Architecture and Civil Engineering, University of Wuppertal, North-Rhein Westfalia, Germany
[2]Nottingham Transportation Engineering Centre, Faculty of Engineering, University of Nottingham, Nottinghamshire, UK

Corresponding Author:
Address correspondence to Dmytro A. Mansura: dmytro.mansura@uni-wuppertal.de

(*7–10*). Neglecting the aerodynamic drag, the following causes are commonly identified:

- Tire macro-distortional losses (texture-independent);
- Tire micro-distortional losses (texture-dependent);
- Pavement macro-distortional losses.

These mechanisms result in energy dissipation either in the tire/tread compound or the pavement structure, offsetting forward force by asymmetrically spreading contact tread compression, where the leading edge of a contact patch is more markedly deformed than the trailing edge. The energy being dissipated in the pavement is tied with wave propagation along, across, and inside the road structure that brings a deflection asymmetry under the tire. Recent computations (*8*) have shown that stiffness of the uppermost pavement layer is much more significant than that of underlying layers, meaning that a stiffer upper pavement saves energy. However, replacement of a flexible pavement with a rigid alternative may decrease pavement RR by about 3 N per truck tire at 40 kN vertical load (*8*). Using a set of falling weight deflectometer (FWD) measurements, Ullidtz compared them with these numerical estimates (*11*). The results matched quite well at lower deflections, but deviated at larger deflections. Other techniques (e.g., Chupin (*9*)) have been applied to analyze the pavement-related energy losses, supporting the main findings.

The other two losses take place within the tire. Emergence of macro-distortional energy dissipation is connected with compressive, tangential (micro-slippage and shear-induced inertia), and wave-induced inertial deformations. According to Michelin (*3*), due to rubber viscosity tread compression and longitudinal shearing contribute the largest amount of energy loss. These texture-independent losses can also be increased by standing waves, centrifugal, and to a lesser degree Corriolis forces.

On the other hand, stone indentations into visco-elastic tread compound induce micro-distortional energy loss. These contact distortions produce local compression which incorporates a visco-elastic and to a smaller extent an inertial component that affect the moment balance in the tire–pavement contact area. Spacing between adjacent aggregates, their size and shape all influence the force level and distribution, having a consequent effect on RR.

Numerous experimental and computational techniques have been employed to study RR. With the help of empirical tests, a clear correlation has been observed between RR and pavement roughness at different scales (*7*). However, disadvantages are that empirical in-situ testing (e.g., coastdown) takes all energy dissipation mechanisms into account, whereas the laboratory method (drum test) either ignores micro-distortional and pavement-induced effects or includes both macro- and micro-distortional tire energy losses. Regarding other set-ups, a double-wheel test rig has been developed, which measures the contact pressures and indentations per single asperity, but which has not been used to predict dissipated energy (*12*). These limitations have been in part tackled by modeling an entire wheel or a tread pattern that interacts with the road surface.

As far as stone indentation influences are concerned, these have been considered with respect to structural vibration, noise emission, wear resistance, and friction (*13*, *14*). Meanwhile most analytical, numerical, and semi-empirical models have omitted surface roughness and associated interfacial penetrations, emphasising macro-distortional RR (*15*, *16*). Those models that include tread–texture interaction have been built to analyze contact stress distribution, but have ignored energy consumption calculations (*12*, *13*, *17*). A few analyses have been carried out to estimate energy losses due to indenting texture (*18–20*), but not all of these have clearly differentiated micro-distortional from tire macro-distortional losses. For instance, Hoever determined that texture-dependent visco-elastic contributions to RR may amount to 0.1% and 0.75% of total tire load level, for a smooth (stone mastic asphalt) and a rough (surface dressing) texture (*19*), respectively, which, by and large, aligns with experimental findings that show greater fuel use on surfaces with coarser stone packing (*7*, *21*).

Until now an in-depth texture-dependent RR analysis is still lacking, and there is therefore a need to investigate individual stone effects, namely size/shape/spacing, discerning the visco-elastic and inertial components of micro-distortional RR. So far none of the existing simulations has examined these individual terms, frequently representing indenters as a series of 2-D/3-D linear or non-linear spring-dampers, or through spectral images derived from scanning real-life surfaces.

A reduction of total RR can only be achieved if tire-related characteristics are modified along with the pavement properties. In this respect, wider tires with shorter sidewalls enable lower RR. At the same time, such tire structural configuration leads to aquaplaning. To decrease tire-related RR (non-texture related) without affecting other performance, a new tall/narrow tire size concept has been suggested (*22*). A methodology that is capable of leading to the design of optimized tire treads, tire material compounds, and road texture for decreased RR, noise, splash, and spray while maintaining adequate drainage, skid resistance (dry, wet, ice, and snow), and ride quality for different operating conditions is therefore badly needed.

The aim of this paper is to suggest an efficient computational multi-scale model for quantification of the micro-distortional RR caused by macro-texture asperities of wavelength 0.5 mm to 50 mm. The model consists of macro-scale (whole tire) and micro-scale (tread–stone contact) simulations and has been developed utilizing ABAQUS software. Numerical RR predictions have then been validated/adjusted by laboratory testing. Furthermore, application of the testing has enabled an evaluation of the energy consumption with variable textures, explanations of which are given below.

Numerical Analysis

Macro-Scale Model

A 175 SR14 slick radial Yokohama tire was adopted from ABAQUS open source to conduct the macro-scale 3-D analysis (*23*). The main purpose of this model was to obtain indentation and release rates to be used in the micro-scale simulation. To model tire free rolling on a smooth and rigid surface, the analysis was broken down into five sequential simulations, namely: inflation, half and full tire stationary footprint, braking/traction, and steady-state rolling. Each successive simulation is dependent on the results transferred from the previous one, and this is performed by means of a built-in transfer capability.

Tire macro-distortion was assessed employing ABAQUS/Implicit (Standard) Steady-State Transport Analysis in a mixed arbitrary Lagrangian eulerian (ALE) formulation, where the deformation is characterized by a Lagrangian method while describing the rigid body rotation in an Eulerian framework. In this analysis the tire was subject to 3300 N load and inflated to a pressure of 200 kPa. To account for large deformation, material, and boundary condition nonlinearities, it was necessary to set the NGEOM function.

Viscous effects, changing geometry, and local instabilities led to a non-convergence problem. To eliminate computational instabilities, a number of techniques was used, namely application of finer mesh, various increment sizes, and global and local stabilization options. After experimentation the time step specified in computation was 1.0 sec; the initial and analysis time increments were taken as 0.001 and 0.01 secs, respectively. The slip tolerance value was set to 0.02 as more relaxed (larger) values help avoid convergence difficulties. The total numbers of elements and nodes in the model were 7,255 and 13,548, respectively. To achieve an efficient runtime and converged solution, a finer mesh was incorporated in the contact region covering 40° of arc and a coarser mesh covering the remaining 320°, as shown in Figure 1, both being discretized with general linear hybrid elements (C3D8H and C3D6H) appropriate for incompressible material. To enforce the constraints strictly, a default hard penalty stiffness contact was adopted with friction defined in terms of a Coloumb law.

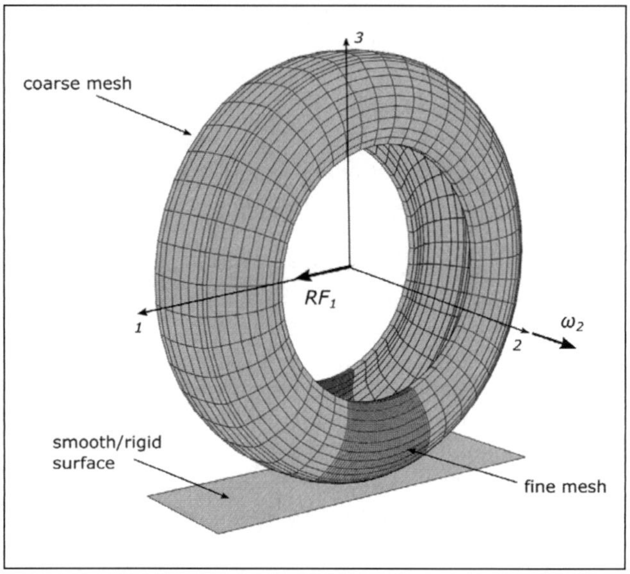

Figure 1. A representation of macro-scale tire model mesh.

The means of arriving at free rolling conditions lay in defining through trial and error the wheel angular velocity (ω_2) at the point where the moment balance about the rim center becomes nearly zero (where achievable); macro-distortional RR is then outputted as the reaction force (RF_1) in the horizontal direction. In this case, free rolling conditions were found at 72.32 rad/sec angular velocity for a vehicle speed of 80 km/h. Computations of the average RR for the ABAQUS tire matched the results reported in the literature for this tire (*24*) in spite of imperfect convergence of RR moment about the hub point.

Tread Compound Properties. Visco-elastic properties of tread compound were represented in ABAQUS in terms of an assembly of rheological Maxwell elements (springs and dashpots). Description of elastic response was achieved by a reduced polynomial (Neo-Hookean) hyperelastic model applying the HYPERELASTIC capability, where just one parameter was used, being calculated as half of the shear modulus. The contribution of the viscous behavior was idealized by including the VISCOELASTIC capability, which comprises an N-term Prony series expansion of the dimensionless relaxation modulus in the time domain. Consideration of a range of temperatures was based on application of the Williams-Landel-Ferry (WLF) empirical time-temperature superposition principle with a reference temperature of 55°.

Micro-Scale Model

A dynamic 3-D tread–stone contact (Figure 2*b*) was analyzed using the ABAQUS/Explicit solver to derive micro-distortional RR estimates based on contact forces between a cubic visco-elastic tread block and a rigid hemispherical stone. Initially, determination of contact forces was attempted by applying ABAQUS/Implicit. However, due to difficulties in achieving a converged solution and an inefficient computation, preference was, in the end, given to the use of ABAQUS/Explicit. The solver adopts central difference integration, solving for node acceleration first and then for velocity and displacement from the previous increment (*23*). In this scheme, to keep a stable solution throughout the analysis, a large number of small time increments are automatically adjusted, selection of which is prescribed by the shortest crossing time of the pressure wave through the smallest element dimension in the mesh. The simulation takes account of three main interaction phases, namely indentation of the asperity into the tread rubber (i.e. loading), holding and release (i.e. unloading or snap-out). In order to construct a simplified and physically reasonable dynamic interaction model, the following set of approximations have been made:

- The tread block is assumed to consist of homogeneous visco-elastic rubber with no steel/textile interlayer reinforcement;
- Texture indents the tread block in the vertical direction; no tangential movement is imposed;
- Texture aggregates have been idealized with hemispherical form and rigid properties (210 GPa modulus; $\nu = 0.3$) with no stone-related roughness (micro-texture);
- Adhesion forces are neglected;
- Rubber temperature is maintained constant;
- Tire belt vibration is excluded;
- Macro-distortion pre-stress (lateral stretching and twisting) of a tread block prior stone-induced deformation is eliminated;
- Duration of interaction phases corresponds to a contact length appearing under the action of a 3300 N tire load rolling at 20, 40, 60, 80, and 100 km/h.

A variety of bituminous mixes are available, all having different macro-texture packing and texture depth. Aggregate gradation, stone shape (angular, cubic, etc.) and maximum stone size are recognized as underlying factors affecting the texture of the surface course. For texture properties, smooth (e.g. 0/8 mm stone mastic asphalt) and rough (e.g. surface dressing) types of pavement can be distinguished. Since inclusion of a realistic surface replica in the analysis would not allow the separation of interlinked influences of stone size and shape, and because gathering such data is expensive, the indenter was assumed to be of a hemispherical form. As shown by Greenwood and others (*12*, *25*), representing surface chippings in terms of hemispheres is usually an acceptable simplification. It is also worth mentioning that the analysis of micro-distortional losses was based on a single stone (Figure 2*b*), which was found to be 5.3% higher compared with a scenario with nine neighboring asperities as presented in Figure 2*a*. However, this represents a negligible impact on the fundamental character of RR, taking the simplified nature of the model into account.

Meshing, Boundary/Contact Conditions and Analysis Steps. In this configuration, a stationary 45 × 45 mm tread block (i.e. a piece of a tire surface) was used, while the stone penetrated vertically into the elastomeric block. This arrangement of boundary conditions (BCs) enabled consistent normal stresses to be captured in comparison with a fixed stone-movable block combination. The surface and sides were left free to move (Figure 2*c*).

To replicate real-life interactions, three analysis steps were simulated: loading, hold ($V_{hold} = 0$), and unloading. The BCs of the stone were described in terms of velocity, which was extracted from the macro-scale model for a series of nodes before and after the vertical velocity becomes zero in the contact area. By assuming that distances between the smooth surface of the macro-scale model pavement and tire surface nodes correspond to indentations caused by texture packing, loading/unloading rates have been generated for a range of tire translational speeds. Durations of time phases have been calculated by summing up the time intervals needed for passage through all internode distances in order to reach a given indentation.

As illustrated in Figure 2*b*, the block consisted of three discretization layers: bottom (in contact with the stone), middle, and top, with a finer mesh in the bottom layer, applying 40,500 8-node linear brick C3D8R continuum elements with reduced integration and hourglass control. After optimization trials, the selected mesh option gave a good trade-off between accuracy and analysis speed.

In addition, the introduction of mass scaling was found to decrease computational time. However, overly high mass scaling induced significant oscillations compared with the unscaled case. Therefore, by varying mass scaling levels, an optimum value was found to obtain efficient computational speed without sacrificing the accuracy/continuity of contact forces. In addition, to avoid computational problems, a linear soft pressure-overclosure relationship (i.e. contact stiffness) was established at the tread-indenter interface. Numerical overlap between contacting bodies can thus be inhibited. An optimized high, but computationally stable, level of contact stiffness was then chosen for all analysis steps. A constant friction of 0.7 was adopted, since this was not observed to substantially change contact forces.

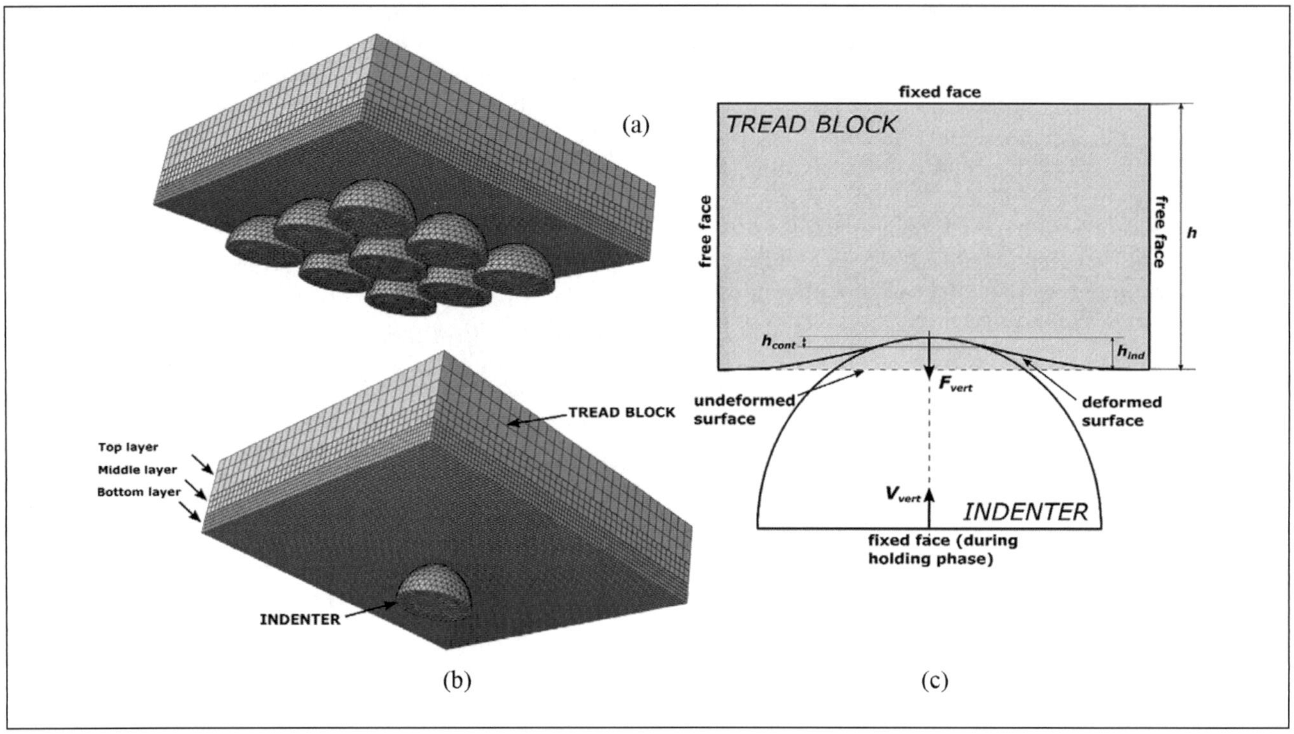

Figure 2. A representation of (*a*) multi-indentation tread–stone model, (*b*) the adopted mesh and tread–stone configuration, and (*c*) a schematic sketch of the FE analysis with indicated boundary conditions, where V_{vert}, h, F_{vert}, h_{ind}, h_{cont} are vertical velocity, tread height, contact force, indentation depth, and depth of indentation contact zone.

Methodology to Quantify Micro-Distortional RR. Predictions of micro-distortional RR were estimated from the energy losses for single stones, derived from contact forces and indentations for a given loading; these losses were scaled to give values for the whole wheel. The results from this approach seem to more closely approximate the effects of protruding stones on RR compared with alternative moment-based procedure, explained in (*26*). Examination of varied texture packings could be accomplished by changing the spacing between indenters along and across the contact allowing both rough and smooth surfaces to be studied.

Description of Packed Indenters Loading Test (PILT)

Adjustment and qualitative confirmation of numerical micro-distortional dissipated energy were achieved with the aid of a packed indenters loading test (PILT) utilizing an Instron 8801 servo-hydraulic machine under load control. Both cored samples and idealized packed spherical surfaces were investigated. In the PILT test, a specimen was placed on a base platform and on top of it a tread pad inserted with the tread pattern facing downwards (Figure 3).

The testing objective was to load the tread pad–texture combination to measure stone penetration into the rubber compound and the force induced. The penetration and force were measured by an internal transducer mounted within a load actuator and a load cell, respectively. The test was conducted at ambient room temperature of 20° C, exposing the specimen to the action of 50 sinusoidal loads at a frequency of 20 Hz. Application of high-frequency cycling was constrained by the limits of the acquisition system and the hydraulic machine. Calculations of energy lost per pad during the 25th cycle were then carried out for a range of tread pads, surfaces, and loading levels. Further, to identify micro-distortional RR per tire, computed energy loss was multiplied by the number of pads in one meter.

Qualitative Confirmation and Adjustment of Computational Predictions

To qualitatively confirm the FE model, numerical computations were compared against the PILT results for a spherical texture of 15.9 mm diameter, both with a tightly packed arrangement and at a spacing of 22 mm. The results from the micro-scale analysis showed good qualitative agreement with the experiments in terms of

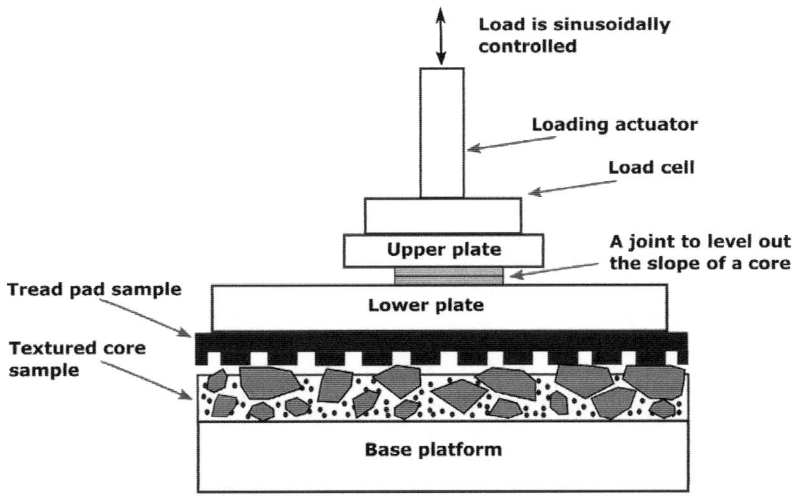

Figure 3. Schematic sketch of PILT set-up.

micro-distortional RR and stone indentation as a function of stone spacing. It was found computationally that an increase in stone spacing from 15.9 mm to 22 mm increased both micro-distortional RR and stone indentation by 1.47 and 1.46 times, respectively. These rates matched the PILT data for the same textures quite well. However, in order to carry out a quantitative comparison between the model and PILT data, the impact of rubber bending between spheres, differences in rubber properties, in loading rates and in contact areas all need to be taken into consideration.

An adjustment was also made to convert numerical energy losses for spherical textures to those of conventionally paved surfaces. For this an adjustment factor was employed, which is the ratio between the energy lost for a realistic surface and that for an idealized surface. Stone mastic asphalt (SMA) and hot rolled asphalt (HRA) was tested (PILT) in this work. These ratios amounted to 0.56 and 0.64, respectively, being averaged to account for four loading levels, three rubber types and two artificial manufactured surfaces. These adjustments were only approximate since the PILT loading rate was significantly lower than those applied in the micro-scale modeling to conform to realistic velocities.

Results and Discussion

The effect of stone packing on the contact forces and RR for a range of hemispherically packed textures was assessed using the numerical approach described above. As presented in Figure 4*a*, indentation of a single 5 mm radius hemisphere by 1 mm into a tread block clearly shows that the tread–stone model distributes compressive forces in the expected visco-elastic fashion.

Figure 4*a* shows that the presence of damping forces generates asymmetry in the normal forces, whereby the peak force is reached at the beginning of the contact area (i.e. the asperity is fully impressed), followed by a gradual force reduction until the beginning of the release phase where the stone disconnects from the rubber surface. A comparison between these computational predictions and results obtained by Boere and Liu (*12*, *18*) showed a qualitative agreement between them, confirming the general validity of the technique.

It was found that, at the lower loading rate relating to 20 km/h, the rubber has a longer time to relax leading to about 13.33% less force compared with faster loading rates, which is a function of rubber-related stiffening at higher frequencies. In addition, as can be seen in Figure 4*a*, the normal forces are slightly noisier, being accompanied by distinct ripples at the end of the loading phase and throughout the holding and unloading phases for higher velocities. The ripples are due to two factors: the stiffer rubber and rigid fixing of the base of the block both restrict block movement giving rise to reflective waves within the block.

It should be noted that a larger stone size resulted in a greater contact force, which is justifiable since more resistance would accumulate from the material. Further, removal of rubber density in the model enabled inertial forces to be separated from visco-elastic and frictional forces. In contrast to expectations, observations indicated that inertia makes a negligible contribution to the contact forces and it is thus of secondary importance

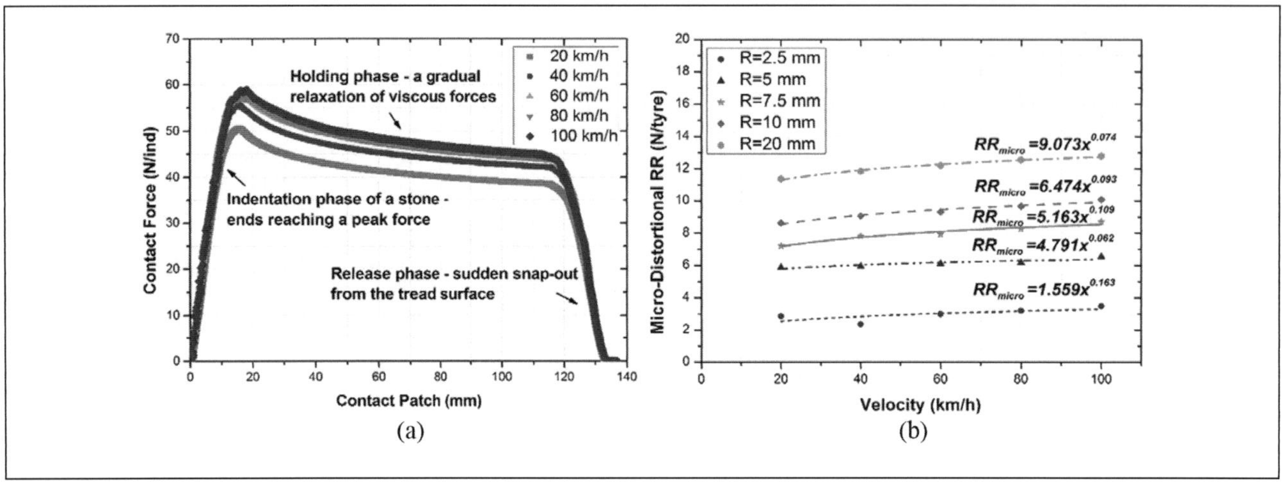

Figure 4. (*a*) Spread of forces in the tire–pavement contact along the contact patch for 5 mm radius indenter at five translational velocities and (*b*) relationship between micro-distortional RR and velocity at a wheel load of 3300 N for a range of densely arranged hemispheres.

with respect to energy loss. In fact, a reduced density scenario led to a very slight increase in the force.

Calculations of micro-distortional RR for a tire under 3300 N load are illustrated in Figure 4*b*. Regardless of the texture size/packing, the micro-distortional energy losses rise according to a power function as the wheel velocity increases. This is again attributed to the stiffening of the compound; larger indentations appear at a lower velocity producing a lower contact force, while generation of a smaller dent (rubber becomes stiffer) is expected for larger loading rates causing greater forces. For instance, a tight packing of 7.5 mm indenters induced a texture-dependent RR of 7.2 N/tire at 20 km/h and 8.7 N/tire at 100 km/h with corresponding contact distortions reducing from 943μm to 882 μm, as can be observed in Figure 5*a*. Overall, for a speed increase from 20 km/h to 100 km/h the variation in RR remained between 11% and 20% for all hemispheres.

Regarding texture packing, a coarser texture caused a larger RR compared with a finer texture, as shown in Figure 4*b*. Increase of hemisphere radius from 2.5 mm to 20 mm magnified micro-distortional RR by a factor of 3.6, from 3.5 N/tire to 12.8 N/tire for a velocity of 100 km/h. From a qualitative point of view, the tendency for RR to increase with texture depth (assuming the hemisphere radius correlates with the texture depth) is in agreement with the results derived from the coastdown method carried out by Hammerström et al. (*7*).

Widening the distance between indenters but keeping the same radius of 7.5 mm also showed that a less dense packing results in greater energy consumption. When varying longitudinal spacing alone, the RR rose from 8.2 N/tire for a tight 15 × 15 mm stone pattern to 10.9 N/tire for a looser 20 × 15 mm arrangement. It can be noted that the increase in energy losses from the 15 × 15 mm to 20 × 15 mm pattern appeared to be of approximately same level as for the velocity effect examined under a similar pressure condition. Furthermore, a rise in the RR of 36% was found when the wheel load increased from 1300 N to 4300 N, from 5.5 N/tire to 8.6 N/tire, when rolling over a tight 15 × 15 mm texture pattern. With 18 × 15 mm and 20 × 15 mm arrangements the difference in energy lost amounts to 50% for the same load increase. The less steep increase in the micro-distortional RR for the 15 × 15 mm pattern appears to be a feature of denser texture patterns.

In general, this is related to a larger compression into the compound by the coarser surfaces and is demonstrated in Figure 5*a*, where a hemisphere radius increase from 5 mm to 10 mm increased contact indentation from 635 μm to 1084 μm. A similar tendency is found for more sparsely packed surfaces. Up to 10 mm radius hemispheres, the indentation levels seem approximately to match the realistic range for asphalt surfaces found experimentally by PILT testing and they agree with the trends and order of magnitude of estimates reported in the literature (*27*).

To differentiate the effects of rubber material, stiff and soft tires were analyzed. The results obtained showed that the stiffer tire, as expected, resulted in less energy loss than the softer tire, on average, imposing 6.1 N/tire and 6.7 N/tire of RR, respectively, over the 20 to 100 km/h speed range. It should be borne in mind that rubber viscosity and stiffness can play a significant role, potentially affecting contact displacement, forces, and fuel losses.

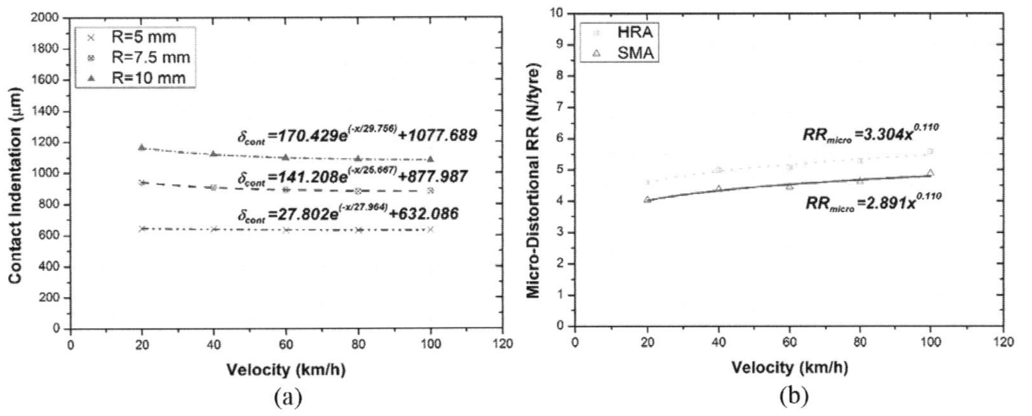

Figure 5. (*a*) Dependence between contact indentation and velocity for three tightly packed texture combinations at 3300 N tire loading and (*b*) relationship between micro-distortional RR and velocity at a wheel load of 3300 N for two bituminous surfaces.

However, inconsistent results were found for the effect of temperature (°C) variation at 3,300 N load and with tight hemispherical packing of 5 mm radius. At 25° C tire temperature, representing the stiffest case, the micro-distortional energy dissipation decreased as the speed increased, while at temperatures from 45° to 85° there was an increase or a small reduction. The largest effect occurred at 35°, inducing 5.9 N/tire RR, although this differed only marginally from the 45° and 55° scenarios. A temperature of 35° had the greatest effect until 40 km/h, after which the highest losses occurred at 45°. The lowest RR was found at 85° until 60 km/h. The origin of the disordered results is assumed to stem from molecular chain realignment in the rubber under the simultaneous action of frequency and temperature effects.

By adjusting the hemispherical surface of 7.5 mm radius it was predicted that the rougher/looser texture of HRA would induce greater RR as compared with the smoother/tighter texture of SMA. Specifically, the mean values of RR for HRA and SMA for a range of velocities were 5.2 N/tire and 4.5 N/tire, respectively, which is in good agreement with the order of magnitude for various asphaltic surfaces computed by Hoever (*19*). In comparison with the total RR, including macro-distortional RR of 42.5 N/tire calculated by the macro-scale FE model, these represent 10.9% and 9.7%, respectively. In general, the findings from the PILT experiment qualitatively confirmed that HRA texture gives rise to more energy loss than SMA texture, as well as being consistently/expectedly higher for the softer tread pad.

Assuming 50,000 car passes in a day on a major highway, the energy lost due to HRA and SMA surfaces would equate to 1.04 MJ/m and 0.91 MJ/m, respectively. This means that, taking 40 years as the whole life of the road, these vehicles would burn an extra 434 L/m and 380 L/m of fuel, which is roughly equivalent to emitting 998 kg and 874 kg of CO_2, respectively. The implication is that purely by switching from the rougher to the smoother surface it would save approximately 54 L/m of fuel or reduce by 124 kg the tailpipe CO_2 emissions for cars with 3.3 kN wheel load. Although the predictions are not directly comparable, this is four times larger than the savings deduced to originate from lowering pavement-induced RR as presented by Thom et al. (*28*) for a traffic flow of 3000 commercial vehicles at 40 kN wheel load, marking it as a substantial source of pavement-related energy loss.

Conclusions

A 3-D multi-scale numerical model has been developed to calculate tire–pavement contact forces, stone indentations, and micro-distortional RR. The estimates of texture-related energy loss were experimentally confirmed and adjusted by use of laboratory testing. The influence on micro-distortional RR of six parameters was studied. Velocity was found to increase lost energy by a factor of 1.1 to 1.2 from 20 km/h to 100 km/h, which was observed to equate approximately to a change from a softer to a stiffer tread compound (i.e. this varies from rubber to rubber). A more significant effect on energy loss was identified due to increasing load level. However, the major effect was related to texture size, increasing the energy dissipation by a factor of 3.6 by replacing a finer with a coarser texture. Varying the spacing between stones led to about 1.3 times difference between dense and loose packings. Results for a range of rubber temperatures gave inconsistent trends, but only slight influence.

A comparison between two bituminous surfaces, HRA and SMA, showed that the smoother texture of SMA may save 54 L/m of fuel or lower tailpipe emissions by around 124 kg of CO_2 over 40 years on a heavily trafficked highway. The pavement surface texture was noted to be much more influential on energy loss compared with the pavement structural properties reported in the literature.

Broadly, the model and the experiment can already be implemented into a life-cycle assessment tool for evaluation of different road surfaces. However, in this study, only car tires have been examined; it is necessary to extend the research further to account for truck tires (i.e. the level of loading, rubber properties), for which losses are expected to be higher. An optimization of pavement texture, taking account of drainage, skid resistance, noise, and vehicle maneuverability, is the next logical step, representing a potential means to advance infrastructure sustainability by lowering environmental cost and financial expense.

Author Contributions

The authors confirm contribution to the paper as follows: study conception and design: D. Mansura, N. Thom, H. Beckedahl; data collection: D. Mansura; analysis and interpretation of results: D. Mansura, N. Thom; draft manuscript preparation: D. Mansura. All authors reviewed the results and approved the final version of the manuscript.

References

1. *Reducing CO_2 Emissions from Passenger Cars*. European Commission, EU Action, 2017. https://ec.europa.eu/clima/policies/transport/vehicles/cars_en. Accessed March 6, 2017.
2. *Regulations for Greenhouse Gas Emissions from Passenger Cars and Trucks*. US Environmental Protection Agency. https://www.epa.gov/regulations-emissions-vehicles-and-engines/regulations-greenhouse-gas-emissions-passenger-cars-and. Accessed March 6, 2017.
3. Société de Technologie Michelin. *The Tire: Rolling Resistance and Fuel Savings*. Michelin, France, 2003.
4. Pettinari, M., B. Schmidt, B. Bo Jensen, and O. Hededal. *New Surface Layers with Low Rolling Resistance Tested in Denmark. Asphalt Pavements*, Taylor & Francis, London, 2014, pp. 323–332.
5. Bandivaker, A. *Evaluating the Impact of Advanced Vehicle and Fuel Technologies in US Light Duty Vehicle Fleet*. PhD thesis, MIT, 2008.
6. Tires and Passenger Vehicle Fuel Economy. *Transportation Research Board Special Report 286*, National Research Council of the National Academies, Washington, D.C., 2006.
7. Hammerström, U., R. Karlsson, and H. Sürensen. *Road Surface Effects on Rolling Resistance-Coastdown Measurements with Uncertainty Analysis in Focus*. Research Report, Deliverable 5a. VTI, Sweden, 2008.
8. Lu, T. *The Influence of Pavement Stiffness on Vehicle Fuel Consumption*. PhD thesis, University of Nottingham, UK, 2010.
9. Chupin, O., J.-M. Piau, and A. Chabot. Evaluation of the Structure-Induced Rolling Resistance (SRR) for Pavements Including Viscoelastic Material Layers. *Journal of Materials and Structures*, Vol. 46, No. 4, 2013, pp. 683–696.
10. Holmberg, K., P. Andersson, and A. Erdermir. Global Energy Consumption due to Friction in Passenger Cars. *Tribology International*, Vol. 47, 2012, pp. 221–234.
11. Ullidtz, P. *Pavement Deflection and Energy Loss*. Green Road Infrastructure Workshop, Denmark, 2014.
12. Liu, S., M. Sutcliffe, and W. Graham. Prediction of Tread Block Forces for a Free-Rolling Tire in Contact with a Rough Road. *Wear*, Vol. 282–283, 2012, pp. 1–11.
13. Pinnington, R. Tire-Road Contact Using a Particle-Envelope Surface Model. *Journal of Sound and Vibration*, Vol. 332, 2013, pp. 7055–7075.
14. Sridharan, K., and R. Sivaramakrishnan. Dynamic Behaviour of Tire Tread Block. *American Journal of Engineering and Applied Science*, Vol. 5, 2012, pp. 119–127.
15. Hall, D., and J. Moreland. Fundamentals of Rolling Resistance. *Journal of Rubber Chemistry and Technology*. Vol. 74, 2001, pp. 525–539.
16. van der Steen, R. *Enhanced Friction Modelling for Steady-State Rolling Tires*. PhD thesis, Eindhoven University of Technology, Netherlands, 2010.
17. Kozhevnikov, I., D. Duhamel, H. Yin, J. Cesborn, and D. Anfosso-Ledee. A New Algorithm For Computing the Indentation of a Rigid Body of Arbitrary Shape on a Visco-Elastic Half-Space. *International Journal of Mechanical Sciences*, Vol. 50, 2008, pp. 1194–1202.
18. Boere, S., I. Arteaga, A. Kuijpers, and H. Nijmeijer. Tire/Road Interaction Model for the Prediction of Road Texture Influence on Rolling Resistance. *International Journal of Vehicle Design*. Vol. 65, 2014, pp. 202–221.
19. Hoever, C. *The Simulation of Car and Truck Tire Vibrations, Rolling Resistance and Rolling Noise*. PhD thesis, Chalmers University of Technology, Sweden, 2014.
20. Arteaga, I., and H. Nijmijer. Visco-Elastic Contact Model for the Prediction of Tire/Road Contact Forces and Rolling Resistance. *Proc., 18th International Congress on Sound and Vibration, ICSV*, Rio de Janeiro, 2011.
21. DeRaad, L. *The Influence of Road Surface Texture on Tire Rolling Resistance*. SAE Technical Paper 780257. Society of Automotive Engineers, Troy, MI, 1978.
22. Vennebörger, M., C. Strübel, B. Wies, and K. Wiese Low Fuel Consumption Tires for Passenger Cars with CO_2 Emission. *ATZ Worldwide*, Vol. 115, No. 7–8, 2013, pp. 16–20.
23. *ABAQUS On-line Documentation*. Dassault Systèmes, Providence, RI, 2014.
24. Behnke, R., and M. Kaliske. Computation of Energy Dissipation in Visco-Elastic Materials at Finite Deformation. *Proceedings of Applied Mathematics and Mechanics*, Vol. 13, 2013, pp. 159–160.
25. Greenwood, J., and J. Wu. Surface Roughness and Contact: An Apology. *Meccanica*, Vol. 36, 2001, pp. 617–630.

26. Mansura, D., N. Thom, and H. Beckedahl. A Novel Multiscale Numerical Model for Prediction of Texture-Related Impacts on Fuel Consumption. *Tire Science and Technology*, Vol. 45, No. 1, 2017, pp. 55–70.
27. Integrated Tire and Road Interaction – Development of a Tire-Road Friction Model. *Specific Targeted Research or Innovative Project*. FP6-PL-0506437, Deliverable 5.1., Institute for Road and Traffic Engineering, Technical University of Aachen, 2006.
28. Thom, N., T. Lu, and T. Parry. "Fuel Consumption Due to Pavement Deflection under Load." Paper presented at the 2nd International Conference on Sustainable Construction Materials and Technologies, Ancona, Italy, June 2010.

The Standing Committee on Pavement Surface Properties and Vehicle Interaction (AFD90) peer-reviewed this paper (18-00907).

Article

TRR
JOURNAL OF THE TRANSPORTATION RESEARCH BOARD

Transportation Research Record
2018, Vol. 2672(40) 440–451

DOI: 10.1177/0361198118774165
journals.sagepub.com/home/trr
SAGE

Pavement Friction Modeling using Texture Measurements and Pendulum Skid Tester

Ahmad Alhasan[1], Omar Smadi[1], Georges Bou-Saab[2], Nacu Hernandez[3], and Eric Cochran[3]

Abstract

Pavement frictional behavior affects pavement performance in terms of vehicle safety, fuel consumption, and tire wear. Comprehending and interpreting pavement friction measurements is a challenging task, because of friction sensitivity to several uncontrollable factors. These factors include: pavement surface conditions, such as the type and thickness of contaminants and fluids on the surface and their interaction with friction forces; and the device operating conditions, such as sliding speed, material properties and geometry of the rubber slider used, and operating temperature. Despite the efforts to describe and quantify the impact of varying conditions on pavement friction, which ultimately will allow for a better harmonization of friction measurements, there is a need to better understand the link between the surface texture and physical friction measurements. In this paper, Persson's friction model is used to analyze and understand the impact of surface texture on frictional behavior of dry pavement surfaces. The model was used to analyze 18 test locations, which were compared with the dry kinetic coefficients of friction (COF) estimated using a British pendulum tester (BPT). The results show that Persson's friction model could predict the COF estimated from the BPT results with relatively high accuracy. In addition, the model could provide a profound explanation of the frictional forces mechanism. Finally, it was found that the mean profile depth (MPD) cannot provide a full picture of the frictional behavior. However, combining MPD with the Hurst exponent, texture measurements can potentially provide a full physical explanation of the frictional behavior for road surfaces.

Pavement frictional behavior (i.e., skid resistance) affects different aspects of pavement performance, including vehicle safety (*1–6*). Several studies have reported the significant impact of skid resistance on dry and wet fatal crash rates (*7–11*). In 2008, more than 19,000 fatalities occurred in roadway departure crashes in the United States. Poor roadway conditions caused by inclement weather—especially wet pavement—have been identified as a major contributing factor in this type of crash (*1*, *12*). De León Izeppi et al. have utilized continuous pavement friction measures in developing standard safety performance functions (*11*). In their study, it was concluded that savings from crash cost avoidance through friction enhancement can be considerable. Moreover, Long et al. have proposed thresholds for skid resistance to achieve a target crash reduction (*12*). In their study, the benefit/cost analysis showed that thresholds on skid resistance will result in a significantly higher benefit cost ratio.

Skid resistance develops as a result of the hysteresis and adhesion mechanisms at the tire–pavement interface (*13*, *14*). The hysteresis component contributes to the total frictional force through the viscoelastic deformation of rubber caused by the road surface roughness. This is a bulk property of the rubber and is referred to as internal friction (*15*). In addition, the adhesion component develops as a result of the shear frictional stresses at the contact area between the tire and the road resurface. This component develops as a result of chemical and dispersive interactions at the molecular level, such as the energy dissipation resulting from the bonding, stretching, and debonding cycles described in previous studies (*14*). In both components, the resulting friction forces, and the

[1]Civil, Construction and Environmental Engineering, Center for Transportation Research and Education, Iowa State University, Ames, IA
[2]Civil, Construction and Environmental Engineering, Iowa State University, Ames, IA
[3]Department of Chemical & Biological Engineering, Iowa State University, Ames, IA

Corresponding Author:
Address correspondence to Ahmad Alhasan: aalhasan@iastate.edu

kinetic coefficient of friction (COF) accordingly, depend on the contact mechanism between the tire and road surface. Persson (*16*) has proposed a new contact model that was verified in several studies (*15*). In his model, the contact area and the kinetic COF depend on multiple factors including the sliding velocity and rubber temperature during the slide (*14*, *17*).

Achieving consistent pavement friction measurements is a challenging task, because of the sensitivity of pavement friction to several varying factors. These factors include: pavement surface conditions, such as the type and thickness of contaminants and fluids on the surface and their interaction with friction forces; and the device operating conditions, such as the sliding speed, the material properties and geometry of the rubber, and the operating temperature (*18*). Kogbara et al. have reviewed various devices and testing procedures to measure pavement skid resistance (*18*). In their study, skid resistance devices can be classified based on three operating principles: longitudinal friction coefficient (LFC), sideway force coefficient, (SFC), and sliders or stationary measuring systems. Many devices follow the first principle (i.e., LFC), which entails the creation of a controlled sliding process of the tire through the application of a braking force. Depending on the braking mode and slip ratio, LFC devices can be subdivided into three categories: locked wheel, fixed slip, and variable slip devices (*1*).

Devices based on SFC measure the skid resistance using a wheel set at an angle, which allows the decomposition of friction measurements into parallel and transverse components. Despite the advantages SFC devices bring, the relative speed between the rubber and the road surface is lower than LFC devices, and thus the system produces low-speed measurements.

Under the third operating principle, sliders or stationary measuring systems require traffic control to conduct the tests. This limitation constrained the use of such devices to specialized investigations at locations of interest. Among these devices are the British pendulum tester (BPT) and the dynamic friction tester. Despite their simplicity and the traffic control limitation, these devices provide the ability to investigate the exact location using multiple systems. The tests can include high-resolution texture scanners that can reach microtexture. Because of the mentioned advantages and the fact that BPT is a very well-established test, it has been deployed in many research studies (*19–21*).

Pavement surface texture is a controlling factor affecting the skid resistance. Surface texture is divided on a scale that ranges from less than 0.5 mm (microtexture), to 0.5–50 mm (macrotexture), to 50–500 mm (megatexture), and to 500 mm to 50 m (unevenness). Most studies relate low-speed skid resistance measurements to the pavement microtexture, which is believed to affect the adhesion component (*22*, *23*), while macrotexture is typically related to friction at high speeds (*13*), which is affected by the hysteresis component of skid resistance (*15*, *24*). In a previous study, Persson indicated that the adhesion component is negligible for surfaces contaminated with dust or fluids (*14*). Furthermore, Lorenz et al. showed that the adhesion component exhibits significance at very low speeds (*15*). In another study, Liu et al. found that test surface macrotexture significantly affected BPT results (*25*).

Several researchers attempted to estimate pavement skid resistance from surface texture measurements (*26–28*). Ergun et al. developed a new model to predict the kinetic COF from surface texture (*26*). The model correlates COF as a function of slip speed to multiple macro- and microtexture coefficients. Lorenz et al. used pavement surface texture scans to predict the kinetic COF at low speed for three asphalt pavement surfaces (*15*). In their study, Persson's friction model (*14*) was used to predict the hysteresis component of the COF. Similarly, Ueckermann et al. presented a concept of contactless skid resistance measurements (*29*). The concept is based on estimating the kinetic COF from Persson's model using texture measurements acquired from an optical measuring system. The study reported very strong correlations (i.e., $R^2 = 0.97$) between the estimated and measured friction coefficients at 60 km/h (~40 mph).

Different devices and operating conditions, such as temperature and speed, will lead to a vastly different estimation of skid resistance for the same pavement surface (*18*, *30*). In an effort to harmonize the skid resistance measurements, the Permanent International Association of Road Congress (PIARC) developed the international friction index (IFI) to compare different friction measurement methods (*31*). IFI incorporates the impact of testing speed through the speed gradient coefficient deduced from macrotexture measurements; this coefficient shifts the friction measurement to an operating speed of 60 km/h (~40 mph). Moreover, PIARC included calibration constants in the final golden standard equation to eliminate the device's effect on the measurements. Flintsch et al. have shown that the model developed by PIARC did not produce harmonious results among the devices included in the study and used by the consortium members of the Virginia Smart Road Rodeo for the multiple tested surface s(*30*).

Despite the large number of studies relating pavement surface texture to skid resistance, there are limited studies showing strong correlations (*18*, *27*, *28*), especially for field measurements. There is a need for a more consistent approach capable of relating texture measurements to skid resistance at various speeds and operating conditions. This paper presents the results of utilizing Persson's friction model to estimate the friction coefficients of 18 texture scans acquired using a high-resolution laser texture scanner. The results were coupled with BPT results in an effort to define a new harmonization approach, especially as the BPT is a widely accepted and well-established simple test for lab studies or localized testing.

Methods

Data Acquisition and Testing

Pendulum Skid Resistance Tester. Field tests were performed using a Cooper pendulum skid resistance tester (Figure 1). Eighteen tests were performed on ten pavement sections, two hot mix asphalt (HMA) surfaces, two Portland cement concrete (PCC) surfaces, and six high-friction surface treatment (HFST) surfaces. Three test points were randomly chosen for the HMA and PCC sections, and one point on each HFST section. At each point wet and dry tests were performed using the BPT. Only dry tests are reported in this study, owing to the modeling complexity in wet conditions. The wet tests were performed in accordance to ASTM E303. Dry tests were performed in accordance to the procedure described in the user manual. According to the ASTM standards and the user manual, the tests were conducted after sweeping the test surface of any dust and repeated five times; the last four repetitions were then used to estimate the average British pendulum number (BPN). The device (Figure 1) consists of a pendulum rotating about a spindle attached to a vertical pillar. At the end of the tubular arm, a head of known mass is fitted with a rubber slider 6 by 25 by 76 mm (0.24 by 0.98 by 2.99 in). The pendulum is released from a horizontal position so that it strikes the sample surface at an approximate constant velocity 10 km/h (~2.78 m/s) and approximately 0.21

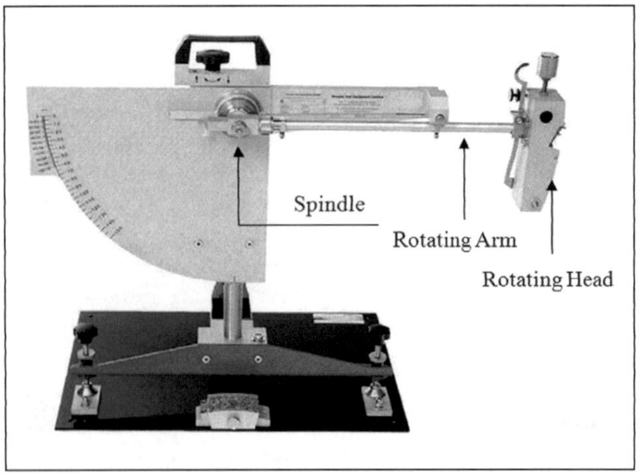

Figure 1. The British pendulum tester (*34*).

MPa (~30 psi) contact pressure. The contact path traveled by the head after striking the surface is between 124 and 127 mm (4.88 and 5 in). Originally, BPT was developed to yield a BPN equal to 100 × COF (*32*). However, there is a slight difference between the intended relation and the true measurements. A more precise relation between the BPN and COF is described in Equation 1 (*33*).

$$COF = \left(\frac{110}{BPN} - \frac{1}{3}\right)^{-1} \quad (1)$$

Surface Texture Measurement. At the same test points, surface texture scans were acquired using the laser texture scanner (LTS) model 9400HD from Ames Engineering. Using the LTS 9400HD, the scanned area for all test points was 70 × 100 mm (2.76 × 3.94 in) as shown in Figure 2. The scanner acquires single point measurements along 405 line scans with a spacing of 6.35×10^{-3} mm (0.25×10^{-3} in) and a total length of 100 mm (3.94 in). The spacing between the line scans was set to 0.17 mm (6.7×10^{-3} in). This allows capturing macrotexture and a sufficient part of the microtexture of the scanned area in the longitudinal direction. The scanner uses red laser with a dot size of 25×10^{-3} mm at the center of range, with a 22° triangulation angle at the center of range. The acquired points are then processed in a triangulation analysis algorithm to produce a 3D point cloud of the surface texture.

Rubber Slider Characterization. Implementing Persson's friction model requires identifying the viscoelastic modulus master-curve of the rubber in contact with the pavement surface. To characterize the mounted rubber slider, TRL (55) specimens were cut from the sliders and tested in a strain controlled Ares-G2 rheometer. Linear dynamic mechanical analysis (DMA) was performed using a 10 mm (0.39 in) 3-point bending fixture. The strain level in the test was set at 0.6%; the frequency sweep ranged from 0.15 Hz to 15 Hz, and the testing temperature ranged from −50°C to 200°C. Using the time–temperature superposition principle, the curves from the frequency sweeps at each testing temperature were shifted to construct the master-curve at a reference temperature 20°C, with a wide frequency range as shown in Figure 3. It should be mentioned that the master-curve could only reach 7.0×10^{7} Hz, which can limit the speeds included in the analysis and simulations, as explained in the following sections.

Analysis and Modeling

Surface Texture Characterization. The friction model used in this paper, Persson's friction model, requires estimating

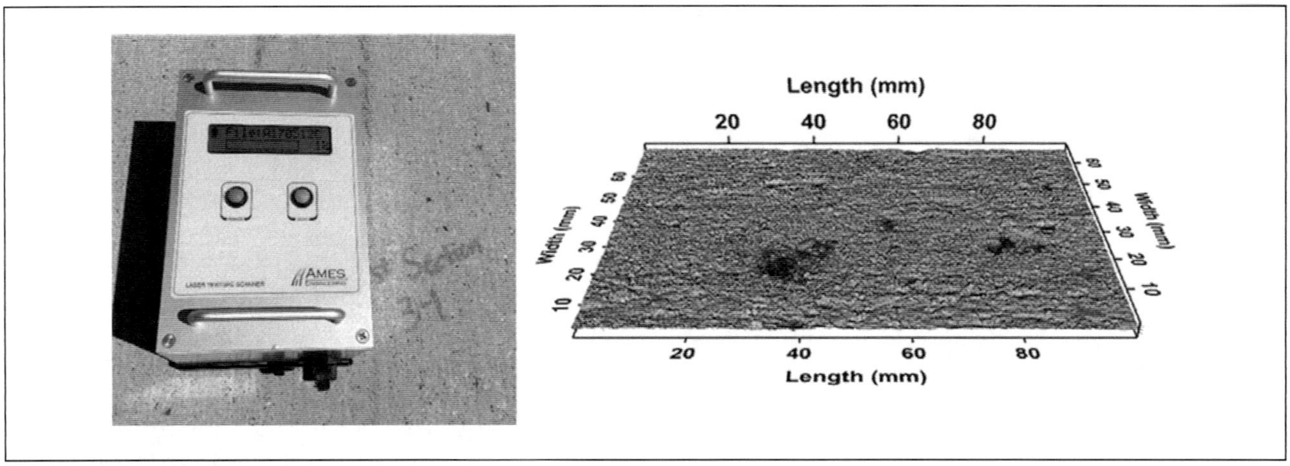

Figure 2. LTS 9400HD and a 3D view of the scanned surface.

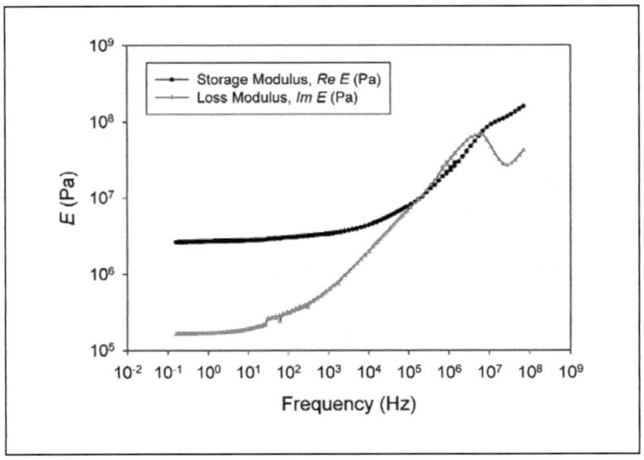

Figure 3. Master-curve of the TRL (55) mounted rubber slider.

the power spectral density (PSD) of the pavement texture scans. PSD can be estimated in a two-dimensional wavenumber plane $q = (q_x, q_y)$; the two-dimensional PSD is then averaged radially about the origin to calculate a one-dimensional function C^{iso}. Alternatively, the PSD in this study was estimated for each line in the scan separately, then averaged at each wavenumber. The average PSD was then used in Equation 2 to estimate the equivalent C^{iso} value (*35*).

$$C^{\text{iso}} \approx \frac{\pi}{q} C^{\text{average}-1\text{D}}(q) \tag{2}$$

The hand-held LTS device had a minimum spacing between point of 6.35×10^{-3} mm (0.25×10^{-3} in), thus the maximum wavenumber reached in the PSD was 73.13 cycle/mm (1.86×10^3 cycle/in). To implement Persson's friction model, PSD at higher wavenumbers should be characterized. Previous studies have shown that pavement surface texture at the macro–micro texture levels is a self-affine fractal surface (*15*). Based on this observation, the PSD can be extrapolated to higher wavenumbers to cover sufficient range for friction analysis. The PSD of self-affine fractals normally consists of two regions (*14*), the constant PSD starting from the lower bound wavenumber q_L to the roll off wavenumber q_0, the second region follows the form presented in Equation 3. In Equations 3 and 4, H is the Hurst exponent, and a higher exponent corresponds to smoother surface texture; q is the wavenumber, k is a constant, h_i is the elevation of a measured point i relative to the average plane; and n is the total number of measured points, $n \approx 6,379,560$ for the LTS 9400 settings used in this study.

$$C(q) \approx \frac{H}{2\pi}\left(\frac{h_0}{q_0}\right)^2 \left(\frac{q}{q_0}\right)^{-2(H+1)} = kq^{-2(H+1)} \tag{3}$$

where

$$h_0 \approx \frac{2}{n}\sum_{i=1}^{n} h_i^2 \tag{4}$$

In addition to Persson's friction model, the pavement texture is summarized using the traditional mean profile depth (MPD). In this study the average MPD values were estimated using the software tool which accompanied the LTS. The MPD is calculated in accordance with the procedure described in ASTM E1845 – 15. Following the standard procedure, the scan is filtered using a 2.5 mm 3rd order Butterworth lowpass filter. Then a least square fit is used to establish a mean line, which is then subtracted from the data to produce a profile with zero mean. Afterwards, the 100 mm lines were divided into two 50 mm segments. Finally, the average of the two peak values from both segment define MPD for each line in the scan. The average MPD corresponds to the average MPD for all the lines in a scan.

Friction Model. Persson's friction model is used to estimate the COF for the tested surfaces from the texture scans. In his model, Persson derived the frictional forces from the work done by the surface asperities to deform the sliding viscoelastic medium (i.e., rubber). This mechanism is responsible for the hysteresis component. Furthermore, Persson has devised a new multiscale contact model to describe the true contact area A between a rubber body and a randomly rough surface (*14*, *15*). The contact model is described in Equations 5–7. From the equations, It can be seen that the true contact area is only a portion of the nominal surface area A_0, where skidding takes place. For the BPT the nominal contact area is equal to the width of the rubber slider multiplied by the path length, which is approximately equal to 76 mm × 125 mm = 9500 mm^2.

$$A_1 = A(q_1) = P(q_1)A_0 \qquad (5)$$

where

$$P(q) = \frac{2}{\pi}\int_0^{\infty} \frac{\sin x}{x} e^{-x^2 G(q)} dx = erf\left(\frac{1}{2\sqrt{G(q)}}\right) \qquad (6)$$

$$G(q) = \frac{1}{8}\int_{q_0}^{q} dq q^3 C(q) \int_0^{2\pi} d\phi \left|\frac{E(qv\cos\phi)}{(1-\upsilon^2)\sigma_0}\right|^2 \qquad (7)$$

q_0 is the roll off wavenumber, q_1 is the cutoff wavenumber, erf is the error function, $C(q)$ is the PSD, E is the rubber dynamic modulus, v is the sliding speed, υ is Poisson's ratio of the rubber ~0.45, and σ_0 is the contact normal pressure between the rubber slider and the tested surface ~0.21 MPa (*32*). From the equations it can be seen that the contact area is a function of speed, surface roughness, and rubber characteristics. Using the contact model described above, the COF owing to the hysteresis component is defined (*14*, *15*) according to Equation 8.

$$COF_{hyst} = \mu_{hyst} \approx \frac{1}{2}\int_{q_0}^{q_1} dq q^3 C(q) S(q) P(q) \int_0^{2\pi} d\phi \cos\phi \, \mathrm{Im} \frac{E(qv\cos\phi)}{(1-\upsilon^2)\sigma_0} \qquad (8)$$

where

$$S(q) = \gamma + (1-\gamma)P(q)^2, \gamma \approx 1/2 \qquad (9)$$

Im is the imaginary part of the dynamic modulus of the rubber, which corresponds to the loss modulus. To include the impact of adhesion, Lorenz et al. have proposed deriving the adhesion component based on actual measurement (*15*). In their approach, the COF owing to hysteresis is predicted using Equation 8, then the difference between the hysteresis component and the total measured coefficient is the adhesive component. In Equation 10, τ_f is the friction shear stress, which describes the adhesion between the rubber and pavement surface.

$$\mu = \mu_{\mathrm{hyst}} + \mu_{\mathrm{adh}} = \mu_{\mathrm{hyst}} + \frac{\tau_f A_1}{\sigma_0 A_0} \qquad (10)$$

In a previous study, Persson indicated that the adhesion component is negligible for surfaces contaminated with dust or fluids (*14*). Furthermore, Lorenz et al. showed that the adhesion component exhibits significance at low speeds (*15*). For the mentioned reasons, and the lack of test data at varying speeds, the adhesion component is excluded at this level of analysis. It should be mentioned, the exclusion of adhesion component does not correspond to neglecting the microtexture in the analysis; in this model both friction components are based on the macrotexture and microtexture. The integration boundaries q_0 and q_1 are free parameters that affect the final results significantly. In this study, q_0 is defined as the point where the PSD deviates from the linear trend significantly as the wavenumber decreases. The cutoff wavenumber can be defined as the wavenumber that results in a cumulative root-mean square slop (h'_{rms}) equal to 1.3. This definition was introduced previously by Lorenz et al. (*15*). In the current study the cutoff wavenumber was chosen according to the mentioned definition except for HFST, where the defined cutoff wavenumber is multiplied by a hundred.

Results and Discussion

Surface Texture Characterization

The PSD curves obtained for all scanned surfaces demonstrated a fractal behavior. The PSD functions in log-log PSD curves followed a straight line trend after reaching the roll off wavenumber (q_0). The slope of the line relaters to the Hurst exponent, H, as described in Equation 3. Figure 4, shows three sample scans acquired on HMA surfaces with the log-log PSD curves in the right panel. The color intensity scale in the scans is proportional to the elevation relative to the mean plane of the scan. Figure 4*a* shows a valley dominant surface, where many points fall below the mean plane.

Figure 5 presents three sample scans acquired on PCC surfaces with the log-log PSD curves in the right

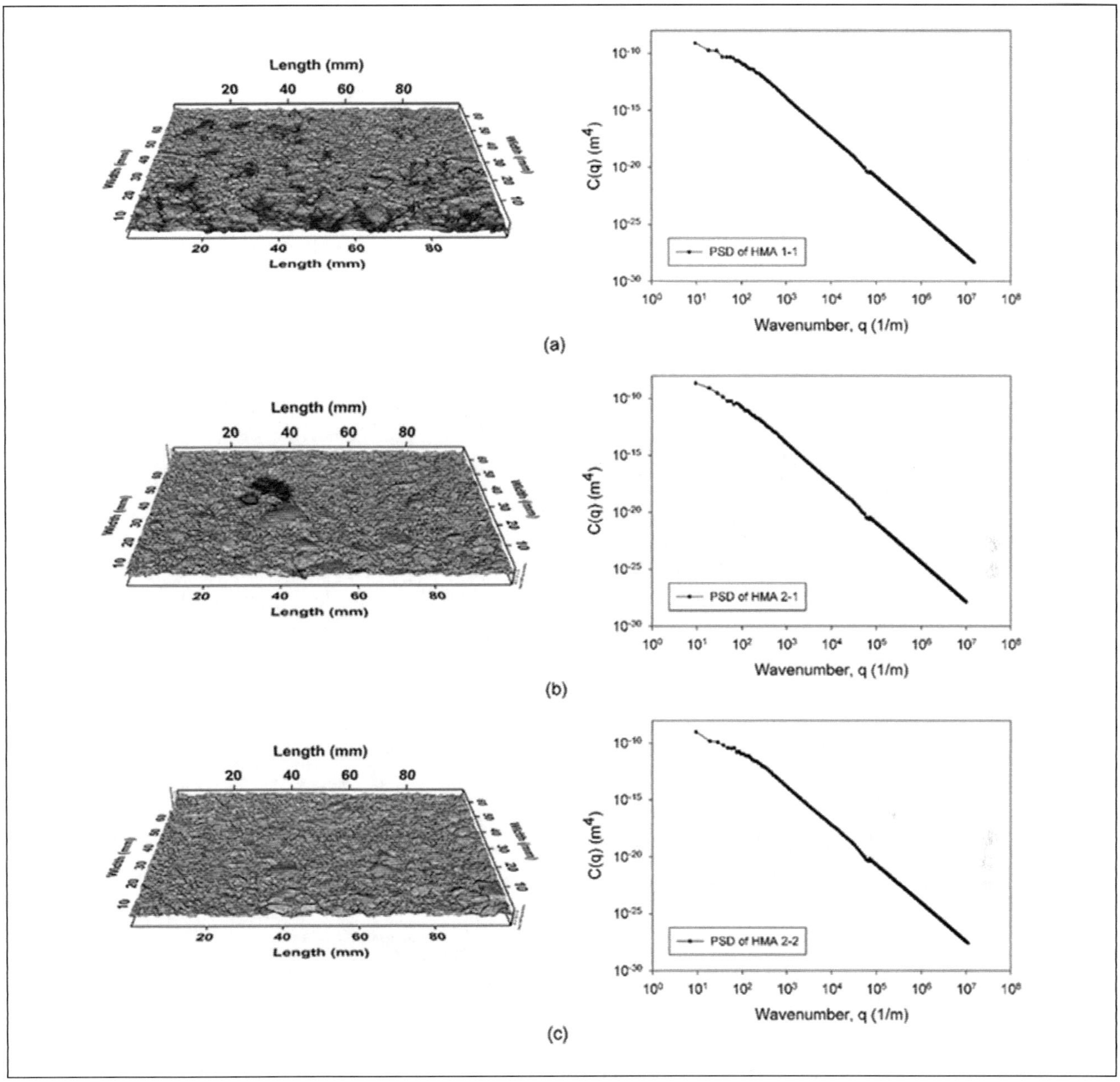

Figure 4. Sample 3D view of scanned HMA surfaces (left panel) and their corresponding PSD (right panel) for (a) test patch 1 on section HMA 1, (b) test patch 1 on section HMA 2, and (c) test patch 2 on section HMA 2.

panel. From the Figures it can be seen that PCC surfaces have less prominent features compared with HMA surfaces. On the other hand, examining the PSD curves, it can be noticed that PCC curves had a lower Hurst exponent, which corresponds to a rougher fractal surface trend. Rougher fractal trends indicate the sharpness of the small features on a surface such as edges, and does not relate to the size of large and smooth features. This discrepancy in understanding the texture behavior between traditional texture analysis and fractal based analysis is significant.

Figure 6 shows three sample scans acquired on HFST surfaces with the log-log PSD curves in the right panel. From the Figures it can be seen that HFST surfaces have more texture features at both microtexture and macrotexture levels. Furthermore, examining the PSD curves, it can be noticed that HFST curves have similar Hurst exponents to PCC sections, owing to the similar

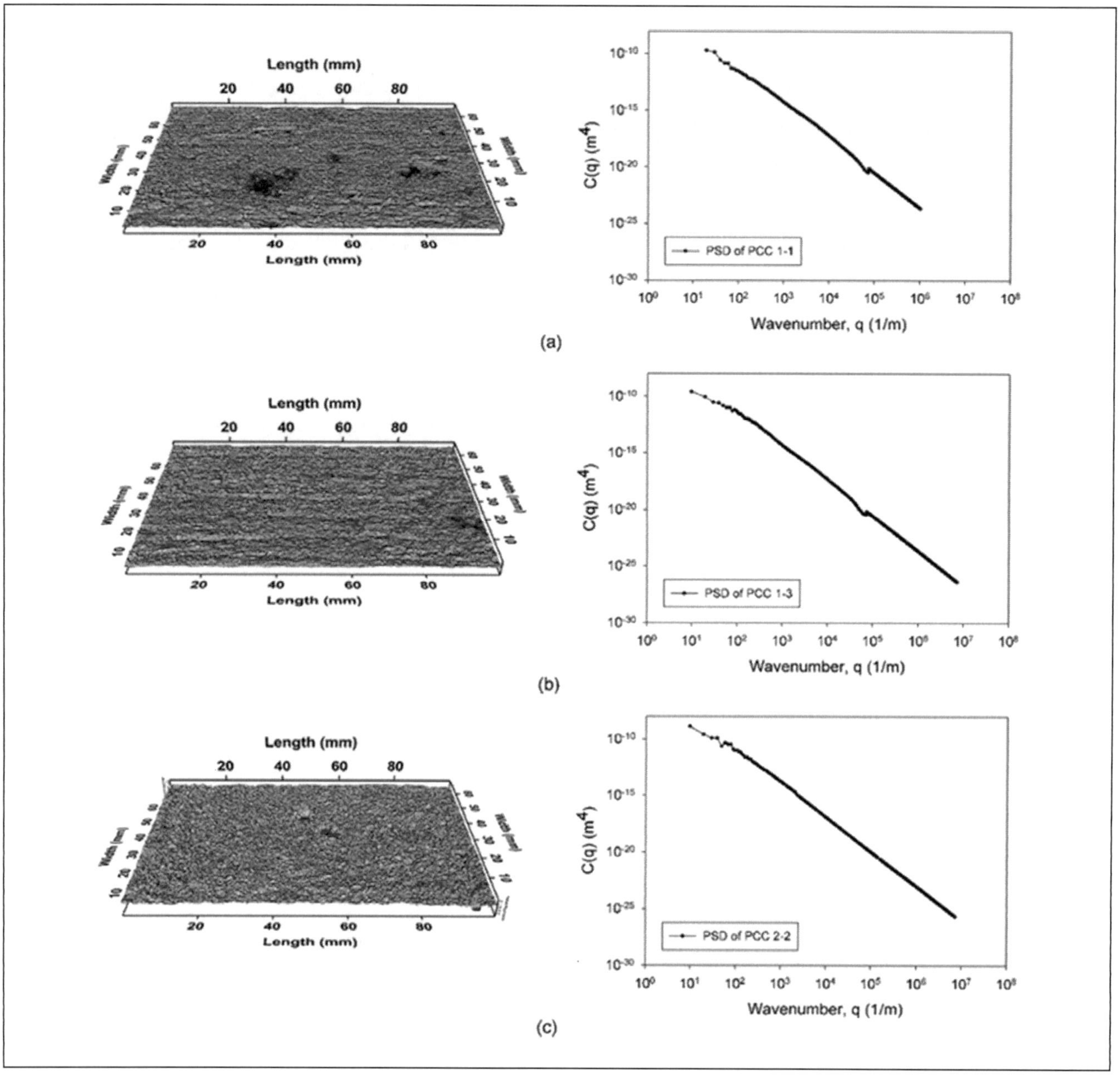

Figure 5. Sample 3D view of scanned PCC surfaces (left panel) and their corresponding PSD (right panel) for (a) test patch 1 on section PCC 1, (b) test patch 3 on section PCC 1, and (c) test patch 2 on section PCC 2.

sharpness in the features and spikes on the surface, yet higher magnitudes. The higher magnitudes are reflected in the higher MPD, which results in higher COF.

Friction Measurements and Modeling

Table 1 summarizes all the average measured BPN and the standard deviation based on four repetitions at each test location. In compliance with ASTM E303, the maximum standard deviation value was 1 BPN. It can be noticed that overall there is no significant difference between the dry COF values for HMA and PCC surfaces. However, the overall MPD is different between the two surface types. On the other hand, HFST results in a significantly higher COF compared with HMA and PCC surfaces, while having a similar Hurst exponent to PCC surfaces. This behavior indicates that MPD solely does not provide a full explanation of the relation between

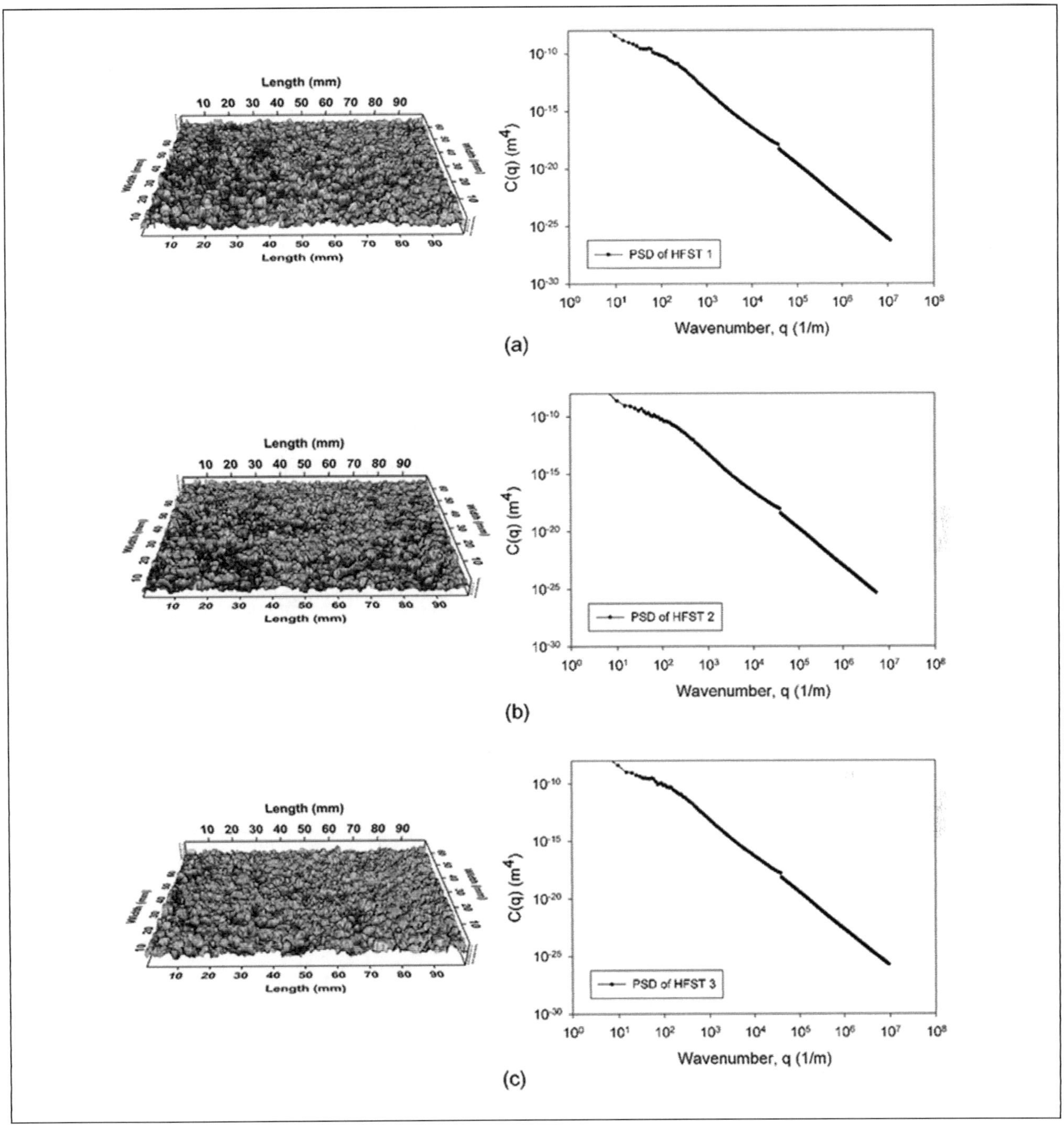

Figure 6. Sample 3D view of scanned HFST surfaces (left panel) and their corresponding PSD (right panel) for (a) test section 1, (b) test section 2, and (c) test section 3.

texture measurements and skid resistance. Alternatively, the friction behavior can be related to the MPD and Hurst exponent, which provides a balanced definition combining the two forms of texture roughness to explain the source of skid resistance.

Table 1 shows that the cutoff wavenumber (q_l) is lower for PCC surfaces as compared with HMA and HFST surfaces. From Table 1 and Figure 7*a* and *b*, it can be seen that compared with the MPD and COF estimated using the BPT, there is an improved agreement

Table 1. Summary of British Pendulum Test Results and Texture Analysis and Modeling

Test ID	MPD (mm)	Hurst exponent (H)	BPN average	BPN standard deviation	COF_{BPN}	q_0 (1/m)	q_1 (10^6/m)	$COF_{Persson}$
HMA 1-1	0.66	0.72	100.00	0.00	1.31	350.00	2.92	1.14
HMA 1-2	0.66	0.72	83.75	0.96	1.02	160.00	2.37	1.07
HMA 1-3	0.52	0.74	90.50	0.58	1.13	230.00	3.90	1.07
HMA 2-1	0.67	0.74	85.25	0.50	1.04	150.00	4.96	1.03
HMA 2-2	0.60	0.70	76.00	0.82	0.90	140.00	1.30	1.07
HMA 2-3	0.68	0.68	88.75	0.96	1.10	117.00	1.07	1.09
Average	0.63	0.72	87.38	0.64	1.08	191.17	2.75	1.08
PCC 1-1	0.31	0.54	80.50	0.58	0.97	150.00	0.13	0.93
PCC 1-2	0.45	0.55	92.25	0.96	1.16	70.00	0.56	1.19
PCC 1-3	0.43	0.57	95.25	0.50	1.22	160.00	0.72	1.21
PCC 2-1	0.34	0.59	84.50	1.00	1.03	130.00	0.30	1.08
PCC 2-2	0.59	0.54	100.75	0.96	1.32	10.00	0.18	1.12
PCC 2-3	0.46	0.54	95.50	1.00	1.22	160.00	0.25	1.17
Average	0.43	0.55	91.46	0.83	1.15	113.33	0.36	1.12
HFST 1	1.74	0.62	104.50	1.00	1.39	221.00	3.46	1.35
HFST 2	1.80	0.59	107.00	0.00	1.44	122.00	2.60	1.36
HFST 3	1.77	0.60	95.00	0.00	1.22	135.00	1.00	1.31
HFST 4	1.68	0.62	107.50	0.58	1.45	130.00	4.50	1.33
HFST 5	1.68	0.61	95.75	0.96	1.23	122.00	3.82	1.33
HFST 6	1.86	0.68	110.00	0.00	1.51	122.00	2.80	1.39
Average	1.75	0.60	103.29	0.42	1.37	142.00	3.03	1.35

between the COF estimated using Persson's model and the COF estimated using the BPT. This observation supports the conclusion that surfaces with different Hurst exponents will have a different trend as a function of MPD. Thus, both roughness definitions should be combined to get a better understanding of the texture impact on skid resistance.

Figure 8 shows the COF curves estimated using Persson's model as function of the sliding speed for all 18 test locations. It should be mentioned that the curves are truncated owing to the limited extent of the master-curve, where the maximum frequency reached was 7.05×10^7 Hz. In the model, the frequency at which the modulus components are used is defined as: $\omega = qv\cos\phi$, thus for a cutoff wavenumber q_1, the maximum speed to be used in the analysis is $7.05/q_1 \times 10^7$ m/s. The curves shown in Figure 8 indicate that the COF friction ramps up from zero to reach a maximum COF and drops back to lower values. Other studies have shown that the curve drops back to zero at much higher speeds. It should be mentioned that the COF estimated from the model is a kinetic COF and thus at lower speeds the static COF might be dominant. The mechanism could be approximated by a step motion, where the rubber stays in static state until it breaks the potential boundary of static friction and moves incrementally at infinitesimal steps relating to rolling resistance. This average movement results in a very slow speed, and accordingly a zero kinetic COF.

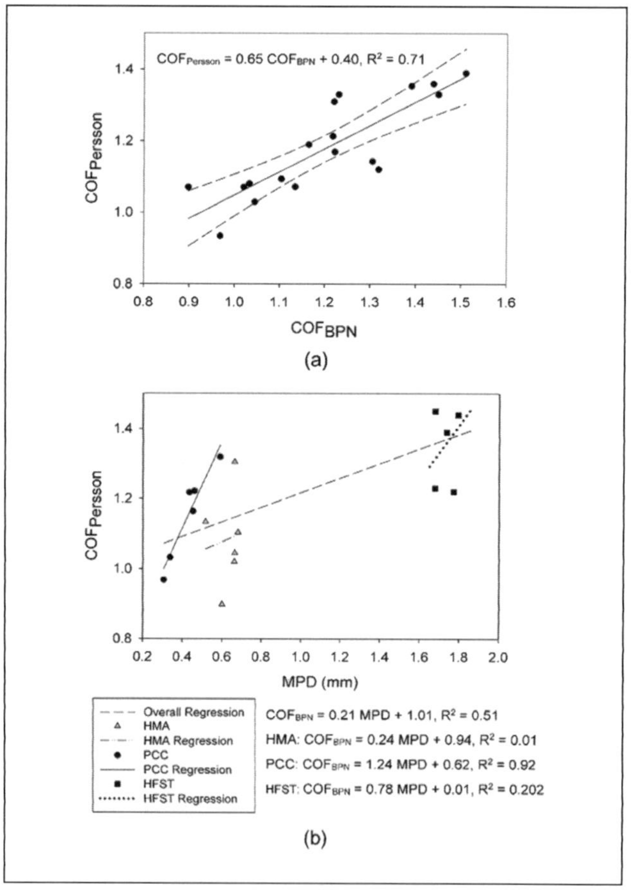

Figure 7. COF Obtained using the British pendulum versus (*a*) COF estimated from Persson's model and (*b*) the MPD.

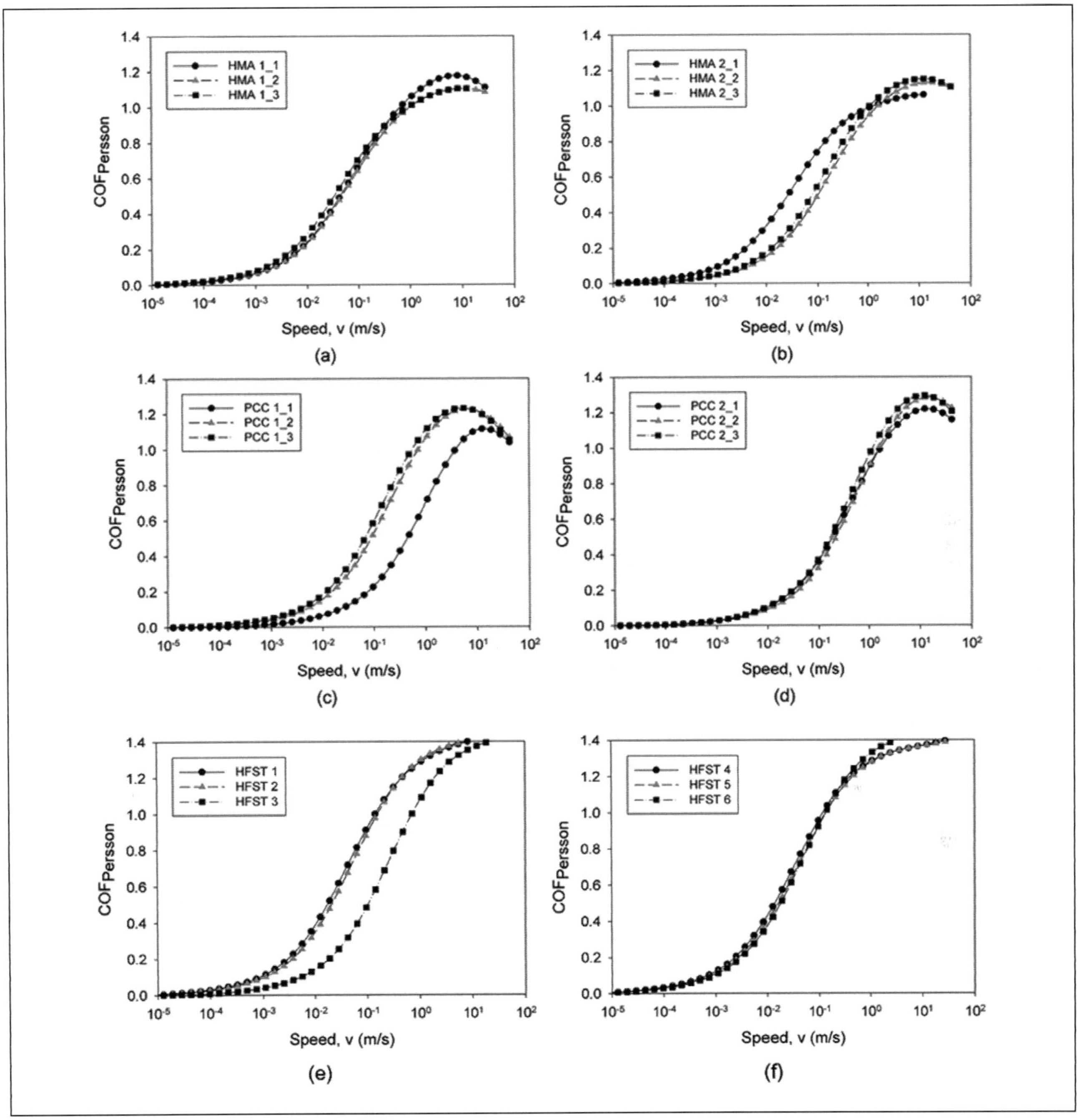

Figure 8. COF predicted using Persson's model as a function of speed for the three test patches on (a) section HMA 1, (b) section HMA 2, (c) section PCC 1, (d) section PCC 2; and the tests on (e) the first three HFST sections, and (f) the last three HFST sections.

Moreover, Figure 8 shows that the peak COF is reached at different speeds for different surfaces. Therefore, defining a universal speed at which all friction measurements peak is not applicable. These curves can provide a valuable tool to derive a universal and consistent harmonization approach. Since the model accounts for the properties of rubber used in the friction device and the testing speed, one can derive the COF for a universal testing condition from the measurements acquired using a single friction device.

Summary and Conclusion

This paper introduced an approach to implement a theoretical friction model, Persson's friction model, to predict the kinetic COF for a BPT from pavement surface texture scans. The field measurements included in the study are: British pendulum measurements, acquired at 18 test points for HMA and PCC pavements, and HFST surfaces; and high-resolution laser texture scans, acquired at the same 18 locations. The key findings from this study are:

- Persson's friction model could predict the COF derived from the British pendulum tester results with high accuracy.
- Persson's friction model provides a promising approach to define a universal harmonization procedure.
- Fractal definition of roughness provides a different understanding compared with the traditional approach using the MPD.
- MPD cannot provide a full picture explaining the frictional behavior. However, combining MPD with the Hurst exponent can potentially provide a full physical explanation of the impact of surface texture on skid resistance.
- Despite the promising results in this study, there is still a need for further investigation to build a general model with a consistent approach for parameter tuning, especially the cutoff wavelength.

Acknowledgments

The authors would like to thank Mark Leichty from Ames Engineering Inc. for providing the laser texture scanners and technical support. The authors would also like to thank Dr. Christopher Williams for arranging for the DMA test.

Author Contributions

The authors confirm contribution to the paper as follows: study conception and design: Ahmad Alhasan, Omar Smadi; data collection: Ahmad Alhasan, Georges Bou-Saab, Nacu Hernandez, Eric Cochran; analysis and interpretation of results: Ahmad Alhasan, Omar Smadi, Georges Bou-Saab, draft manuscript preparation: Ahmad Alhasan, Omar Smadi. All authors reviewed the results and approved the final version of the manuscript.

References

1. Henry, J. J. *NCHRP Synthesis 291: Evaluation of Pavement Friction Characteristics*. TRB, National Research Council, Washington, D.C., 2000.
2. Ivey, D. L., L. Griffin, J. R. Lock, and D. Bullard. *Texas Skid Initiated Accident Reduction Program*. Publication TX-92/910-1F. Texas Department of Transportation, 1992.
3. Flintsch, G., K. McGhee, E. de León Izeppi, and S. Najafi. The Little Book of Tire Pavement Friction. *Pavement Surface Properties Consortium*, 1st ed., 2012.
4. Lowne, R. W. The Effect of Road Surface Texture on Tyre Wear. *Wear*, Vol. 15, No. 1, 1970, pp. 57–70.
5. Cardoso, F. A., A. L. De Aragão Costa, and D. K. Tanaka. Durability Performance of Tire Tread Rubber Compounds as a Function of Road Pavement. *Proc., International Brazilian Conference on Tribology*, Rio de Janeiro, Brazil, 2010. pp. 375–387.
6. Ma, B., H.-G. Xu, Y. Chen, and M.-Y. Lin. Evaluating the Tire Wear Quantity and Differences Based on Vehicle and Road Coupling Method. *Advances in Mechanical Engineering*, Vol. 9, No. 5, 2017, p. 1687814017700063.
7. McGovern, C., P. Rusch, and D. Noyce. *State Practices to Reduce Wet Weather Skidding Crashes*. Federal Highway Administration, U.S. Department of Transportation, Washington, D.C., 2011.
8. Merritt, D. K., C. A. Lyon, and B. N. Persaud. *Evaluation of Pavement Safety Performance*. Report No. FHWA-HRT-14-065. Federal Highway Administration, Washington, D.C., 2015.
9. Wong, S. Effectiveness of Pavement Grooving in Accident Reduction. *ITE Journal*, Vol. 60, No. 7, 1990, pp. 34–37.
10. Najafi, S., G. W. Flintsch, and A. Medina. Linking Roadway Crashes and Tire–Pavement Friction: A Case Study. *International Journal of Pavement Engineering*, Vol. 18, No. 2, 2017, pp. 119–127.
11. de León Izeppi, E., S. W. Katicha, G. W. Flintsch, and K. K. McGhee. Pioneering Use of Continuous Pavement Friction Measurements to Develop Safety Performance Functions, Improve Crash Count Prediction, and Evaluate Treatments for Virginia Roads. *Transportation Research Record: Journal of the Transportation Research Board*, 2016. 2583: 81–90.
12. Long, K., H. Wu, Z. Zhang, and M. Murphy. *Quantitative Relationship between Crash Risks and Pavement Skid Resistance*. Publication FHWA/TX-13/0-6713-1. Texas Department of Transportation, 2014.
13. Hall, J., K. L. Smith, L. Titus-Glover, J. C. Wambold, T. J. Yager, and Z. Rado. *NCHRP Web-Only Document 108: Guide for Pavement Friction*. TRB, National Research Council, Washington, D.C., 2009.
14. Persson, B. N. J. Theory of Rubber Friction and Contact Mechanics. *The Journal of Chemical Physics*, Vol. 115, No. 8, 2001, pp. 3840–3861.
15. Lorenz, B., Y. Oh, S. Nam, S. Jeon, and B. Persson. Rubber Friction on Road Surfaces: Experiment and Theory for Low Sliding Speeds. *The Journal of Chemical Physics*, Vol. 142, No. 19, 2015, p. 194701.
16. Persson, B. N. Contact Mechanics for Randomly Rough Surfaces. *Surface Science Reports*, Vol. 61, No. 4, 2006, pp. 201–227.
17. Persson, B. N. J. Rubber Friction: Role of the Flash Temperature. *Journal of Physics: Condensed Matter*, Vol. 18, No. 32, 2006, p. 7789.

18. Kogbara, R. B., E. A. Masad, E. Kassem, A. Scarpas, and K. Anupam. A State-of-the-art Review of Parameters Influencing Measurement and Modeling of Skid Resistance of Asphalt Pavements. *Construction and Building Materials*, Vol. 114, 2016, pp. 602–617.
19. Subhi, M. B., and R. Farhad. Changes in Asphalt Pavement Friction Components and Adjustment of Skid Number for Temperature. *Journal of Transportation Engineering*, Vol. 131, No. 6, 2005, pp. 470–476.
20. Asi, I. M. Evaluating Skid Resistance of Different Asphalt Concrete Mixes. *Building and Environment*, Vol. 42, No. 1, 2007, pp. 325–329.
21. Araujo, V. M. C., I. S. Bessa, and V. T. F. Castelo Branco. Measuring Skid Resistance of Hot Mix Asphalt Using the Aggregate Image Measurement System (AIMS). *Construction and Building Materials*, Vol. 98, 2015, pp. 476–481.
22. Kassem, E., A. Awed, E. Masad, and D. Little. Development of Predictive Model for Skid Loss of Asphalt Pavements. *Transportation Research Record: Journal of the Transportation Research Board*, 2013. 2372: 83–96.
23. Wu, Z., B. King, C. Abadie, and Z. Zhang. Development of Design Procedure to Predict Asphalt Pavement Skid Resistance. *Transportation Research Record: Journal of the Transportation Research Board*, 2012. 2306: 161–170.
24. Dunford, A. *Friction and the Texture of Aggregate Particles Used in the Road Surface Course*. PhD thesis, University of Nottingham, 2013.
25. Liu, Y., T. F. Fwa, and Y. S. Choo. Effect of Surface Macrotexture on Skid Resistance Measurements by the British Pendulum Test. *Journal of Testing and Evaluation*, Vol. 32, No. 4, 2004, pp. 304–309.
26. Ergun, M., S. Iyinam, and A. F. Iyinam. Prediction of Road Surface Friction Coefficient Using Only Macro- and Microtexture Measurements. *Journal of Transportation Engineering*, Vol. 131, No. 4, 2005, pp. 311–319.
27. Khasawneh, M. A., and R. Y. Liang. Correlation Study between Locked-Wheel Skid Trailer and Dynamic Friction Tester. Presented at 87th Annual Meeting of the Transportation Research Board, Washington, D.C., 2008.
28. Meegoda, J. N., and S. Gao. Evaluation of Pavement Skid Resistance Using High Speed Texture Measurement. *Journal of Traffic and Transportation Engineering (English Edition)*, Vol. 2, No. 6, 2015, pp. 382–390.
29. Ueckermann, A., D. Wang, M. Oeser, and B. Steinauer. Calculation of skid resistance from texture measurements. *Journal of Traffic and Transportation Engineering (English Edition)*, Vol. 2, No. 1, 2015, pp. 3–16.
30. Flintsch, G., E. de León Izeppi, K. McGhee, and J. Roa. Evaluation of International Friction Index Coefficients for Various Devices. *Transportation Research Record: Journal of the Transportation Research Board*, 2009. 2094: 136–143.
31. Wambold, J., C. Antle, J. Henry, and Z. Rado. *PIARC (Permanent International Association of Road Congress) Report, International PIARC Experiment to Compare and Harmonize Texture and Skid Resistance Measurement*. C-1 PIARC Technical Committee on Surface Characteristics, France, 1995.
32. Giles, C., B. Sabey, and K. Cardew. Development and Performance of the Portable Skid-Resistance Tester. *Rubber Chemistry and Technology: September 1965*, Vol. 38, No. 4, 1962, pp. 689–699.
33. Clarke, J., K. Hallas, R. Lewis, S. Thorpe, G. Hunwin, and M. Carré. Understanding the Friction Measured by Standardised Test Methodologies used to Assess Shoe-Surface Slip Risk. *Journal of Testing and Evaluation*, Vol. 43, No. 4, 2014, pp. 723–734.
34. Cooper Technology. *Pendulum Skid Resistance Tester*. https://cooper.co.uk/shop/aggregate-testing/pendulum-skid-tester/. Accessed July 17, 2017.
35. Jacobs, T., T. Junge, and L. Pastewka. Quantitative Characterization of Surface Topography Using Spectral Analysis. *arXiv preprint arXiv:1607.03040*, 2016.

The Standing Committee on Pavement Surface Properties and Vehicle Interaction (AFD90) peer-reviewed this paper (18-00976).

Article

TRR
JOURNAL OF THE TRANSPORTATION RESEARCH BOARD

Transportation Research Record
2018, Vol. 2672(40) 452–462

DOI: 10.1177/0361198118797455
journals.sagepub.com/home/trr
SAGE

Interconversion of Locked-Wheel and Continuous Friction Measurement Equipment (CFME) Friction Measurements

Silvia Barrantes Quiros[1], Gerardo W. Flintsch[1,2], Edgar de León Izeppi[1], and Kevin K. McGhee[3]

Abstract

As transportation agencies are adopting proactive pavement friction management using continuous friction measurement equipment (CFME), the need for a method that allows interconversion between the traditional locked-wheel skid testers (LWSTs) becomes important to assure continuity with past practices. This paper evaluates several conversion methods based on the International Friction Index (IFI) approach and proposes an alternative method that allows predicting the measurements of the LWST using measurements with the Sideway-Force Coefficient Routine Investigatory Machine (SCRIM). The investigation is based on data collected using a SCRIM and two LWSTs on a controlled pavement test facility. The results suggest that a conversion based only on the speed adjustments (FR60) is the most effective method to predict the LWST measurements from SCRIM measurements. The coefficient of determination and average absolute error are comparable to those using the full IFI F60 formula but do not require static reference measurements. The study also showed that the three tested devices produced appropriate repeatability as computed using the limit of agreement at the three tested speeds and on a wide range of surfaces.

Pavement friction is a critical factor in road safety, especially under wet conditions. Towards Zero Deaths (TZD) is an international strategy to reduce fatalities and serious injuries that has the goal of reducing the number of fatalities caused by car accidents around the world in half by 2020 (*1*). Because of the multiple factors involved in crashes, namely the driver, the vehicle, and the road, transportation agencies can only influence the provision of a safer road. In particular, pavement infrastructure with good friction can help reduce crashes by allowing vehicles to stop and maneuver with more control of the vehicle. Thus, it is important that a comprehensive TZD strategy include pavement friction management.

The friction between the pavement and a vehicle's tire is affected by temperature, seasonal changes, pavement condition, and the equipment used for the measurement. This surface characteristic should be taken into account in the design process and monitored during the service life of the road to assure good performance. One of the approaches to reducing crashes is to implement periodic network-level friction measurements. In the U.S., the equipment that has been traditionally used to monitor friction is the locked-wheel skid tester (LWST); however, other equipment has been considered. Several recent projects have experimented with continuous friction measurement equipment (CFME), such as fixed slip, sideway force, and modified locked-wheel testers. However, understanding how to compare pavement surface friction measurements taken with different devices has been a challenge (*2, 3*).

Globalization initiatives that require the use of common roads in adjacent countries (or states) have led to a desire for the harmonization of friction data with different equipment. One of the first attempts to do this by the World Road Association (PIARC) introduced the concept of the International Friction Index (IFI). IFI is a method based on texture properties and friction tests that tries to normalize measurements taken with different equipment. However, this method has been tested and the results have not always been harmonious among all devices (*4*).

The need for a method that allows the comparison of measurements taken with different devices, in different types of pavement, and different conditions, has brought

[1]Center for Sustainable Transportation Infrastructure, Virginia Tech Transportation Institute, Blacksburg, VA
[2]Via Department of Civil and Environmental Engineering, Virginia Polytechnic Institute and State University, Blacksburg, VA
[3]Virginia Transportation Research Council, Charlottesville, VA

Corresponding Author:
Address correspondence to Silvia Barrantes Quiros: silvia20@vt.edu

a new challenge to the pavement industry. This paper proposes an alternative method to the IFI standard that allows measurements taken by the Sideway-Force Coefficient Routine Investigation Machine (SCRIM) to predict LWST values. This is important because it allows the experience gained through emerging proactive pavement management friction programs that use the SCRIM to be integrated with past experience with the LWST.

Objective

The purpose of this study was to compare pavement friction measurements using LWST and CFME and propose an effective method for interconversion between the measures of both systems. The IFI approach was tested and simplified to propose a formula to predict the LWST friction numbers using SCRIM measurements. Data collected on a controlled facility (Texas Transportation Institute [TTI] RELLIS campus) were used to test the various approaches.

Background

Pavement Friction. The friction in the pavement surface, also known as skid resistance, is defined as the force developed between the pavement surface and a tire when skidding or cornering on the pavement surface. The friction coefficient (μ) is obtained by dividing the frictional force by the vertical load on the testing tire. It represents the interaction between the vehicle tire and the pavement without considering the demand for friction resulting from the tangential and/or transversal accelerations. Friction is a significant safety issue that plays a critical role in the reduction of accidents on wet pavements (*5*).

Texture. Friction is influenced by several pavement surface characteristics, mainly microtexture and macrotexture. According to PIARC, pavement texture can be classified according to its wavelength in roughness/unevenness, megatexture, macrotexture, and microtexture (*6*). Megatexture and roughness are not considered significant for safety. Macrotexture, on the other hand, gives the properties needed to allow the runoff of the water squeezed between the pavement surface and the tire, enhancing the contact between them. This characteristic affects the braking capacity in wet conditions and the sound emissions in the tire–pavement interface. Microtexture provides the direct contact and adhesion between the tire and the pavement, so it is directly related to the friction. It depends on the aggregate characteristics and their susceptibility to polishing produced by the contact with the tire.

Friction and Macrotexture Measuring Methods and Equipment. Most states in the U.S. use the locked-wheel trailer, with ribbed, smooth, or both tires, to determine a friction number. The measurement of the macrotexture is less common, but some agencies are starting to measure it. Although most agencies report using the sand patch method, the use of the portable and high-speed laser devices is becoming popular (*7*). Table 1 describes the equipment used for the collection of the friction and macrotexture data for this paper.

Friction Models

The relationship between the pavement friction, texture, and testing speed has been investigated for many years. Leu and Henry first predicted the relationship between friction and skid speed considering the micro- and macrotexture of pavement (*10*). In 1992, Henry proposed a revised formula (known as the Penn State model) (*11*), which relates friction with speed through the following formula:

$$F(S) = F_o * e^{\frac{-S}{S_o}} \tag{1}$$

where,

$F(S)$ = friction measured with an equipment at speed "S";

F_0 = non-dimensional constant that depends on microtexture and is the friction at skid speed of 0;

S = skid speed in km/h; and

S_o = constant with speed units that depends on macrotexture; its value varies between 20 and 600 km/h.

The model was subsequently modified so that the reference speed would be 10 km/h instead of 0 as follows:

$$F(S) = F_{10} * e^{\frac{S-10}{S_0}} \tag{2}$$

where,

F_{10} = friction value obtained for a skid speed of 10 km/h.

The values for F_{10} and S_0 were obtained based on measurements with different equipment using smooth and ribbed tires; however, the data did not present a good correlation in all cases (*12*).

In the 1980s PIARC conducted an international experiment to compare and harmonize the measurements of texture and skid resistance. The result of that experiment was what we know today as the IFI. The experiment involved 16 countries and 47 measurement systems, and evaluated 54 different sections in Spain and Belgium (*13*). The experiment also determined initial harmonization constants for the different equipment used in the experiment. As shown in the following equation, the reference speed was changed to 60 km/h:

$$FR60 = FRS * e^{\frac{s-60}{S_p}} \tag{3}$$

where,

$FR60$ = friction at the skid speed of 60 km/h;

FRS = friction value at S km/h;

s = slip speed; and

Table 1. Equipment Used to Collect Data

Method	Description	Standard
Circular texture meter (CT meter)	The CT meter consists of a displacement laser on an arm that rotates, so the displacement sensor follows a circular trajectory with a diameter of 284 mm, where it collects a detailed pavement profile. The software of this equipment reports directly the mean profile depth (MPD) and root mean square (RMS).	ASTM E2157
Dynamic friction tester (DFT)	The DFT is composed of a 350 mm-diameter, rotating disc with three sliders made of rubber. The disc is propelled by a motor while suspended over a pavement surface until the tangential speed gets to 90 km/h. Water is then applied by the equipment to the pavement surface and the disc is lowered to the surface (*8*). The torque generated while the disc descends is used to calculate the friction as a function of the speed. The friction force and the speed as the disk slows to stop are saved in a file, and the results are reported at speeds of 20, 40, 60, and 80 km/h (*9*).	ASTM E1911
LWST	The LWST records the friction force when a trailer wheel is locked on the wet surface while the vehicle is moving at a constant speed. The equipment consists of a trailer with two wheels, and it locks one at a time. The trailer includes a water distribution system connected to a water tank placed on the pulling vehicle. When the vehicle reaches the desired speed, the water system sprays water before the testing wheel and then the braking system is activated. The wheel is kept blocked for around 1 second and average SN over that period is recorded (*9*). The LWST can use two different types of tires: smooth and ribbed. Ribbed tire measurements are known to be less sensitive to pavement macrotexture and water film depth than smooth tires (*6*).	ASTM E274
SCRIM	The SCRIM is equipment was originally developed by TRL in England and is one of the most used in Europe. It uses an oblique wheel system to determine the transversal friction coefficient in terms of sideway-force coefficient reported as a SCRIM reading. The equipment uses a smooth tire that is mounted at 20° to the direction of travel. A water tank on the vehicle wets the pavement just before the tire and the test tire has its own weight and suspension. The equipment measures the sideways force and computes sideway-force friction coefficient, which is typically reported as the average for each continuous 10 m section.	ASTM WK40015 (under development)

s_p = speed constant, which has speed units and is related to macrotexture.

Other experiments, such as HERMES (*2*) and Tyrosafe (*3*) have also tried to harmonize measurements taken with the different friction measuring equipment to a common scale. However, a general harmonization has proven to be very challenging, although some organizations have had better results for their specific equipment (*9*).

Research Methods

Data Collection

The data used for this study were collected as part of a Federal Highway Administration (FHWA) sponsored project managed by the National Pavement Surface Consortium led by the Virginia Transportation Research Council (VTRC). TTI and the Virginia Tech Transportation Institute (VTTI) collected friction measurements on a controlled facility at the TTI RELLIS Campus in September 2016 (*14*). The objective of this study was to compare the measurements from the SCRIM and the LWST. The facility consists of an old runway. Three lanes have been modified for testing friction equipment. It includes nine sections with different materials, as shown in Figure 1.

Measurements were taken on PAD 1 (Portland cement concrete), PAD 2 (Jennite flush seal), PAD 5 (rounded gravel hot mix), PAD 6 (rounded gravel chip seal), and PAD 7 (lightweight aggregate chip seal). A second set of measurements was taken on the second wheel path of PAD 2, providing, as a result, measurements on six sections. On each section, eight runs were conducted at 30, 40, and 50 mph using the SCRIM and two LWSTs, a reference unit owned by TTI and another unit provided by the Texas Department of Transportation (TxDOT). Both devices used a smooth tire (ASTM E524). In summary, 24 measurements were done with each of the three different devices.

The CT meter and DFT were used to take the reference measurements. Seven stations were defined along each PAD and three runs were done on each station. An

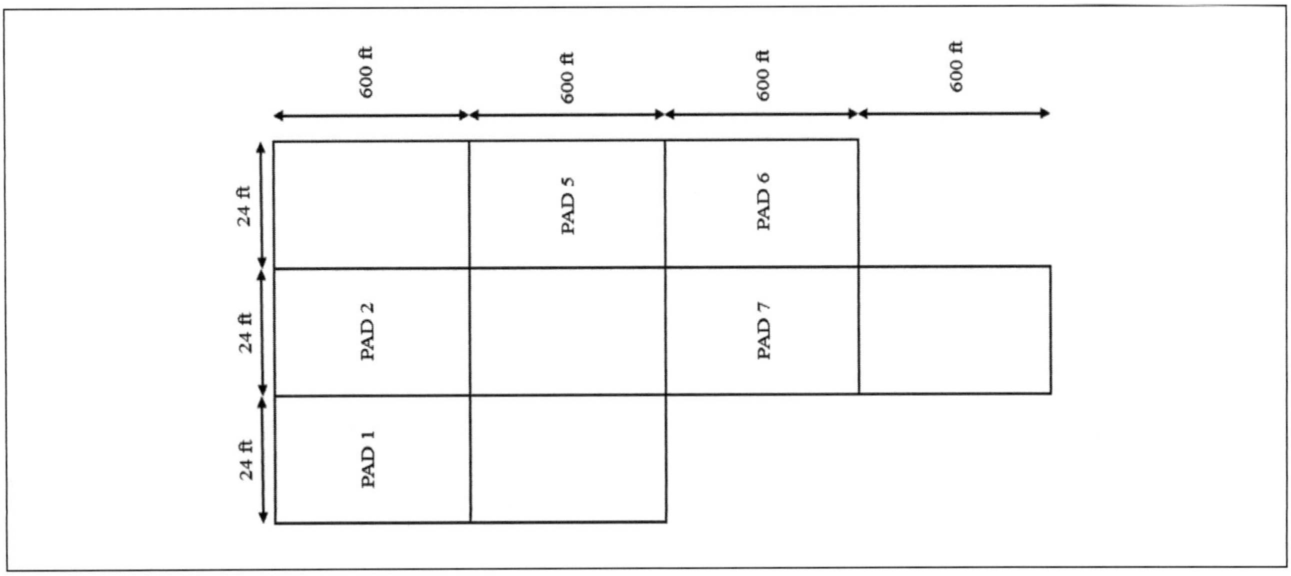

Figure 1. TTI runway.

average of the 21 measurements was obtained in each PAD to use as reference to calculate the S_p and the A and B constants.

Data Analysis Methods

Equipment Comparison using Limits of Agreement (LOA). This method allows the comparison of two measurements taken with different devices or methodologies to show the agreement of the measurements (*15*). This method was first applied in 2012 to compare pavement friction measurements (*16*).

The first step in the LOA analysis is to compute the repeatability coefficient for each device. The data collected are used in an analysis of variance (ANOVA) to obtain the variance (or mean squares) needed as the variance factor to calculate the repeatability for each device. The repeatability, r, for each device is calculated using the following equation:

$$r = \sqrt{s_1^2} * \sqrt{2} * 1.96 \quad (4)$$

where,

s_1^2 = mean square calculated with the ANOVA analysis for the first group of data.

The LOA method then compares the differences (d) between the mean of the measurements taken with one device and the mean measurement of other device. Simply put, LOA is a comparison that can be calculated as $d \pm 2s$, where s is the standard deviation (*17*). The equation used to calculate the LOA for each comparison was:

$$s_c = \sqrt{s_D{}^2 + f_1 * s_1{}^2 + f_2 * s_2{}^2} \quad (5)$$

where,

$s_D{}^2$ = variance of the difference between the measurements of each device;

$s_1{}^2$ and $s_2{}^2$ = variance corresponding to the data set for Device 1 and 2; and

f_1 and f_2 = factors for each device that depends on the number of runs and are defined as follows:

$$f_i = \left(1 - \frac{1}{m_i}\right) \quad (6)$$

m_i = number of runs for the device i.

Data Pre-Processing and Slipping Speed Adjustments. Three methods were investigated to convert data measured with the SCRIM to LWST friction number. All of them focus on bringing the measurements to a common reference slip speed to be able to compare the data. The first method is the direct application of the ASTM International Friction Index or IFI (*18*). The others are derivations (simplifications) of the IFI to adjust for speed. The following subsections describe data preprocessing and the three methods in more detail.

LWST Data Processing

The LWST measurements were filtered to remove the highest and lowest values, except for a few cases where either the two highest or the two lowest measurements were removed (*14*).

SCRIM Data Processing

After data were exported using the equipment software to an Excel file, the measurements were adjusted to the target speed using the manufacturer-recommended equations and the data were filtered to remove outliers.

International Friction Index

ASTM E 1960 "Standard Practice for Calculating International Friction Index of a Pavement Surface" (*18*) normalizes how to obtain the IFI value from measurements of friction and macrotexture. The IFI is composed of two parameters, F60 and S_p. The first value is the friction of the wet pavement at 60 km/h, and the second is the speed parameter that describes the speed dependence of the friction.

IFI uses reference macrotexture (CT meter) and friction (DFT) measurements to normalize the measurements done with different equipment. It was developed to provide a standard comparison of pavement friction measurements around the world (*4*). Through this index, it is possible to compare measurements obtained with different equipment. The procedure followed to calculate the IFI is described as follows:

1. Calculate the constant S_p that is related to the macrotexture, which can be determined using linear regression as a function of the MPD in mm:

$$Sp = a + (b * MPD) \tag{7}$$

where,
a and b = constants that vary according to the equipment used to take the measurement of macrotexture (for this paper the equipment used was the CT meter, a = 14.2 and b = 89.7).

2. Calculate the F60 parameter for the equipment used to determine the friction (FRS), in this case the SCRIM and the LWST, the slipping speed (S), which is the testing speed for the LWST and 0.34 times the testing speed for the SCRIM ($sin(20^o)$ = 0.34), and the S_p determined using the macrotexture measurements using Equation 3.
3. Compute the value of F60 using the following equation:

$$F60 = A + B * FR60 \tag{8}$$

where,
A and B = constants that can be calculated using linear regression, where the independent variable is the F60 obtained with measurements taken using the DFT and the dependent variable is the FR60 using the FRS obtained from the equipment in question.

The IFI is expressed by the pair (F60, S_p). These two numbers were determined for the SCRIM and both LWSTs on each section of the TTI facility (*14*).

Conversion using F60

With the objective to simplify the comparison between the friction measurements, the F60 formulas for the SCRIM and the LWST were matched to obtain an equation to convert measurements taken with the SCRIM to values comparable to those obtained using the LWST:

$$F60_{SCRIM} = F60_{LWST} \tag{9}$$

$$A_{SCRIM} + B_{SCRIM} * FR60_{SCRIM} = A_{LWST} + B_{LWST} * FR60_{LWST} \tag{10}$$

$$A_{SCRIM} + B_{SCRIM} * FRS_{SCRIM} * e^{\frac{s_{SCRIM}-60}{Sp}} = A_{LWST} + B_{LWST} * FRS_{LWST} * e^{\frac{s_{LWST}-60}{Sp}} \tag{11}$$

$$FRS_{LWST} = \frac{A_{SCRIM} - A_{LWST} + B_{SCRIM} * FRS_{SCRIM} * e^{\frac{s_{SCRIM}-60}{Sp}}}{B_{LWST} * e^{\frac{s_{LWST}-60}{Sp}}} \tag{12}$$

Predicted LWST measurements were computed using Equation 12 and compared with the direct measured values.

Conversion using FR60

Another comparison was made by matching FR60 instead of F60. The reason for using this formula was to avoid the need to take measurements with a third device, like the DFT, as it is needed for establishing A and B to compute the F60. The FR60 formulas for the SCRIM and the LWST were matched with the objective of obtaining an equation that would allow converting measurements taken with the SCRIM to values comparable to those obtained using the LWST:

$$FR60_{SCRIM} = FR60_{LWST} \tag{13}$$

$$FRS_{SCRIM} * e^{\frac{s_{SCRIM}-60}{Sp}} = FRS_{LWST} * e^{\frac{s_{LWST}-60}{Sp}} \tag{14}$$

$$FRS_{LWST} = \frac{FRS_{SCRIM} * e^{\frac{s_{SCRIM}-60}{Sp}}}{e^{\frac{s_{LWST}-60}{Sp}}} \tag{15}$$

Predicted LWST measurements were computed using Equation 15 and compared with the direct measured values.

Evaluation of Results. The evaluation used the FR60 and F60 formulas (Equations 12 and 15) to predict the FRS that would be measured by an LWST using the FRS measured with the SCRIM. The predicted values were then compared with the FRS actually determined with the LWST. The comparison was done using linear regression, with the FRS measured as the independent variable and the FRS predicted as the dependent variable. In an ideal case, the regression will result in a 45° equality line with a coefficient of determination, R^2, of 1. The methods were compared based on goodness-of-fit statistics (R^2 and the average absolute error of the predictions) and considerations regarding the need to conduct reference measurements (which requires using additional devices).

Results

Equipment Comparison (LOA)

Table 2 shows the repeatability of the three devices computed using the LOA calculated using Equation 4. The repeatability seems to be good and similar for all three devices.

The table also presents the results from the LOA for pairwise comparisons at the three tested speeds for the three devices studied. The reproducibility between the two LWSTs is good and close to the repeatability of each individual device. However, the measurements taken with both LWSTs are ± 10 units approximately to the SCRIM measurements with a 95% level of confidence at the three speeds. This suggests that the measurements are different. This was expected because the two devices are measuring different frictional properties at very different slipping speeds. Although the testing speed is the same for the three devices, the actual slipping speed for the SCRIM is lower (approximately one-third) than the LWST. These results reinforce the need for speed corrections.

Figure 2 shows the results of the LOA comparison between the TxDOT LWST and SCRIM with the reference to the TTI LWST. The plot shows the LOA for individual measurements, and individual points represent the difference between the average measurements for each device, at each speed, on each pad. The plot shows no bias for the TxDOT device, except at 50 mph, probably as a result of difficulties in maintaining the high speed over the short sections. It is also interesting to note that the LOA band gets wider as speed increases. The figure also shows an average negative bias of 15 friction numbers of the SCRIM and that the bias is dependent on the Skid Number (SN), as in general measurements (and the LOA bands) show a downward trend. This supports the necessity of converting all the measurements to a reference sliding speed as presented in the following sections.

Equipment Interconversion using the IFI (F60)

To calculate the speed constant (S_p), the MPD was measured using a CT meter. Note that the values for different speeds are slightly different because they were computed over the skidding length of the LWST, which varies with speed. Table 3 shows the results obtained for each section and speed. For the calibration of the friction testers, a DFT was used as the friction reference. The FR60 and F60 shown in the first part of Table 4 were calculated as discussed in the Methods section.

Table 2. Repeatability for Devices Used to Collect Data

Equipment	Speed (mph)	Repeatability	LOA ± SCRIM	LOA ± TxDOT
TTI LWST (reference)	30	1.4	10.5	2.6
	40	1.3	9.5	3.6
	50	3.5	10.8	4.2
TxDOT LWST	30	1.1	10.9	-
	40	2.5	9.3	-
	50	4.5	9.0	-
SCRIM	30	3.3	-	-
	40	2.9	-	-
	50	1.3	-	-

The IFI constants A and B (presented in Table 4) were determined using linear regression of the values of FR60 of the equipment in question and the F60 calculated for the DFT. They were then used to calculate the F60 for each equipment using Equation 9 (*14*).

Direct Comparisons using F60

After the calculation of the IFI values for each device, and the calculation of the constants A and B, the next step was to use Equation 12 to predict the values of friction that would be measured by the LWSTs using the SCRIM SR values. The results are presented in the second part of Table 4 for the comparison of the SCRIM with reference to the TTI LWST.

Prediction of LWST Measurement using FR60

Trying to further simplify the prediction of the measurements and reduce the number of variables, the next step was to predict the friction values from the LWST using Equation 15. The results for the TTI LWST are presented in the third part of Table 4. That same table shows the difference between the predicted and the measured data. In Pad 2 there is a significantly greater difference than for the other sections, which could be caused by the irregularities in that specific pavement section.

Prediction of LWST Measurement using FR60 and Texture from SCRIM

To take advantage of the capacity of the SCRIM to measure macrotexture and simplify the prediction of the

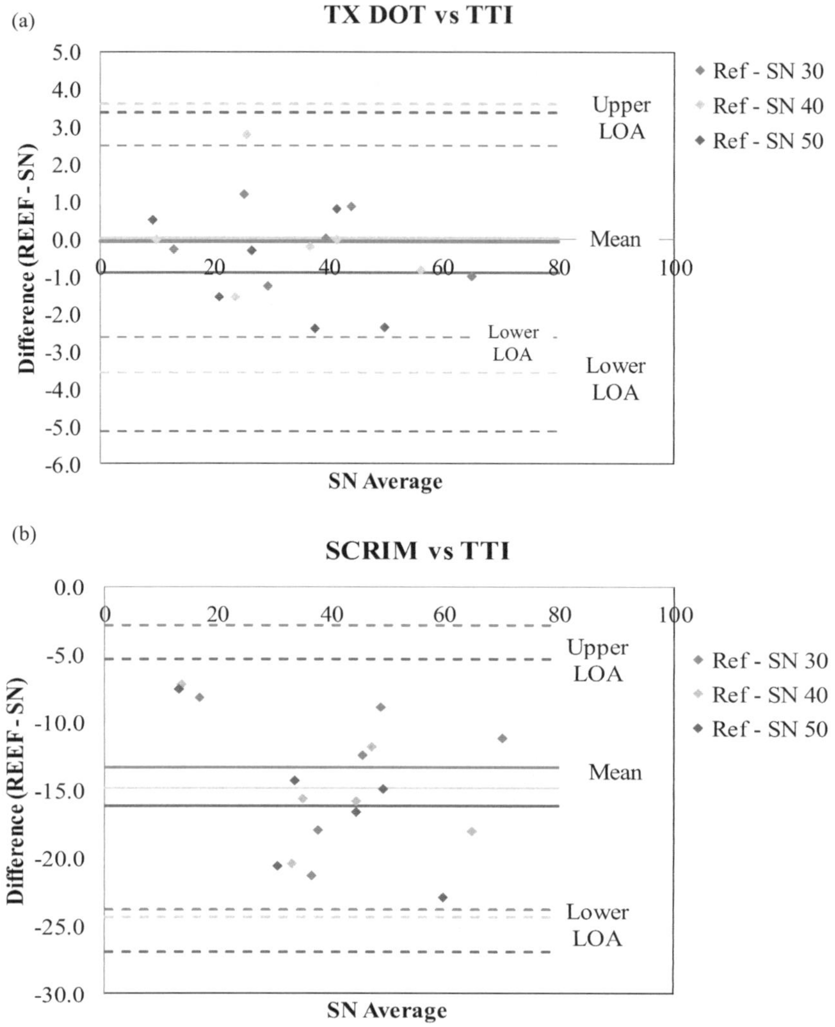

Figure 2. Results of LOA analysis of comparison between devices: (*a*) TxDOT vs. TTI LWST comparison; (*b*) SCRIM vs. TTI LWST comparison.

Table 3. Average Macrotexture (MPD) and Computed Speed Constants (after Fernando et al. [*14*])

Speed	30 km/h		40 km/h		50 km/h	
Section	Avg. MPD (mm)	Sp	Avg. MPD (mm)	Sp	Avg. MPD (mm)	Sp
Pad 1	0.64	71.81	0.65	72.36	0.66	73.28
Pad 2	1.37	137.29	1.42	141.80	1.29	129.79
Pad 2A	0.82	87.85	0.84	89.17	0.84	89.13
Pad 5	2.35	225.39	2.47	235.46	2.76	261.53
Pad 6	1.91	185.63	1.98	192.18	1.92	186.54
Pad 7	1.36	136.09	1.40	139.71	1.46	145.52

LWST measurements, the prediction of the LWST measurements was done using the FR60 formula and the texture measured with the SCRIM. This way the prediction would require only one device, the SCRIM. The third part of Table 4 shows the results of the prediction for the TTI LWST.

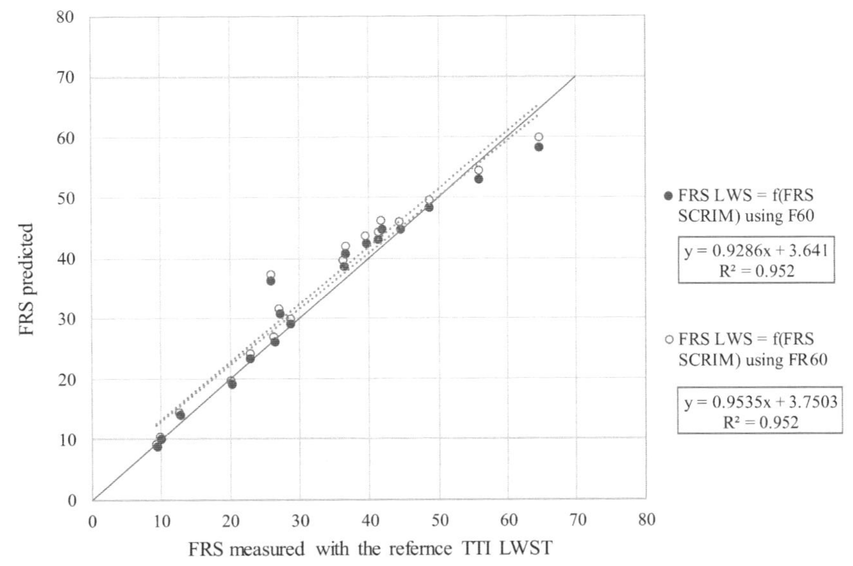

Figure 3. Comparison of measured and predicted friction measurements for TTI LWST.

Analysis and Discussion

Comparison of the Predictions

Figure 3 compares the SN measured with the LWST from TTI and the one predicted based on the SCRIM measurements using the two methods discussed. Using the FR60 or the F60, the coefficient of determination, R^2, is the same; this is expected because of the linear nature of the IFI relationships. It should be noted that each dot in the figure represents the average of several measurements in the same section and differences among individual measurements will certainly have a higher variability. Very similar results were obtained when comparing the SCRIM with the TxDOT LWST.

Both prediction methods have good prediction ability, with a coefficient of determination, R^2, greater than 90%. However, the F60 formula involves an intermediate step that requires measurements with the reference DFT and CT meter. On the other hand, the FR60 formula does not require taking reference friction measurements using the DFT measurements, thus offering a practical alternative.

To simplify the interconversion even further, one additional approach was tested: replacing the CT meter macrotexture measurements with those taken using the SCRIM texture sensor. Figure 4 shows the comparison of the prediction of the LWST measurements using the FR60 formula and the texture measured with the SCRIM. As shown, the R^2 is very close to the one obtained in the previous cases using the texture measured with the CT meter, suggesting very little change in predicting ability.

Discussion

The repeatability of the three devices is in the same order of magnitude, ranging between 1.8 and 3.7 for the three speeds evaluated. The LOA analysis shows that the measurements of the SCRIM are consistently higher than the measurements of the LWST. On average, the SCRIM has a difference of 15 friction numbers from the mean when compared with the LWSTs.

To use Equations 13 and 16 to predict the friction that would be measured with the LWST using the SCRIM measurements, it was necessary to calculate the S_p using the CT meter and the F60 from the DFT to complete the standard procedure for the IFI. According to ASTM E1960, the pavements should have profile depths for the range $0.25\,mm < MPD < 1.5\,mm$ and friction values for the range $0.3 < DFT_{20} < 0.90$. Some of the sections of pavement used for this study have macrotexture and friction characteristics that are out of these ranges. However, all the values measured include a wider range of friction and macrotexture values and improve the relationship.

The results obtained with the methods proposed are encouraging in the search for a way to convert friction measurements. In order to recommend a specific method, several variables were considered: the number of devices required to get the measurements needed to apply the method, the number of variables in the formula applied, and goodness-of-fit statistics (R^2 and the average of the absolute value of the differences between the FRS measured and predicted).

Table 5 presents a summary of the results for the methods tested for the prediction of the friction

Table 4. Average Friction Values Measured, IFI Calculations, and Analysis Results

		IFI calculation												Comparison of TTI LWST SN40 using F60 formula			Comparison using FR60 formula and CT meter macrotexture			Comparison using FR60 formula and SCRIM macrotexture		
		DFT			TTI LWST			TxDOT LWST			SCRIM											
					A = 0.085, B = 0.615			A = 0.089, B = 0.598			A = 0.077, B = 0.599											
Speed (mph)	Section	Avg. SN20	FR60	F60	Avg. SN40	FR60	F60	Avg. SN40	FR60	F60	AVG SR(50)	FR60	F60	Avg. SN meas.	SN pred.	Difference	Avg. SN meas.	SN pred.	Difference	Avg. SN meas.	SN pred.	Difference
30	Pad 1	0.36	0.21	23	29	0.24	23	30	0.25	24	47	0.25	23	28.7	29.2	−0.5	28.7	30.0	−1.3	29.9	25.6	4.4
	Pad 2	0.32	0.24	25	26	0.24	23	25	0.23	22	47	0.34	28	25.8	36.4	−10.6	25.8	37.4	−11.6	24.6	37.9	−13.3
	Pad 2A	0.2	0.12	17	13	0.11	15	13	0.11	16	21	0.13	15	12.6	14.1	−1.5	12.6	14.5	−1.9	12.9	14.9	−2.0
	Pad 5	0.35	0.3	30	44	0.42	34	43	0.41	33	53	0.44	34	44.3	44.9	−0.6	44.3	46.1	−1.9	43.4	45.2	−1.8
	Pad 6	0.34	0.27	28	39	0.37	31	39	0.37	31	52	0.41	32	39.4	42.5	−3.1	39.4	43.7	−4.3	39.4	42.5	−3.1
	Pad 7	0.71	0.53	47	64	0.59	45	65	0.6	45	76	0.55	41	64.4	58.3	6.1	64.4	59.9	4.5	65.4	60.4	5.0
40	Pad 1	0.36	0.2	23	23	0.24	23	24	0.26	24	43	0.26	23	22.9	23.5	−0.7	22.9	24.2	−1.3	24.4	19.7	4.7
	Pad 2	0.33	0.25	26	27	0.28	26	24	0.25	24	43	0.33	27	27.1	30.9	−3.8	27.1	31.7	−4.6	24.3	31.7	−7.4
	Pad 2A	0.21	0.13	18	10	0.1	15	10	0.1	15	17	0.11	14	9.8	10.2	−0.5	9.8	10.5	−0.8	9.8	9.8	−0.1
	Pad 5	0.36	0.31	30	41	0.42	34	41	0.42	34	53	0.45	35	41.2	43.1	−1.9	41.2	44.3	−3.1	41.2	43.4	−2.2
	Pad 6	0.37	0.3	30	36	0.37	31	37	0.37	31	52	0.43	33	36.5	40.9	−4.4	36.5	42.0	−5.5	36.7	39.6	−3.0
	Pad 7	0.76	0.57	50	56	0.57	44	57	0.58	44	74	0.56	41	55.8	53.1	2.7	55.8	54.5	1.2	56.6	54.1	2.5
50	Pad 1	0.34	0.19	22	20	0.26	25	22	0.28	26	41	0.26	23	20.2	19.3	0.9	20.2	19.8	0.4	21.7	16.0	5.7
	Pad 2	0.33	0.24	26	26	0.31	27	27	0.31	27	41	0.32	27	26.3	26.3	0.0	26.3	27.0	−0.7	26.6	27.9	−1.3
	Pad 2A	0.23	0.14	19	9	0.12	16	9	0.11	15	17	0.12	15	9.2	9.0	0.2	9.2	9.2	0.0	8.7	9.3	−0.6
	Pad 5	0.37	0.32	31	42	0.45	36	41	0.44	35	57	0.5	38	41.6	45.0	−3.3	41.6	46.2	−4.5	40.8	45.1	−4.3
	Pad 6	0.37	0.3	30	36	0.4	33	39	0.43	35	53	0.44	34	36.2	38.7	−2.5	36.2	39.8	−3.6	38.6	36.8	1.8
	Pad 7	0.75	0.57	50	49	0.56	43	51	0.58	44	71	0.57	42	48.5	48.4	0.1	48.5	49.7	−1.2	50.9	48.0	2.9
Average of absolute differences																2.41			2.9			2.82

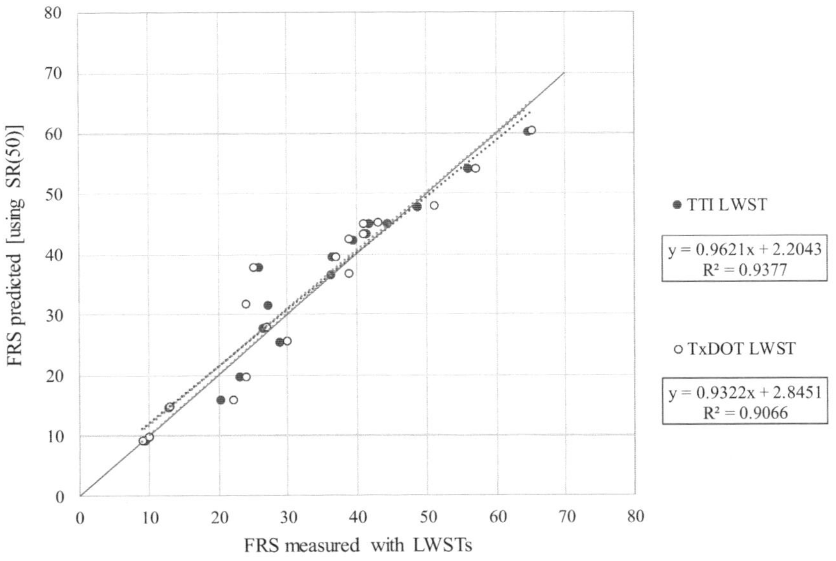

Figure 4. Comparison of predicted FRS using FR60 formula and texture from the SCRIM with measured one.

Table 5. Summary of Comparison of Methods Used

Equipment used for texture	Method	Number of devices required	Variables	Average of absolute value of the difference: TTI	TX DOT
CT meter	F60	3	Macrotexture, speed, SR measured with the SCRIM, previous comparison with DFT (A and B)	2.41	2.93
	FR60	2	Macrotexture, speed, SR measured with the SCRIM	2.90	3.13
Texture of SCRIM	F60	2	Macrotexture, speed, SR measured with the SCRIM, previous comparison with DFT (A and B)	2.71	3.60
	FR60	1	Macrotexture, speed, SR measured with the SCRIM	2.82	3.66

measured with the LWST. The table shows that the four combinations result in similar average absolute errors. Furthermore, the average absolute error is in the same order of magnitude as the equipment's repeatability. As the conversion based only on the speed adjustments (FR60) using the SCRIM macrotexture values does not require reference measurements with static equipment, it is recommended for implementation.

Conclusions

After the comparison done between the measurements of friction taken with the LWST (TTI and TxDOT) and the SCRIM on the same road sections, it can be concluded that:

- All three tested devices produced appropriate repeatability as computed using the LOA, producing measurements within less than $\pm$ 1.1 to 4.5 friction numbers with a 95% level of confidence at the three tested speeds and on a wide range of surfaces.
- The conversion based only on the speed adjustments (FR60) seems to provide the most effective method to predict the LWST measurements from SCRIM measurements. The coefficient of determination and average absolute error are comparable to those using the IFI F60 formula but static reference measurements are not required.

Although the range of surface fiction and texture is rather wide, these conclusions were based on a limited number of tests on a few experimental sections under controlled conditions; thus, it is recommended that the approach be verified using production data collected on a variety of road and surface types.

Acknowledgments

The data used for this paper were collected by TTI and VTTI in a collaborative effort under a study funded by the FHWA. E. Fernando, D. Arrington, R. and Zimmer (from TTI) and B. Hobbs (of VTTI) participated in the experiment design and data collection. Dr. James Wambold oversaw the field experiment and provided guidance during the testing.

Author Contributions

The authors confirm contribution to the paper as follows: study conception and design: GWF, KKM, EDLI; data collection: SBQ; analysis and interpretation of results: SBQ, GWF, EDLI; draft manuscript preparation: SBQ. All authors reviewed the results and approved the final version of the manuscript.

References

1. American Association of State Highway and Transportation Officials. *A Policy on Geometric Design of Highways and Streets.* American Association of State Highway and Transportation Officials, Washington, D.C., 2011.
2. Descornet, G. The HERMES Project. *Proc., 5th Symposium on Pavement Surface Characteristics Conference*, SURF 2004 (CD-ROM), World Road Association, Paris, France, 2004.
3. Vos, E., and J. Groenendijk. *EU Project TYROSAFE, Deliverable D05: Report on Analysis and Findings of Previous Skid Resistance Harmonization Research Projects.* FEHRL, Brussels, Belgium, 2009. http://www.transport-research.info/sites/default/files/project/documents/20120406_001904_79331_TYROSAFE_WP2_D05PreviousSkidResistanceProjects_v2.0_20090225.pdf.
4. Flintsch, G. W., E. de León Izeppi, K. K. McGhee, and J. Roa. Evaluation of the International Friction Index Coefficients for Various Devices. *Transportation Research Record: Journal of the Transportation Research Board*, 2009. 2094: 136–143.
5. Noyce, D., H. Bahia, J. Yambo, and G. Kim. *Incorporating Road Safety into Pavement Management: Maximizing Asphalt Pavement Surface Friction for Road Safety Improvements.* Midwest Regional University Transportation Center, Madison, WI, 2005.
6. Flintsch, G. W., E. de León Izeppi, K. K. McGhee, and S. Najafi. *The Little Book of Tire Pavement Friction.* Unpublished manuscript, 2012.
7. Smith, K. L., R. L. Larson, G. Flintsch, and E. de León Izeppi. *Development and Demonstration of Pavement Friction Management Programs: Theoretical Relationship of Vehicle-Tire-Pavement Interactive and Skid Crashes.* Federal Highway Administration, Washington, D.C., 2011.
8. Choubane, B., H. Lee, C. Houlzchuher, P. Upshaw, and N. Jackson. Harmonization of Texture and Friction Measurements on Florida's Open and Dense Graded Pavements. *Transportation Research Record: Journal of the Transportation Research Board*, 2012. 2306: 122–130.
9. Flintsch, G. W., E. de León Izeppi, J. Izeppi, Roa, K. K. McGhee, and M. Swanlund. Evaluation of the International Friction Index Coefficients for Various Devices. *Proc., 6th Symposium on Pavement Surface Characteristics, SURF 2008*, Portoroz, Slovenia, October 20–22, 2008.
10. Leu, M., and J. Henry. Prediction of Skid Resistance as a Function of Speed from Pavement Texture Measurements. *Transportation Research Record: Journal of the Transportation Research Board*, 1978. 666: 7–13.
11. Henry, J. *Evaluation of Pavement Friction Characteristics.* National Cooperative Highway Research Program, Transport Research Board, Washington, D.C., 2000.
12. Wallman, C., and H. Astrom. *Friction Measurement Methods and the Correlation between Road Friction and Traffic Safety.* Swedish National Road and Transport Research Institute, Linköping, Sweden, 2001.
13. Wambold, J., C. Antle, J. Henry, Z. Rado, G. Descornet, U. Sandberg, M. Gothié, and S. Huschek. *International PIARC Experiment to Compare and Harmonize Skid Resistance and Texture Measurements.* Publication 01.04. PIARC, Paris, France, 1995.
14. Fernando, E., D. Arrington, and R. Zimmer. Comparative Evaluation of Locked-Wheel and SCRIM Testers. Presented at 97th Annual Meeting of the Transportation Research Board, Washington, D.C., 2018.
15. Bland, J., and D. Altman. Statistical Methods for Assessing Agreement between Two Methods of Clinical Measurement. *Lancet*, Vol. 1, No. 8476, pp. 307–310.
16. de León Izeppi, E. D., G. W. Flintsch, and K. K. McGhee. Limits of Agreement Method for Comparing Pavement Friction Measurements. *Transportation Research Record: Journal of the Transportation Research Board*, 2012. 2306: 188–195.
17. Bland, J., and D. Altman. Measuring Agreement in Device Comparison Studies. *Statistical Devices in Medical Research*, Vol. 8, 1999, pp. 135–160.
18. ASTM, E 1960. *Standard Practice for Calculating International Friction Index of a Pavement Surface.* ASTM International, West Conshohocken, PA, 2015. www.astm.org.

The Standing Committee on Pavement Surface Properties and Vehicle Interaction (AFD90) peer-reviewed this paper (18-04991).

Sage Publications USA
sagepub.com

ISBN 978-1-5108-8248-5